Social and Behavioral Sciences

Sports and Recreation

FINITE MATHEMATICS

AN APPLIED APPROACH

THIRD EDITION

FINITE MATHEMATICS

AN APPLIED APPROACH

THIRD EDITION

Paula Grafton Young
Salem College

Todd Lee
Elon University

Paul E. Long
University of Arkansas

Jay Graening
University of Arkansas

PEARSON
Addison
Wesley

Boston   San Francisco   New York
London   Toronto   Sydney   Tokyo   Singapore   Madrid
Mexico City   Munich   Paris   Cape Town   Hong Kong   Montreal

Publisher: Greg Tobin
Acquistions Editors: William Hoffman, Carter Fenton
Managing Editor: Karen Guardino
Executive Project Manager: Christine O'Brien
Editorial Assistant: Mary Reynolds
Editorial Assistant: Sarah Santoro
Production Supervisor: Julie LaChance
Marketing Manager: Pamela Laskey
Marketing Assistant: Heather Peck
Senior Manufacturing Buyer: Evelyn Beaton
Senior Prepress Supervisor: Caroline Fell
Media Producer: Lynn Blaszak
Senior Software Editors: Mary Dougherty, David Malone
Compositor: WestWords, Inc.
Cover Design: Barbara T. Atkinson
Interior Design: Silvers Design/Barbara T. Atkinson

Cover Photo: Precision skiers in "Powder 8" competition. © Gary Brettnachev/Getty Images, Inc.

For permission to use copyrighted material, grateful acknowledgment is made to the copyright holders on page 685, which is hereby made part of this copyright page.

Library of Congress Cataloging-in-Publication Data

Finite math : an applied approach.—3rd ed. / Paula Grafton Young . . . [et al.].
    p. cm.
Rev. ed. of: Finite mathematics / Paul E. Long, Jay Graening. 2nd ed. c1997.
Includes bibliographical references and index.
 ISBN 0-321-17334-1
 1. Mathematics. I. Young, Paula G. II. Long, Paul E. Finite mathematics.

 QA39.3.F56 2003
 510—dc21
                                    2002043790

Printed in the United States of America.

2 3 4 5 6 7 8 9 10-RRD-06 05 04

THIS IS DEDICATED TO:

Dr. Feldman, Dr. Barnard, and Dr. Pierce,
who taught us to be mathematicians,

Our Elon and Salem students,
who taught us to be mentors,

J.R. Boyd, who taught us to be teachers,

Our Mothers, who taught us to be caring humans.

Contents

Preface

About This Book

Finite Mathematics: An Applied Approach, Third Edition, like its previous editions, strives to provide meaningful and accessible contemporary mathematics to students in a variety of disciplines. A major goal of this text is to serve those students who need a terminal mathematics course. The text also provides a solid mathematical foundation that will prepare students for quantitatively oriented courses in disciplines such as business or social statistics, economics, finance, and applied calculus. We, the authors, have the goal of writing a solid, sensible text that will aid both student and instructor. We use a conversational writing style and a minimal amount of technical symbolism in order to make the text more readable and the topics less intimidating. However, great care is taken in the wording of definitions and assertions; basic rigor is never sacrificed.

We are, first and foremost, teachers; we know students who take finite mathematics. Although they will face specific uses of higher mathematics in their future endeavors, most of these students will not be continuing in the physical or mathematical sciences. Many will already have built up a fairly significant aversion to mathematics. In both level and tone, this text reflects our understanding of the student. On the one hand, we do not trivialize the material by being clownish or whimsical; much of the power of mathematics comes from the care taken in what is said. On the other hand, we do not belittle the students' difficulty by using terse explanations, taking self-obliging romps in the obscure aspects of mathematics, or employing words like "obviously" carelessly. Rather, we maintain an honest respect for both the student and the mathematics.

Finite mathematics is often a terminal course. Its purpose is to provide students with a variety of mathematical tools that are significant in the "real world." The guiding philosophy of this course is that what is lost in the mastery that comes with depth is realized in the maturity that comes with breadth. Finite mathematics is a natural course to include in today's curriculum, in which such a heavy emphasis is placed on direct applicability. The text revels in this philosophy and strives to frame as much of the mathematics as possible in the context of the physical universe. However, we do not confuse "real world problems" and *real* problems, and we refuse to blur this distinction with the students. *Real* problems are hard . . . really, really hard. For instance, *real* linear programming problems can have hundreds or even thousands of variables, and simplex methods are not used on them. Is this an argument against the use of "real world" applied problems? No, because such problems are wonderful practice for the tough, but valuable, skill of translating back and forth between the contextual and mathematical. Each of the text's sections ends with a large variety of problems, of many different applications. But the applications are never allowed to obscure the purpose of the problem. The nature of the exercises reflects our belief that students need lots of practice with basic text problems. The examples are constructed to facilitate the students' success with the exercises. Further, at the end of each chapter, we provide *In-Depth Applications* to help students gain insights as to how their growing collection of basic skills can be used in the *real* world.

Finite mathematics is also taught as a precursor to the traditional applied or business calculus course. Most experienced teachers will agree that a terminal mathematics course and a prerequisite mathematics course are different creatures. Where the terminal course has a certain degree of freedom in the areas explored and notation used, a prerequisite course must take great care to provide a firm grounding in basic concepts to be built upon and to use standard notation that is employed in later courses. This distinction only reaffirms the value of the careful, straightforward approach used in this text. We do not construct special notation or techniques that come from personal favorite classroom approaches. Both the student and teacher can rely on the text to provide the basic building blocks for further courses. The explanations, the examples, and the exercises are all written with this aim in mind.

Over the years, finite mathematics has expanded; from institution to institution, the variety of material that is included under this course name is almost boundless. There are two ways in which a text can be of service in dealing with this phenomenon. The first is that chapters can be expanded to cover multiple aspects of the topic at hand. Although this feature allows individual teachers to find the threads they most prefer when covering a topic, we feel that it makes for a bloated and confusing text. We keep the chapters and sections reasonable in size and focus on basic approaches to individual topics. This choice allows us to take the time to go through multiple examples that are pertinent to the progressing story. The text can then be used as a foundation on which any class can be built, as the instructor has the freedom to build and explore upon it as she or he sees fit.

The second way in which a text can address the topical growth of finite mathematics is to add smaller chapters on special topics. This text has always provided such chapters, while staying mindful of the overall size of the book. The inclusion of these topics via separate chapters allows the majority of classes to work through the standard core without distraction, while those classes which go adventuring off the beaten path have a significant set of interesting choices. We shall speak more about this shortly.

What has changed in this new edition? All of the expected changes have been made; data sets have been increased and updated, names and situational problems reflect the world in which we live more accurately, and minor errors have been corrected. Several deeper changes have also been made, but we have not lost sight of the strengths of the previous edition. The breadth and depth of technology have increased, but the independence of the material from the technology is firmly maintained. Color has been incorporated, but only to enhance understanding and engagement. Chapters have been rearranged and reconstructed, but only to facilitate flow and flexibility. At every stage of the revision process, at each decision of change, we have striven to maintain the simplicity, friendliness, and dependability for which the original text was so appreciated by both teachers and students.

Technology

Not only have the topics in finite mathematics changed, but also have the required technologies. We are of the firm belief that technology should be used to enhance the learning of material, not to replace the understanding of the mathematical principles behind it. Therefore, we are careful to use technology—both graphing calculators and spreadsheets—when it is pedagogically enlightening and not merely interesting. A strength of the technology-oriented portions of the text is that they are not just add-ons or last minute insertions. We have a great deal of experience with using technology sensibly in the classroom, and we have tried to use that experience (and good sense) when including technology throughout the revision process.

The use of graphing calculators was successfully included in the previous edition; the use of spreadsheets is new to this edition. Spreadsheet technology is a powerful tool for learning, but that is not the only reason it has been incorporated into the text. It is a fundamental fact that in this day and age, knowledge of spreadsheet packages will be a required lifelong skill for many of our students. For many institutions, this bonus is too much to ignore.

Finite mathematics was taught long before calculators and laptops appeared in the classroom. This text will support the natural use of technology in the classroom, but it absolutely does not require it. Where technology is useful, helpful examples are given. In most all of the examples and exercises, technology can be used, but rarely does it make the problem trivial, and almost never is it absolutely required. The teacher who forbids the use of technology will not find this text restrictive, while classes that use technology constantly will find the natural inclusion equally liberating. However, we have a special place in our hearts (as is expressed in our text) for those teachers who are looking to use technology moderately and who need help in accomplishing that aim.

Course Flexibility

It is our belief that no single ordering of the table of contents will satisfy the needs of every student at every institution. Therefore, we have made the chapters as independent of each other as possible. A course designed to meet quantitative literacy requirements and not designed to prepare students for future quantitatively oriented courses could follow Chapters 1, 2, 3, 6, 7, 8, and 10. Some instructors may wish to begin with Chapter 7 and proceed through Chapters 8, 9, and 10 before progressing to other chapters. Still others, at institutions that offer courses in basic statistics, may wish to omit Chapters 7 through 10 entirely. The following diagram outlines the interdependence of chapters, so that each institution may develop its own ordering of topics:

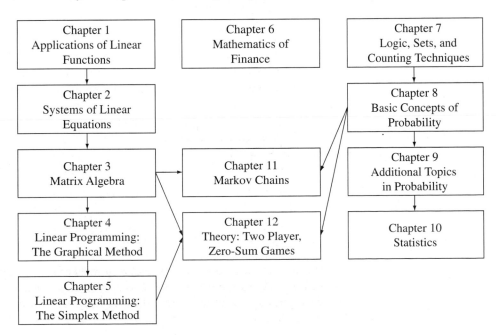

Organization and Content Changes

- In Chapter 2, the connection between the elimination method and Gauss–Jordan elimination has been rewritten to improve clarity.

- Linear programming is divided into two separate chapters. Chapter 4 focuses on graphical methods; Chapter 5 on algorithmic methods.

- Chapter 6, "Mathematics of Finance," has been moved to the center of the text. The formulas have been rewritten to more closely match those used in finance courses.

- In Chapter 7, a new section on logic and a separate section on truth tables have been written, with discussions about equivalence, connectives, and quantifiers included.

- The "count what you do not want" strategy in Chapter 7 has been incorporated throughout the sections on counting, rather than appearing as a separate topic at the end of the chapter.

- Chapter 10, "Statistics," has been completely revised to improve readability and to include more problems involving real data.

- In Chapter 11, a section looking at Markov chains as mathematical models has been added, and new applications, such as epidemiological models, have been included.

- Appendices reviewing algebra and introducing spreadsheets have been added.

Features New to This Edition

In the process of revising this text, we have made some changes designed to improve the learning process for the students and to aide instructors as they adapt the text to their own teaching styles:

- Color has been added to examples in order to clarify the steps used in solving problems. In numerous examples, key quantities are color matched with their corresponding variables in the various equations, figures, and mathematical constructs being used.

- New art has been added, including photographs and updated, color-enhanced graphs. This artwork will improve the clarity of the data being presented and will engage the student.

- Spreadsheet technology has been added to acknowledge both required life skills for the students and current pedagogy.

- Use of the graphing calculator is updated to the TI-83 Plus® model.

- Detailed instructions are provided in the text, rather than strictly in supplements, for technology when its use is appropriate and instructive.

- New exercises and examples using real data from a variety of sources have been added.

- An *In-Depth Application,* reflecting current issues and real data, has been added to the end of 11 of the 12 chapters. This section can be used in a variety of ways, such as for group projects or extra-credit assignments. Each such application incorporates the mathematics developed in the chapter and asks a series of questions related to a timely, interesting topic, such as sports ranking systems (Chapter 3) and message routing through the Internet (Chapter 12).

Hallmark Features

Chapter Problems

Each chapter begins with a relevant problem that can be addressed by the material in the chapter. Each of these problems is completely solved later in the chapter.

Technology Opportunities

Each chapter proceeds with a brief discussion about the uses of graphing calculators and spreadsheets for the topics about to be covered. Detailed instructions for using the TI-83 Plus® and MS Excel® are incorporated into examples near the end of each relevant section to allow instructors flexibility regarding when and how to use the technology they deem most appropriate for their students.

Applications

Applications to a variety of fields are found throughout the examples and exercises. An applications index is provided at the front of the text for reference. Effort has been taken to include applications from many different disciplines, including accounting, the biological sciences, communication and media studies, computers, economics, education, finance, fine arts, the health sciences, management, marketing, physical science, and sports and recreation. Many applications of general interest are also included. Current data from many sources, including governmental agencies and private organizations, are used in the construction of these applications.

Examples and Exercises

As discussed earlier, we have invested a great deal of time and effort to provide an extensive progression of examples in each section. The exercises continue on the same themes introduced by the examples. This connection helps to alleviate some of the more common frustrations college mathematics students can have. Students need a large amount of practice on a large variety of examples; this text provides just that.

Historical Notes

Throughout the text, we have included historical notes, usually in the form of biographical sketches, related to the material being discussed. These notes add a sense of "humanness" to the mathematics. They can be used to foster class discussions or as starting points for writing assignments.

Chapter Summaries and Chapter Tests

Each chapter ends with a summary of the keywords and topics covered, as well as a set of review exercises. The sample tests included at the end of each chapter can also be used to test the student's knowledge of the material covered. The answers to all questions in the sample tests are presented at the back of the text.

Supplements

For the Instructor

The *Instructor's Solutions Manual* (ISBN 0-321-17337-6) contains complete, worked-out solutions for all exercises in the text.

NEW! MyMathLab.com Ideal for lecture-based, lab-based, and on-line courses, MyMathLab provides students and instructors with a centralized point of access to the wide variety of multimedia resources available with their Addison-Wesley textbook. The pages of the actual text are loaded into MyMathLab, and as students work through a section of the on-line text, they can link directly to supplementary resources (such as tutorial software, interactive animations, and audio and video clips) that provide instruction, exploration, and practice beyond what is offered in the printed book. MyMathLab generates personalized study plans for students and offers them unlimited practice exercises in areas in which they need improvement. Instructors can use MyMathLab to create, edit, and assign homework and tests and track all student work in an on-line grade book. Extensive course-management capabilities, including a host of communication tools for course participants, are provided to create a user-friendly and interactive on-line learning environment.

For more information, visit our Web site at www.mymathlab.com, or contact your Addison-Wesley sales representative for a product demonstration (for qualified adopters only).

TestGen4.0 and QuizMaster3.0 let you view and edit test-bank questions, transfer them to tests, and print them in a variety of formats. The program also offers many options for organizing and displaying test banks and tests. A built-in random-number and test generator makes TestGen-EQ ideal for creating multiple versions of tests and provides more possible test items than do printed test-bank questions. Powerful search and sort functions let the instructor easily locate questions and arrange them in the preferred order. Users can export tests as text files so that they can be viewed with a Web browser. In addition, tests created with TestGen can be used with QuizMaster, which enables students to take exams on a computer network. QuizMaster automatically grades the exams, stores results on disk, and allows the instructor to view or print a variety of reports for individual students, classes, or courses.

For the Student

The *Student's Solutions Manual* (ISBN 0-321-17338-4) contains solutions to all odd-numbered section and Chapter Summary exercises and all Sample Test Items and Algebra Review exercises.

The *Graphing Calculator Manual* (ISBN 0-321-17335-X) provides detailed information on using some of the more popular graphing calculator models to work through examples in the text. Listings of programs for graphing calculators are also included.

The *Excel Spreadsheet Manual* (ISBN 0-321-17336-8) provides detailed information on using the Excel spreadsheet program to work through examples in the text.

MathXL is an on-line testing, homework, and tutorial system that uses algorithmically generated exercises correlated to the text. Instructors can assign tests and homework provided by Addison-Wesley or create and customize their own tests and homework assignments. In MathXL, instructors can also track their students' results and tutorial work in an on-line grade book. Students can take chapter tests and receive personalized study plans based on their test results. The study plan diagnoses weaknesses and links students to the areas they need to study and be retested on. Students can work unlimited practice exercises and receive tutorial instruction in areas in which they need improvement. MathXL can be packaged with new copies of this text. Please contact your Addison-Wesley representative for details (for qualified adopters only).

Addison-Wesley Tutor Center: Free tutoring is available to students who purchase a new copy of the third edition of *Finite Mathematics, An Applied Approach* when bundled with an access code. The Addison-Wesley Math Tutor Center is staffed by qualified math instructors who provide students with tutoring on examples in the text and on any exercise with an answer in the back of the book. Tutoring assistance is provided by toll-free telephone, fax, e-mail, and whiteboard technology—which allows tutors and students to actually see the problems worked while they "talk" in real time over the Internet. This service is available five days a week, seven hours a day. For more information, contact your Addison-Wesley sales representative (for qualified adopters only).

Acknowledgments

We thank the following instructors for their contributions in reviewing portions of this text:

Barry A. Balof, Dartmouth College

Pat Bassett, Palm Beach Atlantic University

Rajanikant Bhatt, Texas A&M University

Stuart Boersma, Central Washington University

Miriam Castroconde, Irvine Valley College

Beth Cole, Northwestern State University

Ginny Crisonino, Union County College

Sujay Datta, Northern Michigan University

Donald Davis, Lehigh University

John DeValcourt, Cabrillo College

Michael Divinia, San Jose City College

William Foley, Southwestern College

Gary Frick, College of Southern Maryland

James Jones, Richland Community College

Rita Kolb, Baltimore County Community College, Catonsville

Michael Kovcholovsky, University of Oregon

Michael LaValle, Rochester Community College

Teresa Leyba, South Mountain Community College

C.G. Mendez, University of Michigan

Edward Miller, Lewis–Clark State College

Linda M. Neal, Southern Methodist University

Glenn Pfeifer, Community College of Southern Nevada

Ken Price, University of Wisconsin, Oshkosh

Rama Rao, University of North Florida

Thomas Riedel, University of Louisville

Daphne Rossiter, Mesa Community College

Marilyn Schiermeier, Meredith College

Terry Shell, Santa Rosa College

Darren J. Smith, Montgomery College, Germantown

Robert Stark, University of Delaware

Alexander Suciu, Northeastern University

Stefania Tracogna, Arizona State University

Sharon Vestal, Missouri Western State College

Marlene Will, Spalding University

Neal Wilsey, College of Southern Maryland

Daniel Wilshire, Penn State University, Altoona College

Putting together a textbook is no small feat, and we were lucky to have an outstanding team of talented people on our side at every step of the way. We thank Paul Long and Jay Graening for giving us the green light to revise the text; we hope that this edition is worthy of the trust that they bestowed upon us. Each of us is grateful to our respective institutions for granting us sabbaticals during the revision process. Because of their years of experience with previous editions of this text, the insights and comments from Deborah Harrell, Wenzhi Sun, and Ann Marie Ziegler were valued and welcomed. We also thank Richard Snelsire, Michael Cummings, and Steven Wagner, who offered us their advice about "real problems" in the business world. And, of course we appreciate the patience, support, and enthusiasm of our families, friends, and colleagues who have cheered us on from the beginning.

Our sincere thanks go to our first editor at Addison-Wesley, Laurie Rosatone, for believing in our vision of melding a sensible text with sensible technology. We greatly appreciate Pat McCutcheon at WestWords and Beverly Fusfield at Techsetters for their efforts in bringing the words and art to life. The proofreading and accuracy checking that Elka Block, Frank Purcell, and Jackie Miller did was helpful, as was the work done by Tim Mogill and Deana Richmond as they checked our answers. We thank David Dubriske for putting together the solutions manuals and lending his critical eye to the exercises. Most of all, we thank our entire team at Addison-Wesley: Bill Hoffman, Christine O'Brien, Greg Tobin, Mary Reynolds, Sarah Santoro, Julie LaChance, Barbara Atkinson, Joe Vetere, Beth Anderson, Pam Laskey, Heather Peck, and Lynne Blaszak. They have made our writing experience interesting, relaxed, and a growth experience for both of us.

Paula Grafton Young
Todd Lee

Applications of Linear Functions

PROBLEM

If this trend continues, how many millions of tons of waste will be recycled in the year 2010?

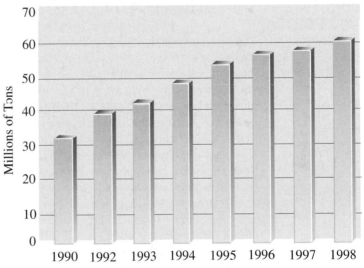

Recovered Waste in the United States
U.S. Census 2000

Millions of Tons

SOURCE: U.S. Bureau of the Census: *Statistical Abstract of the United States*, 2000.

CHAPTER PREVIEW

The central theme of this chapter is the modeling of various phenomena with linear functions. We begin by reviewing the rectangular coordinate system and the graphing of functions. Then the point–slope, slope–intercept, and general-form equations of a straight line are developed. This leads to the concept of a linear function, whose graph is a straight line, and its use in modeling sales, costs, the book value of an asset depreciated by the straight-line method, the decline of fish population as a function of pollution, the growth of municipal solid waste, and so on. The break-even point and the equilibrium supply and demand quantities are then studied as applications of a pair of linear functions. Finally, the concept of a least-squares regression line is introduced as an application of solving a system of two equations in two unknowns and is shown to occur in numerous fields of study.

Technology Opportunities

One of the most powerful uses of graphing calculators is to graph and explore different properties of functions. Using the [TRACE] command, one can estimate different points along a graph. Other built-in commands find specific values of a function, as well as the intersection points of multiple graphs. Later in this chapter, a graphing calculator will be invaluable in displaying scatter plots of data and in computing linear regression equations.

A spreadsheet package is the perfect environment for creating tables of values of functions and for displaying pairs of points that are the solutions of equations. Spreadsheet software also can be readily used for displaying data points along with linear regression curves.

Section 1.1   The Cartesian Plane and Graphing

The **real number line** is a graphic model of the real number system. It consists of a horizontal line with an arbitrary point labeled with the number 0, called the **origin.** A point to the right of the origin is marked to correspond to the number 1, with positive numbers then scaled accordingly to the right of the origin, whereas negative numbers are scaled to the left of the origin. (See Figure 1.) Each real number then corresponds to exactly one point on the real number line, and, conversely, each point on the real number line corresponds to exactly one real number.

The **rectangular coordinate system,** also called the **Cartesian plane** after its inventor, René Descartes, provides a convenient way to graphically represent points in the plane as *ordered pairs of real numbers.* Such a coordinate system is formed by two real number lines intersecting at right angles, as shown in Figure 2. The horizontal number line is usually called the *x*-**axis,** while the vertical number line is usually called the *y*-**axis.** The point of intersection of the two lines is called the **origin,** and the axes separate the plane into four regions, called **quadrants.**

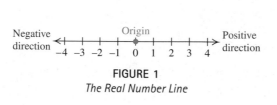

FIGURE 1
The Real Number Line

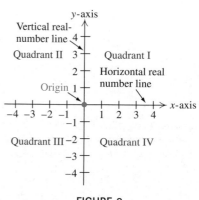

FIGURE 2
The Cartesian Plane

HISTORICAL NOTE

RENÉ DESCARTES (1596–1650)

■ René Descartes (dā-kärt') was a French mathematician and philosopher who is credited with inventing the planar coordinate system that we use today. It is only one of many mathematical and scientific discoveries attributed to Descartes. Among his most important mathematical contributions was the combining of algebra and geometry into a single subject, known as *analytic geometry*. Also known as the "Father of Analytic Philosophy," he is known for the statement, "Je pense, donc je suis." ("I think, therefore I am.")

Each point in the plane can now be matched with exactly one **ordered pair** of real numbers (x, y) called **coordinates of the point;** conversely, each ordered pair of real numbers (x, y) corresponds to exactly one point in the plane. The **x-coordinate** represents the directed distance from the y-axis to the point, and the **y-coordinate** represents the directed distance from the x-axis to the point.

EXAMPLE 1

Plotting Points in the Cartesian Plane

Plot (locate) the points $(1, 3)$, $(-2, 4)$, $(4, 0)$, and $(-3, -2)$ in the Cartesian plane.

SOLUTION To plot the point $(1, 3)$, envision a vertical line through 1 on the x-axis and a horizontal line through 3 on the y-axis. The intersection of these two lines represents the point $(1, 3)$, as shown in Figure 3. The other points are plotted in a similar way.

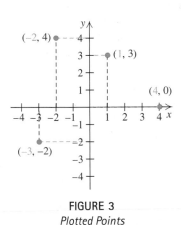

FIGURE 3
Plotted Points ■

The beauty of Descartes' rectangular coordinate system is that it allows us to visualize relationships between two variables. [See Figure 4(b) of Example 2.]

EXAMPLE 2

An Application of the Rectangular Coordinate System

Data collected by the U.S. National Highway Traffic Safety Administration show the following number of fatalities per 100,000 people per year from motor vehicle accidents for the years 1980 through 1998, as shown on the following page. Show these data on a rectangular coordinate system.

SOLUTION Let x represent the years 1980 through 1998, and let y represent the number of deaths per 100,000 people. Because there are many years between 0 and 1980 for which

Year	Deaths
1980	22.48
1982	18.97
1984	18.77
1986	19.19
1988	19.26
1990	17.88
1992	15.39
1994	15.64
1996	15.86
1998	15.34

SOURCE: U.S. National Highway Traffic Safety
Administration: *Traffic Safety Facts 1998.*

no data are shown, it is customary to indicate these omissions graphically by showing a break in the *x*-axis, as in Figure 4(a). The same applies to omissions on the *y*-axis, Plotted points are sometimes connected by line segments, as in Figure 4(b), to more clearly depict any trends that are present.

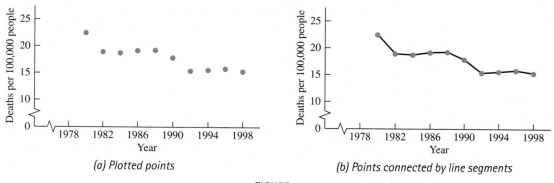

(a) Plotted points   (b) Points connected by line segments

FIGURE 4
Motor Vehicle Deaths per 100,000 People per Year

The relationship between two variables may sometimes be stated by an **equation,** some examples of which are as follows:

$$y = 3x + 2$$
$$x^2 + y^2 = 9$$
$$C = 24x + 1800$$ *C is the cost, in dollars, of making x clock radios.*
$$R = 48x.$$ *R is the revenue gained by selling x ladies' blouses.*

Consider the first of these equations, $y = 3x + 2$. If $x = 1$, for instance, then $y = 3(1) + 2 = 5$. The point $(1, 5)$ is called a **solution** of the equation and may be plotted in the rectangular coordinate system. Another solution is $(0, 2)$, because the substitution of $x = 0$ into the equation results in $y = 2$. There are infinitely many such solutions, and when they are displayed in the rectangular coordinate system, they constitute the **graph** of the equation. This is the link between *algebra* (the equation) and *geometry* (the representation of the solution points in the rectangular coordinate system) that Descartes created with his analytic geometry.

The Graph of an Equation

The **graph** of an equation is the geometric representation of all solutions of the equation in the rectangular coordinate system.

One method of graphing an equation is called the **plotting method.** It consists of the following three steps:

The Plotting Method of Graphing an Equation

The **plotting method** of graphing an equation consists of these three steps:

1. Make a representative table of solutions of the equation.
2. Plot the solutions as ordered pairs in the rectangular coordinate system.
3. Use the representative points to sketch the graph. (*Note:* This does not mean "Connect the dots"!)

EXAMPLE 3

Graphing an Equation

Graph the equation $y = 3x + 2$.

SOLUTION Representative solutions can be efficiently displayed in a table with the headings x, y, and (x, y), as shown below. To find a solution (x, y), choose any number x and then substitute that chosen value into the equation $y = 3x + 2$ to find the corresponding value of y. A good sample of solutions will be found by choosing some positive and some negative values of x. Once the table is constructed, we are ready to plot these solutions in the rectangular coordinate system. The points appear as shown in Figure 5. Finally, we look for a pattern among the points just plotted to determine how a smooth curve through them will appear. In this case, the points appear to be in a straight line, so we connect them accordingly.

x	y	(x, y)
-3	-7	$(-3, -7)$
-2	-4	$(-2, -4)$
-1	-1	$(-1, -1)$
0	2	$(0, 2)$
1	5	$(1, 5)$
1.5	6.5	$(1.5, 6.5)$
2	8	$(2, 8)$

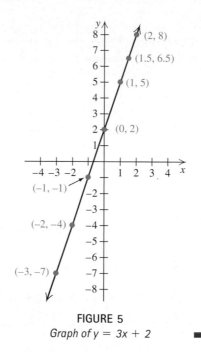

FIGURE 5
Graph of y = 3x + 2 ■

Section 1.2 shows how to identify equations that have straight lines as their graphs. Once they are identified, we need only find *two solutions* (points on the line) to graph the equation. Example 4 demonstrates the plotting method for a different type of equation.

EXAMPLE 4

Sales: Graphing an Equation

The Brite Lite Company hired an economist to study the relationship between the number of flashlights the company could expect to sell and the price it charges for each. The economist discovered that sales S (in hundreds of flashlights per month) could be approximated by the equation

$$S = -0.5p^2 + 32,$$

where p (in dollars) is the price charged for each flashlight. Graph this equation to get a visual picture of the relationship between sales and price.

SOLUTION For this equation, it is convenient to let p play the role of x and S the role of y. Because p (price) cannot be negative, we start seeking solutions of the equation with $p = 0$. Some values of p for which $S = -0.5p^2 + 32$ can be readily calculated and are shown in the table to the left of Figure 6. We find that S drops off rather fast as p increases, until $p = 8$ gives $S = 0$. In other words, when the flashlights are priced at $8, no sales are expected to be made. We may *not* use p values greater than 8, because they would give negative sales. A sketch of the graph is shown in Figure 6.

p	S	(p, S)
0	32	$(0, 32)$
2	30	$(2, 30)$
4	24	$(4, 24)$
6	14	$(6, 14)$
8	0	$(8, 0)$

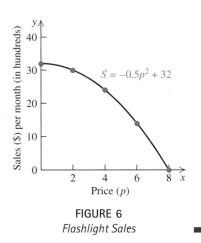

FIGURE 6
Flashlight Sales

NOTE One may argue that it is unrealistic to consider a price of $0 or a price for which there would be zero sales. When using mathematical models, we should always be careful about when the model is realistic and when it is not. This is especially true when using the model in extreme cases, such as a price of $0. Notice that we did not use the model for negative prices or sales, as that definitely makes no sense.

In Example 4, the equation $S = -0.5p^2 + 32$ tells us that the number of sales *depends on* the price charged for each flashlight. In mathematical language, we would say that the sales are a **function** of the price and would denote that fact by the equation $S = f(p)$, read "S equals f of p." In particular, the relationship between sales and the price may be written as

$$f(p) = -0.5p^2 + 32.$$

Such functional notation provides a convenient way to denote sales for various values of p without actually calculating the right side of the equation. For example, $f(4)$ denotes the sales (in hundreds) when the price is $4, $f(7)$ denotes the sales (in hundreds) when the price is $7, and so on. The interval $[0, 8]$ of possible values of p is called the **domain** of the function $S = f(p)$. The formal definition of a function using the variables x and y of the Cartesian plane is given next.

Defining a Function

A **function** is a rule f that assigns, to each value of a variable x, one and only one value $f(x)$, read as "f of x." The set of permissible values of x to which the rule f applies is called the **domain** of the function. The **graph** of a function consists of all ordered pairs $(x, f(x))$ in the rectangular coordinate system.

To graph a function (rule) f that is given in equation form, such as $f(x) = 2x - 5$, it is often more convenient to set $y = f(x) = 2x - 5$ and graph the equation $y = 2x - 5$, rather than use functional notation exclusively. In considering $y = f(x)$, x is called the **independent variable** and y is called the **dependent variable** (since its value depends upon the choice of x).

EXAMPLE 5

Dissolving Salt: Using Functional Notation

Sixteen grams of table salt are stirred into a container of hot water to dissolve. After x minutes, the amount of salt that remains undissolved is given by the function $f(x) = (x - 4)^2$.

(a) How many grams of salt remain undissolved after 1 minute? After 2 minutes?

SOLUTION The answers are found by computing

$$f(1) = (1 - 4)^2 = (-3)^2 = 9 \text{ grams}$$

and

$$f(2) = (2 - 4)^2 = (-2)^2 = 4 \text{ grams}.$$

(b) Find the domain of the function f. In other words, what values of time x are we permitted to use in $f(x) = (x - 4)^2$ to obtain grams of undissolved salt?

SOLUTION Time starts at $x = 0$—the moment salt is put into the container. At that time, the amount of undissolved salt is $f(x) = (0 - 4)^2 = (-4)^2 = 16$ grams. The values of x then increase until x becomes 4, at which time $f(4) = (4 - 4)^2 = 0$ grams of salt undissolved. That is, 4 minutes later, all of the salt has been dissolved. Therefore, a realistic domain for this function is the interval $[0, 4]$.

(c) Graph the function $f(x) = (x - 4)^2$ for the values of x in its domain.

SOLUTION We start by setting $y = (x - 4)^2$ and construct a table of solutions of the given equation using only values of x in the interval $[0, 4]$. Selected values of x and the corresponding values of y are shown in the table to the left of Figure 7.

x	$y = f(x)$	(x, y)
0	16	$(0, 16)$
1	9	$(1, 9)$
2	4	$(2, 4)$
3	1	$(3, 1)$
4	0	$(4, 0)$

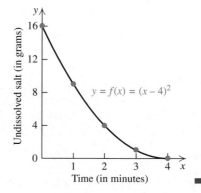

FIGURE 7
Relationship of Time and Undissolved Salt

We note that more suggestive letters than f may be used to denote functions. In Example 5, the notation $y = U(x)$ or $y = S(x)$, for instance, might have been used to denote the undissolved amount of salt. Furthermore, we might want to use t, rather than x, for time. To avoid confusion, be sure to differentiate between the letter denoting the function or rule, say, f, and the functional value, $f(x)$.

As demonstrated in Figures 6 and 7, it is not necessary to use the same scale on the horizontal and vertical axes when graphing functions.

Exercises 1.1

In Exercises 1 through 8, plot the given points on the same set of axes in the rectangular coordinate system.

1. $(3, 5), (6, 1), (-2, -1), (-1, -2)$

2. $(4, 4), (0, 6), (0, -2), (-2, 0)$

3. $(5, 0), (-7, 1), (4, -3)$

4. $(6, 0), (0, 6), (-2, 5), (4, 2)$

5. $(-\frac{1}{2}, 5), (4.5, 2.5), (-1, -4), (0, 0.5)$

6. $(3, -5), (\frac{3}{4}, 4), (-3, -2), (0, 5)$

7. $(4, 0), (4, \frac{1}{2}), (0, 0), (4, -0.5)$

8. $(-3.5, 0.5), (-2, -3), (0, 4.5), (3, -4)$

In Exercises 9 and 10, write the coordinates of each point shown.

9.   **10.**

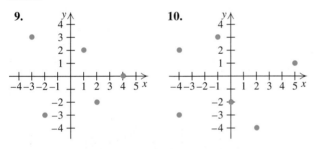

In Exercises 11 through 17, the data were taken from a U.S. Census Bureau publication.* Plot the data in the rectangular coordinate system, and connect the points with line segments.

11. **Exports** Dollar amounts (in billions) of goods and services exported from the United States:

Year	Dollars (billions)
1997	935.0
1998	932.7
1999	957.4
2000	1065.7

*SOURCE: U.S. Bureau of the Census: Statistical Abstract of the United States, 2000.

12. **Civilian Labor Forces** A sample from the last 30 years of the female and male civilian labor forces (in millions) in the United States (ages are over 16 years):

Female (millions)	Male (millions)
31.5	51.2
45.5	61.5
56.8	69
60.9	71.4
64.9	74.5

13. **Gestation Periods versus Life Span** The gestation period (days) and the average life span (years) for several large mammals:

Animal	Gestation (days)	Life Span (years)
Bison	285	15
Elephant (African)	660	35
Hippopotamus	238	41
Moose	240	12
Rhinoceros (Black)	450	15
Tiger	105	16

SOURCE: *The World Almanac and Book of Facts 2000,* World Almanac Education Group, Inc.

14. Newspapers A sample from the last 20 years of the morning and evening daily English newspapers in the United States:

Morning	Evening
387	1388
482	1220
559	1084
656	801
736	760

15. Smokers The percentage of the U.S. population 18 years or older and who smoke cigarettes:

Year	Smokers (percent)
1965	42.2
1974	37.2
1979	33.5
1985	30.1
1990	25.3
1995	24.7
1998	24.1

16. College Enrollment Percentage of the U.S. high school graduates who enrolled in college:

Year	Enrollees (percent)
1989	59.6
1990	59.9
1991	62.4
1992	61.7
1993	62.6
1994	61.9
1995	61.9
1996	65.0
1997	67.0
1998	65.0

17. Household Income The median income per U.S. household:

Year	Median Income (dollars)	Year	Median Income (dollars)
1987	26,061	1993	31,241
1988	27,225	1994	32,264
1989	28,906	1995	34,076
1990	29,943	1996	35,492
1991	30,126	1997	37,005
1992	30,636	1998	38,885

In Exercises 18 through 25, use the plotting method to graph each of the given equations of a straight line.

18. $y = 2x - 3$ **19.** $y = x - 4$

20. $y = -2x$ **21.** $y = 5x$

22. $2x - y = 6$ **23.** $2x + y = 7$

24. $x + 2y = 4$ **25.** $-x + y = 5$

In Exercises 26 through 31, use the plotting method to graph each of the given functions.

26. $f(x) = 3x - 1$ **27.** $f(x) = -2x$

28. $f(t) = -t + 4$ **29.** $P(t) = t^2 - 3$

30. $Q(p) = 2p - 4$ **31.** $Z(p) = p - 1$

In Exercises 32 through 37, use a graphing calculator to graph each of the given functions.

32. $y = x^2$ **33.** $y = x^2 + 1$

34. $y = x^2 - 2$ **35.** $y = x - 6$

36. $y = x^3$ **37.** $y = (x - 1)^3$

38. A certain pollutant is being removed from a small lake. Let t be the number of days from the beginning of the removal process and $f(t) = 50(1 - t/120)$ be the number of parts per million of this pollutant in the water. Find the domain of the function and sketch its graph throughout the domain.

39. Let x be the number of miles, in thousands, that a particular tire is driven and $f(x) = 2(1 - x/40)$ be the depth of the tread in centimeters. Find the domain of this function and sketch its graph throughout the domain.

40. Let p be the price, in dollars, of a particular DVD and $S(p) = 25 - p^2$ be the number sold, in thousands, each month by the ABC Department Store. Find the domain of this function and sketch its graph throughout the domain.

41. Let p be the price, in hundreds of dollars, of a particular brand of stereo and $S(p) = 30(16 - p^2/4)$ be the number sold each year by Andrew's Department Store. Find the domain of this function and sketch its graph throughout the domain.

42. A particular state has a pollution control unit consisting of five inspectors who monitor municipal waste facilities. Each inspection costs the state $120 and requires one inspector to spend an entire day on an inspection. If x is the number of inspections during a given workweek and $C(x) = 500 + 120x$ is the total cost (in dollars) to the state of those inspections, find the domain of this function and sketch its graph throughout the domain.

43. The Hatha Yogurt store employs eight students, each of whom is on call to work a four-hour shift each of the five weekdays. If x is the number of student shifts needed each week and $C(x) = 80 + 40x$ is the store owner's cost (in dollars) for those x shifts, find the domain of this function and sketch its graph throughout the domain.

Section 1.2 Equations of Straight Lines

In this section, we exhibit the various forms of an equation whose graph is a line. Such equations are known as **linear equations.** The term *line* will always mean *straight line*.

All lines, except those parallel to the y-axis (vertical lines) have a **slope,** which is a numerical value that measures the "steepness" of the line. To obtain the slope of a given nonvertical line, let $P(x_1, y_1)$ and $Q(x_2, y_2)$ be two distinct points on the line. (See Figure 8.) As we move along the line from P to Q, the change in the x-coordinate is given by

$$\text{change in } x\text{-coordinate} = x_2 - x_1 = \Delta x \qquad \Delta x \text{ is read "delta x"}$$

and the corresponding change in the y-coordinate is

$$\text{change in } y\text{-coordinate} = y_2 - y_1 = \Delta y.$$

The slope of the line is then defined to be the ratio $\Delta y / \Delta x$ of these changes. A feature that distinguishes a nonvertical line from all other graphs in the plane is that the ratio $\Delta y / \Delta x$ is always the same number m, no matter which point is labeled P or Q or where the points are selected on the line. This implies that Δx and Δy may be thought of as directed changes in their respective variables: Δx positive to the right, negative to the left; Δy positive upward, negative downward. (See Figure 9.) The only note of caution needed in calculating the slope from given coordinates is that the order of subtraction in the numerator must be the same as that in the denominator:

$$m = \frac{y_2 - y_1}{x_2 - x_1} = \frac{\Delta y}{\Delta x}; \qquad m = \frac{y_1 - y_2}{x_1 - x_2} = \frac{-\Delta y}{-\Delta x} = \frac{\Delta y}{\Delta x}; \qquad m \neq \frac{y_2 - y_1}{x_1 - x_2}.$$

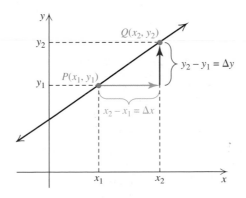

FIGURE 8
Coordinate Difference between Points

The Slope of a Line

The slope m of the nonvertical line passing through the points (x_1, y_1) and (x_2, y_2) is given by the equation

$$m = \frac{y_2 - y_1}{x_2 - x_1} = \frac{\Delta y}{\Delta x}.$$

(For a nonvertical line, $x_2 - x_1 \neq 0$.)

Since $m = \Delta y / \Delta x$, we see that if $\Delta x = +1$, then $m = \Delta y / \Delta x = \Delta y / +1 = \Delta y$. In other words, if the change in x is one unit to the right, then the change in y is precisely the slope of the line. In light of this property, the slope may be defined as the **change in y per unit change in x.**

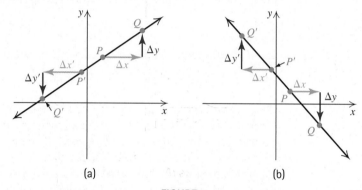

(a)   (b)

FIGURE 9
For a Particular Line, the Ratio $\Delta y / \Delta x$ Is Always the Same.

The Slope as Unit Change

For a particular line with slope m and any point P on that line, if we move to another point on the line for which $\Delta x = 1$, then $\Delta y = m$. In other words, **the slope is the change in y per unit change in x.**

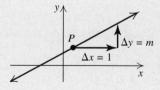

EXAMPLE 1
Finding the Slope of a
Line through Two Points

(a) Find the slope of the line through the points $(2, 4)$ and $(5, 6)$.

SOLUTION Using the definition of the slope, we find that

$$m = \frac{\Delta y}{\Delta x} = \frac{6 - 4}{5 - 2} = \frac{2}{3}. \qquad \text{Choose } (x_1, y_1) = (2, 4)$$

Observe in Figure 10 that the line rises when viewed from left to right. This is a characteristic of all lines with **positive slope.**

(b) Find the slope of the line through the points $(2, 5)$ and $(6, 3)$.

SOLUTION The definition of the slope tells us that

$$m = \frac{\Delta y}{\Delta x} = \frac{3 - 5}{6 - 2} = \frac{-2}{4} = -\frac{1}{2}.$$

Notice in Figure 11 how the line falls when viewed from left to right. This is a characteristic of all lines with **negative slope.**

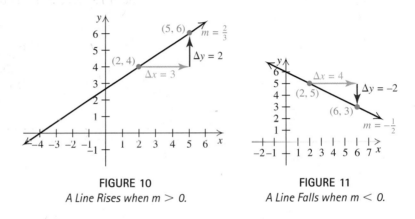

FIGURE 10
A Line Rises when m > 0.

FIGURE 11
A Line Falls when m < 0.

(c) Find the slope of the line through the points $(3, 4)$ and $(6, 4)$.

SOLUTION The slope is

$$m = \frac{\Delta y}{\Delta x} = \frac{4 - 4}{6 - 3} = 0.$$

All horizontal lines have *slope* 0, as suggested in Figure 12.

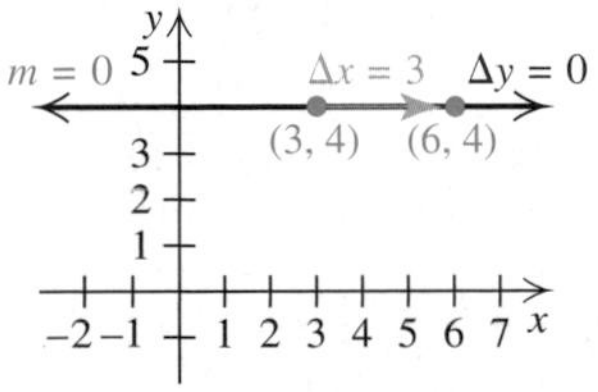

FIGURE 12
Horizontal Lines Have Slope 0. ▬

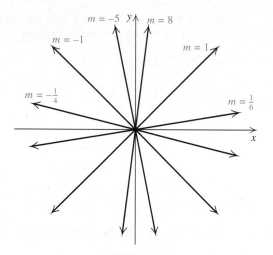

FIGURE 13
Slopes Indicate Steepness of Lines.

The slope of a line measures the steepness of the line when viewed from left to right. As shown in Figure 13, the greater the **absolute value** (i.e., its magnitude, regardless of whether it is positive or negative) of the slope, the steeper the line.

Two points on a vertical line have the same *x*-coordinate, so the calculation of *m* is not defined, because division by 0 is not defined.

Vertical and Horizontal Lines

The slope of a **vertical line** is undefined. The slope of a **horizontal line** is 0.

The **point–slope form** of the equation of a line may be found if a fixed point (x_1, y_1) on the line is given and the line has a slope *m* that is known. The equation is found by considering any point (x, y) on the line other than (x_1, y_1) and noting that

$$\frac{y - y_1}{x - x_1} = m \quad \text{or} \quad y - y_1 = m(x - x_1).$$

The Point–Slope Form of a Line

If the line has slope *m* and passes through the point (x_1, y_1), then the **point–slope form** of an equation of the line is $y - y_1 = m(x - x_1)$.

EXAMPLE 2
Using the Point–Slope
Form of a Line

Find an equation of the line through $(3, 5)$, with slope $m = \frac{3}{4}$.

SOLUTION The information needed for the point–slope form is given. Substituting $x_1 = 3$, $y_1 = 5$ and $m = \frac{3}{4}$ into this equation gives

$$y - 5 = \frac{3}{4}(x - 3)$$

$$4(y - 5) = 4 \times \frac{3}{4}(x - 3)$$ Multiply both sides by 4 to eliminate fraction

$$4y - 20 = 3(x - 3)$$
$$4y - 20 = 3x - 9$$
$$-3x + 4y = 11.$$ ■

EXAMPLE 3

Using the Point–Slope Form of a Line

Write an equation of the line through the points $(2, 3)$ and $(4, 7)$.

SOLUTION First, we find the slope:

$$m = \frac{7 - 3}{4 - 2} = \frac{4}{2} = 2.$$

Second, choose either of the given points, since both will give the same equation. With $(2, 3)$, the point–slope form of a line gives

$$y - 3 = 2(x - 2)$$
$$y - 3 = 2x - 4$$
$$y = 2x - 1.$$

The reader should verify that using the point $(4, 7)$ gives the same equation. (See Figure 14.)

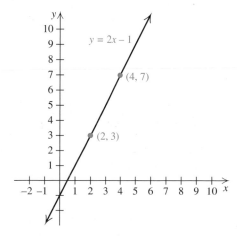

FIGURE 14
An Equation of a Line through Two Points ■

The intended use of the equation of a line often dictates the equation's final appearance. Sometimes it might be more advantageous to write the final form solved for y, as was done in Example 3: $y = 2x - 1$. At other times, the preferred appearance might be to put the terms involving x and y on the left and the constant on the right, as was done in Example 2: $-3x + 4y = 11$. Often, the final choice is a matter of taste. Nonetheless, different forms of an equation are equivalent to each other, and each is a special case of the **general form** $Ax + By = C$, where A and B are not both zero. The equations $-3x + 4y = 11$ and $3x - 4y = -11$ are both in general form. This shows that the general form is not unique.

The General Form of an Equation of a Line

An equation of the form

$$Ax + By = C$$

in which A, B, and C are constants and A and B are not both zero is called the **general form** of an equation of the line.

If the graph of a function intersects the y-axis at a point $(0, b)$, the number b is called the **y-intercept.** Similarly, if the graph of a function intersects the x-axis at a point $(a, 0)$, the number a is known as the **x-intercept.** The **slope–intercept form** of a line may be found if we know the slope m of the line and the y-intercept b of the graph of the line. This form of an equation may be derived from the point–slope form, since a point $(0, b)$ on the line is known:

$$y - b = m(x - 0)$$
$$y - b = mx$$
$$y = mx + b. \qquad \text{The slope–intercept form}$$

The Slope–Intercept Form of a Line

If the slope of a line is m and the y-intercept is b, then the **slope–intercept form** of an equation of a line is $y = mx + b$.

EXAMPLE 4
Using the Slope-Intercept Form of a Line

Find an equation of the line with y-intercept $b = 3$ and slope $m = 5$, and draw the graph of the equation.

SOLUTION The given information readily allows us to write the equation in the slope–intercept form:

$$y = mx + b$$
$$y = 5x + 3.$$

When $x = 0$, the equation shows that $y = 3$, which verifies that the point $(0, 3)$ is on the graph. At the point $(0, 3)$, if the x-coordinate is changed 1 unit to the right, the y-coordinate will change 5 units upward to return to the line, because the slope is 5. These two facts may be used to construct the graph of $y = 5x + 3$, as shown in Figure 15.

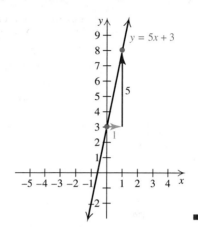

FIGURE 15
*Graphing by Means of the Slope and
the y-Intercept*

EXAMPLE 5

*Finding the Slope and
y-Intercept of a Line*

Find the slope and y-intercept of the line whose equation is $2x + 5y = 7$.

SOLUTION Solving this equation for y will give the slope–intercept form of the line:

$$5y = -2x + 7$$
$$y = -\frac{2}{5}x + \frac{7}{5}.$$

Now we can see that the slope is $-\frac{2}{5}$ and the y-intercept is $\frac{7}{5}$. ■

EXAMPLE 6

*Finding the x- and
y-Intercepts of a Line*

Find the x- and y-intercepts of the line whose equation is $4x - 3y = 6$.

SOLUTION The line will intersect the y-axis at the point on its graph whose x-coordinate is 0. Substituting $x = 0$ into the given equation results in $-3y = 6$, or $y = -2$. This tells us that the y-intercept of the line is -2, as shown in Figure 16.

The line will intersect the x-axis at the point on its graph whose y-coordinate is 0. If we now let $y = 0$ in the given equation, the result is $4x = 6$, or $x = \frac{6}{4} = \frac{3}{2}$. It follows that the x-intercept is $\frac{3}{2}$, as shown in Figure 16.

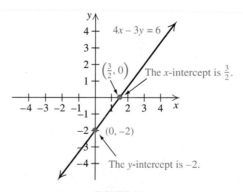

FIGURE 16
Finding the Intercepts of a Line ■

Summarizing the process used to find the intercepts, we see that

to find the y-intercept, let $x = 0$ and solve for y, and

to find the x-intercept, let $y = 0$ and solve for x.

Two lines in the same plane are parallel if they never meet, however far they are extended. The definition of the slope of a line then implies that two distinct lines are parallel only when they have the same slope or when they both have no slope (i.e., they are both vertical). Conversely, two distinct lines have the same slope only when they are parallel.

EXAMPLE 7

Parallel Lines

Find an equation of the line that contains the point $(4, 6)$ and is parallel to the line given by $x - 2y = -4$. Write the answer in general form.

SOLUTION The line we seek must have the same slope as the line given by $x - 2y = -4$. (See Figure 17.) Solving this equation for y gives

$$-2y = -x - 4, \quad \text{or} \quad y = \frac{1}{2}x + 2,$$

from which we conclude that the slope is $\frac{1}{2}$. The point–slope form may now be used to write the desired equation:

$$y - 6 = \frac{1}{2}(x - 4)$$

$$2y - 12 = x - 4 \qquad \text{Multiply both sides by 2}$$

$$-x + 2y = 8 \quad \text{or} \quad x - 2y = -8.$$

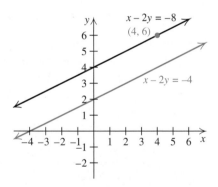

FIGURE 17

The Equation of a Line Parallel to a Given Line

We have already seen that a horizontal line has slope 0. If such a line passes through the point (h, k), the equation of the line is

$$y - k = 0(x - h)$$

$$y - k = 0$$

$$y = k.$$

In other words, the defining property of a horizontal line is that every point on the line has the same y-coordinate.

Even though the slope of a vertical line is undefined, every vertical line has an equation. A vertical line passing through the point (h, k) has the defining property that every point on that line has x-coordinate h. Therefore, the line can be described by the equation $x = h$.

EXAMPLE 8

Equations of Horizontal and Vertical Lines

(a) Find an equation of the horizontal line through the point $(6, -4)$.

SOLUTION The equation is $y = -4$. [See Figure 18(a).]

(b) Find an equation of the vertical line through the point $(7, 4)$.

SOLUTION The equation is $x = 7$. [See Figure 18(b).]

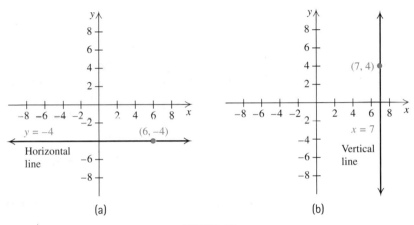

(a)   (b)

FIGURE 18
Equations of Horizontal and Vertical Lines

A summary of the different forms of linear equations is given next.

Various Forms of Linear Equations

Description	Equation
Point–slope form	$y - y_1 = m(x - x_1)$
Slope–intercept form	$y = mx + b$
General form	$Ax + By = C$
Horizontal line through (h, k)	$y = k$
Vertical line through (h, k)	$x = h$

An equation of a line that is not vertical can be expressed in any one of three forms: point–slope, slope–intercept, and general. In particular, the slope–intercept form $y = mx + b$ expresses y as a function of x. Such functions are called **linear.**

Linear Functions

A function is **linear** if it can be written in the form $f(x) = mx + b$, where m and b are real numbers.

EXAMPLE 9

A Linear Cost Function

Suppose that the cost (in dollars) of making x ladies' blouses of a certain brand is given by the linear function $C(x) = 15x + 1800$.

(a) What is the cost of making 80 of these blouses?

SOLUTION The cost is found by replacing x with 80 in the function $C(x) = 15x + 1800$. This gives

$$C(80) = 15 \times 80 + 1800$$
$$= 1200 + 1800$$
$$= 3000.$$

The cost of making 80 blouses is $3000.

(b) If $7800 has been spent making the blouses, how many blouses were made?

SOLUTION We are given that $C(x) = 7800$, and we are to find x. Therefore,

$$7800 = 15x + 1800$$
$$6000 = 15x$$
$$x = 400.$$

If 400 blouses are made, $7800 will be the incurred cost.

(c) Graph the function $C(x)$.

SOLUTION We may visualize the graph as ordered pairs of the form $(x, C(x))$, or we may simply set $y = C(x)$ and consider the graph as the usual ordered pairs (x, y), where

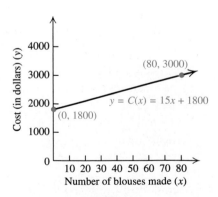

FIGURE 19
Costs for Making Ladies' Blouses

$y = 15x + 1800$. In any event, we need only two such pairs to determine the graph that is a line, as shown in Figure 19. An easy one is found by letting $x = 0$, which gives the pair $(0, 1800)$. The pair $(80, 3000)$ was found in Example 9(a). Note that x cannot be negative. ∎

Linear functions are studied more extensively in the remainder of this chapter.

Exercises 1.2

In Exercises 1 through 10,
(a) Find the slope, if it exists, of the line through the given points.
(b) Check whether your answer in part (a) is reasonable by graphing the points and the associated line.

1. $(4, 5), (6, -1)$
2. $(3, -1), (-2, -1)$
3. $(\frac{1}{3}, 5), (-1, 5)$
4. $(0, 0), (-2, -1)$
5. $(\frac{7}{8}, 3.7), (1, 5.8)$
6. $(1.6, 250), (2.9, 123)$
7. $(3, -2), (3, 5)$
8. $(6, 3), (7, 8)$
9. $(1, 1), (3, 3)$
10. $(-4, 2), (6, 2)$

In Exercises 11 through 21, do the following with the given properties:
(a) Find both the slope–intercept form and a general form of the equation of the line.
(b) Graph the line.

11. Through the point $(2, 5)$ with slope -2
12. Through the point $(-4, 6)$ with slope 4
13. Through the point $(0, 5.7)$ with slope $\frac{2}{3}$
14. Through the point $(4, 0)$ with slope -12
15. Through the points $(3, 4)$ and $(6, -1)$
16. Through the points $(\frac{1}{2}, 4)$ and $(-3, 2)$
17. Through the points $(0.3, 1)$ and $(-2, 3)$
18. Through the points $(10, 500)$ and $(20, 700)$
19. Through the points $(17, 300)$ and $(23, 150)$
20. Having x-intercept 3 and y-intercept -2
21. Having x-intercept 1.6 and y-intercept 4.3

In Exercises 22 through 25, find the equations of both the vertical and the horizontal lines through the given point.

22. $(3, -6)$
23. $(-\frac{1}{2}, 4)$
24. $(-5.7, 200)$
25. $(1.2, -8.6)$

In Exercises 26 through 31, write equations of three different lines, each of which is parallel to the given line, and graph all four lines on the same set of axes.

26. $y = 5x$
27. $4x - 7y = 6$
28. $y = -3x + 7$
29. $x = 5$
30. $y = 7$
31. $-2x + 4y = -3$

In each of Exercises 32 through 35, find an equation of the line that passes through the given point and is parallel to the line for which the equation is given.

32. $(3, 5); x - 2y = 6$
33. $(1.3, 2); y = -3x - 7$
34. $(\frac{1}{2}, -3); 2x - 4y = 7$
35. $(0, -3.5); 2x + 3y = 6$

In Exercises 36 through 41, do the following:
(a) Write equations of three different lines, each of which has the same y-intercept as the given line.
(b) Write equations of three lines, each of which has the same x-intercept as the given line.

36. $y = 2x + 2$
37. $y = -x - 4$
38. $3x + 2y = 4$
39. $x - 2y = 5$
40. $y = 1.3x - 0.6$
41. $7x + 6y = 0$

For each of Exercises 42 through 48, answer the following questions: If (x, y) is a point on the graph of the line and
(a) $\Delta x = 1$
(b) $\Delta x = 2$
(c) $\Delta x = 5$
find Δy.

42. $y = 5x + 1$
43. $y = 3x - 4$
44. $y = -6x + 8$
45. $y = -2x - 5$
46. $x + 2y = 8$
47. $2x + 5y = 7$
48. $x - 2y = 4$

49. **Number of Subscribers** During its first few weeks in operation, a new "advertisements-only" weekly paper described its number of subscribers with the function $S(t) = 800t + 6000$, where t represents the number of weeks in operation.

    (a) How many subscribers did the paper have after 5 weeks of operation? after 20 weeks?

    (b) Are subscriptions increasing each week? If so, by how many?

(c) State how the slope and *y*-intercept of the graph of $y = S(t)$ is related to the number of subscribers.

(d) Graph the given subscriber equation $y = S(t)$ to visually show the number of subscribers from the start of operations forward.

(e) How many weeks will it take for subscriptions to reach 18,000?

50. Sales Sales by the Haworth Motorcycle Company are described by the function $S(t) = 120t + 600$, where *t* is time in years, measured from $t = 0$ as the year 1995, and sales are in thousands of dollars.

(a) What were sales in 1995? in 2000? What is the estimate for sales in 2010?

(b) Are sales increasing each year? If so, by how much?

(c) State how the slope and the *y*-intercept of the graph of $y = S(t)$ are related to sales.

(d) Graph the given sales equation $y = S(t)$ to visually show sales from 1995 forward.

(e) In what year does the graph show that sales will reach $3 million?

51. Population A county agent takes samples of the grasshopper population in late summer and concludes that the population is approximated by the function $P(t) = -2t + 100$, where *t* is measured in days, with $t = 0$ being August 31, and $P(t)$ represents thousands of grasshoppers per acre.

(a) What is the predicted grasshopper population on September 15? on September 20?

(b) Is the grasshopper population increasing or decreasing each day? by how many?

(c) State how the slope and the *y*-intercept of the graph of $y = P(t)$ are related to the number of grasshoppers.

(d) Graph the equation $y = P(t)$ for all applicable values of *t*.

(e) How many days will it take for the grasshopper population to reach 20,000 per acre?

52. Monitoring Pollution The periodic monitoring of a particular water pollutant in Randleman Lake led to the function $P(t) = -800t + 18,000$, where *t* is time, measured from $t = 0$ as the year 1997, and the pollutant is measured in parts per million.

(a) How many parts per million of this pollutant were in the water in 1999? in 2000? What is the estimate for 2006?

(b) Is the pollutant decreasing each year? If so, by how much?

(c) State how the slope and the *y*-intercept of the graph of $y = P(t)$ is related to the pollutant.

(d) Graph the equation $y = P(t)$ for all applicable values of *t*.

(e) According to this model, when will the pollutant be completely removed from the water?

53. Temperature Scales The Fahrenheit and Celsius temperature scales are related by the equation $C = \left(\frac{5}{9}\right)F - \left(\frac{160}{9}\right)$, in which Celsius temperatures are given as a function of Fahrenheit temperatures.

(a) For each rise of 1 degree in Fahrenheit temperature, what is the corresponding rise in the Celsius temperature?

(b) If the Fahrenheit temperature is 80 degrees, what is the Celsius temperature?

(c) Solve the given equation for *F* to obtain an equation that gives the Fahrenheit temperature as a function of the Celsius temperature.

(d) For each rise of 1 degree in the Celsius temperature, what is the corresponding rise in the Fahrenheit temperature?

54. Tire Tread Depth The maker of brand X tires claims that, for well-maintained cars, the tread wear on brand X tires is a function of the number of miles driven and that the function is $T(x) = -x/80,000 + 1/2$, where *x* is the number of miles the tire has been driven and $T(x)$ is given in inches.

(a) Graph the equation $y = T(x)$ for all applicable values of *x*.

(b) What is the tread depth on a new tire?

(c) How many miles of wear will eliminate the tread on one brand X tire?

55. Determining Height Anthropologists have found that the height of a human can be found from a linear function if the length *x*, in centimeters, of the bone from the elbow to the shoulder is known. The linear function is

$M(x) = 2.9x + 70.6$ for a male and $W(x) = 2.8x + 71.5$ for a female.

(a) Graph $y = M(x)$.

(b) If the bone in question measures 46 centimeters for a male, what is the height of that person?

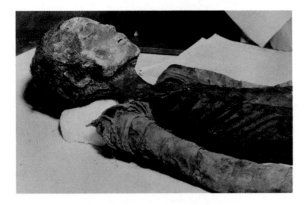

(c) If a female is 180 centimeters tall, what is the length of the bone in question?

56. Median Age From 1970 through 1998, the median age of U.S. men at their first marriage is given by the linear function $A(x) = 0.125x + 21.6$, where x is the number of years beyond 1970.

(a) Graph the given function.

(b) What was the median age of first-married U.S. men in 1982?

(c) When will the median age of first-married U.S. men reach 30 years? 40 years? Do you think that this is a realistic model over a long period of time?

57. Sales A study of records by the Ace Cola Company determined that the number of cases (in thousands) of a particular soft drink that it sells could be estimated by the linear function $S(x) = 0.91x + 28.96$, where x is the number of years past 1995.

(a) Estimate sales for 2006.

(b) In what year should sales be 50,000 cases if this trend is continued?

(c) Graph the given function.

58. Depreciating a Car The "blue-book" value of a particular make of car two years old or older is approximated by the linear function $V(x) = -1333x + 12,666$, where $V(x)$ is the value in dollars when x is the age of the car in years, for $x \geq 2$.

(a) Graph the function.

(b) If you bought a four-year-old car of this make, how much should you expect to pay?

(c) How old will this make of car be before its value is $8000?

Section 1.3 Linear Modeling

When we can describe a phenomenon (e.g., in the physical, business, social, or educational world) with a function, we say that we have constructed a **mathematical model** of that phenomenon. Sometimes these models are exact; at other times they are rather crude approximations. There are many real-world situations in which the function is linear; in those cases, we call the model a **linear model.**

EXAMPLE 1

Modeling a Fish Population

A study of the fish population within a mile of a chemical plant on Spring River showed that when 10 tons per day of pollutants were dumped into the river, the fish population was 30,000. A few years later, when 20 tons per day were being dumped into the river, the fish population had declined to 24,000. In solving the problems presented in (a) and (b), assume a linear relationship between the number of tons of pollutants and the number of fish.

(a) Find an equation that will predict the fish population for any number of tons of pollutants being dumped into the river.

SOLUTION The assumption of a **linear relationship** means that the graph of the equation relating the variables will be a line. For such a relationship, a good rule is to let y be the variable to be predicted—in this case, the number of fish; x will represent the number of tons of pollutants. With this notation, the given information then tells us two points in the rectangular coordinate system: $(10, 30{,}000)$ and $(20, 24{,}000)$. The requested equation will be that of the line through these two points. First, the slope of this line is

$$m = \frac{24{,}000 - 30{,}000}{20 - 10} = -600.$$

The slope represents the *average decline in number* of fish *per ton* of pollutant introduced into the river, as indicated in Figure 20. Now, if we choose the point $(10, 30{,}000)$ and use $m = -600$, the point–slope formula gives the relationship we seek upon solving for y:

$$y - 30{,}000 = -600(x - 10)$$
$$y - 30{,}000 = -600x + 6000$$
$$y = -600x + 36{,}000.$$

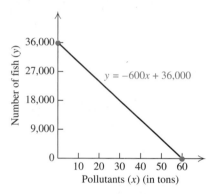

FIGURE 20
Relationship of Pollutants to
Fish Population

The equation $y = -600x + 36{,}000$ gives y (the fish population) as a linear function of x (tons of pollutants).

(b) Use the function $y = -600x + 36{,}000$ to predict the fish population if 25 tons of pollutants per day are dumped into the river.

SOLUTION The answer is

$$y = -600(25) + 36{,}000 = 21{,}000 \text{ fish.}$$

At 25 tons of pollutants per day, the fish population is estimated to be 21,000. ■

EXAMPLE 2
The Price of
Single-Family Homes

According to U.S. Census Bureau data, the median price of a new, privately owned single-family home in the United States was $84,300 in 1985 and has been increasing rather linearly to $159,800 in 1999. (See Figure 21.)

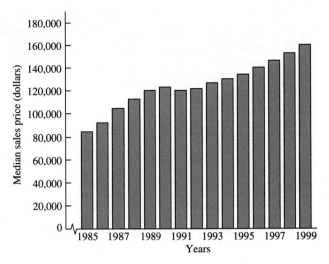

FIGURE 21
Median Sales Price for New Private Homes
SOURCE: U.S. Bureau of the Census:
Statistical Abstract of the United States, 2000.

(a) Use these data to find the average increase in median price per year from 1985 to 1999.

SOLUTION The average is found by calculating the difference in the two median prices and then dividing by the number of years between 1985 and 1999:

$$\frac{159{,}800 - 84{,}300}{1999 - 1985} = \frac{75{,}500}{14} \approx \$5392.$$

(b) Using the slope from part (a) and the data for 1985, find a linear function that will approximate the median price of a new home throughout the period.

SOLUTION Let x be the number of years from 1985 forward, with $x = 0$ representing 1985. Let y be the median cost of housing. Since the slope measures the change in y (housing costs) per unit (year) change in x, part (a) tells us that the slope of our linear function is $5392. Furthermore, when $x = 0$, $y = \$84{,}300$ (the cost in 1985). The slope–intercept form of a line can then be applied to give the linear function $y = 5392x + 84{,}300$. ▬

Linear Cost Models

In the short run, the costs involved in manufacturing a product can be broken down into **fixed costs** and **variable costs.** As the name implies, fixed costs do not change and might include such things as design costs, production machinery costs, and building costs. The firm pays these costs even if it produces no output. Variable costs, on the other hand, are costs that vary according to the production level. They include such things as the cost of labor, raw-material costs, and maintenance costs. The short-run **total costs** are then

Total Costs = Variable Costs + Fixed Costs.

In certain relatively simple situations, short-run variable costs may be calculated by multiplying the **direct cost per item produced,** considered to be the cost of labor and raw materials, by the number of items produced. For example, if the direct costs per item were $50 and x items were produced, the variable costs would be $50x$ dollars. We confine our attention to variable costs such as these because they lead to a linear cost function:

$$\text{Total} = (\text{Direct Cost per item}) \cdot x + \text{Fixed Costs}.$$

EXAMPLE 3

A Linear Cost Function

A new graphing calculator is being readied for production. The company has fixed costs of $80,000, and company officials estimate that direct costs will be $60 per calculator. Find a linear cost function that will give the total costs for producing x calculators.

SOLUTION The company has an initial investment (fixed cost) of $80,000 before production starts. Each calculator manufactured adds $60 to the company costs, so that, upon making x calculators, a variable cost of $60x$ dollars is added to the fixed cost the company has incurred. If $C(x)$ represents the total production cost to the company for making x calculators, then $C(x) = 60x + 80,000$ dollars. ■

At any given production level, knowledge of the extra production costs incurred when changing that level is vital in making management decisions. As a measure of those extra costs, economists define the **marginal cost** as the *change* in total costs associated with producing one additional unit. In Example 3, for instance, if 100 calculators have been made, the marginal cost is $60, because the change in total costs that results from making the 101st calculator is

$$C(101) - C(100) = 60 \times 101 + 80,000 - [60 \times 100 + 80,000]$$
$$= 6060 + 80,000 - 6000 - 80,000$$
$$= 60.$$

Be careful with the negation

This result should come as no surprise: The slope of the graph of any linear function gives the vertical change in the graph per unit change in x; thus, **the value of the slope is always the marginal cost.** Figure 22 gives a visual picture.

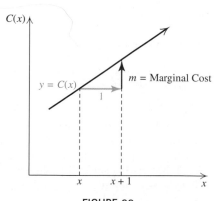

FIGURE 22
Additional Cost of Making Another Item

Linear Cost Functions

A **linear cost function** has the form

$$C(x) = mx + b,$$

where the slope m is known as the **marginal cost** as well as the **direct cost** per item, mx is the **variable cost,** and b is the **fixed cost.**

EXAMPLE 4
Marginal Cost versus
Average Cost

A car rental firm charges $50 per day plus 30 cents per mile to rent a particular make of automobile.

(a) Find a linear cost function that will give the total rental cost per day as a function of the number of miles driven.

SOLUTION The fixed costs are $50, while the direct costs are 30 cents, or $0.30, per mile. If we let x be the number of miles driven during the day, the cost per day for renting this car is $C(x) = 0.30x + 50$ dollars.

(b) If the car has been driven 100 miles, what is the total rental cost?

SOLUTION The total cost is $C(100) = 0.30 \times 100 + 50 = 80$. The total rental cost of driving the car 100 miles is $80.

(c) If 1 more mile is driven, what is the additional rental cost?

SOLUTION The additional cost is the marginal cost, 30 cents.

(d) If the car has been driven 150 miles, what is the additional rental cost for driving 1 more mile?

SOLUTION Again, the answer is 30 cents, the marginal cost.

(e) What is the average rental cost per mile for driving the car 100 miles?

SOLUTION The average rental cost is the total rental cost of driving 100 miles, divided by 100, That is, $C(100)/100 = (0.30 \times 100 + 50)/100 = 80/100 = 0.80$. The average rental cost per mile of driving the car 100 miles is 80 cents per mile. ■

The Average Cost

If $C(x)$ is a cost function, the **average cost** per item for x items is given by

$$\frac{C(x)}{x}.$$

EXAMPLE 5
Finding a Linear Cost
Function, Based on
Limited Information

Suppose that a particular manufacturing process has fixed costs of $2000 and it is known that 50 items cost a total of $6000. Find a linear cost function that models the cost for making x items.

SOLUTION Because the fixed cost is $2000 and the function must be linear, the function must be of the form $C(x) = mx + 2000$, where m is yet to be found. However, we are given that $C(50) = \$6000$, so, upon substituting 50 for x in the equation $C(x) = mx + 2000$, we get

$$C(50) = m \times 50 + 2000$$
$$6000 = 50m + 2000,$$

so that $50m = 4000$, or $m = 80$. Therefore, the function we seek is $C(x) = 80x + 2000$.

To check the validity of the function, substitute $x = 0$ and $x = 50$ into the equation to see whether 2000 and 6000, respectively, are obtained. This is indeed the case, so we can be confident that our function is a reasonable model. ■

An analysis similar to that presented in Example 5 may be done to find the fixed cost when the marginal cost is given instead.

Straight-Line Depreciation

As machines and equipment wear out or become obsolete, businesses take into account the value lost each year over the useful lives of these items. This lost value is called **depreciation.** Depreciation may be calculated in several ways, in accordance with standard accounting procedures and tax laws. One widely used method, and perhaps the simplest, is that of **straight-line depreciation.** Straight-line depreciation assumes that the amount depreciated each year is always the same (i.e., a graph of the value of the object being depreciated is a straight line).

EXAMPLE 6

Straight-Line Depreciation

Kulbeth's cabinet shop bought a new saw for $6000. The shop estimates that the useful life of the saw will be 10 years, at which time its **salvage value** will be $1000. The difference between the purchase price and the salvage value, called the **net cost,** is the *amount to be depreciated:*

$$\$6000 - \$1000 = \$5000.$$

The straight-line method of depreciation assumes that, over a given period, the net cost will be depreciated the *same amount* each year. Therefore, the **annual depreciation** is

$$\frac{\text{net cost}}{\text{useful life}} = \frac{\$5000}{10} = \$500 \text{ per year.}$$

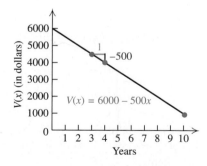

FIGURE 23
Straight-Line Depreciation

If x represents the number of years after the purchase of the saw, the **book value** (the purchase price minus the total depreciation) of the saw is

$$V(x) = 6000 - 500x.$$

Figure 23 illustrates this straight-line depreciation. Notice that the line stops at the point $(10, 1000)$. ■

Straight-Line Depreciation

The **salvage value** of an item is the value, if any, at the end of the item's useful life. The **net cost** of an item is the difference between the purchase price and the salvage value. If N is the net cost of an item, n years is the item's useful life, and D is the item's annual depreciation, then $D = N/n$ is depreciated *each year* under the **straight-line method.** If P is the purchase price, then, after x years, the value V of the item is given by the equation $V(x) = P - Dx$.

Supply and Demand

Economists assert that, all other things being equal, the *supply* of a given item or commodity is a function of its price. In particular, it is assumed that, as the price increases, producers will be willing to supply more of the item, so that the supply increases as well. Just the opposite is true of *demand:* As the price increases, fewer people will want to buy the item. In this text, we are particularly interested in situations in which both the supply and the demand are modeled by functions that are linear.

EXAMPLE 7

Linear Supply and Demand Functions

The marketing department of SuperMart stores estimates that the supply, in dozens, of a particular line of ladies' belts over the next year is given by

$$S(p) = 4p,$$

where p is the price per belt, in dollars. The demand D, in dozens, is estimated to be

$$D(p) = 140 - 10p.$$

(a) What is the demand if the belts are priced at $6 each?

SOLUTION The answer is $D(6) = 140 - 10 \times 6 = 80$ dozen.

(b) At what price will the demand become zero?

SOLUTION We set $0 = 140 - 10p$ and solve for p: $p = \$14$. ■

When the demand for a commodity exceeds the supply, a shortage occurs; if the supply exceeds the demand, a surplus occurs. Of great interest to economists is the point at which the supply equals the demand; that is, there is neither a shortage nor a surplus of the

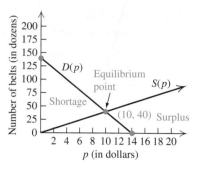

FIGURE 24
Supply and Demand Functions

commodity. When this occurs, the system is said to be in **equilibrium.** The **equilibrium point** is the point where the graphs of $y = S(p)$ and $y = D(p)$ intersect (Figure 24). We will see how to find this point in the next section.

NOTE Economists typically graph supply and demand functions so that price (p) corresponds to the vertical axis and quantity $(S(p)$ or $D(p))$ corresponds to the horizontal axis.

Exercises 1.3

In Exercises 1 through 6, let the starting number represent the number of yogurt outlets a firm owns when starting a business, followed by the resulting expansion or contraction of outlets for a number of years thereafter. Decide which of these can be represented by a linear function. For those that are linear, find the function.

1. Start with 10, and double in number for each year thereafter.

2. Start with 200, and decline by 10 each year thereafter.

3. Start with 500, and increase by 20 each year thereafter.

4. Start with 100, and decrease by half each year thereafter.

5. Start with 60, and show no change thereafter.

6. Start with 10, and increase by 20% each year thereafter.

In Exercises 7 through 12, a cost function is given. For each exercise, do the following:
(a) Find the marginal cost.
(b) Find the fixed cost.
(c) Find the total cost for 20 items.
(d) Find the average cost of producing the first 20 items, 100 items, and 200 items.

7. $C(x) = 3x + 20$

8. $C(x) = 50x + 3000$

9. $C(x) = 3.2x + 1680$

10. $C(x) = 5.8x + 2000$

11. $C(x) = 1.6x + 5000$

12. $C(x) = 200x + 10,000$

13. For the cost function $C(x) = 8x + 75$, sketch a graph of $y = C(x)/x$ and $y = $ (marginal cost) on the same set of axes, and make a comparison of the average cost per item and the marginal cost as production increases.

14. For the cost function $C(x) = 14x + 125$, sketch a graph of $y = C(x)/x$ and $y = $ (marginal cost) on the same set of axes, and make a comparison of the average cost per item and the marginal cost as production increases.

In Exercises 15 through 18, find a linear cost function, using the given information.

15. With fixed costs of $4000, 20 items cost a total of $10,000.

16. With fixed costs of $3500, 22 items cost a total of $5000.

17. With a marginal cost of $25, 40 items cost a total of $4000.

18. With a marginal cost of $80, 30 items cost a total of $6000.

In Exercises 19 through 23, use the given data to find a linear cost function that will give the total cost of producing x units of the item described.

19. Ten refrigerators cost $2000; 15 refrigerators cost $2700.

20. Twenty radios cost $600; 35 radios cost $900.

21. Fifty ladies' coats cost $1000; 60 coats cost $1200.

22. Twelve microwave ovens cost $1400; 20 ovens cost $2250.

23. Fifteen sweepers cost $900; 30 sweepers cost $1560.

24. Water Usage Data from the U.S. Census Bureau* indicated that in 1995 the average person in the United States consumed 192 gallons of water each day. Using this fact, find a linear function that will give the total number of gallons y of water used by x people.

25. Skipping Breakfast If 25% of all Americans regularly skip breakfast, find a linear function that will indicate the number of people y who skip breakfast out of a group of x people.

26. Inventory Reduction A camera store has 150 cameras in stock on January 1 and sells them at the rate of 4 per day. (Do not consider the possible arrival of new shipments of cameras.)

(a) Find a linear function that gives the number of cameras in stock on January 1 and thereafter.

(b) Graph the function found in Exercise 26(a).

(c) State how the slope and y-intercept of your graph are related to the camera inventory.

(d) How many days will it take to reduce the stock to 90 cameras?

27. Enrollment Growth State University had an enrollment of 12,000 in 2000 and has experienced a growth of 600 students per year ever since.

(a) Find a linear function that gives the enrollment of State University for 2000 and beyond.

(b) Graph the function found in Exercise 27(a).

(c) State how the slope and y-intercept of your graph are related to enrollment.

(d) How many years will it take for State's enrollment to reach 20,000 at this rate?

28. Advertising Revenue Advertising revenue for a local newspaper was $500,000 in 2000, when the city's population was 60,000. The paper estimates that advertising revenue rises $250 for each population increase of 1000.

*SOURCE: U.S. Bureau of the Census: *Statistical Abstract of the United States, 2000.*

(a) Find a linear function that will give advertising revenue as a function of the population of the city.

(b) State how the slope and the y-intercept of the graph of the function found in Exercise 28(a) are related to advertising revenue.

(c) If the population increases by 3500 from its 2000 figure, what will the expected advertising revenue be?

29. Automobile Depreciation A new car initially costs $25,000 and loses value at the rate of $3000 each year.

(a) Write a linear function that gives the value of this car in terms of the years after its initial purchase.

(b) State how the slope and the y-intercept of the function found in Exercise 29(a) are related to the value of the car.

(c) How many years will it be before the value of the car will be $5000? $1500?

(d) Graph the function found in Exercise 29(a).

30. Voter Decline In 2000, a small town had 22,000 voters. Voter apathy has created a decline of 500 voters in each election since then.

(a) Write a linear function that will give the number of voters in terms of the number of elections since 2000.

(b) How many voters were there for the sixth election after 2000?

(c) According to the data, how many elections will be held before the number of voters declines to 12,000?

31. Cost of Manufacturing The Sleepmor Manufacturing Co. makes clock radios. A study of short-run costs indicates that fixed costs are $2000, and the direct cost is $20 per clock radio.

(a) What linear cost function gives the total cost for manufacturing x clock radios? Graph this function.

(b) What is the total cost for manufacturing 150 clock radios?

(c) If 200 clock radios have been made, what is the additional cost of making the 201st clock radio?

(d) What is the average cost per clock radio for making 200 clock radios?

(e) What is the average cost per clock radio for making 500 clock radios? for making 1000? for making 2000?

32. Cost of Manufacturing A maker of fax/modems has short-run fixed costs of $75,000 and a direct cost of $80 for each fax/modem.

(a) What linear cost function gives the total cost for manufacturing x fax/modems? Graph this function.

(b) What is the total cost for manufacturing 150 fax/modems?

(c) If 150 fax/modems have been made, what is the additional cost of making the 151st clock radio?

(d) What is the average cost per clock radio for making 300 fax/modems? for 500 fax/modems?

33. **Monitoring Pollution** A state pollution control office estimates that the fixed costs incurred in monitoring waste treatment plants are $25,000 each year, and the cost of each on-site inspection is $1500.

(a) Find a linear cost function in terms of the number x of on-site inspections made each year.

(b) What is the total cost of 50 on-site inspections?

(c) What is the additional cost of making the 51st on-site inspection?

(d) What is the average cost per inspection of making 50 on-site inspections? 100 on-site inspections? 200 on-site inspections?

(e) Graph the function found in Exercise 33(a).

34. **Income Tax Bracket**

(a) What does it mean when your income puts you "in the 28% federal income tax bracket"? Suppose that your taxable income is $25,000 in a given year in which the first $17,850 is taxed at the rate of 15% and the amount over $17,850, but less than $43,150, is taxed at the rate of 28%. Accountants would then say that you are "in the 28% tax bracket" and would refer to the 28% figure as a "marginal tax rate."*

(b) Compute the total tax owed on the $25,000 income. Note that it is quite different from 28% of $25,000.

35. **Sales Income** Assume that you are offered a job paying $1500 per month plus 4% of gross sales.

(a) Find a linear function that will give your monthly income in terms of sales.

(b) Graph the function found in Exercise 35(a).

(c) If you make sales of $20,000 during a given month, what will your income be for that month?

(d) If you make sales of $10,000 at some point during the month, how much additional income will you gain if you make another sale for $1?

(e) What dollar amount in sales will yield a monthly income of $3000?

*This means that, above a certain level, 28 cents of each taxable dollars is paid as income tax.

36. **Sales Related to Unemployment** The Sharper View TV store found that the number of TV sets it sold was a function of the unemployment rate in its community. The store determined that, when the unemployment rate was 2%, it could sell about 400 sets per month and that, for each 1% gain in the unemployment rate, sales fell about 50 sets per month.

(a) Find a linear function that will give the number of sets sold each month as a function of the unemployment rate.

(b) Graph this function.

37. **Apartment Occupancy** An apartment building manager estimates that if the rental rate is $700 per month per apartment, all 80 apartments in the building will be rented. However, if the rent is raised to $800 per month, only 60 of the apartments will be rented. Assume that there is a linear relationship between rent per month and occupancy (i.e., number of apartments rented).

(a) Find a linear function that will predict the occupancy for a given monthly rent, and graph this function.

(b) Check your answer in Exercise 37(a) for accuracy.

(c) Use the function found in Exercise 37(a) to estimate the occupancy if the rent is established at $760 per month.

38. **Parking Fees** When a parking lot charged $3 per car per day, the attendants usually parked 60 cars. When the manager raised the fee to $4 per car, they parked only 50 cars per day.

(a) Find a linear function that gives the number of cars parked each day as a function of the charge per car.

(b) Use the function found in Exercise 38(a) to estimate the number of cars that will be parked if the fee is set at $3.75 per car.

(c) Use the function found in 38(a) to determine how much each car should be charged if the operators of the lot want to have 80 cars parked each day.

(d) According to the function found in 38(a), what is the minimum amount that can be charged per car so that no (zero) cars will be parked there? Do you think this amount is realistic?

(e) If parking were free in this lot, how many cars could be parked there, according to the function found in 38(a)? Do you think this number is realistic?

39. Sales Trends In 1998, SuperMart retail stores had sales of $2.7 million, and in 2000, sales were $3.6 million. Assume that the trend in sales is linear.

(a) Find an equation of a line that will estimate sales (in millions of dollars) in succeeding years.

(b) Using the line found in Exercise 39(a), predict sales in 2005.

(c) Based on the line found in Exercise 39(a), when will sales reach the $10 million mark?

(d) Do you think that the line found in Exercise 39(a) will give a good estimate of sales in the year 2010? In the year 2025?

40. Infant Mortality Rates Health statistics* show that infant mortality rates (per thousand live births) in the United States were 9.2 in 1990, 7.6 in 1996, and 7.2 in 1998.

(a) Assuming a linear decline in infant mortality rates from 1990 through 1998, find an equation of a line that will approximate the mortality rate in a given year during this period. Graph the equation of the line found.

(b) Tell how the slope of the line found in Exercise 40(a) relates to the mortality rate.

(c) Check the accuracy of your line for 1996. What do you predict that the rate will be in 2005?

41. Resource Recovery According to U.S. Census Bureau data,* the amount of paper and paperboard waste generated (in millions of tons) in the United States in 1960 was 5.4, and this amount increased in a rather linear fashion to 38.5 in 1997.

(a) Assuming the growth in waste to be linear, find an equation of a line that will approximate the amount generated for any given year from 1960 to 1997.

(b) Tell how the slope of the line found in Exercise 41(a) relates to the generation of paper and paperboard waste.

(c) The amount generated in 1996 is listed as 38.0 (million tons). How accurate is your equation at this point?

*SOURCE: U.S. Bureau of the Census: *Statistical Abstract of the United States,* 2000.

(d) What is your estimate for the amount generated in 1998? Compare this amount with the actual value reported, 38.2 million tons.

42. Dental Costs According to Social Security Administration statistics,* dental costs per person in the United States were $165 per year in 1995 and have increased rather linearly to $192 in 1998.

(a) Assuming linear growth, find an equation of a line that will approximate the cost of dental service for any year during this period. Graph the line.

(b) The actual figure for 1997 was $184. Check your equation for accuracy at this point.

(c) Relate the slope of the line found in Exercise 42(a) to the cost of dental services.

43. Solid Waste Increase The following data from the U.S. Census Bureau* show (in millions of tons) how the generation of municipal solid waste has increased:

Year	1980	1985	1989	1990	1995	1998
Gross Waste	151.5	164.4	191.4	205.2	211.4	220.2

(a) Find the average growth per year.

(b) From Exercise 43(a), find a linear function that approximates the waste (in millions of tons) generated in any given year on or beyond 1980. Check the accuracy of your function for the year 1995.

(c) Use the function found in Exercise 43(b) to approximate the number of tons of waste that will be generated in 2010.

(d) At this rate, how many years will it take before 300 million tons of waste per year will be generated?

*SOURCE: U.S. Bureau of the Census: *Statistical Abstract of the United States,* 2000.

44. Aging Population The following data from the U.S. Census Bureau* show (in millions) the number of Americans at least 85 years old:

Year	1980	1985	1988	1990
Americans age 85 and over	2.2	2.7	2.9	3.0

Year	1991	1992	1995	2000
Americans age 85 and over	3.2	3.3	3.7	4.3

(a) Assuming linear growth between 1980 and 2000, find the average growth per year (in millions) of persons age 85 and over.

(b) Use the data for the years 1980 and 2000 to obtain a linear growth function for persons age 85 and over. Check for accuracy in 1990 and 1995.

(c) Assuming that the linear trend in Exercise 44(b) continues into the future, in what year will the number of persons age 85 and over reach 6 million?

45. Depreciating a Building A building is purchased for $100,000 and is to be totally depreciated by the straight-line method over a 20-year period. (Assume no salvage value.)

(a) What is the amount to be depreciated each year?

(b) Find a linear function that expresses the book value of the building as a function of years from the date of purchase of the building.

(c) Graph the function found in Exercise 45(b).

46. Depreciating a Machine A copying machine is purchased for $12,500 and is to be totally depreciated by the straight-line method over a five-year period. (Assume no salvage value.)

(a) What is the amount to be depreciated each year?

(b) Find a linear function that expresses the book value of this machine as a function of years from the date of purchase of the machine.

(c) Graph the function found in Exercise 46(b).

47. Depreciating a Machine A particular machine costs $80,000 when purchased new and has a salvage value of $10,000 20 years later. Straight-line depreciation is

*SOURCE: U.S. Bureau of the Census: *Statistical Abstract of the United States,* 2000.

applied to the net value of the machine over a 20-year period.

(a) How much is depreciated each year?

(b) Write a linear function that represents the book value of the machine during its lifetime.

(c) Graph the function found in Exercise 47(b).

48. Depreciating a Building An apartment building is purchased for $160,000. Of this amount, $30,000 is considered the cost of the land on which the building sits. Excluding the cost of the land, the purchase is to be depreciated over a 17-year period by the straight-line method.

(a) What is the amount to be depreciated each year?

(b) Find a linear function that gives the book value of the purchase as a function of the number of years after purchase of the building.

(c) Check the accuracy of the function found in Exercise 48(b).

(d) Graph the function found in Exercise 48(b).

49. Depreciating a Building An apartment building is bought for $550,000, of which $50,000 is estimated to be the cost of the land. The total cost minus the cost of the land is to be depreciated by the straight-line method over a period of 25 years.

(a) How much is depreciated each year?

(b) Write a linear function that represents the book value of the building during the 25-year period.

(c) Graph the function found in Exercise 49(b).

50. Depreciating a Computer The Lewis Bank and Trust Co. buys a computer for $90,000 and plans to use it for eight years. The company estimates that at that time the salvage value will be $7500. On the accounting books, the company is depreciating the computer by the straight-line method over the eight-year period.

(a) What is the amount to be depreciated each year?

(b) Find a linear function that gives the book value of the computer as a function of the number of years after purchase.

(c) Check the accuracy of the function by computing the book value at the time of purchase and again after eight years.

(d) Graph the function.

For Exercises 51 and 52, answer these questions about the given supply and demand functions:

(a) *What is the demand when p = 20? Is there a surplus or a shortage at this price?*

(b) What is the price when the demand is 100? Is there a surplus or a shortage at this price?

(c) Graph $S(p)$ and $D(p)$ on the same set of axes, and identify the regions where shortages and surpluses exist.

51. $S(p) = 6p$
$D(p) = 160 - 6p$

52. $S(p) = 4p + 1$
$D(p) = 148 - 3p$

53. Demand versus Price When the price of soybeans is \$4 per bushel, the demand is 12 million bushels. When the price is \$6 per bushel, the demand is 9 million bushels.

(a) Use the preceding data to find a linear demand function for soybeans (in millions of bushels).

Use the model found in Exercise 53(a) to answer the remaining parts of this exercise.

(b) What is the demand when the price is \$5.25 per bushel?

(c) If the demand for soybeans is 7 million bushels, what is the price?

(d) If soybeans are given away, what does the model in Exercise 53(a) say that the demand will be? Do you think that this demand is realistic?

54. Demand versus Price A market research firm determines that the demand for a particular product, in terms of purchases per thousand population, is 15 when the product is priced at $p = \$20$, but only 5 when the product is priced at \$60.

(a) Find a linear demand function for this product in terms of the price p.

Use the model found in Exercise 54(a) to answer the remaining parts of this exercise.

(b) If the product were priced at \$30, what would be the demand?

(c) At what price is the demand 0?

(d) According to the linear demand function found in part (a), if the product were given away, what would the demand likely be?

Section 1.4 Two Lines: Relating the Geometry to the Equations

We have seen that every linear equation of the general form $Ax + By = C$ has a line in the plane as its graph. Any ordered pair of real numbers (x, y) that is a simultaneous solution of two or more linear equations is a **point of intersection** of the lines forming their graphs. For a pair of lines in the plane, one and only one of the following three statements is true:

I. The two lines intersect in exactly one point.
II. The two lines do not intersect at all (i.e., they are parallel).
III. The two lines are actually the same line (i.e., they have the same graph).

The objectives of this section are (a) to discover, from equations of the lines, which of the foregoing three statements apply and, if the lines intersect, (b) to find the coordinates of the point or points of intersection. This process is referred to as "solving a system of two equations in two unknowns."

The pairs of lines given by the following equations are examples of each of the classifications I, II, and III.

$$\left\{\begin{array}{r} x + 2y = 5 \\ 3x + y = 5 \end{array}\right\} \qquad \left\{\begin{array}{r} 2x - y = 3 \\ 4x - 2y = -9 \end{array}\right\} \qquad \left\{\begin{array}{r} 3x + 2y = 4 \\ 6x + 4y = 8 \end{array}\right\}$$

Example Equations (a)   Example Equations (b)   Example Equations (c)

EXAMPLE EQUATIONS (a)

I. If the slopes of the two lines are not equal, the lines must intersect in exactly one point. Example Equations (a) fall into this classification because the slope of the first line $(y = -\frac{1}{2}x + \frac{5}{2})$ is $-\frac{1}{2}$ and the slope of the second line $(y = -3x + 5)$ is -3. (See Figure 25.)

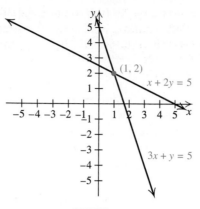

FIGURE 25
Lines That Intersect in One Point ▬

EXAMPLE EQUATIONS (b)

II. If the slopes of the two lines are equal, but the y-intercepts are different, then the two lines are parallel, but distinct. Thus, there are no points of intersection. Example Equations (b) fall into this classification because, when each equation is put into slope–intercept form, $y = 2x - 3$ and $y = 2x + \frac{9}{2}$, the slope of both lines represented is 2, but the y-intercept of the first is -3 while that of the second is $\frac{9}{2}$. (See Figure 26.) ▬

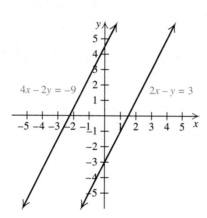

FIGURE 26
Lines That Have No Point in Common
(Parallel Lines)

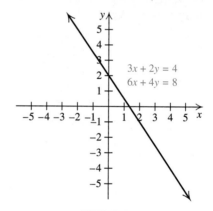

FIGURE 27
Lines That Have Every Point in Common
(Lines with the Same Graph)

EXAMPLE EQUATIONS (c)

III. If the slopes of the two lines are equal, and the lines have the same y-intercept, then the two lines are identical. Thus, every point on one line is also on the other, so there are infinitely many points of intersection. Example Equations (c) fall into this classification because the slopes of both lines represented $\left(y = -\frac{3}{2}x + 2\right)$ are $-\frac{3}{2}$ and the y-intercepts are both 2. (See Figure 27.) ▬

The general problem arising from these three cases is as follows:

> Given two equations of lines, determine whether the lines intersect at exactly one point, intersect at no point, or are the same line.

The two methods of solution considered are the **method of substitution** and the **method of elimination.** Both are demonstrated on each of the pairs in Example Equations (a), (b), and (c), to show how they apply to the three classifications I, II, and III.

The Method of Substitution

This method involves solving for one of the variables in either equation and then substituting the value found into the other equation to obtain one equation with only one unknown (often described as an "equation in one unknown").

The Method of Elimination

This method involves multiplying one of two original equations by a nonzero number to produce a third equation, which, when added to the other original equation, produces a final equation in one unknown.

EXAMPLE 1

Pairs of Lines in
Classification I
(One Point in Common)

Apply the two methods to the Example Equations (a):

$$x + 2y = 5 \tag{a.1}$$
$$3x + y = 5 \tag{a.2}$$

We already know that the two lines represented have exactly one point in common.

SOLUTION

(a) First, we apply the *method of substitution* to this system of equations in order to obtain the point of intersection. In the given system, it is easy to solve for x in the first equation, (a.1):

$$x + 2y = 5 \tag{a.1}$$
$$x = 5 - 2y. \tag{a.3}$$

Now, we substitute this value into Equation (a.2) to obtain the following equations:

$$3x + y = 5$$
$$3(5 - 2y) + y = 5 \qquad \text{Substitute } x = 5 - 2y$$
$$15 - 6y + y = 5$$
$$-5y = -10$$
$$y = 2.$$

Finally, we substitute $y = 2$ into either of the original equations or into Equation (a.3) to find the corresponding value of x:

$$x = 5 - 2y \tag{a.3}$$
$$x = 5 - 2(2) = 1.$$

Therefore, the point of intersection is $(1, 2)$. (See Figure 25.)

SOLUTION

(b) The same point of intersection may be obtained by applying the *method of elimination* to the system of equations. One option is to multiply Equation (a.1) by -3 so that when the resulting equation is added to the second equation, x is eliminated, leaving one equation in the unknown y:

$$x + 2y = 5 \qquad \textbf{(a.1)}$$
$$3x + \ y = 5 \qquad \textbf{(a.2)}$$

$-3(\text{Equation a.1})$
$$\begin{aligned} -3x - 6y &= -15 \\ 3x + \ y &= 5 \end{aligned} \qquad \textbf{(a.2)}$$
$$\begin{aligned} -5y &= -10 \\ y &= 2. \end{aligned}$$

When $y = 2$ is substituted into Equation (a.1) (either original equation will do), we get $x = 1$. Again, the point of intersection is $(1, 2)$.

An alternative process would have been to multiply Equation (a.2) by -2 and then add the result to Equation (a.1) to eliminate y and obtain one equation in the variable x. ■

We now examine how the two solution methods apply to a system in Classification II (i.e., the system has no solution).

EXAMPLE 2
Pairs of Lines in
Classification II
(Parallel Lines)

Apply the two methods to Example Equations (b):

$$2x - \ y = 3 \qquad \textbf{(b.1)}$$
$$4x - 2y = -9. \qquad \textbf{(b.2)}$$

As mentioned previously, we already know that the two lines represented have no point in common.

SOLUTION

(a) The *method of substitution* is applied first. An examination of the two given equations leads to the conclusion that it is easiest to solve Equation (b.1) for y. Taking that approach, we get

$$2x - y = 3 \qquad \textbf{(b.1)}$$
$$y = 2x - 3.$$

Upon substituting this value for y into Equation (b.2), we get

$$\begin{aligned} 4x - 2y &= -9 \qquad \textbf{(b.2)} \\ 4x - 2(2x - 3) &= -9 \qquad \text{Substitute } y = 2x - 3 \\ 4x - 4x + 6 &= -9 \\ 6 &= -9. \end{aligned}$$

The last statement is false (it is a contradiction), which is the signal that the two lines have no point of intersection. (See Figure 26.) Such a system is called **inconsistent.**

SOLUTION

(b) Second, we apply the *method of elimination* to the two given equations. If we multiply Equation (b.1) by -2 and add the results to Equation (b.2), we get

$$2x - y = 3 \qquad \textbf{(b.1)}$$
$$4x - 2y = -9 \qquad \textbf{(b.2)}$$

$-2(\text{Equation b.1}):$
$$-4x + 2y = -6$$
$$\underline{4x - 2y = -9} \qquad \textbf{(b.2)}$$
$$0 = -15.$$

Again, the last statement is a contradiction, which means that the two lines have no point of intersection. ▬

EXAMPLE 3

Pair of Lines in
Classification III
(Same Line)

Apply the two methods to Example Equations (c):

$$3x + 2y = 4 \qquad \textbf{(c.1)}$$
$$6x + 4y = 8. \qquad \textbf{(c.2)}$$

From the previous discussion, we already know that these two equations represent the same line.

SOLUTION

(a) First, we apply the *method of substitution.* Solving Equation (c.1) for y gives

$$3x + 2y = 4 \qquad \textbf{(c.1)}$$
$$2y = -3x + 4$$
$$y = -\frac{3}{2}x + 2.$$

Then, upon substitution into Equation (c.2), we get

$$6x + 4y = 8$$
$$6x + 4\left[-\frac{3}{2}x + 2\right] = 8$$
$$6x - 6x + 8 = 8$$
$$8 = 8.$$

The resulting equation, $8 = 8$, called an **identity,** is a perfectly correct mathematical statement. It does not tell us what the solutions are, but it does signal that one of the equations is unnecessary and does not contribute to the solution. (The graph shown in Figure 27 is the geometric justification for this statement.) Such a system of equations is known as a **dependent system;** in the case of two equations and two unknowns, this happens only when one equation is a multiple of the other. This means that, when graphed, the two resulting lines coincide; hence, all points on one line are also on the other. As a consequence, only one of the equations is needed to determine the form of how the infinitely many solutions appear. (Again, see Figure 27.) A common method of denoting these points of intersection is to take one of the equations, say, the first,

$$3x + 2y = 4, \tag{c.1}$$

and then solve it for one of the variables, say y, to get

$$y = -\frac{3}{2}x + 2.$$

Then the solution is written as

$$x = \text{any number}$$
$$y = -\frac{3}{2}x + 2.$$

If x is assigned any numerical value, we can solve for a corresponding y value, so that (x, y) is a solution to both of the original equations and, hence, is a point of intersection. For instance, if $x = 2$, then $y = -1$, so that $(2, -1)$ represents a point on both lines. If $x = 0$, then $y = 2$, so $(0, 2)$ represents another point of intersection, and so on. In mathematical language, the variable in the solution that can be any number (x in this case) is called a **free variable** (or a **parameter**).

SOLUTION

(b) Second, we apply the *method of elimination* to this pair of equations. If we multiply Equation (c.1) by -2 and add the results to Equation (c.2), the resulting system becomes

$$3x + 2y = 4 \tag{c.1}$$
$$6x + 4y = 8 \tag{c.2}$$

$$-2(\text{Equation c.1}): \qquad -6x - 4y = -8$$
$$\underline{6x + 4y = 8} \tag{c.2}$$
$$0 = 0.$$

This again shows that the system is dependent. This time, if the first equation is solved for x rather than y, the solutions take on the following appearance:

$$y = \text{any number}$$
$$x = -\frac{2}{3}y + \frac{4}{3}.$$

It is a matter of taste or convenience as to which variable is to be considered free. ∎

Eliminating a variable between two linear equations may always be done by multiplying one of the original equations by a particular nonzero number and then adding the resulting equation to the other original equation. This was done in Examples 1(b), 2(b), and 3(b), because the respective multipliers were not difficult to find. Example 4 shows how it may be done for *any* pair of linear equations.

EXAMPLE 4

Using the Elimination Method

Use elimination to solve:

$$2x - 3y = -16 \qquad \textbf{(1)}$$
$$5x + 4y = 29. \qquad \textbf{(2)}$$

SOLUTION After some study of these equations, we conclude that the variable x can be eliminated by multiplying Equation (1) by $-\frac{5}{2}$ and adding the result to Equation (2):

$$(-5/2) \text{ Equation (1):} \qquad -5x + \frac{15}{2}y = 40$$
$$\text{Equation (2):} \qquad \underline{5x + 4y = 29}$$
$$\frac{23}{2}y = 69.$$

To ease the burden of selecting an appropriate multiplier in cases such as this, we show how this selection may be done in a two-stage process. First, multiply Equation (1) by $\frac{1}{2}$ to make 1 the coefficient of x:

$$(1/2) \text{ Equation (1):} \qquad x - \frac{3}{2}y = -8 \qquad \textbf{(1a)}$$
$$\text{Equation (2):} \qquad 5x + 4y = 29. \qquad \textbf{(2)}$$

Now compare the coefficients of x in Equations (1a) and (2): Equation (1a) should be multiplied by -5 and then added to Equation (2) to eliminate x from the latter equation:

$$(-5) \text{ Equation (1a):} \qquad -5x + \frac{15}{2}y = 40$$
$$\text{Equation (2):} \qquad \underline{5x + 4y = 29}$$
$$\frac{23}{2}y = 69.$$

The value of y is then found to be $y = 69(\frac{2}{23}) = 3 \cdot 2 = 6$. The substitution of $y = 6$ into Equation (1), or a similar elimination of y between Equations (1) and (2), gives $x = 1$. ∎

The accuracy of points of intersection obtained by either of the preceding methods should always be checked by substituting the coordinates back into the *original equations* of the lines. If the coordinates actually represent a point on both lines, an equality should result from both equations. This applies even for equations in Classification III.

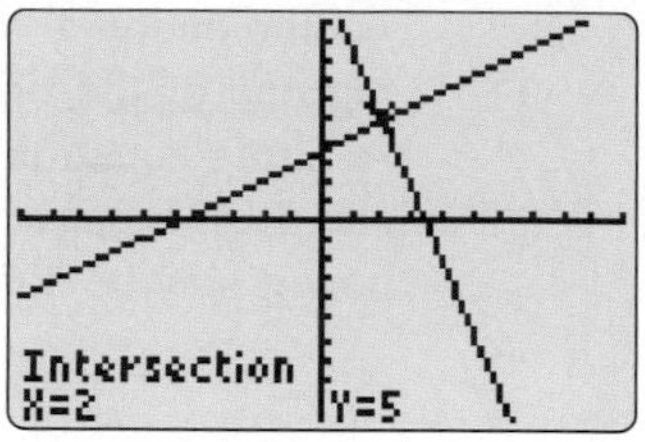

FIGURE 28
Solving a System by Means of
a Graphing Utility

Certain models of graphing calculators may be used to display not only the graphs of a pair of equations, but also any points of intersection, to several decimal places. Shown in Figure 28 is such a display for the linear equations

$$3x - 4y = -14 \quad \text{or} \quad y = \frac{3}{4}x + \frac{14}{4}$$

and

$$7x + 2y = 24 \qquad y = -\frac{7}{2}x + 12.$$

A knowledge of the three classifications of solutions we have discussed is essential in interpreting calculator and computer results, especially in cases in which either no solution exists or infinitely many solutions exist.

Break-Even Analysis

Now let's look at a real-world situation in which the point of intersection of two lines has a significant impact on business decisions. In starting any new manufacturing process, the initial costs are almost always more than the revenue derived from initial sales of the product being manufactured. An example is the introduction of a new automobile. Of great interest in such processes is the number n_0 that needs to be made so that the total revenue from sales finally matches the total costs incurred.

If a number x less than n_0 is made, the total cost $C(x)$ will be greater than the total revenue $R(x)$, so a *loss* results. By contrast, if a number x greater than n_0 is made, then the total cost $C(x)$ is less than the total revenue $R(x)$, so a gain results. When exactly n_0 are made, $R(n_0) = C(n_0)$, so that neither a gain nor a loss results; that is, we **break even.** The number n_0 is known as the **break-even quantity,** while the point where the graphs of $y = C(x)$ and $y = R(x)$ intersect is known as the **break-even point.**

The Break-Even Point

For a cost function $C(x)$ and revenue function $R(x)$, the number n_0 for which $R(n_0) = C(n_0)$ is known as the **break-even quantity,** and the point of intersection of the graphs of $y = R(x)$ and $y = C(x)$ is known as the **break-even point.**

The **profit** from a manufacturing process is related to the revenue and the cost in the following way.

Profit

Profit is defined as revenue minus cost. If $R(x)$ represents the revenue function, $C(x)$ represents the cost function, and $P(x)$ represents the profit function, then

$$P(x) = R(x) - C(x).$$

The profit is always zero at the break-even quantity n_0. The reason is that, at n_0 the revenue equals the cost. Therefore,

$$P(n_0) = R(n_0) - C(n_0) = 0.$$

EXAMPLE 5

Break-Even Analysis

The Allis Manufacturing Company makes clock radios. The company has determined that its short-run fixed costs are $2000 and its direct cost for each clock radio is $20. Suppose that x clock radios are made and each is sold for $30.

(a) Find the total cost-function.

SOLUTION The total cost function is $C(x) = 20x + 2000.$

(b) Find the revenue function.

SOLUTION If x clock radios are sold for $30 each, then $R(x) = 30x.$

(c) Find the profit function.

SOLUTION The profit is the revenue minus the cost. Therefore,

$$\begin{aligned}
P(x) = R(x) - C(x) &= 30x - [20x + 2000] \\
&= 30x - 20x - 2000 \\
&= 10x - 2000.
\end{aligned}$$

(d) Find the break-even quantity and the break-even point.

SOLUTION The break-even quantity is found by setting $R(x) = C(x)$ and solving for x:

$$\begin{aligned}
30x &= 20x + 2000 \\
10x &= 2000 \\
x &= 200.
\end{aligned}$$

The company will break even when it makes and sells 200 clock radios. We lack only the y-coordinate for the break-even point. Substituting $x = 200$ into $R(x) = 30x$ gives $R(200) = 30(200) = \$6000$. The break-even point is then $(200, 6000)$. When the company makes and sells 200 clock radios, both its revenue and its cost will be $6000 (so, its profit will be $0). (See Figure 29.)

(e) Graph $y = C(x)$, $y = R(x)$, and $y = P(x)$ on the same set of axes.

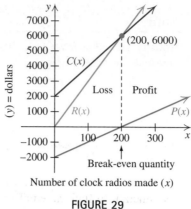

FIGURE 29
Break-Even Analysis

SOLUTION The graphs are shown in Figure 29. Notice that, at $x = 0$, there is a loss of $2000. This is the fixed cost in the clock radio operation. Also, at the break-even quantity $x = 200$, the profit is $0, as we expect. For $0 < x < 200$, the profit is negative (a loss occurs), and for $x > 200$, a positive profit is recorded (a gain occurs). ■

Equilibrium Point

In the previous section, we discussed linear supply and demand models. We noted that, at the point at which the supply of a commodity equals the demand for it, there is neither a shortage nor a surplus of the commodity, and the system is said to be in **equilibrium.** The **equilibrium point** is the point where the graphs of $y = S(p)$ and $y = D(p)$ intersect. The p-coordinate of that point is known to economists as the **equilibrium price,** and the y-coordinate is called the **equilibrium quantity.** The point can be found by solving the system of equations $y = S(p)$ and $y = D(p)$ with the method of substitution discussed in this section.

The Equilibrium Point

If $D(p)$ is a demand function and $S(p)$ is a supply function for a given commodity in terms of its price p, then the **equilibrium price** is that value p_0 of p for which

$$D(p_0) = S(p_0).$$

The **equilibrium quantity** is the value of $D(p_0)$ (or $S(p_0)$). The **equilibrium point** is the ordered pair $(p_0, D(p_0))$ (or $(p_0, S(p_0))$).

EXAMPLE 6

Finding the Equilibrium Point

The marketing department of Super Mart Stores estimates that the supply (in dozens) over the next year of a particular line of ladies' belts is given by $S(p) = 4p$, where p is the price of the belt in dollars. The demand (in dozens) is estimated to be $D(p) = 140 - 10p$.

(a) What is the equilibrium price?

SOLUTION We set $S(p) = D(p)$ and solve for p:

$$4p = 140 - 10p \qquad S(p) = 4p, D(p) = 140 - 10p$$
$$14p = 140$$
$$p = 10.$$

Supply will equal demand at the price of $10 per belt.

(b)   What is the equilibrium quantity?

SOLUTION Now that we have the equilibrium price from part (a), we can substitute it into either function to find the equilibrium quantity. The fastest way to find the equilibrium quantity in this case is to substitute $p = 10$ into the supply function $S(p)$:

$$S(10) = 4 \times 10 = 40.$$

The equilibrium quantity is then 40 dozen belts. Our results are summarized graphically in Figure 30.

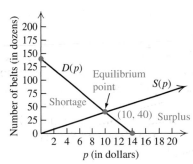

FIGURE 30
Equilibrium Price and Quantity   ▬

One of the later uses for finding a point of intersection of two lines in this text occurs when the coordinates of "corner points" of a region bounded by lines are needed. The next example illustrates this.

EXAMPLE 7

An Application of
Intersection Points

Find the coordinates of the corner points of the shaded region shown in Figure 31.

SOLUTION Of the four corner points, the three located on the coordinate axes are easy to find. Counterclockwise from the upper left, they are $(0, 3)$, $(0, 0)$, and $(2, 0)$. That leaves only the one marked A to be found. To find its coordinates, we first need the equations of the lines that intersect at point A. The line through $(0, 3)$ and $(5, 0)$ has slope

$$m = \frac{y_2 - y_1}{x_2 - x_1} = \frac{0 - 3}{5 - 0} = -\frac{3}{5},$$

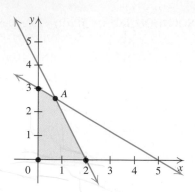

FIGURE 31
Corner Points of a Region

and since the y-intercept is 3, the slope–intercept form of a line may be used to obtain its equation:

$$y = -\frac{3}{5}x + 3. \tag{1}$$

The line through $(0, 4)$ and $(2, 0)$ has slope

$$m = \frac{0 - 4}{2 - 0} = -2,$$

and since the y-intercept is 4, the slope–intercept form of a line may again be used to obtain the equation of this line:

$$y = -2x + 4. \tag{2}$$

Substituting Equation 1 into Equation 2 results in the following series of equations:

$$-\frac{3}{5}x + 3 = -2x + 4$$

$$2x - \frac{3}{5}x = 4 - 3 \qquad \textit{Move like terms together}$$

$$\frac{7}{5}x = 1$$

$$x = \frac{5}{7}.$$

Finally, substituting this value into Equation 2 shows that $y = -2\frac{5}{7} + 4$, or $y = \frac{18}{7}$. Therefore, the point A has coordinates $\left(\frac{5}{7}, \frac{18}{7}\right)$. ▪

Even though there are other methods for solving systems of two linear equations in two unknowns (i.e., finding their point[s] of intersection), the two demonstrated in this section are among the easiest to use. Furthermore, either method may be used to solve any such system.

Exercises 1.4

In each of Exercises 1 through 6, decide, by examining the slope and y-intercept of each pair of equations, which of the classifications the resulting lines fall into: one point in common (I), no point in common (II), or all points in common (III). Do not solve any of these pairs of equations.

1. $4x + 6y = 9$
$x - 3y = 1$

2. $x + 3y = 5$
$2x + 6y = 10$

3. $x - 5y = 8$
$3x - 15y = 3$

4. $2x - y = 4$
$4x - 2y = 8$

5. $-x + 5y = 8$
$x + 5y = 3$

6. $-x + 5y = 8$
$-x + 5y = 3$

In each of Exercises 7 through 24, do the following:
(a) Determine how many solutions, if any, the system has.
(b) Find the point or points of intersection, if any exist, by an appropriate method.

7. $2x + y = 3$
$x - y = 2$

8. $x + 2y = 4$
$2x - 2y = 3$

9. $2x + y = 3$
$x + 0.5y = 1.5$

10. $x + 3y = 2$
$2x + 6y = 4$

11. $x + 2y = 0$
$x - y = 0$

12. $2x - y = 0$
$x + y = 0$

13. $x + 2y = 0$
$2x + 4y = 0$

14. $x + 2y = 0$
$2x + 4y = 1$

15. $x + y = 3$
$2x + 2y = 4$

16. $x - y = 5$
$2x - 2y = 1$

17. $0.5x = 6$
$x + 3y = 4$

18. $y = 5$
$2x - y = 6$

19. $0.5x + 2y = 3$
$x - y = 1$

20. $x + 0.6y = 0$
$x - 2y = 1$

21. $0.2x - y = 0$
$x + 0.6y = 0,$

22. $0.4x - 0.3y = 6$
$x + y = 0$

23. $x - 2y = 3$
$2x - y = 6$

24. $0.2x + 0.7y = 0$
$x - y = 0$

In Exercises 25 through 28, use a graphing calculator or computer to help solve the given systems.

25. $2x + 3y = 150$
$1.8x + 3.5y = 167$

26. $y = 3x$
$y = 3x + 0.1$

27. $0.3x - 0.4y = 0.8$
$6x - 8y = 16$

28. $0.3x - 0.4y = 16$
$x + 0.5y = 21$

In each of Exercises 29 through 34, do the following:
(a) Find the break-even quantity.
(b) Find the break-even point.
(c) Graph $R(x)$ and $C(x)$ on the same set of axes.

29. $R(x) = 50x$
$C(x) = 20x + 900$

30. $R(x) = 0.5x$
$C(x) = 0.3x + 20$

31. $R(x) = 20x$
$C(x) = 10x + 2500$

32. $R(x) = 6x + 10$
$C(x) = 2x + 90$

33. $R(x) = 15x$
$C(x) = 8x + 490$

34. $R(x) = 5x + 12$
$C(x) = 3x + 116$

In each of Exercises 35 through 38, do the following:
(a) Find the break-even quantity.
(b) Find the break-even point.
(c) Find the profit function.
(d) Graph the revenue, cost, and profit functions on the same set of axes.

35. $R(x) = 3x$
$C(x) = 2x + 12$

36. $R(x) = 50x$
$C(x) = 30x + 2000$

37. $R(x) = 20x$
$C(x) = 12x + 490$

38. $R(x) = 60x$
$C(x) = 30x + 1800$

39. Revenue and Costs The Good CD Shop sells only CD's for stereo systems. The shop has a monthly overhead of $8000 (fixed costs) and an average direct cost of $8.36 per CD. Each CD sells for an average of $10.80. Let x be the number of CD's sold each month.

(a) If x CD's are sold during the month, find linear functions for the cost, revenue, and profit.

(b) Find the number of CD's that must be sold each month in order to break even.

(c) Graph $C(x)$, $R(x)$, and $P(x)$ on the same set of axes.

(d) If 800 CD's are sold during the month, will a profit or a loss be incurred? What if 2000 are sold?

40. Revenue and Costs The Friends of Owls Club is planning a picnic. The committee members in charge estimate that they will spend a total of $400 on rental of a pavilion, decorations, and so on. They also that they will spend $7.50 per person for food and drinks. They plan to charge each person who attends $10. Let x be the number of people who attend the picnic.

(a) What is the cost function in terms of x?

(b) What is the revenue function in terms of x?

(c) How many people must attend the picnic in order for the committee to break even?

(d) What is the profit function in terms of x?

(e) Graph the cost, revenue, and profit functions on the same set of axes.

For the supply and demand functions given in Exercises 41–44,
(a) Find the equilibrium price p_0.
(b) Find the equilibrium quantity $S(p_0) = D(p_0)$.
(c) Graph the supply and demand functions on the same set of axes, and label the equilibrium point.

41. $D(p) = 15 - 0.5p$; $S(p) = p + 12$

42. $D(p) = 12.5 - 0.25p$; $S(p) = 0.33p + 9.67$

43. $D(p) = 400 - 10p$; $S(p) = p + 180$

44. $D(p) = 148 - 3p$; $S(p) = 4p + 1$

45. Supply and Demand The Sun Software Company produces tutorial software for finite mathematics. The supply and demand functions for this software are $S(p) = p - 12$ and $D(p) = 50 - 0.5p$, where p is in dollars and $S(p)$ and $D(p)$ are measured in hundreds of units.

(a) Find the equilibrium price for the software.

(b) Find the equilibrium quantity for the software.

46. Supply and Demand Dixon's Manufacturing makes musical instruments. The supply and demand functions for kettledrums made by the company are $S(p) = 0.75p - 120$ and $D(p) = 475 - 0.35p$, where p is in dollars and $S(p)$ and $D(p)$ are in single units. Find the equilibrium point for the kettledrums.

47. Buying and Renting A car rental firm buys a new car for $20,000 and estimates the cost for maintenance, taxes, insurance, and depreciation at 40 cents per mile. The firm charges $50 per day plus 50 cents per mile to rent the car. Let x be the number of miles the car is rented during the first year, and assume that the car is rented for 160 days.

(a) What is the cost function for owning and renting the car, in terms of x?

(b) What is the revenue function in terms of x?

(c) How many miles must the car be driven in order for the company to break even during the first year?

(d) Graph the cost and revenue functions on the same set of axes for $0 \le x \le 150,000$ during the first year.

48. Tool Rental A rental firm has purchased a new ditchdigger for $36,000 and expects that maintenance costs will average 75 cents per actual hour of use of the machine.

The company plans to rent the ditchdigger for $200 per day plus $5 per hour of actual use. Assume that the ditchdigger will be rented for 150 days during the first year. Let x be the number of hours of actual use the ditchdigger gets during that first year.

(a) What is the cost function for purchasing and renting the machine, in terms of x?

(b) What is the revenue function in terms of x?

(c) How many hours must the machine be used for the rental firm to break even during the first year?

(d) Graph the cost and revenue functions on the same set of axes.

49. Revenue and Costs The Greensboro Bottling Co. is planning to introduce a new soft-drink container that holds 3 liters. The company estimates the first-year fixed costs for setting up the new production line at $50,000, and the direct costs for each bottle will be $1.50. The sales department estimates that 60,000 bottles can be sold during the first year at $2 per bottle.

(a) Find the linear cost function $C(x)$ that will give the total costs for selling x bottles of the drink. Also, find the revenue function $R(x)$.

(b) Find the profit function, and graph the cost, revenue, and profit functions on the same set of axes.

(c) If the sales department is correct in having estimated sales of 60,000 bottles during the first year, will a profit or a loss occur that year?

(d) How many bottles must be sold to break even?

(e) What are the total cost and the average cost per 3-liter bottle if 20,000 bottles are produced? What are they if 40,000 bottles are produced?

(f) If 20,000 bottles have been produced, what is the additional cost of making the 20,001st bottle?

50. Manufacturing Revenue and Costs The Hi-Tech Computer Company is planning to introduce a new graphics calculator. For the first year, the company estimates that its fixed costs to set up the production line will be $100,000. Hi-Tech also estimates that the direct costs for making each calculator will be $50 each. The company expects to sell the calculators for $75 each.

(a) Find the linear cost function that gives the total cost of producing x calculators.

(b) Find the revenue function, and use the concept of "marginal" (previously applied to "marginal cost") to state the marginal revenue.

(c) Find the profit function, and graph the cost, revenue, and profit functions on the same set of axes.

(d) How many calculators must be made and sold for the company to break even?

(e) What is the total cost of making 25,000 calculators? If 25,000 have been made, what is the additional cost of making the 25,001st calculator?

(f) What is the average cost per calculator of making 25,000 calculators?

51. Printing Costs The High Point Printing Company has two methods of printing the orders it receives. One is a computer method, which costs $300 to set up and then costs $25 to print 1000 copies. The other is an offset process, which costs $500 to set up and then costs $20 to print 1000 copies.

(a) Let x be the number to be printed in thousands. Find the linear cost function for each of the two methods.

(b) Graph both functions on the same set of axes.

(c) Find the quantity for which the costs are the same.

(d) Assuming equal quality, which method would be most advantageous for the company to use to print an order of 30,000 copies? Of 70,000 copies?

52. Car Revenue and Costs A new-car dealer has overhead costs (fixed costs) of $80,000 per month and average direct costs of $23,000 per car sold by the dealer (including the cost of the car). New cars are sold for an average price of $25,000 per car. Let x be the number of new cars sold each month by the dealer.

(a) Find the linear cost function $C(x)$ that represents the total cost of selling x new cars; also, find the revenue function and the profit function for selling x new cars.

(b) Find the number of new cars the dealer must have sold each month to break even.

(c) Graph $C(x)$, $R(x)$, and $P(x)$ on the same set of axes.

(d) If 30 new cars are sold during the month, will a profit or a loss be incurred by the dealer? What if 50 cars are sold?

53. Comparing Service Costs A new stereo of brand A costs $500 and comes with a five-year "no cost to you" guarantee at no additional cost. Brand B stereo of similar quality costs $350, but comes with an optional service contract, which costs $35 per year and will cover all repair costs.

(a) Let x be the number of years after purchase of one of the stereos, and assume that the service contract is purchased on brand B. Find the linear function that gives the cost of ownership for each stereo.

(b) Graph the two cost functions on the same set of axes.

(c) Find the number of years required for the two costs to be equal.

(d) Based on an ownership period of about five years, how would you decide which stereo has the lower cost?

54. Banking Costs The cost of a student account at the Piedmont Bank is $6 per month, with no fees for automatic teller machine (ATM) transactions. The cost for a similar amount at the Atlantic Bank is $3 per month, and each ATM transaction costs 50 cents.

(a) Let x be the number of ATM transactions per month. Find the linear cost function for having an account at each of the banks in terms of the number of ATM transactions per month.

(b) Graph the cost functions on the same set of axes.

(c) Find the number of ATM transactions for which the account costs are the same.

(d) Based on ATM usage, tell how you would decide which account would give lower overall account costs.

55. Equal Sales In 2000, A-Mart had sales of $36.3 million, and the company has had a constant growth of $1.1 million in sales each year since. B-Mart had sales of $16.7 million in 2000 and has had a constant growth of $2.3 million in sales each year since. If these trends continue, when will B-Mart Sales overtake those of A-Mart?

56. Equal Populations In 2000, Harrisonville had a population of 6254. The town has been gaining population at an average rate of 420 persons per year ever since. Smithville had a population of 17,213 in 2000 and has been losing population at an average rate of 330 persons per year ever since. If these trends continue, in what year will the two towns have the same population?

57. Recycling In 2000, the city of Rumford recycled 80 tons of wastepaper. The amount has been increasing at an average rate of 2 tons per year ever since. Galeton recycled 38 tons of wastepaper in 2000, but the amount has been increasing at the rate of 3 tons per year ever since. If these trends continue, what year will the two cities recycle the same amount of paper?

In Exercises 58 through 63, find the equation of each line that forms a boundary of the shaded region. Then find the coordinates of each corner point of the shaded region.

58.

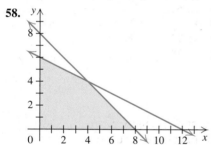

59.

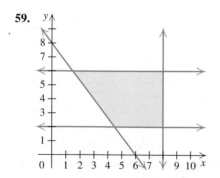

60.

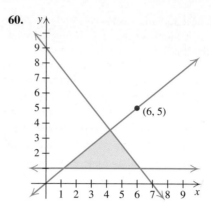

61.

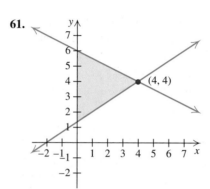

62.

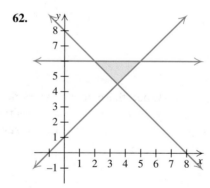

63.

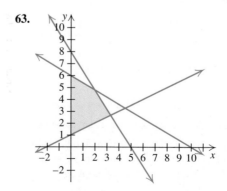

Section 1.5 Regression and Correlation

In Section 1.3, we found equations of lines passing through two points in which the two points represented observed or estimated data. We then saw how those lines could be used, within limits, for prediction purposes. Realistically, however, we would be reluctant to assume, on the basis of just two observations, that the trend of the actual data was linear. If realistic predictions were desired, we would need more evidence of such a trend. Therefore, we usually collect data (observations) over as wide a range as is feasible to begin an investigation of just how, if at all, the two variables are interrelated. Once collected, these data are plotted in a rectangular coordinate system to obtain a **scatter diagram.** Sometimes, the scatter diagram shows a linear tendency, as in Figure 32.

EXAMPLE 1

A Scatter Diagram
Showing a Linear
Tendency

An industrial psychologist gave an aptitude test to 15 new employees in a manufacturing plant and checked on their productivity a few weeks later. The results of the test scores and the corresponding production units for each of these employees were as follows:

$$x = \text{aptitude score}, \; y = \text{production per day}$$

x	y	x	y	x	y	x	y
25	60	60	80	30	62	73	88
80	102	55	76	38	71	40	73
68	85	70	86	50	73	40	68
35	70	45	71	58	78		

When these points are plotted in the plane, the scatter diagram of Figure 32 shows that a straight line can be positioned to provide a reasonable representation of the given data. In such a case, we say that a **linear correlation** between the x and y variables exists.

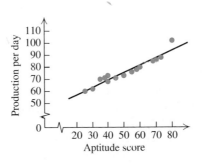

FIGURE 32
A Scatter Plot Showing a Linear Tendency

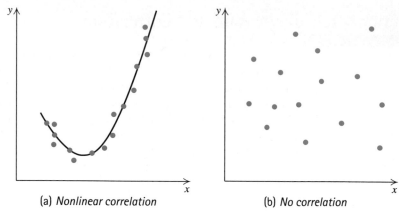

(a) *Nonlinear correlation*   (b) *No correlation*

FIGURE 33
Two Nonlinear Patterns

At other times, the scatter diagram of collected data may show a pattern that is not linear or may even show no pattern at all. Figure 33 shows two such relationships: (a) nonlinear correlation and (b) no correlation.

The general study of how two or more variables are interrelated is called **correlation** analysis; the word *correlation* means "related together."

NOTE Two things being "related" does not imply that one causes the other. Unfortunately, the difference between correlation and causality is as subtle as it is important. For example, the number of television sets sold per month by a store may have a high correlation with the number of refrigerators sold per month, but TV sales do not necessarily affect the sales of refrigerators. Instead, both sales are probably highly linked to local economic prosperity.

After looking at a scatter diagram, we ask two questions: (1) Can we find a mathematical relationship between x and y? and (2) If so, how strong is the relationship? The first question leads us to try to find an equation that expresses the relationship between the two variables. We can choose from many equations, but the simplest is the linear equation, the graph of which is a straight line, so that is where we concentrate our efforts. Once we decide to try to find a linear function that represents the relationship between our two variables, we encounter the problem of how to get the "best" one—that is, the equation that best describes the relationship.

The most widely accepted way to get a best fit to the data is by using the **method of least squares.** The method works on this principle: A given data point (x_k, y_k), where x_k and y_k are constants, differs in the y-coordinate from the point on a line with x-coordinate x_k by a distance $d_k = y_k - (mx_k + b)$. We say that the best fit occurs when the sum of the squares of these distances is as small as possible (a minimum).

To be more specific, consider Figure 34, which shows four points that represent data collected. As m and b vary, the line $y = mx + b$ changes; therefore, the vertical distances d_1, d_2, d_3, and d_4 will change. This means that the sum $S = d_1^2 + d_2^2 + d_3^2 + d_4^2$ will change as well. The method of least squares gets its name from the fact that we want to find the particular values of m and b so that *the sum S of the squared distances is the least (smallest) possible.* Advanced mathematics (calculus) is needed to find the m and b such that $y = mx + b$ and the data have a minimum sum of squares, S. The equations that give the "best" m and b are shown next; the mathematical shorthand symbol $\sum$ means "sum. "

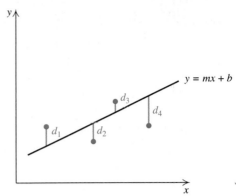

FIGURE 34
m and b Are Selected so as to Minimize
$$S = d_1^2 + d_2^2 + d_3^2 + d_4^2$$

The Least-Squares Regression Line

A linear function $y = mx + b$ is the **least-squares regression line** for the data points

$$(x_1, y_1), (x_2, y_2), \ldots (x_n, y_n)$$

in a scatter diagram if m and b are solutions of the system of equations

$$nb + \left(\sum x\right)m = \sum y \qquad\qquad (1)$$

$$\left(\sum x\right)b + \left(\sum x^2\right)m = \sum xy \qquad\qquad (2)$$

where $\sum x$ = sum of all x values in data points,
   $\sum y$ = sum of all y values in data points,
   $\sum xy$ = sum of products of both coordinates,
   $\sum x^2$ = sum of squares of all x values,
   n = number of data points.

Once Equations 1 and 2 have been solved for m and b, the line $y = mx + b$ so obtained will be known as the **least-squares regression line.** The word *regression* enters the picture to indicate the fact that, once the equation describing the scatter diagram is known (see Figures 35 and 36), that equation may be used to predict the y values by "regressing" on, or literally "going back to," known relationships between the x and y values.

HISTORICAL NOTE

SIR FRANCIS GALTON (1822–1911)

▄▄ The idea of finding the best-fitting line via the least-squares method has been around since the days of Gauss and LaPlace. It was not until Sir Francis Galton, a controversial anthropologist and a cousin of Charles Darwin, presented "Regression towards Mediocrity in Hereditary Stature" to the British Anthropological Institute (1885) that the process became known as *regression*. Galton's use of the word was intended to show that the children of tall people tend to be slightly shorter than their parents and the children of short people tend to be slightly taller than their parents, thereby "regressing" toward a mean height. After his address and subsequent publication of the piece in *Nature*, "linear regression" was used to refer to the method of least squares.

EXAMPLE 2

Finding a Least-Squares
Regression Line

Find the least-squares regression line for these data: $(1, 6), (2, 4), (3, 3), (4, 1)$.

SOLUTION The scatter diagram in Figure 35 suggests that a line will express the relationship between the x and y values rather well. To find m and b in the equation $y = mx + b$, first use the data to organize a table such as the following:

	x	y	x^2	xy
	1	6	1	6
	2	4	4	8
	3	3	9	9
	4	1	16	4
Sum	10	14	30	27

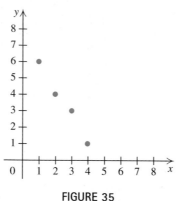

FIGURE 35
Scatter Diagram for Linear Regression

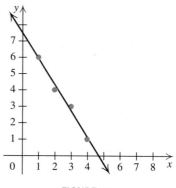

FIGURE 36
Regression Line and Data Points

Thus $\sum x = 10$, $\sum y = 14$, $\sum x^2 = 30$, and $\sum xy = 27$.

Then, because $n = 4$, Equations 1 and 2 become $4b + 10m = 14$ and $10b + 30m = 27$, respectively. Using one of the methods discussed in Section 1.4 (i.e., substitution or elimination) now shows that the solution to this system is $b = 7.5$ and $m = -1.6$, so the desired equation is

$$y = -1.6x + 7.5.$$

Figure 36 shows the scatter diagram of Figure 35, together with the regression line. Note that the regression line does not pass through any of the points in the scatter diagram; this is usually the case. ■

EXAMPLE 3

Finding a Regression
Line for Prediction

Ms. Shrestha is head of advertising sales at a TV station. As part of her sales pitch to other car dealers, she has collected the following data relating advertising spots per week to the number of sales Car Mart made during the week:

$x =$ Number of ads run in a week, $y =$ number of cars sold that week

x	y	x	y
5	15	20	32
15	20	0	10
12	19	9	18
		18	25

Organizing the given data as shown in the following table is helpful in finding the two desired equations in m and b:

x	y	x^2	xy
5	15	25	75
15	20	225	300
12	19	144	228
20	32	400	640
0	10	0	0
9	18	81	162
18	25	324	450
Sum 79	139	1199	1855

In this case, $n = 7$, so the resulting equations are

$$7b + 79m = 139$$

and

$$79b + 1199m = 1855.$$

The solution (using two-decimal accuracy) is $m = 0.93$ and $b = 9.35$. The linear regression line may now be written as $y = 0.93x + 9.35$. This line, along with the scatter diagram, is shown in Figure 37.

(a) According to the regression line, how many car sales could be expected if the ad were run 30 times during the week?

SOLUTION The answer is $y = 0.93(30) + 9.35 \approx 37.25$, or about 37 cars. (See Figure 37.)

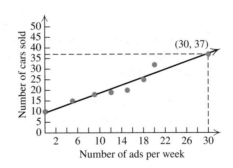

FIGURE 37
Predicting with a Regression Line

(b) If an advertising car agency sells 40 cars during the week, how many times did the sales ad run during the week, as predicted by the regression line?

SOLUTION The answer is found by solving $40 = 0.93x + 9.35$ for x. The result is $x = 32.93$, or about 33 times. ■

CAUTION When using a regression line for prediction purposes, it usually is not wise to stray too far outside the limits of the data gathered. Making predictions outside of the data's domain is called *extrapolation* and must be used with extreme caution.

If a least-squares regression line is graphed on the same set of axes as a scatter diagram, we can visually judge the "goodness" of fit of the line to the data. There is, however, a widely accepted numerical way to judge the goodness of fit, regardless of whether we have such a graph. The technique involves calculating a number, denoted by r, called the **correlation coefficient.** The correlation coefficient is *always* a number between -1 and 1, inclusive. Once calculated, the correlation coefficient works this way: The closer r is to $+1$ or -1, the better the least-squares line fits the data, whereas values closer to 0 (zero) indicate a poor fit. As shown in Figure 38, if $r = +1$ or -1, there is a perfect linear relationship between the x and y values; that is, all points lie on the least-squares line.

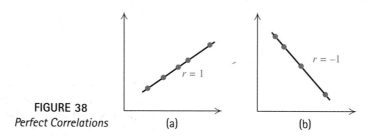

FIGURE 38
Perfect Correlations   (a)   (b)

As shown in Figure 39(a), for r **between 0 and 1** (excluding 0 and including 1), the x and y values have a *positive* correlation. That is, small x values are associated with small y values, and large x values are associated with large y values.

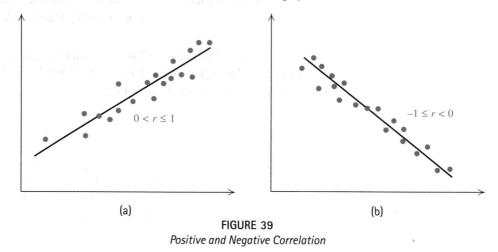

(a)   (b)

FIGURE 39
Positive and Negative Correlation

As shown in Figure 39(b), for r **between -1 and 0** (including -1 and excluding 0), the x and y values have a *negative* correlation. That is, small x values are associated with large y values, and large x values are associated with small y values.

If $r = 0$, there is no apparent linear relationship between the x and y values in the scatter diagram. Figure 40 shows such a diagram. In general, the sign of r indicates the sign of the slope of the linear regression. The magnitude of r is an indication of how well the linear regression "fits" the data.

The interpretation of r will vary, depending on the field of application. In the physical sciences, with experiments closely controlled, r values with absolute values between 0.90 and 1 indicate a high correlation; in the social sciences and educational fields, r values with absolute values between 0.70 and 1 indicate a high to very high correlation. Thus, the correlation coefficient must be judged within the context of the application. The formula used to find r follows:

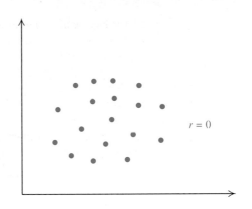

FIGURE 40
Zero Correlation Coefficient

The Correlation Coefficient

If $y = mx + b$ is the linear regression line for the points (x_1, y_1), (x_2, y_2), ... (x_n, y_n), then a measure of how well y fits is given by the **correlation coefficient**

$$r = \frac{n(\sum xy) - (\sum x)(\sum y)}{\sqrt{n(\sum x^2) - (\sum x)^2}\sqrt{n(\sum y^2) - (\sum y)^2}}$$

where n = number of points, $\sum x$ = sum of all x values in data points, $\sum y$ = sum of all y values in data points, and so on.

If r is to be calculated by hand, a table such as the one in Example 3 can be used effectively. Only one column showing the values of y^2 needs to be adjoined to that table to obtain the entries needed in the formula for r:

	x	y	x^2	xy	y^2
	5	15	25	75	225
	15	20	225	300	400
	12	19	144	228	361
	20	32	400	640	1024
	0	10	0	0	100
	9	18	81	162	324
	18	25	324	450	625
Sum	79	139	1199	1855	3059

Substituting the appropriate values from the table into the formula, we calculate

$$r = \frac{7(1855) - (79)(139)}{\sqrt{7(1199) - (79)^2}\sqrt{7(3059) - (139)^2}}$$
$$\approx 0.94.$$

We encourage the use of technology to ease the calculational pain of finding the slope, y-intercept, and correlation coefficient of a least-squares regression line. Most popular spreadsheet packages and graphing calculators will readily calculate these values for entered data, as we illustrate in the next two examples with the use of Excel and a TI-83 Plus.

EXAMPLE 4

Finding a Least-Squares
Regression Line
with Excel

Figure 41 shows how the average number of members per household has been decreasing since 1850. The actual data are as follows:

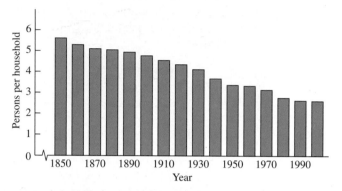

FIGURE 41

Average Number of Persons per Household for Selected Years

Year	1850	1860	1870	1880	1890	1900	1910	1920
Members of Household	5.59	5.28	5.09	5.04	4.93	4.76	4.54	4.34
Year	1930	1940	1950	1960	1970	1980	1990	1999
Members of Household	4.11	3.67	3.37	3.33	3.14	2.76	2.63	2.61

SOURCE: U.S. Bureau of the Census, *Statistical Abstract of the United States,* 1990 and 2000.

Letting t represent the years since 1850, use a least-squares regression line to model the data and find the correlation coefficient.

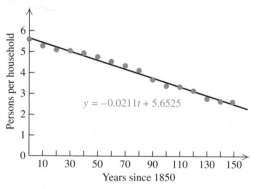

FIGURE 42

Average Number of People in a Household

SOLUTION Both the graph and the actual data show that the decline is linear enough to be reasonably approximated by a linear function. Begin by entering the data into two columns in Excel, with cells A1:A16 containing the values of t, 0 through 149, and cells B1:B16 containing the corresponding number of household members. In any blank cell, type =SLOPE(B1:B16,A1:A16) and press ENTER. The slope of the regression line is then calculated. In another blank cell, type =INTERCEPT(B1:B16,A1:A16) and press ENTER to calculate the y-intercept of the regression line. Finally, type =CORREL (B1:B16,A1:A16) and press ENTER to calculate the correlation coefficient. The results are as follows:

Slope =	-0.0211
Intercept =	5.6525
$r =$	-0.9922

The resulting regression line is then $y = -0.0211t + 5.6525$, where $0 \le t \le 222$. The graph of the line is shown in Figure 42, along with the scatter diagram. The correlation coefficient indicates a strong negative relationship, indicating that as the number of years, t, increases, the number of household members decreases. (The reason for the upper limit, 222, on t is that if $t > 222$, we have $y < 1$, but the number of members per household cannot fall below 1.) ▬

We conclude this section with Example 5, which answers the question posed in the graph at the outset of this chapter.

EXAMPLE 5

Modeling Recovered Waste in the United States since 1990

Use a least-squares regression line and a TI-83 Plus calculator to model the following data:

Year	Recovered Waste (millions of tons)
1990	33.6
1992	40.6
1993	43.8
1994	50.8
1995	54.9
1996	57.3
1997	59.4
1998	62.2

SOURCE: U.S. Bureau of the Census, *Statistical Abstract of the United States,* 2000.

Use your model to predict the number of pounds of waste that will be recovered in the year 2010. A scatterplot of the data, along with the regression line, is shown in Figure 43.

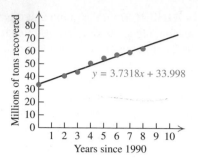

FIGURE 43
Recovered Waste in the United States

SOLUTION If we let $x = 0$ represent the year 1990, then the data, expressed as ordered pairs, take on the appearance $(0, 33.6)$, $(2, 40.6)$, etc. After entering the data into lists L1 and L2 on the TI-83 Plus, we access the LinReg(ax+b) command in the $\boxed{\text{STAT}}$ CALC menu of the TI-83 Plus.0

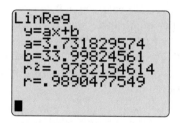

FIGURE 44
*TI-83 Plus Screen Displaying
Linear Regression Information*

(*Note:* If your calculator does not display the value of the correlation coefficient r, then access the CATALOG by pressing $\boxed{\text{2nd}}$ $\boxed{0}$, scroll down to DiagnosticOn, and press $\boxed{\text{ENTER}}$ twice. From now on, the calculator will display the correlation coefficient when the LinReg command is executed.)

As shown in Figure 44, the least-squares regression line for these data is approximately $y = 3.73x + 34$. To predict the amount of waste recovered in the year 2010, first observe that that year is represented by $x = 2010 - 1990 = 20$, and then substitute this value into the regression equation: $y = 3.73 \times 20 + 34 \approx 108.6$ million tons. According to the model, almost 109 million tons of municipal waste will be recovered in the year 2010. ■

Exercises 1.5

In Exercises 1 through 8,

(a) Find the least-squares regression line by setting up and solving the system of two equations and two unknowns that give m and b.

(b) Graph the line and the scatter plot on the same axes. Pay particular attention to Problems 1–4; draw a con-

clusion about the least-squares fit of data containing only two points.

1. $(1, 2), (3, 7)$

2. $(-1, 2), (6, 5)$

3. $(-1, 3), (2, -1)$

4. $(3, 5), (4, 8)$

5. $(1, 1), (2, 4), (3, 8)$

6. $(-1, 2), (1, 0), (5, -5)$

7. $(1, 1), (2, 3), (3, 6), (4, 5)$

8. $(0, 0), (1, -1), (2, -3), (5, -6)$

 In Exercises 9 through 14,

(a) Plot the scatter diagram.
(b) Find and plot the least-squares regression line.
(c) Find the correlation coefficient.

9. $(2, 6), (1, 3), (3, 9)$

10. $(2, 6), (4, 4), (5, 3)$

11. $(3, 8), (5, 3), (6, 7), (8, 2), (9, 4)$

12. $(1, 1), (4, 1), (2, 3), (6, 2), (5, 6), (7, 9)$

13. $(0, 4), (5, 0), (9, 5), (3, 7), (1, 1)$

14. $(3, 10), (4, 2), (6, 8), (7, 3), (8, 5)$

For each of Exercises 15 through 20, do you think there would be a rather high or a rather low correlation between the variables? Estimate the sign of the correlation coefficient.

15. The number of automobile accidents and the number of automobiles on the highways.

16. The heights and the ages of elementary school children.

17. The heights of persons having bank accounts and the amount of money in their bank accounts.

18. The I.Q.'s of college students and their grade-point averages.

19. The ages of cars and their monetary value.

20. Student enrollment in universities and the number of football games won at those universities in 1990.

 In the remaining exercises, use technology where appropriate.

21. Pollutants An experimental study of the environmental impact on black bass in large lakes was conducted. The number of fish (in thousands) within one-half mile of the source of pollution was estimated when various amounts of a particular pollutant were introduced into the water. The following data were collected:

Number of tons of pollutant	20	30	40	50	60	70
Number of fish	40	33	30	26	20	16

(a) Plot the scatter diagram and find the least-squares regression line for these data, using tons of pollutants as the independent variable x.

(b) Find the correlation coefficient.

(c) According to the linear model found in Exercise 21(a), how many fewer fish can be expected for each ton of the pollutant introduced into the water?

(d) Use the line found in Exercise 21(a) to estimate the black bass population if 65 tons of the pollutant were introduced into the lake.

(e) If the black bass population is estimated at 35,000, how many tons of the pollutant can be assumed to have been put into the lake?

22. Estimating Height Anthropologists measured the length of the bone from the elbow to the shoulder of 10 women found in the burial grounds of an ancient civilization. Then the anthropologists measured the respective heights (all in centimeters) of the women. The results were as follows:

Bone length (cm)	Height (cm)
44.0	164.7
42.3	160.0
41.2	157.0
41.8	158.6
40.2	154.1
43.6	162.8
40.6	155.3
38.1	147.2
42.5	160.0
45.0	167.1

(a) Plot the scatter diagram for these data, using bone length as the independent variable x.

(b) Find the least-squares regression line and the correlation coefficient.

23. Reading Readiness A reading-readiness test was administered to a random sample of 10 first-grade students, and the results were compared with the students' I.Q. scores. The following table was compiled:

Person	x = I.Q. score	y = reading readiness score
A	102	45
B	110	50
C	102	48
D	98	38
E	100	40
F	101	45
G	96	42
H	100	45
I	104	39
J	94	38

(a) Plot the scatter diagram for these data.

(b) Find the least-squares regression line, and plot it on the scatter diagram.

(c) Find the correlation coefficient.

24. Aptitude versus Performance An accounting firm gave aptitude tests to all of its new employees during a given year. A year later, the firm compared the test results with the employees' performance ratings. The outcomes for 20 randomly selected employees were as follows:

x = aptitude test score; y = performance rating

x	y	x	y	x	y	x	y
80	85	83	74	79	82	89	86
86	71	93	85	81	90	85	80
87	90	82	85	82	91		
84	75	87	90	92	97		
91	89	72	80	91	95		
60	70	80	91	83	90		

(a) Plot the scatter diagram for these data.

(b) Find the least-squares regression line and the correlation coefficient.

(c) From your findings, do you think that the firm can predict job ratings relatively accurately from the aptitude test scores?

25. Lumber in a Tree Foresters have developed a rule to predict how many board feet of lumber are in a tree: They cut several trees with known diameters and then measure the number of board feet of lumber in each tree. The following is a sample of their findings.

x = diameter (in inches)	17	19	21	23	25
y = volume (in hundreds of board feet)	19	25	33	57	71
x = diameter (in inches)	28	32	38	39	41
y = volume (in hundreds of board feet)	113	123	252	260	293

(a) Find the least-squares regression line and the correlation coefficient.

(b) According to the linear model found in Exercise 25(a), what is the increase in board feet per inch of increase in diameter?

(c) How many board feet of lumber would be produced from a tree that had a diameter of 45 inches?

(d) If a tree produced 25,000 board feet of lumber, what could we assume about its diameter?

26. Wages versus Sales Volume A department store chain collected data from each of its stores for a month. The data on total wages paid to sales personnel and on sales volume are as follows:

Store	Sales wages (thousands)	Sales volume (hundreds of thousands)
A	30.2	3.7
B	52.1	4.8
C	23.6	2.7
D	38.2	3.5
E	45.8	4.7
F	59.2	5.1
G	41.7	3.9
H	62.1	5.4

(a) Plot the scatter diagram, using wages as the independent variable x.

(b) Find the least-squares regression line and the correlation coefficient.

(c) Using the regression line found in Exercise 26(b), determine how many thousands of dollars will be spent in sales wages to support a sales volume of $700,000.

27. Retail Trade Wages Average hourly earnings in the U.S. retail trade industry (in current dollars and constant dollars) are shown in the table at the top of the next page.*

*SOURCE: U.S. Bureau of the Census: *Statistical Abstract of the United States,* 2000.

Year	1980	1985	1990	1995	1999
Current dollars	4.88	5.94	6.75	7.69	9.08
Constant dollars	5.70	5.39	5.07	4.97	5.39

(a) Find the least-squares regression lines that approximate the average hourly earnings in both current dollars and constant dollars for this industry. Find the correlation coefficient in both cases. (Let x be years since 1980.)

(b) Plot both of the regression lines you found in Exercise 27(a) on the same set of axes.

(c) Use the regression lines found in Exercise 27(a) to estimate the difference in current-dollar and constant-dollar average hourly earnings in the year 2005.

28. Cricket Chirps versus Temperature Biologists have studied the relationship between the temperature and the number of chirps made by a cricket. Generally, the warmer the weather is, the more often the cricket chirps. Here are some experimental data collected:

Temperature (in degrees Fahrenheit):	45	52	55	58	60	62
Number of seconds to count 50 chirps:	94	59	42	36	32	32
Temperature (in degrees Fahrenheit):	65	68	70	75	82	83
Number of seconds to count 50 chirps:	26	24	21	20	17	16

(a) Plot the scatter diagram for these data.

(b) Find the least-squares regression line and the correlation coefficient.

(c) If the temperature is 0 degrees Fahrenheit, how many seconds will it take to count 50 chirps, according to the linear model of Exercise 28(b)? Do you believe this?

(d) Use the linear model to estimate the temperature if it takes 1 second to count 50 chirps.

29. Magazine Advertising The number of dollars (in millions) spent advertising drugs and remedies in magazines is shown in the following table for selected years.*

(a) Find the least-squares regression line that will estimate the number of dollars spent on magazine adver-

*SOURCE: U.S. Bureau of the Census: *Statistical Abstract of the United States,* 2000.

Year	1980	1985	1989	1990	1992
Dollars spent	79	135	135	163	279

tising for drugs and remedies. Also, find the correlation coefficient.

(b) Assuming that the trend continues, how many dollars will be spent for this type of advertising in the year 2010?

30. Lead Emissions The following table shows lead emissions in thousands of metric tons for selected years in the United States: *

Year	Lead (thousands of metric tons)
1989	5.47
1990	4.98
1991	4.17
1992	3.81
1993	3.92
1994	4.05
1995	3.93
1996	3.90
1997	3.95
1998	3.97

(a) Find the least-squares regression line that will estimate the lead emissions over the period shown in the table. Also, find the correlation coefficient.

(b) Find the least-squares regression line that will estimate the lead emissions from 1994 through 1998. Also, find the correlation coefficient.

(c) Using the lines found in Exercise 30(a) and (b), determine the year all lead emissions will be eliminated. Which do you judge to be the most accurate calculation?

31. Recycling The table on the next page shows U.S. Census Bureau data* regarding the gross waste generated per year (in millions of tons) in the United States and the amount recovered (e.g., recycled or recovered for energy conversion).

(a) Find the least-squares regression lines that approximate the waste generated and the amount of materials recovered. Give the correlation coefficient in each case.

(b) At the rates shown, will we ever recover even 50% of the waste generated?

*SOURCE: U.S. Bureau of the Census: *Statistical Abstract of the United States,* 2000.

Year	1960	1970	1980	1985
Gross waste generated	87.8	121.9	151.5	164.4
Materials recovered	5.9	8.6	14.5	16.4
Year	1987	1990	1995	1998
Gross waste generated	178.1	205.2	211.4	220.2
Materials recovered	20.1	33.6	54.9	62.2

Maximum Wingspan (mm)	Maximum Caterpillar Length (mm)
115	80
140	90
70	75
100	75
150	130
150	110
135	100

SOURCE: Milne, Lorus, and Margery, *National Audobon Society Field Guide to North American Insects and Spiders* (New York: Alfred A. Knopf, 1995).

(c) Use the functions found in Exercise 31(a) to predict the waste generated and recovered in the year 2010.

32. Moth Measurements The maximum wingspan, in millimeters (mm), of the giant silkworm moth and the maximum length, in mm, of caterpillars of several species of the giant silkworm moth family are given in the table at the top of the next column.

(a) Draw a scatter plot of the data. Do the data appear to have a linear relationship?

(b) Use a least-squares regression line to model the data.

(c) Find the correlation coefficient for the data. Explain what this correlation coefficient represents.

(d) According to the model, what would be the maximum wingspan of a giant silkworm moth species that has a maximum caterpillar length of 150 mm?

(e) The sheep moth (*Hemileuca eglanterina*) has a maximum wingspan of 70 mm and a maximum caterpillar length of 150 mm. What would the model predict for the caterpillar length of this moth?

33. Measure the height and arm length (from shoulder to fingertip) of 10 friends. Then plot a scatter diagram of the data, with length of arm on the *x*-axis and height on the *y*-axis. Find the equation of the regression line and the correlation coefficient. Do you think that your data indicate that a person's height can be predicted with any degree of accuracy by the length of his or her arm? Explain why your correlation coefficient might be considerably different from that of another classmate doing the same experiment.

34. Examine the records of your city for the past several years to find out the annual population. Does the trend appear linear? If so, use the data to obtain a least-squares regression line, and use this line to predict the population 10 years from now. Compare your figures with predictions made by the city. Are the city planners taking into account factors that your linear model does not?

35. Find a source that will show how temperatures cool as the height above the earth increases. Decide whether the data indicate a linear correlation. If so, find the least-squares regression line, and use it to estimate temperatures for heights beyond your data.

Chapter 1 Summary

CHAPTER HIGHLIGHTS

Descartes' *rectangular coordinate system* provides a convenient graphic scheme for representing ordered pairs of real numbers. The *graph* of every linear function is a straight line; conversely, every nonvertical straight line represents a linear function, the equation of which is given by the *point–slope form*, $y - y_1 = m(x - x_1)$, or the *slope–intercept form*, $y = mx + b$. The *general form*, $Ax + By = C$, may be used to represent any line. A pair of lines in the plane may have one point in common, no points in common, or all points in common. To determine which is the case, either the *substitution* or *elimination* method may be used, and either of these methods may also be used to find the point or points of intersection, if any. *Linear functions* are useful in modeling costs, revenue, depreciation, ecology concerns, and many other real-life applications. *Least-squares regression lines* data are collected in various fields and may be used for prediction purposes. The degree of "goodness" of fit between the data and a linear regression line is given by the correlation coefficient.

EXERCISES

1. Calculate the x-intercept and the y-intercept of the line $3x - 5y = 30$.

2. Find the marginal cost and the fixed cost for the function $C(x) = 7x + 430$.

3. Determine the equations of the horizontal line and the vertical line that pass through the point $(-3, 8)$.

4. Calculate the slope of the line $5x + 2y = 13$.

5. If (x, y) is a point on a line whose equation is $y = 3x + 4$, calculate the y-coordinate of the point on this line having x-coordinate $x + 2$.

6. Compute $f(-5)$ for the function $f(x) = 1 + 2x/3$.

7. Write the equation of the line that passes through the origin and is parallel to the line $2y = 5x + 10$.

8. Use the substitution method to find the point or points, if any, that the lines represented by $2x + y = 5$ and $x + y = 7$ have in common.

9. Use the elimination method to find the point or points, if any, that the lines represented by $3x + 6y = 13$ and $5x - 6y = 11$ have in common.

Solve each of the following systems of equations:

10. $5x + 3y = 9$
    $10x + 6y = 10$

11. $5x + 3y = 9$
    $10x + 6y = 18$

12. Find the least-squares regression line for the points $(2, 8)$, $(3, 6)$, $(4, 5)$, and $(5, 9)$. Would the correlation coefficient be nearer to 1, -1, or 0?

13. A manufacturing company has a total cost function given by $C(x) = 7.9x + 2520$.

    (a) Find the total cost of 60 items.

    (b) Find the average cost per item for 60 items.

    (c) Find the additional cost of producing the 61st item.

14. A discount store had sales of $1.8 million in 1990 and $2.7 million in 1992. Assume that the store's sales trend is linear.

    (a) Find a function that will estimate sales in future years.

    (b) Predict the year that sales could be expected to reach $20 million.

15. An appliance company has revenue and cost functions respectively given by $R(x) = 65x$ and $C(x) = 25x + 4000$.

    (a) Find the profit function.

    (b) Find the break-even point.

    (c) Graph the revenue, cost, and profit functions on the same set of axes.

16. Consider the points $(2, 5)$, $(4, 8)$, $(5, 14)$, $(6, 15)$, $(3, 6)$, $(4, 9)$, $(7, 17)$, $(9, 21)$, $(5, 13)$, $(12, 25)$, $(15, 33)$, and $(9, 19)$. Find

    (a) The slope m and the y-intercept b of the least-squares regression line.

    (b) The correlation coefficient r.

    (c) The scatter diagram and graph of the regression line.

17. The following question is similar to one from the Sample Questions for the 2001 PRAXIS *Pre-Professional Skills Test* (PPST):

    Which of the following formulas expresses the relationship between x and y in the foregoing table?

x	0	2	5	6	11
y	4	8	14	16	26

    (a) $y = x + 4$
    (b) $y = 2x + 4$
    (c) $y = 3x - 1$
    (d) $y = 4x$
    (e) $y = 3x - 7$

18. The following question is similar to one from the 1998 *Consumer Duffy CPA Review: Accounting and Reporting* study manual:

Using regression analysis, Clegg Co. graphed the relationship between sales of its most expensive service and its customers' income tax refunds as shown at right.

If there is a strong linear relationship between sales and customers' income tax refunds, which of the following numbers best represents the correlation coefficient for this relationship?

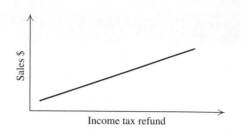

Income tax refund

(a) 0   **(b)** -8.9

(c) 8.9   **(d)** 0.89

(e) -0.89

Chapter 1 Sample Test Items

1. Plot the points $(2, 7)$, $(3, -2)$, $(-5, 0)$, $(-4, -1)$, and $(-2, 6)$ on the same set of axes in the rectangular coordinate system.

2. Graph the functions $y = x - 2$, $y = 2x - 4$, and $y = 3x - 6$ on the same set of axes.

3. Write a linear equation for which the change in y per unit change in x is 3.7 and the y-intercept is -2.8.

4. Write a linear cost function that will represent the total cost of making x items if the total cost is $650 for 50 items and $714 for 70 items.

5. A small town with a population of 2300 has been growing at a rate of about 250 per year. Find a linear function that will give the approximate population x years from now if this trend continues.

6. Find the equation of the vertical line that passes through the point $(-2, 7)$.

7. Find the slope, the x-intercept, and the y-intercept of the line having equation $x - 2y = 9$.

8. Write an equation of a line that has x-intercept -2 and y-intercept 6.4.

9. A restaurant sold 78 luncheon specials when the price was $4.75. When the price was raised to $5.50, 63 specials were sold. Find a linear function that will predict the number of specials sold for a given price.

10. A rancher buys a small tractor for $5600 and plans to depreciate it over 10 years, at which time the value will be $2100. Find a linear function that gives the value of the tractor as a function of the years from the date of purchase.

11. By examining the slopes and y-intercepts, decide whether the equations $2x + 5y = 7$ and $2x - 5y = 7$ determine lines having one point, no points, or all points in common.

12. Solve the system of equations $x + y = 5$ and $2x - y = 12$ by using the substitution method and then the elimination method.

13. Solve the system of equations $0.2x + 0.8y = 3$, and $x + 4y = 15$ by any appropriate method.

14. A graphing calculator company has fixed costs of $40,000 per month and direct sales costs of $73 per graphing calculator. Find the profit function if the calculators sell for an average of $98. What is the break-even point?

15. Betty's Stitchery sells handmade quilts, for which the annual supply and demand functions are $S(p) = 0.03p - 9$ and $D(p) = 17.5 - 0.015p$, respectively. Find the equilibrium price and equilibrium quantity.

16. The Vestmore Company has a mutual fund that returned 8%, 9%, 12%, and 14% over the first four years (starting with year 1).

(a) Plot the scatter diagram for these four years.

(b) Find the least-squares regression line for the data.

(c) Find the correlation coefficient.

(d) Use the least-squares regression line to predict the return on investments for the fifth year.

(e) Predict how many years after the fund started that the return will be 20%.

In-Depth Application

Unemployment and Homicides in the United States, 1980–1999

Objectives

In this extended exercise, you will apply your knowledge of regression analysis and correlation to determine whether unemployment rates and homicide rates in the United States exhibit a linear relationship. In addition, you will learn that the choice of data to include can have a noteworthy impact on the results of your analysis. You will also be asked to interpret what other social and political factors may have contributed to the patterns in unemployment and homicide rates.

The Data

The data on the next page were compiled from *The World Almanac and Book of Facts 2001:*

1. Draw a scatter plot for the unemployment rate in the United States from 1980 to 1999. During what periods did the unemployment rate increase? During what periods did it decrease?

2. Draw a scatter plot for the homicide rate in the United States from 1980 to 1999. During what periods did the homicide rate increase? During what periods did it decrease?

3. Describe any patterns you notice that relate trends in the unemployment rate to trends in the homicide rate.

4. Find the correlation coefficient between the unemployment rate (per 100 population) and the homicide rate (per 100,000 population) for the years 1980 through 1999. Is the correlation positive or negative? Is it a high or low correlation? Does there appear to be a significant relationship between unemployment rates and homicide rates in the United States for the period in question?

(continued)

Year	Unemployed (per 100)	Homicides (per 100,000)
1980	7.1	10.22
1981	7.6	9.83
1982	9.7	9.07
1983	9.6	8.25
1984	7.5	7.91
1985	7.2	7.95
1986	7.0	8.55
1987	6.2	8.26
1988	5.5	8.41
1989	5.3	8.66
1990	5.6	9.42
1991	6.8	9.79
1992	7.5	9.31
1993	6.9	9.51
1994	6.1	8.96
1995	5.6	8.22
1996	5.4	7.41
1997	4.9	6.80
1998	4.5	6.28
1999	4.2	5.70

SOURCE: *The World Almanac and Book of Facts 2000*, World Almanac Education Group, Inc.

5. Find the correlation coefficient for unemployment rates and homicide rates for the period between 1980 and 1988. Is the correlation positive or negative? Is the correlation high or low? Does there appear to be a significant relationship between unemployment rates and homicide rates in the United States for the period 1980–1988?

6. Find the correlation coefficient for unemployment rates and homicide rates for the period between 1989 and 1999. Is the correlation positive or negative? Is the correlation high or low? Does there appear to be a significant relationship between unemployment rates and homicide rates in the United States for the period 1989–1999?

7. Discuss any social, economic, or political factors that may have influenced the different trends in homicide and unemployment during the periods considered in questions 5 and 6. You may wish to consider, for example, who was president of the United States, whether there were any military conflicts involving the nation, or whether there were any significant changes in social programs such as welfare.

Systems of Linear Equations

PROBLEM

A retailer has warehouses in Lima and Canton, from which two stores—one in Tiffin and one in Danville—place orders for bicycles. The store in Tiffin orders 38, and the store in Danville orders 46. Each warehouse has enough bicycles on hand to supply all that have been ordered. Economic conditions dictate that twice as many bicycles be sent from Lima to Danville as from Canton to Tiffin. Find a system of equations whose solutions give possible shipping schedules from the warehouses to the two stores. (See Example 4, Section 2.1, and Example 5, Section 2.3)

CHAPTER PREVIEW

Sometimes, more than one equation is needed to create a mathematical model. In such cases, the model consists of a "system" of equations. Whenever possible, linear equations are used in such systems, because of their simplicity. In this chapter, we first learn how to create mathematical models that involve two or more linear equations, each of which may itself involve several unknowns, such as x, y, or z. Our attention then turns to finding the solutions of such systems. We first write the system in matrix form and then "pivot" on certain elements in the matrix through the use of the Gauss–Jordan elimination method. This two-pronged process recognizes when a given system has one or many solutions—or none at all. The introduction of pivoting here paves the ways for its continued use in later chapters.

Technology Opportunities

Certain computer programs, spreadsheets, and calculators can be used to solve systems of linear equations upon entry of the equations or coefficients of the variables. In using such a tool, knowledge of how it signals the "no solution" case, as well as how, if at all, it displays the "infinitely many solutions" case, is indispensable.

In this chapter, we introduce Excel's Solver to assist with solving systems of linear equations. Solver will be a very useful tool in later chapters as well, so familiarity with its environment now will be beneficial.

Most modern graphing calculators are equipped with a command that places matrices in *reduced row-echelon form*, which will allow us to quickly determine how many solutions, if any, a linear system has.

Section 2.1   Linear Systems as Mathematical Models

Up to this point, linear equations involving only two variables have been considered, and their straight-line graphs have provided visual help with many of the concepts to which they have been applied. The concept of a linear equation may, however, be generalized to any number of variables by following the pattern set forth by linear equations in two variables, namely $ax + by = c$. The left side of this equation consists of a sum of terms, each of which is a constant multiplied by a variable, and the right side is a constant. For example, $3x + 2y - 6z = 7$ is a linear equation in the three variables, x, y, and z, and $w - 4x + 5y - 7z = 9$ is a linear equation in the four variables w, x, y, and z. In general, a linear equation has the form

$$a_1x_1 + a_2x_2 + a_3x_3 + \cdots + a_nx_n = b$$

where $a_1, a_2, a_3, \ldots, a_n$ and b are constants, and $x_1, x_2, x_3, \ldots, x_n$ are variables. Linear equations in three variables have graphs that are planes in a three-dimensional Cartesian coordinate system. Linear equations in four or more variables are too complicated to visualize. Even so, algebraic methods exist for finding solutions of systems of linear equations, regardless of the number of variables involved. One rather straightforward algebraic method is discussed in Section 2.2. Before concentrating on the solutions, however, we need to become proficient at setting up systems that model various situations. Actually, doing that is just a process of translating English sentences into equations, using mathematical symbols. Sometimes it can be done with little effort, while at other times more organization of the facts may be necessary. In the next few examples, a method of organization is set forth that will serve us in working with systems of linear equations, as well as with systems of linear inequalities, which we will encounter in the next chapter.

EXAMPLE 1

A Mathematical Model of a Production Process

A firm produces bargain and deluxe TV sets by buying the components, assembling them, and then testing the sets before shipping. The bargain set requires 3 hours to assemble and 1 hour to test. The deluxe set requires 4 hours to assemble and 2 hours to test. The firm has enough employees so that 390 work hours are available for assembly and 170 work hours are available for testing each week. Use a system of linear equations to model the number of each type of TV set that the company can produce each week while using all of its available labor.

SOLUTION As is typical of this kind of problem, the last sentence asks *how many* of each type of product (in this case, TV sets) should be produced. To start organizing the data presented, put the products "Bargain set" and "Deluxe set" as headings for two columns, along with an x and a y to represent the number of each that will be produced. (See Table 1(a).) Then list the operations necessary to produce these products. In this case, there are two—assembly and testing—which are written in the left column as row headings. (See Table 1(b).) Now reread the problem and fill in the appropriate number for each box, including the numbers that limit each operation. The completed table should look like Table 2.

Bargain Set	Deluxe Set
x	y

(a)

	Bargain Set	Deluxe Set
Number of Each Type of Set	x	y
Assembly Time (hr per set)		
Testing Time (hr per set)		

TABLE 1.   (b)

	Bargain Set	Deluxe Set	Limits on Operations (hr)
Number of Each Type of Set	x	y	
Assembly Time (hr per set)	3	4	390
Testing Time (hr per set)	1	2	170

TABLE 2.

The equations modeling our production process can now be *read as rows* from Table 2: $3x + 4y = 390$, $x + 2y = 170$. The reason this equation holds is that each bargain set requires 3 hours of assembly time, so, to make x units, $3x$ work hours are needed. Similarly, each deluxe set requires 4 hours of assembly time, so, to make y units, $4y$ work hours are needed. Therefore, a total of $3x + 4y$ work hours are needed to produce x bargain sets and y deluxe sets, and we want that to be exactly the available 390 work hours. The reasoning behind the second equation, representing testing work hours, is the same. The resulting system of equations is then

$$3x + 4y = 390$$
$$x + 2y = 170,$$

where x and y are nonnegative. ■

EXAMPLE 2

Modeling a Diet
Problem

A dietitian is to combine a total of 5 servings of cream of mushroom soup, tuna, and green beans, among other ingredients, in making a casserole. Each serving of soup has 15 calories and 1 gram of protein, each serving of tuna has 160 calories and 12 grams of protein, and each serving of green beans has 20 calories and 1 gram of protein. If these three foods are to furnish 370 calories and 27 grams of protein in the casserole, how many servings of each should be used?

SOLUTION This time the foods soup, tuna, and beans head the columns as "products," and the restricted items going across the rows are the total number of servings, the number of calories, and the number of grams of protein. (See Table 3.) Because the total number of servings of these three foods is to be 5, it follows that $x + y + z = 5$. Notice how this fact is reflected in the first row of the table. A reading of the calorie and protein requirements shows how the next two rows of Table 3 are created.

	Soup	Tuna	Beans	Restrictions
Number of Servings of Each Food	x	y	z	5
Calories per Serving	15	160	20	370
Grams of Protein per Serving	1	12	1	27

TABLE 3.

Again, each row with a restriction represents an equation to be satisfied:

$$\begin{aligned}
x + y + z &= 5 \\
15x + 160y + 20z &= 370 \\
x + 12y + z &= 27.
\end{aligned}$$ ∎

EXAMPLE 3

Modeling a Production
Process

Glenda's Woodshop produces dollhouses and magazine racks. Dollhouses require 20 minutes of sawing time, 30 minutes' assembly time, and 10 minutes' sanding time. Magazine racks require 18 minutes of sawing time, 15 minutes' assembly time, and 12 minutes' sanding time. Due to limitations on machinery and work hours, 4 hours per day are devoted to sawing, 5 hours to assembly, and 3 hours to sanding. Find linear equations that will model the number of dollhouses and magazine racks that can be produced each day while fully utilizing the available labor.

SOLUTION There are two products—dollhouses and magazine racks—and three operations—sawing, assembly, and sanding—each restricted by time. As usual, put the products on top as column headers and the restricted operations down the left side, to name the rows. Keeping time in minutes, fill in the boxes by reading the problem.
   The rows in Table 4 give the desired equations:

	Dollhouses	Magazine Racks	Limitations (in minutes)
Number of Units Produced	x	y	
Sawing Time (in minutes per unit)	20	18	$4(60) = 240$
Assembly Time (in minutes per unit)	30	15	$5(60) = 300$
Sanding Time (in minutes per unit)	10	12	$3(60) = 180$

TABLE 4.

$$20x + 18y = 240$$
$$30x + 15y = 300$$
$$10x + 12y = 180. \blacksquare$$

The next example contains a slightly different kind of restriction, but the same format as before may be used to obtain the equations.

EXAMPLE 4

Modeling Shipping Schedules

The problem stated at the outset of this chapter described shipping bicycles from two warehouses, one in Lima and one in Canton, to a store in Tiffin, which ordered 38 bicycles, and a store in Danville, which ordered 46. Each warehouse has enough to supply all that have been ordered, but twice as many are to be shipped from Lima to Danville as from Canton to Tiffin. What equations will describe the possible shipping schedules?

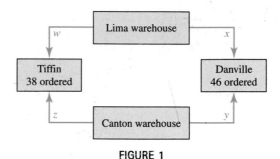

FIGURE 1

SOLUTION Let w, x, y, and z be the number of bicycles shipped from the respective warehouses to the stores, as shown in Figure 1. Because 38 are to be shipped to Tiffin,

$$w + z = 38,$$

and because 46 are to be shipped to Danville,

$$x + y = 46.$$

The requirement that twice as many are to be shipped from Lima to Danville as from Canton to Tiffin means that the x variable is to be twice the z variable; hence, the equation relating x to z is

$$x = 2z \quad \text{or} \quad x - 2z = 0.$$

The preceding equation gives the third row in Table 5.

Number of Bicycles	w	x	y	z	
To Tiffin	1			1	38
To Danville		1	1		46
Relate x, z		1		-2	0

TABLE 5.

This time the equations can be found rather easily, even without the aid of a table. The complete system is

$$
\begin{aligned}
w \qquad\;\; + \;z &= 38 \\
x + y \qquad\;\; &= 46 \\
x \qquad\; - 2z &= 0. \; \blacksquare
\end{aligned}
$$

Once again, the suggested tables will always organize the data for any problem that can be modeled by a linear system of equations, such that the resulting rows give the desired equations. (The tables also will be helpful for setting up spreadsheets.) You may find it useful to describe in words some of the restrictions on the data while the others can be put directly into equation form.

The configuration of numbers obtained in Table 5, from which the equations were written, is an array of numbers consisting of three **rows** (the rows go horizontally) and five **columns** (the columns go vertically). If we insert a zero where each blank space appears and enclose the array with brackets, we obtain the following table:

$$
\text{Row 2} \rightarrow
\begin{bmatrix}
1 & 0 & 0 & 1 & 38 \\
0 & 1 & 1 & 0 & 46 \\
0 & 1 & 0 & -2 & 0
\end{bmatrix}
$$
$$
\underset{\text{Column 4}}{\uparrow}
$$

This table, with brackets enclosing its entries, is called a **matrix.** Among the many applications of matrices is the convenient, useful, and natural representation of *any* system of linear equations, as suggested earlier: Each row represents an equation, and each column, except the last, is used to keep the coefficients of a particular variable aligned. The last column consists of the constants to the right of the *equals* signs. Often, a vertical line is used in place of the equals signs to remind us that the last column is indeed the column of constants to the right of the equals signs. The relation between the equations and the matrix may be represented as follows, where the arrow denotes implication:

$$
\begin{aligned}
w \qquad\;\; + \;z &= 38 \\
x + y \qquad\;\; &= 46 \\
x \qquad - 2z &= 0
\end{aligned}
\quad \rightarrow \quad
\begin{array}{cccc}
w & x & y & z \\
\end{array}
\left[
\begin{array}{cccc|c}
1 & 0 & 0 & 1 & 38 \\
0 & 1 & 1 & 0 & 46 \\
0 & 1 & 0 & -2 & 0
\end{array}
\right]
$$

The matrix to the left of the vertical line is known as the **coefficient matrix** of the original system, for evident reason: The coefficient matrix is the matrix consisting of the coefficients of the variables. The matrix to the right of the vertical line is called the **constant matrix,** since it shows the list of constants from the equations. The entire matrix is known as the **augmented matrix** of the original system. The augmented matrix is the coefficient matrix, augmented by the column of constants. Notice that *all* of the coefficients and constants, even those that are zero, are displayed in the augmented matrix.

EXAMPLE 5

Displaying the
Coefficient, Constant,
and Augmented
Matrices

For the linear system of equations

$$x + 3y + z = 10$$
$$-x \qquad + z = 5$$
$$3x + y \qquad = 0,$$

the coefficient matrix is

$$\begin{bmatrix} 1 & 3 & 1 \\ -1 & 0 & 1 \\ 3 & 1 & 0 \end{bmatrix},$$

the constant matrix is

$$\begin{bmatrix} 10 \\ 5 \\ 0 \end{bmatrix},$$

and the augmented matrix is

$$\left[\begin{array}{ccc|c} 1 & 3 & 1 & 10 \\ -1 & 0 & 1 & 5 \\ 3 & 1 & 0 & 0 \end{array}\right].$$ ■

Each entry (element, number) of a matrix is located where some row and some column intersect, and this leads naturally to a row–column address. In the augmented matrix of Example 5, for instance, the entry -1 is located in the second-row, first-column position, and the entry 10 is in the first-row, fourth-column position. In the second-row, second-column position, there is a zero. In mathematics, it is customary to give the row number first and the column number second when stating the address of a given number in a matrix.

$$\text{First row} \rightarrow \left[\begin{array}{ccc|c} 1 & 3 & 1 & \boxed{10} \\ -1 & 0 & 1 & 5 \\ 3 & 1 & 0 & 0 \end{array}\right]$$

The number 10 is located in the
first-row, fourth-column position

↑
Fourth column

General Notation

A general element in a matrix is denoted by a_{ij} where i is the row number and j is the column number fixing the location of the element. This statement may be shortened to say that a_{ij} is the element in the ith row and jth column in the matrix.

Exercises 2.1

In Exercises 1 through 8, write the coefficient matrix, the constant matrix, and the augmented matrix for each system.

1. $x + 3y = 5$
$2x - y = -4$

2. $-3x + y = 0$
$x + 5y = 6$

3. $x + 3y = 0$
$2x + y - z = 0$
$x - y + z = 0$

4. $2x + 2y - z = 3$
$x - y = 0$
$5x + 3y + z = 5$

5. $w + 2x + 3y + z = 5$
$w + x - y = 3$
$3w + 5x + 2z = -2$

6. $x_1 + x_6 = 2$
$x_2 + x_3 - x_6 = 5$
$x_1 + x_3 + x_4 = -2$
$x_1 + x_2 + x_5 = 7$

7. $5x + 3y + 2z = 0$
$x - 2y = 0$
$3x + y + z = 1$
$x + 4z = 2$

8. $2a + 4b - c = 8$
$a + 3b + c = 6$
$a + b + c = 3$

In Exercises 9–14, rewrite each augmented matrix as a system of equations in x and y or in x, y, and z.

9. $\begin{bmatrix} 2 & 0 & | & 2 \\ 1 & 3 & | & 1 \end{bmatrix}$

10. $\begin{bmatrix} -1 & 3 & | & 1 \\ 4 & 2 & | & 6 \end{bmatrix}$

11. $\begin{bmatrix} 1 & 0 & | & -8 \\ 0 & 1 & | & 7 \end{bmatrix}$

12. $\begin{bmatrix} 0.5 & 0 & | & 2 \\ 0 & 0.5 & | & 4 \end{bmatrix}$

13. $\begin{bmatrix} 1 & 0 & 0 & | & 18 \\ 0 & 1 & 0 & | & 25 \\ 0 & 0 & 1 & | & 30 \end{bmatrix}$

14. $\begin{bmatrix} 2 & 0 & 0 & | & 10 \\ 0 & 1 & 0 & | & 20 \\ 0 & 0 & -5 & | & -15 \end{bmatrix}$

In Exercises 15 through 20, find the system of equations that models the given problem by first making a table like the one suggested in the examples. Do not attempt to solve the system.

15. Investing Emily bought two stocks: Datafix, selling for $30 per share, and Rocktite, selling for $20 per share. She invested a total of $4000. The dividend from Datafix is $2 per share, and the dividend from Rocktite is $1 per share, each year. Emily expects to receive a total of $220 in dividends from the two stocks during a given year. How many shares of each stock did she buy?

16. Training Police Officers A particular police department employs two grades of police personnel: rookies and sergeants. A person at the grade of rookie is to spend 20 hours training and 20 hours on patrol duty each week. A person at the grade of sergeant is to spend 5 hours training and 30 hours on patrol duty each week. The training center can effectively handle 240 person-hours each week, while the department needs at least 440 person-hours each week for patrol duty, for the two grades of personnel. How many persons at each grade does the department have, assuming that the training center operates at full capacity and that the minimum requirements for patrol duty are met?

17. Meeting a Pollution Allowance An electrical plant uses coal and gas to generate electricity. Coal generates 3 megawatt-hours of electricity per ton, while gas generates 4.50 megawatt-hours per thousand cubic feet. Each ton of coal produces 60 pounds of sulfur dioxide per hour, whereas each thousand cubic feet of gas produces only 1 pound of sulfur dioxide per hour. The plant wants to generate electricity at the rate of 183 megawatt-hours, while not exceeding the allowable sulfur dioxide pollution rate of 990 pounds per hour. How many tons of coal and how many thousand cubic feet of gas should the plant burn each hour to generate the desired amount of electricity while meeting the maximum pollution allowance exactly?

18. Advertising A firm wants to start an advertising campaign in newspapers, on radio, and on television to run for one month. The company wants to run a total of 34 ads in these three media. Each time a newspaper ad is run, it costs $100; a radio ad, $80; and a television ad, $400. The company has budgeted $4840 for the entire campaign. Each newspaper ad has an effective rating of 8 points; each radio ad, 3 points; and each television ad, 12 points. The company wants a total of 206 effective rating points from its campaign. Assuming that the entire advertising budget is spent and that the effective rating points are met, how many times should each ad be run?

19. Manufacturing A ballpoint-pen maker produces pens made of wood, silver, and gold. A wooden pen requires 1 minute in a grinder and 2 minutes in a bonder, a silver pen requires $\frac{1}{2}$ minute in a grinder and 2 minutes in a bonder, and a gold pen requires 3 minutes in a grinder and $2\frac{1}{2}$ minutes in a bonder. If there are 200 hours of grinder time and 160 hours of bonder time available each week, how many pens of each type can be produced, assuming that all of the available grinder and bonder time is used?

20. Manufacturing A firm makes standard, deluxe, and superdeluxe model microwave ovens. The assembly times for these ovens are, respectively, 20 minutes, 30 minutes, and 40 minutes. The painting times for the oven are, respectively, 10 minutes, 10 minutes, and 12 minutes. The total production of deluxe and superdeluxe models is to equal that of standard models. If the company has 100 hours of assembly time and 80 hours of painting time each day, and if these times are fully utilized, how many of each type of oven can be made on a daily basis?

For each of the remaining exercises, find a set of equations that models the given problem.

21. Hiring and Training Workers A particular department in a factory hires skilled workers, semiskilled workers, and supervisors. Each skilled worker is paid $12 per hour, each semiskilled worker is paid $9 per hour, and each supervisor is paid $15 per hour. The department is allowed an hourly payroll of $1560 for these three types of workers. The department requires that each worker spend some time in training and safety schooling each week: skilled workers, 3 hours; semiskilled workers, 5 hours; and supervisors, 1 hour. The training center can handle a maximum of 588 person-hours each week, and the department needs 140 skilled and semiskilled workers to meet production schedules. Assuming that the allowable payroll is met, that the training center is fully utilized, and that the department's productions needs are met, how many of each type of worker does the department need?

22. Combining Foods A dietitian wants to combine spinach and lettuce to make a 5-pound salad mixture. The caloric content per pound of spinach and lettuce is, respectively, 80 and 30 calories. The number of units of vitamin A per pound for each are, respectively, 40 and 20 units. If the dietitian wants the total caloric content to be 250 and the total number of units of vitamin A to be 140, how many pounds of each type of food should be used?

23. Making Dog Food A factory makes dog food from four ingredients: A, B, C, and D. These ingredients have respectively, 2, 1, 3, and 1 unit of vitamin A content and, respectively, 4, 6, 2, and 2 units of protein per pound. The company mixes the dog food in 200-pound batches, with 300 units of vitamin A per batch and 600 units of protein per batch. How many pounds of each of the four ingredients—A, B, C, and D—should be used to exactly meet the vitamin and protein requirements per batch?

24. Manufacturing A kitchen appliance manufacturer makes can openers and dough cutters and uses two machines: a press and a riveter. Each can opener requires 0.2 minute in the press and 0.4 minute in the riveter. A dough cutter requires 0.5 minute in the press and 0.3 minute in the riveter. The press can be operated only 3 hours per day and the riveter only 2.5 hours per day. If these two machines are fully utilized, how many can openers and dough cutters can be produced each day?

25. Exporting In a particular developing country, agricultural planners hope to have exports of soybeans, wheat, corn, and barley that will net $10 million from the coming year's crops. The planners estimate that the net profit per acre of these four commodities will be $25, $50, $40, and $30, respectively. They also estimate that the labor needed per acre to plant, harvest, and transport the four commodities will be 5, 6, 8, and 3 persons, respectively. Their country has 50,000 workers available for these operations. Assuming that the estimates are correct and that all 50,000 workers are used, how many acres of each of the four commodities should be planted to meet the planners' goal?

26. Woodworking Charlie's woodshop produces trivets, cutting boards, and bases for pen sets. The only operations involved are sawing and sanding. A trivet requires 10 minutes' sawing time and 5 minutes' sanding time, cutting boards require 5 minutes' sawing time and 5 minutes' sanding time, and pen bases require 10 minutes' sawing time and 15 minutes' sanding time. The per-day limit on sawing is 4 hours and on sanding 5 hours. If the sawing and sanding times are fully utilized, how many of each of the three items can be made each day?

27. **Hiring Workers** Brenda's Drive-In is located in a college town and has rush-hour business during the 5 o'clock hour and again during the 10 o'clock hour. To run her business efficiently, Brenda hires three shifts of workers: one to work from 10:00 A.M. to 6:00 P.M., another from 5:00 to 11:00 P.M., and a third from 10:00 P.M. to 1:00 A.M. Brenda knows from experience that she needs 10 workers during the first- and second-shift overlap and 8 workers during the second- and third-shift overlap. Can Brenda staff her worker needs for each shift if the first and third shifts are to have the same number of workers?

28. **Shipping** A discount store paid $413.28 for a shipment of two kinds of candy bars: Nutty Delite and Chewmoor. The Nutty Delite bars are sold for 50 cents each and the Chewmoor bars for 60 cents each, for a total of $508.40. The Nutty Delite bars cost 42 cents each and the Chewmoor bars cost 52 cents each. How many of each type of bar were in the shipment?

29. **Highway Budget** A state highway department has $120 million to spend on highway construction in three districts: Alma, Bedford, and Cooksville. The cost per mile for construction in the Alma district is $2 million, in the Bedford district $500,000, and in the Cooksville district $1 million. For political reasons, the department needs to build twice as many miles in the Alma district as in the Bedford district. How many miles of highway can be built in each of the three districts while meeting all of the restrictions and using all $120 million?

30. **City Budget** The city of Concord gets a matching grant of $1 million (making a total of $2 million) to spend in three categories: streets, sewers, and parks. The city council decides that the amount spent on parks should equal the total amount spent on streets and sewers and that twice as much should be spent on parks as on sewers. How much money is to be allotted to each of the categories?

31. **Buying Tires** A tire store manager is given $28,500 to buy 600 new tires for the store. Two types of tires are to be purchased: four-ply nylon, costing $40 each, and steel-belted radials, costing $50 each. The manager wants to buy three times as many steel-belted radial tires as four-ply nylon tires. Assuming that all $28,500 is spent, find a system of linear equations that gives the number of each type of tire bought.

32. **Advertising** A company is to run ads in newspapers, on radio, and on television. It wants to run a total of 20 ads in these three media. Each time a newspaper ad is run, it costs $30; a radio ad, $60; and a television ad, $200. The company has budgeted a total of $2400 for the ads. The company wants to run five times as many newspaper ads as television ads. How many times should each ad be run to meet these conditions and utilize all of the advertising budget?

33. **Mixing Gasohol** A gasoline supplier has two large gasohol tanks, one containing 8% alcohol and the other containing 13% alcohol. If 2000 gallons of gasohol with 10% alcohol is needed, how many gallons should be taken from each tank to provide the proper mixture?

34. **Mixing Nuts** A store is filling an order for 60 pounds of a mixture of walnuts, pecans, and peanuts. The mixture will sell for $4.00 per pound. Walnuts normally sell for $3.50 per pound, pecans for $6.00 per pound, and peanuts for $2.25 per pound. The order stipulates that 60% of the nuts are to be walnuts or pecans. If the income from selling the mixture should be the same as that for selling the nuts separately, how many pounds of each type of nut should be used in filling the order?

35. **Mixing Acids** Dr. McKnight, a chemistry professor, wants to make 2 liters of an 18% acid solution by mixing a 21% solution with a 14% solution. How many liters of each type of acid solution should be used to produce the 18% solution?

36. **Election Outcome** A polling firm predicts that 35% of the "yes" vote and 40% of the "no" vote in an upcoming bond election will come from urban areas. A week before the election, supporters of the bond issue estimate that 100,000 votes will be cast in the election, with 37,000 coming from urban areas. Assuming that these predictions and estimates are correct, how will the bond issue fare in the election?

Section 2.2 Linear Systems Having One or No Solutions

In Chapter 1, we saw how two different methods could be used to find points common to a pair of lines or to determine that no such points existed. There, we focused on the geometry and the signals that indicated the classification of a solution, rather than on viewing the methods as just mechanical solutions to systems of two equations and two unknowns. That set the stage for our present work. As it turns out, *every* system of linear equations, regardless of the number of equations or variables involved, falls into one of three categories:

I. There is exactly one solution.
II. There is no solution.
III. There are an infinite number of solutions.

HISTORICAL NOTE

CARL FRIEDRICH GAUSS (1777–1855), WILHELM JORDAN (1842–1899)

The German mathematician Carl Friedrich Gauss is recognized as the greatest mathematician of modern times, the last "complete mathematician." He was connected in some way with nearly every aspect of mathematics during his career, but held number theory in particular in high esteem. Wilhelm Jordan was a German surveyor who presented the elimination method as a way to solve particular surveying problems.

For systems of equations involving three or more variables, it is not as easy to tell by inspection into which category a particular system falls, and the geometry will generally not be available for visual support. So we need a systematic approach that allows us to recognize the relationships between the equations in the system, which, in turn, will tell us why the system falls into a particular category and why the solutions (if any) take on the appearance they do. Only when the nature of solutions to linear systems of equations is understood can we turn the complete task of solution over to technology. Then we can interpret the solution correctly, recognize when particular programs will not solve particular types of systems of linear equations, and know exactly what to seek in a more powerful program or how to proceed manually with a solution if all else fails. Undoubtedly, the most efficient approach to fulfilling the preceding objectives is to use a technique known as the **Gauss–Jordan elimination method,** which is actually an extension of the elimination method discussed in Chapter 1. The objective is to systematically eliminate enough variables so that the solutions, if any, may be found with little or no effort. The Gauss–Jordan method makes use of three basic operations that may be applied to any system of linear equations, each of which *leaves unchanged* any solutions of the original system.

Basic Operations on Any Linear System That Leave the Solutions, if Any, Unchanged

1. Interchange any two equations.
2. Replace any equation by a nonzero multiple of that equation.
3. Replace any equation by that equation plus a multiple of another equation.

NOTE In the discussion that follows, we will begin to develop a method of transforming a system of linear equations into an **equivalent** system with the same solution(s), but for which the solution(s) is (are) relatively easy to determine. Paying careful attention now to the technique and how it works will be crucial in later chapters.

We should think about *why* each of the preceding operations produces a new system of equations with the *same solution(s)* as the original system. Consider the linear system

$$\begin{aligned} x + y &= 3 \\ 2x + y &= 4. \end{aligned}$$

(a)

Using the methods of Section 1.4, we find that the solution of the system is $x = 1$, $y = 2$. Suppose we rewrote the system as

$$\begin{aligned} 2x + y &= 4 \\ x + y &= 3. \end{aligned}$$

(b)

We have applied the first operation: We have interchanged the two equations. But have we changed the solution of the system? No, we have merely changed the order in which we wrote the equations. The solution is still $x = 1$, $y = 2$.

Now suppose we rewrote the original system, (a), as

$$\begin{aligned} -2x + -2y &= -6 \\ 2x + y &= 4. \end{aligned}$$

(c)

A quick inspection verifies that $x = 1$, $y = 2$ is still a solution of the system. The methods of Section 1.4 will also verify that this is the only solution. So, what have we done? We have multiplied the first equation by -2. That is, we have applied the second operation, "Replace any equation by a nonzero multiple of that equation." The system *looks* different, but has the same solution.

Finally, let's multiply the first equation in (a) by -2, add the result to the second equation, and replace the second equation with that result. This would give us

$$\begin{aligned} x + y &= 3 \\ -y &= -2. \end{aligned}$$

The first equation has not changed, but the second looks completely different. We have eliminated x from the second equation, but not from the entire system. Another quick inspection verifies that the solution is still $x = 1$, $y = 2$. Thus, we have applied the third operation, but have not changed the solution of the system.

The next example shows how the foregoing three operations may be used to find the solution of a system of linear equations. Observe carefully the pattern that evolves. We do not randomly choose which operation to use in each step; instead, we have a plan. Our plan is to eventually eliminate all but the first variable from the first equation, all but the second variable from the second equation, and so on, until we have a set of equations that gives the solution of the system—if such a solution exists.

EXAMPLE 1

Illustrating the
Gauss–Jordan Method
on a System of
Equations

Solve the following system of equations by the Gauss–Jordan method, and write the augmented matrix for each equivalent system of equations obtained in the process:

$$\begin{aligned} 2x + y - z &= -2 \\ y + 2x &= 2 \\ x - y + z &= 5 \end{aligned} \qquad \begin{bmatrix} 2 & 1 & -1 & -2 \\ 0 & 1 & 2 & 2 \\ 1 & -1 & 1 & 5 \end{bmatrix}.$$

SOLUTION First, we find an equivalent system that has a 1 as the leading coefficient of the first equation. This may be done a few different ways, but we choose to interchange the first and last equations, giving this equivalent system:

$$\begin{aligned} x - y + z &= 5 \\ y + 2z &= 2 \\ 2x + y - z &= -2 \end{aligned} \qquad \begin{bmatrix} 1 & -1 & 1 & 5 \\ 0 & 1 & 2 & 2 \\ 2 & 1 & -1 & -2 \end{bmatrix}.$$

Next, we replace the last equation by itself plus -2 times the first equation; this will eliminate the x variable in the third equation:

$$\begin{aligned} x - y + z &= 5 \\ y + 2z &= 2 \\ 3y - 3z &= -12 \end{aligned} \qquad \begin{bmatrix} 1 & -1 & 1 & 5 \\ 0 & 1 & 2 & 2 \\ 0 & 3 & -3 & -12 \end{bmatrix}.$$

Now we replace the first equation with itself plus 1 times the second equation:

$$\begin{aligned} x \quad + 3z &= 7 \\ y + 2z &= 2 \\ 3y - 3z &= -12 \end{aligned} \qquad \begin{bmatrix} 1 & 0 & 3 & 7 \\ 0 & 1 & 2 & 2 \\ 0 & 3 & -3 & -12 \end{bmatrix}.$$

To complete our work with the second variable, we replace the last equation by itself plus -3 times the second equation; this eliminates the y variables in the first and third equations:

$$\begin{aligned} x \quad + 3z &= 7 \\ y \quad 2z &= 2 \\ -9z &= -18 \end{aligned} \qquad \begin{bmatrix} 1 & 0 & 3 & 7 \\ 0 & 1 & 2 & 2 \\ 0 & 0 & -9 & -18 \end{bmatrix}.$$

The last equation is now replaced by $-\frac{1}{9}$ times itself; this puts a 1 in the third-row, third-column position:

$$\begin{aligned} x \quad + 3z &= 7 \\ y + 2z &= 2 \\ z &= 2 \end{aligned} \qquad \begin{bmatrix} 1 & 0 & 3 & 7 \\ 0 & 1 & 2 & 2 \\ 0 & 0 & 1 & 2 \end{bmatrix}.$$

Next, we wish to eliminate z from the first and second equations, so we replace the second equation by itself plus -2 times the last equation:

$$
\begin{aligned}
x \quad + 3z &= 7 \\
y \qquad &= -2 \\
z &= 2
\end{aligned}
\qquad
\left[\begin{array}{ccc|c}
1 & 0 & 3 & 7 \\
0 & 1 & 0 & -2 \\
0 & 0 & 1 & -2
\end{array}\right].
$$

Finally, we replace the first equation by itself plus -3 times the last equation; this eliminates the z variables from the first and second equations:

$$
\begin{aligned}
x \qquad &= 1 \\
y \qquad &= -2 \\
z &= 2
\end{aligned}
\qquad
\left[\begin{array}{ccc|c}
1 & 0 & 0 & 1 \\
0 & 1 & 0 & -2 \\
0 & 0 & 1 & 2
\end{array}\right].
$$

This last equivalent system vividly displays the solution. ■

A study of the augmented matrices in Example 1 suggests that **basic row operations on a matrix,** analogous to basic operations on equations, could be defined and used to arrive at the same solution. Such row operations are defined next, using the shorthand notation R_i to represent row i and an arrow ($\rightarrow$) to be read as "replaces," both with the understanding that a "multiple of a row" means to multiply every number in that row by the same constant.

Basic Row Operations on a Matrix

1. Interchange any two rows. $(R_i \leftrightarrow R_k)$
2. Replace any row by a nonzero multiple of itself. $(cR_i \rightarrow R_i)$
3. Replace any row by itself plus a multiple of another row. $(aR_i + R_k \rightarrow R_k)$

Note: The arrow $\rightarrow$ is read as "replaces."

From a matrix point of view, the use of these three operations would give a series of **row-equivalent** matrices, in the sense that each matrix so obtained has rows that represent a different-appearing system of equations, but that *have the same solution as the original system.* At some point, we want to arrive at a matrix that will enable us to easily categorize the system (as one with exactly one solution, no solution, or infinitely many solutions) and allow us to write the solution(s) with minimum effort. We suggest working toward the nice pattern of 0's and 1's exhibited in the last augmented matrix of Example 1. An efficient way to reach this goal is to work *column by column from left to right:*

1. First, use row operations 1 and 2, if necessary, to put a 1 in the first-row, first-column position, and *then* use row operation 3 to rewrite the remaining rows, if necessary, to eliminate all other nonzero entries in the first column.
2. Next, use row operations 1 and 2, if necessary, to put a 1 in the second-row, second-column position, and *then* use row operation 3 to rewrite the remaining rows, if necessary, to eliminate all other nonzero entries in the second column.

3. Continue the pattern as follows:

$$
\begin{bmatrix} 1 & \cdots \\ 0 & \cdots \\ 0 & \cdots \\ 0 & \cdots \end{bmatrix} \rightarrow
\begin{bmatrix} 1 & 0 & \cdots \\ 0 & 1 & \cdots \\ 0 & 0 & \cdots \\ 0 & 0 & \cdots \end{bmatrix} \rightarrow
\begin{bmatrix} 1 & 0 & 0 & \cdots \\ 0 & 1 & 0 & \cdots \\ 0 & 0 & 1 & \cdots \\ 0 & 0 & 0 & \cdots \end{bmatrix} \rightarrow
\begin{bmatrix} 1 & 0 & 0 & 0 & \cdots \\ 0 & 1 & 0 & 0 & \cdots \\ 0 & 0 & 1 & 0 & \cdots \\ 0 & 0 & 0 & 1 & \cdots \end{bmatrix}.
$$

NOTE For some systems, the preceding process will hit "snags." It is precisely these systems that have infinitely many or no solutions. We discuss these later in this section and the next.

In general, a pivot operation consists of the following: the use of row operation 2 to turn any nonzero element in a matrix into a 1, followed by multiple uses of row operation 3 to introduce zeros into all remaining entries of the column in which that element resides. As a memory device, remember that a pivot operation consists of a "one" step followed by several "zero" steps.

The Pivot Operation

The pivot on a nonzero number a in a matrix means to follow these two steps (the number a is called the **pivot element,** the row in which a resides is called the **pivot row,** and the column in which a resides is called the **pivot column**):

1. First, if $a \neq 1$, replace a with 1 by multiplying the pivot row by $1/a$. If $a = 1$, proceed to the next step.
2. If necessary, use row operation 3 repeatedly to replace each nonzero number in the pivot column with a zero by replacing the row in which that nonzero number resides with itself plus an appropriate multiple of the pivot row.

NOTE Once the first step has been completed, the pivot row does not change in appearance throughout the remainder of the pivot operation; only those rows into which zeros are introduced change their appearance.

EXAMPLE 2

Details on How to Perform a Pivot Operation

Pivot on the number located in the first-row, first-column position of this matrix:

$$
\begin{bmatrix} -2 & 1 & 4 & | & 0 \\ 4 & 1 & 8 & | & 3 \\ -3 & 2 & 1 & | & 4 \end{bmatrix}.
$$

SOLUTION According to the previous definitions, row 1 is the pivot row, column 1 is the pivot column, and -2 is the pivot element:

Pivot row $\rightarrow \begin{bmatrix} \boxed{-2} & 1 & 4 & | & 0 \\ 4 & 1 & 8 & | & 3 \\ -3 & 2 & 1 & | & 4 \end{bmatrix}.$ The pivot element is -2.

$\uparrow$
Pivot column

Since the pivot element is not a 1, replace row 1 by $-\frac{1}{2}$ times itself; that is, multiply every number in row 1 by $-\frac{1}{2}$, and use the resulting numbers as the first row of a new matrix, leaving the second and third rows the same as before:

$$-\tfrac{1}{2}R_1 \rightarrow R_1 \quad \begin{bmatrix} ① & -\tfrac{1}{2} & -2 & \big| & 0 \\ 4 & 1 & 8 & \big| & 3 \\ -3 & 2 & 1 & \big| & 4 \end{bmatrix}.$$

Now use the newly appearing pivot row that has a 1 in the pivot position, along with row operation 3, to change the second and third rows so that each has a zero in the first-column position. To accomplish this, the multiplier -4 needs to be applied to the first row, so that when the resulting numbers are added to the second row, a zero is obtained in the first-column position, and, separately, the multiplier 3 needs to be applied to the first row, so that when these resulting numbers are added to the third row, the first-column position also becomes a zero. (This is the beauty of having a 1 in the pivot position; to turn a nonzero number k into zero, multiply the pivot row by $-k$ and add.) We obtain

$$\begin{matrix} \\ -4R_1 + R_2 \rightarrow R_2 \\ 3R_1 + R_3 \rightarrow R_3 \end{matrix} \begin{bmatrix} 1 & -\tfrac{1}{2} & -2 & \big| & 0 \\ 0 & 3 & 16 & \big| & 3 \\ 0 & \tfrac{1}{2} & -5 & \big| & 4 \end{bmatrix}.$$

The pivot operation is now complete. ▬

The next example demonstrates the selection of pivot elements and the details leading to the solution of a particular system of linear equations.

EXAMPLE 3

Solving a System of Equations by Pivoting

Solve the following system of three equations in three variables:

$$\begin{aligned} x \quad\quad + z &= 4 \\ 2y + z &= 7 \\ 2x + y - z &= 1. \end{aligned}$$

SOLUTION The first step is to form the augmented matrix for this system:

$$\begin{matrix} x & y & z & \\ \begin{bmatrix} ① & 0 & 1 & \big| & 4 \\ 0 & 2 & 1 & \big| & 7 \\ 2 & 1 & -1 & \big| & 1 \end{bmatrix}. \end{matrix}$$

Then, working with the first column, we want the number in the first-row, first column position (the circled 1) to be our first pivot element. Since that number is already $+1$, we use appropriate multiples of row 1 to eliminate any other nonzero number in the first column:

$$\begin{matrix} & x & y & z & \\ & \begin{bmatrix} 1 & 0 & 1 & \big| & 4 \\ 0 & ② & 1 & \big| & 7 \\ -2R_1 + R_3 \rightarrow R_3 & 0 & 1 & -3 & \big| & -7 \end{bmatrix}. \end{matrix}$$

The first pivot operation is now complete. Next, working with the second column, we select the element in the second-row, second-column position, which is the circled 2 of the matrix, as our second pivot element. Since that number is not $+1$, the pivot operation is a two-stage process: First we replace the second row by $\frac{1}{2}$ times itself, and then use appropriate multiples of row 2 to eliminate any nonzero numbers in the second column:

$$\frac{1}{2}R_2 \to R_2 \quad \begin{array}{c} x \quad y \quad z \\ \begin{bmatrix} 1 & 0 & 1 & 4 \\ 0 & \textcircled{1} & \frac{1}{2} & \frac{7}{2} \\ 0 & 1 & -3 & -7 \end{bmatrix} \end{array} \quad -1R_2 + R_3 \to R_3 \quad \begin{array}{c} x \quad y \quad z \\ \begin{bmatrix} 1 & 0 & 1 & 4 \\ 0 & 1 & \frac{1}{2} & \frac{7}{2} \\ 0 & 0 & \boxed{-\frac{7}{2}} & -\frac{21}{2} \end{bmatrix} \end{array}.$$

The second pivot operation is now complete. Finally, working with the third column, we selected the circled $-\frac{7}{2}$ in the third-row, third-column position of the last matrix to be our third pivot element. Since that number is not $+1$, the pivot operation is again a two-stage process:

$$-\frac{2}{7}R_3 \to R_3 \quad \begin{array}{c} x \quad y \quad z \\ \begin{bmatrix} 1 & 0 & 1 & 4 \\ 0 & 1 & \frac{1}{2} & \frac{7}{2} \\ 0 & 0 & \textcircled{1} & 3 \end{bmatrix} \end{array} \quad \begin{array}{c} -1R_3 + R_1 \to R_1 \\ -\frac{1}{2}R_3 + R_2 \to R_2 \end{array} \quad \begin{array}{c} x \quad y \quad z \\ \begin{bmatrix} 1 & 0 & 0 & 1 \\ 0 & 1 & 0 & 2 \\ 0 & 0 & 1 & 3 \end{bmatrix} \end{array}.$$

Writing the equations represented in the last matrix gives the solution:

$$x = 1, \qquad y = 2, \qquad z = 3. \quad \blacksquare$$

EXAMPLE 4

Solving a System of Equations by Pivoting

Solve the following system of four equations in three variables:

$$\begin{aligned} x \quad\;\; + z &= 2 \\ y - z &= 4 \\ 2x + y \quad\;\; &= 5 \\ x + y \quad\;\; &= 6. \end{aligned}$$

SOLUTION The first step is to form the augmented matrix for the system:

$$\begin{bmatrix} \textcircled{1} & 0 & 1 & 2 \\ 0 & 1 & -1 & 4 \\ 2 & 1 & 0 & 5 \\ 1 & 1 & 0 & 6 \end{bmatrix}.$$

Then, working with the first column, we want the first-row, first-column element to be our first pivot element. In this case, that element is already a 1. The remaining steps are shown next:

$$\begin{array}{c} -2R_1 + R_3 \to R_3 \\ -1R_1 + R_4 \to R_4 \end{array} \quad \begin{bmatrix} 1 & 0 & 1 & 2 \\ 0 & \textcircled{1} & -1 & 4 \\ 0 & 1 & -2 & 1 \\ 0 & 1 & -1 & 4 \end{bmatrix}.$$

Now we pivot on the element in the second-row, second-column position in the preceding matrix. Here is the result:

$$
\begin{array}{c}
\\
\\
-1R_2 + R_3 \rightarrow R_3 \\
-1R_2 + R_4 \rightarrow R_4
\end{array}
\left[\begin{array}{ccc|c}
1 & 0 & 1 & 2 \\
0 & 1 & -1 & 4 \\
0 & 0 & \boxed{-1} & -3 \\
0 & 0 & 0 & 0
\end{array}\right].
$$

The row of zeros in the fourth row is no cause for alarm. It is a signal that the last equation is a linear combination of the others and contributes nothing to the solution. Such a system is called **dependent.** Finally, we pivot on the -1 in the third-row, third-column position. Because this element is not a $+1$, the pivoting operation will be completed in *two* steps: First, multiply the third row by -1; then turn all other entries in the column into zeros. Here are the details:

$$
\begin{array}{c}
\\
\\
-1R_3 \rightarrow R_3 \\
\\
\end{array}
\left[\begin{array}{ccc|c}
1 & 0 & 1 & 2 \\
0 & 1 & -1 & 4 \\
0 & 0 & \boxed{1} & 3 \\
0 & 0 & 0 & 0
\end{array}\right]
\quad
\begin{array}{c}
-1R_3 + R_1 \rightarrow R_1 \\
1R_3 + R_2 \rightarrow R_2 \\
\\
\\
\end{array}
\left[\begin{array}{ccc|c}
1 & 0 & 0 & -1 \\
0 & 1 & 0 & 7 \\
0 & 0 & 1 & 3 \\
0 & 0 & 0 & 0
\end{array}\right].
$$

The solution of the original system may now be read from the nonzero rows of the last matrix after recalling that the first column represents coefficients of x, the second column coefficients of y, and the third column coefficients of z:

$$
x = -1, \qquad y = 7, \qquad z = 3. \quad \blacksquare
$$

Dependent and Independent Systems of Equations

A system of n equations is called **dependent** if fewer than n of the equations will give the same solution as the original set of equations. Equivalently, at least one of the equations in the system can be obtained by adding multiples of other equations in the system. If a system of equations is not dependent, we call the system **independent.**

Because dependent systems of linear equations in more than two variables are usually difficult or impossible to detect at the outset, one advantage of the pivoting approach is that each equation which is not needed to determine the solution is turned into a row of zeros in the augmented matrix (see Example 4) and is effectively eliminated from consideration. We will see by the end of the chapter that the pivoting approach gives a systematic way to solve *any* system of linear equations.

Sometimes, the orderly progression of pivoting successively down the diagonal can be carried out, but only after an interchange of rows is performed. For instance, if an augmented matrix looks like

$$
\left[\begin{array}{ccc|c}
1 & 3 & 4 & 5 \\
0 & 0 & 2 & 1 \\
0 & 1 & 3 & -2
\end{array}\right],
$$

the next pivot element should be a one in the second-row, second-column position; but there is a zero there, making a pivot operation impossible. However if the second and third rows are interchanged, a 1 will be in the appropriate position and not only can a pivot operation take place, but the interchange won't destroy the work we have done with the first column. The integrity of columns will be preserved on any interchange with rows *below* the desired pivot row.

At other times, it may be impossible to select successive pivot elements down the diagonal and obtain the "ideal" arrangements of 1's and 0's exhibited in the coefficient matrices of Examples 1 and 3. For example, in the matrix

$$\begin{bmatrix} 1 & -4 & 3 & | & 5 \\ 0 & 0 & 2 & | & 3 \\ 0 & 0 & 6 & | & 3 \end{bmatrix},$$

it is impossible to obtain a nonzero entry in the second-row, second-column position upon which to pivot. In such a case, we do the next best thing: seek a pivot element in the second-row, *third-column* position. Now we can pivot on the 2 in that position and proceed with the solution of the original system of equations. The goal is to get as close as possible to the "ideal" arrangement of 1's and 0's in the coefficient matrix. When this is done, the coefficient matrix is said to be in **reduced row-echelon form.**

The Reduced Row-Echelon Form of a Matrix

A matrix is in **reduced row-echelon form** if the following three conditions are satisfied:

1. For each nonzero row, the first nonzero element is a 1.
2. A row with fewer leading zeros is above a row with more leading zeros.
3. The first nonzero element in a row is the only nonzero entry in its column.

EXAMPLE 5

The Reduced Row-Echelon Form of a Matrix

These matrices are all in reduced row-echelon form:

$$\begin{bmatrix} 1 & 0 & 0 \\ 0 & 1 & 0 \\ 0 & 0 & 1 \end{bmatrix} \quad \begin{bmatrix} 1 & 0 & 0 \\ 0 & 1 & 3 \\ 0 & 0 & 0 \end{bmatrix} \quad \begin{bmatrix} 1 & 0 & 0 & 0 & 8 \\ 0 & 0 & 1 & 6 & 7 \end{bmatrix} \quad \begin{bmatrix} 1 & 0 & 0 & 8 \\ 0 & 0 & 1 & 0 \\ 0 & 0 & 0 & 0 \end{bmatrix}$$

None of these matrices are in reduced row-echelon form.

$$\begin{bmatrix} 1 & 0 & 0 \\ 0 & 2 & 3 \\ 0 & 0 & 0 \end{bmatrix} \quad \begin{bmatrix} 1 & 0 & 0 & 4 \\ 0 & 1 & 1 & 9 \\ 0 & 0 & 1 & 0 \end{bmatrix} \quad \begin{bmatrix} 1 & 0 & 0 & 8 \\ 0 & 0 & 0 & 0 \\ 0 & 0 & 1 & 0 \end{bmatrix} \blacksquare$$

The reduced row-echelon form of the coefficient matrix in the next example happens to lead to the conclusion that no solution of the system exists.

EXAMPLE 6

A System of Equations
with No Solution

Solve the following system:

$$2x + y + 4z + 2w = 6$$
$$2z + w = 4$$
$$4z + 2w = 3.$$

SOLUTION

$$\begin{bmatrix} ② & 1 & 4 & 2 & | & 6 \\ 0 & 0 & 2 & 1 & | & 4 \\ 0 & 0 & 4 & 2 & | & 3 \end{bmatrix} \quad \frac{1}{2}R_1 \rightarrow R_1 \begin{bmatrix} 1 & \frac{1}{2} & 2 & 1 & | & 3 \\ 0 & 0 & ② & 1 & | & 4 \\ 0 & 0 & 4 & 2 & | & 3 \end{bmatrix}$$

$$\frac{1}{2}R_2 \rightarrow R_2 \begin{bmatrix} 1 & \frac{1}{2} & 2 & 1 & | & 3 \\ 0 & 0 & ① & \frac{1}{2} & | & 2 \\ 0 & 0 & 4 & 2 & | & 3 \end{bmatrix}$$

$$\begin{matrix} -2R_2 + R_1 \rightarrow R_1 \\ \\ -4R_2 + R_3 \rightarrow R_3 \end{matrix} \begin{bmatrix} 1 & \frac{1}{2} & 0 & 0 & | & -1 \\ 0 & 0 & 1 & \frac{1}{2} & | & 2 \\ 0 & 0 & 0 & 0 & | & -5 \end{bmatrix}.$$

The last row corresponds to the equation $0x + 0y + 0z + 0w = -5$, which has no solution. This means that the entire system has no solution. ∎

Detecting When a System Has No Solution

At any stage in the pivoting process, if a row contains all zeros to the left of the vertical bar and a nonzero number to the right, then the system has **no solution.**

There are several variations of the Gauss–Jordan elimination process that we used as a pivoting process. One variation is to (1) introduce zeros only below the main diagonal of the augmented matrix, (2) find the value of the last variable, and (3) then substitute back to find the values of the other variables. Another variation is to turn the pivot element into some convenient number (not necessarily 1) in order to avoid fractions. Even so, the objective is the same: Identify an equivalent system of equations from which the solutions, if any, can easily be found.

Now that we know about dependent systems and recognize the signal that a system has no solution, we are in a position to take advantage of available technology to do this work for us. Methods for Excel and the TI–83 Plus are outlined next. Other graphing calculators and spreadsheet packages have similar capabilities.

EXAMPLE 7

Using Excel's Solver to
Find the Solution of a
System of Equations

Use Excel's built-in `Solver` to solve the system of equations in Example 3. (Note: `Solver` should be located in the `Tools` menu of Excel. If it is *not*, you will need to locate your original installation disk, select `Add-Ins` from the `Tools` menu, and install the `Solver` program.)

SOLUTION Set up an Excel worksheet that contains as much helpful information as possible, including headings for rows, columns, and even individual cells, much like the tables we set up in the first section of this chapter. To complete this problem with the use of `Solver`, we can set up a worksheet similar to that in Figure 2. Entries in all cells are typed exactly as shown in the figure, except we enter following formulas in cells `B3`, `B4` and `B5`:

Cell	Contents
B3	=B1+F1
B4	=2*D1+F1
B5	=2*B1+D1-F1

	A	B	C	D	E	F
1	x =	0	y =	0	z =	0
2						
3	x+z = 4	0	=	4		
4	2y+z = 7	0	=	7		
5	2x+y-z = 1	0	=	1		

FIGURE 2
Spreadsheet Containing the System of Equations

From the `Tools` menu, select `Solver`. Type in the address of the cell containing the first formula (in this case, cell `B3`) as the "Target Cell." Then check "Value of" and type in the value of the constant for the first equation, which is 4 in this example. The cells we wish to change are the cells next to the labels for *x*, *y*, and *z*, so type `B1,D1,F1` into the box below "By Changing Cells". Our constraints here are the equations themselves, and we have already typed these in. Select `Add`, and type in the name of the cell containing the first formula. Select "=" from the drop-down menu, and then type in the name of the cell containing the constant from the first equation. If your worksheet is set up like Figure 2, then the `Add` window should appear as in Figure 3. If so, click `Add`.

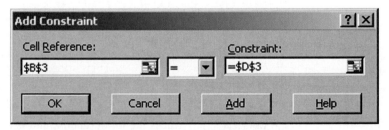

FIGURE 3
Adding Constraints to Solver

Now we need to add the second equation as a constraint, so type in `B4` as the "Cell Reference", select "=" from the drop-down menu, and enter `D4` as the "Constraint". We need to add one more equation, so click `Add` again and repeat the process on the third equation.

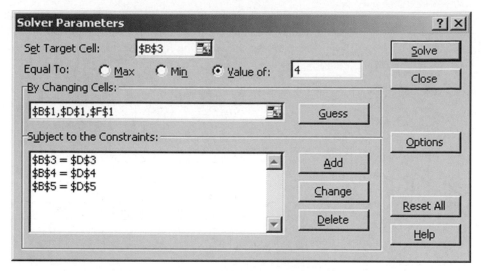

FIGURE 4
Solver Parameters for Example 7

When it is entered, click OK. This will return you to the Solver window, which should now look like Figure 4.

Click Solve. The program will now adjust the values of the variables until all constraints are satisfied, if possible. The top row of the worksheet should now contain the solution of the system of equations: $x = 1$, $y = 2$, and $z = 3$. ∎

NOTE Solver cannot solve systems of equations that have no solution. If you try to use it to solve that type of system, a new window will appear letting you know that Solver was unable to locate a solution of the system.

EXAMPLE 8

Graphing Calculator Method

Use the rref command—the *r*educed *r*ow *e*chelon *f*orm on the TI–83 Plus—to solve the system of equations from Example 3.

SOLUTION We begin by entering the augmented matrix for the system:

$$\begin{bmatrix} 1 & 0 & 1 & | & 4 \\ 0 & 2 & 1 & | & 7 \\ 2 & 1 & -1 & | & 1 \end{bmatrix}$$

Press 2nd x^{-1} to access the MATRX menu, and use the arrow keys to move to the EDIT submenu. We will enter the matrix as [A], so select this matrix name. The matrix has three rows and four columns, so type 3 ENTER 4 ENTER to input these dimensions. Now type in each entry, starting from the first row and pressing ENTER after each. When you finish, QUIT to return to the home screen. Return to the MATRX menu and use the arrow keys to move to the MATH submenu. Scroll down until you find the rref command and select it. Again, access the MATRX menu and select the matrix name, [A].

Press ENTER to see the reduced row-echelon form of our augmented matrix. The result should appear as in Figure 5.

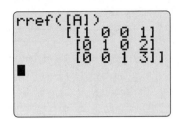

FIGURE 5
TI–83 Plus Screen Showing the
Result of Example 8 ■

Exercises 2.2

In Exercises 1–4, perform the indicated operation(s) on the given augmented matrix.

1. $\begin{bmatrix} 1 & 1 & 1 & | & 2 \\ 0 & 2 & 0 & | & 4 \\ 0 & 1 & 3 & | & 2 \end{bmatrix}; \; -\dfrac{1}{2}R_2 \rightarrow R_2$

2. $\begin{bmatrix} 1 & -2 & 1 & | & 3 \\ 0 & 1 & 2 & | & 1 \\ 0 & 0 & 3 & | & 6 \end{bmatrix}; \; -2R_2 + R_1 \rightarrow R_1$

3. $\begin{bmatrix} 1 & 4 & 0 & | & 5 \\ 0 & ② & 2 & | & 4 \\ 0 & 1 & -1 & | & 0 \end{bmatrix};$ complete the pivot operation on the circled element.

4. $\begin{bmatrix} ④ & 3 & | & -3 \\ 1 & 2 & | & 1 \\ 5 & -1 & | & -2 \\ 9 & -4 & | & 3 \end{bmatrix};$ complete the pivot operation on the circled element.

In Exercises 5–8, use the pivoting process to put the augmented matrix into reduced row-echelon form.

5. $\begin{bmatrix} 1 & 2 & | & 4 \\ 0 & 2 & | & 2 \end{bmatrix}$

6. $\begin{bmatrix} 1 & -3 & | & 2 \\ 0 & 4 & | & -4 \end{bmatrix}$

7. $\begin{bmatrix} 1 & 2 & 1 & | & 7 \\ 0 & 1 & 3 & | & 4 \\ 0 & -1 & 0 & | & 0 \end{bmatrix}$

8. $\begin{bmatrix} 1 & 2 & 5 & | & 0 \\ 0 & 2 & 4 & | & 0 \\ 1 & 0 & 1 & | & 2 \end{bmatrix}$

answer should be
$x = \dfrac{17}{3} \quad y = 0 \quad z = \dfrac{4}{3}$

In Exercises 9 through 26, use the pivoting process to find the solution, if one exists.

9.
$$\begin{aligned} x + y + z &= 4 \\ y - z &= -1 \\ x + 3y + z &= 6 \end{aligned}$$

10.
$$\begin{aligned} 2x + y + 2z &= -5 \\ x - y + z &= -5 \\ x + y \phantom{{}+2z} &= 2 \end{aligned}$$

11.
$$\begin{aligned} y - 3z &= -11 \\ x - y + z &= 5 \\ 3x \phantom{{}+y} + z &= 10 \end{aligned}$$

12.
$$\begin{aligned} y + 5z &= 17 \\ x + 2y - z &= 2 \\ 2x - y + 3z &= 9 \end{aligned}$$

project #2

13. *(circled)*
$$\begin{aligned} 2x + 4y + 2z &= 5 \\ x + y + z &= \tfrac{3}{2} \\ 2x + y + 4z &= 4 \end{aligned}$$

$B3 = D3$

14.
$$\begin{aligned} 3x + 3y + z &= \tfrac{8}{3} \\ x - y + z &= 1 \\ 3x \phantom{{}+y} + z &= -1 \end{aligned}$$

15.
$$\begin{aligned} x + 2y + z &= 3 \\ 3x + 3y + 3z &= 7 \\ 2x + y + 2z &= 1 \end{aligned}$$

16.
$$\begin{aligned} 3x \phantom{{}-y} - z &= 1 \\ 2x - y + z &= 4 \\ x + y - 2z &= 3 \end{aligned}$$

17.
$$\begin{aligned} x + 3y &= 8 \\ 2x - y &= 4 \end{aligned}$$

18.
$$\begin{aligned} 3x - y &= 5 \\ x + 2y &= 0 \end{aligned}$$

19.
$$\begin{aligned} 3x + y &= -1 \\ x - 2y &= 3 \\ 4x - y &= 2 \\ 8x + 5y &= -6 \end{aligned}$$

20.
$$\begin{aligned} 4x - 3y &= -3 \\ x + 2y &= 1 \\ 5x - y &= -2 \\ 9x - 4y &= 3 \end{aligned}$$

21.
$$\begin{aligned} x \phantom{{}+y} + z &= -2 \\ x + y + z &= -2 \\ 3x + 2y + 2z &= -3 \\ 2x + y \phantom{{}+z} &= 2 \end{aligned}$$

22.
$$\begin{aligned} 3x + y \phantom{{}+z} &= 9 \\ x - y + z &= 4 \\ 3x \phantom{{}+y} + z &= 11 \\ 4x - y + 2z &= 15 \end{aligned}$$

23.
$$x + y - 3z = 2$$
$$2x \quad\quad - z = 1$$
$$3x - 2y \quad\quad = 0$$
$$6x - y - 3z = 5$$

24.
$$y + 4z = 7$$
$$2x + y - z = 3$$
$$x - y + z = 1$$
$$3x + y + 5z = 0$$

25.
$$x + y + 3z = 0$$
$$\tfrac{1}{2}x - y - \tfrac{7}{2}z = 0$$
$$x + 2y \quad\quad = 0$$

26.
$$3x - y + z = 0$$
$$x \quad\quad + 3z = 0$$
$$2x + 3y + z = 0$$

In Exercises 27 through 32, use a calculator or computer to find the solution, if one exists.

27.
$$w + x + y - z = 1$$
$$2w + \tfrac{1}{2}x \quad\quad + \tfrac{2}{3}z = \tfrac{5}{12}$$
$$-w - x + \tfrac{1}{5}y \quad\quad = \tfrac{7}{20}$$
$$w + \tfrac{3}{4}x + y + 4z = \tfrac{17}{8}$$

28.
$$\tfrac{2}{5}w + \quad\quad \tfrac{1}{5}y - z = -\tfrac{34}{15}$$
$$-\tfrac{283}{100}w + 2x + y + z = 0$$
$$\tfrac{21}{20}w + \tfrac{2}{5}x + \tfrac{25}{10}y \quad\quad = 2$$
$$\tfrac{3}{10}w + \tfrac{1}{3}x + \tfrac{2}{5}y - z = -\tfrac{28}{15}$$

29.
$$0.2x - 0.82y + 0.03z = -1.96$$
$$0.02x + 0.14y - 0.6z = -2.82$$
$$0.47x + 0.22y + 0.62z = 3.92$$

30.
$$2x + 0.3y + 0.82z = 7.72$$
$$0.82x + 3y - 0.02z = 11.44$$
$$0.57x - y + z = 4$$
$$3.39x + 2y - z = 5$$

31.
$$w - x + 2y + z = 9$$
$$0.3w \quad\quad + 0.52y - z = 1.62$$
$$0.6w + 2x - y \quad\quad = 3$$
$$0.2w + 0.43x + 0.6y + z = 2.15$$

32.
$$2w + 0.6x + 0.3y - z = 2.80$$
$$0.8w + 0.2x + 4y + z = 21.13$$
$$0.36w + 0.4x - 3y \quad\quad = -12.41$$
$$w - x + y + 0.8z = 3.50$$

In the remaining exercises, first set up the system of equations involved, and then solve by using the pivoting process. Use technology where appropriate. You have already set up some of these exercises in Section 2.1.

33. Meeting a Pollution Allowance An electrical plant uses coal and gas to generate electricity. Coal generates 3 megawatt-hours of electricity per ton, while gas generates 4.50 megawatt-hours per thousand cubit feet. Each ton of coal produces 60 pounds of sulfur dioxide per hour, whereas each thousand cubic feet of gas produces only 1 pound of sulfur dioxide per hour. The plant wants to generate electricity at the rate of 183 megawatt-hours, while not exceeding the allowable sulfur dioxide pollution rate of 990 pounds per hour. How many tons of coal and how many thousand cubic feet of gas should the plant burn each hour to generate the desired amount of electricity while meeting the maximum pollution allowance exactly?

34. Hiring and Training Workers A particular department in a factory hires skilled workers, semiskilled workers, and supervisors. Each skilled worker is paid $12 per hour, each semiskilled worker is paid $9 per hour, and each supervisor is paid $15 per hour. The department is allowed an hourly payroll of $1560 for these three types of workers. The department requires that each worker spend some time in training and safety schooling each week: skilled workers, 3 hours; semiskilled workers, 5 hours; and supervisors, 1 hour. The training center can handle a maximum of 588 person-hours each week, and the department needs 140 skilled and semiskilled workers to meet production schedules. Assuming that the allowable payroll is met, that the training center is fully utilized, and that the department's production needs are met, how many of each type of worker does the department need?

35. Computer Printers A large firm needs to purchase several new computer printers to meet its printing needs. The firm is considering buying the Q6, Q10, and Q12 models of a certain brand. The Q6 prints at the rate of 6 pages per minute, has 2 megabytes of memory, and costs $800; the Q10 prints at the rate of 10 pages per minute, has 4 megabytes of memory, and costs $1500; and the Q12 prints at the rate of 12 pages per minute, has 12 megabytes of memory, and costs $3000. Various affected departments in the firm turned in a sum of 3720 pages per hour that they considered adequate for their printing needs, and the treasurer of the firm has set a budget of $11,400 with which to purchase printers. The firm's engineers have decided that printing needs demand a total of 38 megabytes of memory. Annual maintenance costs are estimated at $20, $40, and $120, respectively, for the three models, and the firm has an annual maintenance budget of $380. Assuming that the budgets, printing needs, and memory needs are met exactly, how many of each type of printer should be purchased?

36. Production A kitchen appliance manufacturer makes can openers and dough cutters using two machines: a press and a riveter. Each can opener requires 0.2 minute in the press and 0.4 minute in the riveter. A dough cutter requires 0.5 minute in the press and 0.3 minute in the riveter. The press can be operated only 3 hours per day and the riveter only 2.5 hours per day. If these two machines are fully utilized, how many can openers and dough cutters can be produced each day?

37. Advertising Members of a local charity want to advertise a forthcoming event that they are sponsoring by placing advertisements in the newspapers, on the radio, and on TV. They want to run a total of 24 ads, with three times as many radio ads as TV ads. The newspaper ads have an effective rating of 7; radio ads, 8; and TV ads, 9. The charity wants a total of 188 effective rating points. How many ads should be run in each medium to meet these objectives?

38. Training Personnel The Friendly Secretarial Training Service trains secretaries for three types of office positions: industrial plant (IP), small-business office (SB), and executive secretarial (ES). Each student goes through training in three departments: Typing (T), telephone and miscellaneous (M), and ethics (E). The following table shows the number of units of training each type of secretary receives:

	IP	SB	ES
T	1	1	1
M	0.5	0.5	1.5
E	0.2	0.4	0.8

Each month, typing can deliver 42 units of training; telephone and miscellaneous, 29 units; and ethics, 16 units. For the service members to fully utilize their training capacity, how many of each type of secretary should they admit to their training service each month?

39. Hiring Workers Brenda's Drive-In is located in a college town and has rush-hour business during the 5 o'clock hour and again during the 10 o'clock hour. To run her business efficiently, Brenda hires three shifts of workers: one to work from 10:00 A.M. to 6:00 P.M., another from 5:00 to 11:00 P.M., and a third from 10:00 P.M. to 1:00 A.M. Brenda knows from experience that she needs 10 workers during the first- and second-shift overlap and 8 workers during the second- and third-shift overlap. Can Brenda staff her worker needs for each shift if the first and third shifts are to have the same number of workers?

40. Truck Leasing One day, the Ace Delivery Company leased three trucks, with drivers, to make deliveries. The trucks were driven a total of 6940 miles. Truck A got 12 miles per gallon, and its driver was paid 18 cents per mile. Truck B got 14 miles per gallon, and its driver was paid 22 cents per mile. Truck C got 15 miles per gallon, and its driver was paid 24 cents per mile. The trucks used a total of 500 gallons of gasoline, and the drivers were paid a total of $1519.60. How many gallons did each truck consume during the day?

41. Buying Tires A tire store manager is given $28,500 to buy 600 new tires for the store. Two types of tires are to be purchased: four-ply nylon, costing $40 each, and steel-belted radials, costing $50 each. The manager wants to buy three times as many steel-belted radial tires as four-ply nylon tires. Assuming that all $28,500 is spent, find a system of linear equations that would give the number of each type of tire bought.

42. Investing A bank trust department is to invest $100,000 for a client in three areas: certificates of deposit, earning 8%; junk bonds, earning 12%; and blue-chip stocks, earning 6%. The trust is to earn $8000 each year on these investments, and the client stipulated that three times as much be invested in certificates of deposit and blue-chip stocks as in junk bonds. How much is to be invested in each of the three areas to meet the specified conditions?

43. Dividing an Object Suppose you are to cut a piece of string 160 inches long into two pieces so that one piece is four times as long as the other. How long is each piece of string?

44. Purchasing Copy Machines The Speedy Printing Service is planning on purchasing several new copy machines to meet its customer needs. The company is considering a Model A machine that costs $2000 and prints at the rate of 30 pages per minute, a Model B machine that costs $4000 and prints at the rate of 45 pages per minute, and a Model C machine that costs $8000 and prints at the rate of 60 pages per minute. The firm has budgeted a total of $78,000 for the purchase of these machines and wants to be able to print a total of 45,900 pages per hour. The firm wants to purchase three times as many Model A machines as Model B machines. Assuming that the firm's budget and printing needs are exactly met, how many of each model should be purchased?

45. Production A particular company makes three models of chain saw: A, B, and C. Three operations are involved in the manufacturing process: assembly, painting, and testing. In the assembly operation, model A takes 1 hour,

B takes $1\frac{1}{2}$ hours, and C takes 2 hours. In the painting department, each saw takes $\frac{1}{2}$ hour, and in the testing department, models A and C take $\frac{1}{5}$ of an hour each, and model B takes $\frac{1}{4}$ of an hour. Each week, the company has available 200 hours for assembly, 65 hours for painting, and 26 hours for testing. If all available labor hours are fully utilized, how many of each type of saw can be made each week?

46. Mixing Acids Dr. McKnight, a chemistry professor, wants to make 2 liters of an 18% acid solution by mixing a 21% solution with a 14% solution. How many liters of each type of acid solution should be used to produce the 18% solution?

47. Consolidated Income Company A owns 20% of company B and 10% of company C. Company B owns 30% of company C. Company C owns 5% of company A and 40% of company B. The consolidated income of each company consists of its own net income plus its share of the net income of any other company in which it owns an interest. The net income of each company before intercompany adjustments is as follows:

A:	$1.23 million
B:	$4.6 million
C:	$3.8 million

Find a system of linear equations that will model the consolidated income of each company, and then solve the system.

48. Consolidated Income The Cummings Company owns 30% of the Snelsire Company and 15% of the Johe Company. The Snelsire Company owns 5% of the Cummings Company and 8% of the Esleeck Company. The Esleeck Company owns 12% of the Cummings Company and 2% of the Snelsire Company, while the Johe Company owns 6% of the Cummings Company and 25% of the Snelsire Company. The consolidated income of each company consists of its own net income plus its share of the net income it receives from owning other companies. The net incomes of these companies before intercompany adjustments during a given year are as follows:

Cummings Company:	$5.4 million
Snelsire Company:	$4.8 million
Esleeck Company:	$8.3 million
Johe Company:	$4.0 million

Find a system of linear equations that will model the consolidated income for each of the companies, and then solve the system.

49. Recall that the slope (m) and y-intercept (b) of the least-squares regression line for a set of n data are obtained from the system of equations

$$nb + (\Sigma x)m = \Sigma y$$
$$(\Sigma x)b + (\Sigma x^2)m = \Sigma xy.$$

Solve this system of equations to show that

$$b = \frac{(\Sigma x^2)(\Sigma y) - (\Sigma x)(\Sigma xy)}{n(\Sigma x^2) - (\Sigma x)^2}$$

and

$$m = \frac{n(\Sigma xy) - (\Sigma x)(\Sigma y)}{n(\Sigma x^2) - (\Sigma x)^2}.$$

Section 2.3 Linear Systems Having Many Solutions

In the previous section, we concentrated on systems that had exactly one solution or no solution at all. However, just as in the case of two equations and two unknowns discussed in Chapter 1, larger systems may also have an infinite number of solutions; that is the focus of our attention in this section. With the geometry available in Chapter 1, it was easy to tell at the outset when a system had an infinite number of solutions. However, in three-variable cases, this will usually not be the case, so we rely on the pivoting process to signal when an infinite number of solutions results, and then we use the final matrix to help us display the solutions.

EXAMPLE 1

A System Having an Infinite Number of Solutions

Find the solutions of the system

$$x + 3y + z = 10$$
$$2x + 7y - z = 21.$$

SOLUTION The pivoting process applied to the augmented matrix gives this sequence:

$$\begin{bmatrix} ① & 3 & 1 & | & 10 \\ 2 & 7 & -1 & | & 21 \end{bmatrix} \quad \begin{array}{c} \\ -2R_1 + R_2 \rightarrow R_2 \end{array} \begin{bmatrix} 1 & 3 & 1 & | & 10 \\ 0 & ① & -3 & | & 1 \end{bmatrix}$$

$$-3R_2 + R_1 \rightarrow R_1 \begin{bmatrix} 1 & 0 & 10 & | & 7 \\ 0 & 1 & -3 & | & 1 \end{bmatrix}.$$

The pivoting process has been carried as far as possible. Because the solution is not evident from this matrix, what do we do now? First, we write the equivalent system of equations:

$$x + \quad 10z = 7$$
$$y - 3z = 1.$$

This system of *two* equations and *three* unknowns has what we may call an "extra" unknown, and that is a signal that there will be either an infinite number of solutions or no solution at all. In this case, it is easiest to consider z to be the "extra" unknown and to solve the first equation for x and the second for y, thereby getting solutions for each in terms of the unknown z:

$$x = 7 - 10z$$
$$y = 1 + 3z.$$

When the equations are written this way, we can see that x and y depend upon z. That is, if z is assigned a number, then x and y can be computed so that the three numbers obtained are a solution of the system. *All* solutions may then be stated in the following form:

$$z = \text{any number}$$
$$x = 7 - 10z$$
$$y = 1 + 3z.$$

In particular, if $z = 0$, then $x = 7$, and $y = 1$, so that $x = 7$, $y = 1$, $z = 0$ is a solution. If $z = 2$, then $x = -13$ and $y = 7$; hence, $x = -13$, $y = 7$, $z = 2$ is a solution, and so on. Therefore, this system has infinitely many solutions. ▬

The fact that we are free to assign z any number that we please in the solutions in Example 1 suggests that z may be called a **free variable,** or a **parameter.** We point out that it is not necessary for z to be the free variable; we could just as well have used x as the free variable and solved for y and z in terms of x, or let y be the free variable and solved for x and z in terms of y. Each of these approaches will give exactly the same set of solutions, even though the *appearance* of the solution equations is not the same.

Solutions involving one or more free variables, like the solution in Example 1, may be checked for accuracy just like any other solution. Simply substitute the expression found for x and the expression found for y into each of the original equations, and see whether an equality results:

$$x + 3y + z = 10 \qquad\qquad 2x + 7y - z = 21$$
$$(7 - 10z) + 3(1 + 3z) + z \overset{?}{=} 10 \qquad 2(7 - 10z) + 7(1 + 3z) - z \overset{?}{=} 21 \qquad \textit{Substitute expressions for x and y}$$
$$7 - 10z + 3 + 9z + z \overset{?}{=} 10 \qquad 14 - 20z + 7 + 21z - z \overset{?}{=} 21 \qquad \textit{Simplify}$$
$$10 - 10z + 10z = 10\checkmark \qquad\qquad 21 - 20z + 20z = 21\checkmark$$

EXAMPLE 2

Finding a Parametric Solution

Find the solution of this system of equations:

$$
\begin{aligned}
x - 2y + \ \ z &= 3 \\
2x + \ \ y + \ \ z &= 1 \\
4x - 3y + 3z &= 7
\end{aligned}
$$

SOLUTION The pivoting process used on the augmented matrix for the system results in the following outcomes:

$$
\left[\begin{array}{ccc|c}
① & -2 & 1 & 3 \\
2 & 1 & 1 & 1 \\
4 & -3 & 3 & 7
\end{array}\right]
\begin{array}{c}
\\
-2R_1 + R_2 \to R_2 \\
-4R_1 + R_3 \to R_3
\end{array}
\left[\begin{array}{ccc|c}
1 & -2 & 1 & 3 \\
0 & ⑤ & -1 & -5 \\
0 & 5 & -1 & -5
\end{array}\right]
$$

$$
\begin{array}{c}
\\
\tfrac{1}{5}R_2 \to R_2
\end{array}
\left[\begin{array}{ccc|c}
1 & -2 & 1 & 3 \\
0 & ① & -\tfrac{1}{5} & -1 \\
0 & 5 & -1 & -5
\end{array}\right]
$$

$$
\begin{array}{c}
2R_2 + R_1 \to R_1 \\
\\
-5R_2 + R_3 \to R_3
\end{array}
\left[\begin{array}{ccc|c}
1 & 0 & \tfrac{3}{5} & 1 \\
0 & 1 & -\tfrac{1}{5} & -1 \\
0 & 0 & 0 & 0
\end{array}\right].
$$

Again, we have taken the pivoting process as far as possible. Because there are *two* nonzero rows and *three* variables, there is *one* free variable or parameter. Writing the equivalent equations from the last matrix gives

$$x + \frac{3}{5}z = 1$$

$$y - \frac{1}{5}z = -1,$$

from which it is easiest to solve for x and y in terms of z, so we let z be the free variable and arrive at the parametric solution.

$$x = 1 - \frac{3}{5}z$$

$$y = -1 + \frac{1}{5}z$$

$$z = \text{any number.}$$

Geometrically, the solution of this set of equations is a line that is the intersection of the first two planes in the given system. The last row of zeros in the final matrix means that the line was contained in the plane represented by the third equation; hence, the last plane was not needed to determine the line. The system is **dependent.** ▬

The Gauss–Jordan elimination method, in conjunction with the pivoting process, is very helpful in ultimately revealing the relationships among linear equations that lead to the various types of solutions or in showing that no solution exists. By examining Examples 1 and 2 of this section, can you tell when an infinite number of solutions exist?

When an Infinite Number of Solutions Exist

If the reduced row-echelon form of the augmented matrix has the same number of nonzero rows in the coefficient matrix as in the augmented matrix, and if that number is less than the number of variables, then the system has **an infinite number of solutions.**

A system of linear equations in which the constant term of every equation in the system equals zero is called **homogeneous.** The system in Example 3 is homogeneous. A homogeneous system of linear equations always has at least one solution. Do you see what the solution is?

EXAMPLE 3

Solving a Homogeneous
System of Equations

Solve this homogeneous system of equations:

$$\begin{aligned} x \quad\;\;\; + \;\; z &= 0 \\ x + y + \;\; z &= 0 \\ 2x \quad\;\; + 2z &= 0 \end{aligned}$$

SOLUTION For this particular system, notice that the last equation is twice the first; hence, it does not contribute to the solutions and may be discarded. Solving the first equation for x, we get $x = -z$. Now substitute this value of x into the second equation to obtain $y = 0$. This shows that the solutions are

$$x = -z$$

$$y = 0$$

$$z = \text{any number.} \;\; ▬$$

NOTE For any homogeneous system, one particular solution can always be found by setting all variables equal to zero.

The previous example shows that once we understand the nature of the solutions, it is sometimes possible to obtain them without using the pivoting process, even on problems of substantial size. However, we should stress that the pivoting process applies equally well to *all* systems of linear equations and will guide us to the solution in every case.

A notational scheme for variables that is sometimes used, especially in large systems, is that of "subscripted" variables. The next example shows this notation, along with the fact that sometimes more than one parameter is needed in a solution.

EXAMPLE 4

Subscripted Notation;
More than One
Parameter

Find the solution of the following system of linear equations:

$$x_1 + x_2 - 2x_3 + 4x_4 = 6$$
$$3x_1 + x_2 + 2x_3 \qquad = 5.$$

SOLUTION The pivoting process can be used to advantage here:

$$\begin{bmatrix} ① & 1 & -2 & 4 & | & 6 \\ 3 & 1 & 2 & 0 & | & 5 \end{bmatrix} \quad -3R_1 + R_2 \to R_2 \quad \begin{bmatrix} 1 & 1 & -2 & 4 & | & 6 \\ 0 & ⊘{-2} & 8 & -12 & | & -13 \end{bmatrix}$$

$$-\tfrac{1}{2}R_2 \to R_2 \quad \begin{bmatrix} 1 & 1 & -2 & 4 & | & 6 \\ 0 & ① & -4 & 6 & | & \tfrac{13}{2} \end{bmatrix}$$

$$-1R_2 + R_1 \to R_1 \quad \begin{bmatrix} 1 & 0 & 2 & -2 & | & -\tfrac{1}{2} \\ 0 & 1 & -4 & 6 & | & \tfrac{13}{2} \end{bmatrix}.$$

The system of equations represented in the preceding matrix is

$$x_1 + \quad 2x_3 - 2x_4 = -\frac{1}{2}$$

$$x_2 - 4x_3 + 6x_4 = \frac{13}{2}.$$

This time, there are two equations in four unknowns, resulting in *two* extra variables. It is easiest to solve for x_1 and x_2 in terms of x_3 and x_4, thus making x_3 and x_4 the free variables or parameters:

$$x_1 = -\frac{1}{2} - 2x_3 + 2x_4$$

$$x_2 = \frac{13}{2} + \quad 4x_3 - 6x_4$$

$$x_3 = \text{any number}$$

$$x_4 = \text{any number}.$$

If we let $x_3 = 0$ and $x_4 = 1$, for example, then

$$x_1 = -\frac{1}{2} - 2(0) + 2(1) = \frac{3}{2}$$

and

$$x_2 = \frac{13}{2} + 4(0) - 6(1) = \frac{1}{2}.$$

This means that $x_1 = \frac{3}{2}$, $x_2 = \frac{1}{2}$, $x_3 = 0$, and $x_4 = 1$ is a particular solution of the original system of equations. Each numerical choice of x_3 and x_4 gives a particular solution of the system. ■

EXAMPLE 5

Solving the Shipping Problem Given at the Outset of This Chapter

The problem of shipping bicycles from two warehouses to two stores, as stated at the beginning of this chapter, is described by the equations found in Example 4 of Section 2.1. Those equations are

$$
\begin{aligned}
w \qquad\quad +\ z &= 38 \\
x + y \qquad\ &= 46 \\
x \qquad -\ 2z &=\ 0.
\end{aligned}
$$

If the pivoting process is applied directly to the augmented matrix of this system of equations, the final equivalent matrix, in echelon form, appears as

$$
\left[
\begin{array}{cccc|c}
1 & 0 & 0 & 1 & 38 \\
0 & 1 & 0 & -2 & 0 \\
0 & 0 & 1 & 2 & 46
\end{array}
\right].
$$

(Check your pivoting skills at this point by converting the initial augmented matrix into the preceding one.) The solutions can now be obtained by converting the last matrix to equation form and then considering z to be the parameter. We thus have

$$
\begin{aligned}
w +\ z &= 38 \\
x -\ 2z &= 0 \\
y +\ 2z &= 46,
\end{aligned}
$$

and the solutions become

$$
\begin{aligned}
w &= 38 -\ z \\
x &= \qquad 2z \\
y &= 46 - 2z \\
z &= \text{any number.}
\end{aligned}
$$

In reality, the variables in this particular problem represent numbers of bicycles and cannot be negative. This means that $z \geq 0$, but cannot be so large as to render any of the other variables negative. The first line of the solutions tells us that z cannot be larger than 38, while the third line tells us that z cannot be larger than 23. As a result, z must be an integer from 0 through 23. Now our problem is completely solved, and the solutions give all shipping possibilities:

$$
\begin{aligned}
w &= 38 - z \\
x &= 2z \\
y &= 46 - 2z \\
z &= \text{any integer from 0 through 23.} \quad \blacksquare
\end{aligned}
$$

Later, we will encounter very special systems of equations having an infinite number of solutions, of which the following is a typical example:

$$
\begin{aligned}
x + 2y + u &= 6 \\
3x + y + v &= 8.
\end{aligned}
$$

In dealing with such systems, we are interested, not in all possible solutions, but just in particular ones that may be found by inspection. In this system, if we consider x and y to be free variables (parameters) and set both equal to zero, then the first equation becomes $u = 6$ and the second $v = 8$. Therefore, one solution of this system is $x = 0$, $y = 0$, $u = 6$, and $v = 8$. Suppose that we pivot on the element in the first-row, first-column position of the augmented matrix of the system to obtain

$$
\begin{bmatrix}
1 & 2 & 1 & 0 & 6 \\
0 & -5 & -3 & 1 & -10
\end{bmatrix}.
$$

This matrix represents the following equivalent system of equations:

$$
\begin{aligned}
x + 2y + u &= 6 \\
-5y - 3u + v &= -10.
\end{aligned}
$$

Can you find another particular solution from the preceding matrix? The easiest way is to consider y and u free variables and to set both equal to zero, from which the first equation gives $x = 6$ and the second $v = -10$. A solution is then $y = 0$, $u = 0$, $x = 6$, and $v = -10$.

One final note is in order about the dependent systems encountered in Sections 2.2 and 2.3. An entire row of zeros in an augmented matrix means dependency only within the system. *This fact alone does not, in general, give any clue as to the type of solution the system has; any one of the three classifications might result.* Example 3 of Section 2.2 is a dependent system that has exactly one solution, Example 2 of that section is a dependent system that has an infinite number of solutions, and the following system is a dependent system that has no solution:

$$
\begin{aligned}
x + y &= 2 \\
x + y &= 3 \\
2x + 2y &= 5.
\end{aligned}
$$

(Can you tell why just by visual inspection?)

When using calculators or computers to solve systems of linear equations that have infinitely many solutions, be sure you know how to interpret the output. Realize also that in most cases the output is an *approximation* of the actual answer, because of the round-off involved. Round-off can be so severe at times, that a machine will specify a solution for a system of equations that actually has no solution! Be sure to check such answers for accuracy.

There are also dangers inherent in using technology such as Excel's Solver to find solutions of systems of linear equations that have infinitely many solutions. Solver will more than likely find *one* of those solutions, but will not indicate that there are infinitely more to be found. There are a few ways to find out if Solver's solution is one of many. A simple method is demonstrated in Example 6. You should use this technique *anytime* you apply Solver to a system of linear equations, just to be safe.

EXAMPLE 6

Detecting whether
More than One Solution
Exists

Use `Solver` to find a solution of the system in Example 2, and check to make sure that it is not the only solution.

SOLUTION Figure 6 shows a worksheet that has been set up to contain the information related to the system in Example 2.

	A	B	C	D	E	F
1	x =	0	y =	0	z =	0
2						
3	x-2y+z = 3	0	=	3		
4	2x+y+z = 1	0	=	1		
5	4x-3y+3z = 7	0	=	7		

FIGURE 6
Spreadsheet Containing the System of Equations

We entered the following formulas into cells B3:B5:

Cell	Contents
B3	=B1−2*D1+F1
B4	=2*B1+D1+F1
B5	=4*B1−3*D1+3*F1

Access `Solver` from the `Tools` menu and enter the parameters shown in Figure 7. When you click `Solve`, Excel will display a solution of the system. Recall from Example 2 that this system actually has *infinitely many solutions* and that all of them have the parametric form

$$x = 1 - \frac{3}{5}z$$

$$y = -1 + \frac{1}{5}z$$

$$z = \text{any number.}$$

FIGURE 7
Solver Parameters for Example 6

Substitute the values Excel has provided into these equations to verify that the values do indeed represent a solution of the system.

Now, to check whether there are any other solutions of the system, we can proceed as follows: Choose new values for x, y, and z that are significantly different from the given solution. For instance, if Excel provided the solution $x = 0.7115$, $y = -0.9038$, and $z = 0.4808$, we could try $x = 1000$, $y = 1000$, and $z = 1000$. In other words, we are going to *seed* `Solver` with new starting values for our variables and allow it to find another solution. (See Figure 8.)

	A	B	C	D	E	F
1	x =	1000	y =	1000	z =	1000
2						
3	x-2y+z = 3	0	=	3		
4	2x+y+z = 1	4000	=	1		
5	4x-3y+3z = 7	4000	=	7		

FIGURE 8
Checking Solver for Infinitely Many Solutions

When we run `Solver` again, we will get a new and very different solution: $x = -345.88$, $y = 114.63$, and $z = 578.13$. This indicates that the original solution `Solver` provided is one of many available solutions. `Solver` is not currently equipped with the capability of giving the parametric form of the solutions, but the process we have gone through has shown us that we should complete the problem by means of pivoting, as we did in Example 2. ■

The previous example demonstrates that, while technology can be an efficient time-saver, it cannot substitute for a good understanding of the theory behind solutions of linear systems. A graphing calculator or computer program that can put matrices into reduced row-echelon form will provide us with quick ways to determine the nature of the solutions to any linear systems. Other readily available technology, such as Excel's `Solver`, can reduce the amount of time we spend solving these systems, but cannot always provide us with all the information we need. You should use the technique demonstrated in Example 6 *every time* you apply the `Solver` to a system of equations: Always seed `Solver` with one or two different sets of values for the variables and reexecute it to see whether the same solution is obtained. If a different solution is obtained, then you will know that you have to apply other techniques to find the rest of the solutions.

Exercises 2.3

In Exercises 1 through 16,
(a) *Find the solution in parametric form.*
(b) *Check the answer.*
(c) *Write out three explicit solutions of the system.*

1. $2x - y - 3z = 5$
   $x - 2y + z = 3$

2. $3x + y - z = 0$
   $2x + 3y + 2z = 1$

3. $2x + y - z = 0$
   $x - 3y + z = 1$
   $x + 4y - 2z = -1$

4. $2x - 3y + z = 1$
   $x + y - 3z = 0$
   $4x - y - 5z = 1$

5. $w + 2x + y - 3z = 4$
   $-2w + x + y + z = 2$
   $-w + 3x + 2y - 2z = 6$

6. $2x_1 + 3x_2 + 4x_3 - x_4 = -2$
   $x_1 + 2x_3 = 0$
   $x_1 - x_2 + x_3 + x_4 = 5$
   $x_2 - 2x_3 = 3$

7. $x_1 + 4x_2 + 2x_3 - 3x_4 = 1$
   $3x_1 + x_2 - 2x_3 = -11$
   $x_1 + 2x_2 + x_3 = 5$
   $x_2 - 3x_4 = -8$

8. $2w + x - 3y + z = 8$
   $-w + 2x - y + 2z = 6$
   $3w + 4x - 7y + 4z = 22$
   $-2w + 4x - 2y + 4z = 12$

9. $x + y + z = 3$
   $2x + 2y + 2z = 6$

10. $2x - y + z = 4$
   $6x - 3y + 3z = 12$

11. $2x + 3y + z = 0$
   $x - y + 2z = 0$
   $4x + y + 5z = 0$

12. $3x + 3y - z = 0$
   $4x + 5y - 3z = 0$
   $x + 2y - 2z = 0$

13. $x - z = 0$
   $y + z = 0$
   $x + y = 0$

14. $x = 0$
   $y + z = 0$
   $2y + 2z = 0$

15. $x + z = 0$
   $y = 0$
   $2x + 2z = 0$

16. $x + y = 0$
   $z = 1$
   $x + y + z = 1$

In Exercises 17 through 28, find the solutions, if any exist.

17. $x + 2y - z = 8$
   $2x - y + z = 3$
   $4x + 3y - z = 5$

18. $y + 2z = 5$
   $x + 2y + z = 1$
   $x + 3y + 3z = 2$

19. $w + 2x - 3y + z = 1$
   $4w + 8x - 12y + 4z = 4$
   $2w + 4x - 6y + 2z = 2$

20. $x + y - z = 1$
   $2x + 2y - 2z = 2$
   $3x + 3y - 3z = 3$

21. $2x - 3y + 2z = 3$
   $x + 2y - 3z = 2$
   $3x - y - z = 3$

22. $-w + x + y + 3z = 0$
   $2w + 2x - y - z = 2$
   $w + x + y - 4z = 1$
   $w + 5x + 2y + z = 1$

23. $0.3x + 0.5y + z = 4$
   $5x - 2y + 0.8z = 7$
   $5.6x - y + 2.8z = 15$

24. $6.2x + 3y + 2.7z = 2.3$
   $1.8x - 2.3y - 1.9z = 5.6$
   $9.8x - 1.6y - 1.1z = 13.5$

25. $0.3x + 5.24y - 8.61z = 5.2$
   $1.2x + 20.96y - 34.44z = 8$

26. $4.61w + 2x + 3.5y + z = 8$
   $1.58w + 0.83x - 6.21y + 1.50z = 4$
   $3w + 0.31x + 2y - z = 1.63$
   $10.77w + 3.97x - 6.92y + 3z = 17.63$

27. $w + 2x - 3y + z = 6.3$
   $2w - x + 0.54y + 2.8z = 5$
   $2.83w + 5.72x + y - 3z = 2$
   $12.66w + 15.44x - 2.92y + 2.6z = 32.9$

28. $w + \quad x - y + \quad 2z = 6.2$

$2w - 0.5x + y - 0.46z = 8.1$

In Exercises 29 through 32, solve the problems without using pivoting.

29. $x \qquad - 2z = 0$

$\qquad y \qquad = 0$

$2x \qquad - 4z = 0$

30. $3x - y \qquad = 0$

$\qquad z = 1$

$6x - 2y \qquad = 0$

31. $\qquad y + 2z = 0$

$x \qquad + z = 0$

$2x \qquad + 2z = 0$

32. $x + 4y \qquad = 0$

$\qquad y + z = 0$

$\qquad 3y + 3z = 0$

In Exercises 33 through 36,

(a) Find a specific solution by inspection.

(b) Pivot on the element in the second row and first column, and then find a second solution by inspection.

33. $2x + 2y + u \qquad = 10$

$x + y \qquad + v = 5$

34. $x + 3y + u \qquad = 4$

$x + y \qquad + v = 8$

35. $x - y + u \qquad = 6$

$2x + y \qquad + v = 4$

36. $2x + y + u \qquad = 7$

$3x - y \qquad - v = 10$

37. City Budget The city of Burlington gets a grant for $1 million to spend on various projects. The city's board of directors decides to spend the money on three types of projects: streets, urban renewal, and parks. The board members also decide that the shares going to urban renewal and parks should equal the amount spent on streets. How can the money be allocated to the three types of projects?

38. Highway Construction A state highway department plans to build 100 miles of highways in three districts: A, B, and C. The cost per mile for building highways is $2 million in district A, $500,000 in district B, and $1 million in district C. If $80 million is allocated for construction, can 100 miles of highways be constructed at a total cost of $80 million so that some highway building takes place in each district?

39. Advertising A new no-fat potato chip is being introduced in a small city through a series of advertisements in newspapers, on radio, and on television. A total of 30 advertisements are to run in these media. In addition, twice as many radio ads are to be run as television ads. Finally, the sum of the number of newspaper ads and two times the number of radio ads is to equal the number of ads run plus the number of television ads. How many of each type ad can be run?

40. Housing Construction A construction company is planning a development in which three types of housing—WoodIt, BrickIt, and StoneIt—are to be built. The company estimates that the electrician hours required for WoodIt, BrickIt, and StoneIt houses are 30, 60, and 90 hours, respectively, and the plumber hours are 80, 100, and 120 hours, respectively. If subcontractors can furnish 1080 electrician hours and 1920 plumber hours per year, and these hours are

fully utilized by the contractor, how many of each type of house can be built in a year?

41. Manufacturing A maker of computer printers has three models: the QT20, the QT40, and the QT60. Each QT20 model has 2 megabytes of memory: each QT40, 4 megabytes; and each QT60, 6 megabytes. To get the best prices on memory, the company has an agreement to buy a total of 500 megabytes of memory each week. It requires 3 hours to assemble each QT20 model, 4 hours to assemble each QT40 model, and 5 hours to assemble each QT60 model. The company has 560 worker assembly hours available each week. Each QT20 requires 5 minutes of test time; each QT40, 8 minutes; and each QT60, 11 minutes. The company has available 1060 minutes of testing time each week. Assuming that all available assembly and testing time is used each week, how many of each model of printer can the company assemble?

42. Investments An investor has $100,000 to invest in government bonds, mutual funds, and money market funds. The investor wants to invest as much in government bonds as the total in mutual funds and money market funds. If government bonds earn 7%, mutual funds earn 5%, and money market funds earn 5%, and if the investor expects to earn $6000 in interest during the first year, how much should be invested in each type of investment?

43. Order Fulfillment A company with two warehouses receives orders for 30 and 18 microwave ovens, respectively, from two stores, as shown in the following diagram:

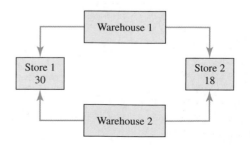

Twice as many ovens are to be shipped from Warehouse 1 to Store 1 as from Warehouse 2 to Store 2. Assuming that the warehouses have ample stock to supply the needs of both stores, find and then solve the equations that model the possibilities of meeting these orders. State three possibilities for supplying the requested microwave ovens.

44. Supplying Orders A company with two warehouses receives orders for TV sets from three stores, as indicated in the diagram at the right.

Find and then solve the system of equations that models this supply problem. State three possibilities for supplying the stores with the requested TV sets.

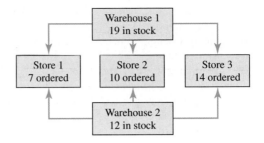

Chapter 2 Summary

CHAPTER HIGHLIGHTS

Linear systems of equations are useful mathematical models for many processes and situations. *Tables* can be helpful in organizing given information. *Augmented matrices,* the *pivoting process,* and the *row-echelon* form of a matrix together provide a convenient, systematic method for solving systems of equations. For systems having an infinite number of solutions, *free variables* or *parameters* can be used to give a general description of all of the solutions.

EXERCISES

In Exercises 1 and 2, write the coefficient matrix, constants matrix, and augmented matrix for each system.

1. $\begin{aligned} 3x \quad\;\; - z &= 0 \\ y \qquad\quad &= 7 \\ 6x \quad\; - 2z &= 0 \end{aligned}$
2. $\begin{aligned} x + y \qquad &= 0 \\ y - 3z &= 4 \\ 5y + 2z &= 8 \end{aligned}$

In Exercises 3 through 5, use the pivoting process to put each matrix into reduced row-echelon form. Indicate whether the corresponding system of equations has one solution, no solution, or infinitely many solutions.

3. $\begin{bmatrix} 1 & 2 & 1 & 0 \\ 3 & 0 & -1 & 2 \\ -1 & 2 & 1 & -2 \end{bmatrix}$
4. $\begin{bmatrix} 1 & 2 & 1 & 0 \\ 3 & 0 & 1 & 2 \\ 2 & -2 & 0 & -1 \end{bmatrix}$

5. $\begin{bmatrix} 1 & 2 & 1 & 0 \\ 3 & 0 & 1 & 2 \\ 2 & -2 & 0 & 2 \end{bmatrix}$

In Exercises 6 through 13, find the solution, if one exists, using the pivoting process.

6. $\begin{aligned} x + 2y &= 7 \\ 3x + y &= 4 \end{aligned}$
7. $\begin{aligned} \tfrac{1}{2}x - 3y &= 5 \\ 2x - 12y &= 18 \end{aligned}$

8. $\begin{aligned} x + 2y \qquad &= 9 \\ 4x - y + z &= 7 \\ 2x - 5y + z &= -11 \end{aligned}$
9. $\begin{aligned} x + 2y + z &= 0 \\ 3y + 5z &= 1 \\ 5x - 4y - 10z &= 12 \end{aligned}$

10. $\begin{aligned} 3x + 2y - z &= -1 \\ 5x \qquad + 7z &= 5 \\ -3y + 4z &= 6 \\ x - 6y + 3z &= 13 \end{aligned}$
11. $\begin{aligned} x - 2y &= 13 \\ 3x + y &= 4 \\ 2x - y &= 11 \\ 4x + 2y &= 2 \end{aligned}$

12. $\begin{aligned} 5x + 2y + z &= 14 \\ x + 2y + 3z &= 6 \\ 2x \qquad - z &= 4 \end{aligned}$
13. $\begin{aligned} x \qquad + 4z &= 7 \\ y \qquad &= 5 \\ 2x \quad\; + 7z &= 13 \end{aligned}$

14. Write a system of equations for which the augmented matrix

$$\begin{bmatrix} 3 & 0 & -2 & 7 \\ -1 & 8 & 5 & 0 \\ 0 & 2 & 1 & -4 \end{bmatrix}$$

would be a model. Solve the system.

15. A student spends a total of 36 hours a week studying history, English, psychology, and accounting. She spends as much time studying history and psychology together as she does studying English and accounting together. The amount of time she studies psychology and accounting together is five times as much time as she spends on studying English. The total time she spends on studying psychology and English is 16 hours less than the time she spends on history and accounting together. How much time during the week does she spend on each subject?

16. To make money for graduate school, Ashley charges $15 per hour for tutoring one student mathematics and $30 per hour for teaching trombone to another student. She has only 10 hours per week that she can spend tutoring. In one particular week, she needs a total of $240 for preparing and submitting applications to graduate school. How many hours should she spend that week tutoring mathematics, and how many should she spend teaching trombone?

In Exercises 17–19, rewrite the given augmented matrix as a system of equations in x, y, and z. Then use substitution to find the solution of the system.

17. $\begin{bmatrix} 1 & 2 & -3 & | & 5 \\ 0 & 1 & 1 & | & 4 \\ 0 & 0 & 1 & | & 3 \end{bmatrix}$ **18.** $\begin{bmatrix} 1 & 5 & 0 & | & 8 \\ 0 & 2 & 1 & | & 7 \\ 0 & 0 & 2 & | & -6 \end{bmatrix}$

19. $\begin{bmatrix} 2 & 1 & -3 & | & 16 \\ 0 & 3 & 1 & | & 12 \\ 0 & 0 & 4 & | & 0 \end{bmatrix}$

20. Explain why the operation $0 \times R_1 \rightarrow R_1$ is not allowed in the pivoting process.

21. Two students were discussing solution(s) of linear systems of equations having more equations than unknowns. Martina says that such systems always have an infinite number of solutions, while Johann says that such systems never have a solution. Who, if anyone, is correct? Explain and justify your response with supporting examples or analyses.

 In Exercises 22 and 23, solve the given systems of equations.

22. $\frac{8}{9}x + \frac{3}{4}y - \frac{7}{8}z = 9$

$\frac{4}{5}x + \frac{5}{6}y + \frac{2}{3}z = \frac{12}{5}$

$\frac{3}{4}x + \frac{2}{3}y - \frac{1}{4}z = 3$

23. $x + 0.2y + 1.4z = 2$

$0.3x - 1.4y + 0.5z = 1.4$

$1.2x + 0.8y - 0.2z = 5$

CUMULATIVE REVIEW

24. Calculate the value of the function $f(x) = 2x - 7$, when $x = 29$.

25. Write the equation of a line having y-intercept 13 and that is parallel to the line having equation $3x + 5y = 27$.

26. Find a total linear cost function having direct costs of $42 and fixed costs such that 30 items cost a total of $3000.

27. Solve the system

$$2x + 3y = 7$$
$$3x + 4y = 9,$$

using any appropriate method.

28. Find a function that will give the net value x years from now of a new printing press that costs $80,000, has a scrap value of $12,000, and is depreciated by the straight-line method over a 15-year period.

Chapter 2 Sample Test Items

SAMPLE TEST ITEMS

1. Write the coefficient matrix for the system

$$x \qquad + 2z = -7$$
$$-3x + 5y - 8z = 0$$
$$x + y \qquad = 43.$$

2. Write the augmented matrix for the system

$$3a - b = 5$$
$$-5a + 17b = -1.2.$$

In each of the next three problems, write a system of equations that models the given information.

3. A parking meter contains 213 coins in nickels, dimes, and quarters, having a total value of $21.15. If the number of dimes is twice the number of quarters, how many nickels are in the meter? The distribution of coins is as follows:

	Nickels	Dimes	Quarters	Total (cents)
Number of Each Coin	N	D	Q	213
Value in Cents (per coin)	5	10	25	2115
Relationship between D and Q		1	-2	0

4. A concert performance brings in receipts of $129,673 from the sale of 9073 tickets. If a student ticket costs $12.50 and an adult ticket costs $16.75, how many students and how many adults attended the concert?

5. Steven is catering a reception for a business convention. He plans to serve three different hors d'oeuvres: sweet and sour meatballs, crab cakes, and Chinese dumplings. The recipe for meatballs makes 20 servings, the recipe for crab cakes makes 10 servings, and the recipe for dumplings makes 30 servings. From past experience, Steven knows that people tend to eat twice as many dumplings as crab cakes. Each recipe for meatballs costs $10, each recipe for crab cakes costs $15, and each recipe for dumplings costs $8. He has been asked to prepare a total of 380 servings and can spend $174 on food without going over budget. How many recipes of each type of hors d'oeuvres should Steven prepare?

6. Which of the following matrices are in reduced row-echelon form?

(a) $\begin{bmatrix} 1 & 0 & 0 \\ 0 & 1 & 0 \\ 0 & 1 & 2 \end{bmatrix}$
(b) $\begin{bmatrix} 1 & 0 & 0 \\ 0 & 0 & 1 \\ 0 & 1 & 0 \end{bmatrix}$

(c) $\begin{bmatrix} 0 & 1 & 0 \\ 0 & 0 & 1 \\ 0 & 0 & 0 \end{bmatrix}$
(d) $\begin{bmatrix} 1 & 0 & 0 \\ 0 & 0 & 0 \\ 0 & 0 & 1 \end{bmatrix}$

(e) $\begin{bmatrix} 1 & 0 & 3 \\ 0 & 1 & -7 \end{bmatrix}$

7. Complete the pivot operation on the circled element of the augmented matrix:

$$\left[\begin{array}{ccc|c} ① & 1 & 3 & -6 \\ 0 & 2 & 4 & 10 \\ 5 & -2 & 1 & 0 \end{array}\right]$$

In Exercises 8 and 9, find the solution, if one exists, using the pivoting process:

8. $\begin{aligned} x - 2y - z &= 1 \\ 3x + y + z &= 10 \\ 2x - y - z &= 0 \end{aligned}$

9. $\begin{aligned} 6x - y &= -11 \\ x + 2y &= 9 \\ 5x + y &= 0 \\ 11x + 3y &= 4 \end{aligned}$

10. Use the pivoting process to solve the system of equations that models the situation given in Question 3 of this test.

11. List three explicit solutions of the following system:

$$\begin{aligned} x - 2y + z &= 5 \\ 3x + 2y + z &= 7. \end{aligned}$$

12. Write the general solution in parametric form for this system:

$$\begin{aligned} x + y \quad &= 5 \\ y + z &= 2 \\ x \quad - z &= 3. \end{aligned}$$

13. Find the solution, if one exists:

$$\begin{aligned} x + 3y - 2z &= 5 \\ x + 2y + z &= 7 \\ y - 3z &= 2. \end{aligned}$$

14. Solve the system of equations that models the situation given in Question 4 of this test.

15. Solve the system of equations that models the situation given in Question 5 of this test.

In-Depth Application

Point-Spread Ranking Systems

Objectives

In this project, you will apply your knowledge of systems of linear equations and their solutions to a ranking system. You will be asked to set up a system of equations, solve it, and determine a ranking for the five teams in the National Football League (NFL) American Football Conference (AFC) West Division during the 2000 regular season. (*Note:* The NFL restructured its divisions, beginning with the 2002 season. The Seattle Seahawks are now in the NFC.)

Introduction

One way to rank athletic teams within a league is by looking at the *point-spread ranking*. Suppose two teams, A and B, with point-spread rankings a and b, respectively, play one another. The *point spread* we would expect at the end of the game is $a - b$; that is, $a - b$ is the difference between A's final score and B's final score. If $a - b$ is positive, then we expect team A to defeat team B by $a - b$ points; if $a - b$ is negative, then we expect team B to defeat team A by $b - a$ points. If team A has a higher point-spread ranking, a, than all the other teams in its league, then we might consider team A to be the "number-one" team in that league.

   Methods similar to the preceding can be applied to a variety of situations, not just to athletic teams. For instance, we could apply such a technique to consumer preference surveys that compare several brands of the same product. In the end, the brand with the highest ranking could be considered the consumer's most preferred brand.

(continued)

Team	AFC Wins	AFC Losses	PTS	OPP	Point-ranking
Denver	6	2	217	190	a
Kansas City	5	3	197	161	b
Oakland	5	3	196	157	c
Seattle	3	5	158	199	e
San Diego	1	7	117	178	d

TABLE 6. 2000 AFC WEST REGULAR SEASON STATISTICS

SOURCE: Data: National Football League Statistics, www.nfl.com.

The Data

During the 2000 regular season, the teams within the AFC West in the NFL stood as shown in Table 6.

Each team within the AFC West played each other team twice. When Denver played Kansas City, we would expect that the point spread was $a - b$ both times that they played, for a total of $2(a - b)$. When Denver played Oakland, we would expect that the point spread was $a - c$ both times, for a total of $2(a - c)$. Thus, the total point spread for Denver against all other AFC West teams would be

$$2(a - b) + 2(a - c) + 2(a - d) + 2(a - e) = 8a - 2b - 2c - 2d - 2e.$$

According to the statistics in the table, the total point spread between Denver and its opponents was $217 - 190 = 27$ points. Thus, our first equation is

$$8a - 2b - 2c - 2d - 2e = 27.$$

Similarly, the equation representing Kansas City's total point spread in the AFC West during the 2000 regular season is

$$2(b - a) + 2(b - c) + 2(b - d) + 2(b - e) = 197 - 161,$$
$$-2a + 8b - 2c - 2d - 2e = 36.$$

1. Set up a system of equations, with each equation similar to the ones created for Denver and Kansas City, that represents the total point spread for all teams within the AFC West during the 2000 regular season.

2. Solve the system of equations created in part 1. (The system has infinitely many solutions; write them using e as the free variable.)

3. Once you have the parametric form for the system of equations, you can quickly find any solution you wish by choosing a particular value for the free variable e. Set $e = 0$ and find the corresponding values of a, b, c, and d.

4. For this particular season, explain what $b - c$ means and calculate it. Based on your calculation, when Kansas City and Oakland played during the 2000 regular season, which team most likely won? Compare your answer with the actual results of the two games: Kansas City 17, Oakland 20; and Kansas City 31, Oakland 49.

5. (a) Judging from the data in Table 6, which of the five teams would you expect to be the "best" team in the league? Explain your reasons.

    (b) Recall that the values found in part 3 represent the *point rankings* for each of the teams. List these values in numerical order and rank the teams from first through fifth.

SOURCE: Problems inspired by Minton, Ronald, "A Mathematical Rating System," *UMAP Journal,* 13 (4), pp. 313–334. COMAP, Inc., 1992.

Matrix Algebra

Pollsters find that intense campaigning in an upcoming bond issue election causes people to change the way they vote in the following way: At the beginning of a given week, 75% of those favoring the issue remain in favor a week later, while 25% change to opposing the issue; 60% of those opposing the issue still oppose the issue at the end of the week, while 40% change to supporting the issue. If a group of 100 voters is selected, 50 of whom support the issue and 50 of whom oppose it, predict how many of these voters will favor, and how many will oppose, the issue a week later, two weeks later, and three weeks later. (See Example 7, Section 3.2.)

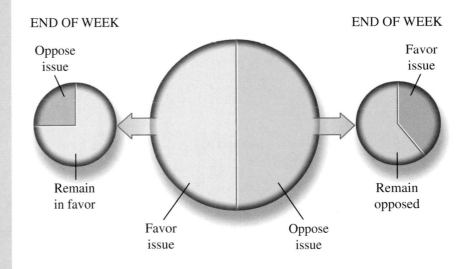

BEGINNING OF WEEK

END OF WEEK

Oppose issue

Remain in favor

Favor issue

Oppose issue

END OF WEEK

Favor issue

Remain opposed

CHAPTER PREVIEW

The notational convenience of using a matrix to represent a system of lin-ear equations is only one of many applications in which matrices can be helpful. In this chapter, we define the addition, subtraction, and multipli-cation of matrices having compatible dimensions and the multiplication of matrices by a real number. The inverse of a matrix and how it may be found (when it exists) rounds out the operations needed to develop and explore a matrix algebra. The rules of operation are compared with those of the algebra of real numbers. The last section of the chapter gives three examples of how the inverse may be applied.

Technology Opportunities

Spreadsheets are ideal for organizing information into rectangular arrays, which are similar to the matrices we will study in this chapter. In particular, a spreadsheet can be very powerful in combining two or more very large sets of data. Most spreadsheet packages also have functions that perform other matrix calculations, such as multiplication.

Graphing calculators usually are preprogrammed to do matrix operations, including addition, multiplication, and finding inverses. Depending on the size of the matrices with which you are dealing, you should choose which technology, if any, to apply in order to reach a solution most efficiently.

Section 3.1 Matrix Addition and Applications

A **matrix** is a rectangular array of numbers. One useful way to classify matrices is by the number of rows and columns they have. If a matrix has m rows and n columns, then we say that the **dimension** ("order," "size") of the matrix is $m \times n$, which is read "m by n." For example,

$$A = \begin{bmatrix} 2 & 3 & 5 \\ 1 & 0 & -2 \end{bmatrix}$$

has a dimension of 2×3, while

$$B = \begin{bmatrix} 2 & 1 \\ 4 & 6 \end{bmatrix}$$

has a dimension of 2×2. The matrix B is an example of a *square matrix*. The matrix A is not square.

One of the useful features of a matrix is that each entry is located at the intersection of some row and some column; this gives a specific location for that entry in terms of a **row-column** designation, called the entry's *position*. As already noted in Chapter 2, the general notation a_{ij} is used to denote the element in the ith row and jth column of a matrix. A general $m \times n$ matrix could then be written as

$$A = \begin{bmatrix} a_{11} & a_{12} & a_{13} & \cdots & a_{1j} & \cdots & a_{1n} \\ a_{21} & a_{22} & a_{23} & \cdots & a_{2j} & \cdots & a_{2n} \\ \vdots & & & & & & \\ a_{i1} & a_{i2} & a_{i3} & \cdots & a_{ij} & \cdots & a_{in} \\ \vdots & & & & & & \\ a_{m1} & a_{m2} & a_{m3} & \cdots & a_{mj} & \cdots & a_{mn} \end{bmatrix}$$

The Dimension of a Matrix

A matrix with m rows and n columns has a **dimension** $m \times n$. A matrix is **square** if the number of rows equals the number of columns.

HISTORICAL NOTE

A HISTORY OF MATRICES

▬▬▬ The idea of a matrix dates at least from its use by Chinese scholars of the Han period (200–100 B.C.) in solving systems of linear equations, although the square arrays themselves were not singled out for attention. It was only in 1850 that James Joseph Sylvester coined the term *matrix*. Soon thereafter, Arthur Cayley (1821–1895) developed the algebra of matrices. He studied mathematics at Trinity College, Cambridge, then became a lawyer, and finally returned to Cambridge in 1863 as a professor of mathematics.

Organizing Information in a Matrix

Often, we are presented with a collection of data or other information that naturally lends itself to being organized into a matrix. Adding appropriate labels to the rows and columns, we can quickly find specific information merely by locating the appropriate address. Several examples follow.

EXAMPLE 1

Organizing Data in a Matrix

Suppose that food A contains 3 ounces of protein, 2 ounces of carbohydrate, and 6 ounces of fat per unit, while food B contains 4 ounces of protein, 5 ounces of carbohydrate, and 2 ounces of fat per unit. Using P, C, and F for protein, carbohydrate, and fat, respectively, this information may be displayed in one of the following two ways, depending on one's preference or the context into which the information falls:

$$
\begin{array}{c}
 \\
\text{Food } A \\
\text{Food } B
\end{array}
\begin{array}{c}
\begin{array}{ccc} P & C & F \end{array} \\
\begin{bmatrix} 3 & 2 & 6 \\ 4 & 5 & 2 \end{bmatrix}
\end{array}
\quad \text{or} \quad
\begin{array}{c}
 \\
P \\
C \\
F
\end{array}
\begin{array}{c}
\begin{array}{cc} \text{Food } A & \text{Food } B \end{array} \\
\begin{bmatrix} 3 & 4 \\ 2 & 5 \\ 6 & 2 \end{bmatrix}
\end{array}
\quad ▬
$$

EXAMPLE 2

Organizing Information in a Matrix

Four stations are linked by communication lines, as shown in Figure 1. Direct communication may take place only in the direction of the arrows. Thus, 3 may communicate directly with 2, but 2 cannot communicate directly with 3.

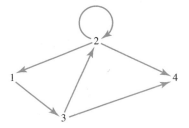

FIGURE 1
Communication Lines

This communications network may be described by the matrix

$$
\begin{array}{c}
 \\
1 \\
2 \\
3 \\
4
\end{array}
\begin{array}{cccc}
1 & 2 & 3 & 4 \\
\left[\begin{array}{cccc}
0 & 0 & 1 & 0 \\
1 & 1 & 0 & 1 \\
0 & 1 & 0 & 1 \\
0 & 0 & 0 & 0
\end{array}\right]
\end{array}
\begin{array}{l}
\text{1 communicates directly with 3} \\
\text{2 communicates directly with 1, 2 and 4} \\
\text{3 communicates directly with 2 and 4} \\
\text{4 does not communicate directly with any other}
\end{array}
$$

where a 1 in the ith row and jth column means that i may communicate directly with j, while a 0 means that it cannot. Such a matrix is called an **incidence matrix.** Unless we have been told otherwise, as we have in this example, we normally assume that a station cannot communicate with itself when we set up an incidence matrix. ■

EXAMPLE 3
Organizing Data in a
Matrix

An appliance store sells brands W, X, Y and Z of television sets, each in a 26-inch, wide-screen, and high-definition model. A beginning inventory matrix on January 1 might appear as follows:

$$
\begin{array}{c}
 \\
\textit{26-inch} \\
\textit{wide-screen} \\
\textit{high-definition}
\end{array}
\begin{array}{cccc}
W & X & Y & Z \\
\left[\begin{array}{cccc}
8 & 6 & 2 & 0 \\
3 & 5 & 5 & 6 \\
1 & 3 & 2 & 4
\end{array}\right] = B.
\end{array}
$$

How many wide-screen sets were in inventory on January 1?

SOLUTION By adding the elements in the second row of matrix B, we see that there were a total of $3 + 5 + 5 + 6 = 19$ wide-screen sets in inventory on January 1. ■

Equivalent Matrices

To bring out the full potential of matrices, we need to develop convenient ways of combining them and manipulating them as entities in their own right. The definitions and rules for doing this make up what is called **matrix algebra.** These rules are the familiar ones of regular algebra, with a few exceptions. To begin our study of matrix algebra, we must define when two matrices are equal.

Matrix Equality

Two matrices are equal if

1. They have the same dimension.
2. The entries in corresponding positions are identical.

From this definition,

$$
\begin{bmatrix}
2 & 3 \\
5 & 7
\end{bmatrix}
\neq
\begin{bmatrix}
2 & 3 & 3 \\
5 & 7 & 5
\end{bmatrix}
$$

because they do not have the same dimension. Also,

$$\begin{bmatrix} 3 & 5 \\ 2 & 1 \end{bmatrix} \neq \begin{bmatrix} 3 & 5 \\ 2 & 7 \end{bmatrix}$$

because the entries in the second-row, second-column positions are not identical. The matrices

$$\begin{bmatrix} 2 & x \\ 1 & 4 \end{bmatrix} \quad \text{and} \quad \begin{bmatrix} 2 & 3 \\ 1 & 4 \end{bmatrix}$$

are equal if and only if $x = 3$.

Scalar Multiplication

Our next step toward developing useful matrix algebra is the idea of multiplying a matrix by a number, a process called **scalar multiplication.** The need for this type of multiplication arises in many situations. Suppose the following matrix S represents monthly sales for two sales representatives in a particular firm:

$$\begin{array}{c} \\ Stecz \\ Plant \end{array} \begin{array}{cc} Jan. & Feb. \\ \begin{bmatrix} 1250 & 1380 \\ 1760 & 420 \end{bmatrix} \end{array} = S$$

If each sales representative earns a 10% commission on monthly sales, what would the matrix C representing monthly commissions look like?

$$\begin{array}{c} \\ Stecz \\ Plant \end{array} \begin{array}{cc} Jan. & Feb. \\ \begin{bmatrix} 0.10(1250) & 0.10(1380) \\ 0.10(1760) & 0.10(420) \end{bmatrix} \end{array} = \begin{bmatrix} 125 & 138 \\ 176 & 42 \end{bmatrix} = C$$

Since each element of the matrix S was multiplied by the same number, we write $0.10S - C$ to indicate that each element of S was multiplied by 0.10.

Multiplying a Matrix by a Number

Let A be a matrix, and let c be any real number. Then the matrix cA is found by multiplying each entry in A by c. Multiplying a matrix by a number is known as **scalar multiplication.**

EXAMPLE 4

Multiplying a Matrix by a Number

Let $A = \begin{bmatrix} 2 & 3 & 0 \\ 1 & -2 & 4 \end{bmatrix}$ and $c = 2$. Then

$$cA = 2A = \begin{bmatrix} 4 & 6 & 0 \\ 2 & -4 & 8 \end{bmatrix}. \quad \blacksquare$$

Matrix Addition

When two matrices with the same dimension contain similar types of information, it may be desirable to add entries together to obtain a new matrix. Consider, for instance, the matrix B from Example 3:

$$\begin{array}{c} \\ 26\text{-}inch \\ wide\text{-}screen \\ high\text{-}definition \end{array} \begin{array}{cccc} W & X & Y & Z \\ \left[\begin{array}{cccc} 8 & 6 & 2 & 0 \\ 3 & 5 & 5 & 6 \\ 1 & 3 & 2 & 4 \end{array}\right] \end{array} = B.$$

This matrix represents the inventory of an appliance store on January 1. Suppose the store places a new order for television sets, represented by the following matrix P, denoting purchases:

$$\begin{array}{c} \\ 26\text{-}inch \\ wide\text{-}screen \\ high\text{-}definition \end{array} \begin{array}{cccc} W & X & Y & Z \\ \left[\begin{array}{cccc} 1 & 1 & 0 & 2 \\ 2 & 0 & 0 & 0 \\ 2 & 1 & 1 & 2 \end{array}\right] \end{array} = P.$$

A natural question might be "What is the inventory after the new purchases are made?" To answer this question, notice that the store has three wide-screen sets of brand W on January 1 and purchases two additional ones. After the order arrives (assuming that no sets have been sold in the meantime), the store would have $3 + 2 = 5$ wide-screen sets of brand W. Working in this manner, we can obtain the matrix E, representing the store's ending inventory:

$$\begin{array}{c} \\ 26\text{-}inch \\ wide\text{-}screen \\ high\text{-}definition \end{array} \begin{array}{cccc} W & X & Y & Z \\ \left[\begin{array}{cccc} 8+1 & 6+1 & 2+0 & 0+2 \\ 3+2 & 5+0 & 5+0 & 6+0 \\ 1+2 & 3+1 & 2+1 & 4+2 \end{array}\right] \end{array}$$

$$= \begin{array}{c} \\ 26\text{-}inch \\ wide\text{-}screen \\ high\text{-}definition \end{array} \begin{array}{cccc} W & X & Y & Z \\ \left[\begin{array}{cccc} 9 & 7 & 2 & 2 \\ 5 & 5 & 5 & 6 \\ 3 & 4 & 3 & 6 \end{array}\right] \end{array} = E.$$

To obtain the new matrix, we have added corresponding entries in matrices B and P. This process is called **matrix addition.**

The Sum of Two Matrices

For matrices A and B of the same dimension $m \times n$, $A + B$ denotes a matrix whose entry in each position is the sum of the entries in corresponding positions of A and B. Matrix addition is not defined for matrices of unequal dimension.

EXAMPLE 5

Adding Matrices

Let

$$A = \begin{bmatrix} 2 & 3 & 5 \\ 0 & 1 & 6 \end{bmatrix}, \quad B = \begin{bmatrix} 2 & 0 & -1 \\ 5 & 3 & 2 \end{bmatrix}, \quad \text{and} \quad C = \begin{bmatrix} 2 & 1 \\ 3 & -2 \end{bmatrix}.$$

Then

$$A + B = \begin{bmatrix} 2+2 & 3+0 & 5-1 \\ 0+5 & 1+3 & 6+2 \end{bmatrix} = \begin{bmatrix} 4 & 3 & 4 \\ 5 & 4 & 8 \end{bmatrix},$$

while $A + C$ and $B + C$ are undefined because the matrices have different dimensions. To calculate $A + 3B$, we have

$$A + 3B = \begin{bmatrix} 2 & 3 & 5 \\ 0 & 1 & 6 \end{bmatrix} + \begin{bmatrix} 6 & 0 & -3 \\ 15 & 9 & 6 \end{bmatrix} = \begin{bmatrix} 8 & 3 & 2 \\ 15 & 10 & 12 \end{bmatrix}.$$

Note that $A + A = 2A$. This means that the two definitions involved yield results that appear like those of regular algebra. ■

Many of the properties of arithmetic carry over to matrix algebra. We will discuss most of these later. One that we can discuss now is the **distributive** property of scalar multiplication. That is, given a scalar c and two matrices A and B that can be added, it follows that $c(A + B) = cA + cB$. This property is demonstrated in part (c) of the next example.

EXAMPLE 6

An Application
Combining Matrix
Addition and
Multiplication of a
Matrix by a Number

An agricultural researcher is observing the growth of soybean, corn, and wheat plants under two fertilizer applications: N and P. Midway through the growing season, the average heights of the plants in each of six plots were recorded in inches as follows:

$$\text{Fertilizer} \begin{array}{c} \\ N \\ P \end{array} \begin{bmatrix} \overset{\text{Soybeans}}{6} & \overset{\text{Corn}}{36} & \overset{\text{Wheat}}{18} \\ 4 & 28 & 20 \end{bmatrix} = A.$$

(a) Use scalar multiplication to convert the average plant height in each plot into feet.

SOLUTION The multiplication of each entry in A by $\frac{1}{12}$ will convert inches to feet. The scalar multiplication is then

$$\frac{1}{12}A = \frac{1}{12}\begin{bmatrix} 6 & 36 & 18 \\ 4 & 28 & 20 \end{bmatrix} = \begin{bmatrix} \frac{1}{2} & 3 & \frac{3}{2} \\ \frac{1}{3} & \frac{7}{3} & \frac{5}{3} \end{bmatrix}.$$

(b) If the average plant heights in each plot are increasing at the rate of 5% per week, use scalar multiplication to calculate the growth in inches during the week past midseason.

SOLUTION The increase in height during the week, in inches, is found by the scalar multiplication $0.05A$:

$$0.05A = 0.05\begin{bmatrix} 6 & 36 & 18 \\ 4 & 28 & 20 \end{bmatrix} = \begin{bmatrix} 0.3 & 1.8 & 0.9 \\ 0.2 & 1.4 & 1.0 \end{bmatrix}.$$

(c) Find a matrix expression that will give the average plant growth in inches for each plot during the second week past midseason.

SOLUTION The growth (in inches) during the first week past midseason was found by looking at the matrix $0.05A$. If we add these growths to the midseason heights found in A, the results will give plant heights at the end of one week past midseason: $A + 0.05A$. Now, applying the 5% growth rate to these heights gives

$$0.05(A + 0.05A) = 0.05A + (0.05)^2A = (0.05 + (0.05)^2)A = 0.0525A$$

for the growth during the second week past midseason. ■

Matrix subtraction combines matrix addition and scalar multiplication.

The Difference of Two Matrices

For matrices A and B of the same dimension, $A - B = A + (-1)B$. This means that the entry in each address of $A - B$ is the difference of the entries in the like addresses of A and B.

EXAMPLE 7
Matrix Subtraction

An investor bought Class A and Class B stock in each of three chemical companies, with dollar values as shown in the following matrix:

$$\begin{array}{c} \\ \text{Ion} \\ \text{Beaker} \\ \text{Atom} \end{array} \begin{array}{cc} \text{Class A} & \text{Class B} \\ \left[\begin{array}{cc} 1500 & 2000 \\ 800 & 1000 \\ 5000 & 3500 \end{array}\right] \end{array} = A.$$

A bear market during the following year saw all of these stocks decline 10% in value. Find a matrix that will give the value of the stocks one year after purchase.

SOLUTION The loss in value may be expressed as $0.10A$, so the actual value of the stocks would be given by $A - 0.10A$:

$$\begin{bmatrix} 1500 & 2000 \\ 800 & 1000 \\ 5000 & 3500 \end{bmatrix} - \begin{bmatrix} 150 & 200 \\ 80 & 100 \\ 500 & 350 \end{bmatrix} = \begin{bmatrix} 1350 & 1800 \\ 720 & 900 \\ 4500 & 3150 \end{bmatrix}. \blacksquare$$

In basic algebra, we understand the importance of the number zero as soon as we learn about subtraction. Given any number a, it is always true that $a - a = 0$. Because we have many different dimensions of matrices, we have many different dimensions for a **zero matrix.** For instance, $\mathbf{0} = \begin{bmatrix} 0 & 0 \\ 0 & 0 \\ 0 & 0 \end{bmatrix}$ is a 3×2 zero matrix. A zero matrix has the property that, when it is added to any matrix A of the same size, the result is A. For example, if $A = \begin{bmatrix} 2 & 5 & 6 \\ 1 & 9 & 3 \end{bmatrix}$ then $\mathbf{0} + A = A$ because $\begin{bmatrix} 0 & 0 & 0 \\ 0 & 0 & 0 \end{bmatrix} + \begin{bmatrix} 2 & 5 & 6 \\ 1 & 9 & 3 \end{bmatrix} = \begin{bmatrix} 2 & 5 & 6 \\ 1 & 9 & 3 \end{bmatrix}.$

The Zero Matrix

A matrix in which the entry in every position is the number zero is referred to as a **zero matrix** and is denoted by **0.**

Properties of Matrix Algebra

The word *compatible* is frequently used to describe matrices between which operations such as addition and subtraction are meaningful. If, for instance, $A + B$ is written, we assume that the two matrices have the same dimension, so that such symbolism actually represents a matrix, and we say that A and B are **compatible** (with respect to addition). With this in mind, matrix addition satisfies the following properties:

Properties of Matrix Addition

Let A, B, and C be $m \times n$ matrices. Then

1. Matrix addition is commutative:

$$A + B = B + A.$$

2. Matrix addition is associative:

$$(A + B) + C = A + (B + C).$$

3. There is a zero matrix, **0,** of dimension $m \times n$, for which

$$A + 0 = 0 + A = A.$$

4. For any number a,

$$a(A + B) = aA + aB.$$

These properties are similar to those for the algebra of real numbers.

EXAMPLE 8

Solving Matrix Equations

(a) Solve this equation for X:

$$X + \begin{bmatrix} 2 & 1 & 3 \\ 5 & 0 & -1 \end{bmatrix} = \begin{bmatrix} 1 & -3 & 2 \\ 1 & 0 & 1 \end{bmatrix}.$$

SOLUTION $X = \begin{bmatrix} 1 & -3 & 2 \\ 1 & 0 & 1 \end{bmatrix} - \begin{bmatrix} 2 & 1 & 3 \\ 5 & 0 & -1 \end{bmatrix} = \begin{bmatrix} -1 & -4 & -1 \\ -4 & 0 & 2 \end{bmatrix}.$

(b) Solve this equation for A:

$$2A - \begin{bmatrix} 1 & 2 \\ 0 & 1 \end{bmatrix} = \begin{bmatrix} 5 & 1 \\ 2 & 1 \end{bmatrix}.$$

SOLUTION $2A = \begin{bmatrix} 5 & 1 \\ 2 & 1 \end{bmatrix} + \begin{bmatrix} 1 & 2 \\ 0 & 1 \end{bmatrix} = \begin{bmatrix} 6 & 3 \\ 2 & 2 \end{bmatrix}$, so that

$$A = \frac{1}{2}\begin{bmatrix} 6 & 3 \\ 2 & 2 \end{bmatrix} = \begin{bmatrix} 3 & \frac{3}{2} \\ 1 & 1 \end{bmatrix}.$$

(c) Solve this equation for X: $3X + 5A - B = 6C$.

SOLUTION $3X = 6C + B - 5A$ so that $X = 2C + \frac{1}{3}B - \frac{5}{3}A$. ■

Inventory Matrices

A typical use of matrices involves inventory tabulations, as seen previously in this section. Usually, such information is stored in computers, perhaps in a spreadsheet and on a much larger scale than that used for our problems here. The general rule of inventory status over a given period is as follows:

Inventory Rule

Let matrices be defined as follows: B is the beginning inventory, P is the purchases, S is the sales, and E is the ending inventory for a given interval. Then $E = B + P - S$.

Whenever three of the four matrices in the preceding equation are known, we may solve for the fourth. Let's look at an example of this rule.

EXAMPLE 9

An Inventory Application

A furniture dealer has four stores—S_1, S_2, S_3, and S_4—in a large city and keeps records about the sales of Early American (E. Am.) and French Provincial (Fr. Prov.) dining room sets in matrix form on a computer. Suppose that the records are as follows:

$$\text{Beginning inventory matrix} = \begin{array}{c} S_1 \\ S_2 \\ S_3 \\ S_4 \end{array} \begin{bmatrix} \overset{\text{E. Am.}}{8} & \overset{\text{Fr. Prov.}}{2} \\ 6 & 9 \\ 12 & 7 \\ 3 & 10 \end{bmatrix} = B,$$

$$\text{Purchase matrix} = \begin{array}{c} S_1 \\ S_2 \\ S_3 \\ S_4 \end{array} \begin{bmatrix} \overset{\text{E. Am.}}{8} & \overset{\text{Fr. Prov.}}{2} \\ 12 & 5 \\ 6 & 10 \\ 9 & 11 \end{bmatrix} = P,$$

$$\text{Ending inventory matrix} = \begin{array}{c} S_1 \\ S_2 \\ S_3 \\ S_4 \end{array} \begin{bmatrix} \overset{\text{E. Am.}}{7} & \overset{\text{Fr. Prov.}}{2} \\ 9 & 6 \\ 10 & 12 \\ 5 & 9 \end{bmatrix} = E.$$

Find the sales matrix.

SOLUTION According to the general rule on inventory, $E = B + P - S$, so that $S = B + P - E$, or

$$S = \begin{bmatrix} 8 & 2 \\ 6 & 9 \\ 12 & 7 \\ 3 & 10 \end{bmatrix} + \begin{bmatrix} 8 & 2 \\ 12 & 5 \\ 6 & 10 \\ 9 & 11 \end{bmatrix} - \begin{bmatrix} 7 & 2 \\ 9 & 6 \\ 10 & 12 \\ 5 & 9 \end{bmatrix} = \begin{bmatrix} 9 & 2 \\ 9 & 8 \\ 8 & 5 \\ 7 & 12 \end{bmatrix}. \quad \blacksquare$$

Transpose of a Matrix

The last operation on a matrix to be mentioned in this section is the transpose of a matrix. We want to show how the transpose is found, so that you are aware of this operation, but we will not see its use until Chapter 5.

The Transpose of a Matrix

The **transpose** of any matrix A, written A^T, is the matrix for which the columns are the respective rows of A. If A is an $m \times n$ matrix, then A^T is an $n \times m$ matrix.

EXAMPLE 10
Finding the Transpose of a Matrix

Find the transpose of the matrix $A = \begin{bmatrix} 3 & 4 & -1 & 7 \\ 2 & 0 & 1 & 0 \\ 1 & 1 & 1 & 2 \end{bmatrix}$.

SOLUTION According to the definition, we start with the first row of A and make it the first column of A^T, then take the second row of A and make it the second column of A^T, and so on. This gives

$$A^T = \begin{bmatrix} 3 & 2 & 1 \\ 4 & 0 & 1 \\ -1 & 1 & 1 \\ 7 & 0 & 2 \end{bmatrix}. \quad \blacksquare$$

The current section has begun the development of an algebra for matrices. The next two sections complete the basic algebraic structure we need.

EXAMPLE 11
Matrix Addition via Spreadsheets

Let $A = \begin{bmatrix} 2 & 6 & 5 \\ 1 & 0 & 2 \end{bmatrix}$ and $B = \begin{bmatrix} -2 & 3 & -5 \\ 1 & 3 & 4 \end{bmatrix}$. Use Excel to find $A + B$.

SOLUTION Set up an Excel worksheet that contains the matrices as shown in Figure 2, along with labels for the individual matrices.

	A	B	C	D	E	F	G
1		A				B	
2	2	6	5		-2	3	-5
3	1	0	2		1	3	4
4							
5							
6		A+B					
7							
8							
9							

FIGURE 2
Setting Up Matrix Addition in a Spreadsheet

Highlight cells A7:C8, so that you have a block of highlighted cells of dimension 2×3, which will be the dimension of $A + B$. While these cells are highlighted, type =A2:C3+E2:G3. Then press Ctrl-Shift-Enter. The result should appear as in Figure 3.

	A	B	C	D	E	F	G
1		A				B	
2	2	6	5		-2	3	-5
3	1	0	2		1	3	4
4							
5							
6		A+B					
7	0	9	0				
8	2	3	6				
9							

FIGURE 3
The Sum of Two Matrices in a Spreadsheet ▬

Graphing calculators have the preprogrammed ability to do matrix algebra, with certain makes and models having more memory than others. Figure 4 shows a graphing calculator screen indicating the calculation of $(2A + B)^T$, where

$$A = \begin{bmatrix} 2 & 6 & 5 \\ 1 & 0 & 2 \end{bmatrix} \quad \text{and} \quad B = \begin{bmatrix} -2 & 3 & -5 \\ 1 & 3 & 4 \end{bmatrix}.$$

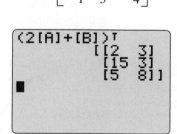

FIGURE 4
Matrix Algebra by Technology

Exercises 3.1

In Exercises 1 through 4, assign the value to the variables so that a matrix equality results.

1. $\begin{bmatrix} t+1 & 5 & 2 \\ 7 & 6 & 0 \end{bmatrix} = \begin{bmatrix} 4 & 5 & 2 \\ 7 & x-3 & 0 \end{bmatrix}$

2. $\begin{bmatrix} 2 & x-4 \\ 3 & 0 \\ 7 & 0 \end{bmatrix} = \begin{bmatrix} 2 & 6 \\ t-1 & 0 \\ 1 & 0 \end{bmatrix}$

3. $\begin{bmatrix} 0 & 0 \\ x & -2 \end{bmatrix} = \begin{bmatrix} 0 & 0 \\ x & -2 \end{bmatrix}$

4. $\begin{bmatrix} x-1 & 4 & 5 \\ 8 & z & 0 \\ y-3 & 0 & 1 \end{bmatrix} = \begin{bmatrix} 2x+3 & 4 & t \\ 8 & z & 0 \\ 2y & 0 & 1 \end{bmatrix}$

In Exercises 5 through 12, use the given matrices A, B, and C to find the indicated expression, if possible. Let

$$A = \begin{bmatrix} 2 & 3 \\ 1 & -1 \end{bmatrix}, \quad B = \begin{bmatrix} 6 & -1 \\ 0 & 2 \end{bmatrix}, \quad \text{and}$$

$$C = \begin{bmatrix} 2 & -3 & 5 \\ 0 & 0 & -2 \end{bmatrix}.$$

5. $-3A + B$

6. $A + 4B$

7. $\frac{1}{3}(A + B)$

8. $\frac{1}{2}(B - A)$

9. $\frac{1}{4}(2A + C)$

10. $A + 2B - C$

11. $-5(C + 3C)$

12. $C - C$

In Exercises 13 and 14, use

$$A = \begin{bmatrix} a & b \\ c & d \end{bmatrix},$$

$$B = \begin{bmatrix} e & f \\ g & h \end{bmatrix},$$

and $\qquad C = \begin{bmatrix} i & j \\ k & l \end{bmatrix}$

in each expression to verify the equality.

13. $A + B = B + A$

14. $(A + B) + C = A + (B + C)$

15. **Seating Passengers** An airline is buying two types of planes: P_1 and P_2. Plane P_1 will seat 30 first-class passengers, 50 tourist-class passengers, and 90 economy-class passengers. Plane P_2 will seat 50 first-class passengers, 60 tourist-class passengers, and 100 economy-class passengers.

(a) Display the preceding information in a 2×3 matrix.

(b) Display the preceding information in a 3×2 matrix.

16. **Advertising** The Linda Morris firm encounters these costs in advertising its computers: $80 per minute per radio ad, $100 per column inch per newspaper ad, and $250 per minute per television ad.

(a) Display the information in a 1×3 matrix.

(b) Display the information in a 3×1 matrix.

17. **College Enrollment** A particular small college has 300 freshmen, 287 sophomores, 250 juniors, and 240 seniors.

(a) Display the preceding information in a 4×1 matrix.

(b) Display the information in a 1×4 matrix.

18. **Communication Outposts** Three military outposts are linked by communication devices as shown in the following graph:

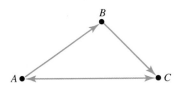

Assume that communication between stations takes place only in the direction of the arrows. Write a matrix that displays this communications network.

19. **Airline Flights** Airline routes through cities *A*, *B*, *C*, *D*, and *E* are shown in the following diagram:

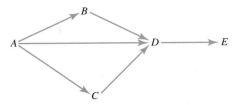

Flights are available in the direction of the arrows only. Assume that each city does not have a flight from itself to itself. Write a matrix that displays this flight network.

A ***dominance relation*** *expresses the fact that, among members of a group, some members exert dominance over other members. Examples are the hierarchy of an office staff, the perceived status among students in a class, and an athletic team winning a sporting contest. A dominance matrix can be built by assigning a row and a column to each member of the group. For any two members, x and y, of the group, we place a 1 in the x row, y column position if x dominates y; otherwise, we place a 0 in that position. Assume that no member of the group dominates him- or herself.*

20. Dominance Relations Among executives Smith, Jones, Barney, and Lemmons, Smith reports to Barney, Jones reports to Barney, and Barney reports to Lemmons. Write a dominance matrix showing these relationships.

21. Dominance Relations The Bears, the Wildcats, the Tigers, and the Cubs play in a basketball tournament. During the first two days of play, the Bears beat the Wildcats and the Tigers, the Tigers beat the Cubs, the Cubs beat the Bears and the Wildcats, and the Wildcats beat the Tigers. Construct a dominance matrix showing these tournament results.

In Exercises 22 through 27, solve each of the matrix equations for the indicated variable.

22. Solve for X: $X + 2B = C$

23. Solve for X: $\frac{3}{4}X - 5T = 3B$

24. Solve for A: $2A + 3B = \mathbf{0}$

25. Solve for B: $\frac{1}{2}A - 3B + C = 2D$

26. Solve for C: $2C + \frac{1}{2}B = 4C$

27. Solve for T: $A - T = 2C + 3T$

In Exercises 28 through 31, solve each matrix equation for X.

28. $X + \begin{bmatrix} 2 & 3 \\ 5 & 1 \end{bmatrix} = \begin{bmatrix} 5 & -1 \\ 2 & 6 \end{bmatrix}$

29. $\begin{bmatrix} 2 & -3 & 1 \\ 4 & 5 & 2 \end{bmatrix} + 2X = -3\begin{bmatrix} -1 & 2 & -1 \\ 1 & 0 & 1 \end{bmatrix}$

30. $\begin{bmatrix} 2 & 1 & -3 \\ 5 & 0 & 0 \\ 1 & -2 & 3 \end{bmatrix} + \frac{1}{3}X = \begin{bmatrix} -1 & -2 & -3 \\ 0 & 1 & 1 \\ 5 & 0 & 2 \end{bmatrix}$

31. $2\begin{bmatrix} 1 & 1 \\ 2 & 0 \\ 3 & -2 \end{bmatrix} + 3X + \begin{bmatrix} 2 & 5 \\ -1 & 3 \\ -4 & -2 \end{bmatrix} = \begin{bmatrix} 0 & 0 \\ 0 & 0 \\ 0 & 0 \end{bmatrix}$

32. Is there a real number x that is a solution of this equation?

$$x\begin{bmatrix} 2 & 3 \\ 4 & 1 \end{bmatrix} + x\begin{bmatrix} 5 & 6 \\ 2 & 2 \end{bmatrix} = \begin{bmatrix} 14 & 18 \\ 12 & 9 \end{bmatrix}$$

33. Is there a real number x that is a solution of this equation?

$$x\begin{bmatrix} 2 & 1 & 1 \\ 0 & 1 & 1 \end{bmatrix} + x\begin{bmatrix} 1 & 2 & 2 \\ 3 & -1 & -1 \end{bmatrix} = \begin{bmatrix} 9 & 9 & 9 \\ 9 & 0 & 0 \end{bmatrix}$$

34. Scalar Multiplication The matrix

$$A = \begin{bmatrix} 20.2 & 50.6 \\ 28.7 & 31.1 \end{bmatrix}$$

gives sales in thousands of dollars. Perform a scalar multiplication that will result in a matrix that displays the sales in dollars.

35. Scalar Multiplication The matrix

$$B = \begin{bmatrix} 0.06 & 0.02 \\ 0.03 & 0.05 \end{bmatrix}$$

gives times in hours. Perform a scalar multiplication that will result in times given in minutes.

Exercises 36 and 37 refer to the following inventory problem:

Inventory During a given week, a small convenience store has a beginning inventory given by the following matrix:

	2% milk	Whole milk	
Quarts	30	20	$= B.$
$\frac{1}{2}$ gal.	18	12	

The sales inventory of the store for the same week is given by the following matrix:

	2% milk	Whole milk	
Quarts	21	18	$= S.$
$\frac{1}{2}$ gal.	22	15	

36. If the ending inventory matrix is $\frac{1}{2}B$, what is the purchase matrix?

37. If the ending inventory matrix is

$$E = \begin{bmatrix} 10 & 8 \\ 12 & 10 \end{bmatrix},$$

what is the purchase matrix?

38. Pollution Control A state pollution-control office is monitoring two waste-treatment plants that discharge into the same watershed. Three of the quantities monitored from plant discharge are nitrates, lead, and oxygen content, measured in parts per million (ppm). At the beginning of the monitoring period, the measurements were as shown in the following matrix:

$$\begin{array}{c} \\ \text{Plant I} \\ \text{Plant II} \end{array} \begin{array}{ccc} \text{Nitrates} & \text{Lead} & \text{Oxygen} \end{array} \\ \left[\begin{array}{ccc} 13 & 17 & 30 \\ 20 & 22 & 50 \end{array}\right] = B.$$

At the end of the monitoring period, six months later, the measurements were as follows:

$$\begin{array}{c} \\ \text{Plant I} \\ \text{Plant II} \end{array} \begin{array}{ccc} \text{Nitrates} & \text{Lead} & \text{Oxygen} \end{array} \\ \left[\begin{array}{ccc} 10 & 15 & 30 \\ 15 & 16 & 45 \end{array}\right] = E.$$

(a) Find the matrix $B - E$.

(b) What do the entries in $B - E$ represent?

(c) Does the matrix $B - E$ show which plant is making the best effort at reducing the aforesaid pollutants in its discharge?

(d) If the trend continues, how many months will it take for plant *A* to entirely eliminate its nitrate discharge?

39. Investing Mr. Jones has two types of IRA retirement accounts: certificates of deposit (CD) and municipal bonds (MB). The accounts are held in two banks: First National and Commercial National. The dollar amounts in each at the beginning of 2003 are given in the following matrix:

$$\begin{array}{c} \\ \text{FN} \\ \text{CN} \end{array} \begin{array}{cc} \text{CD} & \text{MB} \end{array} \\ \left[\begin{array}{cc} 30{,}000 & 25{,}000 \\ 50{,}000 & 40{,}000 \end{array}\right] = A.$$

Assume an 8% appreciation rate on each of these accounts.

(a) Find a matrix expression representing the earnings of the various accounts during 2003.

(b) Compute the matrix expression found in (a).

(c) Find a matrix expression representing the value of each account at the end of 2003.

(d) Compute the matrix expression found in (c).

Exercises 40 through 44 refer to the following problem involving college costs:

College Costs The Ersoffs are comparing the costs per year (in dollars) of sending their daughter to one of two

universities. The costs are shown in the following matrix:

$$\begin{array}{c} \\ \text{Tuition, fees} \\ \text{Living expenses} \end{array} \begin{array}{cc} \text{State} & \text{Private} \end{array} \\ \left[\begin{array}{cc} 4500 & 16000 \\ 5000 & 9400 \end{array}\right] = C.$$

Assume an annual inflation rate of 5%.

40. Find a matrix expression that shows the increases in costs for the second year at each university.

41. Find a matrix expression that represents the actual costs for the second year at each of the universities. Compute this expression.

42. Find a matrix expression that represents the actual costs for the third year at each of the universities. Compute this expression.

43. Find a matrix expression that represents the actual costs for the fourth year at each of the universities. Compute this expression.

44. Find a matrix expression that represents the actual costs of attending four years at each of the universities. Compute this expression.

45. Memory Experiment Three groups of people are tested for their ability to memorize and retain sequences of numbers. The first group had no training, the second group received a short course on memory, and the third group received intensive training on memory. The following table shows the average number of seconds the numbers were retained by each group:

Number of digits in sequence	Group I	Group II	Group III
1–8	14	16	18
9–14	8	11	15
15–18	6	10	13

Construct a matrix showing the percentage by which the performance of the second two groups surpassed the first.

In Exercises 46 through 49, find the transpose of each matrix.

46. $A = \begin{bmatrix} 2 & -3 & 1 \\ 0 & 5 & -2 \end{bmatrix}$

47. $\begin{bmatrix} 1 & 2 \\ 3 & -5 \end{bmatrix}$

48. $C = \begin{bmatrix} 2 \\ -1 \\ 3 \end{bmatrix}$

49. $D = \begin{bmatrix} 1 & 0 \\ 2 & -1 \\ -3 & 2 \\ 4 & 1 \end{bmatrix}$

Section 3.2 Matrix Multiplication and Applications

One of the simplest equations to solve in algebra is $ax = b$, and we know that the solution is $x = \frac{b}{a} = a^{-1}b$, where $a^{-1} = \frac{1}{a}$, provided that $a \neq 0$. A motivation for matrix multiplication comes from the desire to convert a system of linear equations into a similar **matrix equation,** $AX = B$, and then solve for X (under some specific conditions) to obtain a solution of the system of the same form as that in algebra, $X = A^{-1}B$. To this end, our first concern is to define the multiplication of matrices A and X so that the product AX is the matrix B *and* represents an equivalent way of writing the original system of linear equations. An example points the way; it turns out that the process of multiplication is not as easily guessed as it was for addition.

Consider the system

$$2x + 3y = 5$$
$$4x + y = 6.$$

To arrive at the form $AX = B$, the appearance of the system suggests that B should be the 2×1 matrix $\begin{bmatrix} 5 \\ 6 \end{bmatrix}$. The matrix X plays the role of a variable, so it must somehow represent the variables of the system. To do this, define $X = \begin{bmatrix} x \\ y \end{bmatrix}$. Then the only thing left for A is the 2×2 coefficient matrix $A = \begin{bmatrix} 2 & 3 \\ 4 & 1 \end{bmatrix}$. Thus, if $AX = B$, or $\begin{bmatrix} 2 & 3 \\ 4 & 1 \end{bmatrix}\begin{bmatrix} x \\ y \end{bmatrix} = \begin{bmatrix} 5 \\ 6 \end{bmatrix}$, is to hold, then AX must be a 2×1 matrix; furthermore, to represent the given system of equations, it must look like this:

$$AX = \begin{bmatrix} 2x + 3y \\ 4x + y \end{bmatrix} = \begin{bmatrix} 5 \\ 6 \end{bmatrix} = B.$$

That is, $2x + 3y$ represents the single element in the first-row, first-column position, and $4x + y$ represents the single element in the second-row, first-column position. Now notice how each row of A is combined with the column $\begin{bmatrix} x \\ y \end{bmatrix}$ to produce these results:

first row on the left, combined with the column on the right

second row on the left, combined with the column on the right

Do you see how the rows and columns are combined? The rule of combination follows:

$$\left(\begin{matrix} first\ element \\ in\ row \end{matrix}\right) \times \left(\begin{matrix} first\ element \\ in\ column \end{matrix}\right) + \left(\begin{matrix} second\ element \\ in\ row \end{matrix}\right) \times \left(\begin{matrix} second\ element \\ in\ column \end{matrix}\right)$$

This description of combining a row with a column to produce a single number may be thought of as a **row–column product.** On the basis of the preceding pattern, we can now assert a general rule for the row–column product:

The Row–Column Product

Let A be an $m \times n$ matrix, and let B be an $n \times q$ matrix. Then, for a typical row $[a_1 \quad a_2 \quad a_3 \quad \cdots \quad a_n]$ in A and a typical column $\begin{bmatrix} b_1 \\ b_2 \\ b_3 \\ \cdots \\ b_n \end{bmatrix}$ in B, the row–column product, denoted by $\bullet$, gives

$$[a_1 \quad a_2 \quad a_3 \quad \cdots \quad a_n] \bullet \begin{bmatrix} b_1 \\ b_2 \\ b_3 \\ \cdots \\ b_n \end{bmatrix} = a_1 b_1 + a_2 b_2 + a_3 b_3 + \cdots + a_n b_n.$$

The row–column product can be calculated only when the rows of A have the same number of elements as the columns of B.

EXAMPLE 1

Demonstrating the Row-Column Product

Let $A = \begin{bmatrix} 3 & 1 & 2 \\ 1 & -1 & 6 \\ 1 & 2 & 3 \end{bmatrix}$ and let $B = \begin{bmatrix} 5 & 1 \\ 2 & 0 \\ 3 & -1 \end{bmatrix}$.

(a) The row–column product applied to row 2 of matrix A and column 1 of matrix B is

$$[1 \quad -1 \quad 6] \bullet \begin{bmatrix} 5 \\ 2 \\ 3 \end{bmatrix} = 1 \cdot 5 + (-1) \cdot 2 + 6 \cdot 3 = 21.$$

(b) Similarly, row 3 $\bullet$ column 2 $= [1 \quad 2 \quad 3] \bullet \begin{bmatrix} 1 \\ 0 \\ -1 \end{bmatrix}$

$$= 1 \cdot 1 + 2 \cdot 0 + 3 \cdot (-1) = -2. \quad \blacksquare$$

We next define the multiplication of two matrices, using the row–column product.

Defining Matrix Multiplication

Let A be an $m \times n$ matrix and let B be an $n \times q$ matrix. Then the **multiplication of A by B,** denoted by AB, is the $m \times q$ matrix whose entry in the ith row and jth column position is the row–column product of the ith row of A and the jth column of B.

To visualize how this definition of matrix multiplication works, let's return to Example 1. In part (a), we were asked to apply the row–column product to row 2 of A and column 1 of B. According to the definition, this would place the result in the row 2, column 1 position of the product matrix, AB:

$$
\begin{array}{ccc}
A & B & AB
\end{array}
$$
$$
\begin{bmatrix} 3 & 1 & 2 \\ 1 & -1 & 6 \\ 1 & 2 & 3 \end{bmatrix}
\begin{bmatrix} 5 & 1 \\ 2 & 0 \\ 3 & -1 \end{bmatrix}
=
\begin{bmatrix} 23 & 1 \\ 21 & -5 \\ 18 & -2 \end{bmatrix}.
$$

In part (b) of the same example, we were asked to apply the row–column product to row 3 of A and column 2 of B. The result, according to the definition of matrix multiplication, would then be in the row 3, column 2 position of matrix AB:

$$
\begin{array}{ccc}
A & B & AB
\end{array}
$$
$$
\begin{bmatrix} 3 & 1 & 2 \\ 1 & -1 & 6 \\ 1 & 2 & 3 \end{bmatrix}
\begin{bmatrix} 5 & 1 \\ 2 & 0 \\ 3 & -1 \end{bmatrix}
=
\begin{bmatrix} 23 & 1 \\ 21 & -5 \\ 18 & -2 \end{bmatrix}.
$$

EXAMPLE 2

Illustrating Matrix Multiplication

Let $A = \begin{bmatrix} 2 & 3 \\ 5 & 1 \end{bmatrix}$ and $B = \begin{bmatrix} 4 & 2 \\ -1 & 1 \end{bmatrix}$.

(a) Compute AB.

SOLUTION
$$
AB = \begin{bmatrix} 2 & 3 \\ 5 & 1 \end{bmatrix}\begin{bmatrix} 4 & 2 \\ -1 & 1 \end{bmatrix}
$$
$$
= \begin{bmatrix} \text{row 1 of } A \bullet \text{column 1 of } B & \text{row 1 of } A \bullet \text{column 2 of } B \\ \text{row 2 of } A \bullet \text{column 1 of } B & \text{row 2 of } A \bullet \text{column 2 of } B \end{bmatrix}
$$
$$
= \begin{bmatrix} [2 \ \ 3] \bullet \begin{bmatrix} 4 \\ -1 \end{bmatrix} & [2 \ \ 3] \bullet \begin{bmatrix} 2 \\ 1 \end{bmatrix} \\ [5 \ \ 1] \bullet \begin{bmatrix} 4 \\ -1 \end{bmatrix} & [5 \ \ 1] \bullet \begin{bmatrix} 2 \\ 1 \end{bmatrix} \end{bmatrix} = \begin{bmatrix} 5 & 7 \\ 19 & 11 \end{bmatrix}.
$$

(b) Compute BA.

SOLUTION
$$
BA = \begin{bmatrix} 4 & 2 \\ -1 & 1 \end{bmatrix}\begin{bmatrix} 2 & 3 \\ 5 & 1 \end{bmatrix}
$$
$$
= \begin{bmatrix} \text{row 1 of } B \bullet \text{column 1 of } A & \text{row 1 of } B \bullet \text{column 2 of } A \\ \text{row 2 of } B \bullet \text{column 1 of } A & \text{row 2 of } B \bullet \text{column 2 of } A \end{bmatrix}
$$
$$
= \begin{bmatrix} [4 \ \ 2] \bullet \begin{bmatrix} 2 \\ 5 \end{bmatrix} & [4 \ \ 2] \bullet \begin{bmatrix} 3 \\ 1 \end{bmatrix} \\ [-1 \ \ 1] \bullet \begin{bmatrix} 2 \\ 5 \end{bmatrix} & [-1 \ \ 1] \bullet \begin{bmatrix} 3 \\ 1 \end{bmatrix} \end{bmatrix} = \begin{bmatrix} 18 & 14 \\ 3 & -2 \end{bmatrix}.
$$

(c) Compute A^2. (*Note:* $A^2 = AA$, *not* the square of each element of A.)

SOLUTION $A^2 = \begin{bmatrix} 2 & 3 \\ 5 & 1 \end{bmatrix}\begin{bmatrix} 2 & 3 \\ 5 & 1 \end{bmatrix} = \begin{bmatrix} 19 & 9 \\ 15 & 16 \end{bmatrix}.$ ■

Examples 2(a) and 2(b) show that matrix multiplication is *not* commutative; that is, for many matrices A and B, it is often true that $AB \neq BA$.

This computational suggestion might be helpful to you when you are forming the product of two matrices: Take the first row of the matrix on the left, and find its row–column product with *each* column of the matrix on the right. This will give the first row of the product. Next, take the second row of the matrix on the left, and find its row–column product with each column in the matrix on the right. This gives the second row of the product. Next, take the third row of the matrix on the left, and so on.

EXAMPLE 3

Illustrating Matrix Multiplication

Let $A = \begin{bmatrix} 2 & 3 & 4 \\ -2 & 1 & 0 \end{bmatrix}$ and $B = \begin{bmatrix} 1 & 4 & -2 \\ 0 & 1 & 0 \\ 2 & 2 & 1 \end{bmatrix}.$

(a) The respective dimensions of A and B are 2×3 and 3×3. The number of columns, 3, in A is also the number of rows in B, and this must be the case in order for the row–column product to be defined. We have

$$A: 2 \times 3 \qquad\qquad B: 3 \times 3$$

Must match for product

Dimension of product

Therefore, $AB = \begin{bmatrix} 2 & 3 & 4 \\ -2 & 1 & 0 \end{bmatrix}\begin{bmatrix} 1 & 4 & -2 \\ 0 & 1 & 0 \\ 2 & 2 & 1 \end{bmatrix} = \begin{bmatrix} 10 & 19 & 0 \\ -2 & -7 & 4 \end{bmatrix}.$

(b) The product BA does not exist. The reason is that, when the order of B is considered first and the dimension of A second—that is, $B: 3 \times 3$, $A: 2 \times 3$—the middle numbers do not match. If a row–column product were attempted, there would be a mismatch between the elements of any row of B with the elements of any column of A. ■

With matrix multiplication defined, we are now in a position to see how any linear system of equations can be written as an equivalent matrix equation in the form $AX = B$. We must wait until the next section, however, to arrive at the "nice" solution mentioned at the outset of this section.

EXAMPLE 4

A System Written as a Matrix Equation $AX = B$

The system

$$\begin{aligned} -2x + y - z &= 3 \\ x + z &= 2 \\ 5x + y &= -6 \\ x + 2y + z &= 3 \end{aligned}$$

may be written in matrix equation form as

$$\begin{bmatrix} -2 & 1 & -1 \\ 1 & 0 & 1 \\ 5 & 1 & 0 \\ 1 & 2 & 1 \end{bmatrix} \begin{bmatrix} x \\ y \\ z \end{bmatrix} = \begin{bmatrix} 3 \\ 2 \\ -6 \\ 3 \end{bmatrix}.$$

$$4 \times 3 \qquad 3 \times 1 \qquad 4 \times 1 \quad \blacksquare$$

EXAMPLE 5

An Application of
Matrix Multiplication

Costs, in dollars, for radio (per minute), newspaper (per column inch), and TV (per minute) ads in two cities—Ocala and Hawthorne—are given in the following matrix:

$$\begin{array}{c} \\ \text{Ocala} \\ \text{Hawthorne} \end{array} \begin{array}{ccc} \text{Radio} & \text{Newspaper} & \text{TV} \\ \begin{bmatrix} 30 & 20 & 120 \\ 25 & 18 & 140 \end{bmatrix} \end{array}$$

(a) Ads are run five times in each of these media in Ocala and eight times in each of the media in Hawthorne. Display a matrix product that will give the total amount spent on radio, newspaper, and TV ads.

SOLUTION

$$[5 \quad 8] \begin{array}{c} \text{Radio} \quad \text{Newspaper} \quad \text{TV} \\ \begin{bmatrix} 30 & 20 & 120 \\ 25 & 18 & 140 \end{bmatrix} \end{array} = \begin{array}{ccc} \text{Radio} & \text{Newspaper} & \text{TV} \\ [350 & 244 & 1720]. \end{array}$$

This product matrix gives the totals requested.

(b) Suppose that the radio, newspaper, and TV ads are run 20, 15, and 30 times, respectively, in each of the two cities. Use matrix multiplication to find the amount spent on ads in each of the cities.

SOLUTION

$$\begin{array}{c} \text{Ocala} \\ \text{Hawthorne} \end{array} \begin{bmatrix} 30 & 20 & 120 \\ 25 & 18 & 140 \end{bmatrix} \begin{bmatrix} 20 \\ 15 \\ 30 \end{bmatrix} = \begin{bmatrix} 4500 \\ 4970 \end{bmatrix} \begin{array}{c} \text{Ocala} \\ \text{Hawthorne} \end{array}$$

The entries in the product matrix give the total costs in each city.

(c) Suppose that the radio, newspaper, and TV ads are run 30, 40, and 60 times, respectively, in each of the cities in January, and 20, 12, and 22 times, respectively, in February. Use matrix multiplication to find the total amount spent in each city in January and February.

SOLUTION

$$\begin{array}{c} \text{Ocala} \\ \text{Hawthorne} \end{array} \begin{bmatrix} 30 & 20 & 120 \\ 25 & 18 & 140 \end{bmatrix} \begin{array}{c} \text{January} \ \text{February} \\ \begin{bmatrix} 30 & 20 \\ 40 & 12 \\ 60 & 22 \end{bmatrix} \end{array} = \begin{array}{c} \text{January} \ \text{February} \\ \begin{bmatrix} 8900 & 3480 \\ 9870 & 3796 \end{bmatrix} \begin{array}{c} \text{Ocala} \\ \text{Hawthorne} \end{array} \end{array}$$

The first column in the product matrix gives the costs for ads in January in each of the cities, and the second column gives the costs for ads in February in the two cities. ■

In addition to graphing calculators and various other computer programs, spreadsheets not only offer a preprogrammed way to do matrix multiplication, but also allow the labeling of rows and columns, which is helpful in interpreting the meaning of the entries in the product matrix.

EXAMPLE 6

The MMULT Function of Excel

Complete Example 5(c) using Excel.

SOLUTION Begin by entering appropriate row and column labels for the matrices, and then enter the matrices themselves. The result should appear as in Figure 5.

	A	B	C	D	E	F	G	H
1		Radio	Newspaper	TV			January	February
2	Ocala	30	20	120		Radio	30	20
3	Hawthorne	25	18	140		Newspaper	40	12
4						TV	60	22
5								
6	Product:		January	February				
7		Ocala						
8		Hawthorne						

FIGURE 5
Setting Up Matrix Multiplication in a Spreadsheet

Highlight the 2 × 2 block of cells that will contain the product, which is C7:D8. While the cells are highlighted, type =MMULT(B2:D3*G2:H4) and press Ctrl-Shift-Enter. The result should appear as in Figure 6.

	A	B	C	D	E	F	G	H
1		Radio	Newspaper	TV			January	February
2	Ocala	30	20	120		Radio	30	20
3	Hawthorne	25	18	140		Newspaper	40	12
4						TV	60	22
5								
6	Product:		January	February				
7		Ocala	8900	3480				
8		Hawthorne	9870	3796				

FIGURE 6
The Product of Two Matrices in a Spreadsheet ▬

EXAMPLE 7

Predicting Votes in an Election

The voting problem stated at the outset of this chapter tells how voter opinions change from week to week. Assuming that the stated trends hold, can we predict how these 100 voters will vote week by week?

SOLUTION The answer is "yes." We first construct the matrix T whose first row records how the "for" voters at the beginning of the week will, as a percentage, take a "for" or "against" position by the end of the week; the second row records how the "against" voters at the beginning of the week will, as a percentage, take a "for" or "against" position by the end of the week.

$$
\begin{array}{cc}
 & \begin{array}{cc} \text{For} & \text{Against} \end{array} \\
\begin{array}{c} \text{Beginning} \\ \text{of week} \end{array}\begin{array}{c} \text{For} \\ \text{Against} \end{array} & \left[\begin{array}{cc} 0.75 & 0.25 \\ 0.40 & 0.60 \end{array}\right] = T.
\end{array}
$$

Then, we let A be the following matrix showing the initial position of the 100 voters:

$$
\begin{array}{cc}
 & \begin{array}{cc} \text{For} & \text{Against} \end{array} \\
A = & \left[\begin{array}{cc} 50 & 50 \end{array}\right].
\end{array}
$$

Now, if we consider the product

$$AT = \begin{bmatrix} 50 & 50 \end{bmatrix}\begin{bmatrix} 0.75 & 0.25 \\ 0.40 & 0.60 \end{bmatrix} = \begin{bmatrix} 57.5 & 42.5 \end{bmatrix},$$

the entries in the product will have these meanings:

$$
\begin{array}{ccccccc}
50 & \times & 0.75 & + & 50 & \times & 0.40 & = & 57.5, \\
\text{"For"} & & \text{Still} & & \text{"Against"} & & \text{Change} & & \text{"For" at end} \\
& & \text{"For"} & & & & \text{to "For"} & & \text{of week}
\end{array}
$$

$$
\begin{array}{ccccccc}
50 & \times & 0.25 & + & 50 & \times & 0.60 & = & 42.5. \\
\text{"For"} & & \text{Change to} & & \text{"Against"} & & \text{Still} & & \text{"Against" at} \\
& & \text{"Against"} & & & & \text{"Against"} & & \text{end of week}
\end{array}
$$

Therefore, about 58 voters will be for the issue and 42 will be against it at the end of the week. At the end of the second week, the figures will be

$$\begin{bmatrix} 57.5 & 42.5 \end{bmatrix}\begin{bmatrix} 0.75 & 0.25 \\ 0.40 & 0.60 \end{bmatrix} = \begin{bmatrix} 60.125 & 39.875 \end{bmatrix},$$

or about 60 of the voters in favor and about 40 against. At the end of the third week, we find that

$$\begin{bmatrix} 60.125 & 39.875 \end{bmatrix}\begin{bmatrix} 0.75 & 0.25 \\ 0.40 & 0.60 \end{bmatrix} = \begin{bmatrix} 61.04 & 38.95 \end{bmatrix},$$

or about 61 of the voters will be in favor of the issue, while about 39 will be against it. ■

A matrix of special significance is an identity matrix.

Identity Matrices

An **identity matrix** is any square matrix whose entries arc all 1s in the ith-row, ith-column positions and 0s elsewhere. An identity matrix is denoted by I, regardless of its dimension.

An identity matrix has 1's as entries along the top-left to bottom-right diagonal, and 0's everywhere else.

For example,

$$I = \begin{bmatrix} 1 & 0 \\ 0 & 1 \end{bmatrix}$$

has the dimension 2×2, and

$$I = \begin{bmatrix} 1 & 0 & 0 \\ 0 & 1 & 0 \\ 0 & 0 & 1 \end{bmatrix}$$

has the dimension 3×3. Observe that, if

$$A = \begin{bmatrix} 3 & -5 \\ 2 & 8 \end{bmatrix},$$

then

$$AI = \begin{bmatrix} 3 & -5 \\ 2 & 8 \end{bmatrix} \begin{bmatrix} 1 & 0 \\ 0 & 1 \end{bmatrix} = \begin{bmatrix} 3 & -5 \\ 2 & 8 \end{bmatrix} = A$$

and

$$IA = \begin{bmatrix} 1 & 0 \\ 0 & 1 \end{bmatrix} \begin{bmatrix} 3 & -5 \\ 2 & 8 \end{bmatrix} = \begin{bmatrix} 3 & -5 \\ 2 & 8 \end{bmatrix} = A.$$

The preceding discussion illustrates the important fact that an identity matrix plays the *same role* for *square matrices* in matrix algebra as the number 1 does in ordinary algebra: $AI = IA = A$, just like $1 \times a = a \times 1 = a$ in ordinary algebra.

There are several exceptions to the usual properties of algebra in regard to matrix multiplication. For instance, we saw in Example 2 that matrix multiplication is not commutative; that is, AB is not always equal to BA. However, the properties that we have developed so far in this chapter are enough to provide a structure that is useful in many settings. Some properties of matrix multiplication that do hold are summarized in the box that follows. While these may not all be obvious, they are demonstrated by example in the exercises. (Proofs of all the matrix algebra properties presented so far are generally found in texts on *linear algebra*.)

Properties of Matrix Multiplication

For matrices that are compatible under multiplication, matrix multiplication satisfies the following properties:

1. The associative law holds $(AB)C = A(BC)$.
2. If A is a square $n \times n$ matrix, then $IA = AI = A$.
3. Matrix multiplication distributes over addition; that is,

$$A(B + C) = AB + AC$$

and

$$(B + C)A = BA + CA.$$

The distributive property is particularly useful in matrix algebra because, as in ordinary algebra, it allows us to factor matrix expressions. For example,

$$AB + AC = A(B + C)$$
$$BA + A = (B + I)A.$$

Notice that the position of A in both factorizations is dictated by the fact that matrix multiplication is not commutative. Observe the identity matrix I in the second factorization; I in matrix algebra plays the role of 1 in ordinary arithmetic.

HINT When factoring, multiply the result using the properties of matrix algebra, and make sure you get the original expression.

The commutative property of multiplication is not the only property of the algebra of real numbers that fails to carry over to matrix algebra.

Matrix Multiplication Exceptions to Properties of Real Numbers

1. There are matrices A and B for which $AB \neq BA$. Thus, we say that matrix multiplication is *non-commutative*.
2. There are matrices A, B, and C, where A is not the zero matrix, for which $AB = AC$, yet $B \neq C$. That is, *cancellation* is not always permitted.
3. There are matrices A and B, neither of which is the zero matrix, for which $AB = \mathbf{0}$.

Due to the preceding exceptions, we cannot define matrix division.

EXAMPLE 8

Examples of Matrix Algebra Exceptions

(a) Give an example of matrices A, B, and C for which $AB = AC$, but $B \neq C$.

SOLUTION Let

$$A = \begin{bmatrix} 1 & 0 \\ 0 & 0 \end{bmatrix}, \quad B = \begin{bmatrix} 2 & 6 \\ 8 & 7 \end{bmatrix} \quad \text{and} \quad C = \begin{bmatrix} 2 & 6 \\ 4 & 3 \end{bmatrix}.$$

Then

$$AB = \begin{bmatrix} 2 & 6 \\ 0 & 0 \end{bmatrix} \quad \text{and} \quad AC = \begin{bmatrix} 2 & 6 \\ 0 & 0 \end{bmatrix}, \quad \text{but} \quad B \neq C.$$

(b) Give an example of matrices A and B for which $AB = \mathbf{0}$, but $A \neq \mathbf{0}$ and $B \neq \mathbf{0}$.

SOLUTION Let $A = \begin{bmatrix} 2 & 0 \\ 0 & 0 \end{bmatrix}$ and $B = \begin{bmatrix} 0 & 0 \\ 5 & 1 \end{bmatrix}$. Then $AB = \begin{bmatrix} 0 & 0 \\ 0 & 0 \end{bmatrix} = \mathbf{0}$, but neither A nor B is the zero matrix. ∎

Exercises 3.2

In Exercises 1 through 4, write each linear system as a matrix equation of the form $AX = B$.

1. $x + 2y = 7$
   $3x - y = 6$

2. $3x + 2y - z = 2$
   $x - y + z = 4$

3. $x + y = 2$
   $3x - y = 6$
   $x + 5y = -1$

4. $2x + 3y + z - 5w = 1$
   $x - y + 8w = 2$
   $-x + y + 2z = 3$

In Exercises 5 through 8, write each matrix equation as a system of linear equations.

5. $\begin{bmatrix} 2 & 0 \\ 1 & 5 \\ 4 & 6 \end{bmatrix} \begin{bmatrix} x \\ y \end{bmatrix} = \begin{bmatrix} 3 \\ 5 \\ 2 \end{bmatrix}$

6. $\begin{bmatrix} 2 & 0 & 1 & -1 \\ 1 & 2 & 3 & 5 \end{bmatrix} \begin{bmatrix} x \\ y \\ z \\ w \end{bmatrix} = \begin{bmatrix} 0 \\ 0 \end{bmatrix}$

7. $\begin{bmatrix} 1 & 0 & 1 & 0 & 1 \\ 2 & 0 & 3 & 5 & 1 \\ -1 & 2 & 2 & 2 & 0 \end{bmatrix} \begin{bmatrix} x \\ y \\ z \\ w \\ v \end{bmatrix} = \begin{bmatrix} -2 \\ 3 \\ 6 \end{bmatrix}$

8. $\begin{bmatrix} 2 & 1 & 1 \\ 0 & 2 & 3 \\ 5 & 1 & -2 \end{bmatrix} \begin{bmatrix} x \\ y \\ z \end{bmatrix} = \begin{bmatrix} 1 \\ -1 \\ 3 \end{bmatrix}$

In Exercises 9 through 22, find each of the indicated matrices, if possible, for the given A, B, C, and D.

$$A = \begin{bmatrix} 2 & 3 \\ 1 & -2 \end{bmatrix}, \quad B = \begin{bmatrix} 1 & 0 \\ 2 & 3 \end{bmatrix}, \quad C = \begin{bmatrix} 2 & 1 \\ 1 & 0 \\ 3 & 2 \end{bmatrix},$$

and $D = \begin{bmatrix} 1 & 3 & 5 & 9 \\ 2 & 0 & 1 & -3 \end{bmatrix}.$

9. AB

10. CD

11. AC

12. DC

13. AD

14. CB

15. $2B + I$

16. $2I - BA$

17. $B^2 + 2I$

18. $I^2 - A^2$

19. A^2

20. $I - 2D$

21. $A^2 + 3B$

22. D^2

23. Is the matrix product $(A - B)(A + B) = A^2 - B^2$? Explain.

24. Use the rules for matrix algebra to find an expression for $(A + B)^2$. (*Hint:* $(A + B)^2 = (A + B) \times (A + B)$, and use the distributive law.)

In Exercises 25 through 28, use

$$A = \begin{bmatrix} 3 & 1 \\ 2 & 5 \end{bmatrix}, \quad B = \begin{bmatrix} 1 & 0 \\ 2 & 3 \end{bmatrix}, \quad \text{and} \quad C = \begin{bmatrix} 1 & 2 \\ 3 & 4 \end{bmatrix}$$

to verify the given expressions.

25. $(AB)C = A(BC)$

26. $(B + C)A = BA + CA$

27. $A(B + C) = AB + AC$

28. $(AB)^T = B^T A^T$

29. Find matrices A, B, and C, each of dimension 3×3, such that $AB = AC$, but $B \neq C$ and $A \neq \mathbf{0}$.

30. Find matrices A and B, each of dimension 3×3, such that $AB = \mathbf{0}$, but $A \neq \mathbf{0}$ and $B \neq \mathbf{0}$.

In Exercises 31 through 33, use

$$A = \begin{bmatrix} 1 & 2 \\ 3 & 1 \end{bmatrix} \quad \text{and} \quad B = \begin{bmatrix} 2 & 3 \\ 4 & 1 \end{bmatrix}$$

to calculate the indicated quantities.

31. (a) $(2A)B$ **(b)** $A(2B)$ **(c)** $2(AB)$

32. (a) $(-3A)B$ **(b)** $A(-3B)$ **(c)** $-3(AB)$

33. (a) $(kA)B$ **(b)** $A(kB)$ **(c)** $k(AB)$

34. What general conclusion can be drawn from Exercises 31–33?

In Exercises 35 through 42, use the laws of matrix algebra, including the conclusion you drew in Exercise 34, to factor out any common quantities from each expression.

35. $AX + BX$

36. $(AB)C + BC$

37. $AX + 2AY$

38. $AX - X$

39. $3AB + 4CB$

40. $AB + A$

41. $2AX - 3AY$

42. $2BC + 2C$

43. College Costs During Tricia's first year in college, she took 15 credits the first semester and 17 credits the second. She also worked a total of 160 hours the first semester and 130 hours the second. A matrix displaying this information is as follows:

$$\begin{array}{c} \\ \text{Credits taken} \\ \text{Hours worked} \end{array} \begin{array}{cc} \text{1st sem.} & \text{2nd sem.} \\ \left[\begin{array}{cc} 15 & 17 \\ 160 & 130 \end{array}\right] \end{array} = A.$$

If tuition costs $100 per credit and Tricia received $6 per hour at work, find a matrix B so that a multiplication with A will show her income minus tuition each semester. Then do the matrix multiplication.

44. College Costs During her first two years at college, Sara took course hours, plane trips home, and beach weekends, as indicated in this matrix:

$$\begin{array}{c} \\ \text{Hours taken} \\ \text{Trips home} \\ \text{Beach trips} \end{array} \begin{array}{cccc} \text{1st sem.} & \text{2nd sem.} & \text{3rd sem.} & \text{4th sem.} \\ \left[\begin{array}{cccc} 12 & 16 & 15 & 17 \\ 2 & 3 & 1 & 2 \\ 0 & 2 & 1 & 2 \end{array}\right] \end{array} = A.$$

Given that tuition costs $80 per hour, each round-trip plane fare home costs $260, and each beach weekend costs $250, find a matrix B so that a multiplication with the matrix A will give the total cost for each semester for the three activities listed. Then do the matrix multiplication.

45. Nutrition The amounts of fat (F), sodium (S), and protein (P), in units of grams, milligrams, and ounces, respectively, per serving of a soup, a meat, and a vegetable are as follows:

$$\begin{array}{c} \\ \text{Soup} \\ \text{Meat} \\ \text{Vegatable} \end{array} \begin{array}{ccc} F & S & P \\ \left[\begin{array}{ccc} 2 & 830 & 2 \\ 5 & 30 & 8 \\ 0 & 1 & 2 \end{array}\right] \end{array} = A.$$

(a) Given that a person eats 1 serving of soup, 1.5 servings of meat, and 0.73 serving of the vegetable, find a matrix B that, when multiplied with A, will give the total intake of fat, sodium, and protein. Then do the matrix multiplication.

(b) A person on a strict diet for health reasons gets -3 points for each gram of fat, -1.5 points for each milligram of sodium, and 2 points for each ounce of protein. Find a matrix C that, when multiplied with A, will give the total points accumulated for each food. Perform the multiplication.

46. Production The times, in hours, required for assembling, painting, and testing standard and deluxe lawn mowers are as shown in the following matrix:

$$\begin{array}{c} \\ \text{Standard} \\ \text{Deluxe} \end{array} \begin{array}{ccc} \text{Assembling} & \text{Painting} & \text{Testing} \\ \left[\begin{array}{ccc} 2 & 1 & 0.5 \\ 2.5 & 1.2 & 1 \end{array}\right] \end{array} = A.$$

(a) If 80 standard and 100 deluxe mowers are made, find a matrix B that, when multiplied with A, gives the total assembling, painting, and testing time required.

(b) If assembling costs $5 per hour, painting costs $12 per hour, and testing costs $6 per hour, find a matrix C that, when multiplied with A, gives the total cost for making each model of mower.

47. Advertising A particular product was advertised in a newspaper, on the radio, and on television, and the number of men and women influenced by the ads was recorded. The data were as follows:

$$\begin{array}{c} \\ \text{Newspaper} \\ \text{Radio} \\ \text{Television} \end{array} \begin{array}{cc} \text{Men} & \text{Women} \\ \left[\begin{array}{cc} 700 & 1000 \\ 500 & 800 \\ 1500 & 1200 \end{array}\right] \end{array} = A.$$

(a) Suppose that three, five, and two ads in the respective media were run in January, and two, four, and six ads in the respective media were run in June. Write this information in a 2×3 matrix B. Then compute BA, and interpret the entry in each position of the product matrix.

(b) Suppose that the ads were run once each month for three consecutive months. The first time the ads were run, men influenced by the ad spent an average of $10 each, and women influenced by the ad spent an average of $15 each. The second time the ads were run, the figures were $12 and $18, respectively. The third time the ads were run, the figures were $15 and $22, respectively. Record this information in a 2×3 matrix C. Then calculate the product AC, and interpret the entry in each position of the product matrix.

48. Airline Transportation An airline purchases two types of airplanes. Plane P_1 seats 30 first-class passengers, 50 tourist-class passengers, and 90 economy-class passengers. Plane P_2 seats 50 first-class passengers, 60 tourist-class passengers, and 100 economy-class passengers.

(a) Put this information in a 3 × 2 matrix labeled A.

(b) Suppose that five planes of type P_1 and eight planes of type P_2 make flights fully loaded. Put this information in a 2 × 1 matrix labeled B, and calculate the product AB.

(c) To fly between Kansas City and Seattle in January, each first-class passenger is charged $600, each tourist-class passenger is charge $550, and each economy-class passenger is charged $400. In August, these rates change to $800, $650 and $500, respectively. Put this information in a 2 × 3 matrix labeled C. Find the product CA, and interpret the entries in each address.

49. Baking Costs The Better Bread Bakery bakes whole wheat bread and oat bread, with mixing, baking, and packaging times, in hours, as shown.

$$\begin{array}{c} \\ \text{Whole wheat} \\ \text{Oat} \end{array} \begin{array}{ccc} \text{Mix} & \text{Bake} & \text{Package} \\ \begin{bmatrix} 0.03 & 0.05 & 0.01 \\ 0.02 & 0.03 & 0.01 \end{bmatrix} \end{array} = A.$$

Given that an order is received for 200 loaves of whole wheat bread and 300 loaves of oat bread, and the cost of mixing, baking, and packaging is $12, $20, and $1, respectively, per hour, find matrices B and C so that the product BAC will give the total cost (excluding raw materials) for filling the order. Find the total cost.

50. Inventory The Soft Sit Furniture Company has two Stores that stock brands X and Y recliner chairs. The present inventory matrix is as follows:

$$\begin{array}{c} \\ \text{Store 1} \\ \text{Store 2} \end{array} \begin{array}{cc} \text{Brand X} & \text{Brand Y} \\ \begin{bmatrix} 10 & 8 \\ 5 & 12 \end{bmatrix} \end{array} = A.$$

Twenty percent of the inventory in Store 1 and 50% of the inventory in Store 2 is leather. Brand X leather chairs cost $400 each, and brand Y leather chairs cost $520 each. Find matrices B and C so that BAC gives the total cost of the leather chairs in inventory.

Use Example 7 in Section 3.2 as a guide in solving Exercises 50 and 51.

51. Learning Experiment A mouse enters a T-shaped maze and can turn either left, where a shock and a piece of cheese awaits, or right, where just a piece of cheese awaits. Earlier experiments with this maze gave these results: Of those mice that went left on a particular day, 10% went to the left the next day and 90% went to the right. Of those that went to the right on a particular day, 60% went to the right the next day and 40% went to the left. Julie took 100 mice one day and found that 48 went to the left and 52 went to the right. Assuming that the preceding percentages hold, use a matrix product to predict how many of the mice will go left and how many will go right the next day. How many the third day? the fourth day?

52. Social Research A social science researcher studied children whose parents were classified as having high, middle, or low income and compared these classifications with the income levels the children eventually attained in life. The researcher found that, of children whose parents were in the high-income level, 60%, 38%, and 2% eventually were in the high-, middle-, and low-income levels, respectively. Of children whose parents were in the middle-income level, 20%, 60%, and 20% eventually were in the high, middle, and low levels, respectively. Of children whose parents were in the low-income level, 5%, 40%, and 55% eventually were in the high, middle, and low levels, respectively. Use the results of the study and a matrix product to predict how many children would be expected to eventually be in each income level if 40 children from each income level were observed.

53. Construction Fees Estimated fees for architects and structural engineers, as a percentage of total construction cost, for several countries are listed in the following table:

Country	Architects	Structural Engineers
Belgium	7.00%	1.00%
France	4.85%	2.50%
Japan	3.50%	1.00%
Mexico	4.00%	0.80%
United States	4.20%	0.60%

SOURCE: *Hanscomb-Means Report,* May–June 1997

Put the information shown in this table in a 5 × 2 matrix denoted by C. (Remember to convert the given values from percentages to decimals.) Assume that the construction cost of a 2000-square-foot building in each of the countries listed is equivalent to $500,000 (in U.S. dollars) and the cost of a high-rise professional building is $1,000,000 (in U.S. dollars).

(a) Construct a 2×1 matrix B using the costs of the two types of buildings described.

(b) Calculate CB and interpret the meaning of each entry.

54. Executive Reporting Executives A, B, C, D, and E in a particular corporation have a reporting status as shown in the following diagram:

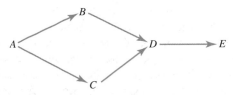

(a) Assuming that no executives report directly to themselves, find a matrix M representing the diagram that is similar to the one in Example 2 in Section 3.1.

(b) Find the matrix M^2, and interpret the entries.

(c) Find the matrix M^3, and interpret the entries.

(d) Find $M + M^2 + M^3$, and interpret the entries.

55. Airline Flights A particular airline has flights between cities A, B, and C, with origin and destination as indicated in the following diagram:

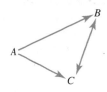

(a) Assuming that there are no direct flights from a city to the same city, construct a matrix M representing the diagram that is similar to Example 2 in Section 3.1, in which a 1 denotes a direct flight from a city on the left to a city on the top.

(b) Find M^2, and discuss how the entries give the number of two-stage flights between cities. Compare these results with the diagram.

(c) Find $M + M^2$, and interpret each entry in the result.

Using matrices $A = \begin{bmatrix} a_{11} & a_{12} \\ a_{21} & a_{22} \end{bmatrix}$, $B = \begin{bmatrix} b_{11} & b_{12} \\ b_{21} & b_{22} \end{bmatrix}$, *and*

$C = \begin{bmatrix} c_{11} & c_{12} \\ c_{21} & c_{22} \end{bmatrix}$, *verify the properties of matrix multiplication given in Exercises 56–58 for* 2×2 *matrices.*

56. $(AB)C = A(BC)$ **57.** $A(B + C) = AB + AC$

58. $(B + C)A = BA + CA$

Section 3.3 The Inverse of a Matrix

We know that any linear system of equations can be written in the equivalent matrix equation form $AX = B$. Now, if there were only some way of "dividing" both sides by A, the solution X of the system could be found. Interestingly, even though *there is no division operation in matrix algebra,* the same result could be obtained by constructing a matrix A^{-1} having the property that $A^{-1}A = I = AA^{-1}$, *if such as matrix exists.* Then multiplication on both sides by A^{-1} on the left of $AX = B$ would give

$$A^{-1}(AX) = A^{-1}B$$

or

$$(A^{-1}A)X = A^{-1}B \qquad \text{Associative Law}$$

so that

$$IX = A^{-1}B \qquad \text{Property of Inverse}$$

and

$$X = A^{-1}B.$$

This would solve the problem if we knew when such a matrix A^{-1}, called the **inverse of A,** could be found for a given A and if we knew just how the construction of A^{-1} took place. Now, we know that only a square matrix A can have an inverse A^{-1}, since the products AA^{-1} and $A^{-1}A$ must both be defined and are both equal to the same identity matrix I. If I has dimension $n \times n$ and $AA^{-1} = I$, then the number of rows in A and the number of columns in A^{-1} must each be n. Similarly, since $A^{-1}A = I$, the number of rows in A^{-1} and the number of columns in A must also each be n. Thus, both A and A^{-1} must be square matrices of the same dimension. To investigate further, we should be aware that the inverse exists only for *some* specific square matrices. To summarize our discussion of inverses so far, only square matrices can have inverses, but not all square matrices do.

The Inverse of a Matrix

Let A be a square $n \times n$ matrix. If there is a square $n \times n$ matrix that, when multiplied on either side by A, produces the identity matrix, then such a matrix is called the **inverse of A** and is denoted by A^{-1}. That is, A^{-1} must have the property that

$$A^{-1}A = I \quad \text{and} \quad AA^{-1} = I.$$

In view of this definition, the solution of the previously mentioned matrix equation $AX = B$ will apply only when the coefficient matrix A is square and then only if the inverse of A exists.

Is it possible for a matrix to have two or more inverses? The answer is "no"; if there is an inverse at all, there is only one. The reason is that if there were two inverses, A^{-1} and, say, Q, of a given matrix A, then $I = A^{-1}A = AA^{-1}$ and $I = QA = AQ$, so that

$$
\begin{aligned}
A^{-1} = IA^{-1} &= (QA)A^{-1} \\
&= Q(AA^{-1}) \qquad \text{Associative Law} \\
&= QI \qquad\qquad \text{Property of Inverse} \\
&= Q.
\end{aligned}
$$

Uniqueness of the Inverse

If a square matrix A has an inverse, then **that inverse is unique.**

EXAMPLE 1

Demonstrating the Property of an Inverse

(a) Let $A = \begin{bmatrix} 2 & 3 \\ 1 & 1 \end{bmatrix}$. Then $A^{-1} = \begin{bmatrix} -1 & 3 \\ 1 & -2 \end{bmatrix}$ because

$$AA^{-1} = \begin{bmatrix} 2 & 3 \\ 1 & 1 \end{bmatrix}\begin{bmatrix} -1 & 3 \\ 1 & -2 \end{bmatrix} = \begin{bmatrix} 1 & 0 \\ 0 & 1 \end{bmatrix} = I$$

and

$$A^{-1}A = \begin{bmatrix} -1 & 3 \\ 1 & -2 \end{bmatrix}\begin{bmatrix} 2 & 3 \\ 1 & 1 \end{bmatrix} = \begin{bmatrix} 1 & 0 \\ 0 & 1 \end{bmatrix} = I.$$

(b) Let $A = \begin{bmatrix} 1 & -1 & 0 \\ 0 & 2 & 1 \\ 1 & 0 & 1 \end{bmatrix}$. Then $A^{-1} = \begin{bmatrix} 2 & 1 & -1 \\ 1 & 1 & -1 \\ -2 & -1 & 2 \end{bmatrix}$ because AA^{-1} and $A^{-1}A$

are both the 3×3 identity matrix. (You should verify these results.)

(c) The matrix $A = \begin{bmatrix} 0 & 0 \\ 3 & 0 \end{bmatrix}$ does not have an inverse. The reason is that there is no

2×2 matrix $\begin{bmatrix} x & y \\ z & w \end{bmatrix}$ such that the product $\begin{bmatrix} 0 & 0 \\ 3 & 0 \end{bmatrix}\begin{bmatrix} x & y \\ z & w \end{bmatrix}$, which is

$\begin{bmatrix} 0 & 0 \\ 3x & 3y \end{bmatrix}$, could equal the identity matrix $\begin{bmatrix} 1 & 0 \\ 0 & 1 \end{bmatrix}$. ∎

One straightforward, systematic way to calculate the inverse of a square matrix, or to detect its nonexistence, is to convert the problem into an equivalent system of equations and use the pivoting process to find the solution. To see how this may be done, consider the matrix $A = \begin{bmatrix} 2 & 3 \\ 1 & 1 \end{bmatrix}$ of Example 1(a). To construct the inverse of A, we need to find a matrix $A^{-1} = \begin{bmatrix} x & y \\ z & w \end{bmatrix}$ such that

$$AA^{-1} = \begin{bmatrix} 2 & 3 \\ 1 & 1 \end{bmatrix}\begin{bmatrix} x & y \\ z & w \end{bmatrix} = \begin{bmatrix} 1 & 0 \\ 0 & 1 \end{bmatrix}$$

or

$$\begin{bmatrix} 2x + 3z & 2y + 3w \\ x + z & y + w \end{bmatrix} = \begin{bmatrix} 1 & 0 \\ 0 & 1 \end{bmatrix}.$$

For the equality between the last two matrices to hold, it follows that

$$2x + 3z = 1$$
$$x + z = 0$$
$$2y + 3w = 0$$
$$y + w = 1.$$

The calculation of A^{-1} has now been reduced to solving two systems of linear equations, each in two unknowns. Notice what happens if each system were to be solved by the pivoting process:

$$\begin{bmatrix} 2 & 3 & | & 1 \\ 1 & 1 & | & 0 \end{bmatrix} \cdots \begin{bmatrix} 1 & 0 & | & x \\ 0 & 1 & | & z \end{bmatrix} \qquad (1)$$

$$\begin{bmatrix} 2 & 3 & | & 0 \\ 1 & 1 & | & 1 \end{bmatrix} \cdots \begin{bmatrix} 1 & 0 & | & y \\ 0 & 1 & | & w \end{bmatrix}. \qquad (2)$$

The coefficient matrix for both systems is exactly the same and is, in fact, the original matrix A. Therefore, the steps in the pivoting process for both systems to reach the final ideal matrix (the identity) would be identical, and furthermore, the values of x, y, z, and w will be in the positions indicated in systems (1) and (2). Instead of doing the work twice, we can do the pivoting for both systems at once by beginning with this slightly adjusted augmented matrix:

$$\begin{bmatrix} 2 & 3 & | & 1 & 0 \\ 1 & 1 & | & 0 & 1 \end{bmatrix} = [A \mid I].$$

Constant column for system 1 ⎯   ⎿ Constant column from system 2

We then do the pivoting process *across entire rows*, until the coefficient matrix becomes the identity, at which point both solutions appear on the right of the vertical bar, like this:

$$\begin{bmatrix} 1 & 0 & | & x & y \\ 0 & 1 & | & z & w \end{bmatrix} = [I \mid A^{-1}].$$

Solution column for system 1 ⎯   ⎿ Solution column from system 2

The actual steps undertaken to accomplish these solutions are as follows:

$$[A \mid I] = \begin{bmatrix} 2 & 3 & | & 1 & 0 \\ 1 & 1 & | & 0 & 1 \end{bmatrix}$$

$$R_1 \longleftrightarrow R_2 \quad \begin{bmatrix} 1 & 1 & | & 0 & 1 \\ 2 & 3 & | & 1 & 0 \end{bmatrix}$$

$$-2R_1 + R_2 \rightarrow R_2 \quad \begin{bmatrix} 1 & 1 & | & 0 & 1 \\ 0 & 1 & | & 1 & -2 \end{bmatrix}$$

$$-1R_2 + R_1 \rightarrow R_1 \quad \begin{bmatrix} 1 & 0 & | & -1 & 3 \\ 0 & 1 & | & 1 & -2 \end{bmatrix} = [I \mid A^{-1}].$$

The part of the last matrix to the right of the vertical bar shows that the solution of System (1) is $x = -1$ and $z = 1$ (the first column) and the solution of System (2) is $y = 3$ and $w = -2$ (the second column), so that the inverse of A is

$$A^{-1} = \begin{bmatrix} -1 & 3 \\ 1 & -2 \end{bmatrix}.$$

It is easy to check the accuracy of A^{-1}: Just make sure that $AA^{-1} = I$ or $A^{-1}A = I$. Either one will do, because it can be mathematically proven that if either of these equalities holds, then so will the other. A check of the inverse just found shows that

$$AA^{-1} = \begin{bmatrix} 2 & 3 \\ 1 & 1 \end{bmatrix} \begin{bmatrix} -1 & 3 \\ 1 & -2 \end{bmatrix} = \begin{bmatrix} 1 & 0 \\ 0 & 1 \end{bmatrix} = I.$$

The mechanics for finding the inverse of any 2×2 matrix are straightforward: Simply form the 2×4 matrix $[A \mid I]$, and then use the pivoting process to transform A into I, if possible. The resulting matrix will then appear as $[I \mid A^{-1}]$. Remember that two systems of equations are being solved simultaneously, and if either system fails to have a solution, then A will fail to have an inverse. Will this process work for 3×3 matrices and even higher dimension square matrices? The answer is "yes" and for exactly the same reasons.

For instance, to find the inverse of $A = \begin{bmatrix} 1 & -1 & 0 \\ 0 & 2 & 1 \\ 1 & 0 & 1 \end{bmatrix}$, the problem becomes one of making sure that

$$AA^{-1} = I \quad \text{or} \quad \begin{bmatrix} 1 & -1 & 0 \\ 0 & 2 & 1 \\ 1 & 0 & 1 \end{bmatrix} \begin{bmatrix} a & b & c \\ d & e & f \\ g & h & i \end{bmatrix} = \begin{bmatrix} 1 & 0 & 0 \\ 0 & 1 & 0 \\ 0 & 0 & 1 \end{bmatrix}.$$

Upon multiplying the two matrices on the left and matching each entry in the product with the corresponding entry in the identity matrix on the right, these three systems of equations result:

$$\begin{array}{lll} a - d \quad\quad = 1 & b - e \quad\quad = 0 & c - f \quad\quad = 0 \\ \quad\quad 2d + g = 0 & \quad\quad 2e + h = 1 & \quad\quad 2f + i = 0 \\ a \quad\quad + g = 0 & b \quad\quad + h = 0 & c \quad\quad + i = 1. \end{array}$$

Because the coefficient matrix in each system is the same (namely, the original matrix A), all three systems can be solved simultaneously:

$$[AI] = \begin{bmatrix} 1 & -1 & 0 & | & 1 & 0 & 0 \\ 0 & 2 & 1 & | & 0 & 1 & 0 \\ 1 & 0 & 1 & | & 0 & 0 & 1 \end{bmatrix} \cdots \begin{bmatrix} 1 & 0 & 0 & | & 2 & 1 & -1 \\ 0 & 1 & 0 & | & 1 & 1 & -1 \\ 0 & 0 & 1 & | & -2 & -1 & 2 \end{bmatrix} = [I \mid A^{-1}].$$

The augmented matrix is shown on the left; row operations yield the last matrix. The reason the inverse is to the right of the vertical bar is that the first column is the solution of System **(1)**, the second column is the solution of System **(2)**, and the third column is the solution of System **(3)**.

The preceding example shows the pattern that develops for any size of square matrix.

Finding the Inverse of a Matrix

Let A be any $n \times n$ matrix. To determine whether A has an inverse,

1. Form the doublewide $n \times 2n$ matrix $[A \mid I]$, where I is the identity matrix of the same dimension as A.
2. Then perform pivot operations on elements on the A side, assuming that rows extend across the entire matrix $[A \mid I]$, until you obtain $[R \mid B]$, where R is in reduced row-echelon form.
3. If $R = I$, then $B = A^{-1}$. If $R \neq I$, then R will contain a row of zeros and A has no inverse.

EXAMPLE 2

Finding the Inverse
of a Matrix

Find the inverse of $A = \begin{bmatrix} 1 & 3 \\ 2 & 1 \end{bmatrix}$.

SOLUTION $[A \mid I] = \begin{bmatrix} 1 & 3 & | & 1 & 0 \\ 2 & 1 & | & 0 & 1 \end{bmatrix}$

$$-2R_1 + R_2 \rightarrow R_2 \quad \begin{bmatrix} 1 & 3 & | & 1 & 0 \\ 0 & -5 & | & -2 & 1 \end{bmatrix}$$

$$-\tfrac{1}{5}R_2 \rightarrow R_2 \quad \begin{bmatrix} 1 & 3 & | & 1 & 0 \\ 0 & 1 & | & \tfrac{2}{5} & -\tfrac{1}{5} \end{bmatrix}$$

$$-3R_2 + R_1 \rightarrow R_1 \quad \begin{bmatrix} 1 & 0 & | & -\tfrac{1}{5} & \tfrac{3}{5} \\ 0 & 1 & | & \tfrac{2}{5} & -\tfrac{1}{5} \end{bmatrix}.$$

Therefore,

$$A^{-1} = \begin{bmatrix} -\tfrac{1}{5} & \tfrac{3}{5} \\ \tfrac{2}{5} & -\tfrac{1}{5} \end{bmatrix}.$$

Verify this result by calculating AA^{-1} and $A^{-1}A$ to see whether I results. ■

EXAMPLE 3

When the Inverse Does
Not Exist

Find the inverse of $A = \begin{bmatrix} 1 & 2 \\ 2 & 4 \end{bmatrix}$.

SOLUTION $[A \mid I] = \begin{bmatrix} 1 & 2 & | & 1 & 0 \\ 2 & 4 & | & 0 & 1 \end{bmatrix}$

$$-2R_1 + R_2 \rightarrow R_2 \quad \begin{bmatrix} 1 & 2 & | & 1 & 0 \\ 0 & 0 & | & -2 & 1 \end{bmatrix}.$$

The reduced row-echelon form of A (the matrix to the left of the vertical bar) is not the identity matrix. Therefore, A has no inverse. From the viewpoint of solutions of systems of equations, the row of zeros to the *left* of the vertical bar means that no solution exists for either of the systems of equations represented. A row of zeros to the left of the vertical bar is *always* a signal that no inverse exists. ■

An alternative method of finding the inverse of a 2 × 2 matrix is presented next. Its validity is shown by using the general method we have been discussing.

Formula for the Inverse of a 2 × 2 Matrix

For the matrix $A = \begin{bmatrix} a & b \\ c & d \end{bmatrix}$, let $D = ad - bc$. If $D = 0$, then A^{-1} does not exist. If $D \neq 0$, then A^{-1} exists and is given by

$$A^{-1} = \frac{1}{D} \begin{bmatrix} d & -b \\ -c & a \end{bmatrix}.$$

To demonstrate the ease of use of this formula, find the inverse of $A = \begin{bmatrix} 3 & 2 \\ 5 & 4 \end{bmatrix}$. The value of D is $D = 3 \cdot 4 - 5 \cdot 2 = 2$. From the formula, we find that

$$A^{-1} = \frac{1}{2}\begin{bmatrix} 4 & -2 \\ -5 & 3 \end{bmatrix} = \begin{bmatrix} 2 & -1 \\ -\frac{5}{2} & \frac{3}{2} \end{bmatrix}.$$

At the outset of the section, we saw that if $AX = B$ is a matrix equation representing a system of linear equations and A has an inverse, then $X = A^{-1}B$. Example 4 shows how the product $A^{-1}B$ gives the solution of a specific system of equations.

EXAMPLE 4

Solving a System by
Using the Inverse of a
Matrix

Convert the system

$$2x + 3y = 1$$
$$x + y = 4$$

into a matrix equation. Then solve the system, using the inverse of the coefficient matrix.

SOLUTION We first rewrite the given system as a matrix equation $AX = B$ and then use the fact that $X = A^{-1}B$ to find the solution. The matrix equation $AX = B$ is

Coefficient Matrix $\rightarrow$ $\begin{bmatrix} 2 & 3 \\ 1 & 1 \end{bmatrix}\begin{bmatrix} x \\ y \end{bmatrix} = \begin{bmatrix} 1 \\ 4 \end{bmatrix}$ $\leftarrow$ Constant Matrix

with $A = \begin{bmatrix} 2 & 3 \\ 1 & 1 \end{bmatrix}$ and $B = \begin{bmatrix} 1 \\ 4 \end{bmatrix}$. Therefore, $X = A^{-1}B$ becomes

$$\begin{bmatrix} x \\ y \end{bmatrix} = \begin{bmatrix} 2 & 3 \\ 1 & 1 \end{bmatrix}^{-1}\begin{bmatrix} 1 \\ 4 \end{bmatrix}.$$

Using the formula for the inverse of a 2×2 matrix, we obtain

$$A^{-1} = \begin{bmatrix} 2 & 3 \\ 1 & 1 \end{bmatrix}^{-1} = \begin{bmatrix} -1 & 3 \\ 1 & -2 \end{bmatrix},$$

so that

$$\begin{bmatrix} x \\ y \end{bmatrix} = \begin{bmatrix} -1 & 3 \\ 1 & -2 \end{bmatrix}\begin{bmatrix} 1 \\ 4 \end{bmatrix} = \begin{bmatrix} 11 \\ -7 \end{bmatrix}.$$

This means that the solution of the original system of equations is $x = 11$ and $y = -7$. Substituting these values for x and y into the original equations will verify the solution:

$$2(11) + 3(-7) = 1$$
$$(11) + (-7) = 4.$$

NOTE We could also solve this system with techniques from Chapter 2, but we present this example to illustrate the process and ideas involved in using a matrix inverse. The process is especially useful if technology is available to invert a matrix and perform matrix multiplication. ■

Various computer programs, spreadsheets, and graphing calculators are prepro-grammed to find the inverse of a matrix. Caution is advised regarding round-off errors, however. Such errors may result in an inverse being found to a matrix that has no inverse. Example 5 demonstrates the Excel commands for calculating a matrix inverse.

EXAMPLE 5

The MINVERSE Function of Excel

Consider the matrix $A = \begin{bmatrix} 2 & 3 \\ 1 & 1 \end{bmatrix}$ from Example 4. Use Excel's MINVERSE function to calculate A^{-1}.

SOLUTION Begin by entering the matrix into a spreadsheet, as in Figure 7.

	A	B	C	D	E
1	**A**			**Inverse**	
2	2	3			
3	1	1			

FIGURE 7
Setting Up a Spreadsheet for Calculating the Inverse of a Matrix

Highlight cells D2:E3, which make up a block of cells with the same dimension as A^{-1} (2 × 2). Type =MINVERSE(A2:B3) and press Ctrl-Shift-Enter. The result should appear as in Figure 8.

	A	B	C	D	E
1	**A**			**Inverse**	
2	2	3		-1	3
3	1	1		1	-2

FIGURE 8
The Result of Executing the MINVERSE Function of Excel ■

Figure 9 shows how a graphing calculator screen might look after the calculation of the inverse and the subsequent matrix multiplication used to solve the system of equations in Example 4.

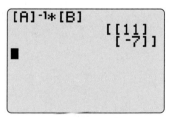

FIGURE 9
Calculator Solution of a System of Equations

The real importance of the inverse of a matrix is in the structure of the matrix algebra. The inverse does in matrix algebra what the reciprocal (division) does in ordinary algebra with real numbers. However, in the real-number system, there is only one number, 0, that does not have an inverse (reciprocal). In matrix algebra, there are many matrices (in fact, many square matrices) that do not have an inverse. One consequence of this difference is that, for real numbers, if $ab = ac$ and $a \neq 0$, then we may cancel the a and obtain $b = c$, while for matrices, $AB = AC$ implies that $B = C$ only when A has an inverse, and then the cancellation is done by multiplying both sides by A^{-1} *on the left*.

The inverse of a matrix has many applications other than those shown in this section. The next section shows three other applications—from cryptography, economics, and manufacturing.

Exercises 3.3

In Exercises 1 through 4, decide whether matrix B is the inverse of A.

1. $A = \begin{bmatrix} 3 & 2 \\ 1 & 1 \end{bmatrix}$ $B = \begin{bmatrix} 1 & -2 \\ -1 & 3 \end{bmatrix}$

2. $A = \begin{bmatrix} 2 & 3 \\ 0 & 3 \end{bmatrix}$ $B = \begin{bmatrix} 2 & \frac{1}{2} \\ 3 & 1 \end{bmatrix}$

3. $A = \begin{bmatrix} 1 & -1 & 0 \\ 2 & 1 & 0 \\ 1 & 1 & 2 \end{bmatrix}$ $B = \begin{bmatrix} 1 & 0 & 0 \\ 2 & 1 & 1 \\ \frac{1}{2} & 3 & -1 \end{bmatrix}$

4. $A = \begin{bmatrix} 1 & 0 & 1 \\ 2 & 0 & 0 \\ 1 & 1 & 1 \end{bmatrix}$ $B = \begin{bmatrix} 0 & \frac{1}{2} & 0 \\ -1 & 0 & 1 \\ 1 & -\frac{1}{2} & 0 \end{bmatrix}$

The inverse of the matrix $A = \begin{bmatrix} 3 & 1 \\ 1 & -1 \end{bmatrix}$ *is* $A^{-1} = \begin{bmatrix} \frac{1}{4} & \frac{1}{4} \\ \frac{1}{4} & -\frac{3}{4} \end{bmatrix}$.

Use this fact to solve the systems of equations in Exercises 5 through 8.

5. $3x + y = -4$
    $x - y = 8$

6. $3u + v = 7$
    $u - v = 1$

7. $3m + n = 0.2$
    $m - n = 0.3$

8. $3r + s = 0$
    $r - s = -5$

In Exercises 9 through 24, find the inverse of each matrix, if it exists.

9. $\begin{bmatrix} 1 & 3 \\ 2 & 5 \end{bmatrix}$

10. $\begin{bmatrix} 2 & 4 \\ 3 & 2 \end{bmatrix}$

11. $\begin{bmatrix} 2 & 3 \\ 1 & \frac{3}{2} \end{bmatrix}$

12. $\begin{bmatrix} 2 & 0 \\ 0 & 1 \end{bmatrix}$

13. $\begin{bmatrix} 0 & 3 \\ 0 & 2 \end{bmatrix}$

14. $\begin{bmatrix} 2 & -1 \\ 0 & 1 \end{bmatrix}$

15. $\begin{bmatrix} -1 & 3 \\ 2 & -4 \end{bmatrix}$

16. $\begin{bmatrix} 1 & 5 \\ 2 & 10 \end{bmatrix}$

17. $\begin{bmatrix} 1 & 0 & 2 \\ 0 & 2 & 2 \\ 1 & 1 & 1 \end{bmatrix}$

18. $\begin{bmatrix} 1 & 2 & 3 \\ 0 & 1 & 1 \\ 2 & 2 & 2 \end{bmatrix}$

19. $\begin{bmatrix} 3 & 1 & 0 \\ 2 & 1 & 1 \\ 5 & 2 & 1 \end{bmatrix}$

20. $\begin{bmatrix} 1 & 1 & 1 \\ 2 & 2 & 2 \\ 3 & 3 & 4 \end{bmatrix}$

21. $\begin{bmatrix} 2 & 1 & 1 \\ 1 & 0 & 1 \\ 1 & 1 & 1 \end{bmatrix}$

22. $\begin{bmatrix} 1 & 0 & 1 \\ 0 & 2 & 1 \\ 2 & 2 & 2 \end{bmatrix}$

23. $\begin{bmatrix} 1 & 0 & 1 & 0 \\ 2 & 2 & 2 & 0 \\ 2 & 0 & 0 & 1 \\ 1 & 0 & 2 & 1 \end{bmatrix}$

24. $\begin{bmatrix} 2 & 1 & 1 & 1 \\ 1 & 1 & 2 & 1 \\ 1 & 2 & 1 & 1 \\ 1 & 1 & 1 & 1 \end{bmatrix}$

25. For $A = \begin{bmatrix} 2 & 3 \\ 1 & -4 \end{bmatrix}$, compute

(a) $I - A$

(b) $(I - A)^{-1}$

26. For $A = \begin{bmatrix} 1 & 2 & 0 \\ 0 & 1 & 1 \\ 1 & 1 & 1 \end{bmatrix}$, compute

(a) $I - A$

(b) $(I - A)^{-1}$

In Exercises 27 through 40, write each system as a matrix equation, and solve each, if possible, using the inverse of the coefficient matrix. Otherwise, use methods from Chapter 2.

27. $x + 3y = 7$
$2x + 4y = -3$

28. $x + 5y = 3$
$2x + 10y = 8$

29. $5x + 2y = 4$
$4x + 2y = -3$

30. $2x + 3y = 8$
$4x + 5y = -2$

31. $x \quad + 2z = 1$
$\quad 2y + 2z = 3$
$x + y + z = 5$

32. $x + 2y + 3z = 4$
$\quad y + z = -2$
$2x + 2y \quad = 3$

33. $2x + y + 2z = -5$
$x - y + z = -5$
$x + y \quad = 2$

34. $\quad y + 5z = 17$
$x + 2y - z = 2$
$2x - y + 3z = 9$

35. $3x + 3y + z = \frac{8}{3}$
$x - y + z = 1$
$3x \quad + z = -1$

36. $3x \quad - z = 1$
$2x - y + z = 4$
$x + y - 2z = 3$

37. $3x - y + z = 0$
$x \quad + 3z = 0$
$2x + 3y + z = 0$

38. $0.2x + 0.82y + 0.03z = -1.96$
$0.02x + 0.14y - 0.6z = -2.82$
$0.47x + 0.22y + 0.62z = 3.92$

39. $2w + 0.6x + 0.3y - z = 2.80$
$0.8w + 0.2x + 4y + z = 21.13$
$0.36w + 0.4x - 3y \quad = -12.41$
$w - x + y + 0.8z = 3.50$

40. $w - x + 2y + z = 9$
$0.3w \quad + 0.52y - z = 1.62$
$0.6w + 2x - y \quad = 3$
$0.2w + 0.43x + 0.6y + z = 2.15$

 In Exercises 41–43,

(a) Set up a system of equations that model the given problem.

(b) Rewrite the systems of equations as the matrix equation $AX = B$.

(c) Use the inverse of matrix A to solve the system of equations.

(Note: Some of these exercises were set up in Section 2.1 of Chapter 2.)

41. **Investing** Emily bought two stocks—Datafix, selling for $30 per share, and Rocktite, selling for $20 per share—for a total of $4000. The dividend from Datafix is $2 per share, and the dividend from Rocktite is $1 per share, each year. Emily expects to receive a total of $220 in dividends from the two stocks during a given year. How many shares of each stock did she buy? (See Exercise 15, Section 2.1.)

42. **Training Police Officers** A particular police department employs two grades of police personnel: rookies and sergeants. A person at the grade of rookie is to spend 20 hours in training and 20 hours on patrol duty each week. A person at the grade of sergeant is to spend 5 hours in training and 30 hours on patrol duty each week. The training center can effectively handle 240 person-hours each week, while the department needs at least 440 person-hours each week for patrol duty for the two grades of personnel. How many persons at each of the two grades does the department have, assuming that the training center operates at full capacity and that the minimum requirements for patrol duty are met? (See Exercise 16, Section 2.1.)

43. **Hiring and Training Workers** A particular department in a factory hires skilled workers, semiskilled workers, and supervisors. Each skilled worker is paid $12 per hour, each semiskilled worker is paid $9 per hour, and each supervisor is paid $15 per hour. The department is allowed an hourly payroll of $1560 for the three types of workers. The department requires that each worker spend some time in training and safety schooling each week: skilled workers, 3 hours; semiskilled workers, 5 hours; and supervisors, 1 hour. The training center can handle a maximum of 588 person-hours each week, and the department needs 140 skilled and semiskilled workers to meet production schedules. Assuming that the allowable payroll is met, the training center is fully utilized, and production needs are met, how many of each type of worker does the department need? (See Exercise 21, Section 2.1.)

44. Refer to Exercise 41. If Emily actually purchased a total of $5000 worth of Datafix and Rocktite stocks, and if she earned $300 in dividends from the two stocks in a given year, how many shares of each stock did she buy?

45. Refer to Exercise 42. Because of city budget cuts, the police department's training center can now handle only 200 person-hours each week. With all other restrictions described in Exercise 42 still operative, how many persons at each of the two grades will the police department be able to have?

46. Refer to Exercise 43. The factory department receives a large order that requires it to hire additional workers to complete. Now the department will be allowed an hourly payroll of $1705, the training center will be able to handle a maximum of 650 person-hours each week, and the department will need a total of 153 skilled and semiskilled workers to complete production. Approximately

how many of each type of worker will the department now be able to employ?

In Exercises 47 through 52, calculate the inverse of each diagonal matrix, and then draw a conclusion from the observed pattern. (A diagonal matrix is a matrix whose only nonzero elements are in a_{ii} positions.

47. $\begin{bmatrix} 3 & 0 \\ 0 & 1 \end{bmatrix}$

48. $\begin{bmatrix} -2 & 0 \\ 0 & 3 \end{bmatrix}$

49. $\begin{bmatrix} 3 & 0 & 0 \\ 0 & -1 & 0 \\ 0 & 0 & 5 \end{bmatrix}$

50. $\begin{bmatrix} -2 & 0 & 0 \\ 0 & 4 & 0 \\ 0 & 0 & 7 \end{bmatrix}$

51. $\begin{bmatrix} 3 & 0 & 0 & 0 \\ 0 & -6 & 0 & 0 \\ 0 & 0 & 2 & 0 \\ 0 & 0 & 0 & 5 \end{bmatrix}$

52. $\begin{bmatrix} 2 & 0 & 0 & 0 \\ 0 & 3 & 0 & 0 \\ 0 & 0 & 4 & 0 \\ 0 & 0 & 0 & 5 \end{bmatrix}$

In Exercises 53 through 58, use the rules for matrix algebra to solve each equation for the indicated variable. Assume that the dimensions are such that matrix multiplication and addition are possible and that inverses exist when needed.

53. Solve for A: $AB = C$

54. Solve for B: $AB = C$

55. Solve for X: $X - AX = B$

56. Solve for Z: $Z + AZ = B$

57. Solve for A: $AB + AZ = C$

58. Solve for Z: $AZ + BZ = D$

In Exercises 59 through 62, use the following matrices to calculate the given expressions:

$$A = \begin{bmatrix} 2 & 3 \\ 1 & 2 \end{bmatrix} \quad \text{and} \quad B = \begin{bmatrix} 3 & 4 \\ 1 & 2 \end{bmatrix}$$

59. (a) $(AB)^{-1}$

  (b) $B^{-1}A^{-1}$

  (c) After comparing the results of parts (a) and (b), state a general rule regarding what you believe about these expressions.

  (d) Use your rule to rewrite $((AB)C)^{-1}$.

60. (a) $(A^2)^{-1}$

  (b) $(A^{-1})^2$

  (c) After comparing the results of parts (a) and (b), state a general rule regarding what you believe about these expressions.

61. (a) $(A^T)^{-1}$

  (b) $(A^{-1})^T$

  (c) After comparing the results of parts (a) and (b), state a general rule regarding what you believe.

62. (a) $(A^{-1})^{-1}$

  (b) After calculating the matrix in part (a), state a general rule regarding what you believe.

63. (a) For matrices A and B such that $AB = 0$, what condition must hold in order for you to conclude that $B = 0$?

  (b) If $AB = 0$, is it possible for both A and B to have inverses?

  (c) If both A and B have inverses, is it possible for $AB = 0$ to hold?

  (d) If $AB = 0$, is it possible that neither A nor B has an inverse?

Section 3.4 More Applications of Inverses

Elementary Cryptography

Cryptography is the study of sending and receiving coded messages. Such messages are **encoded** in such a way that, even if they are intercepted, only the intended receiver should be able to **decode** them.

There are many applications of cryptography that touch our everyday lives. For example, if you have ever made an on-line purchase, your personal information is encoded as it transfers between computers. A bank's automatic teller machine (ATM) must decode certain information we give it to allow access to our accounts. A computer encodes all information entered as binary numbers, does internal work by means of those numbers, and then returns the results in decoded form that we can understand. Encrypted messages are used by the federal government and by the military, during times of both war and peace, to share sensitive information among agencies.

There are many fascinating ways to encode messages. We consider one very simple example, showing how an invertible $n \times n$ matrix A may be used to encode and decode a message. The sender uses A to encode the message, and the receiver uses A^{-1} to decode it. To encode a message, each letter of the alphabet and any other agreed-upon symbols are assigned a positive integer known to both the sender and receiver. The message is then encoded with the use of both these integers and the matrix A and is sent as a string of integers. Finally, the receiver uses this string of integers and A^{-1} to retrieve the original integers and hence the message. Example 1 gives details.

EXAMPLE 1

A Cryptography
Application

Suppose that the sender and receiver agree to the numerical assignment $a = 1, b = 2, \ldots$ $z = 26$, space $= 30$, period $= 40$, and apostrophe $= 60$. (These assignments could be reversed or otherwise scrambled in some way.) The sender has the encoding matrix

$$A = \begin{bmatrix} 1 & 0 & 1 \\ 2 & 0 & 0 \\ 1 & 1 & 1 \end{bmatrix},$$

and the receiver has the decoding inverse matrix

$$A^{-1} = \begin{bmatrix} 0 & \frac{1}{2} & 0 \\ -1 & 0 & 1 \\ 1 & -\frac{1}{2} & 0 \end{bmatrix}.$$

Because A is 3×3, the message will be partitioned into sequences of three symbols each. Suppose that the message to be sent is "MEET ME AT SIX." Then the sequence breakdown, along with the agreed-upon assignment of numbers, is

M	E	E	T		M	E		A	T		S	I	X
13	5	5	20	30	13	5	30	1	20	30	19	9	24	40

The code matrix is constructed by using each sequence as a row:

$$C = \begin{bmatrix} 13 & 5 & 5 \\ 20 & 30 & 13 \\ 5 & 30 & 1 \\ 20 & 30 & 19 \\ 9 & 24 & 40 \end{bmatrix}$$

Now the message will be transmitted as the string of numbers in consecutive rows of CA.

$$CA = \begin{bmatrix} 13 & 5 & 5 \\ 20 & 30 & 13 \\ 5 & 30 & 1 \\ 20 & 30 & 19 \\ 9 & 24 & 40 \end{bmatrix} \begin{bmatrix} 1 & 0 & 1 \\ 2 & 0 & 0 \\ 1 & 1 & 1 \end{bmatrix} = \begin{bmatrix} 28 & 5 & 18 \\ 93 & 13 & 33 \\ 66 & 1 & 6 \\ 99 & 19 & 39 \\ 97 & 40 & 49 \end{bmatrix}.$$

That is, the message is sent as 28, 5, 18, 93, 13, 33, 66, 1, 6, 99, 19, 39, 97, 40, 49. The person receiving the message has the 3×3 matrix A^{-1}, from which he or she will reassemble the message as a matrix having three columns. The receiver also knows that the message is represented by the matrix CA. How does the receiver recover the original matrix C? Just multiply CA on the right by A^{-1}: $(CA)A^{-1} = C$. Thus,

$$(CA)A^{-1} = \begin{bmatrix} 28 & 5 & 18 \\ 93 & 13 & 33 \\ 66 & 1 & 6 \\ 99 & 19 & 39 \\ 97 & 40 & 49 \end{bmatrix} \begin{bmatrix} 0 & \frac{1}{2} & 0 \\ -1 & 0 & 1 \\ 1 & -\frac{1}{2} & 0 \end{bmatrix} = \begin{bmatrix} 13 & 5 & 5 \\ 20 & 30 & 13 \\ 5 & 30 & 1 \\ 20 & 30 & 19 \\ 9 & 24 & 40 \end{bmatrix} = C.$$

Now, equating the number in C with the agreed-upon letters gives

$$\begin{bmatrix} M & E & E \\ T & & M \\ E & & A \\ T & & S \\ I & X & . \end{bmatrix}. \ \blacksquare$$

An Economics Application

Another example of how matrices and their inverses are used is in **input–output analysis.** Briefly, input–output analysis is concerned with the interrelationship among various preselected sectors that make up an economy. Each sector usually requires **input** from all sectors of the economy, including itself. Stated differently, for each unit of output from a given sector, there is an **internal demand** within the system from the other sectors. The **output** of a sector refers to its production. For example, to produce a unit of steel requires machinery from the manufacturing sector, various control devices from the electronics sector, and so on.

HISTORICAL NOTE

WASSILY LEONTIEF

In 1973, the Nobel Memorial Prize in Economic Sciences was awarded to Professor Wassily Leontief (lē-ont'-yēf) of Harvard University. Leontief's work focused on the interrelation of the various sectors of the U.S. economy and made use of input–output matrices. In his original work, Leontief divided the economy into 500 sectors.

Suppose that the economy is divided into the broad sectors of manufacturing (M), electronics (E), and agriculture (A). Then an **input–output matrix** displaying the interrelations among these sectors might appear as

$$
\begin{array}{c}
\text{Output} \\
\begin{array}{ccc}
\text{1 of } M & \text{1 of } E & \text{1 of } A \\
\uparrow & \uparrow & \uparrow
\end{array}
\end{array}
$$

$$
\begin{array}{c}
\text{Input } \begin{array}{c} M \to \\ E \to \\ A \to \end{array}
\left[
\begin{array}{ccc}
0.01 & 0.02 & 0.001 \\
0.10 & 0.01 & 0.02 \\
0.12 & 0.15 & 0.02
\end{array}
\right] = T
\end{array}
$$

with this interpretation: For each unit output of M, 0.01 unit of M itself will be required in the process; for each unit output of E, 0.02 unit of M will be required; for each unit output of A, 0.001 unit of M will be required. This same interpretation may be made for the inputs E and A. Now let x be the total production of M, y the total production of E, and z the total production of A. Then

$$
X = \begin{bmatrix} x \\ y \\ z \end{bmatrix}
$$

is called a **production matrix,** and the matrix product

$$
TX = \begin{bmatrix} 0.01 & 0.02 & 0.001 \\ 0.10 & 0.01 & 0.02 \\ 0.12 & 0.15 & 0.02 \end{bmatrix} \begin{bmatrix} x \\ y \\ z \end{bmatrix} = \begin{bmatrix} 0.01x + 0.02y + 0.001z \\ 0.10x + 0.01y + 0.02z \\ 0.12x + 0.15y + 0.02z \end{bmatrix} \begin{array}{l} \text{total } M \text{ input} \\ \text{total } E \text{ input} \\ \text{total } A \text{ input} \end{array}
$$

is called an **internal demand** matrix because it gives the total input needed from each of the three sectors to produce x, y, and z units, respectively, of M, E, and A.

Now consumer demand enters into the picture. Others will want to purchase some of the output from the various sectors of the economy, and their wants are called the **final demand** on the economy. That demand may be expressed in a matrix with one column:

$$
D = \begin{bmatrix} \text{amount from } M \\ \text{amount from } E \\ \text{amount from } A \end{bmatrix}.
$$

This economy may now be described by the equation

total production = internal consumption by sectors + final demand.

In symbols, the preceding equation can be expressed as $X = TX + D$. In the United States, great sums of money are spent determining what and how much consumers will demand. Usually, a good estimate of the demand is known, so the trick is to find the production matrix X so that the equation $X = TX + D$ holds. Because both T and D are known, solving for X is a matrix algebra problem:

$$
X = TX + D
$$
$$
X - TX = D
$$
$$
(I - T)X = D
$$
$$
X = (I - T)^{-1}D \text{ (if the inverse exists).}
$$

The Output of an Economy

If T is the input–output matrix and D the final demand matrix for a given economy, then the total output for the economy is $X = (I - T)^{-1}D$, assuming that $(I - T)^{-1}$ exists.

The matrix $(I - T)^{-1}$ is known as the "Leontief matrix for the economy," in honor of Wassily Leontief.

EXAMPLE 2

An Input–Output Analysis

Given that the input matrix of an economy that is divided into sectors A and B, is

$$T = \begin{array}{cc} & \begin{array}{cc} \text{Output} \\ A \quad\quad B \end{array} \\ \text{Input } \begin{array}{c} A \\ B \end{array} & \begin{bmatrix} 0.02 & 0.15 \\ 0.18 & 0.10 \end{bmatrix} \end{array}$$

and the final demand is

$$D = \begin{bmatrix} 320 \\ 580 \end{bmatrix} \begin{array}{c} A \\ B \end{array},$$

find the total-output matrix X.

SOLUTION We know that $X = (I - T)^{-1}D$, so

$$X = (I - T)^{-1}D = \left(\begin{bmatrix} 1 & 0 \\ 0 & 1 \end{bmatrix} - \begin{bmatrix} 0.02 & 0.15 \\ 0.18 & 0.10 \end{bmatrix} \right)^{-1} \begin{bmatrix} 320 \\ 580 \end{bmatrix}$$

$$= \begin{bmatrix} 0.98 & -0.15 \\ -0.18 & 0.90 \end{bmatrix}^{-1} \begin{bmatrix} 320 \\ 580 \end{bmatrix}$$

$$= \begin{bmatrix} 1.053 & 0.175 \\ 0.211 & 1.146 \end{bmatrix} \begin{bmatrix} 320 \\ 580 \end{bmatrix}$$

$$= \begin{bmatrix} 438.46 \\ 732.20 \end{bmatrix}$$

Note that once $(I - T)^{-1}$ is known, it is easy to recalculate the total production matrix X if the demand matrix D changes. Therefore, in order to meet both internal and external demand for the economic sectors, sector A should output a total of 438.46 units and sector B should output a total of 732.20 units. Notice, for example, that this implies that $732.20 - 580 = 652.20$ units of sector B are consumed internally, since 580 units are required to meet external demand for that sector. ■

Manufacturing Application

Here is one final example of how the inverse of a matrix can be applied. Think of an assembly process in which a single part sometimes requires other parts for its subassembly. A simple parts diagram might appear as shown in Figure 10. The diagram is interpreted as follows: Part p_2 requires 2 of part p_1, part p_3 requires 1 of part p_1, part p_4 requires 3 of part p_1 and 1 of part p_2, and so on.

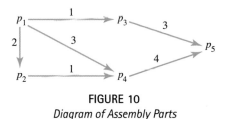

FIGURE 10
Diagram of Assembly Parts

The preceding information can be put into the following **parts matrix:**

$$
P = \begin{array}{c} \\ \text{Input:} \ \begin{matrix} p_1 \\ p_2 \\ p_3 \\ p_4 \\ p_5 \end{matrix} \end{array}
\overset{\begin{matrix} \text{Output:} & p_1 & p_2 & p_3 & p_4 & p_5 \end{matrix}}{\begin{bmatrix} 0 & 2 & 1 & 3 & 0 \\ 0 & 0 & 0 & 1 & 0 \\ 0 & 0 & 0 & 0 & 3 \\ 0 & 0 & 0 & 0 & 4 \\ 0 & 0 & 0 & 0 & 0 \end{bmatrix}}.
$$

Each entry in this matrix represents the number of parts of the kind to the left of the element needed to assemble one part of the kind above the element. The assumption is now made that the total output of these parts is equal to the internal consumption of the assembly process plus the demand outside the system. Therefore, if, as before, we let X be the total-output matrix and D the external-demand matrix, then PX represents the number of parts consumed internally in the assembly process, and $X = PX + D$. Again, D can usually be estimated rather closely, so that the total production X is the sought-after matrix.

Therefore,

$$
\begin{aligned}
X &= PX + D \\
X - PX &= D \\
(I - P)X &= D \\
X &= (I - P)^{-1}D \ \text{(assuming that } (I - P)^{-1} \text{ exists)}.
\end{aligned}
$$

EXAMPLE 3

Using a Parts Matrix

For the assembly process described in Figure 10 and the resulting matrix P, let

$$
D = \begin{bmatrix} 2000 \\ 2500 \\ 3000 \\ 2000 \\ 700 \end{bmatrix} \begin{matrix} p_1 \\ p_2 \\ p_3 \\ p_4 \\ p_5 \end{matrix}. \ \text{Find the production matrix } X.
$$

SOLUTION From the foregoing analysis, $X = (I - P)^{-1}D$, so that, for our situation,

$$X = \left(\begin{bmatrix} 1 & 0 & 0 & 0 & 0 \\ 0 & 1 & 0 & 0 & 0 \\ 0 & 0 & 1 & 0 & 0 \\ 0 & 0 & 0 & 1 & 0 \\ 0 & 0 & 0 & 0 & 1 \end{bmatrix} - \begin{bmatrix} 0 & 2 & 1 & 3 & 0 \\ 0 & 0 & 0 & 1 & 0 \\ 0 & 0 & 0 & 0 & 3 \\ 0 & 0 & 0 & 0 & 4 \\ 0 & 0 & 0 & 0 & 0 \end{bmatrix} \right)^{-1} \begin{bmatrix} 2000 \\ 2500 \\ 3000 \\ 2000 \\ 700 \end{bmatrix}$$

$$= \begin{bmatrix} 1 & -2 & -1 & -3 & 0 \\ 0 & 1 & 0 & -1 & 0 \\ 0 & 0 & 1 & 0 & -3 \\ 0 & 0 & 0 & 1 & -4 \\ 0 & 0 & 0 & 0 & 1 \end{bmatrix}^{-1} \begin{bmatrix} 2000 \\ 2500 \\ 3000 \\ 2000 \\ 700 \end{bmatrix}$$

$$= \begin{bmatrix} 1 & 2 & 1 & 5 & 23 \\ 0 & 1 & 0 & 1 & 4 \\ 0 & 0 & 1 & 0 & 3 \\ 0 & 0 & 0 & 1 & 4 \\ 0 & 0 & 0 & 0 & 1 \end{bmatrix} \begin{bmatrix} 2000 \\ 2500 \\ 3000 \\ 2000 \\ 700 \end{bmatrix} = \begin{bmatrix} 36{,}100 \\ 7300 \\ 5100 \\ 4800 \\ 700 \end{bmatrix}.$$

Therefore, in order to meet both internal and external demand for parts, 36,100 units of p_1, 7300 units of p_2, 5100 units of p_3, 4800 units of p_4, and 700 units of p_5 are required. Notice, for example, that this implies that $36{,}100 - 2000 = 34{,}100$ units of p_1 are consumed during the manufacturing process, since 2000 units of p_1 will be used to satisfy external demand. ▬

Exercises 3.4

In Exercises 1 through 4, use the given encoding matrix A and the given message to

(a) Construct the code matrix after assigning numbers to the letters and symbols in the message, as was done in Example 1.

(b) Compute the matrix the receiver gets.

(c) Find the decoding matrix A^{-1}, and use it to decode the message in a fashion similar to the way the two were used in Example 1.

1. $A = \begin{bmatrix} 1 & 2 \\ -1 & 1 \end{bmatrix}$; message: I HAVE A JOB.

2. $A = \begin{bmatrix} 1 & 0 \\ 1 & 1 \end{bmatrix}$; message: HELP ME.

3. $A = \begin{bmatrix} 1 & 1 \\ 1 & 2 \end{bmatrix}$; message: NO WAY.

4. $A = \begin{bmatrix} 0 & 3 \\ 1 & 1 \end{bmatrix}$; message: FINITE IS FUN.

In Exercises 5 through 8, use matrix A and the encoded message to

(a) First find A^{-1}.

(b) Then use A^{-1} to decode the message.

5. $A = \begin{bmatrix} 1 & 0 & 1 \\ 2 & 0 & 3 \\ 1 & 2 & 1 \end{bmatrix}$; message: 48, 10, 66, 71, 60, 91, 26, 40, 27, 115, 60, 155

6. $A = \begin{bmatrix} 1 & -1 & 0 \\ 0 & 1 & 1 \\ 0 & 0 & 1 \end{bmatrix}$; message: 9, 51, 73, 30, -23, 25, 1, 3, 25, 1, 19, 29, 14, -7, 47

7. $A = \begin{bmatrix} 2 & 0 & 0 \\ 0 & 1 & 0 \\ 0 & 0 & 3 \end{bmatrix}$; message: 14, 12, 45, 4, 1, 36, 60, 23, 3, 36, 13, 27, 28, 7, 90, 6, 15, 42, 40, 9, 42, 42, 5, 57, 80, 30, 90

8. $A = \begin{bmatrix} 4 & 0 & 0 \\ 0 & 1 & 0 \\ 0 & 0 & 1 \end{bmatrix}$; message: 64, 15, 12, 48, 21, 20, 36, 15, 14, 120, 8, 1, 64, 16, 5, 56, 19, 40

In Exercises 9 through 14, answer the following statements about each input–output matrix T and demand matrix D.

(a) *Interpret the entries in each row of T.*

(b) *Compute* $(I - T)^{-1}$.

(c) *Compute the production matrix X.*

(d) *Find the internal consumption.*

9. Input:
Output:

	Steel	Electronics
Steel	0.02	0.10
Electronics	0.15	0.01

$= T$;

$D = \begin{bmatrix} 500 \\ 800 \end{bmatrix}$

10. Input:
Output:

	Agri.	Mfg.
Agri.	0.03	0.20
Mfg.	0.01	0.05

$= T; D = \begin{bmatrix} 1000 \\ 1200 \end{bmatrix}$

11. Input:
Output:

	Mfg.	Electronics
Mfg	0.10	0.20
Electronics	0.15	0.05

$= T$;

$D = \begin{bmatrix} 200 \\ 500 \end{bmatrix}$

12. Input:
Output:

	Service	Mfg.
Service	0.20	0.50
Mfg.	0.10	0.15

$= T; D = \begin{bmatrix} 300 \\ 800 \end{bmatrix}$

13. Input:
Output:

	Agri.	Mfg.	Electronics
Agri.	0.02	0.01	0.01
Mfg.	0.20	0.10	0.10
Electronics	0.15	0.08	0.05

$= T$;

$D = \begin{bmatrix} 500 \\ 1000 \\ 2000 \end{bmatrix}$

14. Input:
Output:

	Service	Mfg.	Electronics
Service	0.40	0.20	0.10
Mfg.	0.01	0.10	0.20
Electronics	0.30	0.20	0.10

$= T$;

$D = \begin{bmatrix} 300 \\ 400 \\ 500 \end{bmatrix}$

In Exercises 15 through 20,

(a) *Construct the parts matrix P.*

(b) *Interpret the entries in each row.*

(c) *Find the production matrix based on P and the given demand matrix D.*

15.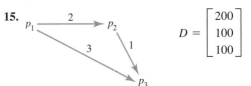
$D = \begin{bmatrix} 200 \\ 100 \\ 100 \end{bmatrix}$

16.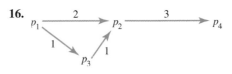

$D = \begin{bmatrix} 200 \\ 300 \\ 400 \\ 100 \end{bmatrix}$

17.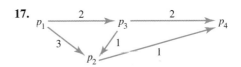

$D = \begin{bmatrix} 200 \\ 80 \\ 100 \\ 50 \end{bmatrix}$

18.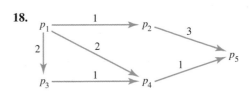

$D = \begin{bmatrix} 100 \\ 200 \\ 300 \\ 100 \\ 50 \end{bmatrix}$

19. $p_1 \xrightarrow{2} p_2 \xrightarrow{2} p_3 \xrightarrow{3} p_4$

$D = \begin{bmatrix} 250 \\ 300 \\ 150 \\ 60 \end{bmatrix}$

20.

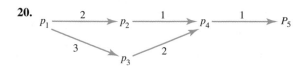

$$D = \begin{bmatrix} 80 \\ 60 \\ 50 \\ 100 \\ 200 \end{bmatrix}$$

In Exercises 21 and 22, each parts matrix represents the number of parts used internally and to make other parts for subassembly in a production process. Construct the parts-arrow diagram (as shown in Exercises 15 through 20) for the subassembly process.

21.

	p_1	p_2	p_3	p_4
p_1	0	2	0	3
p_2	0	0	1	2
p_3	0	0	0	1
p_4	0	0	0	0

22.

	p_1	p_2	p_3	p_4	p_5
p_1	0	2	1	3	1
p_2	0	0	2	0	1
p_3	0	0	0	3	0
p_4	0	0	0	0	2
p_5	0	0	0	0	0

Chapter 3 Summary

CHAPTER HIGHLIGHTS

Matrices provide a convenient structure for organizing and interpreting data. Operations such as multiplying by a number, finding the transpose, determining the inverse, squaring, and cubing can be performed on a single matrix. Operations such as *addition, subtraction,* and *multiplication* use a pair of compatible matrices. Matrix operations follow many of the rules of algebra, but *commutativity* and *cancellation in multiplication* are two notable exceptions. Matrices provide a powerful tool in a variety of areas of applications, including inventories, dominance relations, communication networks, cryptography, input–output analysis, and the production of parts.

EXERCISES

In Exercises 1–11, let

$$A = \begin{bmatrix} 2 & 0 & 1 \\ 3 & -1 & 0 \end{bmatrix}, B = \begin{bmatrix} 1 & -2 \\ 0 & 4 \\ 3 & -1 \end{bmatrix}, C = \begin{bmatrix} 9 & 1 \\ 0 & -1 \end{bmatrix},$$

and $D = \begin{bmatrix} 1 & -2 \\ 3 & 0 \end{bmatrix}$. *Evaluate each expression, if possible.*

1. $C + 3D$ **2.** $\frac{1}{3}C - D$ **3.** B^T **4.** AB **5.** BA

6. BC **7.** CB **8.** D^2 **9.** A^2 **10.** B^{-1} **11.** C^{-1}

12. Solve the matrix equation $X + CX = D$ for X. (Assume that the matrices are compatible and that inverses exist when needed.)

13. Assign values to the variables so that

$$\begin{bmatrix} 27 & -w \\ 2t & 0 \end{bmatrix} = \begin{bmatrix} 3x & 24 \\ 17 & -2v + 1 \end{bmatrix}.$$

14. Find the inverse of the matrix $\begin{bmatrix} 10 & 8 \\ 9 & 7 \end{bmatrix}$.

15. Find the inverse of the matrix $\begin{bmatrix} 2 & 0 & 1 \\ 1 & 3 & 2 \\ 1 & 1 & 0 \end{bmatrix}$.

16. Find the matrices of order 3×3 with $A \neq \mathbf{0}$ and $B \neq \mathbf{0}$ such that $AB = \mathbf{0}$.

17. In the following diagram, airline flights are available through the cities shown in the direction of the arrows only:

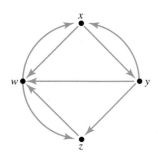

If cities do not have flights from themselves to themselves, write a matrix that displays this flight network.

Use the given coding matrix A and the given message to solve Problems 18 though 21.

$$A = \begin{bmatrix} 0 & 1 \\ 1 & 3 \end{bmatrix}$$ Message: GO PHANTOMS!

18. Construct the code matrix after assigning numbers to the letters and symbols in the message.

19. Compute the matrix the receiver gets.

20. Find the decoding matrix A^{-1}.

21. Decode the message, using A^{-1}.

22. During a given month, a used-book store has a beginning inventory of

	Nonfiction	Fiction	
hardback	327	220	$= B$
paperback	460	112	

and a sales inventory of

	Nonfiction	Fiction	
hardback	54	37	$= S.$
paperback	22	18	

If the ending inventory is

	Nonfiction	Fiction	
hardback	319	245	$= E,$
paperback	460	115	

what is the purchase matrix?

23. Use the inverse of the coefficient matrix, if possible, to solve the following system:

$$\begin{aligned} 2x + y - z &= 3 \\ x + y + 2z &= 5 \\ 3x - y + 2z &= 3. \end{aligned}$$

24. Consider the matrix $A = \begin{bmatrix} 0 & 2 & 3 \\ -1 & 0 & 1 \\ 2 & 1 & 0 \end{bmatrix}$.

(a) Find A^{-1}.

(b) Find $A^2 + 3B$, where

$$B = \begin{bmatrix} -3 & 4 & 6 \\ 2 & 0 & 7 \\ 1 & 9 & 8 \end{bmatrix}.$$

25. Find the inverse of

$$A = \begin{bmatrix} 2 & 4 \\ 1 & 5 \end{bmatrix}.$$

CUMULATIVE REVIEW

26. Find the slope, x-intercept, and y-intercept of the line having the equation $3x + 4y = 12$.

27. Write the equation of the horizontal line that contains the point $(-7, 0)$.

28. Solve the system

$$\begin{aligned} x + 3y &= 7 \\ 3x - 2y &= 9, \end{aligned}$$

using the pivoting process.

29. Find the profit function and the break-even point for a company having a total cost function $C(x) = 60x + 2100$ and a revenue function $R(x) = 80x$.

30. Solve the following system of equations:

$$\begin{aligned} x + y &= 0 \\ y - 7z &= 1 \\ 5y - 35z &= 5 \end{aligned}$$

31. Suppose you believe that you can make 72% on an upcoming Finite Mathematics test with no further study, but estimate that you can increase your test score by 5% for each additional hour of study. Find a linear function that will give your test score in terms of the number of additional hours, t, of study. (Be sure to state the limits on t.)

Chapter 3 Sample Test Items

1. Assign values to the variables so that the matrix

$$\begin{bmatrix} 2 & 0 & 3x \\ t+4 & 7 & 19 \end{bmatrix} \text{ equals the matrix}$$

$$\begin{bmatrix} y-7 & w & 9 \\ 10 & 7 & 3r+1 \end{bmatrix}.$$

In Problems 2–7, evaluate each expression, if possible, using the following matrices:

$$\text{Let } A = \begin{bmatrix} 2 & 0 & 3 \\ -1 & 1 & 0 \\ 0 & 0 & 4 \end{bmatrix}, B = \begin{bmatrix} 1 & 2 & 3 \\ 4 & 5 & 6 \\ 7 & 8 & 9 \end{bmatrix},$$

$$\text{and } C = \begin{bmatrix} -2 & 1 & 4 \\ 6 & 3 & -1 \end{bmatrix}.$$

2. $2A - B$ **3.** $\frac{1}{2}B + C$ **4.** A^T **5.** B^2 **6.** AB **7.** C^2

8. In the following diagrams, three communication stations communicate directly only in the direction of the arrows.

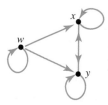

Find a matrix that displays this communication network.

9. Find a linear system of equations corresponding to this matrix equation

$$\begin{bmatrix} 2 & 0 & 1 \\ 3 & -2 & 5 \end{bmatrix} \begin{bmatrix} x \\ y \\ z \end{bmatrix} = \begin{bmatrix} 4 \\ 17 \end{bmatrix}.$$

10. The T-Mart Company has two stores that stock brands X and Y microwave ovens. The current-inventory matrix is

$$\begin{array}{cc} & \text{Store 1} \quad \text{Store 2} \\ \begin{array}{c} \text{Brand X} \\ \text{Brand Y} \end{array} & \begin{bmatrix} 15 & 12 \\ 8 & 3 \end{bmatrix} = A. \end{array}$$

Brand X microwave ovens cost \$175 each, and Brand Y ovens cost \$250 each. Find a matrix B and its product with A so that the resulting matrix gives the total cost of the microwave oven inventory in each store.

11. Decide whether $\begin{bmatrix} 3 & -1 \\ 2 & -1 \end{bmatrix}$ and $\begin{bmatrix} 1 & 1 \\ 2 & 3 \end{bmatrix}$ are inverses of each other.

12. Find the inverse of $\begin{bmatrix} 5 & 7 \\ 2 & 3 \end{bmatrix}$.

13. Find the inverse, if one exists, for the matrix

$$\begin{bmatrix} 1 & 4 & -3 \\ 0 & 1 & 7 \\ 0 & 3 & 20 \end{bmatrix}.$$

14. Solve for X in the matrix equation $XA + B = Y$.

15. Use the fact that

$$\begin{bmatrix} 3 & -1 & 0 \\ 1 & 0 & -1 \\ -1 & 1 & -1 \end{bmatrix}^{-1} = \begin{bmatrix} 1 & -1 & 1 \\ 2 & -3 & 3 \\ 1 & -2 & 1 \end{bmatrix}$$

to quickly solve the system

$$\begin{array}{rcl} 3x - y & & = 2 \\ x & - z & = 0 \\ -x + y - z & & = 3 \end{array}$$

by matrix multiplication.

The following diagram indicates the parts needed to assemble other parts:

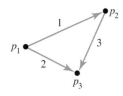

Use the diagram to answer Exercises 16 through 20.

16. Construct the parts matrix P.

17. Interpret the entries in each row of P.

18. Compute $(I - P)^{-1}$.

19. If $D = \begin{bmatrix} 100 \\ 200 \\ 300 \end{bmatrix}$ is the demand matrix, calculate the production matrix X.

20. Find the number of each part consumed internally in the production process.

In-Depth Application

Communication Networks and Powers of Incidence Matrices

Objectives

In this application, you will apply your knowledge of matrices and matrix operations to answer questions about a communication network. More specifically, you will be asked to create an incidence matrix for a network and use matrix multiplication to determine how long it takes for information to pass from one station to another.

Introduction

In Example 2 of Section 3.1, a communication network was introduced, along with its corresponding incidence matrix. In general, a set of objects with arrows between some of them is called a **directed graph.** The objects are called **vertices** and the arrows are called **edges.** If there are n vertices, we can label them from 1 to n. Thus, we can define the elements of the corresponding incidence matrix A by

$$a_{ij} = \begin{cases} 1 & \text{if there is an edge from the } i\text{th vertex to the } j\text{th vertex,} \\ 0 & \text{otherwise.} \end{cases}$$

Such matrices can be used to model the exchange of information among computers within a corporation, transport routes among cities, and communication paths among intelligence agents behind enemy lines.

The Communication Network

Suppose we have a network of five communication stations. The following directed graph represents the way information is passed between the stations:

(continued)

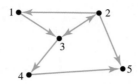

1. Create the incidence matrix A for this network.

2. According to the graph, information can be carried from station 1 to station 5 along several different paths. For instance, information could be passed from station 1 to station 3, from station 3 to station 2, then from station 2 to station 5. We can write this path as $1 \rightarrow 3 \rightarrow 2 \rightarrow 5$. Another path from station 1 to station 5 could be written as $1 \rightarrow 3 \rightarrow 4 \rightarrow 5$. List at least two paths from station 2 to station 4.

3. The **length** of a path in a directed graph is defined as the number of steps required to move from one vertex to another. For instance, the path $1 \rightarrow 3 \rightarrow 2$ has length 2. The incidence matrix A for the network essentially identifies all one-step paths. The matrix A^2 then lists the number of two-step paths from one station to another. Calculate A^2. According to this matrix, how many two-step paths start at station 4 and end at station 2? How many start at station 3 and end at station 5?

4. If matrix A indicates all one-step paths, and matrix A^2 represents the number of two-step paths, what does the matrix $A + A^2$ represent? Calculate this matrix and verify your conclusion.

5. In working with communication networks, it is important to know how many steps are required for information to pass from one station to another. Suppose we wish to pass information from station 1 to station 5. According to matrix A, this cannot be done in one step, and according to matrix A^2, it also cannot be done in two steps (since there are 0's in the first-row, fifth-column entries of each matrix). For the matrix $A + A^2$, identify all pairs of stations for which information *cannot* be passed in *two or fewer* steps.

6. According to the graph, can information be passed from station 1 to station 5 in three steps? If so, what path(s) can be taken?

7. Calculate A^3. What does this matrix represent? How many three-step paths are there that start at station 1 and end at station 5? Compare this answer with your answer to Problem 6.

8. Calculate $A + A^2 + A^3$. Are there any pairs of stations remaining for which information cannot be passed in *three or fewer steps*?

9. Calculate $A + A^2 + A^3 + A^4$. Are there any pairs of stations remaining for which information *cannot* be passed in *four or fewer steps*?

10. Are there any pairs of stations for which information can *never* be passed between them? Compare these pairs with your answer to Problem 9.

11. In general, for a communication network with n stations and incidence matrix A, the entries in the matrix sum $A + A^2 + \cdots + A^k$ represent the number of paths required to pass information from one station to another in k *or fewer* steps. hy do we not need to calculate this sum beyond $k = n - 1$?

Linear Programming: A Graphical Approach

▮ PROBLEM

Two kinds of foods are being mixed to provide certain minimum vitamin requirements. Each food's price and vitamin content per kilogram are indicated. How much of each food should be included in the mixture both to meet all of the minimum vitamin requirements and to minimize costs? (See Example 5, Section 4.3.)

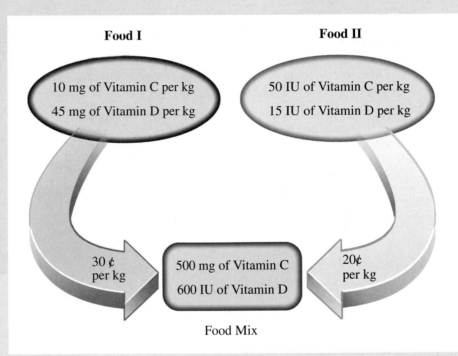

Food I

10 mg of Vitamin C per kg

45 mg of Vitamin D per kg

Food II

50 IU of Vitamin C per kg

15 IU of Vitamin D per kg

30 ¢ per kg

500 mg of Vitamin C

600 IU of Vitamin D

20¢ per kg

Food Mix

CHAPTER PREVIEW

Linear programming is one of the premier applications of modern mathematics. It is used extensively in business, health, agricultural, industrial, and military settings. Linear programming is a technique for finding the maximum or minimum value of a linear function involving many variables—variables that may assume values only in a region restricted by a system of linear inequalities. After a section on the mathematical modeling of this type of problem, a method for a graphical solution to problems involving two variables is explored.

Technology Opportunities

Spreadsheets are helpful for organizing data into tables and calculating function values across various subsets of the data. All of the tables of data and calculations that are used in this chapter can be organized in a spreadsheet. Most graphing calculators also have the ability to create basic tables, but are particularly useful in producing the graphs presented in this chapter.

Modern spreadsheet packages have fairly sophisticated tools for solving linear programming problems; these tools will be explored in the next chapter.

Section 4.1   Modeling Linear Programming Problems

We resume our modeling of linear systems, but instead of using linear equalities exclusively, we now allow **linear inequalities.** The problems modeled here fall under the general classification of **optimization problems.** An optimization problem involves trying to maximize or minimize some specific quantity. For example, we may want to maximize the profit for a business or minimize the pollution output of a city. In this chapter, we discuss special kinds of optimization problems that require the variables to satisfy a linear system (perhaps with both equalities and inequalities). Such problems are known as **linear programming problems,** and their mathematical formulation is called a **linear program.** We begin our study of these topics by exploring several examples that illustrate the general nature of linear programming problems, and linear programs.

HISTORICAL NOTE

LEONID VITAL'EVICH KANTOROVICH (1912–1986)

Although Kantorovich spent much of his professional life as a theoretical mathematician, he is credited with being the first to use linear programming to solve problems in economics. In 1939, he published his work involving linear programming techniques designed to optimize production. He was a joint winner of the 1975 Nobel prize in economics, along with Tjalling C. Koopmans, for work on the theory of optimal allocation of resources.

EXAMPLE 1

Maximizing Profit

A company makes desks in two models: a student model and a secretarial model. Each student model requires 2 hours of woodworking and 3 hours of finishing. Each secretarial model requires 3 hours of woodworking and 5 hours of finishing. The company has a total of 240 work hours available in the woodworking department and a total of 390 work hours available in the finishing department each week. A profit of $20 is made on each student model, and $50 is earned on each secretarial one. Assuming that all desks manufactured are sold, how many of each type of desk should be made each week to maximize the company's profit from these items?

SOLUTION: We are not yet prepared to solve this problem, but we can give its mathematical formulation, or linear program.

The statement of the problem asserts that the woodworking department has a total of 240 work hours *available* each week. This means that the woodworking department may expend any number of hours *up to and including* 240 during the week. The mathematical symbol used to describe this limit is the **inequality sign** $\leq$, read "less than or equal to." The woodworking department's restriction may then be stated as

$$[\text{hours used in woodworking department}] \leq 240.$$

Similarly, the finishing department may use any number of hours *up to and including* 390 hours. This restriction may be stated as

$$[\text{hours used in finishing department}] \leq 390.$$

To describe how these inequalities relate to the production of the two types of desks, we use a table like those in Chapter 2 to organize the given data. The only difference is that now inequalities are involved and the profit is taken into account. Let x represent the number of student models made and y the number of secretarial models. Table 1 gives a systematic organization of all the data.

	Student Desk	Secretarial Desk	Limitations (hrs)
Number Produced (units)	x	y	
Woodworking (hrs/unit)	2	3	≤ 240
Finishing (hrs/unit)	3	5	≤ 390
Maximize Profit ($/unit)	20	50	

TABLE 1.

Just as with equalities, we read the inequalities across the rows:

$$2x + 3y \leq 240$$
$$3x + 5y \leq 390.$$

The first of these inequalities states that if x student desks are made, $2x$ hours of woodworking will be required, and if y secretarial desks are made, $3y$ more hours of woodworking will be required for a total of $2x + 3y$ hours, and the total must be less than or equal to 240 hours. In mathematical symbols, this is precisely the description of the woodworking limitations. Similarly, the second inequality describes the limitations on the finishing department. Because of the linear form of the left side of these two inequalities, they are known as **linear inequalities,** and they are the structural **constraints** on the problem.

The statement expressing the profit P is read from the last row of Table 1 as $P = 20x + 50y$ (in dollars). The function P is known as the **objective function** for this problem. Notice that P is a function of *two* variables: x and y. To maximize P means to find values of x and y that satisfy all of the problem's constraints and make P as large as possible.

One last consideration finishes our mathematical model: Because x and y represent numbers of desks to be made, both must be nonnegative. Using the symbol $\geq$, which is read "greater than or equal to," these conditions may be written $x \geq 0$ and $y \geq 0$ and become part of the system of constraints. In particular, these constraints are called **non-negativity constraints**. The complete mathematical description of the problem may now be stated as the following linear program:

$$\text{Maximize } P = 20x + 50y \quad \} \quad \leftarrow \textit{Objective function}$$
$$\text{subject to } 2x + 3y \leq 240 \quad \}$$
$$3x + 5y \leq 390 \quad \} \quad \leftarrow \textit{Structural constraints}$$
$$x \geq 0, y \geq 0. \quad \} \quad \leftarrow \textit{Nonnegativity constraints} \quad \blacksquare$$

The next linear programming problem asks that the objective function be minimized (made as small as possible) subject to the given constraints.

EXAMPLE 2

Minimizing Weight

A dietitian is to mix two foods, SureGrow and HealthBlast, to meet nutritional requirements. Each package of SureGrow weighs 0.2 pound (lb) and contains 20 milligrams (mg) of vitamin C, 40 international units (IU) of vitamin D, and 3 mg of Vitamin E. Each package of HealthBlast weighs 0.15 lb and contains 5 mg of vitamin C, 80 IU of vitamin D, and 2 mg of vitamin E. The mixture of the two foods is to contain the daily recommended allowance of these vitamins for adult males; thus, we need at least 60 mg of vitamin C, at least 200 IU of vitamin D, and at least 10 mg of vitamin E. How many packages of each food should be mixed to minimize the total weight? (We assume that we can use a fraction of a package if necessary.) Give the linear program of this problem.
SOURCE: Dietary Reference Intakes, National Academy of Science, 1997.

SOLUTION   Again, we are interested only in constructing the mathematical formulation of this problem. The statement that at least 60 mg of vitamin C must be in the mixture means that 60 mg *or more* are to be included. The mathematical symbol $\geq$, read "greater than or equal to," is used to describe this condition. Applying the restrictions stated in the problem, we have

$$[\text{Total mg of vitamin C}] \geq 60,$$
$$[\text{Total IU of vitamin D}] \geq 200,$$
$$[\text{Total mg of vitamin E}] \geq 10.$$

Let x represent the number of packages of SureGrow used and y the number of packages of HealthBlast used. The data from the problem are organized as shown in Table 2.

	SureGrow	HealthBlast	Limitation (mg or IU)
Number of Packages Used (pack)	x	y	
Vitamin C (mg/pack)	20	5	≥ 60
Vitamin D (IU/pack)	40	80	≥ 200
Vitamin E (mg/pack)	3	2	≥ 10
Minimize Weight (lb/pack)	0.2	0.15	

TABLE 2.

In this problem, the total weight W of the mixture is to be minimized. Writing the total weight first and the inequality constraints next, we see that the rows of Table 2 give the following linear program:

$$
\begin{aligned}
\text{Minimize } W &= 0.2x + 0.15y \quad \} \quad \leftarrow \textit{Objective function} \\
\text{subject to } 20x &+ \ 5y \geq 60 \\
40x &+ 80y \geq 200 \quad \} \quad \leftarrow \textit{Structural constraints} \\
3x &+ \ 2y \geq 10 \\
x &\geq 0, y \geq 0. \quad \} \quad \leftarrow \textit{Nonnegativity constraints} \ \blacksquare
\end{aligned}
$$

Some linear programming problems have "mixed constraints," which means that some constraints are inequalities of the form $\leq$, others are of the form $\geq$, and some may even be in the form of an equality.

EXAMPLE 3

Mixed Constraints: Maximizing Image Points

A new line of snack cracker is being introduced by a food company, which has allotted a maximum of $50,000 for advertising in newspapers, on radio, and on television. Research shows that each newspaper ad costs $100 and will cause 200 people to try the new cracker, each radio ad costs $150 and will cause 300 people to try the new cracker, and each television ad costs $300 and will cause 600 people to try the new cracker. The company wants to reach at least 10,000 people through these advertisements and has agreed to run at least twice as many radio ads as television ads. Company image ratings are 4 points for each newspaper ad, 2 for each radio ad, and 3 for each television ad. How can the company meet these objectives while maximizing the total number of company image points? Give the linear program for this problem.

SOLUTION Let x represent the number of newspaper ads, y the number of radio ads, and z the number of television ads.

Table 3 gives the pertinent information for this linear program.

	Newspaper	Radio	Television	Limitations ($ or people)
Number of Ads Purchased (ads)	x	y	z	
Costs ($/ad)	100	150	300	$\leq 50,000$
Exposures (people/ad)	200	300	600	$\geq 10,000$
Agreement		1	-2	≥ 0
Maximize Image Point (pts/ad)	4	2	3	

TABLE 3.

The restriction on the total cost of ads may then be written as

$$100x + 150y + 300z \leq 50,000.$$

The total number of people buying the new cracker is to be at least 10,000, so this restriction becomes

$$200x + 300y + 600z \geq 10,000.$$

The entries for the Agreement row may seem confusing, but note that just one algebra step is needed in order to standardize the entries in that row. The agreement is that the company will have at least twice as many radio ads as television ads. This relationship can be written as $y \geq 2z$, which we can rewrite as $y - 2z \geq 0$—precisely the inequality that is generated from the Agreement row of the table. This new inequality can be interpreted as "The number of radio ads has to be big enough so that when twice the number of television ads is subtracted, the total stays nonnegative."

The company's total number of image points, I, which are to be maximized, may be written as $I = 4x + 2y + 3z$. The mathematical formulation of the problem may now be stated:

Maximize $I = 4x + 2y + 3z$ } ← *Objective function*

subject to $100x + 150y + 300z \leq 50{,}000$ ⎫
$200x + 300y + 600z \geq 10{,}000$ ⎬ ← *Structural constraints*
$y - 2z \geq 0$ ⎭

$x \geq 0,\, y \geq 0,\, z \geq 0.$ } ← *Nonnegativity constraints* ■

EXAMPLE 4

The Mathematical Formulation of a Linear Programming Problem

A firm makes three types of golf bags: economy, deluxe, and superdeluxe. The economy bag requires 15 minutes of cutting, 20 minutes of sewing, and 30 minutes of trimming, and it sells for $80. The deluxe bag requires 15 minutes of cutting, 20 minutes of sewing, and 50 minutes of trimming, and it sells for $90. The superdeluxe bag requires 20 minutes of cutting, 30 minutes of sewing, and 50 minutes of trimming, and it sells for $100. The firm has available 80 work hours of cutting time, 90 work hours of sewing time, and 120 work hours of trimming time each day. A contract from a large discount store requires that the firm make at least 20 economy bags and 25 deluxe bags each day. The company also wants to make an equal number of deluxe and superdeluxe bags. How many bags of each type should be made to meet the stated restrictions and to maximize revenue? Give the mathematical formulation of this problem.

SOLUTION Let x represent the number of economy bags made, y the number of deluxe bags made, and z the number of superdeluxe bags made. Table 4 shows how these data may be organized. (Hours are converted to minutes.) Note that the restriction $y = z$ has been rewritten as $y - z = 0$.

	Economy	Deluxe	Super Deluxe	Limitations (min or bags)
Number of Bags Produced	x	y	z	
Cutting (min/bag)	15	15	20	$\leq 80(60) = 4{,}800$
Sewing (min/bag)	20	20	30	$\leq 90(60) = 5{,}400$
Trimming (min/bag)	30	50	50	$\leq 120(60) = 7{,}200$
Contract (Economy)	1			≥ 20
Contract (Deluxe)		1		≥ 25
Production Equality		1	-1	$= 0$
Maximize Profit ($/bag)	80	90	100	

TABLE 4.

The linear program for this problem is

$$\text{Maximize } P = 80x + 90y + 100z \quad \} \quad \leftarrow \text{Objective function}$$

$$
\begin{aligned}
\text{subject to } \quad 15x + 15y + 20z &\leq 4800 \\
20x + 20y + 30z &\leq 5400 \\
30x + 50y + 50z &\leq 7200 \\
x \qquad\qquad\qquad &\geq 20 \\
y \qquad\qquad &\geq 25 \\
y - \quad z &= 0
\end{aligned}
\right\} \quad \leftarrow \text{Structural constraints}
$$

$$x \geq 0, y \geq 0, z \geq 0. \quad \} \quad \leftarrow \text{Nonnegativity constraints} \blacksquare$$

The preceding four examples illustrate the general nature of a linear programming problem: to maximize or minimize an objective function that is subject to particular constraints. The objective function is always a linear function of several variables, x_1, $x_2 \ldots x_n$, meaning that it has the form

$$f = c_1x_1 + c_2x_2 + \cdots + c_nx_n,$$

where $c_1, c_2, \ldots c_n$ are real numbers, not all of which are zero. All constraints (restrictions) are also linear and may be put into one of these forms:

$$a_1x_1 + a_1x_2 + \cdots + a_nx_n \leq b,$$
$$a_1x_1 + a_2x_2 + \cdots + a_nx_n > b,$$
$$a_1x_1 + a_2x_2 + \cdots + a_nx_n = b.$$

In addition, the nonnegativity conditions apply: $x_1 \geq 0, x_2 \geq 0, x_3 \geq 0, \ldots, x_n \geq 0$.

In the next two sections, we present a graphical method for solving linear programming problems with two variables. In the next chapter, we examine an algebraic method as well.

Exercises 4.1

In Exercises 1 through 4,

(a) *Organize the data given into a table resembling any of Tables 1–4.*

(b) *From the table, formulate the linear programming problem by writing the objective function and constraints.*

1. **Manufacturing** The Fine Line Pen Company makes two types of ballpoint pens: a silver model and a gold model. The silver model requires 1 minute in a grinder and 3 minutes in a bonder. The gold model requires 3 minutes in a grinder and 4 minutes in a bonder. Because of maintenance procedures, the grinder can be operated no more than 30 hours per week and the bonder no more than 50 hours per week. The company makes $5 on each silver pen and $7 on each gold pen. How many of each type of pen should be produced and sold each week to maximize profits?

2. **Inventory** An appliance dealer, Andrea Kurtz, wants to purchase a combined total of no more than 100 refrigerators and dishwashers for inventory. Refrigerators weigh 200 pounds each, and dishwashers weigh 100 pounds each. Suppose that the dealer is limited to a total of 12,000 pounds for these two items. If a profit of $35 on each refrigerator and $20 on each dishwasher is projected, how many of each should be purchased and sold to make the largest profit?

3. **Production** A company makes two calculator models, one designed specifically for financial use and the other designed for scientific use. The financial model contains 10 microcircuits, and the scientific model contains 20. A contract with a supplier of semiconductor chips requires the use of at least 3200 chips each day. A contract with a supplier of the off–on

switches used in both calculators requires the use of at least 300 switches each day. The company would like to produce at least 100 financial models each day. If each financial calculator requires 10 production steps and each scientific calculator requires 12 production steps, how many calculators of each type should be produced to minimize the number of production steps?

4. **Drug Administration** After a particular operation, a patient receives three types of painkillers. Type A rates a 2 in effectiveness, type B rates a 2.5, and type C rates a 3. For medical reasons, the patient can receive no more than three doses of type A, no more than five doses of types A and C combined, and no more than seven doses of types B and C combined during a 24-hour period. How many doses of each type can be safely given to the patient in a 24-hour period with maximum painkilling benefits?

In the remaining exercises, write the mathematical formulation of the linear programming problem, and identify the objective function and all constraints.

5. **Nutrition** A particular salad contains 4 units of vitamin A, 5 units of vitamin B complex, and 2 mg of fat per serving. A nutritious soup contains 6 units of vitamin A, 2 units of vitamin B complex, and 3 mg of fat per serving. If a lunch consisting of these two foods is to have at least 10 units of vitamin A and at least 10 units of vitamin B complex, how many servings of each should be used to minimize the total number of milligrams of fat?

6. **Minimizing Manufacturing Costs** A tire company produces a bias-ply, a radial, and an off-road tire. Each bias-ply tire takes 2 minutes and costs $20 to produce; each radial tire, 6 minutes and $30; and each off-road tire, 10 minutes and $50. Orders the company receives dictate that it must make a total of at least 5000 tires each day, and furthermore, at least 1500 of the total must be radial tires and at least 60 of the total must be off-road tires. The union contract with the company stipulates that at least 400 worker-hours must be used each day on the production lines. How many of each type of tire should be made to meet these conditions and keep costs to a minimum?

7. **Radio Programming** A radio station is planning to air a new talk show, new entertainment show, new stock market show, and new national news show. Market research shows the expected number of listeners to be, respectively, 10,000, 2500, 800, and 600 each time the shows air. The production costs for each time the shows air are, respectively, $1,000, $800, $400, and $300. The weekly budget for production costs is not to exceed $20,000. Advertisers would like to run the talk show at least 5 times each week and the entertainment show at least 3 times each week. The station wants to limit the number of times each show is run

during the week to 8, 5, 3, and 10 times, respectively. How many times each week should the station air each of the programs to maximize the total number of listeners?

8. **Production and Sales** A firm makes a standard and a deluxe model of electric skillet. Each standard model costs $25 and each deluxe model $40 to make. The firm must make at least 100 of the standard model and at least 75 of the deluxe. The firm sells each standard model for $30 and each deluxe model for $75, and it must have a total revenue of at least $2250 from sales of the two models. How many of each model should the firm make and sell to minimize costs?

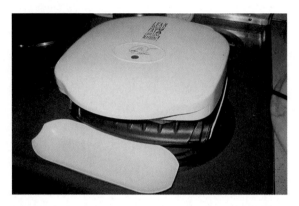

9. **Industrial Mixtures** A road-paving firm has on hand three types of paving material. Each barrel of type A contains 2 gallons of carbon black and 2 gallons of thinning agent and costs $5. Each barrel of type B contains 3 gallons of carbon black and 1 gallon of thinning agent and costs $3. Each barrel of type C contains 3 gallons of carbon black and 1 gallon of thinning agent and costs $4. The firm needs to fill an order for which the final mixture must contain at least 12 gallons of carbon black and at least 6 gallons of thinning agent. How many barrels of each type of paving material should be used to fill this order at minimum expense?

10. **Crop Yield** Members of the Ministry of Agriculture in a particular country are planning next year's rice, wheat, corn, and barley crops. They estimate that each hectare of rice planted will yield 20 cubic meters; each hectare of wheat, 18 cubic meters; each hectare of corn, 15 cubic meters; and each hectare of barley, 22 cubic meters. They would like to raise a total of at least 20 million cubic meters of these four grains. In addition, they would like to raise at least 4 million cubic meters of rice, 5 million of wheat, 6 million of corn, and 2 million of barley. Soil conditions dictate that at least three times as many hectares of corn be planted as rice. The Ministry estimates that it has enough machinery to plant at least 1 million hectares of these four grains and that, for planting, harvesting, and

transportation, each hectare of the grains will require, respectively, six, three, four, and three workers. How many hectares of each type of grain should be planted to meet the requirements of the Ministry of Agriculture and to use the least number of workers?

11. **Investments** An investment club has at most $30,000 to invest in either junk bonds or premium-quality bonds. Each type of bond is bought in $1000 denominations. The junk bonds have an average yield of 12%, and the premium-quality bonds yield 7%. The policy of the club is to invest at least twice the amount of money in premium-quality bonds as in junk bonds. How should the club invest its money in these bonds to receive the maximum return on its investment?

12. **Pollution Control** The Environmental Protection Agency (EPA) has issued new regulations that require the New Molecule Chemical Co. to incorporate a new process to reduce the pollution caused when a particular chemical is produced. The old process releases 2 grams of sulphur and 5 grams of lead for each liter of the chemical produced. The new process releases 1 gram of sulphur and 2.6 grams of lead for each liter of the chemical produced. The company makes a profit of 20 cents for each liter produced under the old method and 16 cents for each liter produced under the new process. The new EPA regulations do not allow for more than 10,000 grams of sulphur and 8000 grams of lead each day. How many liters of the chemical can be produced under the old method and how many under the new method to maximize profits and stay within EPA regulations?

13. **Farming Profits** A farmer has 500 acres available on which to plant wheat, soybeans, or both. The planting costs per acre are $200 for wheat and $350 for soybeans, and the farmer has a total of $25,000 to spend on planting. The anticipated income per acre from wheat is $350 and from soybeans $525. How many acres should be planted in wheat and how many in soybeans to maximize profit?

14. **Order Fulfillment** The Movie Mania Store is about to place an order for VHS tapes and DVD's. The distributor the store orders from requires that an order contain at least 250 items. The prices that Movie Mania must pay are $12 for each tape and $16 for each DVD. The distributor also requires that at least 30% of any order be DVD's. How many tapes and DVD's should Movie Mania order so that its total ordering costs will be kept to a minimum?

15. **Airline Transportation** A charter plane service contracts with a retirement group to provide flights from New York to Florida for at least 500 first-class, 1000 tourist-class, and 1500 economy-class passengers. The company has two types of planes. Type A costs $12,000 per flight and carries 50 first-class, 60 tourist-class, and 100 economy-class passengers. Type B costs $10,000 per flight and carries 40 first-class, 30 tourist-class, and 80 economy-class passengers. How many of each type of plane should be used to minimize flight costs?

16. **Minimizing Production Costs** A firm makes three models of microwave oven: a standard model, a deluxe model, and a superdeluxe model. The firm has contracts that call for making at least 80 standard, at least 50 deluxe, and at least 90 superdeluxe ovens each week, and the total of all these models must be at least 300. The cost is $100 for making a standard model, $110 for a deluxe model, and $120 for a superdeluxe model. How many ovens of each model should be made each week to minimize costs?

17. **Retail Sales** An appliance store sells two brands of television sets: Nitebrite and Solar II. Each Nitebrite sells for $420, and each Solar II sells for $550. The store's warehouse capacity for television sets is 300, and new sets are delivered only once a month. Past records show that customers will buy at least 80 Nitebrite sets and at least 120 Solar II sets each month. How many of each type of set should the store buy each month to maximize revenue?

18. **Political Campaign** The campaign manager of a candidate for a local political office estimates that each 30-minute speech to a civic group will generate 20 votes, each hour spent on the telephone will generate 10, and each hour spent campaigning in shopping areas will generate 40. The candidate wants to make at least twice as many trips to shopping areas as speeches to civic groups and wants to spend at least 5 hours on the telephone. The campaign manager thinks her candidate can win if he can generate a total of at least 1000 votes by the three methods. How can the candidate meet these goals and keep his campaigning time to a minimum?

19. **Nutrition** Gary is going on a low-carbohydrate diet that calls for him to eat two foods at each meal for a week. One unit of Food I contains 5 grams of carbohydrate, 80 calories, and 2 grams of fat. One unit of Food II

contains 7 grams of carbohydrate, 110 calories, and 5 grams of fat. Gary wants each meal to provide at least 400 calories, but no more than 105 grams of carbohydrate. In addition, at least three times as many units of Food I as Food II should be used in each meal for taste appeal. How many units of each food should Gary eat per meal to minimize his intake of fat?

20. Refining A refinery is processing gasoline and fuel oil. The refining process demands that at least 3 gallons of gasoline to each gallon of fuel oil be refined. To meet the demand, at least 4.2 million gallons of fuel oil per day must be refined. The demand for gasoline is no more than 7.6 million gallons per day. If the refinery sells gasoline for 87 cents per gallon and fuel oil for $1.20 per gallon, how many gallons of each of these products should be refined each day to maximize revenues?

Section 4.2 Linear Inequalities in Two Variables

The inequality constraints of a linear programming problem in two variables form a **system of linear inequalities** whose solutions may be displayed as a shaded region in the plane. Learning how to find such a shaded region and, more importantly, its corner points (if any) is the goal of this section. Once that is done, we will have taken the first step toward a graphic solution to a linear programming problem.

A linear inequality in two variables has one of the forms

$$ax + by < c,$$
$$ax + by \leq c,$$
$$ax + by > c,$$
$$ax + by \geq c,$$

where a, b, and c are real numbers and a and b are not both zero. For instance, $2x + y \leq 8$, $x - 3y > 7$, and $x \geq 0$ are all examples of linear inequalities. Should we need them, the usual laws governing inequalities prevail.

Some Laws of Inequalities

1. If $a < b$ and c is any real number, then $a + c < b + c$.
2. If $a < b$ and c is any *positive* number, then $ac < bc$.
3. If $a < b$ and c is any *negative* number, then $ac > bc$.

EXAMPLE 1
Illustrating the Laws of Inequalities

Solve the inequality $x - 2y \leq 6$ for y.

SOLUTION The following steps will isolate y on the left side of the inequality:

$$
\begin{aligned}
x - 2y &\leq 6 \\
-x + (x - 2y) &\leq -x + 6 \qquad &\text{Law (1)} \\
-2y &\leq -x + 6 \qquad &\text{Algebra of real numbers}
\end{aligned}
$$

$$\left(-\frac{1}{2}\right)(-2y) \ge \left(-\frac{1}{2}\right)(-x + 6) \quad \text{Law (3)}$$

$$y \ge \frac{1}{2}x - 3. \qquad \text{Algebra of real numbers} \quad \blacksquare$$

Solutions of a Linear Inequality

The solution of a linear inequality in two variables consists of all ordered pairs (x, y) that, when substituted into the inequality, result in a true statement.

For the inequality $x - 2y \le 6$ of Example 1, the ordered pair $(5, 5)$ is a solution, because the substitution of $x = 5$ and $y = 5$ into the inequality results in

$$5 - 2(5) \le 6,$$

or

$$-5 \le 6, \text{ a true statement.}$$

The ordered pair $(10, 1)$ is not a solution, because, upon substitution, we have

$$10 - 2(1) \le 6$$

or

$$8 \le 6, \text{ a false statement.}$$

How do we identify all such solutions? Actually, an infinite number of solutions usually exist, and they are best understood when displayed graphically in the coordinate plane. To see where such solutions lie, first replace $\le$ with $=$ to obtain the equation $x - 2y = 6$. Now graph the line $x - 2y = 6$ (or its equivalent, $y = \frac{1}{2}x - 3$), as shown in Figure 1(a), by finding the x- and y-intercepts: When $y = 0$, $x = 6$, so that 6 is the x-intercept, and when $x = 0$, $y = -3$, which shows that -3 is the y-intercept. The line divides the plane into two **half-planes**—in this case, an "upper half-plane" and a "lower half-plane"—and is the **boundary line** between them. Now, if a point (x_1, y_1) in the plane is substituted into the equation $y = \frac{1}{2}x - 3$, and the left side is compared with the right side, three possibilities exist:

a. $\quad y_1 = \frac{1}{2}x_1 - 3;\quad$ the point is on the line,

b. $\quad y_1 < \frac{1}{2}x_1 - 3;\quad$ the point is below the line,

c. $\quad y_1 > \frac{1}{2}x_1 - 3;\quad$ the point is above the line.

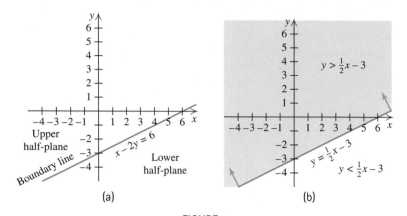

FIGURE 1

The Shaded Region, along with the Boundary Line, Represents the Solution
to $y \geq \frac{1}{2}x - 3$.

Since the inequality $x - 2y \leq 6$ can be rewritten as $y \geq \frac{1}{2}x - 3$, it follows that all points making (a) or (c) true represent the part of the plane satisfying the inequality. Geometrically, this region consists of all points in the plane that lie on or above the line $x - 2y = 6$, as shown in Figure 1(b).

The preceding analysis suggests that when a linear inequality has a solution, its graph will always contain *one* of the *half-planes* determined by the boundary line, but *not* the other. This suggestion may be proved mathematically. Some typical solutions of inequalities are shown in Figure 2. A solid line [as in (a) and (b)] indicates that the boundary line is part of the solution, while a dashed line [as in (c) and (d)] indicates that it is not. Note that for vertical lines, we have left and right half-planes. The fact that one of the half-planes is always part of the solution means that, besides deciding whether to include the boundary line, we need to determine which half-plane should be shaded for the graphic solution. The easiest way to do this is to substitute the coordinates of any point not on the boundary line into the original inequality and then to decide whether that point is a solution of the inequality.

The Graphic Solution of a Linear Inequality

1. Replace the inequality symbol with = to obtain the equation of the boundary line.
2. Plot the line represented by the boundary equation.
3. Select a test point *not* on the boundary line, and substitute it into the original inequality. If the resulting inequality is true, shade in the half-plane in which the test point lies. If the resulting inequality is false, shade in the half-plane opposite where the test point lies.
4. If the inequality is of the form $\leq$ or $\geq$, include the line in the set of solutions. If the inequality is of the form $<$ or $>$, use a dashed line.

NOTE   The quickest and easiest point to use as a test point is the origin $(0, 0)$, provided, of course, that it is not on the boundary line that divides the plane into two half-planes. If $(0, 0)$ is on the line, the next best test point to use is one of the form $(a, 0)$ or $(0, b)$ on one of the axes.

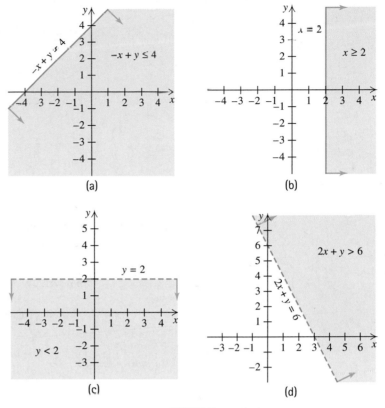

FIGURE 2
Typical Graphic Solutions of Inequalities

EXAMPLE 2

Graphic Solution for a
Boundary Line
Containing the Origin

Shade in the graphic solution of the inequality $3x - y > 0$.

SOLUTION Replace the inequality with an equality to get the boundary line $3x - y = 0$. If we let $x = 0$, we find that $y = 0$ also. Therefore, the boundary line passes through the origin, $(0, 0)$. One more point is needed to graph the line. Suppose we let $x = 1$. Then $3(1) - y = 0$, or $y = 3$. Therefore, $(1, 3)$ is another point on the line. (See Figure 3.) Due to the strict inequality $>$, the line is not part of the solution and is dashed.

Since $(0, 0)$ is on the boundary line $3x - y > 0$, it may *not* be used as a test point. Suppose the point $(4, 0)$ is arbitrarily selected as a test point. The test of this point is as follows:

Test Point	*Substituted into the Original Inequality*
$(4, 0)$	$3(4) - 0 > 0$ or $12 > 0$, which is true.

As a consequence of this test, the half-plane determined by the line $3x - y = 0$ and containing $(4, 0)$ should be shaded. (See Figure 3.) ▬

EXAMPLE 3

Graphic Solution for a
Vertical Boundary Line

Shade in the graphic solution of the linear inequality $3x + 5 \leq 1$.

SOLUTION Replacing the inequality with an equality and simplifying, we obtain $x = -4/3$. This is the equation of the vertical line intersecting the x-axis at $(-4/3, 0)$. We can use the

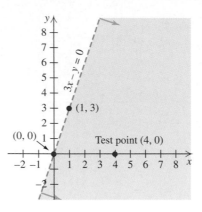

FIGURE 3
A Test Point That Is Not the Origin

origin as a test point, since it is not on the graph of the line. When we do we obtain the following result:

Test Point	*Substituted into the Original Inequality*
(0,0)	$3(0) + 5 \leq 1$ or $5 \leq 1$, which is false.

The test point is to the right of the boundary line and does not solve the inequality; thus, we shade the half-plane to the left of the boundary line (Figure 4). Since the linear inequality includes the equality, we include the boundary line in the graphical solution.

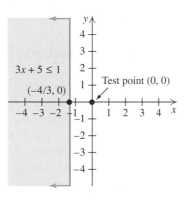

FIGURE 4
*Graphical Solution of Linear
Inequality with a Vertical Boundary* ■

Knowing how to construct the graphic solution of a single inequality leads to the construction of the graphic solution of a system of two or more linear inequalities. In such systems, a solution will consist of points, each of which makes *all* of the inequalities in the system simultaneously true.

EXAMPLE 4

Finding Solutions of a
System of Linear
Inequalities

Find the graphic solution of this system of inequalities:

$$x + 3y \geq 6$$
$$x - y \leq 2$$
$$x \quad\quad \geq 0.$$

SOLUTION First, replace each inequality with an equality to give the boundary equations:

Inequality	Boundary Line
$x + 3y \geq 6$	$x + 3y = 6$
$x - y \leq 2$	$x - y = 2$
$x \geq 0$	$x = 0$ (the y-axis)

Now consider the inequalities, one at a time. Graph $x + 3y = 6$ by using the x-intercept, 6, and the y-intercept, 2. (See Figure 5.) Then substitute $(0, 0)$ as a test point in the inequality $x + 3y \geq 6$, because $(0, 0)$ is not a point on the line $x + 3y = 6$. The result is as follows:

Test Point	Substituted into Original Inequality
$(0, 0)$	$0 + 3(0) \geq 6$ or $0 \geq 6$, which is false.

Therefore, the half-plane *not* containing the origin is to be shaded, as indicated by the arrows attached to the line $x + 3y = 6$ in the figure. Now the second line, $x - y = 2$, is plotted. The x-intercept is 2 and the y-intercept is -2. Again, the test point $(0, 0)$ may be used:

Test Point	Substituted into Original Inequality
$(0, 0)$	$0 - 0 \leq 2$ or $0 \leq 2$, which is true.

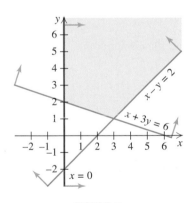

FIGURE 5
*Graphic Solution of a System
of Linear Inequalities*

So the half-plane that *does* contain the origin is to be shaded, as indicated by the arrows attached to the line $x - y = 2$ in Figure 5.

Finally, the inequality $x \geq 0$ has as solutions all points in the plane that have a non-negative x-coordinate (the points on or to the right of the y-axis), as shown in Figure 5. The solutions of the given system are those points in the plane that *simultaneously make all three* inequalities *true,* as indicated by the shaded region in the figure. This collection of points is known as the **feasible region** for the given system. ■

The Feasible Region

For any system of linear inequalities, the set of points that makes all inequalities in the system true simultaneously is called the **feasible region (set of feasible solutions)** for that system. In the plane, the **graph of the feasible region** is found by shading the set of points in the rectangular coordinate system that represents the feasible region.

We suggest that, in graphing the feasible region for a system of two or more inequalities in the plane, you use the less confusing indicator arrows like those shown in Figure 5 to help locate the final graph, rather than shading each individual half-plane involved. Because of the conceptual difficulties involved in drawing three-dimensional figures on paper, no graphs of feasible regions for systems with three or more variables are attempted here.

The feasible region shown in Figure 5 has two **corner points:** (0, 2) and (3, 1). These points belong to the feasible region and are the intersection of two lines that bound the region.

A Corner Point of a Feasible Region

For a feasible region in the plane defined by a system of linear inequalities, a point of intersection of two boundary lines that is also part of the feasible region is called a **corner point (vertex) of the feasible region.**

A corner point is always the intersection of two boundary lines, but not all such intersections are corner points, as is demonstrated in Example 5.

EXAMPLE 5

Finding the Corner
Points of a Feasible
Region

Find the corner points of the feasible region described by

$$x + y \leq 5$$
$$x + 2y \leq 8$$
$$x \geq 0, y \geq 0.$$

SOLUTION The boundary lines for these inequalities are, respectively, $x + y = 5$, $x + 2y = 8$, $x = 0$, and $y = 0$. You should be able to calculate the x- and y-intercepts for each of these lines. (Compare your calculation with Figure 6.) Inserting the test point $(0, 0)$ into each of the first two inequalities shows that we should shade on the origin side of both boundary lines. The nonnegativity constraints *always* tell us to shade in the first quadrant. The feasible region thus obtained is shown in the figure, with the intersection of each pair of boundary lines denoted by one of the letters A, B, C, D, E, or F. According to the previous definition, the point labeled B is a corner point of the feasible region, because it is part of the feasible region and because it is the intersection of the lines $x = 0$ (the y-axis) and $x + 2y = 8$. The substitution of $x = 0$ into $x + 2y = 8$ gives $y = 4$, so the coordinates of B are $(0, 4)$. The point labeled C is also a corner point of the feasible region, because it is the intersection of the lines $x = 0$ and $y = 0$. The coordinates of C are $(0, 0)$. The point labeled D is another corner point of the feasible region, represented by the intersection of $y = 0$ and $x + y = 5$. Substituting $y = 0$ into $x + y = 5$ gives $x = 5$. The

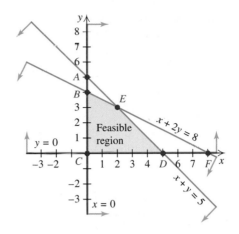

FIGURE 6
*B, C, D, and E Are Corner Points
of the Feasible Region.*

coordinates of D are therefore $(5, 0)$. The corner point E is the intersection of the lines $x + 2y = 8$ and $x + y = 5$. The elimination method introduced in Chapter 1 may be used to obtain the coordinates of E:

$$x + 2y = 8 \quad \textbf{(1)}$$
$$x + y = 5 \quad \textbf{(2)}$$

$-1 \times$ Equation (2)

$$\begin{aligned} x + 2y &= 8 \\ -x - y &= -5 \\ \hline y &= 3. \end{aligned}$$

Substituting $y = 3$ into Equation (2) results in $x + 3 = 5$, so $x = 2$. The coordinates of E are then $(2, 3)$. The points labeled A and F are the intersection of two boundary lines, but are not part of the feasible region; therefore, they are not corner points of the region. ■

NOTE Do not lose track of what the graph of the feasible region represents. For example, in Figure 5, the shaded four-sided region, along with the line segments $\overline{BC}$, $\overline{CD}$, $\overline{DE}$, and $\overline{EB}$, is a "picture" of all the points that simultaneously solve each of the four stated inequalities. You probably can even guess why that set of points is called the feasible region: Those points are the ones that are feasible (usable) if the system of inequalities is the set of constraints for a linear program.

EXAMPLE 6

Finding the Corner
Points of a Feasible
Region

Find the feasible region and identify the corner points for the system.

$$-2x + y \leq 1$$
$$x + y \leq 4$$
$$y \geq 2.$$

SOLUTION The boundary lines are $-2x + y = 1$, $x + y = 4$, and $y = 2$. Compare your calculations of the x- and y-intercepts of each line with the intercepts shown in Figure 7. The test point $(0, 0)$ applied to the first two inequalities shows that we should shade on the origin side of both corresponding boundary lines. The third inequality tells us to shade above the line $y = 2$. The feasible region is shown in the figure.

The corner point A is the intersection of the boundary lines $-2x + y = 1$ and $x + y = 4$. Solving the first equation for y ($y = 2x + 1$) and substituting the value of y into

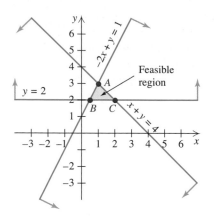

FIGURE 7
*A, B, and C Are Corner Points
of the Feasible Region.*

$x + y = 4$ gives $x + (2x + 1) = 4$, or $x = 1$. When $x = 1$ in $x + y = 4$, $y = 3$, which shows that the corner point A is $(1, 3)$. Corner point B is the intersection of the boundary lines $-2x + y = 1$ and $y = 2$. Substituting $y = 2$ into $-2x + y = 1$ gives $-2x + 2 = 1$, or $x = \frac{1}{2}$. The coordinates of B are $\left(\frac{1}{2}, 2\right)$. Corner point C is the intersection of $x + y = 4$ and $y = 2$. Substituting $y = 2$ into $x + y = 4$ gives $x + 2 = 4$, or $x = 2$. The coordinates of C are $(2, 2)$. ■

Given any linear programming problem in two variables, we are now in a position to graph the feasible region as defined by the constraints of the problem and to find the corner points of that region. As the next section shows, these are the first steps in the graphical solution of a linear programming problem.

Exercises 4.2

In Exercises 1 through 4, determine whether each point is or is not in the feasible region of the given system of inequalities.

1. $2x - 3y \leq 4$

   (a) $(0, 0)$

   (b) $\left(3, \dfrac{2}{3}\right)$

   (c) $(4, 0)$

2. $2x - 3y > 4$

   (a) $(0, 0)$

   (b) $\left(3, \dfrac{2}{3}\right)$

   (c) $(4, 0)$

3. $x + y \leq 2$
     $6x + 5y \leq 11$
         $x > 0$
         $y > 0$

   (a) $(0, 0)$

   (b) $(1, 1)$

   (c) $(1.9, 0)$

4. $2x - y > 2$
         $x > 0$
         $y > 0$

   (a) $(0, 0)$

   (b) $(1, 0)$

   (c) $(2, 1)$

In each of Exercises 5 through 16,
(a) Write the equation of the boundary line.
(b) Select an appropriate test point and use it to identify the half-plane in the feasible region.
(c) Shade in the feasible region.

5. $3x + y \leq 9$ **6.** $x + 2y \geq 6$

7. $2x + 3y < 6$ **8.** $2x + y \geq 10$

9. $-2x + y \geq 8$ **10.** $-x \geq 3$

11. $2x > 6$ **12.** $3y \leq 6$

13. $3x - 2y \leq 12$ **14.** $y < x$

15. $x - y \leq 0$ **16.** $x \geq 0$

In Exercises 17 through 24,
(a) Find the boundary line for each inequality.
(b) Identify the various half-planes represented by attaching arrows to the separating boundary lines; then shade in the feasible region.

(c) *For each corner point of the feasible region, identify the two lines whose intersection is that point, and then solve each system of equations to find the coordinates of the corner point.*

17. $x + 2y \le 6$
$x \ge 0$
$y \ge 0$

18. $x \ge 3$
$y \ge 4$
$x + y \le 10$

19. $x \ge y$
$x \le 2y$
$x \le 3$

20. $x - y \ge 0$
$x \le 3$
$y \ge 1$

21. $x + 3y \le 9$
$3x + y \le 11$
$x \ge 0$
$y \ge 0$

22. $2x + 5y \le 20$
$x + 2y \le 9$
$x > 0$
$y \ge 0$

23. $x + y \le 8$
$2x + y \le 10$
$4x + y \le 16$
$x \ge 0$
$y \ge 0$

24. $x + y \le 11$
$x + 2y \le 18$
$x + 5y \le 20$
$x \ge 0$
$y \ge 0$

In Exercises 25 through 34, graph the feasible region, find the coordinates of the corner ponts, and determine whether the points (0, 0) and (1, 3) are in the feasible region.

25. $2x - y \le 4$
$2x + 3y \le 12$
$x \ge 0, y \ge 0$

26. $2x + y \le 8$
$x + y \le 6$
$x \ge 0, y \ge 0$

27. $x + 3y \le 9$
$x + y \le 7$
$x \ge 0, y \ge 0$

28. $x + 5y \le 25$
$4x - 2y \le 4$
$x \ge 0, y \ge 0$

29. $x \ge 1$
$x \le 4$
$y \ge 0$
$y \le 3$

30. $-2x + y \le 0$
$x + y \le 4$
$y \ge 0$

31. $-2x + y \ge 0$
$x \ge 0$
$y \le 4$

32. $-3x + y \ge 3$
$x \ge 0$
$y \le 5$

33. $x - y \ge 4$
$x \le 6$
$y \ge 0$

34. $x - y \le 6$
$x \ge 6$
$y \ge 0$

In Exercises 35 through 46,
(a) *Give the mathematical formulation of the linear programming problem.*

(b) *Graph the feasible region described by the constraints, and find the corner points. Do not attempt to solve the linear program.*

35. Sales A sales representative covers territory in Iowa and Kansas. Her daily travel expenses average $120 in Iowa and $100 in Kansas. Her company provides an annual travel allowance of $18,000. Her company also stipulates that she must spend at least 50 days in Iowa and 60 days in Kansas per year. If sales average $3000 per day in Iowa and $2500 per day in Kansas, how many days should she spend in each state to maximize sales?

36. Tutoring A college student, Amanda Sherman, tutors her peers in finite math and calculus, limiting herself to no more than 12 hours per week tutoring these two courses. She also limits herself to no more than 8 hours per week tutoring in finite math. If she charges $10 per hour to tutor in finite math and $12 per hour to tutor in calculus, how many hours should she spend tutoring each subject to maximize her income?

37. Nutrition A dietitian is blending two foods for a special diet. Food I has 30 units of vitamin A and 90 units of vitamin B complex per kilogram. Food II has 40 units of vitamin A and 50 units of vitamin B complex per kilogram. The mixture is to have at least 1200 units of vitamin A and at least 2970 units of vitamin B complex. If Food I costs 80 cents per kg and Food II costs $1.10 per kg, how many kilograms of each should be used to minimize costs?

38. Purchasing Costs A concrete company needs at least 4 tons of #2-grade crushed rock and at least 12 tons of #6-grade crushed rock to mix an upcoming order for concrete. The Hardrock Company ships carloads containing 1 ton of #2 rock and 6 tons of #6 rock. The Morstone Company ships carloads containing 2 tons of #2 rock and 4 tons of #6 rock. If each carload from the Hardrock Co. costs $2000 and each carload from the Morstone Co. costs $2300, how many carloads should be ordered from each to minimize costs?

39. Shipping A particular part is shipped each week to assembly plants in Sacramento and Hobbs. The plant in Sacramento needs at least 1000 of these parts each week, and the plant in Hobbs needs at least 1200 each week. It costs $2 to ship each part to Sacramento and $3 to ship each part to Hobbs. The budget dictates that costs for shipping the parts are not to exceed $14,000 each week. The work hours required to ship a part to Sacramento are 0.2 hour and to ship a part to Hobbs are 0.25 hour. How many parts should be shipped to each location if the work hours are to be kept to a minimum?

40. Revenue A firm makes a standard model and a deluxe model of microwave oven. Each standard model requires 10 minutes of painting time, and each deluxe model requires 15 minutes of painting time. The firm has available a total of 6 hours each day for painting purposes. The firm must produce at least 15 standard units and at least 8 deluxe units each day. If each standard model sells for $120 and each deluxe model sells for $150, how many of each should be produced each day to maximize revenue?

41. Revenue The WriteMor Pen Company makes two types of ballpoint pens. A silver pen requires 1 minute in a grinder and 3 minutes in a bonder. A gold pen requires 3 minutes in a grinder and 4 minutes in a bonder. Because of cleaning, adjustments, and maintenance, the grinder can be operated no more than 30 hours per week, and the bonder can be operated no more than 50 hours per week. If each silver pen sells for $30 and each gold pen sells for $80, how many of each should be made each week to maximize revenue?

42. Revenue A gift company employs two workers to make gavels and plaques. To make a gavel, Ms. Jones performs some tasks that take $\frac{1}{2}$ hour; Mr. Smith performs the remaining tasks, which also take $\frac{1}{2}$ hour. To make a plaque, Ms. Jones's part takes $\frac{1}{2}$ hour, while Mr. Smith's part takes 1 hour. Mr. Smith can work no more than 25 hours per week, whereas Ms. Jones can work no more than 40 hours per week. If each gavel sells for $12 and each plaque for $20, how many of each should be made each week to maximize revenue?

43. Production A company makes two calculators: a business model and a scientific model. The business model contains 10 microcircuits, and the scientific model contains 20 microcircuits. The company has a contract that requires it to use at least 320 microcircuits each day, and the manufacturing capacity for business calculators is 50 per day. The company also wants to make at least twice as many business calculators as scientific calculators. If each business model is sold for $50 and each scientific model for $60, how many of each should be made and sold each day to maximize revenue?

44. Purchasing The Psychology Department at State University buys mice and rats for experimental purposes. It buys at least three times as many mice as rats. Each mouse costs $2, each rat costs $5, and the departmental budget dictates that no more than $500 be spent on such purchases. If each mouse can be used for three experiments and each rat for two, how many of each should be purchased to maximize the total number of experiments that can be run?

45. Psychology Survey A fashion magazine has commissioned the Psychology Department at State University to do a survey on the lifestyles of men who have graduated from college within the past 5 years and of men under age 40 years who have never attended college. The interviewing is done by professors and graduate students. A professor spends 30 minutes interviewing a college graduate and 35 minutes interviewing a person who never attended college. A graduate student spends 20 minutes interviewing a college graduate and 45 minutes interviewing a person who never attended college. During a semester, professors have at most 80 hours for interviewing, and graduate students have at most 100 hours for interviewing. How many college graduates and how many men who never attended college should be interviewed during a semester to maximize the total number interviewed?

46. Event Planning During a weeklong state fair, officials want to provide entertainment that will draw crowds to the fair. They find that a top star demands $10,000 for a performance and should draw a crowd of 8000 people. A slightly faded star demands $6000 for a performance and should draw a crowd of 3000 people. Fair officials have no more than $40,000 to spend on such entertainers and would like to draw a total of at least 25,000 people for their performances. Assume that each entertainer performs only once at the fair. If all top stars donated $500 and all slightly faded stars donate $450 of their fees back to community charities, how many top stars and how many slightly faded stars should be brought in so that donations to community charities are maximized?

Section 4.3 Solving Linear Programming Problems Graphically

The constraints listed for a linear programming problem lead naturally to a feasible region for the problem. In the previous section, we learned how to graph and find the corner points of such regions in the coordinate plane. In this section, we use these concepts as the basis for completing the steps involved in graphically solving a linear programming problem.

A typical problem, such as the following one, shows how the graphic solution process unfolds:

$$\text{Maximize } f = 5x + 2y \quad \} \quad \leftarrow \textit{Objective function}$$
$$\text{subject to } x + y \leq 6$$
$$2x - y \leq 6 \quad \} \quad \leftarrow \textit{Structural constraints}$$
$$x \geq 0, y \geq 0. \quad \} \quad \leftarrow \textit{Nonnegativity constraints}$$

The problem is to select the point or points in the feasible region determined by the constraints that will make $f = 5x + 2y$ as large as possible. Because there are an infinite number of points in the feasible region, the task of finding this maximum value of f may seem hopeless; but it is not that difficult, as we shall see in this section. We begin by graphing the feasible region determined by the constraints and finding the coordinates of the corner points. (See Figure 8.)

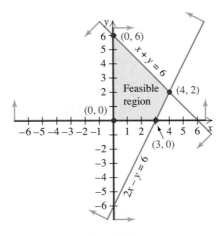

FIGURE 8
Graph and Corner Points
of a Feasible Region

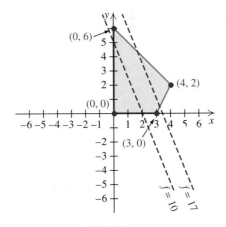

FIGURE 9
Objective Function Values
and the Feasible Region

We now experiment for just a moment. Suppose that the point (2, 0) is arbitrarily selected from the feasible region and f is evaluated at that point. Then $f = 5(2) + 2(0) = 10$. However, (2, 0) is not the only point in the feasible region that gives f a value of 10: There is an entire line, $5x + 2y = 10$, for which each point (x, y) on the line gives

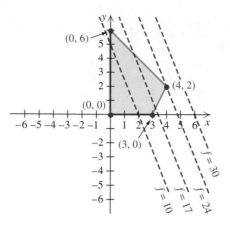

FIGURE 10
f Is Maximized at a Corner Point.

$f = 10$. (See Figure 9.) In particular, points on this line and also in the feasible set meet the constraints of the problem. What if the point $(3, 1)$ were chosen? Then $f = 5(3) + 2(1) = 17$ at that point. Again, any point (x, y) on the line $5x + 2y = 17$ gives f a value of 17. Notice the relationship between the graphs of these lines. The slope–intercept form of each line, namely, $y = -\frac{5}{2}x + \frac{10}{2}$ and $y = -\frac{5}{2}x + \frac{17}{2}$, shows that both lines have the same slope, $-\frac{5}{2}$, and that the second has the larger y-intercept. (See Figure 9.) Now, solving any line $ax + by = c$ in standard form for y, to obtain the slope–intercept form, gives $-a/b$ as the slope and c/b as the y-coordinate of the y-intercept. Therefore, all lines in standard form with left side $5x + 2y$ have the same slope, $-\frac{5}{2}$, and are parallel. Furthermore, if various values of c are assigned to f, then $f = 5x + 2y = c$; the larger c becomes, the higher the y-intercept is on the y-axis. This is shown in Figure 10, in which the larger the value of c, the farther upward and away from the origin the graph of $5x + 2y = c$ is. Now the picture should be clear as to how we find the point in the feasible region that makes f as large as possible: The *last point of the feasible region* touched by the *upward-moving parallel lines* $5x + 2y = c$ is the point that *maximizes f*. In this case, that point is $(4, 2)$, and the value of f at that point is $f = 5(4) + 2(2) = 24$. Our linear programming problem is thus solved.

This example suggests that if a linear objective function f has a maximum value on the feasible region bounded by linear constraints, then such a maximum will occur at a corner point of that region. Now, might that indeed *always* be the case—even for problems asking for the *minimum* value of f? In other words, does every linear programming problem have a solution? The mathematical groundwork for investigating these questions has already been done. One aspect of the answer concerns *bounded feasible regions,* which may be described as those regions for which a positive number a can be selected so that a circle with its center at the origin and radius a contains the entire region. The other aspect concerns the concept of a *convex region,* which means that, if any two points in the region are connected by a line segment, then that segment lies entirely within the region. (See Figure 11.)

The fundamental theorem that is stated next is the cornerstone to finding solutions to linear programming problems.

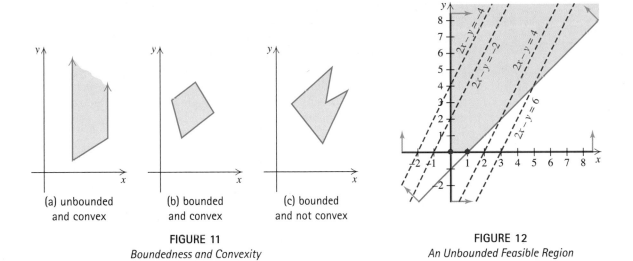

| (a) unbounded and convex | (b) bounded and convex | (c) bounded and not convex |

FIGURE 11
Boundedness and Convexity

FIGURE 12
An Unbounded Feasible Region

The Fundamental Theorem of Linear Programming

If the feasible region for a linear programming problem is nonempty and convex, and if the objective function has a maximum (or minimum) value within that set, then that maximum (or minimum) will always correspond to at least one corner point of the region.

The fundamental theorem does not guarantee a solution to any linear programming problem: it just says that *if* the objective function has a maximum or minimum value in the feasible region, then that value must occur at some corner point of the feasible region. From the discussion of Example 1 of this section, however, we can assert that if the feasible region has at least one point and is *bounded,* then *any* objective function will have both a maximum and a minimum value at a corner point. On the other hand, if the feasible region is *unbounded,* then the existence of a solution depends upon both the nature of the objective function and the shape of the feasible region. To elaborate on this last statement, consider the unbounded region in Figure 12 described by $x - y \leq 1$, $x \geq 0$, and $y \geq 0$ and the following three objective functions:

a. The function $f = 2x - y$ has no maximum or minimum value over this region. (See Figure 12 for a visual justification of this statement.) Along the boundary given by $x - y = 1$ $(y = x - 1)$, $f = 2x - y$ is reduced to $f = x + 1$ by substitution. This function has no maximum; therefore $f = 2x - y$ has no maximum over the entire region. In contrast, along the y-axis $(x = 0)$, $f = 2x - y$ becomes $f = -y$, from which we see that f has no minimum. As a consequence, f has no minimum over the entire region.

b. The function $g = x + y$ has a minimum, but no maximum value over the region. The minimum value is 0 and occurs at the point $(0, 0)$. Inspection leads to the conclusion that no maximum exists.

c. The function $h = x - y$ has a maximum, but no minimum value over the region. To see this, note that the graph of $h = x - y = 1$ is precisely that of the "lower right" boundary of the feasible region. (See Figure 12.) Thus, at any point along this boundary line, the value of h is 1, and that is the maximum value that h may attain. The reason is that if $k > 1$, the graph of $h = x - y = k$ does not intersect the feasible region at all, whereas if $k < 1$, the graph of $h = x - y = k$ always intersects the feasible region; hence, the more negative k becomes, the more negative the value of h becomes, which shows that no minimum exists.

The Existence of Solutions

For a linear programming problem with a nonempty feasible region R and an objective function f,

1. If R is bounded, then f has both a maximum and a minimum value at some corner point of R.
2. If R is unbounded and if f has a maximum or a minimum value, on R, then that value will occur at a corner point. However, the nature of f and the shape of R will determine whether a maximum or a minimum exists.

This existence statement shows us how to graphically solve a linear programming problem. The details are listed next.

How to Solve a Linear Programming Problem Graphically

If a linear programming problem has a solution, the following steps will locate it:

1. Graph the feasible region, as determined by the constraints.
2. Find the coordinates of each corner point of the feasible region.
3. Evaluate the objective function f at each corner point, and select the point (or points) that gives the maximum or minimum, as required.

EXAMPLE 1
Solving a Linear Programming Problem Graphically

Maximize $f = 3x + 4y$ } ← Objective function

subject to $2x + 3y \leq 8$ ⎱
             $x + 3y \leq 9$ ⎰ ← Structural constraints

$x \geq 0, y \geq 0.$ } ← Nonnegativity constraints

SOLUTION First, the feasible region must be graphed and the corner points found. Figure 13 shows the completed graph of the feasible region. The coordinates of all of the corner points except A may be found from the graphing process, where the intercepts were recorded: $B(0, 3)$, $C(0, 0)$, and $D(4, 0)$. Point A is the intersection of the lines whose equations are

$$2x + \ y = 8$$
$$x + 3y = 9.$$

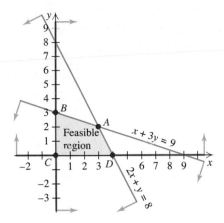

FIGURE 13
*Feasible Region for a Linear
Programming Problem*

To find A, solve the first equation for y; (i.e., $y = -2x + 8$), and then substitute the result into the second:

$$x + 3(-2x + 8) = 9$$
$$x - 6x + 24 = 9$$
$$-5x = -15$$
$$x = 3.$$

Now substituting $x = 3$ into $y = -2x + 8$ gives

$$y = -2(3) + 8 = 2.$$

The coordinates of point A are $(3, 2)$. Finally, evaluating f at each corner point of the feasible region (see Table 5) enables us to determine the maximum value of f.

Corner Point	Value of $f = 3x + 4y$
$A(3, 2)$	17—Maximum
$B(0, 3)$	12
$C(0, 0)$	0
$D(4, 0)$	12

TABLE 5.

These calculations reveal the solution to this problem: The maximum value of f is 17 and occurs when $x = 3$ and $y = 2$ (the corner point $(3, 2)$). ■

EXAMPLE 2

Solving a Linear
Programming Problem
Graphically

Maximize $f = x + 6y$ over the set of constraints given in Example 1.

SOLUTION We already know the corner points, so the tabulation of f at those points is all that remains to be done.

Corner Point	Value of $f = x + 6y$
$A(3, 2)$	15
$B(0, 3)$	18—Maximum
$C(0, 0)$	0
$D(4, 0)$	4

TABLE 6.

Table 6 shows that the maximum value of f is now 18 and occurs when $x = 0$ and $y = 3$. ■

EXAMPLE 3

Solving a Linear Programming Problem Graphically

Maximize $f = 2x + y$ over the same set of constraints as in Example 1.

SOLUTION Again, only the tabulation is necessary to find the maximum.

Corner Point	Value of $f = 2x + y$
$A(3, 2)$	8—Maximum
$B(0, 3)$	3
$C(0, 0)$	0
$D(4, 0)$	8—Maximum

TABLE 7.

From Table 7, this time the solution is 8, and it occurs at two corner points: (3, 2) and (4, 0). In fact, this solution occurs at *any* point on the line segment that joins these two points. The reason is that when the objective function $f = 2x + y$ is set equal to any constant c, the slope of $2x + y = c$ is the same as the slope of the boundary line $2x + y = 8$. This means that the upward-moving parallel lines $2x + y = c$ last touch the feasible region along the entire segment of $2x + y = 8$ between (3, 2) and (4, 0). The fact remains, nonetheless, that the solution still occurs at a corner point of the feasible region. ■

EXAMPLE 4

Using Technology to Generate Tables

Find the maximum and minimum values of the objective function in the linear program.

$$\text{Objective } f = 2x + 5y$$
$$\text{subject to} \quad 2x + 3y \leq 12$$
$$-2x + y \leq 0$$
$$y \geq 2$$
$$x \geq 0.$$

SOLUTION Figure 14 indicates that all three possible intersections of the constraint lines are corner points of the feasible region. Corner A is the intersection of $-2x + y = 0$ and $y = 2$; a quick substitution yields the coordinates (1, 2) for this point. The same technique works for the intersection of $2x + 3y = 12$ and $y = 2$, giving the coordinates (3, 2) for corner B. For corner C, we must do a little more work to find that the intersection of $-2x + y = 0$ and $2x + 3y = 12$ is $\left(\frac{3}{2}, 3\right)$.

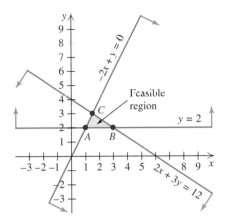

FIGURE 14
Feasible Region Used to Find
Maximum and Minimum Values

Now we will set up a spreadsheet to compute the values of the objective function at each of these points. (Although a spreadsheet is an elaborate tool to use in this situation, the practice of setting up a table will be quite useful later, when we face more difficult problems and require the power of spreadsheets.) Figure 15 illustrates the setup of the table. In Figure 15(a), the column labels and corner points have been entered. In Figures 15(b) and (c), we have typed the objective function into the first cell of the objective function column, being careful to reference the appropriate cells in row 2 for each variable. Finally, in Figure 15(d), the formula in the first cell is copied down to complete the table. As the table demonstrates, the objective function reaches a minimum of 12 at corner $A(1, 2)$ and a maximum of 18 at corner $C(1.5, 3)$.

	A	B	C	D
1	X Values	Y Values	Obj. Func.	
2	1	2		
3	3	2		
4	1.5	3		
5				

(a)

	A	B	C	D
1	X Values	Y Values	Obj. Func.	
2	1	2	=2*A2+5*B2	
3	3	2		
4	1.5	3		
5				

(b)

	A	B	C	D
1	X Values	Y Values	Obj. Func.	
2	1	2	12	
3	3	2		
4	1.5	3		
5				

(c)

	A	B	C	D
1	X Values	Y Values	Obj. Func.	
2	1	2	12	
3	3	2	16	
4	1.5	3	18	
5				

(d)

FIGURE 15
Creating a Spreadsheet for Finding Values of the Objective Function

We now turn our attention to the problem posed at the beginning of this chapter.

EXAMPLE 5

Nutrition Problem
Solved Graphically

A dietitian is blending two foods for a mixture that has a high content of vitamins C and D. Food I has 10 mg of vitamin C per kg of food and 45 International Units (IU) of vitamin D per kg. Food II has 50 mg of vitamin C and 15 IU of vitamin D per kg. The mixture is to have at least 500 mg of vitamin C and 600 IU of vitamin D. If Food I costs 30 cents per kg and Food II costs 20 cents per kg, how many kg of each food should be used to meet these requirements while keeping costs to a minimum?

SOLUTION The constraints are suggested by Table 8, along with the usual condition that the variables be nonnegative.

	Food I	Food II	Limitations (mg or IU)
Mass of Food (kg)	x	y	
Vitamin C (mg/kg)	10	50	≥ 500
Vitamin D (IU/kg)	45	15	≥ 600
Minimize Cost (cents/kg)	30	20	

TABLE 8.

The corresponding linear program is

$$\text{Minimize } Cost = 30x + 20y$$
$$\text{subject to } 10x + 50y \geq 500$$
$$45x + 15y \geq 600$$
$$x \geq 0, y \geq 0.$$

The feasible region is shown in Figure 16, with the corner points labeled A, B, and C. As in previous examples, we compute the coordinates and cost at each point and put their approximate values in a table (Table 9).

Corner Point	Cost
$A(0, 40)$	800
$B(10.71, 7.86)$	478.57 Minimum
$C(50, 0)$	1500

TABLE 9.

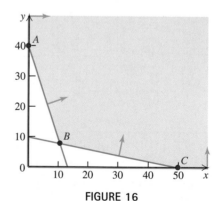

FIGURE 16
Feasible Region for Blending Problem

A minimum cost of 478.57 cents, or about $4.79, is obtained by mixing 10.71 kg of Food I with 7.86 kg of Food II. ■

NOTE The value 478.57 in Example 5 was obtained by using as many decimals as the technology allows for the coordinates of point B; we did not round off the results of any calculations until the end. In general, one should avoid rounding until reaching a final value. Also, remember that even when your technology displays only a few decimal places, it is using many more in its calculations.

You may well have spotted a pattern in all of the examples in this section. Why do we need to look at a graph of the feasible region when all the important intersection points with nonnegative coordinates seem to be fairly easy to guess and check against the objective function? However, not all intersection points of the boundary lines have to be corner points of the feasible region, as the next two examples demonstrate.

EXAMPLE 6

Nonuseful Intersection
Points

Consider again the food mixture problem given in Example 5, but with the added constraint of needing vitamin E as well. Suppose Food I has 20 mg of vitamin E per kg and Food II has 30 mg of vitamin E per kg. Furthermore, we need a total of at least 350 units of vitamin E. The latter condition adds the constraint

$$20x + 30y \geq 350$$

to the linear program given in Example 5. Before creating a new table of values of the objective function, consider the graph of the new feasible region given in Figure 17.

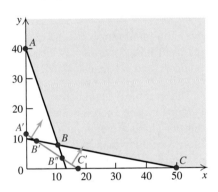

FIGURE 17
The Feasible Region for the Food Mixture Problem, with the Added Constraint of Needing 350 Units of Vitamin E

Notice that the new constraint adds four new intersection points in the first quadrant, but the feasible region (and thus the corner points) remains unchanged! We need to do no more work, as our solution from Example 5 still holds. ▬

EXAMPLE 7

Nonuseful Intersection
Points

Let us explore the food mixture problem given in Examples 5 and 6 one more time. Suppose there is still a need for vitamin E, as stated in Example 6, but now there is a need for a total of at least 600 units, not 350 units. This modifies the corresponding constraint to

$$20x + 30y \geq 600.$$

As before, it is very important to view a graph of the feasible region (Figure 18) before proceeding.

Notice that the region from Example 5 has now gained two new corner points and lost one. The one lost is precisely where our previous minimum cost was found, but it is not a point where the minimum vitamin E requirement is met. Constructing a new table with the four corner points of the feasible region and the corresponding costs (Table 10) will point to a new solution.

Corner Point	Cost
$A(0, 40)$	800
$B'(8.57, 14.29)$	542.86 Minimum
$B''(21.43, 5.71)$	757.14
$C(50, 0)$	1500

TABLE 10.

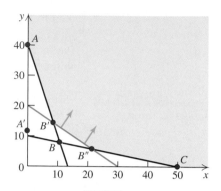

FIGURE 18
The Feasible Region for the Food Mixture Problem, with the Added Constraint of Needing 600 Units of Vitamin E

Now the minimum is 542.86 cents, or about $5.42, which is the cost of mixing 8.57 kilograms of Food I with 14.29 kilograms of Food II. (Again, note that the costs for mixtures at points B' and B'' are calculated with more decimal places than are indicated in the table.) ▬

The graphical approach to solving linear programming problems is limited to two or, perhaps for the energetic and visually perceptive, three variables. This is evidence enough that an algebraic approach is needed, because most real-world problems involve many more variables than two or three. We begin the development of such an approach in the next chapter. However, we keep the graphical theme intact to help us understand why particular techniques are needed.

Exercises 4.3

Exercises 1 and 2 apply to the feasible region shown in the following graph:

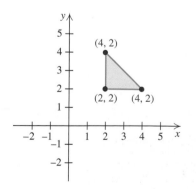

1. (a) If the objective function is $f = 3x + 2y$, graph the lines $3x + 2y = c$, where c takes on the values $-6, 6, 12, 18,$ and 24, superimposed on the feasible region.

(b) On the basis of these lines, where do the maximum and minimum values of f occur? Make a statement about the movement of the parallel lines $f = c$ versus the values of c.

(c) If the objective function is ($g = -f = -3x - 2y,$) graph the lines $-3x - 2y = c$, where c takes on the values $6, -6, -12, -18,$ and -24, superimposed on the feasible region.

(d) On the basis of these lines, where do the maximum and minimum values of $g = -f$ occur? Make a statement about the movement of the parallel lines $g = c$ versus the value of c.

2. (a) If the objective function is $f = 3x - 2y$, graph the lines $3x - 2y = c$, where c takes on the values of $-6, 6, 12, 18,$ and 24, superimposed on the feasible region.

   (b) On the basis of these lines, where do the maximum and minimum values of f occur? Make a statement about the movement of the parallel lines $f = c$ versus the values of c.

   (c) If the objective function is

   $$g = -f = -3x + 2y,$$

   graph the lines $-3x + 2y = c$, where c takes on the values $-6, 6, 12, 18,$ and 24, superimposed on the feasible region.

   (d) On the basis of these lines, where do the maximum and minimum values of $g = -f$ occur? Make a statement about the movement of the parallel lines $g = c$ versus the values of c.

Solve each of the linear programming problems in Exercises 3 through 15 by the graphic method.
(a) Graph the feasible region.
(b) Find the coordinates of the corner points.
(c) Evaluate f at each corner point, and find the extreme value requested.

3. Maximize $P = 4x + 4y$

   subject to $x + 3y \le 30$
   $$2x + y \le 20$$
   $$x \ge 0, y \ge 0.$$

4. Maximize $P = x + 5y$

   subject to $2x + y \le 12$
   $$2x + 5y \le 20$$
   $$x \ge 0, y \ge 0.$$

5. Maximize $f = x + 2y$

   subject to $2x + y \ge 10$
   $$x \le 4$$
   $$y \le 6.$$

6. Maximize $f = 6x + y$

   subject to $x + y \ge 8$
   $$-3x + y \ge 0$$
   $$y \le 6.$$

7. Minimize $C = x - 2y$

   subject to $x \ge 2, x \le 4$
   $$y \ge 1, y \le 5.$$

8. Minimize $C = 4x + 2y$

   subject to $-x + y \le 0$
   $$x + 2y \ge 6$$
   $$y \ge 0.$$

9. Minimize $C = 2x + 3y$

   subject to $9x + 4y \ge 40$
   $$x + 4y \ge 8$$
   $$x \ge 0, y \ge 0.$$

10. Minimize $C = 5x + 6y$

    subject to $4x + y \ge 6$
    $$x + 8y \ge 17$$
    $$x \ge 0, y \ge 0.$$

11. Minimize $C = 5x + 3y$

    subject to $x + y \ge 6$
    $$6x + y \ge 16$$
    $$x + 6y \ge 16$$
    $$x \ge 0, y \ge 0.$$

12. Maximize $P = 3x + 5y$

    subject to $x + 2y < 8$
    $$x + y \le 6$$
    $$4x + 2y \le 16$$
    $$x \ge 0, y \ge 0.$$

13. Maximize $P = 8x + y$

    subject to $x + y \le 6$
    $$x + 3y \le 12$$
    $$3x + y \le 12$$
    $$x > 0, y > 0.$$

14. Maximize $f = 2x + y$

    subject to $x + y \le 5$
    $$x \ge 2$$
    $$y \ge 0.$$

15. Minimize $f = 2x - 5y$

    subject to $x + y \le 6$
    $$x \ge 1$$
    $$y = 2.$$

16. Consider the feasible region defined by the constraints $x \ge 1, y > 1,$ and $y \le 3$.

    (a) Does the objective function $f = 2x + y$ attain a maximum at some point in the feasible region? If so, what is that point, and what is the maximum value of f?

1st Question #2 similar to these two — (handwritten annotation)

(b) Does the objective function $f = 2x + y$ attain a minimum at some point in the feasible region? If so, what is that point, an what is the minimum value of f?

17. Consider the feasible region defined by the constraints $x \geq 1$, $x \leq 4$, and $y \geq 2$.

(a) Does the objective function $P = x + 3y$ attain a maximum at some point in the feasible region? If so, what is that point, and what is the maximum value of f?

(b) Does the objective function $P = x + 3y$ attain a minimum at some point in the feasible region? If so, what is that point, and what is the minimum value of f?

In the remaining exercises, first formulate the linear program, and then solve it by the graphical method. Some of these exercises have already appeared in Section 4.1.

18. **Manufacturing Profits** The Fine Line Pen Company makes two types of ballpoint pens: a silver model and a gold model. The silver model requires 1 minute in a grinder and 3 minutes in a bonder. The gold model requires 3 minutes in a grinder and 4 minutes in a bonder. Because of maintenance procedures, the grinder can be operated no more than 30 hours per week and the bonder no more than 50 hours per week. The company makes $5 on each silver pen and $7 on each gold pen. How many of each type of pen should be produced and sold each week to maximize profits?

19. **Inventory** The NewAge Technical Avenues store is accumulating an inventory of computers and fax machines. The store wants no more than 30 of these two items in the store. On the average, each computer it stocks costs $2000 and each fax machine $800. The store has set an inventory budget constraint of $40,800 for the two items. If the net profit on each computer is $200 and on each fax machine is $100, how many computers and how many fax machines should NewAge Technical Avenues have in inventory to maximize its net profit?

20. **Nutrition** A dietitian is to prepare two foods to meet certain requirements. Each pound of Food I contains 100 units of vitamin C, 40 units of vitamin D, and 10 units of vitamin E and costs 20 cents. Each pound of Food II contains 10 units of vitamin C, 80 units of vitamin D, and 5 units of vitamin E and costs 15 cents. The mixture of the two foods is to contain at least 260 units of vitamin C, at least 320 units of vitamin D, and at least 50 units of vitamin E. How many pounds of each type of food should be used to minimize the cost?

21. **Production** The Traction Tire Company produces two types of tires: four-ply and radial. Each four-ply tire requires 2 minutes to make and costs $20. Each radial tire requires 6 minutes to make and costs $30. Orders received by the company dictate that it must make a total

of at least 4000 tires each day, of which at least 1600 must be radials. The union contract with Traction requires that the company use at least 320 work-hours each day making the two types of tires. How many tires of each type should be made to meet the conditions and minimize costs?

22. **Radio Programming** A radio station is introducing a new talk show and a new gardening show into its programming. The station's marketing consultant estimates that each time the talk show is aired, 20,000 people will listen, while each time the gardening show is aired, 8000 people will listen. The station does not want to air the two shows a total of more than 12 times each week. Furthermore, they want to air the gardening show at least three times, but not more than five times, each week. How many times each week should each show be aired to maximize the number of listeners?

23. **Processing** Two grains—barley and corn—are to be mixed for an animal food. Barley contains 1 unit of fat per pound, and corn contains 2 units of fat per pound. The total number of units of fat in the mixture are not to exceed 12 units. No more than 6 pounds of barley and no more than 5 pounds of corn are to be used in the mixture. If barley and corn each contain 1 unit of protein per pound, how many pounds of each grain should be used to maximize the number of units of protein in the mixture?

24. **Manufacturing** A company makes two calculators: a business model and a scientific model. The business model contains 10 microcircuits, and the scientific model contains 20 microcircuits. The company has a contract that requires it to use at least 320 microcircuits each day. The company also wants to make at least twice as many business calculators as scientific calculators. If each business calculator requires 10 production steps and each scientific calculator requires 12 production steps, how many calculators of each type should be made to minimize the number of production steps?

25. **Manufacturing** A company makes desks in two models: a student model and a secretary model. The company makes a $40 profit on each student model and a $50 profit on each secretary model it sells. Each student model requires 2 hours of woodworking and 3 hours of finishing. Each secretary model requires 3 hours of woodworking and 5 hours of finishing. The company has a total of 240 work-hours available in the woodworking department and a total of 390 work-hours available in the finishing department. Due to new business orders, the company needs to make at least 40 of the secretary model desks during the next month. How many desks of each model should the company make and sell during the next month to maximize its profit?

26. Pollution Control The Environmental Protection Agency (EPA) has issued new regulations that require the New Molecule Chemical Co. to install a new process to help reduce pollution caused in the production of a particular chemical. The old process releases 2 grams of sulphur and 5 grams of lead for each liter of the chemical produced. The new process releases 1 gram of sulphur and 2.6 grams of lead for each liter of the chemical produced. The company makes a profit of 20 cents for each liter produced under the old method and 16 cents for each liter produced under the new process. The new EPA regulations do not allow for more than 10,000 grams of sulphur and 8000 grams of lead to be released each day. How many liters of the chemical should be produced under the old method and how many liters should be produced under the new method to maximize profits and yet stay within EPA regulations?

27. Airline Transportation A charter plane service contracts with a retirement group to provide flights from New York to Florida for at least 500 first-class, 1000 tourist-class, and 1500 economy-class passengers. The company has two types of planes. Type A carries 50 first-class, 60 tourist-class, and 100 economy-class passengers and costs $12,000 to make each flight. Type B carries 40 first-class, 30 tourist-class, and 80 economy-class passengers and costs $10,000 to make each flight. How many of each type of plane should be used to minimize flight costs?

28. Order Fulfillment The Movie Mania store is about to place an order for VHS tapes and DVD's. The distributor it orders from requires that an order contain at least 250 items. The prices Movie Mania must pay are $12 for each tape and $16 for each DVD. The distributor also requires that at least 30% of any order be DVD's. How many tapes and how many DVD's should Movie Mania order so that its costs will be kept to a minimum?

29. Manufacturing A firm makes a standard and a deluxe model of microwave oven. Each oven requires one plastic case, and the firm has 300 cases on hand. Each standard model requires 10 minutes to assemble, and each deluxe model requires 15 minutes to assemble. The orders received dictate that the firm produce at least 120 standard models and at least 150 deluxe models each day. To do this, the company has eight employees in the assembly department, each working an 8-hour day. If the firm makes a profit of $25 on each standard model and $35 on each deluxe model, how many of each model should be made each day to maximize profit?

30. Agriculture Agricultural planners in a certain country are planning next year's rice and wheat acreages. They estimate that each acre of rice planted will yield 30 bushels and each acre of wheat planted will yield 20 bushels. They want to raise at least 4 million bushels of rice and at least 5 million bushels of wheat. They have enough machinery to plant at least one million acres of these two crops combined. For planting, harvesting, and transporting the grains, the planners estimate that it will take six workers for each acre of rice planted and three workers for each acre of wheat planted. How can the planners meet the requirements set forth and use the minimum number of workers?

31. Investments An investment club has at most $30,000 to invest in junk and premium-quality bonds. Each type of bond is bought in $1000-dollar denominations. The junk bonds have an average yield of 12%, and the premium-quality bonds yield 7%. The policy of the club is to invest at least twice the amount of money in premium-quality bonds as in junk bonds. How many of each type of bond should the club buy to maximize its investment return?

32. Agriculture The New Age Protein Company has 500 acres of land available on which to plant wheat and soybeans. The planting costs per acre are $120 for wheat and $200 for soybeans. The company has a budget of $76,000 for planting. Due to soil conditions on the land, no more than 400 acres may be planted in wheat, while exactly 100 acres must be planted in soybeans. If the anticipated income per acre from wheat is $350 and from soybeans is $520, how many acres of each grain should be planted to maximize the company's income?

33. Purchasing A particular psychology department buys mice and rats for experimental purposes. It buys at least three times as many mice as rats. Each mouse costs $2, each rat costs $4, and the total departmental budget for such purchases is $500. Due to space limitations, the number of mice purchased cannot exceed 200, and the number of rats purchased needs to be exactly 25. If each mouse can be used in six experiments and each rat in four, how many of each should be purchased to maximize the number of experiments?

Chapter 4 Summary

A *linear programming problem* is a special kind of real-world problem in which one wants to maximize or minimize a certain quantity while observing *constraints* placed on the variables. The mathematical formulation of such a problem, the *linear program*, has three parts: the *objective function* (a linear function of the variables which represents the quantity that is to be maximized or minimized), the *structural constraints* (a linear system of inequalities), and the *nonnegativity constraints* (a requirement that all of the variables be nonnegative). The solution set of the system of constraints is called the *feasible region*. The objective of a linear program is to find where in the feasible region the objective function has a maximum or minimum value. When there are exactly two variables, the feasible region can be plotted in the coordinate plane. In such a case, if the objective function takes on a maximum or minimum value over the feasible region, then that value must correspond to at least one *corner point* of the feasible region.

EXERCISES

1. For each given point, determine whether or not the point is in the solution set for $x - 2y \geq 4$.

   (a) $(0, 0)$

   (b) $\left(6, \dfrac{1}{2}\right)$

   (c) $(4, 0)$

2. Graph the solution set for the inequality $2x + 3y < 10$.

3. Graph the solution set for the inequality $x > 3$.

4. Graph the solution set for the inequality $y \geq 2x + 7$.

5. Graph the solution set for the inequality $y \leq -3$.

Problems 6 through 9 refer to the following feasible region:

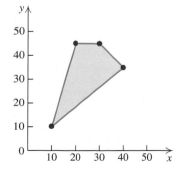

6. Write the system of inequalities that would determine the feasible region.

7. Find all the corner points of the feasible region.

8. Find the maximum of $f = 3x + 2y$ over the feasible region.

9. Find the minimum of $f = 20x + 15y$ over the feasible region.

Problems 10 through 13 refer to the following system of inequalities:

$$5x + 2y \leq 20$$
$$x + 4y \leq 10$$
$$x \geq 0, y \geq 0.$$

10. Graph the feasible region for the given system of inequalities.

11. Find each corner point of the feasible region for the given system of inequalities.

12. Find the maximum of $f = 12x + 25y$ over the feasible region.

13. Find the minimum of $f = 4x + 5y$ over the feasible region.

14. Graphically solve the following linear programming problem:

$$\text{Maximize } f = 50x + 75y$$
$$\text{subject to } 30x + 40y \leq 150$$
$$25x + 20y \leq 100$$
$$x \geq 0, y \geq 0.$$

15. Graphically solve the following linear programming problem:

$$\text{Minimize } C = 3x + 2y$$
$$\text{subject to } 2x + 4y \geq 6$$
$$x + y \geq 2$$
$$x \geq 0, y \geq 0.$$

16. Write a complete word problem expressing a linear programming problem that fits the following system:

$$\text{Maximize } P = 3x + 4y$$
$$\text{subject to } x + 2y \leq 12$$
$$4x + 5y \leq 20$$
$$5x + y \leq 10$$
$$x \geq 0, y \geq 0.$$

CUMULATIVE REVIEW

17. Write an equation of the line that has a slope of 4 and passes through the point $(2, -5)$.

18. Solve the following system of equations, using the substitution method:

$$2x - 3y = 9$$
$$4x + y = -3.$$

19. Solve the following system of equations:

$$2x + 3y - 5z = 0$$
$$4x + 2y + z = 7$$
$$x - 3y = -2.$$

20. Multiply: $\begin{bmatrix} 2 & -3 \\ 4 & 1 \end{bmatrix} \cdot \begin{bmatrix} 1 & -2 \\ 7 & 0 \end{bmatrix}$.

21. Find the transpose of the matrix $A = \begin{bmatrix} 1 & 3 & 0 \\ 2 & -4 & 1 \\ 3 & 2 & -2 \end{bmatrix}$.

Chapter 4 Sample Test Items

1. Graph and shade the feasible region determined by the system of inequalities

$$x + 2y \leq 6$$
$$5x + 2y \leq 10$$
$$x \geq 0$$
$$y \geq 0.$$

2. Graph and shade the feasible region determined by the system of inequalities

$$x \geq 1$$
$$y < 3.$$

3. Find all corner points for the feasible region determined by

$$x + 5y \leq 13$$
$$2x + y \leq 8$$
$$x \geq 0, y \geq 0.$$

4. Find all corner points for the feasible region determined by

$$4x + 2y \geq 100$$
$$20x + 30y \geq 150$$
$$x \geq 0, y \geq 0.$$

5. Solve the following linear programming problem graphically:

$$\text{Maximize } f = 2x + 5y$$
$$\text{subject to } 7x + y \leq 21$$
$$2x + 9y \leq 18$$
$$x \geq 0, y \geq 0.$$

6. Solve the following linear programming problem graphically:

$$\text{Minimize } g = 4x - y$$
$$\text{subject to } 3x + y \geq 6$$
$$3x + 4y \geq 12$$
$$x \geq 0, y \geq 0.$$

7. Test Construction A test is made up of true–false questions worth 3 points each and multiple-choice questions worth 5 points each. You must answer at least 10 true–false questions and at least 6 multiple-choice questions. Because of time constraints, you can answer at most 30 true–false and at most 20 multiple-choice questions. You are allowed to answer no more than 36 questions. State the linear program for this problem and plot the feasible region. How many of each type of question should you answer to maximize your score? What is this maximum score?

8. Sales Commission Javier, a part-time furniture salesman, earns commission from selling finished and unfinished bookcases. He earns a $20 commission from selling each finished bookcase and a $15 commission from selling each unfinished bookcase. He spends about 30 minutes with each customer who wishes to buy a finished bookcase and about 15 minutes with each customer who wishes to buy an unfinished bookcase. If he can work no more than 24 hours per week, and he is expected to sell a total of at least 30 bookcases each week, how many of each type of bookcase should Javier sell in a week to maximize his commission?

CHAPTER **5**

Linear Programming: The Simplex Method

PROBLEM

Refrigerators are to be shipped from warehouses in Wilmar and Lubbock to two stores, one in Phoenixville and one in Mebane, with orders and shipping costs per refrigerator as indicated in the illustration. How many refrigerators should be shipped from each warehouse to minimize costs? (See Section 5.3, Example 8).

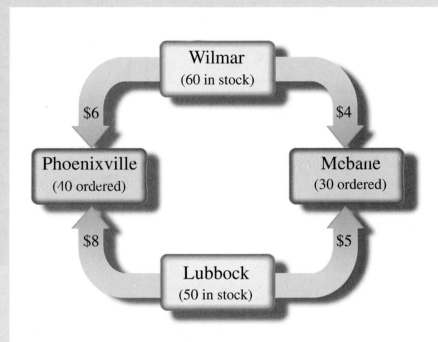

CHAPTER PREVIEW

Practical applications of linear programming problems usually involve many more than just two variables. In fact, problems with dozens of variables are not uncommon. The first algebraic method for solving such problems, called the simplex method (or simplex algorithm), was developed by George Dantzig in 1947. Although more advanced techniques are now used for large-scale problems, the simplex method still makes for an enlightening introduction to algebraic approaches to linear programming. In the first section of this chapter, we explore matrix row operations for moving around the corner points of the feasible region. In the sections that follow, we use the technique to develop the simplex method and apply it to both standard and nonstandard linear programming problems.

Technology Opportunities

Excel has a powerful tool called `Solver` for solving linear programming problems. We have already used the `Solver` in Chapter 2 to solve a system of linear equations; in this chapter, we will see its real power.

Most calculators do not include linear programming solvers, but do have advanced matrix manipulation capabilities, as seen in Chapter 2.

Section 5.1　Slack Variables and Pivoting

In Chapter 4, we explored how to set up linear programming problems and how to solve two-variable linear programming problems graphically. However, most linear programming problems have many more variables than two, and we need a less restrictive, algebraic method for solving these problems. The **simplex method (simplex algorithm)** is a well-known algebraic method for solving a large class of linear programming problems. To be able to use this algorithm, we first must define certain general types of linear programming problems and learn how to navigate around the corner points of the feasible regions of these special problems.

The simplex algorithm is used to solve linear programming problems that already are, or that can be converted to, **standard maximum-type** problems. The following is an example of a standard maximum-type problem:

$$\text{Maximize } f = 2x + y$$
$$\text{subject to } 2x + y \le 8$$
$$x + y \le 5$$
$$x \ge 0, y \ge 0.$$

The Standard Maximum–Type Problem

A linear programming problem is a **standard maximum-type problem** if the following conditions are met:

1. The objective function is linear and is to be *maximized*.
2. The variables are all nonnegative.
3. The structural constraints are all of the form $ax + by + \cdots \le c$, where $c \ge 0$.

The first step toward using the simplex algorithm, and the key to viewing a linear programming problem as an algebraic process, is the conversion of each structural constraint into an equality by the addition of a **slack variable** to the left side. To turn the constraint $2x + y \le 8$ into an equality, for example, we add a variable, s_1, to the left side to create the equality $2x + y + s_1 = 8$. As x and y change, the variable s_1 continually "takes up the slack" to ensure that an equality exists (hence the name). Adding the slack variable s_2 to the second constraint completes a system of equations called **slack equations:**

$$\left. \begin{array}{l} 2x + y + s_1 \quad\ = 8 \\ x + y \quad\quad\ + s_2 = 5 \end{array} \right\} \leftarrow \text{Slack equations}$$

NOTE Slack variables are never added to the nonnegativity constraints in any linear programming problem.

Now think of the first slack equation, $2x + y + s_1 = 8$, as one equation in three variables, whose solutions may be expressed as

$$s_1 = 8 - (2x + y),$$
$$x = \text{any nonnegative number},$$
$$y = \text{any nonnegative number},$$

where x and y are free variables. The last expression for s_1 tells us that

$$\text{if } 2x + y = 8, \quad \text{then} \quad s_1 = 0 \text{ (there is no slack)},$$
$$\text{if } 2x + y < 8, \quad \text{then} \quad s_1 \text{ is positive, and}$$
$$\text{if } 2x + y > 8, \quad \text{then} \quad s_1 \text{ is negative.}$$

That is, for points (x, y) in the first quadrant satisfying $2x + y \leq 8$, the value of the slack variable is either positive or zero. A similar situation prevails for the second slack equality, $x + y + s_2 = 5$. This shows us that, for points in the feasible region where both of the inequalities involved are satisfied, each slack variable has a value that is either positive or zero, as shown in Figures 1 and 2. This conclusion is valid for all linear programming problems.

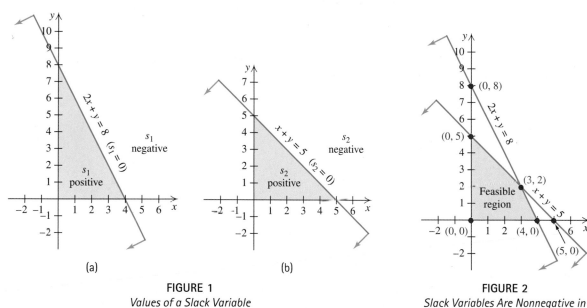

FIGURE 1
Values of a Slack Variable

FIGURE 2
Slack Variables Are Nonnegative in the Feasible Region.

Property of Slack Variables

For any point in the feasible region of a standard maximum-type linear programming problem, the value of each slack variable is nonnegative.

We explore one last bit of geometry connected with slack variables in the next example.

EXAMPLE 1

Coordinates of Points
where Two Boundary
Lines Intersect

Find the values of the variables x and y and the slack variables s_1 and s_2 at each point of intersection for each pair of boundary lines of the graph of the system of linear inequalities

$$2x + y \leq 8$$
$$x + y \leq 5.$$

SOLUTION The x- and y-coordinates of these points have already been demonstrated in Figure 2. Next, substitute each of these points into the slack equations:

$$2x + y + s_1 \quad\ = 8$$
$$x + y \quad\ + s_2 = 5.$$

The first equation gives the value of s_1 and the second equation the value of s_2. (See Table 1.)

Point	First Equation	Second Equation
(0, 8)	$s_1 = 0$	$s_2 = -3$
(0, 5)	$s_1 = 3$	$s_2 = 0$
(0, 0)	$s_1 = 8$	$s_2 = 5$
(4, 0)	$s_1 = 0$	$s_2 = 1$
(5, 0)	$s_1 = -2$	$s_2 = 0$
(3, 2)	$s_1 = 0$	$s_2 = 0$

TABLE 1.

Notice that when a point of intersection lies outside of the feasible region, at least one of the slack variables is *negative,* as shown in Figure 3.

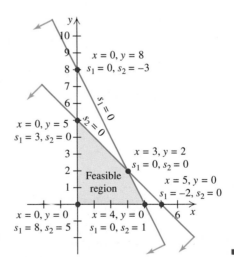

FIGURE 3
*Two of the Four Variables
Are Always Zero.*

Now that we have the geometry of our example in mind, we turn our attention to how particular solutions of the slack equations relate to that geometry. The system of slack equations

$$2x + y + s_1 \quad\ = 8$$
$$x + y \quad\ + s_2 = 5$$

is a system of two equations in four unknowns and, as we saw in Section 2.3, has an infinite number of solutions. Because we have four variables and only two equations in the system, we may choose *two* variables to be free and assign them any values we please. The choice that offers the easiest solution is to let x and y be the free variables and set both equal to zero, so that, by inspection of the system, $s_1 = 8$ and $s_2 = 5$. Notice that this solution, $x = 0$, $y = 0$, $s_1 = 8$, and $s_2 = 5$, corresponds to the values of the variables at the origin in Figure 3. Solutions following this pattern not only are the easiest to find, but also will be seen as the most useful in arriving at the final solution to a linear programming problem by the simplex algorithm. At any stage of the solution process, such a solution relies upon dividing the variables into two groups, each with a special name in the language of linear programming. In one group, each variable is considered to be free and is set equal to zero; this is the group of **nonbasic variables.** The remaining variables, whose values are determined once the nonbasic variables are set equal to zero, are called **basic variables.** The solution is most easily read by first identifying the nonbasic variables in the augmented matrix that corresponds to the system of linear equations then "mentally blocking out" their columns (the equivalent of setting them equal to zero), leaving each row as an equation involving a basic variable only. Observe how this may be done for the above slack equations:

$$
\begin{array}{cccc}
x & y & s_1 & s_2 \\
\end{array}
$$
$$
\left[\begin{array}{cccc|c}
2 & 1 & 1 & 0 & 8 \\
1 & 1 & 0 & 1 & 5
\end{array}\right]. \tag{1}
$$

Nonbasic variables (set = 0) ↑ ↑   ↑ ↑ Basic variables

NOTE See Chapter 2 for a review of augmented matrices and matrix row operations.

To further investigate the nature of basic and nonbasic variables, suppose we take the augmented matrix **(1)** and pivot on the *arbitrarily chosen* element that is circled in the second-row, second-column position. Then we obtain

$$
\begin{array}{cccc}
x & y & s_1 & s_2
\end{array}
\left[\begin{array}{cccc|c}
2 & 1 & 1 & 0 & 8 \\
1 & ① & 0 & 1 & 5
\end{array}\right]
\quad -1R_2 + R_1 \rightarrow R_1 \quad
\begin{array}{cccc}
x & y & s_1 & s_2
\end{array}
\left[\begin{array}{cccc|c}
1 & 0 & 1 & -1 & 3 \\
1 & 1 & 0 & 1 & 5
\end{array}\right]. \tag{2}
$$

Nonbasic variable ↑   ↑ ↑ Basic variables   ↑ Nonbasic variable

Thus, after the pivot, we have a new-appearing system of two equations and four unknowns with infinitely many solutions. Now, which two variables should we choose to set equal to zero (i.e., to be free, or nonbasic) and hence mentally "block out" those columns, so as to most easily read the values of the remaining variables? Let us choose x and s_2 as those nonbasic variables. Once x and s_2 are set equal to zero, the first row represents the equation $s_1 = 3$ and the second row represents the equation $y = 5$. A solution is then

$$x = 0, \quad s_2 = 0, \quad y = 5 \quad \text{and} \quad s_1 = 3.$$

Relate this solution to the point $(0, 5)$ in Figure 3, and observe that two things have happened due to this pivot. First, from the standpoint of the solutions obtained, the pivot operation has "moved" us from the corner point $(0, 0)$ to the adjacent corner point $(0, 5)$. Second, the classification of two variable has changed. y changed from nonbasic to basic and s_2 changed from basic to nonbasic. Such a trade-off will occur every time a pivot operation is performed, which implies that the *number* of variables in each classification *never* changes. This, in turn, implies that there will always be two nonbasic variables whose values are zero, and that is precisely what is needed to identify the intersection of two boundary lines. (See Figure 3.)

The foregoing two augmented matrices show the pattern for identifying basic and nonbasic variables according to the distinctive appearance of their corresponding columns. Nonbasic variables—the free ones to be set equal to zero—are variables whose columns have a "cluttered" look, while basic variables are variables whose columns contain a 1 somewhere, with all other elements being zero.

Basic and Nonbasic Variables

For a system of slack equations represented in an augmented matrix, the **basic variables** are those variables whose column of coefficients contains exactly one positive 1 and all other entries 0. The remaining variables are called **nonbasic variables.**

Continuing our investigation of pivoting and the solutions obtained from classifying basic and nonbasic variables, suppose we now pivot on the arbitrarily chosen element in the first-row, fourth-column position of augmented matrix **(2)**. The result of this pivot (now circled) is

$$
\begin{array}{cccc}
x & y & s_1 & s_2
\end{array}
\left[\begin{array}{cccc|c}
1 & 0 & 1 & \boxed{-1} & 3 \\
1 & 1 & 0 & 1 & 5
\end{array}\right]
\rightarrow
\begin{array}{cccc}
x & y & s_1 & s_2
\end{array}
\left[\begin{array}{cccc|c}
-1 & 0 & -1 & 1 & -3 \\
2 & 1 & 1 & 0 & 8
\end{array}\right]. \qquad \textbf{(3)}
$$

Nonbasic variables ⟶   Basic variables

The basic variables are now y and s_2, while x and s_1 are nonbasic. Letting $x = 0$ and $s_1 = 0$, we see that the first row shows that $s_2 = -3$ and the second that $y = 8$. This is the solution depicted at the point $(0, 8)$ in Figure 3. Even though it is not a corner point of the feasible region, it is the intersection of two boundary lines, $x = 0$ and $s_1 = 0$ $(2x + y = 8)$. This solution coincides with what we have already observed geometrically about any variable: If a slack variable has a negative value, then the point represented lies beyond the bounds of the feasible region.

The Geometry of Pivoting

If a pivot is performed on any *nonzero* element in the augmented matrix of slack equations, and if the nonbasic variables are set equal to zero in the resulting matrix, then the resulting solution is a point where two boundary lines of the feasible region intersect.

We have seen that when we pivot, the resulting solution obtained from the classification of basic and nonbasic variables sometimes gives another corner point of the feasible region and at other times does not. To be useful in solving a linear programming problem, the solution obtained after *every* pivot operation must result in a corner point of the feasible region and not just the intersection of two of its boundary lines. Fortunately, there is a way to select the row in which the pivot resides to make this happen. Let us develop a selection rule using the initial augmented matrix of slack equations for our example, namely,

$$\begin{array}{cccc} x & y & s_1 & s_2 \\ \begin{bmatrix} 2 & 1 & 1 & 0 \\ 1 & 1 & 0 & 1 \end{bmatrix} & & & \begin{matrix} 8 \\ 5 \end{matrix} \end{array}.$$

Suppose that we have decided to pivot arbitrarily on an element in the x-column and we want to be sure that the resulting basic solution represents a corner point of the feasible region shown in Figure 2. The initial solution is the origin, and a pivot in the x direction will take us to either the point (4, 0) or the point (5, 0) of Figure 2. These two points were determined from the graphs of the boundary lines *by dividing each coefficient of x into the corresponding constant term on the right*. Now look at the augmented matrix, and see how the same things is accomplished:

$$\begin{bmatrix} 2 & 1 & 1 & 0 \\ 1 & 1 & 0 & 1 \end{bmatrix} \begin{matrix} 8 \\ 5 \end{matrix}$$

2 into 8 → Quotient: 4

1 into 5 → Quotient: 5

The *smaller* quotient, 4, identifies the corner point (4, 0), and the larger quotient, 5, identifies the noncorner point (5, 0), on the x-axis of Figure 2. We should choose the element in the x column that gives the *smallest quotient*—the 2 in the first-row, first-column position—as the pivot element in order to remain in the feasible region. The actual pivot operation gives this result:

$$\begin{array}{cccc} x & y & s_1 & s_2 \\ \begin{bmatrix} ② & 1 & 1 & 0 \\ 1 & 1 & 0 & 1 \end{bmatrix} & & & \begin{matrix} 8 \\ 5 \end{matrix} \end{array} \rightarrow \begin{array}{cccc} x & y & s_1 & s_2 \\ \begin{bmatrix} 1 & \frac{1}{2} & \frac{1}{2} & 0 \\ 0 & \frac{1}{2} & -\frac{1}{2} & 1 \end{bmatrix} & & & \begin{matrix} 4 \\ 1 \end{matrix} \end{array}.$$

The solution given by the last augmented matrix is $y = 0$, $s_1 = 0$, $x = 4$, and $s_2 = 1$, and this is precisely the corner point we wanted. To reinforce the "smallest quotient" pivot selection, we next show that a pivot on the "1" in the first column *does not* give a corner point of the feasible region:

$$\begin{bmatrix} 2 & 1 & 1 & 0 \\ ① & 1 & 0 & 1 \end{bmatrix} \begin{matrix} 8 \\ 5 \end{matrix} \rightarrow \begin{bmatrix} 0 & -1 & 1 & -2 \\ 1 & 1 & 0 & 1 \end{bmatrix} \begin{matrix} -2 \\ 5 \end{matrix}.$$

The solution, $x = 5$, $y = 0$, $s_1 = -2$, and $s_2 = 0$, reflects the point (5, 0) where two boundary lines intersect, but is not a corner point of the feasible region. (See Figure 3.)

Similarly, entries in the y column give quotients of 8 and 5, from top to bottom. Can you use this information to tell in advance where, in a geometric sense, a pivot on either element in the y column will take the solution?

Clearly, the division of elements in the pivot column into the corresponding constants holds the key to staying in the feasible region. Pivoting where the smallest nonnegative quotient occurs restricts the resulting constants to being non-negative. This, in turn, keeps us on points that are both the intersection of boundary lines *and* on the feasible region. Thus, the resulting points are always corner points of the feasible region. This analysis suggests a smallest-quotient rule. Before verifying the suggestion, however, we look at one more example.

EXAMPLE 2

A Solution after Pivoting

Consider the linear programming problem

$$\text{Maximize } f = 5x + 2y$$
$$\text{subject to } \quad x + y \le 6$$
$$2x - y \le 6$$
$$x \ge 0, y \ge 0.$$

Inserting a slack variable for each appropriate constraint gives this system of slack equations:

$$x + y + s_1 \quad\quad = 6$$
$$2x - y \quad\quad + s_2 = 6.$$

Figure 4 shows the feasible region and the values of all variables where two boundary lines intersect.

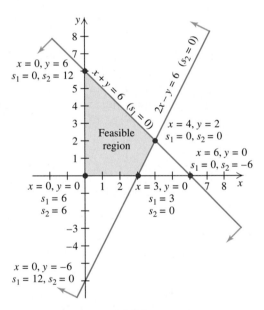

FIGURE 4
Coordinates of Points where
Boundary Lines Intersect

The initial augmented matrix for the system of slack equations is

$$
\begin{array}{cccc}
x & y & s_1 & s_2 \\
\end{array}
$$
$$
\left[\begin{array}{cccc|c}
1 & 1 & 1 & 0 & 6 \\
2 & -1 & 0 & 1 & 6
\end{array}\right].
$$

Upon setting the nonbasic variables x and y equal to zero, we find that this augmented matrix has a solution of $x = 0$, $y = 0$, $s_1 = 6$, and $s_2 = 6$. Suppose we decide to pivot on an element in the y column. Computing the quotients as before gives

$$\begin{array}{c} \overset{\lceil \text{1 into 6} \rceil}{} \\ \begin{bmatrix} 1 & 1 & 1 & 0 & 6 \\ 2 & -1 & 0 & 1 & 6 \end{bmatrix} \\ \overset{}{\lfloor -1 \text{ into 6} \rfloor} \end{array} \begin{array}{l} \text{Quotient: } 6 \\ \text{Quotient: } -6 \end{array}$$

We know in advance that if a pivot is performed on the 1 (which gave the quotient 6), the new solution will give the point $(0, 6)$:

$$\begin{bmatrix} 1 & 1 & 1 & 0 & 6 \\ 3 & 0 & 1 & 1 & 12 \end{bmatrix}, \quad \textit{(Solution: } x = 0, y = 6, s_1 = 0, \text{ and } s_2 = 12; \text{ see Figure 4.)}$$

while if a pivot is performed on the -1 (which gave the quotient -6), the new solution should give the point $(0, -6)$:

$$\begin{bmatrix} 3 & 0 & 1 & 1 & 12 \\ -2 & 1 & 0 & -1 & -6 \end{bmatrix}. \quad \textit{(Solution: } x = 0, y = -6, s_1 = 12, \text{ and } s_2 = 0; \\ \textit{see Figure 4.)}$$

Thus, the pivot on a negative element has resulted in a point that is not in the feasible region. ▄▄

Example 2 illustrates that when choosing a pivot row, division by a negative number should be avoided if the solution determined by setting the nonbasic variables equal to zero is to reflect a corner point of the feasible region after the pivot has been completed.

The following conditions set the stage for our pivoting rule that takes us from one corner point to another corner point.

Assume now that:

(a) We have started with the augmented matrix for a system of slack equations obtained from *any standard maximum linear programming problem,*
(b) We have obtained a feasible *solution* by setting the nonbasic variables equal to zero, and
(c) The current augmented matrix has a *solution that reflects a corner point* of the feasible region.

Then we have the following rule that takes us from one corner point to another:

The Smallest-Quotient Rule

Under the preceding conditions (a), (b), and (c):

1. Select any new pivot column.
2. Divide each positive number in that column into the corresponding number in the constants column of the matrix.
3. Select as the pivot row the row corresponding to the *smallest nonnegative quotient* so obtained. (Zero is included.)

Pivoting on the element where the pivot column and pivot row intersect will always result in a solution that occurs at a corner point of the feasible region. Such a solution is called a **basic feasible solution** to the linear programming problem.

The smallest-quotient rule tells us that "once on a corner point, always on a corner point" of the feasible region.

EXAMPLE 3
Using the Smallest-
Quotient Rule

Pivot on the system of slack equations shown in Example 2, each time using the smallest-quotient rule, where the first pivot column is the second column of the augmented matrix. Then, on the resulting matrix, choose the pivot column to be the first column, and then on *that* resulting matrix, choose the pivot column to be the third column.

SOLUTION The system of slack equations has the following augmented matrix:

$$\begin{array}{cccc} x & y & s_1 & s_2 \end{array}$$
$$\begin{bmatrix} 1 & 1 & 1 & 0 & 6 \\ 2 & -1 & 0 & 1 & 6 \end{bmatrix}.$$

Initially, s_1 and s_2 are basic variables, while x and y are nonbasic variables and equal zero. Letting $x = 0$ and $y = 0$, we see that the first row reads $s_1 = 6$ and the second $s_2 = 6$. This solution represents the origin of Figure 5. Choose the y column as the column from which to select a pivot element. The only appropriate quotient is 6 from the first row, because we do not divide by negative numbers. The row with the smallest nonnegative quotient is the first; hence, pivot on the 1 in the first-row, second-column position:

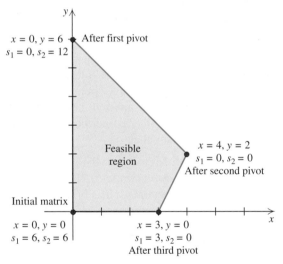

$x = 0, y = 6$
$s_1 = 0, s_2 = 12$ After first pivot

Feasible
region

$x = 4, y = 2$
$s_1 = 0, s_2 = 0$
After second pivot

Initial matrix

$x = 0, y = 0$
$s_1 = 6, s_2 = 6$

$x = 3, y = 0$
$s_1 = 3, s_2 = 0$
After third pivot

FIGURE 5
*Tracing the Movement of Solutions
after Successive Pivots*

$$\begin{array}{cccc} x & y & s_1 & s_2 \end{array}$$
$$\begin{bmatrix} 1 & ① & 1 & 0 & 6 \\ 2 & -1 & 0 & 1 & 6 \end{bmatrix} \quad \text{Quotient: 6} \quad \leftarrow \text{Pivot row}$$

Pivot column

$$1R_1 + R_2 \rightarrow R_2 \begin{bmatrix} 1 & 1 & 1 & 0 & 6 \\ 3 & 0 & 1 & 1 & 12 \end{bmatrix}.$$

The basic feasible solution for the last augmented matrix is $x = 0$, $s_1 = 0$, $y = 6$, and $s_2 = 12$. (See Figure 5.)

From the last matrix, we choose the x column from which to select a pivot element. The selection process, along with the result of the pivot, is as follows:

$$
\begin{array}{cccc}
x & y & s_1 & s_2 \\
\end{array}
$$
$$
\begin{bmatrix}
1 & 1 & 1 & 0 & 6 \\
\boxed{3} & 0 & 1 & 1 & 12
\end{bmatrix}
\begin{array}{l}
\text{Quotient: 6} \\
\text{Quotient: 4} \quad \leftarrow \text{Pivot row}
\end{array}
$$

$\uparrow$ Pivot column

$$
\begin{array}{cccc}
x & y & s_1 & s_2 \\
\end{array}
$$
$$
\tfrac{1}{3}R_2 \to R_2
\begin{bmatrix}
1 & 1 & 1 & 0 & 6 \\
1 & 0 & \tfrac{1}{3} & \tfrac{1}{3} & 4
\end{bmatrix}
$$

$$
\begin{array}{cccc}
x & y & s_1 & s_2 \\
\end{array}
$$
$$
-1R_2 + R_1 \to R_1
\begin{bmatrix}
0 & 1 & \tfrac{2}{3} & -\tfrac{1}{3} & 2 \\
1 & 0 & \tfrac{1}{3} & \tfrac{1}{3} & 4
\end{bmatrix}.
$$

The last augmented matrix gives the basic feasible solution: $s_1 = 0$, $s_2 = 0$, $y = 2$, and $x = 4$. (See Figure 5.)

Finally, on the last matrix, we pivot on an element in the s_1 column. The pivot selection and resulting pivot operation are

$$
\begin{array}{cccc}
x & y & s_1 & s_2 \\
\end{array}
$$
$$
\begin{bmatrix}
0 & 1 & \boxed{\tfrac{2}{3}} & -\tfrac{1}{3} & 2 \\
1 & 0 & \tfrac{1}{3} & \tfrac{1}{3} & 4
\end{bmatrix}
\begin{array}{l}
\text{Quotient: 3} \quad \leftarrow \text{Pivot row} \\
\text{Quotient: 12}
\end{array}
$$

$\uparrow$ Pivot column

$$
\begin{array}{cccc}
x & y & s_1 & s_2 \\
\end{array}
$$
$$
\tfrac{3}{2}R_1 \to R_1
\begin{bmatrix}
0 & \tfrac{3}{2} & 1 & -\tfrac{1}{2} & 3 \\
1 & 0 & \tfrac{1}{3} & \tfrac{1}{3} & 4
\end{bmatrix}
$$

$$
\begin{array}{cccc}
x & y & s_1 & s_2 \\
\end{array}
$$
$$
-\tfrac{1}{3}R_1 + R_2 \to R_2
\begin{bmatrix}
0 & \tfrac{3}{2} & 1 & -\tfrac{1}{2} & 3 \\
1 & -\tfrac{1}{2} & 0 & \tfrac{1}{2} & 3
\end{bmatrix}.
$$

The basic feasible solution of the last augmented matrix is $y = 0$, $s_2 = 0$, $x = 3$, and $s_1 = 3$. (See Figure 5.) ∎

As seen in the previous example, by following the smallest-quotient rule, we are able to "move" the basic variables so that they correspond with corner points of the feasible region. This approach may be quite interesting, but how does it solve a linear programming problem? If, somehow, the augmented matrix could be expanded to include the objective function, then perhaps the selection of pivots could be guided by whether the objective function's value is increased. In other words, the technique would be extremely useful if we could "move" to basic feasible solutions that had larger and larger objective function values. This is exactly what the simplex method does, as we shall see in the next section.

Exercises 5.1

In Exercises 1 through 8, determine which constraints are not in the form appropriate for a standard maximum-type problem.

1. $x + y \leq 3$

2. $3x + 4y \leq -4$

3. $3x - 2y \leq 6$

4. $x - y \geq 0$

5. $4x - 5y \geq 6$

6. $-2x + y \leq 6$

7. $7x + 5y \geq -4$

8. $3y \geq 7$

For the inequality $x + 3y \leq 9$, insert a slack variable s_1, and answer the questions in Exercises 9 through 13 about s_1.

9. What is the value of s_1 if $x = 3$ and $y = 2$?

10. What is the value of s_1 if $x = -2$ and $y = 5$?

11. What is the value of s_1 if $x = 0$ and $y = 0$?

12. What is the range of values of s_1 if $x \geq 0$, $y \geq 0$, and $x + 3y \leq 9$?

13. If s_1 is negative, where must the points (x, y) lie in the plane, relative to the boundary line $x + 3y = 9$?

For the inequality $2x + y \leq 8$, insert a slack variable s_2, and answer the questions in Exercises 14 through 18 about s_2.

14. What is the value of s_2 if $x = 2$ and $y = 1$?

15. What is the value of s_2 if $x = -2$ and $y = 7$?

16. What is the value of s_2 if $x = 0$ and $y = 0$?

17. What is the range of values of s_2 if $x \geq 0$, $y \geq 0$, and $2x + y \leq 8$?

18. If s_2 is negative, where must the points (x, y) lie in the plane, relative to the boundary line $2x + y = 8$?

In Exercises 19 through 24, consider each augmented matrix to be the result of one or more pivots from an initial augmented matrix representing the slack equations of a linear programming problem in standard form. Find the solution in which the nonbasic variables are zero, and decide whether the solution is a corner point of the feasible region for the problem.

19.
$$
\begin{array}{cccc}
x & y & s_1 & s_2 \\
\end{array}
$$
$$
\left[\begin{array}{cccc|c}
2 & 1 & 0 & 3 & 2 \\
1 & 0 & 1 & 5 & 7
\end{array}\right]
$$

20.
$$
\begin{array}{cccc}
x & y & s_1 & s_2 \\
\end{array}
$$
$$
\left[\begin{array}{cccc|c}
1 & -3 & 5 & 0 & 6 \\
0 & 2 & -1 & 1 & 1
\end{array}\right]
$$

21.
$$
\begin{array}{cccc}
x & y & s_1 & s_2 \\
\end{array}
$$
$$
\left[\begin{array}{cccc|c}
2 & 0 & -4 & 1 & -2 \\
0 & 1 & 2 & 0 & 5
\end{array}\right]
$$

22.
$$
\begin{array}{cccc}
x & y & s_1 & s_2 \\
\end{array}
$$
$$
\left[\begin{array}{cccc|c}
1 & 0 & 0 & 2 & -2 \\
0 & -3 & 1 & -1 & 5
\end{array}\right]
$$

23.
$$
\begin{array}{cccccc}
x & y & z & s_1 & s_2 & s_3 \\
\end{array}
$$
$$
\left[\begin{array}{cccccc|c}
2 & 1 & 0 & 1 & 4 & 0 & 2 \\
-1 & 0 & 1 & 1 & 2 & 0 & 1 \\
3 & 0 & 0 & 3 & 0 & 1 & 5
\end{array}\right]
$$

24.
$$
\begin{array}{cccccc}
x & y & z & s_1 & s_2 & s_3 \\
\end{array}
$$
$$
\left[\begin{array}{cccccc|c}
0 & 3 & -1 & 1 & -2 & 0 & 1 \\
1 & 0 & 1 & 0 & -3 & 0 & 5 \\
0 & 1 & 0 & 0 & 4 & 1 & 2
\end{array}\right]
$$

Exercises 25 through 30 apply to this linear programming problem:

$$
\begin{aligned}
\text{Maximize } & f = x + 3y \\
\text{subject to } & x + y \leq 9 \\
& -x + 2y \leq 12 \\
& x \geq 0, y \geq 0.
\end{aligned}
$$

25. Insert slack variables where appropriate in the constraints, and form the system of slack equations.

26. Graph the feasible region, and label the intersection of each pair of half-plane lines with four coordinates.

27. Form the initial augmented matrix from the slack equations, and find the pivot element in the x column so that when the pivot operation is performed on that element, a basic feasible solution results. Pivot on the other element in the x column to confirm your result.

28. Form the initial augmented matrix from the slack equations, and find the pivot element in the y column so that if the pivot operation were performed on that element, a basic feasible solution would result. Pivot on the other element in the y column to confirm your result.

29. Pivot on an element in the x column of the initial augmented matrix of slack equations so that a basic feasible solution results. Then apply the smallest-quotient rule to the y column of the resulting augmented matrix to find a pivot element in the y column so that, upon pivoting, another basic feasible solution results.

30. Pivot on an element in the y column of the initial augmented matrix of slack equations so that a basic feasible solution results. Then apply the smallest-quotient rule to the x column of the resulting augmented matrix to find a pivot element in the x column so that, upon pivoting, another basic feasible solution results.

Exercises 31 through 36 apply to this linear programming problem:

$$\text{Maximize } f = 5x + 2y$$
$$\text{subject to } \quad x + \ y \le 10$$
$$2x - 2y \le \ 8$$
$$x \ge 0, y \ge 0.$$

31. Insert slack variables where appropriate in the constraints, and form the system of slack equations.

32. Graph the feasible region, and label the intersection of each pair of half-plane lines with four coordinates.

33. Form the initial augmented matrix from the slack equations, and find the pivot element in the x column so that when the pivot operation is performed on that element, a basic feasible solution results. Pivot on the other element in the x column to confirm your result.

34. Form the initial augmented matrix from the slack equations, and find the pivot element in the y column so that if the pivot operation were performed on that element, a basic feasible solution would result. Pivot on the other element in the y column to confirm your result.

35. Pivot on an element in the x column of the initial augmented matrix of slack equations so that a basic feasible solution results. Then apply the smallest-quotient rule to the y column of the resulting augmented matrix to find a pivot element in the y column so that, upon pivoting, another basic feasible solution results.

36. Pivot on an element in the y column of the initial augmented matrix of slack equations so that a basic feasible solution results. Then apply the smallest-quotient rule to the x column of the resulting augmented matrix to find a pivot element in the x column so that, upon pivoting, another basic feasible solution results.

Exercises 37 through 42 refer to this linear programming problem:

$$\text{Maximize } P = x + 4y$$
$$\text{subject to } x + \ y \le 8$$
$$x + 2y \le 10$$
$$x \ge 0, y \ge 0.$$

37. Insert slack variables where appropriate in the constraints, and form the system of slack equations.

38. Graph the feasible region, and label the intersection of each pair of half-plane lines with four coordinates.

39. Form the initial augmented matrix from the slack equations, and find the pivot element in the x column so that when the pivot operation is performed on that element, a basic feasible solution results. Pivot on the other element in the x column to confirm your result.

40. Form the initial augmented matrix from the slack equations, and find the pivot element in the y column so that if the pivot operation were performed on that element, a basic feasible solution would result. Pivot on the other element in the y column to confirm your result.

41. Pivot on an element in the x column of the initial augmented matrix of slack equations so that a basic feasible solution results. Then apply the smallest-quotient rule to the y column of the resulting augmented matrix to find a pivot element in the y column so that, upon pivoting, another basic feasible solution results.

42. Pivot on an element in the y column of the initial augmented matrix of slack equations so that a basic feasible solution results. Then apply the smallest-quotient rule to the x column of the resulting augmented matrix to find a pivot element in the x column so that, upon pivoting, another basic feasible solution results.

In each of the Exercises 43 and 44, a set of inequalities describing the feasible region of a standard maximum-type linear programming problem is given. Insert the appropriate slack variables, and sketch the feasible region. Starting at the origin, select successive pivot elements in the resulting augmented matrices so that the respective solutions, with nonbasic variables set equal to zero, will occur at successive corner points, going completely around the feasible region with a clockwise motion and returning to the origin. Then do the same things with a counterclockwise motion.

43. $x + y \le 8$ **44.** $2x - y \le 8$

$3x + y \le 12$ $4x + y \le 22$

$x \ge 0, y \ge 0$ $x \ge 0, y \ge 0$

45. The following initial augmented matrix represents the slack equations for a standard maximum-type linear programming problem:

$$\begin{array}{cccc} x & y & s_1 & s_2 \end{array}$$
$$\begin{bmatrix} 1 & 1 & 1 & 0 & | & 6 \\ 7 & 3 & 0 & 1 & | & 21 \end{bmatrix}.$$

(a) Write the set of all constraints for the problem.

(b) Graph the feasible region.

(c) Perform one pivot on the given matrix that will give a solution at the point (3, 0).

(d) Perform one pivot on the given matrix that will give a solution at the point (0, 6).

(e) Perform two successive pivots on the given matrix so that the last resulting matrix will give the solution at the point $\left(\frac{3}{4}, \frac{21}{4}\right)$.

46. The following initial augmented matrix represents the slack equations for a standard maximum-type linear programming problem:

$$
\begin{array}{cccc}
x & y & s_1 & s_2 \\
\end{array}
$$
$$
\left[\begin{array}{cccc|c}
1 & 1 & 1 & 0 & 4 \\
1 & -1 & 0 & 1 & 2
\end{array}\right].
$$

(a) Write the set of all constraints for the problem.

(b) Graph the feasible region.

(c) Perform one pivot on the given matrix that will give a solution at the point (2, 0).

(d) Perform one pivot on the given matrix that will give a solution at the point $(0, -2)$.

(e) Perform two successive pivots on the given matrix so that the last resulting matrix will give the solution at the point (3, 1).

Section 5.2   The Simplex Algorithm

The simplex algorithm (or simplex method) is an algebraic technique that solves standard maximum-type linear programming problems. In view of the smallest-quotient rule that we established in Section 5.1, all that remains to be done is to involve the objective function in the pivoting procedure on the set of slack equations so that we can actually find the maximum value. To the augmented matrix representing the slack equations, we add a new row at the bottom that represents the objective function in such a way that its value becomes part of any basic feasible solution read from that matrix. Then, finally, we discover the signal that tells us when the maximum has been reached. We now consider a typical problem with two variables and view the associated geometry:

$$
\begin{aligned}
\text{Maximize } f &= 5x + 3y \quad \} \quad \leftarrow \text{Objective function} \\
\text{subject to } 2x + 3y &\le 18 \quad \Big\} \\
2x + y &\le 10 \quad \Big\} \quad \leftarrow \text{Structural constraints} \\
x \ge 0, y &\ge 0. \quad \} \quad \leftarrow \text{Nonnegativity constraints}
\end{aligned}
$$

The feasible region for this problem, shown in Figure 6, has each corner point labeled with its coordinates and its value of f.

HISTORICAL NOTE

GEORGE DANTZIG (1914–)

During World War II, Dantzig worked with the United States Air Force as a combat analyst. He completed his doctorate in mathematics in 1946. In 1947, he developed the simplex algorithm for solving linear programming problems analytically. Dantzig's work had a major impact on the military and on the field of economics, both of which are areas in which optimization of resources is vital.

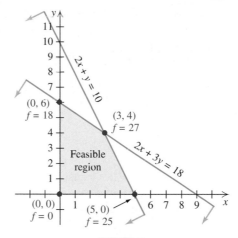

FIGURE 6
*Relating the Simplex Algorithm
to the Feasible Region*

The first step in the simplex algorithm (already noted in the previous section) is the formation of the slack equations.

Step 1 in the Simplex Algorithm

Insert a slack variable into each of the structural constraints.
 The result is this system of slack equations:

$$2x + 3y + s_1 \qquad = 18$$
$$2x + y \qquad + s_2 = 10.$$

Step 2 in the Simplex Algorithm

Rewrite the objective function to match the format of the slack equations, and adjoin the objective function to the bottom of the slack equation system.
 Accordingly, rewrite the objective function $f = 5x + 3y$ as $-5x - 3y + f = 0$ to match the format of the slack equations, and adjoin the function to the bottom of the slack equations in this fashion:

$$2x + 3y + s_1 \qquad\qquad = 18$$
$$2x + y \qquad + s_2 \qquad = 10$$
$$-5x - 3y \qquad\qquad + f = 0.$$

Step 3 in the Simplex Algorithm

Write the augmented matrix for this new system of equations.
 The augmented matrix is called the **initial simplex tableau.** Each number in the bottom row to the left of the vertical bar is called an **indicator.**

In our present problem, the initial simplex tableau is

$$
\begin{array}{ccccc}
\ x & y & s_1 & s_2 & f \\
\end{array}
$$

$$
\left[
\begin{array}{ccccc|c}
2 & 3 & 1 & 0 & 0 & 18 \\
2 & 1 & 0 & 1 & 0 & 10 \\
\hline
-5 & -3 & 0 & 0 & 1 & 0
\end{array}
\right].
$$

$\}\leftarrow$ Slack equations

$\leftarrow$ Objective function

$\uparrow \quad \uparrow \quad \uparrow \quad \uparrow \quad \uparrow$

Indicators

Traditionally, the last equation is separated from the slack equations in the augmented matrix by a horizontal bar, to remind us of the special role that equation plays in the maximization process that unfolds.

The constraints of a standard maximum-type linear programming problem assure us that the origin (all original variables equal zero) will always be in the feasible region; in fact, the origin will always be the **basic feasible solution** to the initial simplex tableau: $x = 0$, $y = 0$, $s_1 = 18$, and $s_2 = 10$. Continuing to read this solution from the last row, we see that $f = 0$. (See Figure 6.)

Step 4 in the Simplex Algorithm

The most negative indicator in the last row of the tableau determines the pivot column. The smallest-quotient rule is applied to that column.

The simplex algorithm provides an orderly scheme for selecting pivot elements. The pivot column is selected in such a way that the column's variable will cause f to increase faster per unit change than with any other column variable. The last equation of our present simplex tableau (the initial one) indicates that $-5x - 3y + f = 0$, or $f = 5x + 3y$, which shows that a unit increase in x will produce an increase of 5 in f, while a unit increase in y will produce an increase of 3 in f. Therefore, the pivot column will be the x column. Notice that this choice is displayed in the simplex tableau as *the most negative indicator* in the last row. We now move up to the slack equations and use the smallest-quotient rule to select the pivot row:

$$
\begin{array}{ccccc}
\ x & y & s_1 & s_2 & f \\
\end{array}
$$

$$
\left[
\begin{array}{ccccc|c}
2 & 3 & 1 & 0 & 0 & 18 \\
② & 1 & 0 & 1 & 0 & 10 \\
\hline
-5 & -3 & 0 & 0 & 1 & 0
\end{array}
\right].
$$

Quotient: 9

Quotient: 5 $\leftarrow$ Pivot row

$\uparrow$

Pivot column

The circled element where the arrow from the pivot row and the arrow from the pivot column meet is the pivot element.

Step 5 in the Simplex Algorithm

Perform the pivot operation on the selected pivot element.

Next, we pivot on the circled element by making it a 1 and *all* other elements in that column a 0, including the -5 in the last row:

$$\frac{1}{2}R_2 \to R_2 \quad \begin{array}{ccccc} x & y & s_1 & s_2 & f \\ \end{array}$$

$$\frac{1}{2}R_2 \to R_2 \quad \left[\begin{array}{ccccc|c} 2 & 3 & 1 & 0 & 0 & 18 \\ 1 & \frac{1}{2} & 0 & \frac{1}{2} & 0 & 5 \\ -5 & -3 & 0 & 0 & 1 & 0 \end{array}\right]$$

$$\begin{array}{c} -2R_2 + R_1 \to R_1 \\ \\ 5R_2 + R_3 \to R_3 \end{array} \quad \begin{array}{ccccc} x & y & s_1 & s_2 & f \\ \end{array}$$

$$\begin{array}{c} -2R_2 + R_1 \to R_1 \\ \\ 5R_2 + R_3 \to R_3 \end{array} \quad \left[\begin{array}{ccccc|c} 0 & 2 & 1 & -1 & 0 & 8 \\ 1 & \frac{1}{2} & 0 & \frac{1}{2} & 0 & 5 \\ 0 & -\frac{1}{2} & 0 & \frac{5}{2} & 1 & 25 \end{array}\right].$$

The basic feasible solution is now $y = 0$, $s_2 = 0$, $x = 5$, $s_1 = 8$, and $f = 25$. Note the corner point with these x, y, and f values in Figure 6. This pivot operation has moved the basic feasible solution from the origin to the point $(5, 0)$ and simultaneously given us the value $f = 25$ at that point!

Now, we know from Figure 6 that the maximum value of f has not yet been reached. But how do we know this from the last matrix? Note that the last row, $-\frac{1}{2}y + \frac{5}{2}s_2 + f = 25$, can be rewritten as $f = 25 + \frac{1}{2}y - \frac{5}{2}s_2$. Because both y and s_2 are zero in the current solution, and neither is permitted to be negative, if we are to stay in the feasible region, any positive change in s_2 would *decrease* the value of f from its present value of 25. On the other hand, any positive change in y will further *increase* the value of f, because y has a positive coefficient. However, this positive coefficient *is displayed as negative* in the format of the augmented matrix, which implies that, from the matrix point of view, *any negative indicator* indicates that further growth in f is possible. Moreover, the most negative of all such indicators identifies the variable giving f the fastest growth per unit increase in that variable. Thus, our next pivot column will be the y column, and, again, the smallest-quotient rule applied to the coefficients in that column will determine the pivot row:

$$\begin{array}{ccccc} x & y & s_1 & s_2 & f \\ \end{array}$$

$$\left[\begin{array}{ccccc|c} 0 & ② & 1 & -1 & 0 & 8 \\ 1 & \frac{1}{2} & 0 & \frac{1}{2} & 0 & 5 \\ 0 & -\frac{1}{2} & 0 & \frac{5}{2} & 1 & 25 \end{array}\right] \begin{array}{l} \text{Quotient: } 4 \quad \leftarrow \text{Pivot row} \\ \text{Quotient: } 10 \\ \\ \end{array}$$

$$\underset{\text{Pivot column}}{\uparrow}$$

Pivoting on the circled element requires two steps:
First,

$$\frac{1}{2}R_1 \to R_1 \quad \begin{array}{ccccc} x & y & s_1 & s_2 & f \\ \end{array}$$

$$\frac{1}{2}R_1 \to R_1 \quad \left[\begin{array}{ccccc|c} 0 & 1 & \frac{1}{2} & -\frac{1}{2} & 0 & 4 \\ 1 & \frac{1}{2} & 0 & \frac{1}{2} & 0 & 5 \\ 0 & -\frac{1}{2} & 0 & \frac{5}{2} & 1 & 25 \end{array}\right],$$

and second,

$$\begin{array}{c} \\ \\ -\frac{1}{2}R_1 + R_2 \rightarrow R_2 \\ \frac{1}{2}R_1 + R_3 \rightarrow R_3 \end{array} \begin{array}{cccccc} x & y & s_1 & s_2 & f \\ \left[\begin{array}{ccccc|c} 0 & 1 & \frac{1}{2} & -\frac{1}{2} & 0 & 4 \\ 1 & 0 & -\frac{1}{4} & \frac{3}{4} & 0 & 3 \\ 0 & 0 & \frac{1}{4} & \frac{9}{4} & 1 & 27 \end{array}\right]. \end{array}$$

This time, the basic feasible solution is $s_1 = 0$, $s_2 = 0$, $x = 3$, $y = 4$, and $f = 27$. Note the point in Figure 6 with these x, y, and f values. This pivot has moved the solution from the point $(5, 0)$ to the point $(3, 4)$ and has simultaneously calculated the new value of f at that point: 27.

Geometrically, we know that the maximum value of f for points in the feasible region has been reached. However, can we tell this from the augmented matrix? The answer is yes, and the proof is found by examining the last row. This time, the equation in the last row is $\frac{1}{4}s_1 + \frac{9}{4}s_2 + f = 27$, which can be rewritten as $f = 27 - \frac{1}{4}s_1 - \frac{9}{4}s_2$. At present, both s_1 and s_2 are 0, but if we pivot further, one will necessarily change from a nonbasic variable to a basic variable. This will result in a positive value for that variable, which will *decrease* the value of f from its present value of 27. Therefore, no further increase in f is possible. The coefficients of these variables, when viewed as indicators in the last simplex tableau, are *positive,* which leads to the conclusion that if no negative indicator appears in the last row of the matrix, then the maximum value of f has been reached.

Step 6 of the Simplex Algorithm

If a negative indicator is present, repeat steps 4 and 5. If no negative indicator is present, the maximum of the objective function has been reached.

Even though it is instructive and interesting to match the geometry with each basic feasible solution whenever possible, the beauty of the simplex algorithm is that it is algebraic in nature and can be performed on any standard maximum-type linear programming problem, regardless of its size. Furthermore, the repetitive nature of the pivoting process is why we call this process an algorithm.

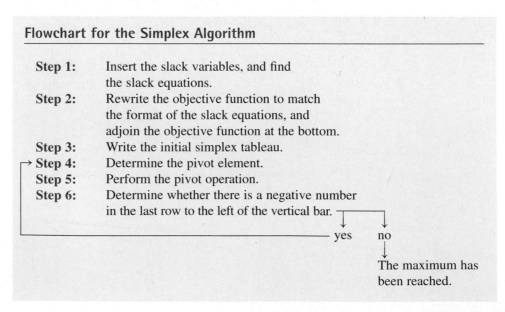

Flowchart for the Simplex Algorithm

Step 1:	Insert the slack variables, and find the slack equations.
Step 2:	Rewrite the objective function to match the format of the slack equations, and adjoin the objective function at the bottom.
Step 3:	Write the initial simplex tableau.
→ **Step 4:**	Determine the pivot element.
Step 5:	Perform the pivot operation.
Step 6:	Determine whether there is a negative number in the last row to the left of the vertical bar.

yes   no

The maximum has been reached.

Just as with our developmental example, this step-by-step procedure always starts with the basic solution at the origin (all variables in the original problem equal zero) and then successively pivots, giving basic feasible solutions at adjacent corner points on the feasible region until the corner that gives the maximum of the objective function is reached.

EXAMPLE 1

Using the Simplex Algorithm

Use the simplex algorithm to solve this problem:

$$\text{Maximize } P = 3x + 4y + z$$
$$\text{subject to } \quad x + 2y + z \le 6$$
$$2x + 2z \le 4$$
$$3x + y + z \le 9$$
$$x \ge 0, y \ge 0, z \ge 0.$$

SOLUTION Following the steps in the flowchart, we form the slack equations first:

$$x + 2y + z + s_1 = 6$$
$$2x + 2z + s_2 = 4$$
$$3x + y + z + s_3 = 9.$$

Now, rewriting the objective function as $-3x - 4y - z + P = 0$ and adjoining it to the bottom of the slack equations, we obtain a new system of four equations with seven unknowns, whose initial simplex tableau is

$$
\begin{array}{ccccccc}
x & y & z & s_1 & s_2 & s_3 & P \\
\end{array}
$$
$$
\left[
\begin{array}{ccccccc|c}
1 & ② & 1 & 1 & 0 & 0 & 0 & 6 \\
2 & 0 & 2 & 0 & 1 & 0 & 0 & 4 \\
3 & 1 & 1 & 0 & 0 & 1 & 0 & 9 \\
\hline
-3 & -4 & -1 & 0 & 0 & 0 & 1 & 0
\end{array}
\right].
$$

← Slack equations

← Objective function

Indicators

Next, the most negative indicator in the last row is -4, showing that the pivot column is the y column. Applying the smallest-quotient rule to the coefficients of y *in the slack equations only* gives quotients of 3 (from the first row) and 9 (from the third row). We cannot divide by 0 in the second row. Because the smallest quotient, 3, was obtained from the first row, the first row will be the pivot row. The resulting pivot element will be the circled 2 in the initial simplex tableau. The completed pivot operation and the preparations for selecting the next pivot element are as follows:

$$
\begin{array}{ccccccc}
x & y & z & s_1 & s_2 & s_3 & P \\
\end{array}
$$
$$
\left[
\begin{array}{ccccccc|c}
\frac{1}{2} & 1 & \frac{1}{2} & \frac{1}{2} & 0 & 0 & 0 & 3 \\
② & 0 & 2 & 0 & 1 & 0 & 0 & 4 \\
\frac{5}{2} & 0 & \frac{1}{2} & -\frac{1}{2} & 0 & 1 & 0 & 6 \\
\hline
-1 & 0 & 1 & 2 & 0 & 0 & 1 & 12
\end{array}
\right].
$$

Quotient: 6

Quotient: 2 ← Pivot row

Quotient: $\frac{12}{5}$

Pivot column

The resulting pivot element is circled in the tableau. Pivoting on that element gives this tableau:

$$
\begin{array}{ccccccc}
x & y & z & s_1 & s_2 & s_3 & P \\
\end{array}
$$

$$
\left[
\begin{array}{ccccccc|c}
0 & 1 & 0 & \frac{1}{2} & -\frac{1}{4} & 0 & 0 & 2 \\
1 & 0 & 1 & 0 & \frac{1}{2} & 0 & 0 & 2 \\
0 & 0 & -2 & -\frac{1}{2} & -\frac{5}{4} & 1 & 0 & 1 \\
\hline
0 & 0 & 2 & 2 & \frac{1}{2} & 0 & 1 & 14 \\
\end{array}
\right].
$$

No negative indicators exist, which means that the maximum for P on the feasible region has been reached. The solution to the linear programming problem is then $P = 14$ when $z = 0$, $s_1 = 0$, $s_2 = 0$, $x = 2$, $y = 2$, and $s_3 = 1$. That is, the maximum value of P on the feasible region is 14 and occurs at the point $(2, 2, 0)$. ■

EXAMPLE 2

Using the Simplex Algorithm

Use the simplex algorithm to solve this problem:

$$
\begin{aligned}
\text{Maximize } & f = 3x + 2y \\
\text{subject to} \quad & x + y \leq 4 \\
& -2x + y \leq 2 \\
& x \geq 0, y \geq 0.
\end{aligned}
$$

SOLUTION The corner points of the feasible region, each labeled with its x, y, and f values, are shown in Figure 7. Inserting slack variables into the first two inequalities and rewriting f as $-3x - 2y + f = 0$ gives this system of equations:

$$
\begin{aligned}
x + y + s_1 &= 4 \\
-2x + y + s_2 &= 2 \\
-3x - 2y + f &= 0.
\end{aligned}
$$

The initial simplex tableau and preparations for selecting the first pivot element are as follows:

$$
\begin{array}{ccccc}
x & y & s_1 & s_2 & f \\
\end{array}
$$

$$
\left[
\begin{array}{ccccc|c}
① & 1 & 1 & 0 & 0 & 4 \\
-2 & 1 & 0 & 1 & 0 & 2 \\
\hline
-3 & -2 & 0 & 0 & 1 & 0 \\
\end{array}
\right].
$$
Quotient: 4 ← Pivot row

↑ Pivot column

Recall that the smallest-quotient rule allows division by *positive* numbers only. The first pivot element is the circled 1 in the first-row, first-column position. The result of this pivot is

$$
\begin{array}{ccccc}
x & y & s_1 & s_2 & f \\
\end{array}
$$

$$
\begin{array}{r}
\\
3R_1 + R_2 \rightarrow R_2 \\
3R_1 + R_3 \rightarrow R_3
\end{array}
\left[
\begin{array}{ccccc|c}
1 & 1 & 1 & 0 & 0 & 4 \\
0 & 3 & 2 & 1 & 0 & 10 \\
0 & 1 & 3 & 0 & 1 & 12 \\
\end{array}
\right].
$$

The fact that no negative indicators appear shows that the maximum value of f on the feasible region has been reached. That maximum value is 12 and occurs when $x = 4$ and $y = 0$. Check this result against Figure 7. ∎

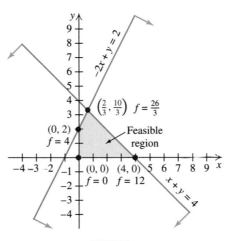

FIGURE 7
A Graphic Illustration of
Basic Feasible Solutions

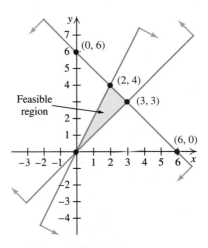

FIGURE 8
Illustrating Why 0 Is Allowed
as the Smallest Quotient

EXAMPLE 3

Using the Simplex
Algorithm

Use the simplex algorithm to solve this linear programming problem:

$$\text{Maximize } f = 2x + y$$
$$\text{subject to } x + y \le 6$$
$$y \le 2x$$
$$x \quad \le y$$
$$x \ge 0, y \ge 0.$$

SOLUTION The graph of the feasible region is shown in Figure 8. Next, the algebra of inequalities can be used to rewrite the second and third constraints so that they satisfy the conditions set forth for a standard maximum-type problem:

$$-2x + y \le 0$$
$$x - y \le 0.$$

Therefore, the constraint system now appears as

$$x + y \le 6$$
$$-2x + y \le 0$$
$$x - y \le 0.$$

Next, inserting the appropriate slack variables and rewriting f gives this system of equations:

$$
\begin{aligned}
x + y + s_1 & & & & = 6 \\
-2x + y & + s_2 & & & = 0 \\
x - y & & + s_3 & & = 0 \\
-2x - y & & & + f & = 0.
\end{aligned}
$$

The initial simplex tableau and preparations for selecting the first pivot element are as follows:

$$
\begin{array}{cccccc}
x & y & s_1 & s_2 & s_3 & f \\
\end{array}
$$

$$
\left[
\begin{array}{cccccc|c}
1 & 1 & 1 & 0 & 0 & 0 & 6 \\
-2 & 1 & 0 & 1 & 0 & 0 & 0 \\
① & -1 & 0 & 0 & 1 & 0 & 0 \\
-2 & -1 & 0 & 0 & 0 & 1 & 0
\end{array}
\right]
\quad
\begin{array}{l}
\text{Quotient: } 6 \\
\\
\text{Quotient: } 0 \leftarrow \text{Pivot row} \\
\end{array}
$$

↑ Pivot column

The smallest-quotient rule of the simplex algorithm states that the smallest *nonnegative* quotient is to be used to select the pivot row. Because 0 is that number in this case, the third row is the pivot row. The result of this pivot is

$$
\begin{array}{cccccc}
x & y & s_1 & s_2 & s_3 & f \\
\end{array}
$$

$$
\left[
\begin{array}{cccccc|c}
0 & ② & 1 & 0 & -1 & 0 & 6 \\
0 & -1 & 0 & 1 & 2 & 0 & 0 \\
1 & -1 & 0 & 0 & 1 & 0 & 0 \\
0 & -3 & 0 & 0 & 2 & 1 & 0
\end{array}
\right]
\quad
\text{Quotient: } 3 \leftarrow \text{Pivot row}
$$

↑ Pivot column

The negative indicator, -3, shows the need for another pivot; according to the smallest-quotient rule, that pivot element will be the circled 2 in the preceding augmented matrix. The pivot operation results in this tableau:

$$
\begin{array}{cccccc}
x & y & s_1 & s_2 & s_3 & f \\
\end{array}
$$

$$
\left[
\begin{array}{cccccc|c}
0 & 1 & \frac{1}{2} & 0 & -\frac{1}{2} & 0 & 3 \\
0 & 0 & \frac{1}{2} & 1 & \frac{3}{2} & 0 & 3 \\
1 & 0 & \frac{1}{2} & 0 & \frac{1}{2} & 0 & 3 \\
0 & 0 & \frac{3}{2} & 0 & \frac{1}{2} & 1 & 9
\end{array}
\right]
$$

The last tableau shows the maximum value of f to be 9 when $x = 3$ and $y = 3$. ∎

EXAMPLE 4

Using the Simplex Algorithm to Solve a Manufacturing Problem

The Hi Tech Company is making three models of computer desks: an executive model, an office model, and a student model. Construction of the executive model requires 2 hours in the cabinet shop, 1 hour in the finishing department, and 1 hour in the crating department. The office model requires 1 hour, 2 hours, and 1 hour, respectively, in these departments, while the student model requires 1 hour, 1 hour, and $\frac{1}{2}$ hour, respectively, in these departments. On a daily basis, the cabinet shop has 16 hours available; the finishing department, 16; and the crating department, 10. If the company realizes a profit of $150 on each executive model, $125 on each office model, and $50 on each student model, how many of each type of computer desk should it make and sell to maximize profits?

SOLUTION Let x, y, and z be the number of executive, office, and student models, respectively, to be produced and sold. The problem may then be translated into a linear programming problem as follows:

$$\text{Maximize } P = 150x + 125y + 50z$$

$$
\begin{aligned}
\text{subject to } 2x +\ & y +\ z \le 16 && \text{(Cabinet hours)} \\
x + 2y +\ & z \le 16 && \text{(Finishing hours)} \\
x +\ & y + \tfrac{1}{2}z \le 10 && \text{(Crating hours)} \\
& x \ge 0, y \ge 0, z \ge 0.
\end{aligned}
$$

After inserting a slack variable for each of the first three constraints and rewriting the objective function as $-150x - 125y - 50z + P = 0$, we may write the initial simplex tableau, along with the notations for finding the first pivot element, as

$$
\begin{array}{ccccccc|c}
x & y & z & s_1 & s_2 & s_3 & P & \\
\textcircled{2} & 1 & 1 & 1 & 0 & 0 & 0 & 16 \\
1 & 2 & 1 & 0 & 1 & 0 & 0 & 16 \\
1 & 1 & \tfrac{1}{2} & 0 & 0 & 1 & 0 & 10 \\
\hline
-150 & -125 & -50 & 0 & 0 & 0 & 1 & 0
\end{array}
$$

Quotient: 8 ← Pivot row
Quotient: 16
Quotient: 10

Pivot column

The smallest-quotient rule implies that the 2 in the first-row, first-column position should be the first pivot element. Upon completion of the pivoting process, the resulting augmented matrix, along with the preparations for finding the next pivot element, appear as follows:

$$
\begin{array}{ccccccc|c}
x & y & z & s_1 & s_2 & s_3 & P & \\
1 & \tfrac{1}{2} & \tfrac{1}{2} & \tfrac{1}{2} & 0 & 0 & 0 & 8 \\
0 & \tfrac{3}{2} & \tfrac{1}{2} & -\tfrac{1}{2} & 1 & 0 & 0 & 8 \\
0 & \textcircled{\tfrac{1}{2}} & 0 & -\tfrac{1}{2} & 0 & 1 & 0 & 2 \\
\hline
0 & -50 & 25 & 75 & 0 & 0 & 1 & 1200
\end{array}
$$

Quotient: 16
Quotient: $\tfrac{16}{3}$
Quotient: 4 ← Pivot row

Pivot column

The negative indicator shows that another pivot is required, and the smallest-quotient rule indicates that the new pivot element will be the $\tfrac{1}{2}$ in the third-row, second-column position. The augmented matrix resulting from the pivoting process is

$$
\begin{bmatrix}
1 & 0 & \tfrac{1}{2} & 1 & 0 & -1 & 0 & 6 \\
0 & 0 & \tfrac{1}{2} & 1 & 1 & -3 & 0 & 2 \\
0 & 1 & 0 & -1 & 0 & 2 & 0 & 4 \\
\hline
0 & 0 & 25 & 25 & 0 & 100 & 1 & 1400
\end{bmatrix}.
$$

Because no negative indicators are present, the maximum for P within the feasible region has been found. The maximum value of P is 1400 (dollars), and it occurs when $x = 6$, $y = 4$, and $z = 0$. That is, to maximize profits under the given constraints, the company should make six executive models, four office models, and no student models. ▬

Important Facts about the Simplex Algorithm

A Surplus in a Given Constraint The values of the slack variables in the solution to Example 4 are $s_1 = 0$, $s_2 = 2$, and $s_3 = 0$. Each of these variables is associated with a particular constraint of the problem: s_1 with the cabinet constraint, s_2 with the finishing constraint, and s_3 with the crating constraint. The fact that s_1 and s_3 are zero means that there is no slack in those constraints. In other words, all of the available resources of the cabinet and crating departments are used when P is maximized. In contrast, $s_2 = 2$ means that there are 2 hours of time in the finishing department that are unused, or are **surplus** to the production process.

The Column Representing the Objective Function Note that the column representing the function to be maximized never changes throughout the solution process, and some textbooks do not bother to put it in the tableau. If you choose not to put it in, just remember that the number in the lower right corner of the tableau is the current value of f.

What If there Is a Tie in Selecting the Pivot Column or Pivot Row? Just use the column or row of your choice. In the rare event that your choice leads to looping within a fixed set of tableaus, go back and try the other choice.

There Is Never a Negative Number in the Rightmost Column above the Horizontal Bar. Due to the nature of a standard maximum-type linear programming problem, the initial simplex tableau can never have a negative number in the rightmost column above the horizontal bar. In addition, if the pivoting rules of this book are correctly followed, there can *never* be a negative number appearing in the rightmost column above the horizontal bar in any matrix obtained from successive pivots on the initial simplex tableau, regardless of how many pivot operations have been performed.

Is It Possible for a Negative Number to Appear in the Lower Right-Hand Corner? Depending on the nature of the objective function, it is possible for a negative number to appear (even in the final solution matrix) in the lower right-hand corner of a matrix associated with the pivoting process. If a negative number appears in the final solution matrix, it means that the maximum value of the objective function at that point is negative. For example, if $f = -2x - 3y$ is to be maximized over a feasible region in the first quadrant that does not contain the origin, then the corner point at which the maximum occurs will have at least one coordinate that is not zero. Hence, when f is evaluated there, a negative maximum value will result, and this value will appear in the lower right-hand corner of the final tableau.

What If No Optimal Solution Exists? No optimal solution exists when, at some point, we are unable to move forward in the selection of the next pivot element. As an example, suppose we ask the simplex algorithm to find the maximum value of $f = x + y$ subject to

$$x \leq 1$$
$$x \geq 0, y \geq 0.$$

(The feasible region is a one-unit-wide vertical strip along the y-axis, unbounded in the y-direction. We already know that there is no maximum for the objective function over this region.) There is only one inequality, $x \leq 1$, that requires a slack variable:

$$x + s_1 = 1.$$

Rewriting the objective function, we get

$$-x - y + f = 0.$$

The initial simplex tableau, the selection of the first pivot element, and the matrix resulting from that pivot are as follows:

$$
\begin{array}{cccc}
x & y & s_1 & f \\
\end{array}
\left[\begin{array}{cccc|c}
\textcircled{1} & 0 & 1 & 0 & 1 \\
-1 & -1 & 0 & 1 & 0
\end{array}\right]
\begin{array}{c} \\ 1R_1 + R_2 \rightarrow R_2 \end{array}
\begin{array}{cccc}
x & y & s_1 & f \\
\end{array}
\left[\begin{array}{cccc|c}
1 & 0 & 1 & 0 & 1 \\
0 & -1 & 1 & 1 & 1
\end{array}\right].
$$

The negative indicator in the second matrix indicates that the maximum for f has not been reached and that at least one more pivot is required. The pivot column must be the y column; however, it is impossible to select a pivot row with the use of the smallest-quotient rule. This is one signal that no maximum exists.

Our last example introduces the use of Excel's Solver for finding the solution to a linear program problem.

EXAMPLE 5

Using Excel to Solve Linear Programming Problems

Consider again the production problem given in Example 4. Use Excel to find the production rates that maximize profit.

SOLUTION First, a spreadsheet should be labeled and filled with pertinent information, as shown in Figure 9. Although this is not necessary, it greatly helps illuminate what is going

	A	B	C	D	E	F
1		**Variables**				
2	Exec Model	x	=			
3	Office Model	y	=			
4	Student Model	z	=			
5		**Objective Function**				
6	Maximize	P=150x+125y+50z				
7		**Structural Constraints**				
8	Cabinet Shop Hours	2x+1y+1z<=16		<=		
9	Finishing Dept. Hours	1x+2y+1z<=16		<=		
10	Crating Dept. Hours	1x+1y+(1/2)z<=10		<=		
11						

FIGURE 9
Initial Labeling of Spreadsheet

on. All of the cell entries up to this point are text, not formulas. It is important to recognize the three main parts of our spreadsheet: the range in which we are going to load the values of the variables (D2:D4), the cell that will contain the objective function's value (C6), and the ranges containing the constraints (C8:C10 and E8:E10).

Now we enter the values and formulas that will be used by Solver to compute the solution. Enter 0 into each of the cells D2, D3, and D4, that being the initial values of x, y, and z (since, initially, those are our nonbasic variables.) In cell C6, we enter the formula for the profit in terms of the variables' cells, namely,

$$=150*D2+125*D3+50*D4.$$

Finally, the constraints are entered; this requires two columns of cells. The first column contains the formulas for the left-hand side of the constraints, while the second column contains the constants on the right-hand side of the constraints.

Cell	Content	Cell	Content
C8	=2*D2+D3+D4	E8	16
C9	=D2+2*D3+D4	E9	16
C10	=D2+D3+.5*D4	E10	10

When this material is entered, the spreadsheet looks like Figure 10.

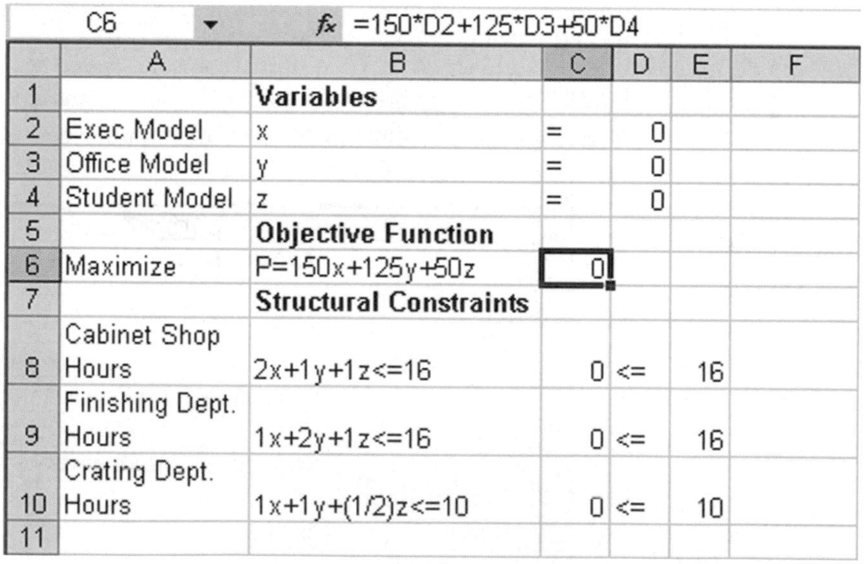

C6	▼	f_x =150*D2+125*D3+50*D4				
	A	B	C	D	E	F
1		**Variables**				
2	Exec Model	x	=	0		
3	Office Model	y	=	0		
4	Student Model	z	=	0		
5		**Objective Function**				
6	Maximize	P=150x+125y+50z	0			
7		**Structural Constraints**				
8	Cabinet Shop Hours	2x+1y+1z<=16	0	<=	16	
9	Finishing Dept. Hours	1x+2y+1z<=16	0	<=	16	
10	Crating Dept. Hours	1x+1y+(1/2)z<=10	0	<=	10	
11						

FIGURE 10
The Initial Values at the Origin of the Feasible Region

We now run the Solver function, which can be found under the Tools menu (Figure 11). The safest first step is to click and confirm the Reset All button, which clears all previous entries. Next, enter the cell ranges exactly as seen in Figure 11. Notice how the addresses needed by Solver are divided into precisely the three areas we set up

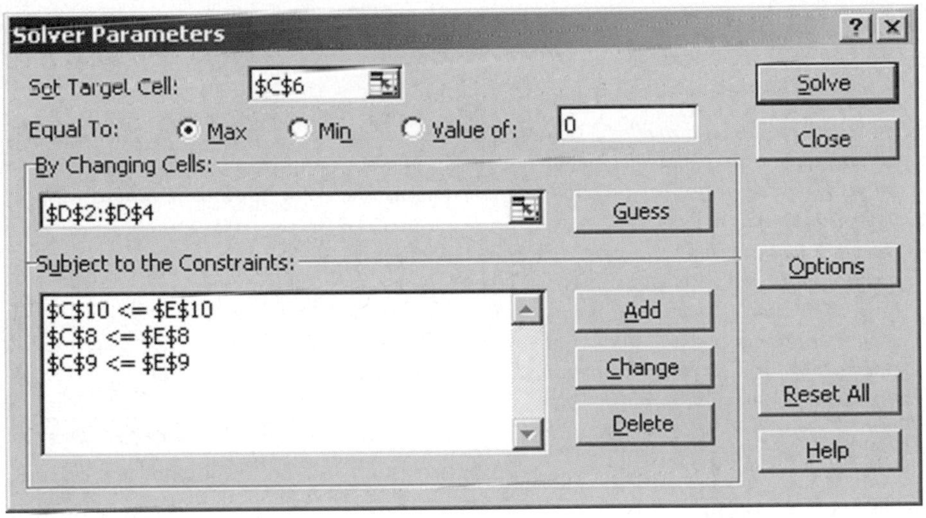

FIGURE 11

The `Solver`, *Ready to Run*

in the spreadsheet. Since this is a maximization problem, we make sure that the correct radio button (`Max`) is selected in the `Equal To` area.

Before clicking the `Solve` button, there is one last step: Select `Options`, and make sure that both `Assume Linear Model` and `Assume Non-Negative` are checked. The first option allows the program to be more efficient in its computations, while the second is the reason there was no need to include the non-negativity constraints in the spreadsheet.

After the `Solver` runs, a pop-up window asks whether the results should be accepted; in this case, they should be. Our final solution can be read from the screen. (See Figure 12.) Even the slack of 2 hours in the finishing department is seen in the constraint range.

	A	B	C	D	E	F
1		**Variables**				
2	Exec Model	x	=	6		
3	Office Model	y	=	4		
4	Student Model	z	=	0		
5		**Objective Function**				
6	Maximize	P=150x+125y+50z	1400			
7		**Structural Constraints**				
8	Cabinet Shop Hours	2x+1y+1z<=16	16	<=	16	
9	Finishing Dept. Hours	1x+2y+1z<=16	14	<=	16	
10	Crating Dept. Hours	1x+1y+(1/2)z<=10	10	<=	10	
11						

FIGURE 12

The Solution to the Production Problem

Exercises 5.2

In each of Exercises 1 through 6, suppose that the matrix shown represents a simplex tableau after a sequence of pivot operations has been performed on an initial tableau. Answer these questions about each matrix:

(a) Does the tableau show that a maximum for the objective function has been reached?

(b) Find the value of f and the point at which that value occurs at this stage of the simplex algorithm.

(c) Does a surplus exist for any constraint? If so, which constraint and how much surplus?

1.
$$\begin{array}{ccccc|c} x & y & s_1 & s_2 & f & \\ 1 & 0 & \frac{2}{3} & \frac{1}{3} & 0 & 5 \\ 0 & 1 & 3 & 0 & 0 & 8 \\ 0 & 0 & 5 & 6 & 1 & 20 \end{array}$$

2.
$$\begin{array}{ccccc|c} x & y & s_1 & s_2 & f & \\ 2 & 0 & 1 & 1 & 0 & 40 \\ 1 & 1 & 0 & 3 & 0 & 60 \\ -1 & 0 & 0 & 5 & 1 & 38 \end{array}$$

3.
$$\begin{array}{ccccc|c} x & y & s_1 & s_2 & f & \\ 1 & \frac{2}{5} & \frac{1}{5} & 0 & 0 & \frac{8}{5} \\ 0 & \frac{5}{6} & \frac{1}{3} & 1 & 0 & 16 \\ 0 & \frac{8}{5} & \frac{6}{5} & 0 & 1 & \frac{80}{5} \end{array}$$

4.
$$\begin{array}{cccccc|c} x & y & z & s_1 & s_2 & f & \\ 1 & 2 & 0 & 2 & 8 & 0 & 8 \\ 0 & 3 & 1 & 3 & 1 & 0 & 20 \\ 0 & 0 & 0 & -1 & 2 & 1 & 50 \end{array}$$

5.
$$\begin{array}{cccccc|c} x & y & s_1 & s_2 & s_3 & f & \\ 1 & 2 & 0 & 0 & \frac{5}{3} & 0 & 6 \\ 0 & -1 & 1 & 0 & 3 & 0 & 8 \\ 0 & 4 & 0 & 1 & \frac{1}{3} & 0 & 2 \\ 0 & -\frac{2}{5} & 0 & 0 & \frac{2}{3} & 1 & 25 \end{array}$$

6.
$$\begin{array}{cccccc|c} x & y & s_1 & s_2 & s_3 & f & \\ 1 & 0 & -\frac{2}{5} & \frac{5}{2} & 0 & 0 & \frac{12}{5} \\ 0 & 1 & 0 & -2 & 0 & 0 & 8 \\ 0 & 0 & \frac{1}{6} & \frac{2}{5} & 1 & 0 & \frac{4}{5} \\ 0 & 0 & \frac{5}{6} & \frac{7}{6} & 0 & 1 & \frac{17}{3} \end{array}$$

In each of Exercises 7 through 22,

(a) Graph the feasible region.

(b) At each corner point of the feasible region, find the value of x, y, f, and each slack variable.

(c) Solve the linear programming problem by the simplex algorithm, and after each pivot identify the basic feasible solution with a corner point of the feasible region.

7. Maximize $f = 8x + 6y$

   subject to $2x + y \le 10$

   $x + y \le 7$

   $x \ge 0, y \ge 0$.

8. Maximize $f = 2x + 3y$

   subject to $x + y \le 8$

   $3x + 5y \le 30$

   $x \ge 0, y \ge 0$.

9. Maximize $f = x + 4y$

   subject to $2x + 5y \le 20$

   $x + 4y \le 12$

   $x \ge 0, y \ge 0$.

10. Maximize $f = 2x + y$

    subject to $x + 3y \le 14$

    $2x + y \le 11$

    $x \ge 0, y \ge 0$.

11. Maximize $f = 5x + y$

    subject to $x + 2y \le 6$

    $4x + 3y \le 120$

    $x \ge 0, y \ge 0$.

12. Maximize $f = 2x + 3y$

    subject to $x + y \le 2$

    $x \le 2$

    $x \ge 0, y \ge 0$.

13. Maximize $f = 2x + y$

    subject to $x + y \le 5$

    $y \le 5$

    $x \ge 0, y \ge 0$.

14. Maximize $f = 3x + 5y$

    subject to $2x + y \le 6$

    $y \le 4$

    $x \ge 0, y \ge 0$.

15. Maximize $f = 3x + 3y$

subject to $\quad x + \quad y \le 9$

$\qquad -x + 2y \le 10$

$\qquad x \ge 0, y \ge 0.$

16. Maximize $f = -x - 3y$

subject to $\quad x + y \le 5$

$\qquad 2x - y \le 4$

$\qquad x \ge 0, y \ge 0.$

17. Maximize $f = -2x - y$

subject to $x - \quad y \le 6$

$\qquad x + 2y \le 8$

$\qquad x \ge 0, y \ge 0.$

18. Maximize $f = 2x + 2y$

subject to $-2x + 3y \le 12$

$\qquad 2x + \quad y \le 6$

$\qquad x \ge 0, y \ge 0.$

19. Maximize $f = 3x + y$

subject to $y \le x$

$\qquad x \le 2$

$\qquad x \ge 0, y \ge 0.$

20. Maximize $f = x + 4y$

subject to $y \le x$

$\qquad x \le 2$

$\qquad x \ge 0, y \ge 0.$

21. Maximize $f = 5x + 2y$

subject to $x \le y$

$\qquad y \le 3$

$\qquad x \ge 0, y \ge 0.$

22. Maximize $f = 3x + y$

subject to $x + y \le 8$

$\qquad x \le 2y$

$\qquad y \le 6$

$\qquad x \ge 0, y \ge 0.$

Use the simplex algorithm to solve each of the linear programming problems in Exercises 23 through 37.

23. Maximize $f = 3x + 2y$

subject to $\quad x + \quad y \le \quad 9$

$\qquad 2x + \quad y \le 12$

$\qquad x + 4y \le 24$

$\qquad x \ge 0, y \ge 0.$

24. Maximize $P = 2x + y$

subject to $4x + \quad y \le 20$

$\qquad x + 2y \le 16$

$\qquad 3x + 2y \le 20$

$\qquad x \ge 0, y \ge 0.$

25. Maximize $f = 2x + y + 3z$

subject to $\quad x + 2y + \quad z \le 25$

$\qquad 3x + 2y + 2z \le 30$

$\qquad x \ge 0, y \ge 0, z \ge 0.$

26. Maximize $P = x + 2y + 4z$

subject to $x \qquad + 2z \le 10$

$\qquad 3y + \quad z < 24$

$\qquad x \ge 0, y \ge 0, z \ge 0.$

27. Maximize $f = 5x + 3y + 4z$

subject to $2x + \quad y + 2z \le 40$

$\qquad 2x + 2y + 3z \le 50$

$\qquad x + 3y + 2z \le 20$

$\qquad x \ge 0, y \ge 0, z \ge 0.$

28. Maximize $f = 3x + 2y + z$

subject to $-x + 2y + \quad z \le 12$

$\qquad 2x + \quad y - \quad z \le 6$

$\qquad x + 2y + 2z \le 8$

$\qquad x \ge 0, y \ge 0, z \ge 0.$

29. Maximize $P = 2x + 2y + z$

subject to $\quad x - 2y + \quad z \le 4$

$\qquad x - 4y + 3z \le 12$

$\qquad 2x + 2y + \quad z \le 17$

$\qquad x \ge 0, y \ge 0, z \ge 0.$

30. Maximize $f = 2x + 3y + z$

subject to $x \qquad + z \le 8$

$\qquad y - z \le 10$

$\qquad x - y - z \le 12$

$\qquad x \ge 0, y \ge 0, z \ge 0.$

31. Maximize $f = 2x + 3y + 3z$

subject to $\quad x \qquad + 2z \le 8$

$\qquad -y + 3z \le 10$

$\qquad -x + \quad y - \quad z \le 12$

$\qquad x \ge 0, y \ge 0, z \ge 0.$

32. Maximize $f = x + 3y + z$

subject to $x + y \leq 6$

$y \leq 4$

$x \leq 5$

$x \geq 0, y \geq 0, z \geq 0.$

33. Maximize $f = x + 2y + 4z$

subject to $y + z \leq 4$

$y \leq 3$

$z \leq 2$

$x \geq 0, y \geq 0, z \geq 0.$

34. Maximize $f = 4x + 3y + 2z + w$

subject to $2x + y + 2w \leq 12$

$-x + 2y + z + 3w \leq 6$

$x + 2z + 3w \leq 15$

$x + y + z + w \leq 10$

$x \geq 0, y \geq 0, z \geq 0, w \geq 0.$

35. Maximize $f = 4x + 4y + 2z + w$

subject to $x + z + 2w \leq 6$

$2x + y + w \leq 12$

$x + 2z + 3w \leq 18$

$x + y + 2z + w \leq 10$

$x \geq 0, y \geq 0, z \geq 0, w \geq 0.$

36. Maximize $f = 2x_1 + 3x_2 - 4x_3 - x_4$

subject to $x_1 + 2x_2 + 3x_3 + x_4 \leq 6$

$x_2 + 2x_3 \leq 5$

$x_1 - x_2 + 2x_4 \leq 4$

$x_1 \geq 0, x_2 \geq 0, x_3 \geq 0, x_4 \geq 0.$

37. Maximize $f = 2x_1 - x_2 + x_3 - 2x_4 - 3x_5$

subject to $2x_1 + x_2 + x_4 + 2x_5 \leq 2$

$4x_1 + 3x_3 + 2x_4 + x_5 \leq 8$

$x_1 + 2x_2 + 4x_3 + x_4 \leq 4$

$x_1 \geq 0, x_2 \geq 0, x_3 \geq 0, x_4 \geq 0, x_5 \geq 0.$

In the remainder of the exercises, find the objective function and the constraints, and then solve the problem by using the simplex method.

38. Manufacturing The Fine Line Company makes silver and gold ballpoint pens. A silver pen requires 1 minute in the grinder and 3 minutes in the bonder. A gold pen requires 3 minutes in the grinder and 4 minutes in the bonder. The grinder can be operated no more than 30 hours

per week and the bonder no more than 50 hours per week. If the company makes 30 cents on each silver pen and 60 cents on each gold pen, how many of each type of pen should be made and sold each week to maximize profits?

39. Manufacturing A bicycle manufacturer makes a 3-speed model and a 10-speed model with two operations: assembly and painting. Each 3-speed bike requires 1 hour to assemble and 2 hours to paint. Each 10-speed bike takes 1 hour to assemble and 1 hour to paint. The assembly operation has 80 hours per week available, and the painting operation has 100 work hours available each week. If the company makes $80 on each 3-speed bike and $60 on each 10-speed bike, how many bicycles should be made and sold each week to maximize profits?

40. Advertising An auto dealer's staffers can run a particular size of newspaper ad in black and white for $200 per printing and in color for $300 per printing. The company has budgeted a maximum of $6000 for such ads. The staffers know that they can run each ad no more than 20 times before it loses its effectiveness. They estimate that each time the black-and-white version is run, it will be read by 3000 people; each time the color version is run, it will be read by 5000 people. How many times should each ad—black and white, and the color—be run to reach the maximum number of readers?

41. Test Construction Suppose that you take a finite-mathematics test that has some true–false questions and some open-ended questions. The true–false questions are worth five points each, and the open-ended questions are worth eight points each. The rules of the test are that no more than 20 true–false questions are to be answered, and the total number of questions answered must be no more than 30. How many of each type of question should you answer correctly to maximize the number of points scored on the test?

42. Manufacturing A maker of microwave ovens is gearing up for a production run of deluxe- and standard-model ovens. A deluxe model requires 3 hours to assemble, and a standard model requires 2 hours to assemble. There are 180 hours of assembly time available. The manufacturer has on hand 50 deluxe-model cabinets and 30 standard-model cabinets for its production run. The company makes a profit of $30 on each deluxe model and $25 on each standard model. How many of each type of oven should be made to maximize profits?

43. Purchasing The Journalism Department at State University wants to buy computers to serve two departmental purposes. For one purpose, brand X computers, which cost $2500 each, are to be used; for the other purpose, brand Y computers, which cost $5000 each, are to be used. The department has space for no more than 15 brand X computers. If the department can spend no more than $100,000 for computers, how many of each brand of computer should be purchased to maximize the total number bought?

44. Manufacturing A confectioner has 600 pounds of chocolate, 100 pounds of nuts, and 50 pounds of fruit in inventory with which to make three types of candy: Sweet Tooth, Sugar Dandy, and Dandy Delite. A box of Sweet Tooth uses 3 pounds of chocolate, 1 pound of nuts, and 1 pound of fruit and sells for $8. A box of Sugar Dandy requires 4 pounds of chocolate and $\frac{1}{2}$ pound of nuts and sells for $5. A box of Dandy Delite requires 5 pounds of chocolate, $\frac{3}{4}$ pound of nuts, and 1 pound of fruit and sells for $6. How many boxes of each type of candy should be made from the available inventory to maximize revenue?

45. Advertising An advertising agency has developed radio, newspaper, and television ads for a particular business. Each radio ad costs $200, each newspaper ad costs $100, and each television ad costs $500 to run. The business does not want the television ad to run more than 20 times, and the sum of the numbers of times the radio and newspaper ads can be run is to be no more than 110. The agency estimates that each airing of the radio ad will reach 1000 people, each printing of the newspaper ad will reach 800 people, and each airing of the television ad will reach 1500 people. If the total amount to be spent on ads is not to exceed $15,000, how many times should each type of ad be run so that the total number of people reached is a maximum?

46. Wildlife Management A state fish-and-game commission is charged with managing an area designated as a wildlife refuge. The commission's members are interested in the populations of deer, quail, and turkeys. Studies indicate that each deer needs 1 unit of food group 1, 2 units of food group 2, and 2 units of food group 3. Each quail needs 1.2 units of food group 1, 1.8 units of food group 2, and 0.6 unit of food group 3. Each turkey needs 2 units of food group 1, 0.8 unit of food group 2, and 1 unit of food group 3. With its management practices, the commission estimates that there will be 600 units of food group 1 available, 900 units of food group 2, and 720 units of food group 3.

If the total number of the three species is to be a maximum, how many of each species will the refuge support?

47. Manufacturing A maker of toaster ovens has three models: a superdeluxe, a deluxe, and a standard model. It uses three operations to produce the ovens: assembly, painting, and testing/packaging. Each superdeluxe model requires 4 hours in assembly, 2 hours in painting, and 1 hour in testing/packaging. Each deluxe model requires 2 hours in assembly, 1 hour in painting, and 1 hour in testing/ packaging. Each standard model requires 1 hour in assembly, 1 hour in painting, and $\frac{1}{2}$ hour in testing/packaging. The manufacturer has 160 work hours available in assembly, 100 in painting, and 60 in testing/packaging each week. If a profit of $20 is made on each superdeluxe model, $16 on each deluxe model, and $12 on each standard model, how many of each model should be made each week to maximize profits?

48. Purchasing The Psychology Department at State University buys mice, rats, and snakes for experimental purposes. Each mouse costs $2, each rat costs $5, and each snake costs $6. The departmental budget requires that no more than $800 be spent on such purchases. Each mouse requires 1 square foot of living space, each rat requires 2 square feet of living space, and the department is limited to a maximum of 90 square feet for these rodents. It wants to purchase no more than 10 snakes and no more than 30 rats. How many mice, rats, and snakes should be purchased by the department to accommodate the maximum number? Will there be surplus money in the budget? Will there be a surplus of living space for the rodents?

49. Event Planning The directors of a state fair want to bring in entertainers to bolster attendance at the fair. They find that a top star demands $10,000, a faded star $6000, and high-quality local talent $3000 for each performance. The directors estimate that 8000 people will attend a top-star performance, 3000 will attend a faded-star performance, and 1200 will attend a high-quality local-talent performance. The total amount to be spent for such entertainers is not to exceed $50,000, and the directors decide to contract for no more than six of them. The amount spent on advertising is

not to exceed $4000, and the directors estimate that the advertising costs for each type of entertainer will be $500, $400, and $250, respectively. Assuming one performance for each entertainer, how many of each type should be contracted to maximize attendance? Will there be any surplus contracting or advertising funds left over? If so, how much?

50. **Investing** An investor has up to $150,000 to invest in any of three areas: stocks, bonds, or a money market fund. The annual rates of return are estimated to be 5%, 9%, and 8%, respectively. No more than $75,000 can be invested in any one of these areas. How much should be invested in each area to maximize annual income? Will there be any of the $150,000 that is not invested in one of these three areas?

51. **Manufacturing** A maker of computer printers has four models: the QP20, the QP40, the QP60, and the QP100. Each QP20 model has 2 megabytes of memory; each QP40, 4 megabytes; each QP60, 6 megabytes; and each QP100, 10 megabytes. In planning the next week's production, the company wants to buy no more than 820 mega-

bytes of memory because of budget constraints, and furthermore, it wants to make at least twice as many QP20 models as QP100 models, at least as many QP40 models as QP60 models, and no more than 20 QP100 models. If the printers sell for $500, $600, $1000, and $2000, respectively, how many of each model of printer should the company assemble during the next week to maximize revenue?

52. **Investments** An investor has $100,000 to invest in government bonds, an aggressive-growth mutual fund, a derivative-laden foreign-bond fund, and a money market fund. The investor wants to invest at least as much in government bonds as the total in the aggressive-growth mutual fund and the foreign-bond fund. Furthermore, no more than $20,000 is to be invested in the money market fund. If government bonds earn 7%, the aggressive-growth mutual fund earns 5%, the foreign-bond fund earns 8%, and the money market funds earn 5%, how much should be put into each of these investments to maximize annual returns?

Section 5.3 Nonstandard and Minimization Problems

There are many useful linear programming problems that do not meet the conditions of the standard maximum-type problem. Examples include any minimization problem, maximization problems in which one or more of the constraints (aside from non-negativity) is of the ≥ type, problems with an equality constraint, and so forth. Such problems fall into the classification of **nonstandard problems.** In this section, we develop the necessary tools for solving all such problems, if a solution exists. First, we turn the problem into a **maximum-type** linear programming problem:

A Maximum–Type Linear Programming Problem

A maximum-type linear programming problem satisfies the following conditions:

1. The objective function is to be maximized.
2. The nonnegativity constraints are present. That is, all variables in the original problem are nonnegative.
3. All structural constraints are of the form $ax + by + \cdots \le c$ (c need not be positive).

Then, if any constant c on the right side of a constraint is negative, we consider an ingenious pivoting scheme that will eventually turn the initial tableau for the maximum-type problem into a tableau to which the simplex algorithm can be applied to finish the solution process.

Notice that the only difference between a maximum-type problem and a standard maximum-type problem is that the requirement that $c \ge 0$ has been dropped.

To accomplish the first step in our plan, we examine the possible conversion procedures needed to change a nonstandard problem into a maximum-type problem.

Constraints of the $\geq$ type may be converted to the desired $\leq$ form by multiplying the inequality by -1, because we know that multiplication by any negative number reverses the direction of an inequality. For example, the inequality $2x + 3y \geq 5$ becomes $-2x - 3y \leq -5$ upon multiplication by -1, and the inequality $5x - 3y + 7z \geq 6$ becomes $-5x + 3y - 7z \leq -6$ upon multiplication by -1.

Equality constraints may be eliminated by solving for one of the variables and then substituting into each remaining structural constraint and the objective function. Example 6 will show the details of how this may be done. An alternative approach is given in Exercises 46 and 47 at the end of the section.

Finally, how do we change a problem whose objective function is to be minimized into an equivalent one whose objective function is to be maximized? The key lies in observing that the *minimum* of an objective function f occurs at *exactly the same point in the feasible region* that the *maximum* of $-f$ does. For example, consider $f = 2x + y$ and $g = -f = -2x - y$ over the arbitrary feasible region shown in Figure 13. In the figure, the line $f = 2x + y = 8$ is the same line as $g = -2x - y = -8$. The figure demonstrates the relationship between f and g: The further downward that f is moved, the *smaller* is its value; the further downward that g is moved, the *larger* is its value. We conclude that the *minimum of f* over the feasible region occurs where the *maximum of $g = -f$* occurs over the same feasible region.

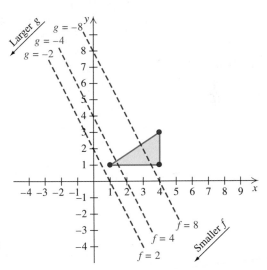

FIGURE 13
The Maximum for g Gives the Minimum for f.

EXAMPLE 1

Converting a Minimization Problem to a Maximum-Type Problem

Convert this problem to a maximum-type problem:

$$\text{Minimize } f = 5x + 3y + z$$
$$\text{subject to } \quad x + 2y + 5z \geq 6$$
$$2x - y + 6z \geq 8$$
$$x \geq 0, y \geq 0, z \geq 0.$$

SOLUTION The structural constraints are converted to $\leq$ by multiplying by -1:

$$-x - 2y - 5z \leq -6$$
$$-2x + y - 6z \leq -8.$$

As noted, the objective function part of the problem—minimize $f = 5x + 3y + z$ —is replaced with maximize $g = -5x - 3y - z$. The maximum-type problem is then written as

$$\text{Maximize } g = -5x - 3y - z$$
$$\text{subject to } -x - 2y - 5z \le -6$$
$$-2x + y - 6z \le -8$$
$$x \ge 0, y \ge 0, z \ge 0. \quad \blacksquare$$

Now that we see how to convert any linear programming problem to a maximum-type problem, why not just proceed with the simplex algorithm? The next example explains why we cannot immediately do that.

EXAMPLE 2

A Maximization
Problem with
Mixed Constraints

$$\text{Maximize } f = 2x + 3y$$
$$\text{subject to } x + y \le 8$$
$$2x + y \ge 10$$
$$x \ge 0, y \ge 0.$$

When some of the inequalities are $\le$ and some are $\ge$, as in this example, the problem is referred to as one with **mixed constraints.** The feasible region is shown in Figure 14, along with the coordinates of all points where two boundary lines intersect. The only thing that prevents our problem from being in the maximum-type format is the constraint $2x + y \ge 10$. Upon multiplying this inequality through by -1, the structural constraints now have the appearance

$$x + y \le 8$$
$$-2x - y \le -10$$

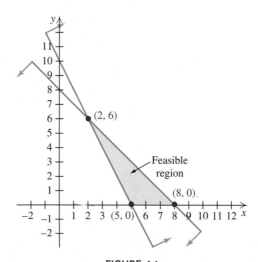

FIGURE 14
The Origin Is Not a Basic Feasible Solution.

and, hence, are in the desired form for a maximum-type problem. We proceed as usual to insert the slack variables and to rewrite the objective function to make it a part of the system of equations on which we pivot:

$$x + y + s_1 \qquad\qquad = 8$$
$$-2x - y \qquad + s_2 \qquad = -10$$
$$-2x - 3y \qquad\qquad + f = 0.$$

The initial simplex tableau may now be written

$$
\begin{array}{ccccc}
x & y & s_1 & s_2 & f \\
\end{array}
$$
$$
\left[
\begin{array}{ccccc|c}
1 & 1 & 1 & 0 & 0 & 8 \\
-2 & -1 & 0 & 1 & 0 & -10 \\
\hline
-2 & -3 & 0 & 0 & 1 & 0
\end{array}
\right].
$$

Up to this point, everything falls nicely into place. However, on reading the solution from the initial tableau in the usual manner ($x = 0$, $y = 0$, $s_1 = 8$, $s_2 = -10$), we realize that this solution does not represent a corner point of the feasible region because the value of one of the variables, s_2, is negative. (See Figure 14.) The fact that our first basic solution does not reflect a corner point of the feasible region means that we cannot employ the simplex algorithm to go from corner point to corner point of the feasible region until the maximum is reached. ■

Since the initial simplex tableau for any linear programming problem has the original variables (not the slack variables) as the first set of nonbasic variables, the first solution obtained by setting these variables equal to zero always gives the origin of the coordinate system as the first solution. Every standard maximum-type problem has the origin as a corner point of the feasible region; hence, the initial tableau always starts with a solution at a corner point of that region. However, when the origin is not in the feasible region, as in Example 2, such a first solution is of no value to us, because we have no rules to guide us in the pivoting process. To deal with this problem, we first give an easy way to recognize when the solution to any simplex tableau does not reflect a corner point of the feasible region. This recognition is based on the fact that at least one variable will have a negative value. (See the tableau of Example 2.)

When a Tableau Does Not Represent a Corner Point

For any simplex tableau that has one or more *negative numbers* in the rightmost column above the horizontal bar, the solution obtained by setting all nonbasic variables equal to zero does *not* represent a corner point of the feasible region.

Next, we need an organized pivoting scheme that will *eliminate these negative numbers*, thereby producing nonnegative entries in the rightmost column above the horizontal bar. At that point in the solution process, the scheme will give a *basic feasible solution*. Once this is accomplished, the appearance of the problem becomes that of a standard maximum-type problem, and the rules of the simplex algorithm, developed in Section 5.2, may be used to complete the maximization process, because those rules will continue to give tableaus whose solutions are basic feasible solutions to the problem. The pivoting scheme we prefer to use in eliminating these negative numbers is a set of rules first devised by Professor J. Conrad Crown in 1982 and later modified slightly by Professor Arthur Hobbs to improve their efficiency. We refer to these rules as **Crown's rules.**

Crown's Rules: Eliminating Negative Numbers in the Rightmost Column above the Horizontal Line in a Simplex Tableau

1. If there is a row above the horizontal bar in the tableau whose *only* negative entry is in the rightmost column, then the problem has no feasible solution and hence cannot have an optimal solution. Stop in such a case. Otherwise,

2. (a) To find the pivot column, locate the first negative number from the top in the rightmost column of the tableau. Then select the most negative entry in that row to the left of the vertical bar. The column in which that negative number resides is the pivot column.*

   (b) To find the pivot row, divide each *positive* entry in the pivot column into the corresponding entry in the rightmost column *only* if that entry is *nonnegative*. If such quotients exist, the row giving the smallest nonnegative quotient is the pivot row. If no such quotients exist, divide each *negative* entry in the pivot column into the corresponding entry in the rightmost column only if that entry is also *negative*. In this case, the row giving the *largest* positive quotient is the pivot row. (If no positive–nonnegative quotient exists, there will always be at least one negative–negative quotient.)

   (c) Pivot as usual, by first turning the pivot element into a one and all other elements in that column into zeros.

3. Repeat steps 1 and 2 until there are no negative entries in the rightmost column above the objective function row.

The pivoting scheme just described assures us that the number of negative entries in the rightmost column above the horizontal bar can never go up; each time the scheme is applied, it either leaves the topmost negative entry in the rightmost column alone (if the pivot row has a zero as its rightmost entry) or makes it "less negative" (including making it, and possibly others, nonnegative), thereby giving it a measure of progress toward a feasible solution. Once these negatives are eliminated, the usual simplex algorithm is applied until there are no negative indicators in the objective function row.

EXAMPLE 3

Using the Crown Pivoting Rules

Complete the solution to the linear programming problem of Example 2.

SOLUTION The initial tableau has already been constructed in Example 2 and appears as follows:

$$
\begin{array}{ccccc}
x & y & s_1 & s_2 & f \\
\end{array}
$$
$$
\left[
\begin{array}{ccccc|c}
\textcircled{1} & 1 & 1 & 0 & 0 & 8 \\
-2 & -1 & 0 & 1 & 0 & -10 \\
\hline
-2 & -3 & 0 & 0 & 1 & 0
\end{array}
\right].
$$

The Crown pivoting rules are used first to eliminate the -10 in the rightmost column above the horizontal bar, which is the first, and in this case only, negative number in this area. (Ignore the indicators until all negatives in the area have been removed.) The most negative number to the left of the vertical bar in the row where -10 resides is -2. Since -2 is in the x column, that column is the pivot column. In the x column, above the horizontal bar, there is only one positive–nonnegative quotient, 1 divided into 8, so it is automatically the smallest such quotient. Since this quotient comes from the first row,

---

* The selection of a pivot column corresponding to *any* negative number in that row to the left of the vertical bar is just as good. Sometimes one choice will be more efficient than another in removing negatives in the rightmost column, but there is no consistently "most efficient" choice. We choose the "most negative" simply to have a specific set of instructions. For pencil-and-paper calculations, investigate by inspection to gain efficiency.

that row is the pivot row. Consequently, the 1 in the first-row, first-column position is the pivot element. The tableau resulting from pivoting on this element is shown next. Notice that this pivot eliminated the one and only negative number in the rightmost column above the bar. This means that a basic feasible solution has been found ($f = 16$ when $x = 8$ and $y = 0$; see Figure 14). Now we check the indicators to see whether further pivoting is required to complete the maximization of the objective function. We have

$$
\begin{array}{ccccc}
x & y & s_1 & s_2 & f \\
\end{array}
$$
$$
\begin{bmatrix}
1 & 1 & 1 & 0 & 0 & 8 \\
0 & \textcircled{1} & 2 & 1 & 0 & 6 \\
0 & -1 & 2 & 0 & 1 & 16
\end{bmatrix}
\quad
\begin{array}{l}
\text{Quotient: 8} \\
\text{Quotient: 6} \quad \leftarrow \quad \textit{Pivot row}
\end{array}
$$

$\uparrow$ Pivot column

Noting a negative indicator, we observe that the resulting quotients used in the smallest-quotient rule imply that the next pivot element is the circled 1 in the second-row, second-column position. The result of that pivot is the following tableau:

$$
\begin{array}{ccccc}
x & y & s_1 & s_2 & f \\
\end{array}
$$
$$
\begin{bmatrix}
1 & 0 & -1 & -1 & 0 & 2 \\
0 & 1 & 2 & 1 & 0 & 6 \\
0 & 0 & 4 & 1 & 1 & 22
\end{bmatrix}.
$$

Because no negative indicators now appear, the maximum of f has been reached and is 22. the maximum occurs when $x = 2$ and $y = 6$. (See Figure 14.)

In this particular problem, had we selected the first pivot column as the one determined by the least negative number in the row containing -10 (the y column), the first pivot element would have been the 1 in the first-row, second-column position. The pivot operation on that element would have turned the -10 into a -2; progress toward a nonnegative entry there would have been made, but not as much as with the pivot we made in this particular case. ∎

EXAMPLE 4

Minimizing an Objective Function

Minimize $C = x + 2y + 4z$
subject to $x + 2y + z \geq 5$
$2x - y \geq 8$
$x + y + 2z \leq 12$
$x \geq 0, y \geq 0, z \geq 0.$

SOLUTION As noted earlier, to turn this problem into a maximum-type problem, we must maximize $g = -C = -x - 2y - 4z$. Upon multiplying the first two constraints by -1, our problem can then be restated as:

Maximize $g = -x - 2y - 4z$
subject to $-x - 2y - z \leq -5$
$-2x + y \leq -8$
$x + y + 2z \leq 12$
$x \geq 0, y \geq 0, z \geq 0.$

After inserting the slack variables and rewriting g in the usual fashion, our initial tableau is

$$
\begin{array}{ccccccc}
x & y & z & s_1 & s_2 & s_3 & g \\
\end{array}
$$

$$
\left[
\begin{array}{ccccccc|c}
-1 & -2 & -1 & 1 & 0 & 0 & 0 & -5 \\
-2 & 1 & 0 & 0 & 1 & 0 & 0 & -8 \\
1 & ① & 2 & 0 & 0 & 1 & 0 & 12 \\
\hline
1 & 2 & 4 & 0 & 0 & 0 & 1 & 0 \\
\end{array}
\right].
$$

The first negative number encountered in the rightmost column is -5, and the most negative number in that row is -2, in the y column. Therefore, that column is our first pivot column. There is one positive–nonnegative quotient in the y column, namely, 1 divided into 12, establishing the third row as the pivot row. Therefore, the first pivot element is the circled one in the third-row, second-column position. The result of using that pivot is

$$
\begin{array}{ccccccc}
x & y & z & s_1 & s_2 & s_3 & g \\
\end{array}
$$

$$
\left[
\begin{array}{ccccccc|c}
1 & 0 & 3 & 1 & 0 & 2 & 0 & 19 \\
-3 & 0 & -2 & 0 & 1 & -1 & 0 & -20 \\
① & 1 & 2 & 0 & 0 & 1 & 0 & 12 \\
\hline
-1 & 0 & 0 & 0 & 0 & -2 & 1 & -24 \\
\end{array}
\right].
$$

The rightmost column now has only one negative number, -20, above the objective function, and the most negative number in that row is -3, in the x column. This establishes that column as our next pivot column. In the x column, there are two positive–nonnegative quotients, with 12 being the smallest. Therefore, the circled 1 in the third-row, first-column position is our next pivot element. The result of using this pivot is

$$
\begin{array}{ccccccc}
x & y & z & s_1 & s_2 & s_3 & g \\
\end{array}
$$

$$
\left[
\begin{array}{ccccccc|c}
0 & -1 & 1 & 1 & 0 & ① & 0 & 7 \\
0 & 3 & 4 & 0 & 1 & 2 & 0 & 16 \\
1 & 1 & 2 & 0 & 0 & 1 & 0 & 12 \\
\hline
0 & 1 & 2 & 0 & 0 & -1 & 1 & -12 \\
\end{array}
\right].
$$

Finally, the Crown pivoting rules have put us on a corner point of the feasible region. The negative indicator shows, however, that the maximum of g has not yet been reached. So the s_3 column is the pivot column, and the smallest-quotient rule tells us that the circled 1 in the first-row, sixth column position is the new pivot element. That pivot gives our final tableau:

$$
\begin{array}{ccccccc}
x & y & z & s_1 & s_2 & s_3 & g \\
\end{array}
$$

$$
\left[
\begin{array}{ccccccc|c}
0 & -1 & 1 & 1 & 0 & 1 & 0 & 7 \\
0 & 5 & 2 & -2 & 1 & 0 & 0 & 2 \\
1 & 2 & 1 & -1 & 0 & 0 & 0 & 5 \\
\hline
0 & 0 & 3 & 1 & 0 & 0 & 1 & -5 \\
\end{array}
\right].
$$

The maximum of g has now been reached and is seen to be -5. This means that the minimum of f, which is the negative of g, is 5. This minimum occurs when $x = 5$, $y = 0$, and $z = 0$. ∎

EXAMPLE 5
When Fixed Costs
Are Involved

A small subcontractor for a large shoe company does the cutting and gluing for walking shoes and jogging shoes. Each pair of walking shoes requires 2 minutes' cutting time and 4 minutes' gluing time. Each pair of jogging shoes requires 3 minutes' cutting time and 2 minutes' gluing time. The subcontractor has two employees who do the cutting and three employees who do the gluing and guarantees that each will work at least 30 hours per week. The cost of materials is $8 for each pair of walking shoes and $10 for each pair of jogging shoes, with an overhead cost of $200. How many pairs of each type of shoe should be made during the week to minimize total material costs?

SOLUTION Let x be the number of walking shoes and y be the number of jogging shoes to be made during the week. Then the constraints are

$$2x + 3y \geq 2(30)(60) = 3600 \text{ (minutes)}$$
$$4x + 2y \geq 3(30)(60) = 5400 \text{ (minutes)},$$

where the employees are working at least 30 hours per week, 60 minutes per hour. The feasible region is shown in Figure 15. The objective function representing the total cost of materials becomes

$$C = 8x + 10y + 200,$$

which we must minimize. This problem can then be stated as a maximum-type problem:

$$\text{Maximize } g = -8x - 10y - 200 \text{ (in dollars)}$$
$$\text{subject to } -2x - 3y \leq -3600$$
$$-4x - 2y \leq -5400$$
$$x \geq 0, y \geq 0.$$

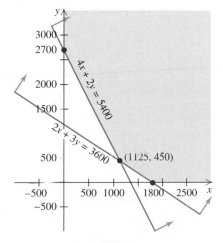

FIGURE 15
Feasible Region for Shoe Production

The initial simplex tableau then becomes

$$
\begin{array}{ccccc}
x & y & s_1 & s_2 & g \\
\end{array}
$$

$$
\left[
\begin{array}{ccccc|r}
-2 & -3 & 1 & 0 & 0 & -3600 \\
-4 & \boxed{-2} & 0 & 1 & 0 & -5400 \\
\hline
8 & 10 & 0 & 0 & 1 & -200
\end{array}
\right]
\quad
\begin{array}{l}
\text{Quotient: } 1200 \\
\text{Quotient: } 2700 \\
\\
\end{array}
$$

Since negative entries appear in the rightmost column, the Crown pivoting rules must be applied. The first number encountered is -3600, and the most negative number in that row to the left of the vertical bar is -3. This means that the y column will be the pivot column. Because only negative–negative quotients exist in that column, the larger quotient determines the pivot row. The first pivot element is then the circled -2 located in the second-row, second-column position. To get a 1 in that position, replace row 2 by $-\frac{1}{2}$ times itself:

$$
-\tfrac{1}{2}R_2 \to R_2 \quad
\begin{array}{ccccc}
x & y & s_1 & s_2 & g \\
\end{array}
$$

$$
-\tfrac{1}{2}R_2 \to R_2 \quad
\left[
\begin{array}{ccccc|r}
-2 & -3 & 1 & 0 & 0 & -3600 \\
2 & 1 & 0 & -\frac{1}{2} & 0 & 2700 \\
\hline
8 & 10 & 0 & 0 & 1 & -200
\end{array}
\right].
$$

Introducing zeros above and below the 1 completes the pivot operation:

$$
\begin{array}{l}
3R_2 + R_1 \to R_1 \\
\\
-10R_2 + R_3 \to R_3
\end{array}
\left[
\begin{array}{ccccc|r}
\boxed{4} & 0 & 1 & -\frac{3}{2} & 0 & 4500 \\
2 & 1 & 0 & -\frac{1}{2} & 0 & 2700 \\
\hline
-12 & 0 & 0 & 5 & 1 & -27{,}200
\end{array}
\right].
\quad
\begin{array}{l}
\text{Quotient: } 1125 \\
\text{Quotient: } 1350 \\
\\
\end{array}
$$

No negative entries appear in the rightmost column above the horizontal bar, which is the signal to return to the regular simplex algorithm rules. The pivot column will be the x column, and the smallest quotient indicates that the first row will be the pivot row. Pivoting on the circled 4 in the first-row, first-column position yields the matrix

$$
\begin{array}{ccccc}
x & y & s_1 & s_2 & g \\
\end{array}
$$

$$
\left[
\begin{array}{ccccc|r}
1 & 0 & \frac{1}{4} & -\frac{3}{8} & 0 & 1125 \\
0 & 1 & -\frac{1}{2} & \frac{1}{4} & 0 & 450 \\
\hline
0 & 0 & 3 & \frac{1}{2} & 1 & -13{,}700
\end{array}
\right].
$$

The absence of negative indicators signals that the maximum of g has been reached and is $-13{,}700$. This means that the minimum cost C is \$13,700 and occurs when 1125 walking shoes and 450 jogging shoes are made during the week. ∎

Sometimes a maximization or minimization linear programming problem has an equality as a constraint. The simplest way to handle such a constraint is to solve for one of the variables in the equality and substitute for it throughout the problem, thereby reducing the number of variables by one. Example 6 gives the details of this operation.

EXAMPLE 6

**A Constraint That Is
an Equality**

Maximize $f = 3x + 2y + 5z$

subject to $\quad x + \;\; y + z \le 8 \qquad\qquad\qquad$ **(1)**

$\qquad\qquad x + \;\; y \quad\;\;\; \le 4 \qquad\qquad\qquad$ **(2)**

$\qquad\qquad\qquad\;\; 2y + z = 6 \qquad\qquad\qquad$ **(3)**

$\qquad\qquad x \ge 0,\, y \ge 0,\, z \ge 0.$

SOLUTION Solving the equality in **(3)** for z, we have $z = 6 - 2y$. Substituting the expression for z into constraint **(1)** gives

$$x + y + z = x + y + (6 - 2y) \le 8$$

or

$$x - y \le 2.$$

Constraint **(2)** does not contain z and therefore is not affected by the substitution process:

$$x + y \le 4.$$

Because $z \ge 0$, we must have $z = 6 - 2y \ge 0$, or $2y - 6 \le 0$, or

$$y \le 3.$$

Finally, using $z = 6 - 2y$, we rewrite the objective function as

$$f = 3x + 2y + 5z = 3x + 2y + 5(6 - 2y) = 3x - 8y + 30.$$

Rewriting the new linear programming problem without the z variable gives

Maximize $f = 3x - 8y + 30$

subject to $\;\; x - y \le 2,$

$\qquad\qquad x + y \le 4$

$\qquad\qquad\quad\; y \le 3$

$\qquad\quad x \ge 0,\, y \ge 0.$

Now rewrite the objective function as $-3x + 8y + f = 30$ and construct the initial tableau:

$$
\begin{bmatrix}
x & y & s_1 & s_2 & s_3 & f & \\
\textcircled{1} & -1 & 1 & 0 & 0 & 0 & 2 \\
1 & 1 & 0 & 1 & 0 & 0 & 4 \\
0 & 1 & 0 & 0 & 1 & 0 & 3 \\
\hline
-3 & 8 & 0 & 0 & 0 & 1 & 30
\end{bmatrix}.
$$

There are no negative numbers in the rightmost column; therefore, the initial tableau has a basic feasible solution, and the Crown pivoting rules are not needed. Consequently, we turn to the rules of the simplex algorithm. The most negative indicator is in the x column;

hence, that column is the pivot column. The smallest-quotient rule then tells us that the first row is the pivot row. After pivoting on the 1 in the first-row, first-column position, we find that the resulting matrix looks like this:

$$
\begin{array}{cccccc}
x & y & s_1 & s_2 & s_3 & f \\
\end{array}
$$

$$
\left[
\begin{array}{cccccc|c}
1 & -1 & 1 & 0 & 0 & 0 & 2 \\
0 & 2 & -1 & 1 & 0 & 0 & 2 \\
0 & 1 & 0 & 0 & 1 & 0 & 3 \\
\hline
0 & 5 & 3 & 0 & 0 & 1 & 36
\end{array}
\right].
$$

Since there are no negative indicators in the last matrix, the maximum has been reached. The maximum value of f is 36 and occurs when $x = 2$, $y = 0$, and $z = 6 - 2y = 6 - 2(0) = 6$. ▪

IMPORTANT NOTE When an equality constraint is present and you solve for a particular variable, do *not* forget to substitute into the nonnegativity constraint of that variable. As was seen in the last example, the nonnegativity constraint on z gave the added structural constraint $y \leq 3$.

EXAMPLE 7

A Problem Having
No Solution

Use the simplex algorithm to solve this linear programming problem:

$$
\begin{aligned}
\text{Maximize } f &= 2x + y \\
\text{subject to } x &\geq 1 \\
x &\leq 2 \\
y &\geq 0.
\end{aligned}
$$

SOLUTION Rewriting the first inequality as $-x \leq -1$, we obtain the slack equations

$$
\begin{aligned}
-x + s_1 \quad\quad &= -1 \\
x \quad\quad + s_2 &= 2.
\end{aligned}
$$

The initial simplex tableau is then

$$
\left[
\begin{array}{ccccc|c}
-1 & 0 & 1 & 0 & 0 & -1 \\
① & 0 & 0 & 1 & 0 & 2 \\
\hline
-2 & -1 & 0 & 0 & 1 & 0
\end{array}
\right].
$$

The Crown pivoting rules indicate that the circled 1 in the second-row, first-column position is the first pivot element. On completing that pivot, we find that the tableau becomes

$$
\left[
\begin{array}{ccccc|c}
0 & 0 & 1 & 1 & 0 & 1 \\
1 & 0 & 0 & 1 & 0 & 2 \\
\hline
0 & -1 & 0 & 2 & 1 & 4
\end{array}
\right].
$$

The negative indicator implies that a pivot is needed in the second column. However, the only choice for a pivot element is a zero, and that is an impossible choice. We conclude that the problem has no solution. Sketch the feasible region for this problem, and discover graphically why this is true. ▪

The problem posed at the outset of this chapter is solved in the next (and last) example of this section. This problem is on an extremely small scale compared with most **transportation problems,** and, in fact, you may be able to solve it without the aid of linear programming. Nonetheless, the technique shown for setting up the constraints can be applied to problems of large scope.

EXAMPLE 8

Solving a
Transportation Problem

The problem stated at the outset of this chapter involves shipping refrigerators at minimum cost from two different warehouses to fill orders from two different stores. If we let x represent the number of refrigerators to be shipped from Wilmar to Phoenixville, then $40 - x$ represents the number that must be shipped from Lubbock to Phoenixville. Likewise, if we let y represent the number of refrigerators to be shipped from Wilmar to Mebane, then $30 - y$ represents the number that must be shipped from Lubbock to Mebane. Figure 16 describes the problem.

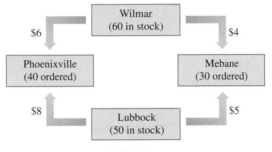

FIGURE 16

The inequalities and objective function may now be constructed from the diagram.

Minimize $f = 6x + 4y + 8(40 - x) + 5(30 - y) = -2x - y + 470$

subject to $x \quad \leq 40$

$\qquad y \leq 30$

$\qquad x + y \leq 60$

$\qquad (40 - x) + (30 - y) \leq 50 \quad \text{or} \quad -x - y \leq 20$

$\qquad x \geq 0, y \geq 0.$

To minimize f, we need to maximize $g = -f$, or $g = 2x + y - 470$. The initial tableau and the subsequent pivots are as follows:

$$
\begin{bmatrix}
1 & 0 & 1 & 0 & 0 & 0 & 0 & 40 \\
0 & \textcircled{1} & 0 & 1 & 0 & 0 & 0 & 30 \\
1 & 1 & 0 & 0 & 1 & 0 & 0 & 60 \\
-1 & -1 & 0 & 0 & 0 & 1 & 0 & -20 \\
\hline
-2 & -1 & 0 & 0 & 0 & 0 & 1 & -470
\end{bmatrix}
\rightarrow
\begin{bmatrix}
1 & 0 & 1 & 0 & 0 & 0 & 0 & 40 \\
0 & 1 & 0 & 1 & 0 & 0 & 0 & 30 \\
\textcircled{1} & 0 & 0 & -1 & 1 & 0 & 0 & 30 \\
-1 & 0 & 0 & 1 & 0 & 1 & 0 & 10 \\
\hline
-2 & 0 & 0 & 1 & 0 & 0 & 1 & -440
\end{bmatrix}
$$

$$
\rightarrow
\begin{bmatrix}
0 & 0 & 1 & \textcircled{1} & -1 & 0 & 0 & 10 \\
0 & 1 & 0 & 1 & 0 & 0 & 0 & 30 \\
1 & 0 & 0 & -1 & 1 & 0 & 0 & 30 \\
0 & 0 & 0 & 0 & 1 & 1 & 0 & 40 \\
\hline
0 & 0 & 0 & -1 & 2 & 0 & 1 & -380
\end{bmatrix}
\rightarrow
\begin{bmatrix}
0 & 0 & 1 & 1 & -1 & 0 & 0 & 10 \\
0 & 1 & -1 & 0 & 1 & 0 & 0 & 20 \\
1 & 0 & 1 & 0 & 0 & 0 & 0 & 40 \\
0 & 0 & 0 & 0 & 1 & 1 & 0 & 40 \\
\hline
0 & 0 & 1 & 0 & 1 & 0 & 1 & -370
\end{bmatrix}
$$

The final tableau shows that the solution is $x = 40$ and $y = 20$, which means that the shipping schedule should look like this:

From Wilmar to Phoenixville, 40 refrigerators,

From Wilmar to Mebane, 20 refrigerators,

From Lubbock to Phoenixville, 0 refrigerators,

From Lubbock to Mebane, 10 refrigerators.

The minimum shipping costs will be $370. ■

Exercises 5.3

In each of Exercises 1 through 16,
(a) Sketch the feasible region, and find the coordinates of the corner points.
(b) Solve the problem with the simplex algorithm, using the Crown pivoting rules if necessary, and record the movement of the corner points for each tableau.
(c) Check the solution by computing the value of f at each corner point of the feasible region.

1. Maximize $f = 2x + y$

   subject to $x + y \leq 6$

   $4x + y \leq 12$

   $x \geq 0, y \geq 0.$

2. Maximize $f = x + 2y$

   subject to $x + y \geq 6$

   $3x + y \leq 9$

   $x \geq 0, y \geq 0.$

3. Maximize $f = 4x + 2y$

   subject to $x + y \geq 5$

   $3x + y \leq 12$

   $x \geq 0, y \geq 0.$

4. Maximize $f = 4x + y$

   subject to $x - y \geq 0$

   $x \quad \leq 4$

   $y \geq 1.$

5. Maximize $f = 2x + 3y$

   subject to $2x - y \leq 0$

   $x \quad \geq 1$

   $y \leq 6.$

6. Maximize $f = 2x - 3y$

   subject to $y \geq x$

   $y \leq 4$

   $y \geq 2$

   $x \geq 0.$

7. Maximize $f = -4x + 5y$

   subject to $x - y \geq 1$

   $x + y \leq 7$

   $x \geq 0, y \geq 0.$

8. Maximize $f = 2x + 3y$

   subject to $x + y \geq 6$

   $2x + y \geq 8$

   $y \leq 8$

   $x \quad \leq 6$

   $x \geq 0, y \geq 0.$

9. Minimize $f = x + 3y$

   subject to $x + y \geq 6$

   $2x + y \geq 8$

   $x \geq 0, y \geq 0.$

10. Minimize $f = 4x + 3y$

   subject to $2x + y \geq 8$

   $x + 2y \geq 6$

   $x \geq 0, y \geq 0.$

11. Minimize $f = x + y$

   subject to $x + y \leq 6$

   $x \quad \geq 2$

   $y \geq 2.$

12. Minimize $C = 5x + y$

subject to $2x + y \geq 8$

$\quad 2x + y \leq 12$

$\quad x \quad\quad \geq 1$

$\quad x \quad\quad \leq 2.$

13. Minimize $C = x - 2y$

subject to $x + y \geq 6$

$\quad x \quad\quad \geq 2$

$\quad x \quad\quad \leq 8$

$\quad\quad y \geq 2$

$\quad\quad y \leq 6.$

14. Minimize $C = -3x + 2y$

subject to $x - 2y \leq 0$

$\quad x \quad\quad \geq 2$

$\quad\quad y \leq 7.$

15. Minimize $f = 2x + y + 10$

subject to $\quad\quad y \leq x$

$\quad 2x - y \geq 6$

$\quad x \geq 0, y \geq 0.$

16. Minimize $f = 2x + y + 18$

subject to $x \geq 2$

$\quad y \geq x$

$\quad y \leq 2x.$

Make use of the Crown pivoting rules and the simplex method to solve Exercises 17 through 30.

17. Minimize $C = 3x + 4y + z + 50$

subject to $x \quad\quad \leq y$

$\quad\quad z \geq 2$

$\quad x \quad\quad \geq 4$

$\quad x + y + z \geq 6$

$\quad\quad y \geq 0.$

18. Minimize $C = 2x - 3y + z + 80$

subject to $x + 2y + z \geq 2$

$\quad\quad z \leq 6$

$\quad x \quad\quad \geq y$

$\quad\quad y \quad\quad \leq 4$

$\quad x \geq 0, y \geq 0, z \geq 0.$

19. Maximize $f = 3x + 2y + 6z$

subject to $x + 2y + z \leq 6$

$\quad\quad y + 2z \leq 4$

$\quad 2x \quad\quad + z \geq 4$

$\quad x \geq 0, y \geq 0, z \geq 0.$

20. Maximize $P = 4x + 5y + z$

subject to $x + 2y + 3z \leq 6$

$\quad x + y + 2z \geq 3$

$\quad 3x + y \quad\quad \geq 5$

$\quad\quad y \quad\quad \geq 1$

$\quad x \geq 0, y \geq 0, z \geq 0.$

21. Maximize $P = 3x + 4y + z$

subject to $3x + 2y + 4z \leq 12$

$\quad x + \quad\quad z \geq 1$

$\quad\quad y \quad\quad \geq 1$

$\quad\quad z \leq 1$

$\quad x \geq 0, y \geq 0, z \geq 0.$

22. Minimize $f = 2x - y + 2z$

subject to $x + 2y \quad\quad \geq 40$

$\quad x + 2y + 3z \geq 60$

$\quad\quad y + z \geq 30$

$\quad x \geq 0, y \geq 0, z \geq 0.$

23. Minimize $C = 4x + 3y + 2z$

subject to $x + 2y + z \geq 6$

$\quad\quad y + 2z \geq 1$

$\quad 2x + y + 2z \leq 4$

$\quad x \geq 0, y \geq 0, z \geq 0.$

24. Minimize $f = x + 2y + 4z$

subject to $x \quad\quad \geq 2$

$\quad x + y + z \geq 6$

$\quad\quad z \geq 3$

$\quad x \geq 0, y \geq 0.$

25. Maximize $f = x + 2y + z$

subject to $x + y + 2z \leq 12$

$\quad 2x + y + z \leq 16$

$\quad x + y \quad\quad = 3$

$\quad x \geq 0, y \geq 0, z \geq 0.$

26. Maximize $P = 2x - 3y + z$

subject to $x + y + 2z \leq 6$

$\quad\quad y \quad\quad \leq 2$

$\quad x \quad\quad \geq z$

$\quad x + y \quad\quad = 4$

$\quad x \geq 0, y \geq 0, z \geq 0.$

27. Minimize $C = 3x + y - 2z$

    subject to $2x - y \quad\quad = 3$

    $\quad\quad x + y - 2z \le 4$

    $\quad\quad x \quad\quad - z \ge 0$

    $\quad\quad y \ge 0, z \ge 0.$

28. Minimize $f = 2x + 3y + z$

    subject to $x + y + z \ge 4$

    $\quad\quad x \quad\quad + 3z \ge 6$

    $\quad\quad x \quad\quad = 1$

    $\quad\quad x \ge 0, y \ge 0, z \ge 0.$

29. Maximize $f = 3x_1 + 4x_2 + x_3$

    subject to $\quad\quad x_2 \quad\quad = 2$

    $\quad\quad x_1 + x_2 + x_3 \le 8$

    $\quad\quad x_2 + x_3 \le 4$

    $\quad\quad x_1 \ge 0, x_2 \ge 0.$

30. Maximize $C = 2x_1 + 3x_2 - x_3 + 100$

    subject to $x_1 \quad\quad \le x_2$

    $\quad\quad x_3 \ge 2$

    $\quad\quad x_2 \quad\quad \ge 4$

    $\quad\quad x_1 + x_2 + x_3 \ge 8$

    $\quad\quad x_1 \ge 0.$

In the remaining exercises, find the mathematical formulation of the problem and then solve by applying the Crown pivoting rules, if necessary, and then the simplex method. Some of these exercises may be familiar to you from earlier sections.

31. **Retail Sales** An appliance store sells two brands of television sets. Each Daybrite set sells for $420, each Noglare set for $550. The store's warehouse capacity for television sets is 300, and new sets are delivered only once each month. Records show that customers will buy at least 80 Daybrite sets and at least 120 Noglare sets each month. How many of each brand should the store stock and sell each month to maximize revenue?

32. **Manufacturing** A firm makes three types of golf bags: A, B, and C. Type A requires 15 minutes to cut, 30 minutes to sew, and 30 minutes to trim and sells for $80. Type B requires 15 minutes to cut, 20 minutes to sew, and 50 minutes to trim and sells for $90. Type C requires 20 minutes to cut, 30 minutes to sew, and 50 minutes to trim and sells for $100. The firm has available 80 work hours for cutting, 90 work hours for sewing, and 120 work hours for trimming each day. A contract from a large discount store

requires that the firm make at least 20 bags of type A and 25 bags of type B each day. How many of each type bag should the firm produce to maximize its revenue?

33. **Political Campaign** The campaign manager of a candidate for a local political office estimates that each 30-minute speech to a civic group will generate 20 votes, each hour spent on the phone will generate 10 votes, and each hour spent campaigning in shopping areas will generate 40 votes. The candidate wants to make at least twice as many shopping area stops as speeches to civic groups and wants to spend at least three hours on the phone. The campaign manager thinks that her candidate can win if he generates at least 1000 votes by the three methods. How can the candidate meet these goals with minimum time spent?

34. **Advertising** A new line of snack cracker is being introduced by a food company that has allotted no more than $50,000 for advertisements in newspapers, on the radio, and on television. Each newspaper ad costs $100 and will cause 200 people to try the new cracker, each radio costs $150 and will cause 300 people to try the cracker, and each television ad costs $300 and will cause 600 people to try the cracker. The company wants to reach at least 10,000 people through these advertisements and has agreed to run at least twice as many radio ads as television ads. How can the company meet these objectives at minimum cost?

35. **Drug Administering** After a certain operation, a patient receives three types of painkiller: A, B, and C. A dose of A rates a 2 in effectiveness, a dose of B rates a 2.5 in effectiveness, and a dose of C rates a 3 in effectiveness. For medical reasons, the patient can receive no more than three doses of A, no more than five doses of A and C combined, and no more than seven doses of B and C combined during a 24-hour period. How many doses of each type should be given to the patient in a 24-hour period in order to maximize the painkilling effect?

36. **Refining Gasoline** A refiner blends high- and low-octane gasoline into three grades for sale to the wholesaler:

regular, premium, and superpremium. The regular grade consists of 30% high-octane and 70% low-octane gasolines, the premium grade consists of 50% of each, and the superpremium grade consists of 70% high-octane and 30% low-octane gasolines. The refiner can sell each gallon of regular for 60 cents, each gallon of premium for 65 cents, and each gallon of superpremium for 70 cents. If there are currently 150,000 gallons of high-octane gasoline and 130,000 gallons of low-octane gasoline in stock, how many gallons of regular, premium, and superpremium should be made if the refiner is to maximize revenues?

37. **Inventory** The InfoAge Communication Store stocks fax machines, computers, and portable CD players. Space restrictions dictate that it stock no more than a total of 100 of these three machines. Past sales patterns indicate that it should stock an equal number of fax machines and computers and at least 20 CD players. If each fax machine sells for $500, each computer for $1800, and each CD player for $1000, how many of each should be stocked and sold for maximum revenues?

38. **Nutrition** A dietitian is to prepare two foods to meet certain requirements. Each pound of Food I contains 100 units of vitamin C, 40 units of vitamin D, and 10 units of vitamin E and costs 20 cents. Each pound of Food II contains 10 units of vitamin C, 80 units of vitamin D, 5 units of vitamin E and costs 15 cents. The mixture of the two food is to contain at least 260 units of vitamin C, at least 320 units of vitamin D, and at least 50 units of vitamin E. If the fixed cost for handling and storage of the two foods is $10, how many pounds of each food should be used to minimize the total cost?

39. **Manufacturing** The Wide-Track Tire Company makes radial, snow, and off-road tires. Each radial tire takes 4 minutes and costs $20 to produce, each snow tire takes 6 minutes and costs $30 to produce, and each off-road tire takes 8 minutes and costs $40 to produce. Orders received by the company dictate that at least 1000 radial, at least 400 snow, and at least 100 off-road tires be made each day. The union contract with Wide-Track requires the use of at least 500 work hours each day. If the fixed costs for production of the tires is $2000, how many of each type of tire should be made each day to minimize total costs?

40. **Radio Programming** Radio station WMJX is adding a talk show, a sports call-in show, and a country-and-western show to its programming lineup. The company's marketing consultant estimates that each time these shows are aired, they will attract, respectively, 5000, 6000, and 3000 listeners. Advertising revenues dictate that, together, the talk show and the sports call-in show can be aired a total of no more than eight times each week and the country and western show no more than six times each week. Furthermore,

the station manager wants to run the country-and-western show twice as many times as the talk show. How many times should each show be aired each week to attract the maximum number of radio listeners?

41. **Processing** Two grains—barley and corn—are to be mixed for an animal food. Barley contains 1 unit of fat per pound and corn contains 2 units of fat per pound. The total number of units of fat in the mixture should not exceed 12. No more than 6 pounds of barley and no more than 5 pounds of corn are to be used in the mixture. If barley and corn each contain 1 unit of protein per pound, how many pounds of each grain should be used to maximize the number of units of protein in the food mixture?

42. **Nutrition** Gary is going on a low-carbohydrate diet that calls for him to eat two foods—Food I and Food II—at each meal for a week. One unit of Food I contains 5 grams of carbohydrate, 80 calories, and 2 grams of fat. One Unit of Food II contains 7 grams of carbohydrate, 110 calories, and 5 grams of fat. Gary wants each meal to provide at least 400 calories, but no more than 105 grams of carbohydrate. In addition, at least three times as many units of Food I as Food II are to be used in each meal for taste appeal. How many units of each food should Gary use per meal to minimize his intake of fat?

43. **Shipping Costs** Find the minimum cost of shipping TV sets from the warehouses to the stores in the following figure:

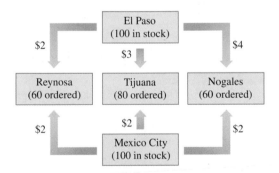

44. **Shipping Costs** The Raw Wood Furniture Company gets an order from its store in Toronto for 20 desks, an order from its store in Montreal for 12 desks, and an order from its store in Winnipeg for 15 desks. The warehouse in Ottawa has 25 desks and the warehouse in Calgary has 34 desks from which these orders are to be filled. The costs to the company for shipping desks from Ottawa to the three stores are $4, $6, and $5, respectively. The costs to the company for shipping desks from Calgary to the three stores are $5, $4, and $5, respectively. How many desks should the company ship from

each warehouse to each store so that shipping costs will be a minimum?

45. Shipping Costs The Pranab Book Company has boxes of a particular textbook in two warehouses: one in Hyderabad with 28 boxes and one in Calcutta with 40 boxes. It gets orders for the text from three bookstores: 20 boxes from New Delhi, 15 boxes from Mumbai, and 10 boxes from Kanpur. The shipping costs per box from Hyderabad to the three locations are $1, $1, and $2, respectively. The shipping costs per box from Calcutta to the three locations are $2, $2, and $1, respectively. How many boxes of books should be shipped from each warehouse to each location so that shipping costs will be minimal?

46. Consider this problem:

$$\text{Maximize } f = 2x + y$$
$$\text{subject to } \qquad y = 1$$
$$x + y \le 2$$
$$x \ge 0, y \ge 0.$$

(a) Graph the feasible region.

(b) Set up the initial tableau, and solve the problem by using the rules shown in this section. Keep track of the geometry of the basic solutions.

(c) Take the initial tableau of part (b), and attempt to solve the problem by using this pivot-selection rule:

Select the pivot column as usual. Then, in that column, divide each element above the horizontal line into the corresponding element in the rightmost column, and select the pivot row to be the one giving the smallest *positive* quotient. Anytime a negative element appears in the rightmost column, use the Crown pivoting rules to eliminate that element. Keep track of the geometry of the basic solutions.

What is happening after several pivots that tells us that the foregoing rule is not desirable?

47. Consider this problem:

$$\text{Maximize } f = 2x + 5y$$
$$\text{Subject to } x \qquad = 3$$
$$x + y \le 4$$
$$x \ge 0, y \ge 0.$$

(a) Graph the feasible region.

(b) Solve the problem by substituting the value of x into each remaining structural constraint. (See Example 6.)

(c) Solve the problem by replacing the equality constraint with two inequality constraints, one using $\le$ and the other using $\ge$. Verify that the answer is the same as in part (b).

Note: This technique can be generalized to any problem involving equality constraints.

Section 5.4 The Dual Problem

Every linear programming problem has associated with it another linear programming problem called its **dual.** If the original problem is one whose objective function is to be maximized, its dual problem is one whose objective function is to be minimized. Similarly, if the original problem is one whose objective function is to be minimized, its dual problem is one whose objective function is to be maximized. As our exploration of the dual problem unfolds, we will see that the final tableau for a dual problem not only gives the solution to the dual problem, but also contains the solution to the original problem.

Here is how the relationship between the original problem and its dual works: For a maximizing problem, every structural constraint must first be written in the $\le$ form. This may require multiplying certain of the inequalities by -1. The dual problem then becomes a minimizing problem with constraints in the $\ge$ form. Analogously, before finding the dual of a minimizing problem, every constraint must first be written in the $\ge$ form. Again, this may require multiplying certain of the inequalities by -1. The dual problem then becomes a maximizing problem with constraints in the $\le$ form. In either case, once the constraints are written in the proper form, the problem is known as the **primal problem.** We summarize these comments with the following chart:

Primal Problem ———————————→ Dual Problem

Maximizing problem	Minimizing problem
Nonnegativity constraints present	Nonnegativity constraints present
Structural constraints written as $\leq$	Structural constraints written as $\geq$

Dual Problem ◄——————————— Primal Problem

We use the sample problem

$$\text{Minimize } f = x + 2y$$
$$\text{subject to } x + \ y \geq 6$$
$$x + 3y \geq 12$$
$$x \geq 0, y \geq 0.$$

to explain how the dual problem is formed and how its final tableau gives the solution to the example problem. Notice that all constraints are already in the $\geq$ form; hence, the original problem is also the primal problem. Figure 17 shows the feasible region in our problem. A few calculations show that the minimum value of f is 9 and occurs when $x = 3$ and $y = 3$. With that in mind, here is how we proceed to create the dual problem associated with this sample problem. First, put the constraints, *without slack variables,* and the coefficients of the objective function into a matrix, as follows:

$$P = \begin{bmatrix} 1 & 1 & 6 \\ 1 & 3 & 12 \\ \hline 1 & 2 & 0 \end{bmatrix}.$$

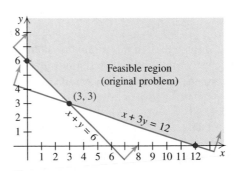

FIGURE 17
*Feasible Region for Original
Minimization Problem*

The matrix P is known as the **primal matrix** for this problem. Next, form the transpose of this matrix, P^T:

$$P^T = \begin{bmatrix} 1 & 1 & 1 \\ 1 & 3 & 2 \\ \hline 6 & 12 & 0 \end{bmatrix} = D.$$

Finally, from D, a standard maximum-type problem, called the **dual problem,** is created by considering the last row to be a new objective function that is to be maximized and the remaining rows as new constraints of the $\leq$ type. Thus, the dual problem becomes

$$\text{Maximize } d = 6u + 12v$$
$$\text{subject to } u + v \leq 1$$
$$u + 3v \leq 2$$
$$u \geq 0, v \geq 0.$$

Notice that we use different variables to avoid confusion.

Figure 18 shows the feasible region for the newly created dual problem. Calculations demonstrate that the *maximum* value of $d = 6u + 12v$ over this region is 9 and occurs at the point where $u = \frac{1}{2}$ and $v = \frac{1}{2}$. Note that this *maximum* value is the same as the *minimum* value of the original objective function in the primal problem.

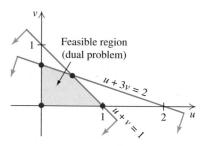

FIGURE 18
Feasible Region for the Dual Problem

Even though the original problem and its dual are quite different in appearance and have very different feasible regions, there is a link between the two, described be a rather deep mathematical theorem known as the **duality theorem:**

The Duality Theorem

Suppose that a primal linear program, with objective function f, and its dual, with objective function d, both have solutions. Then

1. If the primal problem is a maximization problem, the dual will be a minimization problem and max f = min d.
2. If the primal problem is a minimization problem, the dual will be a maximization problem and min f = max d.

We have already observed in a geometric sense that our sample problem agrees with the general statement of the theorem. But there is more: The duality theorem also tells us where to look in the final tableau of the dual problem for the corner points at which these equal optimal solutions occur.

The Duality Theorem, Continued

In solving the dual problem by the simplex algorithm (using Crown's rules first, if necessary), the coordinates of the point at which the optimal solution occurs are read from the final tableau in the usual way. In that same final tableau, the coordinates of the point at which the optimal solution to the primal problem occurs are found in the objective-function row and the slack-variable columns.

To demonstrate the preceding fact, we solve the dual problem from our present example by using the simplex algorithm. First, the slack variables are inserted and the objective function is rewritten as usual:

$$
\begin{aligned}
u + \quad v + s_1 \quad\quad &= 1 \\
u + \quad 3v \quad + s_2 \quad &= 2 \\
-6u \quad -12v \quad\quad + d &= 0.
\end{aligned}
$$

The initial simplex tableau is then found to be

$$
\begin{array}{ccccc}
u & v & s_1 & s_2 & d \\
\end{array}
$$
$$
\left[
\begin{array}{ccccc|c}
1 & 1 & 1 & 0 & 0 & 1 \\
1 & ③ & 0 & 1 & 0 & 2 \\
\hline
-6 & -12 & 0 & 0 & 1 & 0
\end{array}
\right]
\begin{array}{l}
\text{Quotient: } 1 \\
\text{Quotient: } \frac{2}{3} \quad \leftarrow \text{Pivot row}
\end{array}
$$

Pivot column

The first pivot element is the circled 3 in the second-row, second-column position. Completion of that pivot results in this matrix:

$$
\begin{array}{ccccc}
u & v & s_1 & s_2 & d \\
\end{array}
$$
$$
\left[
\begin{array}{ccccc|c}
\frac{2}{3} & 0 & 1 & -\frac{1}{3} & 0 & \frac{1}{3} \\
\frac{1}{3} & 1 & 0 & \frac{1}{3} & 0 & \frac{2}{3} \\
\hline
-2 & 0 & 0 & 4 & 1 & 8
\end{array}
\right]
\begin{array}{l}
\text{Quotient: } \frac{1}{2} \quad \leftarrow \text{Pivot row} \\
\text{Quotient: } 2
\end{array}
$$

Pivot column

The negative indicator, -2, in the last row indicates that the maximum for d has not yet been reached, so another pivot is necessary on the circled $\frac{2}{3}$ in the first-row, first-column position. The result of this pivot is

$$
\begin{array}{ccccc}
u & v & s_1 & s_2 & d \\
\end{array}
$$
$$
\left[
\begin{array}{ccccc|c}
1 & 0 & \frac{3}{2} & -\frac{1}{2} & 0 & \frac{1}{2} \\
0 & 1 & -\frac{1}{2} & \frac{1}{2} & 0 & \frac{1}{2} \\
\hline
0 & 0 & 3 & 3 & 1 & 9
\end{array}
\right].
$$

The maximum value for d on the feasible region of the dual problem has now been reached. It is 9 and occurs when $u = \frac{1}{2}$, $v = \frac{1}{2}$, $s_1 = 0$, $s_2 = 0$. As asserted in the duality theorem, the maximum value of d is the same as the minimum value of the original objective function f. The solution to the dual problem is $u = \frac{1}{2}$ and $v = \frac{1}{2}$. How can we use the dual's final tableau to determine where the solution to the primal problem occurs? Remember that the solution occurred when $x = 3$ and $y = 3$. As promised by the duality theorem, the bottom row of the final tableau of the dual problem has, in the slack-variable columns, the values for x and y, respectively: $x = 3$, and $y = 3$. Now our problem is completely solved from the final tableau of the dual problem; furthermore, this solution is confirmed by the geometry of the situation.

The sample problem just completed suggests one reason that we study the dual problem: If an original minimization problem has all of its constraints of the form $\geq$, then the dual problem becomes a standard maximum-type problem that can be solved by using only the pivoting rules of the simplex algorithm, thereby enabling us to dispense with the Crown pivoting rules. We present one more example of this type before turning to more general minimization and maximization problems.

EXAMPLE 1

Using the Dual to Solve a Minimization Problem

A company makes three types of pencil sharpeners,: NoDull, SharpenIt, and PinPoint. Production costs for each are, respectively, $7, $12, and $10. Contracts for the upcoming month require that the number of NoDull models plus the number of SharpenIt models be at least 200, the number of SharpenIt models be at least 100, and the sum of all three models be at least 500. How many of each model should be scheduled for production next month to minimize production costs?

SOLUTION Let x, y, and z represent the number of NoDull, SharpenIt, and PinPoint models, respectively, to be produced. The problem then can be stated as

$$\text{Minimize } C = 7x + 12y + 10z$$
$$\text{subject to } x + y \quad\ \geq 200$$
$$y \quad\ \geq 100$$
$$x + y + z \geq 500$$
$$x \geq 0, y \geq 0, z \geq 0.$$

The primal and dual matrices are, respectively,

$$P = \begin{bmatrix} 1 & 1 & 0 & 200 \\ 0 & 1 & 0 & 100 \\ 1 & 1 & 1 & 500 \\ \hline 7 & 12 & 10 & 0 \end{bmatrix}; P^T = D = \begin{bmatrix} 1 & 0 & 1 & 7 \\ 1 & 1 & 1 & 12 \\ 0 & 0 & 1 & 10 \\ \hline 200 & 100 & 500 & 0 \end{bmatrix}.$$

The dual problem can be stated from D, as shown in the following:

$$\text{Maximize } d = 200u + 100v + 500w$$
$$\text{subject to } u \quad\ + w \leq 7$$
$$u + v + w \leq 12$$
$$w \leq 10$$
$$u \geq 0, v \geq 0, w \geq 0.$$

Inserting the three required slack variables and rearranging the objective function that is to be maximized, we obtain the initial simplex tableau for our dual problem.

$$
\begin{array}{c}
\begin{matrix} u & v & w & s_1 & s_2 & s_3 & d \end{matrix} \\
\left[\begin{array}{ccccccc|c}
1 & 0 & ① & 1 & 0 & 0 & 0 & 7 \\
1 & 1 & 1 & 0 & 1 & 0 & 0 & 12 \\
0 & 0 & 1 & 0 & 0 & 1 & 0 & 10 \\
\hline
-200 & -100 & -500 & 0 & 0 & 0 & 1 & 0
\end{array}\right].
\end{array}
$$

Our pivot element is the circled 1 in the initial tableau. The result of using that pivot operation and the next pivot element (circled) are shown in the following matrix:

$$
\begin{array}{c}
\begin{matrix} u & v & w & s_1 & s_2 & s_3 & d \end{matrix} \\
\left[\begin{array}{ccccccc|c}
1 & 0 & 1 & 1 & 0 & 0 & 0 & 7 \\
0 & ① & 0 & -1 & 1 & 0 & 0 & 5 \\
-1 & 0 & 0 & -1 & 0 & 1 & 0 & 3 \\
\hline
300 & -100 & 0 & 500 & 0 & 0 & 1 & 3500
\end{array}\right].
\end{array}
$$

The negative indicator in the objective-function row tells us that at least one more pivot operation is needed. The result of that pivot operation is as follows:

$$
\begin{array}{c}
\begin{matrix} u & v & w & s_1 & s_2 & s_3 & d \end{matrix} \\
\left[\begin{array}{ccccccc|c}
1 & 0 & 1 & 1 & 0 & 0 & 0 & 7 \\
0 & 1 & 0 & -1 & 1 & 0 & 0 & 5 \\
-1 & 0 & 0 & -1 & 0 & 1 & 0 & 3 \\
\hline
300 & 0 & 0 & 400 & 100 & 0 & 1 & 4000
\end{array}\right].
\end{array}
$$

The absence of negative indicators in this tableau shows that the maximum for d has now been achieved and is 4000. Therefore, the minimum production costs will be $4000, which will occur when 400 of the NoDull model, 100 of the SharpenIt model, and none of the PinPoint model are made. (These values are read from the objective-function row in the slack-variable columns.) ▬

Since we have the Crown pivoting rules from the previous section, we can exploit the full nature of the dual problem. The next example is a much more general minimization problem.

EXAMPLE 2

Using the Dual to Solve a Minimization Problem

Solve this linear programming problem through the use of its dual problem:

$$\text{Minimize } f = 2x - 3y$$
$$\text{subject to} \qquad y \le 3$$
$$x \qquad \le 4$$
$$x + y \ge 5$$
$$x \ge 0, y \ge 0.$$

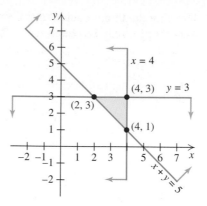

FIGURE 19
The Feasible Region for Example 2

SOLUTION The feasible region is shown in Figure 19.

To obtain the dual problem, we first rewrite the structural constraints of the problem as

$$-y \geq -3$$
$$-x \quad\quad \geq -4$$
$$x + y \geq \quad 5$$

so that the primal matrix becomes

$$P = \left[\begin{array}{cc|c} 0 & -1 & -3 \\ -1 & 0 & -4 \\ 1 & 1 & 5 \\ \hline 2 & -3 & 0 \end{array}\right],$$

from which it follows that

$$P^T = \left[\begin{array}{ccc|c} 0 & -1 & 1 & 2 \\ -1 & 0 & 1 & -3 \\ \hline -3 & -4 & 5 & 0 \end{array}\right] = D.$$

The dual problem may now be written as

$$\text{Maximize } d = -3u + -4v + 5w$$
$$\text{subject to} \quad\quad -v + w \leq \quad 2$$
$$-u \quad\quad + w \leq -3$$
$$u \geq 0, v \geq 0, w \geq 0.$$

The initial tableau for the dual problem is

$$\left[\begin{array}{cccccc|c} u & v & w & s_1 & s_2 & d & \\ 0 & -1 & 1 & 1 & 0 & 0 & 2 \\ \boxed{-1} & 0 & 1 & 0 & 1 & 0 & -3 \\ \hline 3 & 4 & -5 & 0 & 0 & 1 & 0 \end{array}\right].$$

The Crown pivoting rules are needed first to remove the -3 in the rightmost column. According to those rules, the first pivot element will be -1, in the second-row, first-column position. After multiplying the second row by -1 and pivoting on the resulting 1 in the second-row, first-column position, we find that the resulting tableau is

$$
\begin{array}{cccccc|c}
u & v & w & s_1 & s_2 & d & \\
\hline
0 & -1 & ① & 1 & 0 & 0 & 2 \\
1 & 0 & -1 & 0 & -1 & 0 & 3 \\
\hline
0 & 4 & -2 & 0 & 3 & 1 & -9
\end{array}.
$$

The negative above the horizontal bar in the rightmost column has now been removed, so we return to the simplex algorithm rules. The most negative indicator is -2, and the only positive number in that column is the 1 in the first-row, third-column position. That, then, becomes the new pivot position. Upon completion of that pivot, the next and final tableau becomes

$$
\begin{array}{cccccc|c}
u & v & w & s_1 & s_2 & d & \\
\hline
0 & -1 & 1 & 1 & 0 & 0 & 2 \\
1 & -1 & 0 & 1 & -1 & 0 & 5 \\
\hline
0 & 2 & 0 & 2 & 3 & 1 & -5
\end{array}.
$$

The solution to the maximization of d has now been completed. The maximum value for d is -5, which occurs when $u = 5$, $v = 0$, and $w = 2$. According to the duality theorem, the minimum value of f in our original problem is also -5 and occurs when $x = 2$ (from the bottom of the s_1 column) and $y = 3$ (from the bottom of the s_2 column). ■

We now turn to the dual of a problem whose objective function is to be maximized.

EXAMPLE 3

The Dual of a
Maximization Problem

Find and solve the dual problem to this linear programming problem:

$$
\begin{aligned}
\text{Maximize } f &= 2x + 3y \\
\text{subject to} \quad x + y &\le 6 \\
2x + y &\le 8 \\
x \ge 0, \, y &\ge 0.
\end{aligned}
$$

SOLUTION Since the original problem has all of its structural constraints in the $\le$ form, it is also the primal problem. The primal matrix may now be constructed as follows:

$$
P = \begin{bmatrix} 1 & 1 & 6 \\ 2 & 1 & 8 \\ \hline 2 & 3 & 0 \end{bmatrix}
$$

The transpose of this matrix, the dual matrix, is then

$$
P^T = \begin{bmatrix} 1 & 2 & 2 \\ 1 & 1 & 3 \\ \hline 6 & 8 & 0 \end{bmatrix} = D.
$$

7. Solve this linear programming problem by using the simplex algorithm.

In Exercises 8 and 9, a linear programming problem requires you to

$$\text{Minimize } f = 2x + 3y$$
$$\text{subject to } x + y \geq 3$$
$$x + 2y \geq 4$$
$$x \geq 1, y \geq 0.$$

8. Formulate the dual problem.

9. Use the simplex algorithm to solve the dual problem, and find the solution to the primal and dual problems, including the points at which it occurs.

In Exercises 10 through 12, a linear programming problem requires you to

$$\text{Minimize } f = 3x + y$$
$$\text{subject to } x + y \geq 7$$
$$2x + y \geq 10$$
$$x \geq 0, y \geq 0.$$

10. Sketch the feasible region, and find the coordinates of the corner points.

11. Solve this problem by using the Crown pivoting rules.

12. Check your solution to the problem by evaluating f at each corner point of the feasible region.

13. Use the Crown pivoting rules to

$$\text{Maximize } f = 2x + 3y + 5z$$
$$\text{subject to } x + y + 2z \leq 6$$
$$2y + z \geq 4$$
$$2x + z \leq 4$$
$$x \geq 0, y \geq 0, z \geq 0.$$

14. Faculty Salaries To minimize costs, a college mathematics department is analyzing the salaries paid to professors and lecturers. The average annual salary is \$40,000 for professors and \$15,000 for lecturers. Space allocations permit no more than 50 total faculty, but at least 25 must be professors, and no more than 15 can be lecturers. To teach the necessary classes requires at least 35 faculty members. How many professors and lecturers should be employed to minimize total salary costs?

In-Depth Application

Shadow Prices

Objective

In this application, you will investigate an interpretation of the optimal solution for the dual of a linear programming problem.

Introduction

A question one may face when investigating linear programming problems is how to compute changes in the optimal value of the objective function if we change one or more of the constant bounds in the constraints. These rates of change (the change in the objective function's value per unit change in a constraint) are called **shadow prices.** How to compute shadow prices is illustrated here.

The Problem

A retirement group contracts a charter plane service to provide flights from New York to Florida for at least 500 first-class, 1000 tourist-class, and 1500 economy-class persons in the organization. The charter service offers two makes of planes, described in the following table:

	Cost per Flight (thousand dollars)	First-Class Seats	Tourist-Class Seats	Economy-Class Seats
Plane A	12	50	30	100
Plane B	10	40	60	80

(continued)

How many flights of each plane should the retirement group charter if it wants to minimize costs?

1. Set up and solve this linear programming problem.

2. Take note of how much the minimum cost is to charter the flights. Also, record the solution points for both the primal and dual problem. (*Hint:* The duality theorem tells how you can get the solution points for the primal and dual problem from either the primal or dual final tableau.)

   Notice that the primal solution point has two coordinates, while the dual solution has three coordinates. The primal solution corresponds to the number of flights of each type of plane the retirement group should charter both to meet the occupancy requirements and to minimize cost. (As with many "real" problems, the mathematics gives a decimal number for these solutions, while reality forces whole numbers. Ignore this fact for the time being.) Is there an interpretation for the three coordinates of the dual solution?

3. Compute the number of available first-class seats under the optimal solution, and compare it with the number of first-class seats actually needed. If the retirement group were to add one more passenger to the first-class group, would the optimal cost go up?

4. Increase the number of tourist-class seats needed by one (i.e., to 1001), and solve the problem again. By how much did the cost increase? Now go back to the original problem, add one to the economy-class seat constraint and solve again. How much did the cost increase this time?

5. Compare both your answers to (3) and (4) and the coordinates with the original dual solution. Make a conjecture as to why the dual-solution coordinates are called shadow prices.

6 Mathematics of Finance

Sarah is to repay a $10,000 home improvement loan at 8.7%, with payments of $206.13 made monthly. After one year of making payments, Sarah gets a raise in salary and decides that she can afford to pay $250 per month instead. How long will it take Sarah to finish paying off the loan? (See Example 6, Section 6.3.)

CHAPTER PREVIEW

This chapter deals with facts that everyone should know when investing or borrowing money. We begin with the concepts of simple and compound interest and how they compare. We then move to *annuities,* a term describing the accumulation of funds when periodic payments are made into an account and, once deposited, begin to earn interest immediately. The meaning of the consumer term APR (annual percentage rate) is then explored. Next, a method for calculating payments on monthly consumer loans and how to calculate the length of a loan, assuming a predetermined payment level, are considered. Realistic interest rates quoted by lenders are often given over a wide range of decimal percentages, and for that reason, calculators with a natural logarithm key are frequently suggested for use in this chapter.

Technology Opportunities

A calculator with the capability of exponentiation, a logarithm key, and an e key is essential to completing many of the calculations in this chapter. The TI-83 Plus calculator also has a *time value of money* (TVM) solver which simplifies many of the calculations that occur in later sections. Spreadsheets, too, have built-in formulas for completing these calculations. In addition, specialized financial calculators are available, some of which are widely used by finance professionals. Care should be taken with any of these tools, since it is important to know exactly how to enter given information into it. In general, money that is paid "out of pocket," such as an investment or a payment, is entered as a negative value; money that is received, such as a loan amount or the future value of an investment, is entered as a positive value.

Section 6.1 Simple and Compound Interest

If you put money into a savings account, have a certificate of deposit at a financial institution, or buy a government bond or municipal bond, you have, in effect, agreed to lend money to that entity. Put differently, that entity is borrowing from you. If, instead, you borrow money to buy a car, a computer, or a house or to start a business, then a financial institution has agreed to lend you money. If either case, the amount of money lent (or borrowed) is called the **principal.** The principal may be thought of as the **present value** of the money at the time it is lent or borrowed. The person or institution that does the borrowing always agrees to pay a fee for the privilege of borrowing or using the money for a specified length of time. Such a fee is called **interest.** Interest falls into two general classifications—**simple interest** and **compound interest**—and there is a significant difference between the two.

EXAMPLE 1
Calculating Simple Interest

If Mr. Stiener puts $3000 into a savings account that pays 5% simple interest, how much money will be in the account at the end of one year if no withdrawals or additions are made?

SOLUTION As usual, the interest is stated in terms of the percentage of the principal *per year.* The actual interest itself is calculated as

$$I = \$3000 \times 0.05 = \$150.$$

Therefore, the amount in the account at the end of the year will be $3000 + $150 = $3150. ■

EXAMPLE 2
Calculating Simple Interest

The Bank of Benton charges 16% simple interest on its MasterCard accounts. If you charged $500 on such a card and were assessed interest for one month on that amount, how much would the interest be?

SOLUTION The term "16% simple interest" means that the interest for borrowing the money is calculated at the rate of 16% of the $500 for an entire year. If the money is borrowed for only one month, then you should be charged only $\frac{1}{12}$ of the yearly amount. Therefore, the interest would be

$$I = \$500 \times 0.16 \times \frac{1}{12} \approx \$6.67. \quad ■$$

EXAMPLE 3

Calculating Simple Interest

Suppose that you borrowed $5000 from your grandfather for three years at the rate of 7% simple interest. How much would you owe your grandfather at the end of the three years?

SOLUTION The rate of 7% simple interest means that the borrower agrees to pay 7% of the $5000 principal *each year* as a rental fee for the use of the money. Because the money is borrowed for three years, the interest will be

$$I = \$5000 \times 0.07 \times 3 = \$1050.$$

Therefore, the total accumulated amount to be repaid at the end of the three-year period is $A = \$5000 + \$1050 = \$6050.$ ■

The total amount A that will be repaid (or accumulated by the lender) is sometimes known as the **future value** of the loan. Software, such as MS Excel, and calculators with financial functions often use the abbreviation FV or Fv for this value.

Simple Interest

Let P = principal (amount borrowed or lent), or present value
  r = interest rate *per year*, as a decimal
  t = length of loan, measured *in years*.

Then the **simple interest I** on the loan is

$$I = Prt.$$

The **total accumulated amount A** (or future value) to the person making such a loan—hence, the total amount repaid by the person borrowing the money—is

$$A = P + I = P + Prt = P(1 + rt).$$

EXAMPLE 4

Comparing Financing Options

Jeff is planning to buy a new car for $17,000. The car dealership offers three financing plans, none of which requires a down payment: 6.9% simple interest for five years; 3.9% simple interest for four years; and 0% interest for three years. How much would Jeff's monthly payments be under each financing option?

SOLUTION To calculate Jeff's monthly payment under each financing option, we must first calculate the total amount accumulated under each plan. Regardless of the financing option Jeff chooses, his principal is $P = \$17,000$. If he chooses the 6.9% simple-interest plan for five years, then $r = 0.069, t = 5$, and the future value of the loan will be

$$A = P(1 + rt) = \$17,000(1 + 0.069 \times 5) = \$17,000(1.345) = \$22,865.$$

Since Jeff will be required to make monthly payments for five years, he will make a total of $12 \times 5 = 60$ payments. Thus, each monthly payment under the 6.9% financing option will be

$$\$22,865 \div 60 \approx \$381.08.$$

If Jeff chooses the 3.9% financing option for four years, then $r = 0.039, t = 4$, and the future value of the loan will be

$$A = P(1 + rt) = \$17,000(1 + 0.039 \times 4) = \$17,000(1.156) = \$19,652.$$

Under this plan, Jeff will make a total of 12 × 4 = 48 monthly payments, so each payment is equal to

$$\$19,652 \div 48 \approx \$409.42.$$

Finally, if Jeff chooses the 0% financing plan for three years, then no interest is accrued. In this case, Jeff would make 12 × 3 = 36 monthly payments, each in the amount of

$$\$17,000 \div 36 \approx \$472.22. \quad \blacksquare$$

Simple-interest calculations are often used to determine the interest portion of each monthly payment made to pay back a loan. In such a case, the interest is calculated on a daily basis on the new balance or the principal established at that time. Interest rates on savings accounts, by contrast, are almost always given in terms of *compound* interest. An easy example will show how compound interest differs from simple interest. Suppose that $1000 is put into an account that pays 10% *compounded annually*. At the end of the first year, the interest will be $100, so the account balance is $1100. If the account is left untouched for a second year, the account earns 10% of the *new principal,* $1100. Therefore, the total amount accumulated at the end of the second year is

$$\$1100 + (\$1100)(0.10) = \$1100 + \$110 = \$1210.$$

This is the idea of **compound interest**—the *earning of interest on interest.* What is the amount accumulated at the end of the third year? In this case, during the third year, the account would earn 10% interest on the new principal of $1210 established at the end of the second year. The total accumulated would then be

$$\$1210 + (\$1210)(0.10) = \$1210 + \$121 = \$1331.$$

The compounding period of one year used here is only one of many possible terms. Table 1 shows some of the more common compounding terms advertised by financial institutions.

Type of Interest	Number of Interest Periods per Year	Length of Each Interest Period
Compounded Annually	1	1 year
Compounded Semiannually	2	6 months
Compounded Quarterly	4	3 months
Compounded Monthly	12	1 month
Compounded Daily	365 (or 360)	1 day

TABLE 1. SOME COMMON COMPOUNDING PERIODS

What do such rates mean? Here are some examples: If rates are advertised as 6% compounded quarterly, the interest rate *per quarter* is $\frac{6\%}{4} = 1.5\%$, and this rate is applied to the total in the account at the end of each quarter. If rates are advertised as 8% compounded monthly, the interest rate *per month* is $\frac{8\%}{12} \approx 0.67\%$, and this rate is applied to the total in the account at the end of each month.

Now suppose that we have invested a particular principal P at an annual interest rate of r and that interest is compounded m times per year. Then interest is applied at a rate of $i = r/m$ per interest period. If the account is left untouched for t years, then the total number of interest periods is $n = mt$. At the end of the first interest period, the accumulated amount is

$$A_1 = P + Pi$$
$$= P(1 + i).$$

At the end of the second interest period, the accumulated amount is

$$A_2 = A_1 + A_1 i$$
$$= P(1 + i) + P(1 + i)i \qquad \text{Substitute expression for } A_1.$$
$$= P(1 + i)(1 + i) = P(1 + i)^2. \qquad \text{Factor out } P(1 + i).$$

At the end of the third interest period, the accumulated amount is

$$A_3 = A_2 + A_2 i$$
$$= P(1 + i)^2 + P(1 + i)^2 i \qquad \text{Substitute expression for } A_2.$$
$$= P(1 + i)^2(1 + i) = P(1 + i)^3. \qquad \text{Factor out } P(1 + i)^2.$$

Following this general pattern gives the accumulated amount at the end of the nth interest period:

The Compound Interest Formula

Let P = principal
t = number of years
m = number of interest periods per year
r = annual interest rate, compounded m times per year, *as a decimal*
$i = \dfrac{r}{m}$ = interest rate *per period*
$n = mt$ = total number of interest periods.

Then the future value after n interest periods is

$$A = P\left(1 + \frac{r}{m}\right)^{mt} = P(1 + i)^n.$$

Example 5 demonstrates how to use the compound interest formula.

EXAMPLE 5

Using the Compound-Interest Formula

Jamal puts $500 into a savings account that pays 4.2% interest compounded quarterly. If he makes no additional payments to the account and does not withdraw any money from it for five years, what is the future value of the account?

SOLUTION We begin by identifying the values needed by the compound interest formula. Jamal invests a principal of $P = \$500$, at an annual interest rate of $r = 0.042$, with $m = 4$ interest periods per year, for $t = 5$ years. Thus, $i = r/m = \frac{0.042}{4} = 0.0105$ and $n = mt = 4 \times 5 = 20$. Therefore, the future value of Jamal's savings account is

$$A = P(1 + i)^n = 500(1 + 0.0105)^{20} = 500(1.0105)^{20} \approx \$616.16. \quad \blacksquare$$

Both Excel and the TI-83 Plus have built-in features for calculating the future value of an investment or loan. The next two examples demonstrate how this is accomplished. When using either tool, care should be taken to input values correctly. If money is taken out of someone's pocket and placed into an account, used to make a payment, or invested, then the dollar value is assumed to be *negative* by both Excel and the TI-83 Plus.

EXAMPLE 6

Using Excel to Calculate the Future Value of an Investment

Suppose $1000 is invested at 4.5% interest. What is the future value of the investment at the end of 5 years if the interest is

a. compounded annually?

b. compounded quarterly?

SOLUTION

a. The Excel function FV calculates the future value of an investment. From any blank cell, select Insert, then Function, and choose the Financial function category. Scroll through the list to locate the FV function. Highlight this function and click OK. A pop-up menu (see Figure 1) will appear.

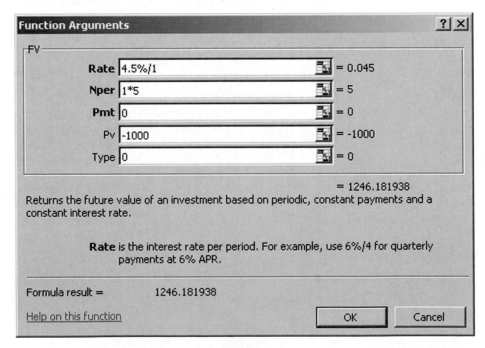

FIGURE 1
The Future Value Pop-Up Menu from Excel, with Values from Example 6(a)

For the Rate, we enter the value of i, which is $r/m = \frac{0.045}{1}$. For Nper, the number of interest periods, we enter the value of n, which is $mt = 1 \times 5$. Since we will not be making any additional payments into the account, we enter 0 for Pmt. Because we have invested the principal (i.e., we have *paid* the principal into the account), we enter −1000 as the present value, Pv. Finally, we enter 0 for Type since we want interest paid at the end of each interest period. When we click OK, the future value of the investment, $1246.18, is displayed in the spreadsheet cell.

b. Beginning in another blank cell, access the FV function as in part (a). When the pop-up menu appears, we will enter the values shown in Figure 2, since interest is compounded quarterly.

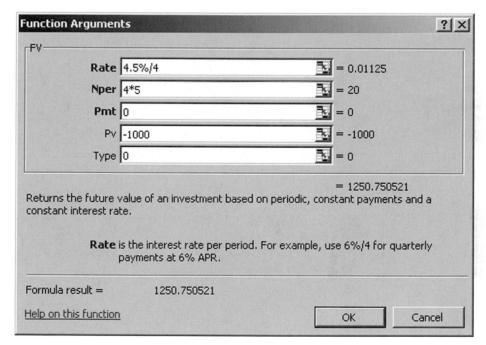

FIGURE 2
The Future Value Pop-Up Menu from Excel, with Values from Example 6(b)

We enter $i = \frac{0.045}{4}$ for Rate, $n = 4 \times 5$ for Nper, 0 for Pmt, −1000 for Pv, and 0 for Type. When we click OK, the future value of the investment, $1250.75, is displayed in the spreadsheet cell. ■

EXAMPLE 7

Using the TVM Solver on the TI-83 Plus to Calculate Future Value.

Suppose $1000 is invested at 4.95% interest. What is the future value of the investment at the end of 5 years if the interest is

a. compounded annually?

b. compounded quarterly?

SOLUTION Access the TVM Solver on the TI-83 Plus by pressing [APPS] [ENTER]. We must enter values for all of the variables listed, *except* the one in which we are interested. So, for this example, we must enter values for all variables except FV.

a. We enter 1×5 for N, the number of payment periods, and then 4.95 for I%, which is the annual interest rate *as a percentage*. Since we have placed our principal *into* an account, we enter -1000 as the present value, PV. We will not be making any additional payments into the account, so we enter 0 for PMT. Because we do not know the future value of the investment and will eventually be solving for it, we enter 0 for FV. Since interest is compounded annually in this example, we enter 1 for P/Y, which represents the number of payments made each year, and 1 for C/Y, which represents the number of times interest is compounded per year. Finally, we highlight END to indicate that interest is paid at the end of each interest period. (See Figure 3.)

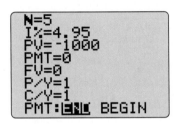 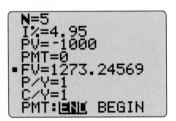

FIGURE 3
The TVM Solver *of the TI-83 Plus, with Values from Example 7(a)*

Now, use the arrow keys to move the cursor beside FV and press [ALPHA] [ENTER] to execute the SOLVE command. The value of FV is updated, indicating that the future value of the account will be approximately $1273.25.

b. Since most of the values required for the TVM Solver for this example already have been entered, we need only modify the values of N, P/Y, and C/Y to reflect the fact that interest is compounded quarterly. Thus, we change the value of N to 4×5 and the values of P/Y and C/Y to 4. (See Figure 4.)

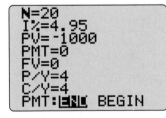 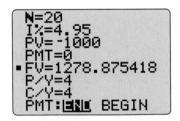

FIGURE 4
The TVM Solver *of the TI-83 Plus, with Values from Example 7(b)*

After moving the cursor beside FV, we press [ALPHA] [ENTER] to execute the SOLVE command. The value of FV is updated, indicating that the future value of the account will be approximately $1278.88. ■

What effect does increasing the number of interest periods have on the future value of an investment? In Example 8, we will explore this question.

EXAMPLE 8

Using the Compound Interest Formula

If $100 is invested at 8%, how much is in the account at the end of 10 years if the interest is

(a) Simple?

(b) Compounded annually?

(c) Compounded semiannually?

(d) Compounded quarterly?

(e) Compounded daily?

SOLUTION

(a) $A = P(1 + rt) = \$100.000(1 + 0.08(10)) = \$100.00(1 + 0.8)$

$$= \$100.00(1.8) = \$180.00.$$

(b) $A = P(1 + i)^n = P\left(1 + \dfrac{r}{m}\right)^{mt} = \$100.00(1 + 0.08)^{10} = \$100.00(1.08)^{10}$

$$\approx \$215.89.$$

(c) $A = P(1 + i)^n = P\left(1 + \dfrac{r}{m}\right)^{mt} = \$100.00\left(1 + \dfrac{0.08}{2}\right)^{2(10)} = \$100.00(1.04)^{20}$

$$\approx \$219.11.$$

(d) $A = P(1 + i)^n = P\left(1 + \dfrac{r}{m}\right)^{mt} = \$100.00\left(1 + \dfrac{0.08}{4}\right)^{4(10)} = \$100.00(1.02)^{40}$

$$\approx \$220.80.$$

(e) $A = P(1 + i)^n = P\left(1 + \dfrac{r}{m}\right)^{mt} = \$100.00\left(1 + \dfrac{0.08}{365}\right)^{365(10)} \approx \$222.53.$ ∎

NOTE For the examples in this section and throughout this chapter, we have used a spreadsheet or calculator for our computations. If your answers do not agree precisely, it may be due to round-off error or the number of decimal places your calculator or spreadsheet is set to carry, especially with numbers raised to very large powers. In all cases, accuracy is greater if you wait until the end to round, rather than rounding at intermediate steps.

Example 8 shows that the greater the number of compounding period per year, the greater is the return (applied to the same principal). Could we indefinitely increase the amount earned in a 10-year period if the number of compounding periods continued to grow?

Table 2 examines what happens to a principal of $1, invested for one year at an annual interest rate of 100%, when the number of compounding periods increases.

Interest Type	Number of Interest Periods per Year	Future Value of $1 after 1 year
Compounded Annually	1	$2.000000
Compounded Semiannually	2	$2.250000
Compounded Quarterly	4	$2.441406
Compounded Monthly	12	$2.613035
Compounded Daily	365	$2.714567
Compounded by the Hour	8760	$2.718127
Compounded by the Minute	525,600	$2.718279
Compounded by the Second	31,536,000	$2.718282

TABLE 2. THE EFFECT OF INCREASING THE NUMBER OF INTEREST PERIODS ON A $1 INVESTMENT

It may surprise you to see that the future value of our principal seems to "level off" to about $2.72 as the number of interest periods increases. The number that our future value is approaching appears in many different applications of mathematics. It is called e (named for Leonhard Euler) and is an *irrational number*, meaning that its decimal representation is non–terminating and non–repeating. Using methods from calculus, we can show that, as the number of compoundings per year, m, increases without bound, our formula for calculating the future value of an investment, $A = P(1 + i)^n = P(1 + r/m)^{mt}$, approaches $A = Pe^{rt}$. In this case, we say that interest is *compounded continuously*. This means that, at each instant of time, the investment grows in proportion to the amount present at that instant. The amount to which such an investment P grows after t years is given next.

Continuous Compounding

Let P = principal
   r = rate of interest **compounded continuously**
   t = time in years.

Then the accumulated amount after t years is

$$A = Pe^{rt},$$

where

$$e = 2.718. \ldots$$

Example 9 continues the comparison made in Example 8.

EXAMPLE 9

Compounding Money Continuously

 If $100 is invested at 8% compounded continuously, how much will be in the account at the end of 10 years?

SOLUTION Substituting into the formula for continuous compounding gives

$$A = \$100 \cdot e^{0.08(10)} = \$100 \cdot e^{0.8}.$$

To approximate this value with the TI-83 Plus, type $\boxed{1}\boxed{0}\boxed{0}\boxed{\times}$, press $\boxed{\text{2nd}}\boxed{\text{LN}}$, and then enter the exponent 0.8. Press $\boxed{\text{ENTER}}$ to see that $\$100 \cdot e^{0.8} \approx \$222.55.$ ∎

To approximate the value in Example 9 with Excel, in any blank cell, type =100*EXP(0.8) and press Enter. It may be surprising to note that after rounding to two decimals, $100 grows only slightly faster over a 10-year period at 8% when compounded continuously than when compounded daily ($222.53). In other words, continuous compounding is better, but not that much better. The comparison between simple interest and compound interest given in Examples 7 and 8 may also be shown graphically. Notice that simple compounding is *linear;* that is, $A = P(1 + rt) = P + Prt$ is linear in the variable t, provided that P and r remain fixed. The A-intercept is P and the slope of the line is Pr. By contrast, compound interest is exponential in nature. That is to say, the variable t is an exponent in the formula $A = P\left(1 + \frac{r}{m}\right)^{mt}$. Figure 5 is a graphic comparison between simple and compound interest.

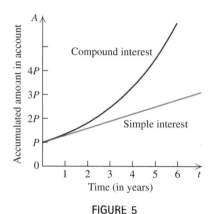

FIGURE 5

Comparing the Growth of Principal between Simple and Compound Interest

Solving for Other Values in a Compound Interest Formula

Solving for some variables in finance formulas requires logarithms, which we may obtain from our calculators. The solution of a commonly used equation is shown next, as a brief review of needed algebra skills. (See Appendix I for additional information about logarithms.)

An Algebra Refresher

To solve the equation $a^n = b$ for n, where a and b are positive real numbers, calculate

$$n = \frac{\ln(b)}{\ln(a)} = \frac{\log(b)}{\log(a)},$$

where $\ln(b)$ is the logarithm of b to the base e and $\log b$ is the logarithm of b to the base 10. The solutions comes from a law of logarithms, which states that if $a^n = b$, then $n \ln(a) = \ln(b)$.

EXAMPLE 10

How Fast Will
Money Double?

 How long will it take an amount of money P to double if the interest rate is 6% compounded quarterly?

SOLUTION The question asks, how many interest periods n will it take for P to reach the accumulated amount $2P$? The compound interest formula $A = P(1 + i)^n$ tells us that

$$2P = P\left(1 + \frac{0.06}{4}\right)^n,$$

and we are to solve for n. After dividing each side by P, we have

$$2 = (1.015)^n,$$

and from the algebra refresher,

$$n = \frac{\ln(2)}{\ln(1.015)} \approx 46.56$$

interest periods, which translates into $\frac{46.56}{4} = 11.64$ years. ∎

EXAMPLE 11

Using the Compound
Interest Formula

What amount of money should be deposited now in an account paying 8% compounded quarterly so that 10 years later the account balance will be $12,000? (Assume no further withdrawals or additions to the account.)

SOLUTION The number of interest period is $n = mt = 4 \times 10 = 40$, and the interest rate per period is $i = \frac{0.08}{4} = 0.02$. The compound interest formula then tells us that

$$\$12{,}000 = P(1 + 0.02)^{40}.$$
$$= P(2.2080397).$$

Solving for P gives $P \approx \$5{,}434.68$. ∎

EXAMPLE 12

Interest Needed to
Double Money

 What rate of interest, compounded annually, should you receive for an amount of money P to double in 10 years?

SOLUTION This time we want to find the annual interest rate r for which P will become $2P$ when $n = 10$. That is, we are to solve

$$2P = P(1 + r)^{10}$$

for r. To do this, first cancel P on both sides to obtain

$$2 = (1 + r)^{10},$$

and then take the tenth root of both sides, getting

$$2^{1/10} = 1 + r$$

from which it follows that

$$r = 2^{1/10} - 1$$
$$= 2^{0.10} - 1$$
$$\approx 1.0717735 - 1$$
$$= 0.0717735,$$

or about 7.18%. ■

EXAMPLE 13

The Battle
against Inflation

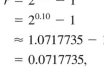 If money is losing value at the rate of 5% annually because of inflation, how long will it take a dollar, now assumed to be worth 100 cents, to be worth 80 cents?

SOLUTION Instead of increasing in value, the principal $P = \$1$ is now *losing* value at the rate of 5%, compounded annually. Thus, after the first year, the accumulated value for an amount P is $P - 0.05P = P(1 - 0.05)$; after the second year, the accumulated value is $P(1 - 0.05) - P(1 - 0.05)0.05 = P(1 - 0.05)^2$, and so on. In particular, for $P = \$1$, we seek n such that

$$\$0.80 = \$1.00(1 - 0.05)^n$$

or

$$0.80 = 0.95^n.$$

Using our algebra refresher, we find that

$$n = \frac{\ln(0.80)}{\ln(0.95)} \approx 4.35 \text{ years.} ■$$

Example 13 points out that even though our investment may earn a handsome return, inflation reduces its spending power. Example 14 shows another factor affecting our investment or savings return.

EXAMPLE 14

Taxes

Suppose an investment of $5000 earns 8% compounded quarterly for a period of one year. In computing your income taxes, you find that the applicable federal rate is 28% and the applicable state rate is 7%. How much of the return on your investment do you get to keep?

SOLUTION The return on your investment will be

$$\$5000\left(1 + \frac{0.08}{4}\right)^4 - 5000 = \$412.16.$$

The federal tax on this amount will be

$$\$412.16(0.28) = \$115.40$$

and the state tax will be

$$\$412.16(0.07) = \$28.85.$$

Therefore, the total income tax due on the return is $115.40 + $28.85 = $144.25. It follows that you get to keep

$$\$412.16 - \$144.25 = \$267.91. \quad \blacksquare$$

Problems similar to the last several examples can be completed with the TVM Solver of the TI-83 Plus or with one of the built-in financial functions in Excel. Some of these are demonstrated in the last two examples of this section.

EXAMPLE 15

Using the TI-83 Plus to Calculate Present Value

Complete Example 11, using the TVM Solver of the TI-83 Plus.

SOLUTION In Example 11, we wished to find the amount of money that, when deposited in an account paying 8% interest compounded quarterly, would result in an account balance of $12,000 in 10 years. In the TVM solver, we enter 4*10 = 40 for the number of payment periods, N; 8 for the annual percentage rate, I%; 0 for the periodic payment, PMT, since no additional money will be added to the account; 12000 for the future value of the account, FV; and 4 for both P/Y and C/Y, since interest is compounded quarterly. Move the cursor next to PV, the present value of the account, and press [ALPHA] [ENTER] to access the SOLVE command. (See Figure 6.) We need to deposit $5,434.68 in the account today in order to attain a balance of $12,000 in 10 years. Note that the amount appears as a negative value in the TVM Solver, since that amount of money must be taken out of pocket.

FIGURE 6
Calculating Present Value with the TVM Solver ■

In general, when using the TVM Solver on the TI-83 Plus, enter all the known information related to the financial problem, remembering that money taken out of pocket is to be entered as a negative number. Then move the cursor to the unknown value and access the SOLVE command.

EXAMPLE 16

Using Excel to Calculate the Number of Interest Periods

How many interest periods will it take $1200 to double if the interest rate is 4.5% compounded quarterly? Approximately how many years is this number equivalent to?

SOLUTION From any blank cell in a worksheet, access the Nper function, which calculates the number of periods desired. A dialog box similar to Figure 7 appears. Enter 4.5%/4 for the periodic interest Rate, as a percent; 0 as the payment amount, Pmt, since no additional payments will be made into the account; −1200 for the present value, Pv; and 2400 as the future value of the account, Fv. Click OK and the number of periods, 61.96, or approximately 62 periods. This corresponds to 62/4 = 15.5 years.

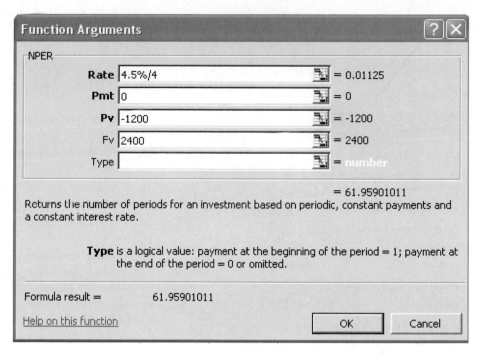

FIGURE 7
Excel's NPER Function for Calculating Required Number of Interest Periods

Excel has many built-in financial functions. To see what is available, choose `Insert`, then `Function`, and then select `Financial` from the drop-down menu. As you scroll through the list of available functions, a brief description of their purpose is displayed at the bottom of the dialog window.

Exercises 6.1

 A calculator or spreadsheet will be helpful in these exercises.

In Exercises 1 through 10, find the amount that will be accumulated in each account under the conditions set forth.

1. A principal of $2000 is accumulated with 7.5% interest compounded monthly for 4 years.

2. A principal of $5000 is accumulated with 10% interest compounded annually for 8 years.

3. A principal of $3000 is accumulated with 6% interest compounded quarterly for 12 years.

4. A principal of $7000 is accumulated with 9.5% interest compounded annually for 15 years.

5. A principal of $6000 is accumulated with simple interest of 12% for 1 year.

6. A principal of P is accumulated with 6% interest compounded quarterly for 5 years.

7. A principal of $800 is accumulated for 12 years

   (a) At 7% simple interest

   (b) At 7% compounded quarterly

   (c) At 7% compounded monthly

8. A principal of $4000 is accumulated for 8 years

   (a) At 9% compounded quarterly

   (b) At 9% compounded monthly

   (c) At 9% compounded daily

9. A principal of $1 is accumulated for 10 years

   (a) At 12% compounded semiannually

   (b) At 12% compounded quarterly

   (c) At 12% compounded daily

10. A principal of $10,000 is accumulated for 4 months

    (a) At 14% simple interest

    (b) At 14% compounded monthly

    (c) At 14% compounded daily

11. Graph, on the same set of axes, the accumulated value for each of the first five years if $5000 earns

    (a) simple interest of 10%

    (b) 10% interest compounded quarterly

12. Graph, on the same set of axes, the accumulated value for each of the first five years if $10,000 earns

    (a) simple interest of 6%

    (b) 6% interest compounded monthly

In Exercises 13 through 18, find the rate of interest required to achieve the conditions set forth.

13. $A = \$20,000$; $P = \$8000$; $t = 10$ years; interest is compounded annually

14. $A = \$12,000$; $P = \$3500$; $t = 6$ years; interest is compounded annually

15. $A = \$5000$; $P = \$1250$; $t = 12$ years; interest is compounded quarterly

16. $A = \$6500$; $P = \$2300$; $t = 5$ years; interest is compounded quarterly

17. $A = \$7500$; $P = \$2500$; $t = 6$ years; interest is compounded semiannually

18. $A = \$3000$; $P = \$1500$; $t = 8$ years; interest is compounded semiannually

In Exercises 19 through 24, find the number of interest periods required to achieve the conditions set forth.

19. $A = \$3500$; $P = \$1200$; interest is 8% compounded semiannually

20. $A = \$5000$; $P = \$2400$; interest is 6% compounded semiannually

21. $A = \$6000$; $P = \$1000$; interest is 12% compounded quarterly

22. $A = \$30,000$; $P = \$4800$; interest is 15% compounded quarterly

23. $A = \$50,000$; $P = \$6000$; interest is 10% compounded annually

24. $A = \$60,000$; $P = \$10,000$; interest is 9% compounded annually

In Exercises 25 through 30, find the principal P required to achieve the stated conditions.

25. $A = \$5000$; rate is 6% compounded annually for a period of 10 years

26. $A = \$12,000$; rate is 9% compounded annually for a period of 15 years

27. $A = \$30,000$; rate is 12% compounded quarterly for a period of 10 years

28. $A = \$35,000$; rate is 6% compounded quarterly for a period of 12 years

29. $A = \$3000$; rate is 7.5% compounded monthly for a period of 10 years

30. $A = \$10,000$; rate is 9% compounded monthly for a period of 8 years

31. **Investing** If $5000 is invested at the rate of 7% compounded semiannually, what will be the value of the investment 10 years from now, assuming no withdrawals?

32. **Investing** If $10,000 is invested at the rate of 10% compounded quarterly, what will be the value of the investment 15 years from now, assuming no withdrawals?

33. **Investing** Dr. Bishop is saving money to send his children to college. How much will he need to invest now at 8% compounded quarterly if he wants the accumulated value of the investment to be $20,000 in 12 years?

34. **Investing** Ramero plans to buy a new car three years from now. Rather than borrow at that time, he plans to invest part of a small inheritance at 7.5% compounded semiannually to cover the estimated $6000 trade-in difference. How much does he need to invest if he starts investing now?

35. **Investing** The Tanners have received an $8000 gift from one of their parents to invest in their child's college education. They estimate that they will need $20,000 in 12 years to achieve their educational goals for their child. What interest rate compounded semiannually would the Tanners need to achieve this goal?

36. **Investing** When she took early retirement, Nancy received $50,000 from a savings plan she had established with her company while she was still working. She plans to open her own business and work for herself for a few years before retiring completely. She estimates that she needs $75,000 to start the business she has in mind, and she plans to take three years to find the right location. What interest rate compounded quarterly would Nancy need to achieve this goal?

37. **Tripling Your Money** How many years will it take for a sum of money to triple if the interest rate is 8% compounded quarterly?

38. **Doubling Your Money** How many years will it take for $4000 to double if the interest rate is 10% compounded annually?

39. Growth of Money How many years will be required to turn $5000 into $8000 if the interest rate 7.5% compounded semiannually?

40. Growth of Money How many years will be required to turn $10,000 into $25,000 if the interest rate is 8% compounded quarterly?

Checking Accounts Exercises 41 and 42 refer to the following regulations: Banking rules require that a specified percentage of funds in checking accounts be set aside as a reserve and not be used for loans. If a bank states that it is paying interest on available funds, then it is paying interest only on the percentage of your account that is outside the reserve requirements. Assume this reserve to be 15%.

41. Suppose that your checking account pays interest at the rate of 4.5% annually on the monthly minimum balance of available funds. If your minimum balance this month was $1200, how much interest was credited to your account?

42. Suppose that your checking account pays interest at the rate of 5% annually on the monthly minimum balance of available funds. If your minimum balance this month was $2000, how much interest was credited to your account?

43. Cost of Living If the cost of living is expected to increase at the rate of 4% compounded annually each year for the next five years, what will the cost then be of goods priced at $50 now?

44. Cost of Living If the cost of living is expected to increase at the rate of 6% compounded annually each year for the next 10 years, what will the cost then be of goods priced at $100 now?

45. Inflation If the purchasing power of the dollar is expected to decrease at the rate of 5% each year, how long will it take before the purchasing power is cut in half?

46. Inflation If the purchasing power of the dollar is expected to decrease at the rate of 8% each year in the foreseeable future, how long will it take before the purchasing power is reduced to three-fourths of its current value?

47. Retirement What income will a couple that is comfortably retired now on $25,000 per year need 10 years from now to maintain the same standard of living if the inflation rate is assumed to be 5% per year?

48. Retirement What income will a couple that is comfortably retired now on $45,000 per year need eight years from now to maintain the same standard of living if the inflation rate is assumed to be 6% per year?

49. Cars and Inflation If a new car costs $20,000 now, and the inflation rate is assumed to be 5% per year, what will a new car cost eight years from now?

50. Homes and Inflation If the median price of a new home is $110,000 now, and the inflation rate is expected to be 5% per year, what will the median price of a new home be 10 years from now?

51. Construction Costs A contractor building a new eight-story office building estimates that the cost of each floor will be 20% more than the cost of the floor just below it. If the cost of the first floor is $250,000, what

(a) Will the fourth floor cost?

(b) Will the eighth floor cost?

(c) Will the entire building cost?

52. Rate of Return on an Investment If Joan bought a lot for $18,000 and sold it five years later for $24,500, what was her percentage rate of return on the investment if interest was compounded annually?

Section 6.2 Ordinary Annuities

An **annuity** is a sequence of equal payments, each made at equally spaced time intervals. When the interest on the payments is compounded at the same time the payment is made, the annuity is called an **ordinary annuity.** We consider only ordinary annuities here. The following are examples of ordinary annuities:

   a sequence of equal annual payments into an IRA account;

   a sequence of equal monthly payments to pay off a car loan;

   a sequence of equal monthly payments to pay off a house loan;

   a sequence of equal monthly payments into an account to be used later for a college education.

When making payments into an annuity account, we are naturally interested in the total accumulated in the account after a given number of payments. This total is called the **future value** of the annuity, and it may be calculated by a formula that makes use of the compound-interest formula. For example, suppose that we plan to put $500 into an account at the end of each year for five years, assuming that interest is 8% compounded annually.

0	1	2	3	4	5	Years
	$500	$500	$500	$500	$500	

The following table shows how each of the $500 amounts earns interest compounded annually.

Payment Number	Amount of Payment	Number of Years at 8% Interest	Accumulated Value
1	$500	4	$500(1 + 0.08)^4$
2	$500	3	$500(1 + 0.08)^3$
3	$500	2	$500(1 + 0.08)^2$
4	$500	1	$500(1 + 0.08)^1$
5	$500	0	$500

TABLE 3. AMOUNT OF INTEREST EARNED ON EACH PAYMENT

The future value A of the account is the sum of the entries in the last column of the table:

$$A = \$500(1.08)^4 + 500(1.08)^3 + 500(1.08)^2 + 500(1.08) + 500$$
$$= \$500\left[(1.08)^4 + (1.08)^3 + (1.08)^2 + (1.08) + 1\right].$$

Notice the pattern set out in the last line. Could you now write an expression for the future value of this annuity after the twelfth payment? If you write

$$A = \$500\left[(1.08)^{11} + (1.08)^{10} + (1.08)^9 + \cdots + 1\right],$$

you are correct! Following these guidelines, if an amount p were deposited into an account at the end of each interest-bearing period, and if the interest rate were i *per period,* then the future value A of the annuity at the end of the nth payment would be

$$A = p\left[(1 + i)^{n-1} + (1 + i)^{n-2} + (1 + i)^{n-3} + \cdots + 1\right].$$

The sum inside the square brackets has a very special form. It is called a **geometric series,** and an algebraic formula gives its sum:

$$A = p\left[\frac{(1 + i)^n - 1}{i}\right].$$

(See Appendix I for an explanation of a geometric series.)

Future Value of an Ordinary Annuity

Let p = amount of each payment made at the end of each period
  t = number of years
  m = number of periods per year
  r = annual interest rate, compounded m times per year, as a decimal
  $i = \dfrac{r}{m}$ = interest rate per period
  $n = mt$ = total number of periods (payments).

Then the **future value** of the annuity is the sum of all payments and interest earned on those payments and is found by the formula

$$A = p\left[\frac{\left(1 + \dfrac{r}{m}\right)^{mt} - 1}{\dfrac{r}{m}}\right] = p\left[\frac{(1 + i)^n - 1}{i}\right].$$

EXAMPLE 1

Calculating Retirement Funds

Ms. O'Brien sets up an individual retirement account (IRA) with a local bank. She plans to put the limit of $2000 into the account each year. The account will earn 7.5% guaranteed annual compound interest. (Such accounts are free of taxes until withdrawal at retirement.) If she starts the account at age 30 and retires at age 65, what amount will have accumulated in the account?

SOLUTION This an ordinary annuity with $p = \$2000$, $i = 0.075$, and $n = 35$. The amount accumulated in the account at the end of 35 years, including the thirty-fifth payment, is given by the future value:

$$A = \$2000\,\frac{(1 + 0.075)^{35} - 1}{0.075}$$

$$\approx \$2000(154.25160)$$

$$\approx \$308,503.21. \qquad \blacksquare$$

Example 1 vividly shows the power of compound interest over a long period. Ms. O'Brien has contributed $2000 \times 35 = \$70,000$ to the account, but will enjoy a retirement fund of over $300,000.

When large expenses are foreseen in the future by families or businesses, they sometimes elect to set aside a sum of money in an interest-bearing account on a regular basis to cover those expenses. For example, a family might want to set aside a certain amount each month for 16 years to have $80,000 for their child's college expenses, or a business might want to set aside a certain amount each year to accumulate $50,000 to buy a piece of equipment 10 years from now. In either case, we say that the family or business is setting up a **sinking fund.** In a *sinking fund,* we seek to find the amount p to deposit each period to achieve the stated financial goal in the desired length of time. We assume that the *deposits* will be made at the *end of each interest-paying period,* and we know the future value, the interest rate, and the time needed to achieve our goal. Thus, all of the quantities in

$$A = p\left[\frac{(1 + i)^n - 1}{i}\right]$$

are known, except for p. Solving for p gives the amount of each payment, as shown next.

Payments to a Sinking Fund

Let A = the future value desired in the fund
  t = number of years
  m = number of periods (payments) per year
  r = annual interest rate, compounded m times per year, as a decimal
  $i = \dfrac{r}{m}$ = interest rate per period
  $n = mt$ = total number of periods (payments).

Then the amount to be deposited at the end of each interest period in order to achieve A is

$$p = A\frac{\left(\dfrac{r}{m}\right)}{\left(1 + \dfrac{r}{m}\right)^{mt} - 1} = A\frac{i}{(1 + i)^n - 1}.$$

NOTE Many books use the letter S rather than A for the sum of the terms giving the total accumulated value. In addition, because the fraction

$$\frac{(1 + i)^n - 1}{i}$$

is used so often, some books use the notation $s_{\overline{n}|i}$ for this quantity. It is read "s sub n angle i."

EXAMPLE 2
Finding the Payments to
a Sinking Fund

Suppose that Dr. Pate wants to save $10,000 over a period of three years to buy a fishing boat. His bank will pay 8% compounded semiannually on an account into which he will place his payments. If the payments are made semiannually, how much should each payment be for Dr. Pate to achieve his savings goal?

SOLUTION This is a sinking-fund problem, with A = $10,000, $i = \frac{0.08}{2} = 0.04$ per period, and $n = 3 \times 2 = 6$. Therefore, the payments should be

$$p = \$10,000\left[\frac{0.04}{(1 + 0.04)^6 - 1}\right]$$
$$\approx \$10,000(0.1507619)$$
$$\approx \$1507.62.$$

Suppose that an interest-bearing account already has an amount P in it and that equal payments p are made into the account periodically, as in an ordinary annuity. What will the accumulated value of the account be after n periods? By the compound interest formula, the amount P will grow to $P(1 + i)^n$, and, in addition, the annuity portion will grow to

$$A = p\left[\frac{(1 + i)^n - 1}{i}\right].$$

Therefore, the total accumulated amount will be that given by the ordinary annuity formula.

The Ordinary Annuity Formula

Let P = the initial amount in an account
$\quad p$ = amount deposited at the end of each period
$\quad t$ = number of years
$\quad m$ = number of periods per year
$\quad r$ = annual interest rate, compounded m times per year, as a decimal

$\quad i = \dfrac{r}{m}$ = interest rate per period

$\quad n = mt$ = total number of periods (payments).

Then the total amount accumulated in the account after n periods is

$$A = P\left(1 + \frac{r}{m}\right)^{mt} + p\left[\frac{\left(1 + \dfrac{r}{m}\right)^{mt} - 1}{\dfrac{r}{m}}\right] = P(1 + i)^n + p\left[\frac{(1 + i)^n - 1}{i}\right].$$

This formula encompasses two of the major formulas that we have already developed: the compound interest formula (let $p = 0$) and the formula for the future value of an annuity (let $P = 0$).

EXAMPLE 3

Using the Ordinary-
Annuity Formula

Suppose that an account paying 6% compounded semiannually now has $10,000 in it and its owner plans to add $500 at the end of each six-month period for the next eight years. What will be the accumulated value of the account at the end of eight more years?

SOLUTION The ordinary-annuity formula may be used to find the accumulated value. We have $P = \$10{,}000$, $i = \frac{0.06}{2} = 0.03$, $n = 16$, and $p = \$500$. Therefore,

$$S = \$10{,}000(1 + 0.03)^{16} + \$500\frac{(1 + 0.03)^{16} - 1}{0.03}$$

$$\approx \$10{,}000(1.604706) + \$500(20.1568813)$$

$$\approx \$26{,}125.50.$$

Just as an annuity has a future value, it also has a *present value*. Suppose that we have an annuity of n payments of p dollars each, with an interest rate of i per payment period. The **present value** of such an annuity is defined to be *single deposit of P dollars*, earning the same rate of interest for n periods, that will generate the same accumulated value as the annuity. That is, we want to find the value of P in the formula

$$P(1 + i)^n = p\left[\frac{(1 + i)^n - 1}{i}\right]$$

or

$$P = p\left[\frac{(1 + i)^n - 1}{i(1 + i)^n}\right] = p\left[\frac{1 - (1 + i)^{-n}}{i}\right].$$

Present Value of an Ordinary Annuity

For an annuity with
   p = amount of each payment made at the end of each period
   t = number of years
   m = number of periods per year
   r = annual interest rate, compounded m times per year, as a decimal
   $i = \dfrac{r}{m}$ = interest rate per period
   $n = mt$ = total number of periods (payments).

Then the **present value** of the annuity is

$$P = p\left[\frac{1 - \left(1 + \dfrac{r}{m}\right)^{-mt}}{\dfrac{r}{m}}\right] = p\left[\frac{1 - (1 + i)^{-n}}{i}\right].$$

Here is an alternative way to look at the present value of an annuity: Suppose that we put P dollars into the interest-bearing account, *withdraw p dollars each period*, and redeposit the p dollars into a new account earning the same interest rate. Then, after n periods, the new account would have the accumulated value of the annuity, which means that the original account starting with P dollars must now be exhausted! In other words, the *present value of an annuity is the **single** deposit that will finance the annuity.*

EXAMPLE 4

Using the Present Value
of an Annuity

Mr. and Mrs. McKnight want to put a sum of money into an account so that when their daughter starts college so that they can withdraw $1000 from the account each month for four years. They will make no further payments into the account once their daughter starts college. Assume that their account will earn interest at the rate of 8% compounded monthly. How much should be in the account when their daughter starts college?

SOLUTION This question can be restated to ask, "What is the present value of an annuity with $p = \$1000$, $i = \frac{0.08}{12}$, and $n = 48$?" The present-value formula gives the amount needed as

$$P = \$1000 \frac{1 - \left(1 + \dfrac{0.08}{12}\right)^{-48}}{\dfrac{0.08}{12}}$$

$$\approx \$40,961.91. \quad \blacksquare$$

EXAMPLE 5
Small-Business Application

Appliance City has just sold a refrigerator to a family that, after making a small down payment, agrees to pay $40 per month for a period of three years. Like most small businesses, Appliance City sells its loan to a bank so that the company can receive its money up front. If interest rates are 10% compounded annually, how much should Appliance City expect to receive from the bank for the loan?

SOLUTION The question asks for the present value of the loan. In this case, $p = \$40$, $i = \frac{0.10}{12}$ (because the payments are made monthly), and $n = 36$. Substituting these numbers into the present-value formula gives

$$P = 40\left[\frac{1 - \left(1 + \dfrac{0.10}{12}\right)^{-36}}{\dfrac{0.10}{12}}\right] \approx \$1239.65. \quad \blacksquare$$

Exercises 6.2

In Exercises 1 through 6, find the future value of each of the annuities with the given interest rate per period.

1. $p = \$500$, $n = 30$, $i = 0.025$

2. $p = \$1200$, $n = 24$, $i = 0.04$

3. $p = \$700$, $n = 20$, $i = 0.01$

4. $p = \$3000$, $n = 40$, $i = 0.015$

5. $p = \$8000$, $n = 20$, $i = 0.05$

6. $p = \$10,000$, $n = 12$, $i = 0.03$

In Exercises 7 through 12, a sinking fund is being established, the goal of which is to reach $100,000. Calculate the payments in each case, where i is the interest rate per period.

7. $n = 20$, $i = 0.025$

8. $n = 30$, $i = 0.01$

9. $n = 16$, $i = 0.02$

10. $n = 18$, $i = 0.04$

11. $n = 22$, $i = 0.03$

12. $n = 26$, $i = 0.05$

13. Saving for the Future Nanette plans on putting $500 in an account on her daughter's birthday, beginning with the first one and continuing through age 16. If the account pays 8% compounded annually, how much should be in the account on Nanette's daughter's 16th birthday?

14. Saving for the Future Timothy has $10,000 in an account now and plans to add $1000 to the account at the end of each quarter from now on. If the account is assumed to pay 6% compounded quarterly, how much will be in it at the end of eight years?

15. Investing for College Elena's son will enter college 16 years from now. At that time, she would like to have $20,000 available for college expenses. For that purpose, her bank will set up an account that pays 7% compounded quarterly. If she makes payments into the account at the end of each quarter, what must Elena's payments be to achieve her goal?

16. Business Investment A freight-hauling firm estimates that it will need a new forklift in six years. The estimated cost of the vehicle is $40,000. The company sets up a sinking fund that pays 8% compounded semiannually, into which it will make semiannual payments to achieve the goal. Calculate the size of the payments.

17. Investing for Retirement Eduardo is a 40-year-old individual who plans to retire at age 65. Between now and then, $2000 is paid annually into his IRA account, which is anticipated to pay 5% compounded annually. How much will be in the account upon Eduardo's retirement?

18. Investing for Retirement At the end of each year, Jeremy invests $1000 in a company retirement plan in which the employer matches the employee's contribution. If the plan pays 8% compounded annually and Jeremy will retire in 30 years, what will be the total accumulated value of the account?

19. Comparing Investments Which of the following investments earns more?

 (a) At the end of each year for 20 years, $2000 is invested into an account that pays 7% compounded annually

 (b) At the end of each quarter for 20 years, $1800 is invested into an account that pays 6.5% compounded quarterly

20. Planning for Retirement Compare the accumulated value of IRA accounts in which $2000 is invested at the end of each year at an interest rate of 6% compounded annually if

 (a) You start at age 25 and retire at age 65.

 (b) You start at age 35 and retire at age 65.

 (c) You start at age 40 and retire at age 65.

 (d) You start at age 50 and retire at age 65.

21. Planning for Retirement Compare the accumulated value of IRA accounts into which $500 is invested at the end

of each quarter at an interest rate of 6% compounded quarterly if

 (a) You start at age 25 and retire at age 60.

 (b) You start at age 30 and retire at age 60.

 (c) You start at age 35 and retire at age 60.

 (d) You start at age 40 and retire at age 60.

22. Planning for Retirement If you are 30 years of age now and if you plan to retire at age 60 with an IRA account having a total accumulated value of $300,000, how much would you have to invest at the end of each year if the account

 (a) Paid 6% compounded annually?

 (b) Paid 8% compounded annually?

23. Planning for Retirement Katy wants to have enough in her retirement accounts so that, upon retirement, she can withdraw $500 each month for the next 20 years. Assuming that her accounts will earn an average of 7% compounded monthly, what sum of money should she have in her accounts upon retirement?

24. Savings The Brewsters are saving for their daughter's college days. They would like to be able to withdraw $800 each month from their account for five years once their daughter starts college. Assuming that their account will earn interest at the rate of 9% compounded monthly, what sum of money should the Brewsters have in the account when their daughter starts college?

25. Savings The Meek brothers are planning a trip around the world. They hope to work some as they go, but believe that they should have accessible $800 per month so they can live in relative comfort for the eight months they plan to be gone. How much should they have in an account earning 6% compounded monthly when they leave so that they can withdraw the desired $800 each month for eight months?

26. Savings A small business will start paying a $5000 franchising fee beginning next year. If a fund earning 8% compounded annually is set aside from which payments will be made, what should be the size of the fund to guarantee annual payments of $5000 for the next seven years?

27. Lottery Earnings If you won a lottery that paid $5000 each year for the next 10 years, and interest rates were 10% compounded annually, what would be the present value of your prize?

28. Present Value of a Loan Suppose you lend a friend money, and she begins repaying you at the rate of $150 per month. Suppose the loan still has three years of payments left and you would like to sell it and receive your money up front. If interest rates are 8% compounded annually, what is a fair price for your loan?

29. **Present Value of a Loan** Appliance City sells refrigerators on credit to customers. It just sold a refrigerator for payments of $65 each month for a period of three years. Like most small businesses, Appliance City sells its loan to the bank to have cash available now. If interest rates are 9% compounded annually, how much should the company expect to receive for its loan?

30. **Present Value of a Loan** The Albright Tractor Company has just sold a tractor to a farmer who financed a portion of the purchase for payments of $1500 quarterly for a period of four years. If interest rates are 12% compounded annually and the company sells the loan to the bank, how much should Albright expect to receive?

31. Suppose that Bronwyn deposits $100 at the end of each month into an account that pays 5.25% compounded monthly. After three years, she puts the accumulated amount into a certificate of deposit paying 7.5% compounded quarterly for two-and-a-half years. After the two-and-a-half years are up, she puts the money into a 90-day certificate of deposit paying 6.8% annually. When the certificate matures, how much money has Bronwyn accumulated?

32. Erin deposits $200 at the end of each month into an account that pays 6.25% compounded monthly. After two years, she puts the accumulated amount into a 90-day certificate of deposit that pays 7.5% annually. Upon maturity of the certificate, she puts the accumulated amount into a three-year certificate of deposit paying 8% compounded quarterly. When this last certificate matures, how much money will Erin have accumulated?

Section 6.3 Consumer Loans and the APR

The Truth in Lending Act passed by the U.S. Congress in 1969 created a common reference rate that lenders must state to protect consumers from various and sometimes confusing ways in which interest rates are quoted. That common reference rate is called the **effective interest rate** or, more commonly, the **annual percentage rate,** abbreviated APR. The APR is a statement of the interest rate that the loan earns *compounded annually.* Put differently, the APR is the *simple interest rate* that would be needed to produce the same income as the compounding procedure over a one-year period. Any rate, annual or otherwise, can easily be converted to an APR. For instance, if an interest rate is quoted as being 6% compounded quarterly, then the compound-interest formula tells us that $1 will grow to

$$\$1\left(1 + \frac{0.06}{4}\right)^4 = \$1(1.015)^4 \approx \$1(1.06136) = \$1.0614$$

in one year. Therefore, the quoted rate of 6% compounded quarterly, sometimes called the **nominal rate,** yields

$$\$1.0614 - \$1 = \$0.0614 \text{ interest}$$

for one year. This is the same yield as a simple interest rate of 6.14% applied to $1 for a year:

$$0.0614(\$1) = \$0.0614 \text{ interest.}$$

Therefore, the APR is 6.14%. The same pattern of reasoning on a $1 investment allows us to arrive at a general rule for finding the APR:

The Annual Percentage Rate

Let r = the decimal equivalent of a quoted nominal rate,
  m = the number of compounding periods per year,
  $i = \dfrac{r}{m}$ = the periodic interest rate.

Then the **effective rate of interest** or the **annual percentage rate** is given by

$$APR = \left(1 + \frac{r}{m}\right)^m - 1 = (1 + i)^m - 1.$$

The intended purpose of the APR is to allow a convenient comparison of rates.

EXAMPLE 1

Calculating the APR

 Which investment earns the most?

(a) A savings account paying 6.25% compounded monthly

(b) A government bond paying 6.5% compounded semiannually

(c) A certificate of deposit paying 6.3% compounded quarterly

SOLUTION When each rate is converted to the APR, we will immediately know which is the better investment.

(a) $APR = (1 + i)^m - 1 = \left(1 + \dfrac{0.0625}{12}\right)^{12} - 1 \approx 0.06432$, or 6.43%.

(b) $APR = (1 + i)^m - 1 = \left(1 + \dfrac{0.065}{2}\right)^{2} - 1 \approx 0.06606$, or 6.61%.

(c) $APR = (1 + i)^m - 1 = \left(1 + \dfrac{0.063}{4}\right)^{4} - 1 \approx 0.06450$, or 6.45%.

These calculations show that (b) is slightly better than either of the other two investments. ■

Both spreadsheets and the TI-83 Plus include features that calculate the APR. To calculate the APR in Example 1(a) with Excel, we type the formula =EFFECT(0.0625,12) into a blank cell and press Enter. In general, Excel's EFFECT function requires the nominal interest rate as a *decimal*, followed by a comma and the number of interest periods per year.

To calculate the APR with the TI-83 Plus, from the home screen, access the Eff function in the APPS Finance menu, then type 6.25,12), and press ENTER. In general, the Eff function of the TI-83 Plus requires the nominal interest rate as a *percentage*, followed by a comma and the number of interest periods per year.

If you bought a new car and needed to borrow $20,000 for 60 months at a 10.5% APR, do you know how much the monthly payments would be? If you borrowed $5000 at 12% APR and wanted to pay it back at the rate of $100 per month, do you know how long it would take? Of course, the auto dealers or bankers are expected to tell you the answers to these questions. They will either look up the answers in a specially prepared table or find them upon keying the pertinent information into a calculator or computer. In this section, we intend to look behind the scenes to gain some insight into the logic of how such calculations are made.

Most consumer loans (personal, automobile, or home loans) are required to be repaid with monthly payments. In reality, however, most of us pay a little early or a little late, and this affects the interest that is being charged on the loan. The last payment on the loan then may be a little more or a little less than the others, to adjust for the interest charges. Methods of calculating interest on loans are set forth by the U.S. Comptroller of the Currency and, in some cases, by state law. One of the most common arrangements is that of paying interest on the *unpaid balance* of the loan. For instance, suppose that a $1000 personal loan is taken out at the rate of 12% simple interest applied to the monthly unpaid balance. The loan is to be repaid in six equal monthly payments of $172.55. During the first month, the borrower has use of the entire $1000 and therefore should be charged $\frac{1}{12}$ of one year's interest, or

$$\$1000 \times 0.12 \times \frac{1}{12} = \$10.$$

So the first payment of $172.55 includes $10 interest, and the remainder, $162.55, is applied to the principal. The *new loan* or *unpaid balance* is now $1000 − $162.55 = $837.45. For the second payment, this unpaid balance is multiplied by the factor $\left(\frac{1}{12}\right)(0.12) = 0.01$ to get the interest ($8.37) portion of the payment. The entire financial picture for this loan can be pictured in Table 4. Notice that a portion of each payment is interest, and the remainder is applied to the outstanding balance of the loan. The interest portion of each payment decreases over time, while the principal portion increases.

Payment Number	Amount	Interest	Applied to Principal	Unpaid Balance
1	$172.55	$10.00	$162.55	$837.45
2	$172.55	$8.37	$164.18	$673.27
3	$172.55	$6.73	$165.82	$507.45
4	$172.55	$5.07	$167.48	$339.97
5	$172.55	$3.40	$169.15	$170.82
6	$172.53	$1.71	$170.82	$0.00

TABLE 4. AN AMORTIZATION SCHEDULE

A table such as Table 4 is called an **amortization table** or **amortization schedule.** There are many software programs available to compute amortization tables, and most personal finance software, such as Quicken® or MS Money®, have the capability of creating them as well. We can also utilize the functionality of spreadsheets to create amortization tables, as is demonstrated in Example 2.

EXAMPLE 2
Creating an
Amortization Schedule
with Excel

Suppose John and Emma have a 15-year mortgage at a nominal rate of 7.2% with a beginning principal of $72,000 and monthly payments of $655.23. Create an amortization table with Excel to demonstrate how much of each monthly payment will be applied to interest and how much will be applied to principal during the first two years.

SOLUTION To create an amortization table with Excel, begin by setting up column headings such as those shown in Table 4. Next, format all but the first column so that values are displayed as currency. In the next row, we enter the payment number, 1, and then the amount of the monthly payment, $655.23. Then enter the amount of interest charged to the unpaid balance by typing in the formula =72000*0.072/12. Now we enter the amount of the first payment actually applied to the principal, which is the total monthly payment minus the interest payment. We input this amount with the formula =B2-C2. Finally, we enter the amount of principal remaining after the first payment, which is the original principal minus the amount of the first payment applied to the principal. This is accomplished with the formula =72000-D2.

Once we set up the second row, we can fill the remaining rows to complete the table for the first two years. We will use the following formulas:

Cell	Contents
A3	2
B3	655.23
C3	=E2*0.072/12
D3	=B3-C3
E3	E3-D3

Now we highlight the third row and drag to fill so that the results for 24 months (24 payments) are displayed, as in Table 5.

Payment Number	Amount	Interest	Applied to Principal	Unpaid Balance
1	$655.23	$432.00	$223.23	$71,776.77
2	$655.23	$430.66	$224.57	$71,552.20
3	$655.23	$429.31	$225.92	$71,326.28
4	$655.23	$427.96	$227.27	$71,099.01
5	$655.23	$426.59	$228.64	$70,870.38
6	$655.23	$425.22	$230.01	$70,640.37
7	$655.23	$423.84	$231.39	$70,408.98
8	$655.23	$422.45	$232.78	$70,176.20
9	$655.23	$421.06	$234.17	$69,942.03
10	$655.23	$419.65	$235.58	$69,706.45
11	$655.23	$418.24	$236.99	$69,469.46
12	$655.23	$416.82	$238.41	$69,231.05
13	$655.23	$415.39	$239.84	$68,991.21
14	$655.23	$413.95	$241.28	$68,749.92
15	$655.23	$412.50	$242.73	$68,507.19
16	$655.23	$411.04	$244.19	$68,263.01
17	$655.23	$409.58	$245.65	$68,017.35
18	$655.23	$408.10	$247.13	$67,770.23
19	$655.23	$406.62	$248.61	$67,521.62
20	$655.23	$405.13	$250.10	$67,271.52
21	$655.23	$403.63	$251.60	$67,019.92
22	$655.23	$402.12	$253.11	$66,766.81
23	$655.23	$400.60	$254.63	$66,512.18
24	$655.23	$399.07	$256.16	$66,256.02

TABLE 5. AN AMORTIZATION TABLE FOR THE FIRST TWO YEARS OF A 15-YEAR MORTGAGE

Rather than computing interest monthly on the unpaid balance of a loan, most lending institutions charge interest on a daily basis.

EXAMPLE 3

The Interest Portion of a Monthly Payment

Sue has a car loan from her credit union at a rate of 10.5%, for which her payments are $235 per month. The interest is computed on a daily basis on the unpaid balance of the loan. If the loan balance after her last payment was $8647.50 and Sue makes her next payment 32 days later, how much of the $235 is paid toward interest?

SOLUTION The interest rate per day is $i = \frac{0.105}{365}$. This rate is applied to the balance of $8647.50 for 32 days, giving

$$(\$8647.50)\left(\frac{0.105}{365}\right)(32) = \$79.60. \quad \blacksquare$$

When we borrow money for a certain length of time, our immediate concern is the amount of each periodic payment. Suppose that we investigate how to arrive at the amount of each monthly payment calculated for the $1000 loan shown in Table 4. The line of thought we employ leads us to the general formula for doing so.

First, consider the transaction from the viewpoint of the lender, who lets you borrow $1000 for six months for a rental fee of 1% per month. The lender is then entitled to an accumulated value of $1000(1 + 0.01)^6 = \$1061.52$ by the compounded interest law, assuming the lender would have invested the $1000 had he not loaned it to you. Now consider the transaction from the viewpoint of the borrower. When the borrower makes a monthly payment, think of that payment as an investment in an account that earns interest at the rate of 1% per month. After making the sixth payment, the total accumulated in the account must be $1061.52, to which the lender is entitled. However, this is just an annuity in which p is unknown and the future value is $1061.52.

Therefore, we must have

$$\$1061.52 = p\frac{(1 + 0.01)^6 - 1}{0.01},$$

or

$$\$1061.52 = p(6.152015),$$

so that

$$p = \frac{\$1061.52}{6.152015} = \$172.55.$$

This thought process and the accompanying calculations serve as a model for a general formula from which monthly payments (or any other periodic payments) to retire a loan may be calculated. If an amount P is borrowed at i percent per period for n periods, and if n periodic equal payments are to be made to retire the loan, then the lender is entitled to a total accumulation of $P(1 + i)^n$ at the end of n periods. The borrower thinks of a like amount gained through an annuity of n equal payments of p dollars each, earning interest at the rate of i percent compounded at the end of each period, for a total of

$$p\frac{(1 + i)^n - 1}{i}.$$

Equating these two quantities and solving for p gives

$$P(1 + i)^n = p\frac{(1 + i)^n - 1}{i} \quad \text{or} \quad p = \frac{iP(1 + i)^n}{(1 + i)^n - 1}.$$

Amount of Payments to Retire a Loan

Let P = amount borrowed
   t = number of years in the life of the loan
  m = number of periods (payments) per year
   r = annual interest rate, compounded m times per year, as a decimal
   $i = \dfrac{r}{m}$ = interest rate per period
  $n = mt$ = total number of periods (payments).

Then the amount of each payment needed to pay off the loan is

$$p = \frac{\left(\dfrac{r}{m}\right)P}{1 - \left(1 + \dfrac{r}{m}\right)^{-mt}} = \frac{iP}{1 - (1 + i)^{-n}}.$$

EXAMPLE 4
Calculating Payments
for a Loan

Mr. and Mrs. Kalman bought a lot for $40,000 on which they plan to build a new house in the near future. The banker required a $5000 down payment and financed the remainder at 6% compounded semiannually, with equal payments to be made semi-annually for eight years. What are the Kalmans' semiannual payments?

SOLUTION In this case, $P = \$35,000$, $i = \frac{0.06}{2} = 0.03$, and $n = 16$. Therefore, the Kalmans' semiannual payments will each be

$$P = \frac{0.03(\$35,000)}{1 - (1 + 0.03)^{-16}} \approx \$2786.38. \quad \blacksquare$$

Even though you may not know exactly how the finance company will figure the monthly payments when an APR is quoted, a close approximation may be obtained by dividing the APR by 12 and using the resulting percentage as the compounding rate.

EXAMPLE 5

Estimating the Monthly
Payments for a Loan

Joe has just finished college; upon landing his first job, he decides to buy a new car. He thinks that after trading in his junker, he will still need to finance $13,000 at the current 10.9% APR advertised for 60 months. Find Joe's approximate monthly payment.

SOLUTION We assume that interest will be charged monthly on the unpaid balance of the loan, so that

$$i = \frac{0.109}{12} \approx 0.00908333$$

is to be applied to the monthly balance each month. The amount-of-payments formula then tells us that the monthly payments will be

$$p = \frac{0.00908333(\$13,000)}{1 - (1 + 0.00908333)^{-60}} \approx \$282. \blacksquare$$

If the amount of payments formula is solved for $(1 + i)^n$, we get

$$(1 + i)^n = \frac{p}{p - Pi} \quad \text{or} \quad n = \frac{\ln\left(\dfrac{p}{p - Pi}\right)}{\ln(1 + i)}.$$

Thus, if we know p, P, and i, we can calculate the number of payments necessary to retire the loan if a solution of the preceding equation exists. (A solution exists when $p > Pi$.)

Number of Payments Needed to Retire a Loan

Let P = amount of each payment made at the end of each period
p = amount of each payment
m = number of periods per year
r = annual interest rate, compounded m times per year, as a decimal
$i = \dfrac{r}{m}$ = interest rate per period

Then the **number of payments** needed to retire the loan is

$$n = \frac{\ln\left(\dfrac{p}{p - P\left(\dfrac{r}{m}\right)}\right)}{\ln\left(1 + \dfrac{r}{m}\right)} = \frac{\ln\left(\dfrac{p}{p - Pi}\right)}{\ln(1 + i)}.$$

We are now ready to solve the problem posed at the outset of this chapter.

EXAMPLE 6

Calculating the Number
of Payments Needed to
Retire a Loan

Sarah is to repay a $10,000 loan at a nominal rate of 8.7%, with 60 payments of $206.13 made monthly. After one year of making payments, Sarah gets a raise in salary and decides that she can afford to pay $250 per month instead. How long will it take Sarah to finish paying off the loan?

SOLUTION First, we must calculate the unpaid balance of Sarah's $10,000 loan. Table 6 summarizes the calculations:

Payment Number	Amount	Interest	Applied to Principal	Unpaid Balance
1	$206.13	$72.50	$133.63	$9866.37
2	$206.13	$71.53	$134.60	$9731.77
3	$206.13	$70.56	$135.57	$9596.20
4	$206.13	$69.57	$136.56	$9459.64
5	$206.13	$68.58	$137.55	$9322.09
6	$206.13	$67.59	$138.54	$9183.55
7	$206.13	$66.58	$139.55	$9044.00
8	$206.13	$65.57	$140.56	$8903.44
9	$206.13	$64.55	$141.58	$8761.86
10	$206.13	$63.52	$142.61	$8619.25
11	$206.13	$62.49	$143.64	$8475.61
12	$206.13	$61.45	$144.68	$8330.93

TABLE 6. CALCULATING SARAH'S UNPAID BALANCE AFTER 12 PAYMENTS

As we see from the table, Sarah still owes $8330.93 of her $10,000 loan after one year. Using the formula for calculating the number of payments required to retire a loan, we get

$$n = \frac{\ln\left(\dfrac{p}{p - Pi}\right)}{\ln(1 + i)} = \frac{\ln\left(\dfrac{250}{250 - 8330.93(0.00725)}\right)}{\ln(1 + 0.00725)} \approx \frac{\ln 1.318560133}{\ln 1.00725} \approx 38.28 \text{ payments.}$$

Therefore, if Sarah increases her monthly payments from $206.13 to $250, she will pay off her loan in around 38 more months, 10 months before she otherwise would have. (It would take her 48 more months to pay off the loan at the original monthly payments of $206.13.) ∎

Figure 8 shows how to set up the TVM Solver of the TI-83 Plus in order to complete Example 6. Notice that 0 has been entered as the future value of the loan, since Sarah wishes to pay off the loan. Notice also that the present value has been entered as a positive number and that the amount of Sarah's monthly payment has been entered as a negative number.

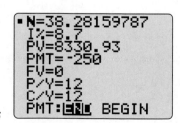

FIGURE 8
Using the TVM Solver *to Complete Example 6*

EXAMPLE 7

Paying Off the Mortgage

The Cochrans obtained a loan of $87,500 to purchase a new house. The nominal loan rate is 9.8%, with payments to be made at the end of each month for 30 years.

(a) How much are the Cochrans' monthly payments?

SOLUTION The solution is found by using the amount-of-payments formula. In this case, $P = \$87,500$, $i = \frac{0.098}{12} \approx 0.00816667$, and $n = 360$. The amount of each payment is then

$$p = \frac{(0.00816667)(\$87,500)}{1 - (1.00816667)^{-360}} \approx \$754.98.$$

(b) Assuming that timely payments are made throughout the life of the loan, what is the total amount (principal plus interest) paid for the loan?

SOLUTION The total will be $\$754.98(360) = \$271,792.80$.

(c) Suppose that, after learning of the monthly payment amount, the Cochrans decide they can afford to pay an extra $50 each month toward paying off their loan. Again, assuming timely payments, what total will they now pay for their loan?

SOLUTION Payments of $804.98 are now being made to retire a loan of $87,500. The number-of-payments formula shows that

$$n = \frac{\ln\left(\dfrac{\$804.98}{804.98 - (\$87,500)(0.00816667)}\right)}{\ln(1.00816667)} \approx 268.84 \text{ months.}$$

Rounding up to 269 months, we find that the total amount paid is $269(\$804.98) = \$216,539.62$. (Compare this with the amount in part (b)).

(d) After making timely payments for 20 years on this loan at the rate of $754.98 per month, how much is the unpaid balance?

SOLUTION After 20 years, 240 payments have been made toward retiring the loan. If we think of the remaining 120 payments as an annuity, then its present value is equal to the unpaid balance of the loan. Using the present value formula of Section 6.2, we find that

$$P = \$754.98 \frac{1 - \left(1 + \dfrac{0.098}{12}\right)^{-120}}{\dfrac{0.098}{12}} \approx \$57,611.96. \quad \blacksquare$$

Exercises 6.3

In Exercises 1 through 8, convert the given interest rate to the APR.

1. 6.5% compounded quarterly

2. 8% compounded monthly

3. 4% compounded semiannually

4. 12% compounded daily

5. 16% compounded monthly

6. 10% compounded quarterly

7. 9% compounded daily

8. 7% compounded monthly

In Exercises 9 through 14, calculate the APR and determine which rate will result in the most interest per year.

9. (a) 5% compounded semiannually

   (b) 4.8% compounded quarterly

   (c) 4.6% compounded monthly

10. (a) 10.5% compounded quarterly

    (b) 11% compounded daily

    (c) 10.6% compounded semiannually

11. (a) 6.3% compounded monthly

    (b) 6.5% compounded quarterly

    (c) 6.8% compounded annually

12. (a) 12% compounded annually

    (b) 11.8% compounded quarterly

    (c) 11.9% compounded semiannually

13. (a) 8% compounded semiannually

    (b) 8.2% compounded annually

    (c) 6% compounded monthly

14. (a) 3% compounded quarterly

    (b) 2.5% compounded daily

    (c) 3.1% compounded semiannually

15. **Credit Card Charges** If a credit card charges 1% per month on the unpaid balance, what APR is the company charging?

16. **Credit Card Charges** If a gasoline credit card charges 0.8% per month on the unpaid balance, what APR is the company charging?

In Exercises 17 through 24, find the payment for the given length of time, amount, and interest rate.

17. $P = \$20,000$; $i = 3\%$ per month; monthly payments for 6 years

18. $P = \$18,000$; $i = 4\%$ per quarter; quarterly payments for 10 years

19. $P = \$40,000$; $i = 2\%$ per month; monthly payments for 20 years

20. $P = \$30,000$; $r = 8\%$ compounded semiannually; semi-annual payments for 25 years

21. $P = \$12,000$; $r = 11.9\%$ compounded monthly; monthly payments for 60 months

22. $P = \$15,000$; $i = 0.0833\%$ per month; monthly payments for 36 months

23. $P = \$22,000$; $r = 10\%$, compounded quarterly; quarterly payments for 12 years

24. $P = \$80,000$; $r = 9.3\%$, compounded annually; annual payments for 10 years

25. Suppose that your credit union charges 9.5% simple interest on the unpaid balance of your loan for each day since that last payment. The monthly payments are $326, and you made your last payment 35 days ago, at which time the unpaid balance was $9643.68. What portion of your next payment will be interest?

26. Suppose that your house loan charges 12.3% simple interest on the unpaid balance of your loan for each day since the last payment. The last payment was made 28 days ago, when the loan balance was $56,743.54. How much of the payment that you are about to make will be interest?

27. **Borrowing Money** Suppose that you borrow $12,500 for the purchase of a new car at 9.8% APR for 48 months. What is the approximate amount of your monthly payment?

28. **Repaying a Loan** A business borrowed $50,000 at 12% per year on the unpaid balance. The loan is to be repaid in 10 equal payments, beginning at the end of the first year. Find the amount of each payment.

29. **Borrowing Money** An appliance dealer advertises a particular brand of refrigerator for $800, which may be financed at 7.9% APR for 48 months. As usual, the payments are to be made in equal monthly installments. Find the approximate amount of each payment.

30. **Borrowing Money** A department store advertises a stereo for $1500, which it will finance for 60 months at 8% APR. Payments are to be made monthly, and all are to be of equal amount. Find the approximate amount of each payment.

31. **Borrowing Money** A farmer buys a new tractor for $25,000, of which his banker finances $20,000 for eight years at 11% per year on the unpaid balance. Assume that the farmer repays the loan in equal annual installments. What is the amount of each payment?

32. **Borrowing Money** Jared buys a new motorcycle for $8500 from The Fastrack Motorcycle Co., which agrees to finance 80% of the purchase at 9% APR for a period of 60 months. What is the approximate size of each of Jared's monthly payments?

33. **Borrowing Money** The Bigger Wave Boat Co. sells George a new fishing boat for $9200. The company agrees to finance 90% of the purchase at 8.5% simple interest, computed on the unpaid monthly balance, for a period of 72 months.

(a) What is the amount of each of George's monthly payments?

(b) Assuming timely payments of the amount found in Exercise 33(a), what is the unpaid balance of the loan after the 24th payment?

34. Borrowing Money Pat and Maxine decide to buy a lot on which they plan to build a new house. The lot cost $28,500, and the bank agrees to finance 80% of the cost at 9% compounded semiannually for a period of eight years. Payments on the loan are to be made in equal semiannual installments.

(a) Find the amount of each payment.

(b) Assuming timely payments of the amount found in Exercise 34(a), what is the unpaid balance after the 12th payment?

35. Borrowing Money Diane decides to buy a small farm near the edge of town for $95,000. The bank agrees to finance 70% of the cost at 10% compounded quarterly for a period of 10 years. Payments on the loan are to be made in equal quarterly installments.

(a) What is the amount of each payment?

(b) Assuming timely payments of the amount found in Exercise 35(a), what is the unpaid balance after the 20th payment?

In Exercises 36 through 42, a couple gets financing for 90% of the $130,000 purchase price of a house at the rate of 9.5% on the monthly unpaid balance.

(a) Find the amount of the monthly payments to repay the loan only (excluding taxes and insurance).

(b) Find the total amount paid to the finance company (the monthly payment multiplied by the number of payments) for each of the following repayment periods.

36. The loan is repaid in 20 years.

37. The loan is repaid in 25 years.

38. The loan is repaid in 30 years.

39. The loan is repaid in 35 years.

40. The loan is repaid in 40 years.

41. The loan is repaid in 45 years.

42. The loan is repaid in 50 years.

Home Mortgages *In Exercises 43–49, Allan finances $125,000 toward the purchase of a new home through a 15-year mortgage. Create an amortization table for the first year of the mortgage if the interest rate applied to the monthly unpaid balance is*

43. 9.25%   **44.** 7.33%

45. 7.95%   **46.** 7.60%

47. 8.20%   **48.** 6.97%

49. 4.75%

Chapter 6 Summary

CHAPTER HIGHLIGHTS

Mortgages, car loans, savings accounts, certificates of deposit, and many other financial and consumer instruments involve the lending or borrowing of money. The amount of money involved is called the *principal,* and the fee charged can be *simple interest* or *compound interest* (interest earned on both the principal and the interest). *Ordinary annuities* and *sinking funds* involve sequences of equal payments made at equally spaced intervals, with interest paid at the time of the payments. Ordinary annuities and sinking funds have a predetermined goal amount or *future value.* A common reference rate, the *annual percentage* rate *(APR),* has been established to help consumers determine the true cost of loans and to aid them in comparing rates.

EXERCISES

Find the accumulated value if $2000 is invested for three years at 7% under the following schemes:

1. Simple interest

2. Interest compounded annually.

3. Interest compounded quarterly.

4. Interest compounded semiannually.

Find the number of interest periods if $1500 accumulates to $3000 at 8% compounded

5. Quarterly.   **6.** Daily.

Find the principal if the accumulated amount is $5000 over 10 years, at 9% compounded

7. Monthly.   **8.** Semiannually.   **9.** Annually.

Find the future value of an annuity with

10. $i = 0.01$, $n = 18$, and $p = \$1000$.

11. $i = 0.03$, $n = 40$, and $p = \$5000$.

Calculate the payments for a sinking fund having a goal of $40,000, with

12. $i = 0.02$ and $n = 30$.   **13.** $i = 0.04$ and $n = 12$.

Convert the given interest rate to the APR:

14. 8% compounded daily

15. 6% compounded quarterly

16. 7% compounded monthly

17. 10% compounded semiannually

18. Thi is planning to put $125 per month into an ordinary annuity paying 7.2% interest. How much will she have available for her daughter's college education 15 years later?

19. Approximate your monthly payment if you borrow $14,000 for a new car at 6.9% APR for three years.

20. Calculate the amount of the monthly payment necessary, based on the best available current interest rates, for you to have $1 million in a sinking fund by age 65. (First guess or estimate the amount, and then compare the estimate with your calculated amount.)

21. Analyze the given interest rates, and decide which gives the greatest yield:

(a) 8% compounded annually

(b) 7.9% compounded semiannually

(c) 7.8% compounded quarterly

(d) 7.7% compounded monthly

(e) 7.6% compounded daily

(f) 7.5% compounded continuously

22. Given that $12,000 is borrowed for 60 months at 8.6% APR, estimate the monthly payments, assuming that the interest is charged on a

(a) Monthly basis

(b) Daily basis

CUMULATIVE REVIEW

23. Find the inverse of the matrix

$$\begin{bmatrix} 1 & 0 & 3 \\ 2 & 4 & 6 \\ 0 & -2 & 1 \end{bmatrix}$$

24. Use the simplex algorithm to

$$\text{Maximize } f = 5x + 4y$$
$$\text{subject to } x + 2y + z \le 10$$
$$x + 3y \quad\ \le\ 6$$
$$y + z \le\ 5$$
$$x \ge 0, y \ge 0, z \ge 0.$$

25. Write an equation for the line with slope 3 and y-intercept 2.

Chapter 6 Sample Test Items

In Exercises 1 through 4, find the accumulated value under the given conditions.

1. The principal is $1200, with simple interest of 6% for five years.

2. The principal is $3000, with 7% interest compounded semiannually for six years.

3. The principal is $8000, with 9% interest compounded monthly for eight months.

4. The principal is $5000, with simple interest of 10% for eight months.

In Exercises 5 and 6, find the number of interest periods required to meet the given condition.

5. $A = \$7000$; $P = \$2400$; the interest is 8% compounded semiannually.

6. $A = \$25,000$; $P = \$3000$; the interest is 10% compounded annually.

In Exercises 7 and 8, find the principal required to achieve the given conditions.

7. $A = \$20,000$; the rate is 12% compounded monthly for 7 years.

8. $A = \$30,000$; the rate is 6.5% compounded quarterly for 12 years.

9. Find the number of years it will take to double your money if the interest rate is 8% compounded daily.

In Exercises 10 and 11, find the future value of each annuity for the given interest rate per period.

10. $p = \$800$; $n = 20$; $i = 0.05$.

11. $p = \$2000$; $n = 30$; $i = 0.02$.

In Exercises 12 and 13, a sinking fund is being established with a goal of \$200,000. Calculate the payments if i is the interest rate per period:

12. $n = 24$; $i = 0.03$.   13. $n = 20$; $i = 0.04$.

14. Eli Whitney Co. sold a big-screen television to a customer for payments of \$120 per month for three years. If the company sells its loan to a bank whose interest rates are 8% compounded annually, how much should Whitney receive for the loan?

In Exercises 15 and 16, convert each given interest rate to the APR:

15. 7% compounded quarterly

16. 9% compounded daily

17. Find the monthly payment for a loan if $P = \$50,000$, $i = 0.2\%$ per month on the unpaid balance, and the payments are for 15 years.

18. Find the number of years required to pay off a \$70,000 mortgage on a house at a 9.4% interest rate compounded monthly if payments of \$800 are made toward the loan.

In-Depth Application

Credit and Debt Consolidation

Objectives

In this application, you will apply your knowledge of compound interest and the time required to retire a debt in order to determine how best to consolidate debts and reduce interest payments.

(continued)

Introduction

In the past several years, the availability of credit cards and the ease of applying for credit have caused many American citizens to find themselves in financial trouble. Credit card companies offer many options for new applicants; usually included are low introductory interest rates, low monthly payments, and "cash advance" options. Without reading the fine print in credit applications, one can easily become locked in a spiral of debt.

The Data

Suppose that Suzanne and Steven recently got married and have five credit cards between them. The balances, annual interest, credit available, and minimum monthly payments are shown in the following table, after this month's payments have been made:

Company	Current Balance	Annual Interest (compounded monthly)	Available Credit	Minimum Payment
Hometown Bank	$3421.84	18.5%	$6578.16	$144
Trinity Credit Services	$981.13	21.5%	$4018.87	$75
AmeriCredit	$6201.22	7.9%	$2798.78	$150
Fray's Department Store	$1102.15	18.0%	$3897.85	$91
Splurge Credit	$821.79	4.9%	$4178.21	$45

In an effort to reduce their current credit card debt, Suzanne and Steven agree not to make any additional purchases with their credit cards. Assume that all minimum monthly payments are applied first to monthly interest and then to the principal from current balances.

1. (a) Calculate Suzanne and Steven's total current credit card debt.
   (b) Calculate their total minimum monthly payments.

2. Calculate the amount of each monthly payment that will be applied to interest for each account at the end of each month over the next year.

3. Calculate the amount of each monthly payment that will be applied to principal for each account at the end of each month over the next year.

4. If Suzanne and Steven continue to make the minimum monthly payments shown and do not make any additional purchases with any of the five credit cards, how many months will it take to retire each debt?

Hometown Bank sends Suzanne and Steven an offer to transfer balances from other credit cards at a fixed annual rate of 4.5% for one year. In other words, any credit card balances they transfer from other accounts to their Hometown Bank account will accrue interest at a rate of 4.5% annually for one year, and then any remaining balance from these transfers will begin to accrue interest at the 18.5% rate. The *current* balance of $3421.84 will continue to accrue interest at the annual rate of 18.5%. Assume that all monthly payments are applied first to the interest on balance transfers, then to the interest on the previous balance, next to the principal from balance transfers, and finally to the principal from previous balances.

5. If Suzanne and Steven decide to transfer the entire balance of their Trinity Credit Services account to their Hometown Bank account,
   (a) How much of their minimum monthly payment of $144 will be applied to total interest each month over the next year?
   (b) How much of their minimum monthly payment of $144 will be applied to principal each month over the next year?
   (c) How much of the balance transfer will remain as principal at the end of one year?
   (d) How much of the previous balance will remain as principal after one year?
   (e) How long will it take to retire this loan after one year?

(continued)

6. If Suzanne and Steven decide to transfer the entire balance of their Fray's Department Store account to their Hometown Bank account,
   (a) How much of their minimum monthly payment of $144 will be applied to total interest each month over the next year?
   (b) How much of their minimum monthly payment of $144 will be applied to principal each month over the next year?
   (c) How much of the balance transfer will remain as principal at the end of one year?
   (d) How much of the previous balance will remain as principal after one year?
   (e) How long will it take to retire this loan after one year?

7. After consulting a credit counselor, Suzanne and Steven decide to be more creative as they take advantage of Hometown Bank's balance transfer offer. Since the offer applies only to new balance transfers, they first transfer their entire Hometown Bank balance to their AmeriCredit account. Next, they transfer their new AmeriCredit balance to their Hometown Bank account, so that the entire amount is eligible for the balance transfer offer.
   (a) How much of their minimum monthly payment of $144 will be applied to total interest each month over the next year?
   (b) How much of their minimum monthly payment of $144 will be applied to principal each month over the next year?
   (c) How much of the balance transfer will remain as principal at the end of one year?
   (d) How long will it take to retire this loan after one year?

8. Which of the actions described in Problems 5, 6, and 7 is the most beneficial, financially, to Suzanne and Steven? Is there a better action that can be taken in response to Hometown Bank's balance transfer offer? Justify your answer with information such as differences between total monthly payments, differences between monthly interest payments, and differences in length of time required to retire the loans.

Logic, Sets, and Counting Techniques

PROBLEM

The Galax city council is composed of 10 members, of which 6 are Democrats and 4 are Republicans. A committee of 4 council members is to be formed so that there is at least one representative of each political affiliation. How many committees can be formed under this restriction? (See Example 10, Section 7.7.)

CHAPTER PREVIEW

We begin this chapter with a brief study of the basic concepts of logic. Connectives, such as *and, or,* and *not,* will be used to translate sentences into symbolic statements. The connection between statements of *implication* and *disjunction,* as well as other logically *equivalent* statements will be examined with an eye toward rewriting statements in a more convenient way. *Truth tables* will be employed to continue our study of logically equivalent statements. We will use *Venn diagrams* to organize information about *sets,* including the *complement* of a set and the *union, intersection,* and *difference* of two sets. Finally, with the building blocks provided by our study of logic and sets, we will reach the main objective of the chapter: to utilize *counting techniques* such as *permutations* and *combinations.* The techniques developed here will be invaluable in our later study of probability.

Technology Opportunities

Calculators can be used effectively to compute factorials, combinations, and permutations. Common calculator commands for these calculations are an exclamation point (!) for factorials, nCr for combinations, and nPr for permutations.

Spreadsheet packages also have built-in functions for calculating factorials, combinations, and permutations.

Section 7.1 Logic

To comprehend successfully the concepts and exercises in our upcoming work in counting, probability, and statistics, you must be able to interpret the mathematical meanings of key words such as *and, or, not, at least,* and *at most.* In more complicated statements, you may come across words and phrases like *every* and *there exists.* For example, suppose a survey of college freshmen and sophomores yielded these data:

	Taking Geology	**Not Taking Geology**
Freshmen	20	15
Sophomores	18	10

Do you know how many of these students are described by each of the following two sentences?

The student is a freshman *and* is taking geology. (There are 20.)

The student is a sophomore *or* is taking geology. (There are 48.)

These key words are also used in defining operations between sets and in rephrasing statements in a logically equivalent way to make them more useful. We will study their use as set operations in later sections; for now, we concentrate on their use in logic. Who makes the ground rules for the use of these words? How do we know when the rephrasing of a statement is logically equivalent to the original statement? The answer to these questions is found in the subject of logic, and we will use that viewpoint to obtain a basic understanding of the key words needed to carry us through counting techniques (and beyond). The subject of logic is a vast one, but we will confine our attention only to the ideas that are directly needed for our mathematical developments. We begin by exploring the concept of a statement and how statements are combined to form compound sentences.

By a **statement,** we mean a declarative sentence that is either true or false, and not both, or an open sentence that can be made true or false by substituting for the pronoun or variable in the sentence. Following are some examples of statements:

There are five positions on a basketball team. (True)

Every student wears glasses. (False)

December has 31 days. (True)

$2 + 3 = 7$. (False)

$x + 3 = 5$. (True for $x = 2$; false otherwise.)

She is the president of the First Federal Bank. (True for one person; false for all others.)

No dog has a tail. (False)

From such statements, we may form compound sentences by the use of **connectives.** Two important and often-used connectives are *and* and *or.* To illustrate, let p and q represent the following statements:

p: The student is a freshman.

q: The student is taking geology.

Then the **conjunction** of p and q is constructed as follows: "The student is a freshman *and* is taking geology."

The following is the special logical symbolism used for conjunction:

Conjunction of p and q: $p \wedge q$ (Read p **and** q).

A compound sentence using the connective *or* is known as the **disjunction** of p and q and is formed in the following manner:

The student is a freshman *or* is taking geology.

The special logical symbolism for disjunction is as follows.

Disjunction of p and q: $p \vee q$ (Read p **or** q).

In logic, there are two ways of interpreting *or.* Consider the following two statements:

p: You must do your mathematics homework or study for your history exam.

q: For dinner, you may have a hamburger or a pizza.

In statement p, it is certainly acceptable to perform both tasks, but the statement requires that you do at least one of them. However, in statement q, it is likely that you are being given a choice between two dinner options and that you will not have both a hamburger and a pizza for dinner. The interpretation of *or* in statement p is known as an **inclusive or** (we "include," or allow, the possibility of both parts of the disjunction occurring). The *or* used in statement q is known as an **exclusive or** (we "exclude" the possibility of both parts of the disjunction from occurring). As with most multiple word usages, one usually must take which *or* is being used from the context. However, in mathematics, we always assume that the word *or* is inclusive unless otherwise stated. If we wish to indicate that the word *or* is intended to be exclusive, we add the phrase "but not both" to our statement. As an example, consider the following restatement of q:

q': For dinner, you may have a hamburger or a pizza, but not both.

This restatement makes it clear that the use of the word *or* is intended to be *exclusive*.

The word *not* is also considered a connective and, for a statement *p*, is called the **negation** of *p*. It works just as you think it should:

p: The student is a freshman.

The negation of *p*: The student is *not* a freshman.

There is also a logical symbolism for negation:

Negation of *p*: ∼*p* (Read **not** *p*).

We will study the concept of logically *equivalent* statements in the next section. However, it is useful at this point to note that the statements ∼(∼*p*) and *p* are logically the same statement. For instance, consider the statement

p: The truck is green.

Then the statement ∼*p* says "The truck is not green," implying that the truck is some other color. Finally, the statement ∼(∼*p*) asserts "The truck is *not* not green." This is not a grammatically correct sentence in the English language, but we can interpret what it is implying. If ∼*p* indicates that the truck is some color other than green, then ∼(∼*p*) says that the truck is *not* some color other than green. That is, the truck is green, which means that statement *p* actually holds.

The statements ∼(∼*p*) and *p* are logically equivalent.

The three connectives just introduced—*and, or,* and *not*—serve as building blocks for most of our needs in this textbook, with the understanding that variations in the use of English may sometimes give alternative wording. For example, the use of "but . . . not" has the same meaning as "and . . . not." (See Example 2.) These connectives and others form a logical basis for a formal mathematical language.

EXAMPLE 1

Illustrating Connectives and Symbols

Consider the following statements:

p: The shirt was made by the Ace Company.

q: The shirt has a flaw.

Our three connectives can be used to create the following statements:

p ∧ *q*: The shirt was made by Ace **and** it has a flaw.

p ∨ *q*: The shirt was made by Ace **or** it has a flaw.

∼*p*: The shirt was **not** made by Ace.

~q: The shirt does **not** have a flaw.

~$p \land q$: The shirt was **not** made by Ace **and** it has a flaw.

$p \land$ ~q: The shirt was made by Ace **and** it does **not** have a flaw.

~$p \land$ ~q: The shirt was **not** made by Ace **and** it does **not** have a flaw.

~$p \lor q$: The shirt was **not** made by Ace **or** it has a flaw.

$p \lor$ ~q: The shirt was made by Ace **or** it does **not** have a flaw.

~$p \lor$ ~q: The shirt was **not** made by Ace **or** it does **not** have a flaw. ∎

EXAMPLE 2

Translating Sentences
to Symbols

Consider the following statements:

p: The computer has a DVD drive.

q: The Internet is accessible.

We now show each of the following sentences in symbolic form:

The computer has a DVD drive and the Internet is accessible: $p \land q$.

The Internet is not accessible: ~q.

The computer has a DVD drive or the Internet is accessible: $p \lor q$.

The computer has a DVD drive, but the Internet is not accessible: $p \land$ ~q. ∎

Implications

Perhaps the most powerful type of statement in mathematics is the **implication.** Sometimes known as *if–then* statements, implications assert that when one statement is true, another statement is also true. For example, you may wake up one morning and say, "If the sun comes out today, then I will go to the park." You are asserting that the sun's coming out will ensure that you will go to the park. It is important to notice that nothing is implied on the chances of a cloudy day: In such a case, you may or may not go to the park. Writing the compound statement symbolically is fairly straightforward:

p: The sun comes out today.

q: I will go to the park.

The special symbolism for "if p, then q" or "p **implies** q" is $p \Rightarrow q$. We refer to the first statement, p, as the **hypothesis** and the second statement, q, as the **conclusion.**

p implies q: $p \Rightarrow q$ (Read "**if** p, **then** q").

EXAMPLE 3

Interpreting
Implications

Consider the statements

p: You eat little green apples.

q: You get a stomachache.

Then

1. $p \Rightarrow q$: If you eat little green apples, then you get a stomachache.
2. $q \Rightarrow p$: If you get a stomachache, then you eat little green apples.
3. $p \Leftrightarrow q$: You get a stomachache if and only if you eat little green apples.

The first statement only gives the results of eating little green apples; it does not say what will happen if you leave the apples alone. The second statement asserts that you eat little green apples whenever you get a stomachache. It does not rule out the fact that you may well eat little green apples all the time, regardless of whether you have a stomachache. The third statement is really an assertion of both statements (1) and (2). The "if and only if" phrase indicates that p is true exactly when q is true. ■

In the last statement in Example 3, we see an "if and only if" phrase $p \Leftrightarrow q$, which is symbolic shorthand for saying both $p \Rightarrow q$ and $q \Rightarrow p$ at the same time. When we have "p if and only if q," we say that p is *logically equivalent* to q. This is the second time we have spoken about logically equivalent statements. Earlier, we said that the statement p is equivalent to the statement $\sim(\sim p)$. In the next section, we will directly address the idea of logical equivalence, but for now your intuition for equivalence should serve you well.

Implications are really a very special type of connective in disguise. Consider again the earlier mentioned implication "If the sun comes out today, then I will go to the park." As described in Example 3, this statement does not assert what might happen if the sun does *not* come out today; it merely says what must happen if the sun *does* comes out today. There are three things that can happen without violating this implication:

(1) The sun comes out today and I go to the park.

(2) The sun does not come out today and I go to the park.

(3) The sun does not come out today and I do not go to the park.

Notice that statement (1) is the guaranteed result of going to the park because the sun does come out. By contrast, the statement "The sun comes out today and I do not go to the park" violates the implication and is not in our list of possibilities. Statements (2) and (3) illustrate the fact that our implication says *nothing* when the hypothesis "the sun comes out today" does not hold true.

Let us define the hypothesis and conclusion symbolically:

p: The sun comes out today.

q: I will go to the park.

The foregoing three possibilities can now be written as follows:

(1′) $p \wedge q$.
(2′) $\sim p \wedge q$.
(3′) $\sim p \wedge \sim q$.

Now, consider the statement $\sim p \vee q$, which symbolically represents the sentence "The sun does not come out today or I go to the park." What events could occur to make this statement true? The use of the word *or* here is inclusive, so we again have three possibilities: only q is true, both $\sim p$ and q are true, and only $\sim p$ is true.

($1''$) $\sim p$ is false and q is true: The sun comes out today and I go to the park.

$(p \wedge q)$

($2''$) $\sim p$ is true and q is true: The sun does not come out today and I go to the park.

$(\sim p \wedge q)$

($3''$) $\sim p$ is true and q is false: The sun does not come out today and I do not go to the park.

$(\sim p \wedge \sim q)$

These three possibilities that make the statement $\sim p \vee q$ true are precisely the possibilities that do not refute the implication $p \Rightarrow q$. Thus, the statements "If the sun comes out today, then I will go to the park" and "The sun does not come out today or I will go to the park" are logically equivalent.

We can give the above argument for general statements p and q and obtain the following theorem:

> The statement "If p, then q" is equivalent to "not p or q." Symbolically,
>
> $$(p \Rightarrow q) \Leftrightarrow (\sim p \vee q).$$

We shall end the section with a discussion of several **quantifiers.** Many times, quantifiers such as *all* and *at least* are used to specify the number of objects being considered. For instance, there is a big difference between saying "*All* dogs are brown" and saying "*Some* dogs are brown."

At Least, At Most

In previous chapters, we occasionally encountered the terms *at least* and *at most* in rather straightforward numerical settings. For example, if we say "At least 20 TV sets are to be made," then we know immediately that any number that is 20 or larger will satisfy the statement. In some situations that we will encounter later in the text, however, a translation of *at least* into a logically equivalent statement using the connectives *or* and *and* will be very helpful. The following examples illustrate what we mean.

EXAMPLE 4
Translating an At Least Statement

Translate the following sentence into one using the connectives *or* and *and*:

From a random sample of six calculators, at least four are in good working order.

SOLUTION The given sentence means that, from among the six calculators, four or more are in good working order. This, in turn, can be restated in terms of *or* as

"Four are in good working order, or five are in good working order, or six are in good working order."

To be even more specific, one last sentence will tell the full story:

"(Four are in good working order, and two are not), or (five are in good working order, and one is not), or (all six are in good working order)."

The last translation is the most specific and is also the most useful in counting and probability problems. Notice that with this breakdown, no two of the component parts separated by an *or* can happen simultaneously (more about this later). ■

EXAMPLE 5

Translating an *At Most*
Statement

Translate the following sentence into one using the connectives *or* and *and*: Among five students, at most two are taking psychology.

SOLUTION The given sentence means that none, one, or two of the students are taking psychology. To be more precise,

"None of the five are taking psychology, or one is taking psychology and four are not, or two are taking psychology and three are not." ▄

Notice in the preceding examples that when *at least* is being applied to a numerical value, it can be translated as "greater than or equal to." For example, the sentence

"Our team won **at least** three basketball games."

can be translated as

"The number of games our team won is **greater than or equal to** three."

Similarly, the *at most* quantifying phrase can be rewritten using *less than or equal to*.

We would also like to be able to write the negations of sentences containing *at least* or *at most*.

EXAMPLE 6

Negation of *At Least*
and *At Most*

(a) State the negation of the following sentence:

"Our team won at least three basketball games."

SOLUTION The given sentence says that the total number of games our basketball team won this year was a number *greater than or equal to* three. The negation of that sentence is

"Our team did not win at least three games."

We can say this negation without using *not* by using connectives:

"Our team won two, one, or no games."

A shorter way of saying this is

"Our team won at most two basketball games."

(b) State the negation of the sentence

"The speedometer read at most 70 miles per hour."

SOLUTION The given statement says that, at a specific moment, the speedometer's value was less than or equal to 70 mph. The negation of this is that the speedometer's reading was greater than 70 mph, or

"The speedometer read at least a little more than 70 miles per hour." ▄

In both cases in Example 6, the negation of one phrase can be rewritten in terms of the other by changing the "boundary value" (e.g., "not at least *three* games" becomes "at most *two* games"). This was trickier to do in the speedometer example, since speed can take on a continuous range of values; thus, "not at most *70* mph" becomes "at least *a little over 70* mph.

All, Some

Many times we want to restrict the number of objects we are considering, without giving a specific number. Suppose, in Example 6, that we did not remember the specific number of games the team had won, but we did remember the team winning. To handle this situation, we could have said, "This year our team won *some* of its games." In another example, suppose a child says, "*All* dogs are brown." Then, although the child probably has no idea of the number or the colors of different dogs that are living, the statement includes every single one of them. In both examples, there are numerous ways of saying the sentences while maintaining their precise meaning.

EXAMPLE 7
Alternatives to *All* and *Some*

(a) Rewrite the following sentence in three different ways without using the quantifier *some:*

"Some of the flowers are red."

SOLUTION There are several different quantifiers that are equivalent to *some*. In everyday contexts, words like *some* or *few* imply different numerical ranges, but in mathematics we assume that *some* means only that there exists at least one of the objects in question. Accordingly, we can rewrite the given sentence in any of the following ways:

"There is at least one flower that is red."

"There exists a flower that is red."

"Not all of the flowers are a color other than red."

(b) Rewrite the following sentence in three different ways without using the quantifier *all:*

"All mathematics problems are challenging."

SOLUTION In rewriting this sentence, we need to express the idea that every single mathematics problem is challenging. Each of the following sentences captures that idea:

"Every mathematics problem is challenging."

"There are zero mathematics problems that are not challenging."

"There does not exist a mathematics problem that is not challenging." ■

We finish this section with examples of how to negate sentences with these types of quantifiers.

EXAMPLE 8
Negation of *All* and *Some*

(a) State the negation of the sentence

"All students study diligently."

SOLUTION One way of looking at this sentence is to say that there are zero students who do not study diligently. Thus, the negation would be that there is at least one student that does not put forth a diligent effort in his or her studies:

"There exists a student that does not study diligently."

(b) State the negation of the sentence

"Some skateboards have 53 wheels."

SOLUTION The negation is fairly straightforward: We want to create a sentence that says that the number of skateboards with 53 wheels is zero. As a matter of fact, we can say this explicitly:

"The number of skateboards with 53 wheels is zero."

A shorter way of saying it, without mentioning the number of skateboards, is

"No skateboard has 53 wheels."

Notice that the following sentence is *not the negation of the given sentence:*

"Some skateboards do not have 53 wheels."

Why does this sentence fail to negate the given one? Because stating that there are examples of skateboards that do not have 53 wheels does not rule out the possibility of some skateboards having 53 wheels. We leave this example by examining another way of phrasing the negation that is technically correct, but can be misleading:

"All skateboards do not have 53 wheels."

Although the preceding sentence, taken literally, says that every skateboard has a number of wheels different from 53, it fails to convey this meaning in everyday speech. Most people would view the last statement as saying that not all skateboards have 53 wheels—in other words, only some skateboards do not have exactly 53 wheels. If we insist on using the quantifier *all* in our negation, it would be much clearer to say,

"All skateboards have a number of wheels different from 53." ▬

(Note: We must never forget that nonmathematical sentences often are ambiguous, so great care must be taken in interpreting them.)

Exercises 7.1

Consider the following statements:

      p: The student is a female.

      q: The student is a sophomore.

In Exercises 1 through 6, rewrite the indicated sentences in English.

1. $p \vee q$ **2.** $p \wedge q$ **3.** $\sim p \vee q$

4. $p \wedge \sim q$ **5.** $p \Rightarrow q$ **6.** $\sim q \Rightarrow \sim p$

Consider the following statements:

      p: The hat is red.

      q: The belt is black.

In Exercises 7 through 12, rewrite the indicated sentences in English.

7. $p \vee q$ **8.** $p \wedge q$ **9.** $\sim p \vee q$

10. $p \wedge \sim q$ **11.** $p \Rightarrow q$ **12.** $\sim q \Rightarrow \sim p$

Rewrite each of the sentences in Exercises 13 through 16 as disjunctions.

13. If the card is a five, then the card is a diamond.

14. If the physician is skilled, then the surgery will be successful.

15. The student's test grades will improve if she does her homework.

16. The dog will be clean if it is bathed.

Consider the following statements:

p: Ninety percent of all cataract surgeries are successful.

q: Eighty percent of all kidney transplants are successful.

In Exercises 17 through 20, rewrite each of the sentences in symbolic form.

17. Ninety percent of cataract surgeries and 80% of all kidney transplants are successful.

18. Ten percent of all cataract surgeries fail, or 80% of all kidney transplants are successful.

19. Ten percent of all cataract surgeries fail, and 20% of all kidney transplants fail.

20. Ninety percent of all cataract surgeries are successful, and 20% of all kidney transplants are unsuccessful.

Consider the following statements:

p: All of the coins are silver.

q: Some of the marbles are purple.

In Exercises 21 through 24, rewrite each of the sentences in symbolic form.

21. All of the coins are silver, and some of the marbles are purple.

22. Some of the coins are not silver.

23. At least one of the coins is not silver, or none of the marbles are purple.

24. Every coin is silver, and at least one marble is purple.

In Exercises 25 through 28, rewrite each of the sentences by removing the quantifiers and using connectives. (Use Examples 4 and 5 as a guide.)

25. In a group of five students, at least four are taking a mathematics course.

26. In a collection of eight calculus books, at least five have computer support.

27. In a box of six shirts, at most three have flaws.

28. In a case of 24 soda cans, at most 4 have been opened.

In Exercises 29 through 38, write the negation of each sentence.

29. The car is red or has a CD player.

30. The modem is fast, or the scanner will reproduce color.

31. The blouse is not blue, and the belt is black.

32. The I.D. number begins with the digit 2, and the employee is not computer literate.

33. At least one of the two marbles is red.

34. At least one of the two students is majoring in history.

35. At least two of the three students graduated with honors.

36. At least two of the three cars have an automatic transmission.

37. At most three of the four marbles are green.

38. At most one of the three students is on the basketball team.

Section 7.2 Truth Tables

In the previous section, we encountered three instances of *equivalent* statements: The statement $\sim(\sim p)$ was said to be *equivalent* to the statement p; the "if and only if" statement $p \Leftrightarrow q$ was interpreted as p is *equivalent* to q, and the implication $p \Rightarrow q$ was said to be *equivalent* to the disjunction $\sim p \vee q$. By developing what we mean by the *truth* of a statement, and by utilizing **truth tables,** we can more clearly see what we mean by *equivalent* in these and other situations.

Because we are dealing only with statements that are true or false and not both, each such statement has exactly two possible **truth values,** T or F, and of these, only one may describe the statement. The following simple table displays the truth values of some English-language statements:

Statement	True Value
All Toyotas are red.	F
3 + 5 = 8	T
Some Toyotas are red.	T

A sentence using any two such statements p and q will always have four possible pairs of truth values to consider:

p	q
T	T
T	F
F	T
F	F

The truth value of a particular sentence involving p, q, and the connectives we have discussed will then be determined for each possible combination. The basic building blocks in such a process are the truth values assigned to $p \wedge q$, $p \vee q$, and $\sim p$. The last would be assigned truth values just as you think it should: If p is true, then $\sim p$ is false, while if p is false, then $\sim p$ is true:

p	$\sim p$	Examples of Assigning Truth Values to $\sim p$
T	F	p: 3 + 5 = 8
		$\sim p$: 3 + 5 ≠ 8
F	T	p: All Toyotas are red.
		$\sim p$: Not all Toyotas are red. *or*
		$\sim p$: Some Toyotas are not red. *or*
		$\sim p$: There is at least one Toyota that is not red.

The accepted way of assigning truth values to $p \wedge q$ is given in the next table, along with an example of each case:

p	q	$p \wedge q$	Examples of Assigning Truth Values to $p \wedge q$
T	T	T	Chicago is in Illinois, and Dallas is in Texas.
T	F	F	Chicago is in Illinois, and Dallas is in Alaska.
F	T	F	Chicago is in Wyoming, and Dallas is in Texas.
F	F	F	Chicago is in Wyoming, and Dallas is in Alaska.

That is, the entire sentence $p \wedge q$ (p and q) is true when *both* p and q are *true;* otherwise, the entire sentence is false. In other words, for the sentence $p \wedge q$ to be true, p and q must be simultaneously true.

The accepted way of treating the disjunction $p \vee q$ is shown next, with accompanying examples to help us grasp why the truth values are assigned as they are.

p	q	$p \vee q$	Examples of Assigning Truth Values to $p \vee q$
T	T	T	Two is an even number or 5 is odd.
T	F	T	Two is an even number or 5 is even.
F	T	T	Two is odd or 5 is odd.
F	F	F	Two is odd or 5 is even.

That is, the entire sentence $p \vee q$ (p or q) is true when *either* p is true or q is true or when *both* are true; otherwise the entire sentence is false. Stated differently, $p \vee q$ is true when *at least* one of its two components is true; otherwise, $p \vee q$ is false. Put still differently, p or q is false if both p and q are false; otherwise, the entire sentence is true.

NOTE Throughout the remainder of this book and, indeed, throughout all of mathematics and statistics, the words *and* and *or* are used as logical connectives with truth values as just defined.

With these building blocks, we are now ready to construct a truth table for any sentence that uses p, q, and the connectives $\wedge$, $\vee$, and $\sim$.

EXAMPLE 1

Constructing
a Truth Table

Make a truth table that exhibits the truth values of $(p \wedge q) \vee \sim p$.

SOLUTION Some intermediate steps will prove useful in reaching our final goal:

p	q	$p \wedge q$	$\sim p$	$(p \wedge q) \vee \sim p$
T	T	T	F	T
T	F	F	F	F
F	T	F	T	T
F	F	F	T	T

The table shows that the given sentence is false when p is true and q is false, but is true for the other possible combinations of T and F. ■

An important application of truth tables is to help us decide when two sentences are logically equivalent.

Logically Equivalent Sentences

Two sentences are **logically equivalent** if they have identical truth values. The notation $p \equiv q$ is used to denote the fact that p and q are logically equivalent.

The usefulness of the concept of logical equivalence lies in the fact that we can often translate a given sentence into one that is easier to understand.

EXAMPLE 2

Showing the
Equivalence of
Sentences

Show, by means of a truth table, that p and $\sim(\sim p)$ are equivalent.

SOLUTION The statements p and $\sim(\sim p)$ are equivalent, as shown by the first and third columns of the following truth table:

p	$\sim p$	$\sim(\sim p)$
T	F	T
F	T	F

EXAMPLE 3

Showing the Equivalence of Statements

Show that $\sim(p \wedge q) \equiv \sim p \vee \sim q$ by the use of a truth table.

SOLUTION The following truth table demonstrates that the given statements are equivalent:

p	q	$p \wedge q$	$\sim(p \wedge q)$	$\sim p$	$\sim q$	$\sim p \vee \sim q$
T	T	T	F	F	F	F
T	F	F	T	F	T	T
F	T	F	T	T	F	T
F	F	F	T	T	T	T

The fourth and last columns are identical and therefore show the desired equivalence. Note that columns 3, 5, and 6 are inserted solely for the purpose of making the two desired columns easier to complete.

In the previous section, we presented a theorem concerning an equivalent way of writing an implication; the following example proves the theorem, using a truth table.

EXAMPLE 4

Using Truth Tables to Verify Logical Equivalence

Show that $p \Rightarrow q \equiv \sim p \vee q$ by the use of a truth table.

SOLUTION The truth table for the two statements can be completed as follows:

p	q	$p \Rightarrow q$	$\sim p$	$\sim p \vee q$
T	T	T	F	T
T	F	F	F	F
F	T	T	T	T
F	F	T	T	T

The third and fifth columns are identical, showing the desired equivalence.

DeMorgan's Laws of Logic

Example 3 actually demonstrates the equivalence of the first of two very important equivalences known as **DeMorgan's laws of logic.**

DeMorgan's Laws of Logic

Let p and q be logical statements. Then

$$1.\ \sim(p \wedge q) \equiv \sim p \vee \sim q.$$
$$2.\ \sim(p \vee q) \equiv \sim p \wedge \sim q.$$

DeMorgan's laws demonstrate how negations can be "distributed" across connectives.

HISTORICAL NOTE

AUGUSTUS DEMORGAN (1806–1871)

Augustus DeMorgan, an Englishman, was among the first mathematicians who became aware of the structure of algebra—the commutative laws of addition and multiplication, the associative laws, the distributive law, and so on.

EXAMPLE 5

Using DeMorgan's Laws

Each of the following uses one of DeMorgan's laws of logic to translate a given sentence into a logically equivalent one.

(a) What is the negation of "The fax machine is busy, and the copier is jammed."

SOLUTION "The fax machine is not busy, or the copier is not jammed."

(b) What is the negation of "The first ball is red, or the second ball is not green"?

SOLUTION "The first ball is not red, and the second ball is green."

(c) What is the negation of "The shirt has a flaw and did not come from Belk's department store"?

SOLUTION "The shirt does not have a flaw or came from Belk's." ▄

Exercises 7.2

In Exercises 1 through 6, make a truth table for each of the given sentences.

1. $p \vee \sim q$
2. $p \vee (q \wedge \sim p)$
3. $\sim p \Rightarrow \sim q$
4. $\sim p \Rightarrow (q \wedge p)$
5. $\sim(\sim p \vee q)$
6. $\sim p \wedge (\sim q \vee q)$

In Exercises 7 through 9, use a truth table to show the given equivalences.

7. $\sim(p \vee q) \equiv \sim p \wedge \sim q$
8. $p \vee q \equiv q \vee p$
9. $p \wedge (q \vee r) \equiv (p \wedge q) \vee (p \wedge r)$

In Exercises 10 through 13, use truth tables to determine which of the pairs of sentences are equivalent.

10. $p \vee \sim q : \sim p \wedge q$
11. $p \wedge q : \sim p \Rightarrow \sim(q \wedge p)$
12. $\sim p \wedge q : \sim(p \vee \sim q)$
13. $q \wedge (\sim p \vee q) : \sim(\sim p \vee q)$

In Exercises 14 through 22, decide what truth value (T or F) to assign to each sentence.

14. If an animal is a dog, then it is a cocker spaniel.
15. The earth is not bigger than the moon.
16. $2 + 3 = 5$, and January has 35 days.
17. $2 + 3 = 5$, or January has 35 days.
18. $2 + 3 \neq 7$, or Social Security numbers have nine digits.
19. Some college students drink coffee, and all college students own automobiles.
20. Some college students drink coffee, or all college presidents are female.
21. All mountain bikes are red, or there have been at least two male presidents of the United States.
22. All heart transplants are successful, and *The New York Times* is read by millions.

Consider the following statements:

p: The die shows an odd number.

q: The die does not show a five.

In Exercises 23 through 27, apply DeMorgan's laws of logic to each sentence and then rewrite it in English.

23. $\sim(p \wedge q)$

24. $\sim(p \vee q)$

25. $\sim(p \vee \sim q)$

26. $\sim(\sim p \wedge q)$

27. $\sim(\sim p \vee \sim q)$

28. At the beginning of this section, we indicated that we were dealing only with statements that are either true or false, but not both. You might have wondered whether there were any statements that could somehow be both true and false at the same time. To answer this question, consider the statement "This sentence is false," symbolized by p.

(a) Assume that p is true. In your own words, describe what this assumption says about the statement.

(b) Assume p is false. In your own words, describe what this assumption says about the statement.

(c) Is statement p a true statement or a false statement?

29. A **tautology** t is a sentence whose only possible truth value is T. Give an example of a tautology in symbolic form (e.g., using p, q, $\vee$, $\wedge$, $\sim$), and verify that it is always true through the use of a truth table.

30. A **contradiction** c is a sentence whose only possible truth value is F. Give an example of a contradiction in symbolic form (e.g., using p, q, $\vee$, $\wedge$, $\sim$), and verify that it is always false by using a truth table.

31. The **contrapositive** of $p \Rightarrow q$ is defined to be the implication $(\sim q) \Rightarrow (\sim p)$.

(a) Make a truth table showing that an implication and its contrapositive are equivalent.

(b) List three mathematical theorems, and state the contrapositive of each.

Section 7.3 Sets

The language of sets provides a common way of expressing ideas in a universal notation that permeates all fields of mathematics. In addition, sets provide us with a link to logic and a visual aid in expressing the thoughts of counting and probability.

A **set** is a well-defined collection of distinct objects. The term *well-defined* means that some property, rule, or description will allow us to determine whether a given object belongs to the set. The following are examples of well-defined sets:

The set of people in the United States who voted in the 2000 presidential election.

The set of letters in the Greek alphabet.

The whole numbers between 0 and 10, inclusive.

These sets are not well defined:

The set of small numbers.

The set of young people in Boston.

The objects that make up a set are called **elements** of the set, and we write $a \in S$ to denote the fact that the element a belongs to the set S. If a is not an element of S, we write $a \notin S$. If the elements of a set can be listed, braces $\{\ \}$ are usually used to enclose them. Sometimes, a set may have no elements at all, in which case, we call the set

empty. The symbol for an empty set is $\varnothing$ (with *no* braces around it) or braces { } with no symbol within.

<u>EXAMPLE 1</u>

Set Notation

(a) Let A denote the set of integers from 0 to 5, inclusive. Then $A = \{0, 1, 2, 3, 4, 5\}$, where $3 \in A$, but $8 \notin A$.

(b) Let B represent the set of positive integers. Then $B = \{1, 2, 3, 4, \ldots\}$ is a common way of expressing a set that has a definite pattern, but cannot be listed in its entirety. Another way is to write $B = \{n : n$ is a positive integer$\}$, where the colon is read "such that"; B is the set of all n such that n is a positive integer.

(c) Let C be the set of integers between $\frac{1}{2}$ and $\frac{3}{4}$. There are no integers in this range of numbers, so we write $C = \varnothing$. ■

If it happens that each element of one set, say, A, is also an element of another set, B, then we write $A \subseteq B$ and say that A is a **subset** of B.

Definition of Subset

Let A and B be sets. Then $A \subseteq B$ if, and only if, each element of A is also an element of B.

The concept of a subset now allows us to define the concept of **set equality.**

Set Equality

Let A and B be sets. Then $A = B$ if, and only if, $A \subseteq B$ and $B \subseteq A$. In other words, two sets are equal if they contain exactly the same elements.

For example, $A = \{1, 2, 3\}$ is a subset of $B = \{0, 1, 2, 3, 4, 5\}$, so we could write $A \subseteq B$. In this case, B is not a subset of A, because there is at least one element of B that is not an element of A, so $A \neq B$. As another example of a subset, the set consisting of the months whose names begin with a "J," namely, January, June, and July, is a subset of the set of the 12 months of the year.

It is generally convenient to regard all sets in a particular discussion to be subsets of a single larger set U called the **universal set.** For example, sets of numbers may come from the universal set of real numbers, and sets of letters may come from the universal set called the English alphabet. The idea of a universal set allows us to create an algebra of sets that results in properties paralleling those of logic.

Earlier, we mentioned another special set—$\varnothing$—the empty set. Notice that any set with no elements is equal to the empty set. For example, the set of all two-ton canaries is equal to the empty set, and so is the set of all living saber-toothed tigers. Thus, the set of all two-ton canaries is equal to the set of all living saber-toothed tigers. Regardless of how freaky the description, the result is the same: a set with no elements.

The empty set also has the interesting property that it is a subset of every set! Seeing why this is true can be difficult, but reread the definition of a subset and think about it for a moment. Given any set A, is it true that each element in $\varnothing$ is also an element in A? Yes, it is true, since the empty set has no elements that could fail to be in A. There is another way of looking at this: Ask yourself what would have to be true for the empty set *not* to be a subset of a set A.

Intersection and Union

When dealing with multiple sets, there are certain combinations that are particularly interesting. For example, consider the set A of all your classmates and the set B of all your friends. One new set that may be of interest is the set of all people who are *both* your classmates *and* your friends. This set is called the **intersection** of A and B, denoted by $A \cap B$. The intersection of two or more sets is the set that contains all of the mutually "shared" elements.

The Intersection of Sets

Let A and B be sets. Then the **intersection** $A \cap B$ is the set of all elements that belong to both A and B. That is,

$$A \cap B = \{x : (x \in A) \text{ and } (x \in B)\}.$$

Continuing with the previous example, you may be interested in the set of people who are *either* your classmates *or* your friends (remember that we are using the inclusive *or,* which means that we are including people who are both your classmates and your friends). This set is called the **union** of A and B, denoted by $A \cup B$. The union of two or more sets contains all of the elements that are in at least one of the sets.

The Union of Sets

Let A and B be sets. Then the **union** $A \cup B$ is the set of all elements that belong to either A or B. That is,

$$A \cup B = \{x : (x \in A) \text{ or } (x \in B)\}.$$

A convenient way of visualizing intersections and unions is the **Venn diagram,** shown in Figure 1. A Venn diagram is constructed by drawing a rectangle to represent the universal set U and then drawing two or more circles within the rectangle, each representing a set under discussion. You must make sure that the circles overlap in every possible combination to allow for all possible intersections. The diagram is completed by shading the part or parts representing the given set expression. For example, for two sets A and B, you would draw two overlapping circles, one for A and one for B. The overlapping area would be shaded to represent those elements in the intersection $A \cap B$. We would shade all of both disks to represent the union $A \cup B$. Both cases are illustrated in Figure 1.

HISTORICAL NOTE
JOHN VENN (1834–1923)

Venn diagrams are named after the English logician John Venn, who was a lecturer of moral sciences at Cambridge University. His diagrams for representing the relationships between sets became the accepted standard and are still used today. Such diagrams are also important in the study of logic.

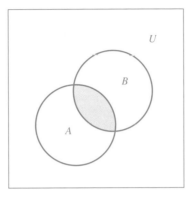

FIGURE 1a
The shaded portion represents $A \cap B$.

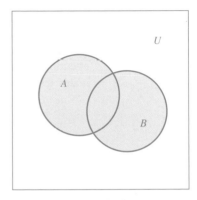

FIGURE 1b
The shaded portion represents $A \cup B$.

EXAMPLE 2

Illustrating the Intersection and Union

Let $A = \{0, 1, 2, 3\}$, $B = \{2, 3, 5\}$, and $C = \{5\}$ be subsets of $U = \{0, 1, 2, 3, 4, 5\}$. Then

$$A \cap B = \{2, 3\}$$
$$A \cup B = \{0, 1, 2, 3, 5\}$$
$$A \cap C = \varnothing$$
$$B \cup C = \{2, 3, 5\} = B$$
$$B \cap C = \{5\} = C$$
$$A \cup C = \{0, 1, 2, 3, 5\}. \quad \blacksquare$$

Notice that the intersection of two sets will always be a subset of each of the sets ($\varnothing$ is a subset of every set), and the union will always contain each of the sets. In fact, you can think of the union as taking all of the elements of one of the sets and then adjoining to them all the remaining elements of the other set. Observe that we do not list a given element in a set twice, and the order in which we list the elements of a set is immaterial.

Complement and Difference

As you should start to see, when we begin with simple sets, many more complicated sets can be naturally created. In the case of the set A of all your classmates, we could consider the set of all people who are *not* your classmates. This set is called the **complement** of A, denoted by A'. Notice that for this particular new set to make any sense, we must consider a universal set U. If the universal set is the set of all students, then the complement of A is the set of all students who are not your classmates.

The Complement of a Set

Let A be a subset of a universal set U. Then the **complement** A' is the set of all elements of U that are not in A. That is,

$$A' = \{x : (x \in U) \text{ and } (x \notin A)\}.$$

Sometimes we want to restrict the elements of one set to those that lie outside another set. For instance, you may only want to consider your friends (B) who are not your classmates (A). This new set is called the **difference** of B and A, denoted by $B - A$. Pay close attention to the order of the sets, as the difference $A - B$ would be the set of your classmates who are not your friends.

The Difference of Sets

Let A and B be sets. Then the **difference** $B - A$ is the set of all elements of B that are not in A. That is,

$$B - A = \{x : (x \in B) \text{ and } (x \notin A)\}.$$

In Figure 2, we see that sets created by complements or differences can be illustrated by Venn diagrams.

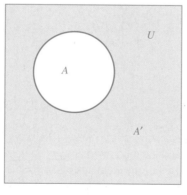

FIGURE 2a
The shaded portion represents A′.

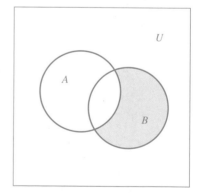

FIGURE 2b
The shaded portion represents B − A.

EXAMPLE 3
Illustrating Set
Operations

Let $A = \{1, 2, 4\}$, $B = \{0, 1, 2, 3\}$, $C = \{4, 5\}$, and $U = \{0, 1, 2, 3, 4, 5, 6, 7, 8\}$. Then

$$A' = \{0, 3, 5, 6, 7, 8\}$$
$$A - B = \{4\}$$
$$B - A = \{0, 3\}$$
$$A - U = \varnothing$$
$$A \cap B = \{1, 2\}$$
$$A \cup B = \{0, 1, 2, 3, 4\}$$
$$B \cup C' = \{0, 1, 2, 3, 6, 7, 8\}$$
$$B \cap C = \varnothing$$
$$C - B = \{4, 5\} = C$$
$$B - C = \{0, 1, 2, 3\} = B. \quad \blacksquare$$

Sets and Logic

You may have sensed by now that there is a strong connection between logic and sets, at least as far as we have developed them. To bring out this connection, suppose that P is the set of elements in a universal set U that makes statement p true, and suppose that $Q \subseteq U$ is the set that makes statement q true. The sets P and Q are then known as **truth sets** for p and q, respectively. Sentences constructed using p, q, and the connectives we have studied also have truth sets. For example, the set of elements in $P \cap Q$ would make $p \wedge q$ true. The following table gives the translation between some logical connectives and the corresponding set operations.

Statement	Truth Set
And: $p \wedge q$	Intersection: $P \cap Q$
Or: $p \vee q$	Union: $P \cup Q$
Negation: $\sim p$	Complement: P'
$p \wedge \sim q$	$P \cap Q' = P - Q$

EXAMPLE 4

Illustrating the
Connection between
Logic and Sets

Roll a single six-sided die, and count the number of pips showing on the top side. The number of pips showing is an element of the set $\{1, 2, 3, 4, 5, 6\}$. Consider the following statements:

> p: The die shows an even number.
>
> q: The die shows a number greater than 4.

The truth set for p is $P = \{2, 4, 6\}$, and that for q is $Q = \{5, 6\}$ (see Figure 3). The conjunction p *and* q is then

> $p \wedge q$: The die shows a number that is even and is greater than 4.

This statement has the truth set

$$P \cap Q = \{6\}.$$

The sentence describing the disjunction, p *or* q, is

> $p \vee q$: The die shows a number that is even or is greater than 4.

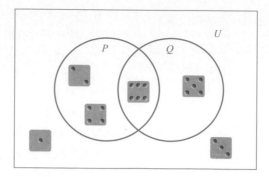

FIGURE 3
Different Possibilities for the Roll of a Die

This statement has the truth set

$$P \cup Q = \{2, 4, 5, 6\}.$$

The negation

$\sim p$: the die shows an odd number.

This statement has the truth set

$$P' = \{1, 3, 5\}.$$

Finally, consider

$\sim p \wedge q$: The die does not show an even number and does show a number greater than 4.

The truth set for this statement is

$$P' \cap Q = \{5\}. \blacksquare$$

A study of Example 4 illustrates how we are naturally and, quite correctly, led to use a logical sentence structure and its truth set representation interchangeably. In fact, most of the time, we just refer to translating a sentence into set form or vice versa, without mentioning the truth-set connection. The process is sometimes made easier through the use of a Venn diagram, as shown in the next two examples. As is often the case, the intersection or union of more than two sets is involved.

General Unions and Intersections

For the sets $A_1, A_2, \ldots A_n$, the **intersection**

$$A_1 \cap A_2 \cap \ldots \cap A_n = \{x : x \text{ belongs to every set}\},$$

while the **union**

$$A_1 \cup A_2 \cup \ldots \cup A_n = \{x : x \text{ belongs to at least one of the sets}\}.$$

EXAMPLE 5

Shopping Survey: From
Logic to Sets

A survey was made of all shoppers who checked out in a particular lane of Harps IGA grocery store one day between the hours of 2:00 and 7:00 P.M. Among those shoppers, let C represent those who bought cinnamon, P those who bought pepper, and S those who bought salt. In each of the following exercises, use a Venn diagram to identify the set of shoppers, and then use the symbols C, P, and S, along with $\cup$, $\cap$, $'$, and $-$ to express the set.

(a) Shoppers who bought cinnamon, but not pepper.

SOLUTION The Venn diagram shown in Figure 4 identifies the shoppers in question.

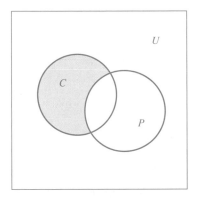

FIGURE 4
$C - P$

The shaded set may be expressed as

$$C - P = C \cap P'.$$

(b) Shoppers who bought cinnamon and pepper, but not salt.

SOLUTION The Venn diagram shown in Figure 5 identifies the shoppers in question.

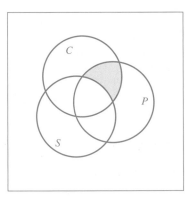

FIGURE 5
$(C \cap P) - S$

The shaded region may be expressed as

$$(C \cap P) - S = (C \cap P) \cap S'.$$

(c) Shoppers who bought at least one of the three products.

SOLUTION The Venn diagram shown in Figure 6 identifies the shoppers in question.

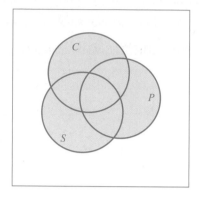

FIGURE 6
$C \cup P \cup S$

The shaded region may be expressed as

$$C \cup P \cup S.$$

(d) Shoppers who bought exactly one of the three products.

SOLUTION The Venn diagram shown in Figure 7 identifies the shoppers in question.

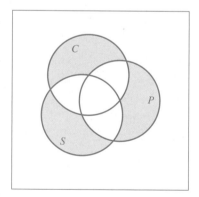

FIGURE 7

The shaded region may be expressed in more than one way:

$$[C \cap (P \cup S)'] \cup [P \cap (C \cup S)'] \cup [S \cap (C \cup P)']$$
$$= [C \cap (P' \cap S')] \cup [P \cap (C' \cap S')] \cup [S \cap (C' \cap P')]. \blacksquare$$

EXAMPLE 6

Interpreting Sets in English

Suppose we reach into a toolbox and grab a tool. Some of the tools in the toolbox are wrenches and some of the tools have a rubber grip. If, of all the tools in the toolbox, W is the subset of wrenches and R is the subset of tools with a rubber grip, write an English sentence describing each of the following possibilities:

(a) $W \cap R$

SOLUTION We grabbed a wrench with a rubber grip.

(b) $(W \cup R)'$

SOLUTION We grabbed a tool that was not a wrench, and it did not have a rubber grip.

(c) $W' \cup R$

SOLUTION We did not grab a wrench or the tool did have a rubber grip. ▬

Exercises 7.3

For the sets $A = \{1, 2, 3\}$, $B = \{2, 3, 5\}$, $C = \{6, 7\}$, and $U = \{0, 1, 2, 3, 4, 5, 6, 7\}$, find the elements of the sets described in Exercises 1 through 10.

1. A'
2. $A \cup B$
3. $\varnothing'$
4. $B \cap C$
5. $A' \cap B$
6. $A \cup B'$
7. $(A \cup C)'$
8. $A' \cup B'$
9. $A - B'$
10. $C - A$

For the sets $A = \{a, b, c\}$, $B = \{c, d\}$, $C = \{e\}$, and $U = \{a, b, c, d, e, f\}$, find the sets described in Exercises 11 through 20.

11. B'
12. $A' \cup B$
13. $A \cap B$
14. $B \cap C$
15. U'
16. $(A \cap B')'$
17. $B' \cup C$
18. $(A' \cup B')'$
19. $A - C$
20. $A - B$

In each of Exercises 21 through 28, shade in the part of a Venn diagram representing the given set.

21. $A \cup B'$
22. $(A \cup B)'$
23. $A' \cap B'$
24. $(A \cup B) - C$
25. $(A \cap B) - C$
26. $A - (B \cap C)$
27. $A' \cap B$, if $A \subseteq B$
28. $A \cup (B' \cap C)$, if $A \subseteq B$

Answer Exercises 29 through 34 as true or false. For each false statement, make up sets that demonstrate its falsity.

29. A is always a subset of $A \cup B$.
30. A is never a subset of $A \cap B$.
31. $A \cap B$ is always a subset of $A \cup B$.
32. $A \cup B$ is never a subset of A.
33. $A \cap \varnothing = \varnothing$
34. If $A \neq \varnothing$, $A \cup \varnothing = \varnothing$

Selecting Numbers Let $U = \{1, 2, 3, \ldots, 16\}$ be the universal set and P be the set of those numbers which make the following statement true:

p: The number is odd.

Let Q be the set of those numbers which make the following statement true:

q: The number is divisible by 3.

Write each of Exercises 35 through 40 in sentence form.

35. $P \cap Q$
36. $(P \cap Q)'$
37. $P \cup Q'$
38. $P' \cap Q$
39. $P - Q$
40. $Q' - P$

Let the universal set U be the set of people currently living in your hometown. Let G represent all of those people who make the following statement true:

p: The person wears glasses.

Let F represent all of those people who make the following statement true:

q: The person is over 40 years of age.

Write each of Exercises 41 through 46 in sentence form.

41. F'　　　　**42.** $G' \cup F$　　　　**43.** $G \cap F$

44. $G' - F$　　　**45.** $G \cap F'$　　　**46.** $(G \cup F)'$

Subscription Survey Let the universal set U be the set of households in Elgin, let G be the set of households in Elgin that subscribe to the *Gazette,* and let T be the set of households in Elgin that subscribe to the *Tribune.* In each of Exercises 47 through 54,

(a) Use the given set symbols to construct a Venn diagram identifying the three types of household.

(b) Use the given set symbols, along with ∩, ∪, ′, or −, to write a set expression identifying the following households:

47. Households in Elgin that do not subscribe to the *Gazette.*

48. Households in Elgin that subscribe to both the *Gazette* and the *Tribune.*

49. Households in Elgin that subscribe to the *Gazette,* but not to the *Tribune.*

50. Households in Elgin that subscribe to neither of the two newspapers.

51. Households in Elgin that do not subscribe to the *Gazette,* but do subscribe to the *Tribune.*

52. Households in Elgin that subscribe to exactly one of the two newspapers.

53. Households in Elgin that subscribe to neither the *Gazette* nor the *Tribune.*

54. Households in Elgin that subscribe to the *Gazette* or do not subscribe to the *Tribune.*

Course Enrollment Let the universal set U be the set of students enrolled in a given finite math class, let E be the set of those students currently enrolled in an English class, H be the set of those students currently enrolled in a history class, and G be the set of those students currently enrolled in a geology class. In each of Exercises 55 through 62,

(a) Use the given set symbols to construct a Venn diagram identifying the various sets of students.

(b) Use the given set symbols, along with ∩, ∪, ′, or −, to write a set expression identifying the following sets of students:

55. Students who are currently enrolled in English and history, but not geology.

56. Students who are currently enrolled in English, but not in history or geology.

57. Students who are currently enrolled in geology or English, but not history.

58. Students who are currently enrolled in all four courses.

59. Students who are currently enrolled in exactly one of the four classes.

60. Students who are currently enrolled in none of the four classes.

61. Students who are currently enrolled in at least one of the four classes.

62. Students who are currently enrolled in at least two of the four courses.

Section 7.4　Application of Venn Diagrams

Identifying a region in a Venn diagram to represent a particularly complicated set expression may prove to be taxing. A method that eliminates the need to cross-shade component parts required to reach the final representation is shown in Example 1.

EXAMPLE 1

Constructing a Venn Diagram

Use a Venn diagram to represent $A \cap (B \cup C')$.

SOLUTION First, draw three overlapping circles in a universal set U. Next, label each separate region with a number, letter, or symbol, as shown in Figure 8(a). (We prefer to label with numbers.) Now, record the numbers that make up each of the regions A, B, and C' involved in the given expression $A \cap (B \cup C')$. Then record the numbers that identify regions in any parts of the expression, such as $(B \cup C')$, that would be used as building blocks in constructing the expression. (Our list is shown next.)

Set	Region Labels
A	1, 2, 3, 4
B	3, 4, 5, 6
C'	1, 4, 5, 8
$B \cup C'$	1, 3, 4, 5, 6, 8

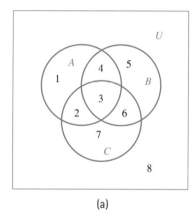 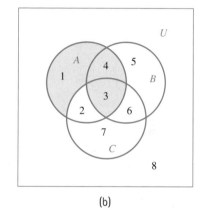

(a) (b)

FIGURE 8
Construction of a Venn Diagram: (a) Labeling the Regions; (b) Shading the Regions

Finally, $A \cap (B \cup C') = \{1, 3, 4\}$, which identifies the regions to be shaded, as shown in Figure 8(b). ■

DeMorgan's Laws of Sets

Venn diagrams provide a quick, easy way to indicate whether two set statements represent the same set.

EXAMPLE 2
Using Venn Diagrams to Show Set Equality

Use Venn diagrams to show that $(A \cap B)' = A' \cup B'$.

SOLUTION First, draw a Venn diagram, labeling each region as indicated in Figure 9. Then, as was done in Example 1, find the final regions to be shaded for each of the expressions in question.

$(A \cap B)'$	
Set	Region Labels
A	1, 2
B	2, 3
$A \cap B$	2
$(A \cap B)'$	1, 3, 4

$A' \cup B'$	
Set	Region Labels
A	1, 2
B	2, 3
A'	3, 4
B'	1, 4
$A' \cup B'$	1, 3, 4

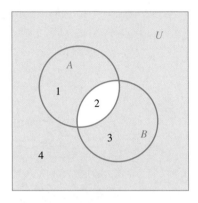

FIGURE 9
Using Venn Diagrams to Show Set Equality

The last line in each column shows that the same region should be shaded, which demonstrates that $(A \cap B)' = A' \cup B'$. ▬

The equality shown in Example 2 is one of two that are known as **DeMorgan's laws of sets.** Notice how they parallel DeMorgan's laws of logic.

DeMorgan's Laws of Sets

Let A and B be sets. Then

$$1. \ (A \cup B)' = A' \cap B'.$$
$$2. \ (A \cap B)' = A' \cup B'.$$

The preceding are but two of many equalities among set expressions. A partial list of some of the more prominent ones is shown next for sets A, B, and C in a universal set U. Note the similarity between these equalities and the usual ones of algebra.

Some Properties of Set Algebra

Let A, B, and C be sets. Then the following properties hold:

Commutative laws:

1. $A \cap B = B \cap A$ 2. $A \cup B = B \cup A$

Associative laws:

1. $A \cap (B \cap C) = (A \cap B) \cap C$ 2. $A \cup (B \cup C) = (A \cup B) \cup C$

Distributive laws:

1. $A \cap (B \cup C) = (A \cap B) \cup (A \cap C)$
2. $A \cup (B \cap C) = (A \cup B) \cap (A \cup C)$

DeMorgan's laws of sets:

1. $(A \cap B)' = A' \cup B'$ 2. $(A \cup B)' = A' \cap B'$

Set difference: $A - B = A \cap B'$
Double complementation: $(A')' = A$

Counting Problems

Venn diagrams are also used in counting problems. A counting problem is one that asks, "How many?" and the answer is always one of the whole numbers, $0, 1, 2, 3, \ldots$. For a finite set A, the notation $n(A)$ is used to denote the number of elements in the set. For instance,

If $A = \{a, b, c\}$, then $n(A) = 3$.

If $B = \{0, 1, 2, 3, 4, 5, 6\}$, then $n(B) = 7$.

If S is the set of cards in a standard deck of cards, then $n(S) = 52$.

For $\varnothing$, $n(\varnothing) = 0$.

The next example shows how a Venn diagram may be used to help count the number of elements in some set expressions.

EXAMPLE 3

Hospital Admission: Using a Venn Diagram to Help with Counting

A hospital admissions office checked the records of 40 recently admitted patients and found that 24 had major medical coverage (M), 15 had catastrophic medical coverage (C), and 6 had both types of coverage. Use a Venn diagram to find the number of these patients who had neither type of coverage.

SOLUTION First draw a Venn diagram with two circles, one labeled M and the other C. Then label each separate region of the diagram with the number of patients that have the particular type of insurance represented by that region. Start the labeling process by considering the intersection $M \cap C$. The data tell us that 6 patients had both types of coverage, which means that $n(M \cap C) = 6$. So we label the intersection with a 6, as shown in Figure 10(a). Now we look back at the data and read that 24 had major medical coverage, which means that a *total* of 24 should be represented in the M circle. There are already 6 in this circle, so $24 - 6 = 18$ should be used to count the remainder, $M - (M \cap C)$, as shown in Figure 10(b). The data also tell us that a total of 15 should be represented in the C circle. There are already 6 in this circle, so $15 - 6 = 9$ should be used to count the region $C - (M \cap C)$, as shown in Figure 10(c). Finally, the total in our universal set is 40, which means that the remaining uncounted region outside the two circles should be labeled with $40 - (9 + 6 + 18) = 40 - 33 = 7$, as shown in Figure 10(d). Now we can answer the question posed at the outset of this example and any other counting question about these data by directly examining the Venn diagram in Figure 10(d).

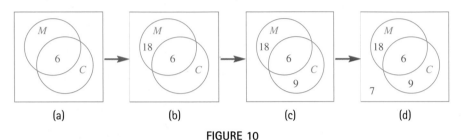

(a)   (b)   (c)   (d)

FIGURE 10
Sequence of Steps in Labeling a Venn Diagram for Counting Purposes

(a) How many of these patients had neither of the two types of insurance? The answer is the number in the region outside the two circles: 7.

(b) How many of these patients had major medical, but not catastrophic, insurance? The answer is the number in the region represented by $(M - C)$: 18.

(c) How many of these patients had major medical or catastrophic insurance? The answer is the number in $M \cup C$: $18 + 6 + 9 = 33$. ∎

EXAMPLE 4

Shopping Survey:
Counting with Venn
Diagrams

One hundred customers of a discount grocery store were asked about their purchases of candy, apples, and popcorn during the previous month. Thirty-two said they had purchased candy, 48 had purchased apples, 35 had purchased popcorn, 15 had purchased both candy and apples, 7 had purchased both candy and popcorn, 5 had purchased both apples and popcorn, and 2 had purchased all three foods. A Venn diagram representing these purchases may be constructed similar to that in Example 3. Here, we start with the intersection of all three circles and build outward to obtain the numbers shown in Figure 11.

(a) Among these customers, how many purchased candy or popcorn, but not apples?

SOLUTION In symbols, Question (a) asks for $n[(C \cup P) - A]$, and the answer from Figure 11 is found to be $12 + 5 + 25 = 42$.

(b) For these customers, how many purchased none of these foods?

SOLUTION The answer is 10.

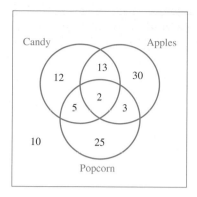

FIGURE 11
Grocery Store Purchases

(c) For these customers, how many purchased exactly two kinds of food?

SOLUTION The answer is $5 + 3 + 13 = 21$. ▬

A useful counting formula arises for $n(A \cup B)$ that we can employ now as well as in later sections. To justify this formula and its special case, consider the Venn diagrams in Figure 12. From Figure 12(a), $n(A) = 5$ and $n(B) = 7$. However, $n(A) + n(B) = 12$ *includes* the count of the two elements in the intersection *twice*. Therefore, *excluding* $n(A \cap B)$ once will leave each element in $A \cup B$ counted exactly once. It follows that

$$n(A \cup B) = n(A) + n(B) - n(A \cap B).$$

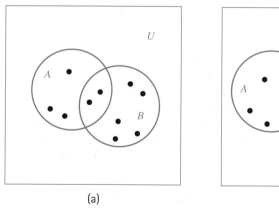

(a)   (b)

FIGURE 12
Developing a Counting Formula

Figure 12(b) shows the special case when $A \cap B = \emptyset$. Because $n(A \cap B) = 0$, the formula reduces to $n(A \cup B) = n(A) + n(B) - n(A \cap B) = n(A) + n(B)$.

In general, the method for counting the size of the union of sets is known as the inclusion–exclusion principle.

Important Counting Formula: The Inclusion–Exclusion Principle

For any two sets A and B,

$$n(A \cup B) = n(A) + n(B) - n(A \cap B),$$

or equivalently,

$$n(A \text{ or } B) = n(A) + n(B) - n(A \text{ and } B).$$

The inclusion–exclusion principle may be restated as

"When counting, 'or' means 'add,' but don't double-count any outcome."

We note that, when a Venn diagram is used for counting purposes, the construction we have demonstrated is designed specifically to eliminate any double-counting of elements, so there is no particular need for a formula there. However, as we move forward with other counting techniques in which the word "or" is involved, the use of the inclusion–exclusion principle, or its equivalent restatement, will preclude any double-counting of outcomes.

EXAMPLE 5

Using the Inclusion–Exclusion Principle

If $n(A) = 26$, $n(B) = 32$, and $n(A \cup B) = 47$, what is $n(A \cap B)$?

SOLUTION Any time we know three of the four terms in the inclusion–exclusion principle, we can algebraically solve for the fourth. Direct substitution of the given data into

$$n(A \cup B) = n(A) + n(B) - n(A \cap B)$$

gives

$$47 = 26 + 32 - n(A \cap B)$$

so that

$$47 = 58 - n(A \cap B)$$

or

$$n(A \cap B) = 58 - 47 = 11. \quad \blacksquare$$

EXAMPLE 6

Quality Control: Using the Word "or" in Counting

A check of 90 lightbulbs revealed the following data:

Bulb Type	Defect	No Defect
60-W bulbs	3	20
100-W bulbs	5	22
150-W bulbs	7	33

(a) How many bulbs are 60 W or defective?

SOLUTION Consider one set to be the 60-W bulbs and the other to be the defective bulbs. We are asked to find $n(60 \text{ W or defective})$. From the chart, the meaning of "or," and the fact that no bulb is to be counted twice, we have

$$n(60 \text{ W or defective}) = 3 + 20 + 5 + 7 = 35$$

without having to remember a formula.

Of course, the inclusion–exclusion counting formula gives the same result when we realize that three bulbs lie in both sets:

$$n(60 \text{ W or defective}) = n(60 \text{ W}) + n(\text{defective}) - n(60 \text{ W and defective})$$
$$= 23 + 15 - 3 = 35.$$

(b) Find $n(60 \text{ W or } 150 \text{ W})$.

SOLUTION Since no bulb can be both 60 W and 150 W,

$$n(60 \text{ W or } 150 \text{ W}) = n(60 \text{ W}) + n(150 \text{ W})$$
$$= 23 + 40 = 63. \quad \blacksquare$$

Occasionally, it is more efficient to count what we do *not* want, rather than to count what we want. A technique for doing that is outlined in our last example of this section.

EXAMPLE 7

Counting What We Do Not Want

One hundred Fleetmart shoppers were surveyed, and the following results were obtained:

	Visit Fleetmart Less than Once per Week	Visit Fleetmart Once per Week	Visit Fleetmart More than Once per Week
Spend $50 or less per visit	21	17	15
Spend more than $50 per visit	2	41	4

Of those shoppers surveyed, how many visit the store at least once per week?

SOLUTION There are two ways to solve this problem. First, we could add together all of the numbers of shoppers in the last two columns of the table, since those columns represent the shoppers who visit Fleetmart at least once per week:

$$17 + 41 + 15 + 4 = 77.$$

However, we can solve the problem in a more efficient way by recalling that there is a total of 100 shoppers, and the only ones we do **not** wish to count are those in the first column. Thus, we can subtract the numbers of shoppers in the first column from the total number and obtain the count we wanted:

$$100 - 21 - 2 = 77.$$

We call this method "more efficient" because it requires fewer arithmetic steps. $\quad \blacksquare$

The preceding procedure is a specific application of the inclusion–exclusion principle in which we are dividing the universal set into a subset and its complement. In Example 7, the universal set U is the set of 100 Fleetmart shoppers. We are interested in the subset W of Fleetmart shoppers who visit at least once per week. Our approach uses the facts that $U = W \cup W'$ and $W \cap W' = \varnothing$. Thus, the inclusion–exclusion principle gives $n(W \cup W') = n(W) + n(W') - n(W \cap W') = n(W) + n(W') - 0$. This gives us the ability to compute the size of the desired outcome set with a single subtraction $n(W) = n(U) - n(W')$. We summarize this statement with the **complement principle.**

Important Counting Formula: The Complement Principle

For any set A and universal set U,

$$n(A) = n(U) - n(A'),$$

or equivalently,

$$n(A) = n(\text{universal set}) - n(\text{not } A).$$

We will use the complement principle repeatedly when computing the number of times that something is *not* true is easier than computing when it is true.

Exercises 7.4

In Exercises 1 through 10, shade in a Venn diagram to represent each set.

1. $(A' \cup B) \cap C$

2. $(A \cap B) \cap C'$

3. $(A \cup B)' \cap C'$

4. $A' \cap (B \cup C')$

5. $(A \cap B')'$, if $A \cap B = \varnothing$

6. $A \cup (B' \cap C)$, if $A \subseteq B$

7. $(A \cap B) \cup C'$

8. $[(A \cup B)' \cap C] - B$

9. $(A - C) \cup (B \cap C')$

10. $B - [(A \cup B)' - C]$

In Exercises 11 through 14, use a Venn diagram to verify the set equalities.

11. $(A \cup B)' = A' \cap B'$

12. $A \cap (B \cup C) = (A \cap B) \cup (A \cap C)$

13. $[A' \cap (B \cup C)] - B = (A \cup B)' \cap C$

14. $[(A \cup B) - C] - B = A - (B \cup C)$

In Exercises 15 through 20, use a Venn diagram to decide which pairs of sets are equal.

15. $A \cap B'$; $A - B$

16. $A' \cap B$; $A \cup B'$

17. $(A \cup B)'$; $A' \cap B$

18. $(A \cup B) - C$; $(A \cap C') \cup (B \cap C')$

19. $C - (A \cup B)$; $(C \cap A) \cup (C \cap B)$

20. $B - (A \cup C)$; $(B \cap A') \cup (B \cap C')$

Use the counting formula $n(A \cup B) = n(A) + n(B) - n(A \cap B)$ to answer Exercises 21 through 26.

21. Find $n(A \cup B)$ if $n(A) = 8$, $n(B) = 12$, and $n(A \cap B) = 3$.

22. Find $n(A \cup B)$ if $n(A) = 6$, $n(B) = 3$, and $A \cap B = \varnothing$.

23. Find $n(A)$ if $n(A \cup B) = 8$, $n(B) = 5$, and $n(A \cap B) = 2$.

24. Find $n(B)$ if $n(A) = 6$, $n(A \cup B) = 8$, and $n(A \cap B) = 1$.

25. Find $n(A \cap B)$ if $n(A \cup B) = 12$, $n(A) = 7$, and $n(B) = 9$.

26. Find $n(A \cap B)$ if $n(A \cup B) = 20$, $n(A) = 12$, and $n(B) = 6$.

27. **Reading Proficiency** Thirty senior students at a state university were asked about their reading proficiency in French and German. Twelve said they could read French,

5 said they could read German, and 2 said they could read both languages.

(a) Draw and label a Venn diagram to represent this survey numerically.

In (b) through (f), rewrite each question in set notation, and then answer the question, using the Venn diagram constructed in (a). How many of these students read

(b) French, but not German?

(c) French or German?

(d) At least one of these two languages?

(e) Exactly one of these two languages?

(f) Neither of these two languages?

(g) Rework Exercise 27(c), using the counting formula $n(A \cup B) = n(A) + n(B) - n(A \cap B)$.

28. Marketing Survey A food store surveyed 50 of its shoppers as to whether they had purchased bacon or steak during the past week. The results were 30 had purchased bacon, 20 had purchased steak, and 8 had purchased both bacon and steak.

(a) Draw and label a Venn diagram to numerically represent this survey.

In (b) through (f) rewrite each question in set notation, and then answer the question, using the Venn diagram constructed in (a). How many of these shoppers bought

(b) Bacon, but not steak?

(c) Bacon or steak?

(d) At least one of these two meats?

(e) Exactly one of these two meats?

(f) Neither of these two meats?

(g) Rework Exercise 28(c), using the counting formula $n(A \cup B) = n(A) + n(B) - n(A \cap B)$.

In each of the remaining exercises, use a Venn diagram to help answer the counting question.

29. Marketing Survey In a particular city, 80 households were asked these two questions:

> Q1: Do you own a Chevrolet?
>
> Q2: Do you own at least two cars?

Twenty-five households answered "yes" to Question 1, 52 answered "yes" to Question 2, and 18 answered "yes" to both questions. How many of these households

(a) Answered "yes" to at least one of these questions?

(b) Answered "yes" to exactly one of these questions?

(c) Answered "no" to both questions?

(d) Own a Chevrolet, but do not own more than one car?

30. Demographic Survey In a particular school district, 90 families were asked these two questions:

> Q1: Do you have children attending public kindergarten?
>
> Q2: Do you have children in grades 1 through 5 attending public school?

Thirty answered "yes" to Question 1, 50 answered "yes" to Question 2, and 10 answered "yes" to both questions.

(a) How many answered "yes" to at least one of these questions?

(b) How many answered "no" to both questions?

(c) How many have no children in grades 1 through 5 attending public school?

(d) What percentage of these families had children in public kindergarten or in public school grades 1 through 5?

31. Marketing Survey There were 150 respondents to a survey about ownership of radios and color TV sets. They were asked these questions:

> Q1: Do you own a color TV set?
>
> Q2: Do you own at least two radios?

There were 110 people who answered "yes" to Question 1, 90 who answered "yes" to Question 2, and 120 who answered "yes" to Question 1 or Question 2.

(a) How many of these people answered "yes" to both questions?

(b) How many of these people answered "no" to both questions?

(c) How many of these people answered "yes" to Question 1, but "no" to Question 2?

(d) What percentage of the respondents own at least two radios, but no color TV set?

32. Enrollment Survey In a history class, 30 students were asked about other courses they were enrolled in. Twenty said they were in math, 22 in English, and 26 in either math or English.

(a) How many students are in both math and English?

(b) How many are in neither?

(c) How many are in math, but not English?

(d) What percentage of these students are in math, but not English?

33. Marketing Survey Ninety customers of a discount store were asked about their purchases during the previous month. Twenty said they had purchased books, 45 had purchased film, 38 had purchased jewelry, 15 had purchased both books and film, 8 had purchased both books and jewelry, 6 had purchased both film and jewelry, and 3 had purchased all three articles. How many of these people had purchased

(a) Books or film, but not jewelry?

(b) Books and film, but not jewelry?

(c) Jewelry, but not books or film?

(d) At least one of these three articles?

(e) Exactly one of the three articles?

34. Subscription Survey A poll of 100 executives showed that 45 subscribe to *The Wall Street Journal,* 25 to *Business Week,* 8 to both *The Wall Street Journal* and *Barron's,* 12 to *Barron's* and *Business Week,* 6 to *The Wall Street Journal* and *Business Week,* 2 to all three publications, and 22 to none of the publications. How many of these executives

(a) Subscribe to at least one of these publications?

(b) Subscribe to *Business Week,* but not to *The Wall Street Journal* or *Barron's?*

(c) Do not subscribe to *Barron's?*

(d) Subscribe to exactly two of these publications?

(e) Subscribe to *Business Week* or *The Wall Street Journal,* but not to *Barron's?*

(f) Subscribe to *Barron's* and *The Wall Street Journal,* but not to *Business Week?*

35. Political Survey In the town of Wilmar, 150 residents were asked to answer the following questions:

Q1: Are you in favor of building the proposed East–West expressway?

Q2: Are you in favor of the loop expressway?

Q3: Are you in favor of the proposed city bond issue?

Sixty-three answered "yes" to Question 1, 82 answered "yes" to Question 3, 15 answered "yes" to Questions 1 and 2, 42 answered "yes" to Questions 1 and 3, 40 answered "yes" to Questions 2 and 3, 12 answered "yes" to all three questions, and 27 answered "no" to all three questions. From among these people, how many

(a) Answered "yes" to Question 2 or Question 3?

(b) Answered "yes" to at least one of the questions?

(c) Answered "yes" to exactly one of these questions?

(d) Favored the East–West expressway, but did not favor the other two projects?

36. Political Survey The following questions were asked of a random sample of 1000 people:

Q1: Do you favor nationalized health care?

Q2: Do you favor higher Social Security taxes?

Q3: Do you favor more reliance on private retirement systems?

Of these people, 553 answered "yes" to Question 1, 224 answered "yes" to Question 2, 438 answered "yes" to Question 3, 87 answered "yes" to Questions 1 and 2, 102 answered "yes" to Questions 1 and 3, 93 answered "yes" to Questions 2 and 3, and 50 answered "yes" to all three questions. What percentage of these people

(a) Favor nationalized health care or higher Social Security taxes?

(b) Do not favor any of these three ideas presented?

(c) Favor nationalized health care, but not increased Social Security taxes?

37. Cognition Study A psychologist took 20 monkeys and spent a week trying to teach them to ride a tricycle, another week teaching them to do chin-ups, and yet another week teaching them to shoot a basketball into a low hoop. The results were as follows:

Two could be taught to do all three activities.

Four could be taught to ride a tricycle and shoot a basketball.

Three could be taught to do chin-ups and shoot a basketball.

Five could be taught to ride a tricycle and do chin-ups.

Nine could be taught to ride a tricycle.

Eight could be taught to shoot a basketball.

Five could not be taught to do any of these activities.

How many of these monkeys were taught to

(a) Do chin-ups, but could not be taught to ride a tricycle?

(b) Do at least two of these activities?

How many could *not* be taught to

(c) Ride a tricycle or shoot a basketball?

38. Automobile Options A Chevrolet dealer has 160 new cars for sale on her lot. Among these cars,

Fifty have four-cylinder engines.

Eighty have a tilted steering wheel.

Thirty have power windows.

Forty-two have four-cylinder engines and a tilted steering wheel.

Eighteen have four-cylinder engines and power windows.

Fifteen have all three features.

Sixty-five have none of the features.

How many of these cars have

(a) A tilted steering wheel and power windows?

(b) At least one of these features?

(c) Exactly two of these features?

39. Marketing Survey A student was hired to stand near the spice section of a supermarket from 5:00 P.M. to 8:00 P.M. on a particular day to count the number of people who bought spices and to make special note of the number of people who bought cinnamon or pepper. The student reported that 60 people bought a spice of some kind, 50 bought cinnamon, 40 bought pepper, and 5 bought both cinnamon and pepper. Should the student be believed?

40. Skill Survey A plant supervisor reported that, of the 20 people being supervised, 16 could operate a lathe, 12 could operate a grinder, 8 could operate a bender, 2 could operate a lathe and a grinder, 4 could operate a lathe and

a bender, 5 could operate a grinder and a bender, and 1 could operate all three machines. Should the supervisor be believed?

41. Unemployment Unemployment data for a small town in Virginia are as follows:

	Employed	Unemployed
Male	60	10
Female	40	5

How many of these people are

(a) Male or unemployed?

(b) Female and employed?

42. Advertising Expenses The Bureau of the Census's *Statistical Abstract of the United States* for the year 2000 shows these amounts (in millions of dollars) spent on newspaper advertising:

Year	Nationally (millions of $)	Locally (millions of $)
1997	5016	36,654
1998	5402	38,840
1999	5942	40,640

According to the data, how many millions of dollars were spent on advertising

(a) During 1998 or 1997?

(b) During 1997?

(c) Nationally during 1997 or 1998?

(d) Nationally or locally in 1999?

Section 7.5 The Multiplication Principle

Knowing how to count is one of the keys to being able to compute many of the probabilities discussed in Chapter 8. To *count* means to find the number of elements in a set, even in sets that are too large to explicitly write down every element. It is the large sets (such as the set of possible licenses plates or the set of possible picks in a lottery) that drive the need for counting techniques. We have already seen that Venn diagrams may be applied to specific types of counting problems. However, other counting problems require more sophisticated techniques, some of which we will explore in the remainder of this chapter. Regardless of the counting techniques applied,

1. A counting problem always asks, "How many?"
2. The answer is always found to be one of the *whole numbers:* 0, 1, 2, 3, ...

Here are some examples of counting problems that are unlike any we have seen up to this point, but that we should be able to answer after studying this section:

How many different Social Security numbers are possible?

How many different telephone numbers can be given the area code 870?

In how many ways can a president and a vice president be selected from among 12 people?

How many ways can the sum of eight be rolled on a pair of dice?

One way to solve these or any other counting problems is to make a set of all of the objects to be counted and then simply count them. However, as you can see from the preceding examples, this technique is limited in its application because of the extremely long lists that can occur. Even so, one helpful suggestion on any counting problem is to list a few of the items to be counted so that it is clear exactly what is, and what is not, to be counted. Beyond that, we need to know techniques that will give us a count of the number in the set without actually viewing the entire set. Some examples in which we *can* write down the entire set of items to be counted may help set the stage for the techniques of this section.

EXAMPLE 1
Viewing the List and
Then Finding a Formula

A box contains three slips of paper, *a* is written on one, *b* on another, and *c* on the third. A slip is drawn, the letter on it recorded, and the slip is *not* returned to the box. A second slip is then drawn and the letter recorded. Each sequence of two letters so selected is known as an **outcome** from this activity. How many different outcomes are there?

SOLUTION The set of all possible outcomes may be listed as

$$A = \{(a, b), (a, c), (b, a), (b, c), (c, a), (c, b)\}.$$

A **tree diagram** like the one shown in Figure 13 gives a visual description of the selection process leading to the six possible outcomes. From left to right, there are six paths through the diagram, and each of these paths corresponds to the ordered pair listed at the end of the path. Since each ordered pair is an outcome, there are six possible outcomes.

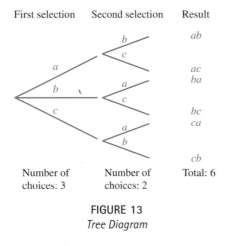

First selection	Second selection	Result

Number of
choices: 3

Number of
choices: 2

Total: 6

FIGURE 13
Tree Diagram

The tree diagram helps explain how we may obtain the total number of outcomes in terms of the number of choices available for each of the two successive draws: there are three choices of slips for the first draw and two choices of slips for the second draw. (The first slip was not returned to the box.) Thus,

$$n(A) = (\text{Number of Choices on First Draw}) \times (\text{Number of Choices on Second Draw})$$
$$= 3 \times 2 = 6$$

gives the correct result. Note that the product correctly counts the interchange of two letters in an outcome as a new and different outcome; for example, (a, b) and (b, a) are different outcomes. Thus, the *order* in which the letters appear is *important*. We say that each outcome is an **ordered arrangement** of the elements in the set. ▄

In Example 1, each selection comes from the same set $\{a, b, c\}$, but once a letter is removed, we are not allowed to use it again in future selections. This procedure is known as selection **without replacement** or selection when **no repetitions are allowed.** In the next example, we will consider the possibility of **repeating** a selection, and later we will deal with sets with several elements that are **indistinguishable.**

EXAMPLE 2
Viewing the List and
Then Finding a Formula

A coin is tossed three times. An **outcome** from this activity is an ordered sequence of three symbols: a T for tails or H for heads to represent the result of the first toss, a T or H to represent the result of the second toss, and a T or H to represent the result of the third toss. How many outcomes are possible?

SOLUTION Recording the possibilities for successive tosses of the coin gives this set:

$$A = \{HHH, HHT, HTH, THH, TTH, THT, HTT, TTT\}.$$

Therefore, there are a total of eight possible outcomes for this activity. As in Example 1, we may obtain this total by thinking in terms of the number of choices available for each of the three tosses making up an outcome: There are two choices (H or T) for the first toss, two choices for the second toss, and two for the third toss, from which the product

$$n(A) = (\text{Number of Choices on First Toss}) \times (\text{Number of Choices on Second Toss})$$
$$\times (\text{Number of Choices on Third Toss})$$
$$= 2 \times 2 \times 2 = 8$$

gives the correct result. Note that the product correctly counts the interchange of two different letters in an outcome as a new and different outcome. ▄

In Example 2, we may view the selections as coming from the set $\{H, T\}$ of two different objects, an H and a T, and, once a letter is selected, it is returned to the set for possible selection again in the same outcome. (For example, *HHT* or *TTT*.) We refer to this as selection **with replacement** or selections when **repetitions are allowed.**

EXAMPLE 3
Viewing the List and
Then Finding a Formula

A certain model of hair dryer has three blower-speed settings—B_1, B_2, and B_3—and two heat settings—H_1 and H_2. How many settings does this dryer have for styling purposes?

SOLUTION The possible settings (outcomes) can easily be listed:

$$B_1H_1 \qquad B_2H_1 \qquad B_3H_1$$
$$B_1H_2 \qquad B_2H_2 \qquad B_3H_2$$

The list tells us that there are six settings for styling purposes. Why do we not put H_1B_1 (and so on) in the list? Because it is *not* a different dryer setting. That is, changing the order in which the individual settings are selected does *not* result in a new dryer setting. Even so, the total number may be obtained by multiplying the respective number of choices, as was done in Examples 1 and 2:

(Number of Choices for Blower Settings) $\times$ (Number of Choices for Heat Settings)
$$= 3 \times 2 = 6. \quad \blacksquare$$

Think of making selections from two *different sets:* the blower speeds in one set and the heat settings in another set. Select from the set of blower speeds first, and then select from the set of heat settings. The product $3 \cdot 2 = 6$ gives all possible settings for selections made in that order. A tree diagram for the outcome set of this selection process is shown in Figure 14. Notice that the leading branches of the tree are labeled with blower settings *only,* reflecting a selection from just *one* of the two sets. The attached branches reflect selections made from a *different* set, namely, the heat settings. A path along two consecutive branches results in the blower setting shown at the end of the second branch. Thus, following along any one path, from beginning to end, gives exactly one element of the outcome set.

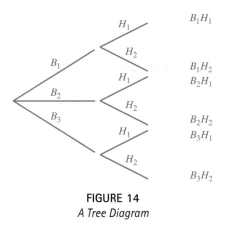

FIGURE 14
A Tree Diagram

Our tree could have been drawn by depicting heat selections first and blower settings second, still giving $2 \cdot 3 = 6$ different paths or dryer settings.

The **multiplication principle** demonstrated in Examples 1, 2, and 3 is a very general and powerful counting technique. It counts *all possible distinct ordered outcomes,* each consisting of successive selections from one or more sets.

> ## The Multiplication Principle
>
> If an outcome consists of k successive selections with n_1 choices for the first selection, n_2 choices for the second selection, . . . , and n_k choices for the kth selection, then the size of the set A of distinct ordered outcomes is $n(A) = n_1 \cdot n_2 \cdot n_3 \cdot \cdots \cdot n_k$.

NOTE Sets are always considered to have different-appearing, or **distinguishable,** elements, unless otherwise noted. Thus, the two a's in $\{a, a, b, c\}$ are considered **indistinguishable** (more on this in the next two sections). Direct application of the multiplication principle *will not* correctly count all ordered outcomes in sets with indistinguishable elements.

EXAMPLE 4

Using the Multiplication Principle

How many telephone numbers can be put on a 555-*xxxx* exchange?

SOLUTION We have the set of ten digits—0, 1, 2, 3, . . . , 9—from which to choose. Because a telephone number may have repeated digits, we assume that the selection process takes place with replacement; that is, once a digit is selected, we still may use it again. The multiplication principle tells us that the total number of four-digit numbers possible is

(Number of Choices for First Digit) $\times$ (Number of Choices for Second Digit)

$\times$ (Number of Choices for Third Digit) $\times$ (Number of Choices for Fourth Digit)

$$= 10 \times 10 \times 10 \times 10 = 10{,}000. \quad\blacksquare$$

EXAMPLE 5

Using the Multiplication Principle

How many four-digit personal identification numbers (like those used for ATM cards) are possible if zero cannot be used as the first digit and no digit may be repeated?

SOLUTION Because no repetition of digits in the identification number is allowed, the selection process is one without replacement. Since the first digit cannot be zero, there are nine choices for that digit. For the second digit, we may now use zero or any digit not used in the first selection; accordingly, there are nine choices for the second digit. For the third digit, there are eight choices; for the fourth digit, seven choices. The multiplication principle then tells us that the total number of different personal identification numbers is

(Number of Choices for First Digit) $\times$ (Number of Choices for Second Digit)

$\times$ (Number of Choices for Third Digit) $\times$ (Number of Choices for Fourth Digit)

$$= 9 \times 9 \times 8 \times 7 = 4536. \quad\blacksquare$$

Up to this point, all of our selections that went into forming an outcome have rather naturally gone from left to right. The multiplication principle does not require that any particular order be followed, however. We are allowed to start in any digital slot that is most convenient.

EXAMPLE 6

Using the Multiplication Principle

Three different novels and two different technical manuals are to be arranged on a shelf. In how many ways can they be arranged if a novel is to occupy the middle position?

SOLUTION It is most convenient to select a novel for the middle position first, and there are three choices from which to do so. We then may select from the four remaining books to occupy any one of the four remaining positions. Once selected and put on the shelf, a book cannot be used again in the selection process; that is, we are selecting without replacement. Here are the number of choices involved:

$$(\text{Second Selection}) \times (\text{Third Selection}) \times (\text{First Selection})$$
$$\times (\text{Fourth Selection}) \times (\text{Fifth Selection})$$
$$= 4 \times 3 \times 3 \times 2 \times 1 = 72. \blacksquare$$

EXAMPLE 7

Using the Multiplication Principle

Three boys and two girls are to be seated in a row of chairs to have a picture taken. In how many ways can they be seated for the picture

(a) If there are no restrictions?

SOLUTION Think of five slots in a row, each representing a chair, and then start denoting the successive choices for the number of people available for the respective chairs. (We may start this process at any position; however, because we normally read from left to right, it is customary to begin with the position on the left.) There are five choices for seating a person in the leftmost chair. Once the person on the left is seated, there are four choices remaining for the second seat (once seated, a person cannot be seated again), and so forth. Therefore, there are

$$5 \cdot 4 \cdot 3 \cdot 2 \cdot 1 = 120$$

possible seating arrangements (outcomes).

(b) If boys are to sit in the end chairs?

SOLUTION There are three choices for seating a boy in the left end seat and then two choices for seating a boy in the right end seat. The remaining three people may be seated with no restrictions. Thus, there are $3 \cdot 3 \cdot 2 \cdot 1 \cdot 2 = 36$ different ways. Figure 15 below may give some help.

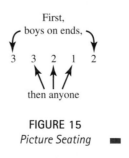

FIGURE 15
Picture Seating ■

EXAMPLE 8

Using the Counting Principle with Inclusion–Exclusion

Three boys and two girls are to be seated in a row. In how many different ways can this be done

(a) If the boys are to sit next to each other and the girls are to sit next to each other?

SOLUTION To meet the seating conditions of the problem,

> A: The boys could be seated in the first three seats on the left, followed by the girls on the right, *or*

> B: The girls could be seated in the first two seats on the left, followed by the boys on the right.

Our outcome set is $C = A \cup B$. Recall that the inclusion–exclusion formula for counting may be stated as "*or* means add, but don't double-count any outcome." In this case, no arrangement (outcome) in A is found in B, leading us to conclude that there will be no double-count when the number of possibilities for A are added to those for B. In symbols,

$$n(C) = n(A \cup B) = n(A) + n(B) - n(A \cap B) = n(A) + n(B).$$

Therefore, the total number of arrangements (outcomes) is

$$n(C) = (3 \cdot 2 \cdot 1 \cdot 2 \cdot 1) + (2 \cdot 1 \cdot 3 \cdot 2 \cdot 1) = 2(3 \cdot 2 \cdot 1 \cdot 2 \cdot 1) = 24.$$

Figure 16 shows how we arrive at this product.

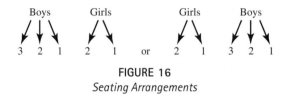

FIGURE 16
Seating Arrangements

(b) If a boy is seated on the left end or a girl is seated on the right end?

SOLUTION In some seating arrangements, a boy is seated on the left and a girl on the right. By the inclusion–exclusion principle, we must be careful not to double-count any arrangements. Therefore,

$n(\text{boy on left } or \text{ girl on right}) = n(\text{boy on left}) + n(\text{girl on right}) - n(\text{boy on left } and \text{ girl on right})$

Boy first,   Girl first,   Boy first, girl second,

$= 3 \cdot 4 \cdot 3 \cdot 2 \cdot 1 \quad + \quad 4 \cdot 3 \cdot 2 \cdot 1 \cdot 2 \quad - \quad 3 \cdot 3 \cdot 2 \cdot 1 \cdot 2$

then anyone   then anyone   then anyone

$= 72 + 48 - 36 = 84.$ ∎

EXAMPLE 9

Using the Multiplication Principle

An experiment in a psychology laboratory involves luring a rat through a maze from A to D, with various enticements in each tunnel. How many different paths (outcomes) can the rat take through the maze, shown in Figure 17?

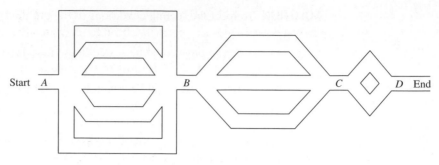

FIGURE 17
Paths through a Maze

SOLUTION Think of the maze as consisting of three different sets of paths: one set from *A* to *B*, one set from *B* to *C*, and one set from *C* to *D*. We have five choices from the first set of paths, three choices from the second set of paths, and two choices from the third set of paths. The multiplication principle then tells us that there are

$$5 \cdot 3 \cdot 2 = 30$$

different paths through the maze. Observe that this product does *not* count (and correctly so) the paths from right to left, because they are the same as the paths from left to right. This is so because we are making our selections from different sets. ▬

EXAMPLE 10

Using the Multiplication Principle

Suppose that a license plate is made up of two letters followed by five digits. How many different such plates can be made?

SOLUTION Because no restrictions are given, we assume that license plates such as AA–00000 are permitted. We are to select letters first and then digits. Think of the 26 letters in the alphabet in a first box (set) and the ten digits 0, 1, 2, 3, 4, 5, 6, 7, 8, 9 in a second box.

There are $26 \cdot 26$ ways of selecting two letters, with replacement, from the first box. For each of these ways, there are $10 \cdot 10 \cdot 10 \cdot 10 \cdot 10$ ways of selecting five digits, with replacement, from the second box. The multiplication principle states that the total number of different license plates is the product of $26 \cdot 26$ and $10 \cdot 10 \cdot 10 \cdot 10 \cdot 10$, or

$$26 \cdot 26 \cdot 10 \cdot 10 \cdot 10 \cdot 10 \cdot 10 = 67{,}600{,}000. \quad ▬$$

It is *very important* to recognize that the multiplication principle counts as a new outcome the interchange of two different elements in a given outcome when selections are made from the same set. This knowledge is used throughout the remainder of the chapter and carries forward into our study of probability.

Our final example of the section demonstrates combining the multiplication principle with the complement principle.

EXAMPLE 11

Counting What We Do Not Want

Suppose that seven-digit telephone numbers are to be created so that all of the numbers contain at least one zero. How many such telephone numbers can be created?

SOLUTION To solve this problem, we begin by counting the size of the universal set of all possible seven-digit telephone numbers. Since there are 10 possible digits,

$$n(\text{Telephone Numbers}) = 10 \cdot 10 \cdot 10 \cdot 10 \cdot 10 \cdot 10 \cdot 10 = 10{,}000{,}000.$$

That is, there are 10 million different possible seven-digit telephone numbers. Now count how many of those we *do not* want; in other words, count how many telephone numbers do not contain the digit 0. (Notice that this set is the complement of the set of telephone numbers that contain at least one zero). Since 0 is excluded from the possible choices of digits, we see that there are nine choices for each digit in the telephone number. Thus,

$$n(\text{Telephone Numbers That Do Not Contain Zero}) = 9 \cdot 9 \cdot 9 \cdot 9 \cdot 9 \cdot 9 \cdot 9 = 4{,}782{,}969$$

different telephone numbers can be created without using the digit 0. If we now apply the complement principle, we will have the count we want:

$$n(\text{Telephone Numbers that Contain at Least One 0})$$
$$= n(\text{Telephone Numbers}) - n(\text{Telephone Numbers We Do Not Want})$$
$$= 10{,}000{,}000 - 4{,}782{,}969 = 5{,}217{,}031.$$

There are 5,217,031 possible seven-digit telephone numbers that contain at least one zero. ∎

Exercises 7.5

How many possible outcomes are there for each of the activities described in Exercises 1 through 4?

1. Roll a single die.

2. Draw one card from a standard deck of cards.

3. Record a score on a 10-point quiz.

4. Have a birthday in July.

In Exercises 5 through 11, write down the entire list to be counted, and then count the items in the list. Assume that there are no restrictions unless indicated.

5. A coin is tossed four times. How many possible outcomes are there?

6. Two-digit numbers are to be made up from the digits 2, 3, and 4. How many such numbers can be made if repetition of digits is allowed?

7. Two-digit numbers are to be made up from the digits 0, 1, 2, and 3. How many such numbers can be made if 0 cannot be the first digit and repetition of digits is allowed?

8. Each of three slips of paper has just the number 1, 2, or 3 written on it. A slip is drawn, the number recorded, and the slip laid aside. Now a second slip is drawn and the number recorded. How many distinct outcomes are there?

9. A secretary types letters to three people and then types the corresponding envelopes. In how many ways can the letters be put in the envelopes so that exactly two are in the wrong envelopes?

10. If there are three roads from Goshen to Gravette and two roads from Gravette to Tontitown, how many ways are there to go from Goshen to Tontitown through Gravette?

11. Botany Experiment A botanist wants to test three varieties of soybeans under two levels of fertilizer and two types of insecticides. How many experimental plots are needed?

In the remaining exercises, use any appropriate technique that has been presented to obtain the count requested.

12. Electing Officers In how many ways can a president, a secretary, and a treasurer be elected from among 10 people?

13. Social Security Numbers How many Social Security numbers are there if

(a) There are no restrictions?

(b) No number can begin with zero?

(c) Neither of the first two digits can be zero?

14. Telephone Numbers How many seven-digit telephone numbers can be put on the 501 area code if

(a) There are no restrictions?

(b) No telephone number can begin with zero?

(c) No telephone number can begin with zero and the last digit must be odd?

15. In how many ways can five boys and three girls be seated in a row if

(a) There are no restrictions?

(b) Boys and girls are seated alternately?

(c) Girls must be seated in both end seats?

16. In how many ways can five boys and three girls be seated in a row if

(a) Boys and girls are seated alternately?

(b) Boys sit together and girls sit together?

(c) One of the girls, Sue, must be seated on the left end?

17. In how many ways can eight books be arranged on a shelf if

(a) There are no restrictions?

(b) One of the books, *Of Mice and Men,* must be displayed on the left end?

18. How many three-digit numbers can be constructed from the digits 2, 3, 4, 5, 6, and 7 if

(a) There are no restrictions?

(b) The numbers constructed must all be even?

(c) The numbers constructed must all be even and no digit may be repeated?

19. How many four-digit numbers can be constructed if

(a) There are no restrictions? (Assume that 0 can be the first digit.)

(b) Zero cannot be the first digit and no digit can be repeated?

(c) The first digit is even, no digit can be repeated, and each number constructed must be odd?

20. Class Schedules If three subjects are available for an 8 A.M. class and five other subjects for a 10 A.M. class, how many class schedules are possible for these two classes?

21. A coin is tossed twice. How many possible outcomes are there? How many if the coin is tossed five times? Seven times?

22. Serial numbers assigned to calculators by a manufacturer have J, H, or T as a first symbol, to indicate the plant in which the calculator was made. There then follow one of the numbers 01, 02, . . . , 12, to indicate the month in which the calculator was made, followed in turn by four digits. How many different serial numbers are possible?

23. If a firm has 600 employees, what is the smallest number of digits that an employee I.D. number can have if each I.D. number has x digits? What is the smallest number if the firm has 12,000 employees?

24. In how many ways can four of nine monkeys be arranged in a row for a genetics experiment?

25. Quiz Responses In how many ways can a 10-question true–false quiz be answered?

26. Test Responses In how many ways can a 10-question multiple-choice test be answered if there are four choices for each question?

27. In how many ways can a couple select a bathtub if a tub is made of one of three materials and each material comes in two colors?

28. How many different car license plates can be made using three letters followed by three digits if

(a) There are no restrictions?

(b) No letter can be repeated?

(c) No letter can be repeated and zero cannot be used as the first digit?

29. How many different car license plates can be made using two letters followed by four digits if

(a) There are no restrictions?

(b) No letter can be repeated?

(c) The last digit must be a 4 and no digit can be repeated?

30. Radio Call Letters How many four-letter radio station call letters are there if

(a) Each must begin with a K or a W?

(b) Each must begin with a K or a W and no letter may be repeated?

(c) Each must begin with a K or a W, no letter may be repeated, and the last letter must be a Z?

31. Radio Call Letters How many four-symbol radio station call signs are there that use either all letters or one letter followed by three digits if

(a) The first letter must be a K or a W?

(b) The first letter must be a K or a W and no digit is repeated?

(c) The first letter must be a K or a W and no letter or digit may be repeated?

32. A microwave oven dealer has in stock three models, each in five colors. In how many ways can a customer select a microwave oven to purchase?

33. **Race Results** Thirty runners enter a race in which first-, second-, third-, and fourth-place winners are recognized. In how many ways can these runners place in these four positions?

34. How many three-letter codes are there if

(a) There are no restrictions?

(b) The letter z may not be used as the first letter?

(c) The letter z may not be used as the first letter and no letter may be repeated?

35. In how many ways can a 40-member football team select a captain and a cocaptain?

36. **Batting Orders** The manager of an amateur baseball team has selected the nine players for an upcoming game. How many batting orders are possible if

(a) There are no restrictions?

(b) The pitcher must be last, the first baseman must bat sixth, and the catcher must bat second?

37. In how many different ways can four of eight people be seated in four chairs?

38. In how many different ways can five of nine people be seated in five chairs?

39. A coin is tossed three times. How many outcomes have

(a) A head on the first or third toss?

(b) A head on the first toss or a tail on the second toss?

40. Three-letter code words are to be made from the letters a, b, c, and d, with no repetition of letters allowed. How many three-letter code words are possible if

(a) The first letter is an a and the last letter a b?

(b) The first letter is an a or the last letter a b?

41. Four-digit identification numbers are being made, with repetition of digits within the numbers allowed. How many such identification numbers are possible if

(a) Each must begin with a zero and end with an eight?

(b) Each must begin with a zero or end with an eight?

42. **Computer Programming** In certain programming languages, an identifier is a sequence of characters in which the first character must be a letter in the English alphabet and the remaining characters may be either a letter or a digit. For identifiers of length five, how many

(a) Begin with an h and end with a 6?

(b) Begin with an h or end with a 6?

43. From among eight computers, how many samples of three can be selected

(a) Without replacement, where the order of selection is important?

(b) With replacement, where the order of selection is important?

44. From among 10 computer disks, how many samples of 4 can be selected

(a) Without replacement, where the order of selection is important?

(b) With replacement, where the order of selection is important?

Section 7.6 Permutations

Counting problems sometimes involve the product of consecutive positive integers, beginning with 1 and continuing to some fixed integer n. To save massive amounts of writing when n is large, the mathematical shorthand notation **$n!$,** read **"n factorial,"** is used to denote such a product. For example, $4! = 1 \cdot 2 \cdot 3 \cdot 4 = 24$, and $6! = 1 \cdot 2 \cdot 3 \cdot 4 \cdot 5 \cdot 6 = 720$.

n Factorial

Let n be a positive integer. Then the product of the integers from 1 through n is denoted by $n!$ and read "n factorial." That is,

$$1 \cdot 2 \cdot 3 \cdot \cdots \cdot n = n!$$

By definition, $0! = 1$.

EXAMPLE 1

Using Factorial Notation

In how many ways can eight children be seated in a row for a school picture?

SOLUTION According to the multiplication principle, the answer is

$$8 \cdot 7 \cdot 6 \cdot 5 \cdot 4 \cdot 3 \cdot 2 \cdot 1 = 8!$$
$$= 40,320. \quad \blacksquare$$

In Example 1, when two children in the row exchange seats, the resulting new arrangement is said to be a **permutation** of the former arrangement. In that sense, our interpretation of the word "permutation" is the same as the dictionary meaning: To *permute* is to *interchange*. In general, then, we may think of successively selecting a first object, then a second object, then a third object, etc., as giving an **ordered outcome,** whereas the selection of these same objects in a different order gives a permutation of the first outcome. Stated differently, a **permutation** is any **ordered arrangement of objects on a line,** and any reordering of those objects on the line is a *new permutation.* Accordingly, one counting problem would be to count the number of different permutations of a given set of objects. In Example 1, for instance, there are 8! = 40,320 different permutations, or different orderings, of the eight children when all are seated in a row.

In carrying these thoughts further, we might ask about the number of permutations of the letters in a word—for example, "batter." Some of the permutations are batter, abttre, and tarebt. Notice that, for any ordered arrangement, the interchange of the two t's *does not* give a new permutation. We say that the two t's are *indistinguishable.* This is in contrast to Example 1 involving children: People are, of course, *distinguishable.* Evidently, the counting process for determining the number of permutations in which indistinguishable objects are present will be different from that in which all objects are distinguishable. It seems logical to begin with the latter case, in which all objects are distinguishable, because the multiplication principle will count the number of permutations as it did in Example 1. In fact, we take several aspects of Example 1 to begin our study of permutations.

All Distinguishable Objects

Selection Rules for Permutations

1. All objects are selected from the *same set S.*
2. All objects in S are considered *distinguishable.*
3. Successive selections from S are made *without replacement.*
4. Successive selections are considered to be an ordered arrangement.

In selecting elements (objects) from a given set S, it is not necessary that *all* objects in the set be in the ordered arrangement that serves as a permutation. Example 2 illustrates this idea with a set of letters from which the permutations can actually be written down and then counted.

EXAMPLE 2

Counting Permutations

Find the number of permutations of the letters in the set $S = \{a, b, c, d\}$, when selected two at a time.

SOLUTION In this case, the set A of the ordered arrangements of pairs of elements from S is

$$A = \{ab, ba, ac, ca, ad, da,$$
$$bc, cb, bd, db,$$
$$cd, dc\}.$$

There are 12 such permutations, so we say "There are 12 permutations of four things taken two at a time," and we use the notation $P(4, 2)$ to denote this fact.

The multiplication principle gives this same result: There are four choices for the first selection, and, for each of these, there are three choices for the second selection. Therefore, $n(A) = P(4, 2) = 4 \cdot 3 = 12$. ■

EXAMPLE 3

Computing the Number of Permutations

In how many ways can the positions of nursing coordinator, insurance benefits coordinator, and records coordinator at Alamance Regional Medical Center be filled from among six internal candidates, all of whom are qualified to fill any of the three positions?

SOLUTION The positions must be filled in some order. Suppose that the nursing coordinator is chosen first, then the insurance benefits coordinator, and, finally the records coordinator. The order in which people are selected for these positions is important in the sense that if Otto is selected for nursing and Carlos for insurance, then the selection of Carlos for nursing and Otto for insurance is a new outcome of selections, or a permutation of the first outcome. Of course, once a person is selected for a particular position, he or she cannot be put back into the pool of people eligible for the remaining positions. The fact that we are to select three people successively, without replacement, from a set of six people in which the order of selection is important provides the conditions for a permutation. Therefore,

$$(\text{Choices for Nursing}) \times (\text{Choices for Insurance}) \times (\text{Choices for Records})$$
$$= 6 \times 5 \times 4 = 120.$$

This is the same as asking for the number of ordered arrangements of six people taken three at a time. In the language of permutations, we are saying "the number of permutations of six things taken three at a time" is 120; this count is denoted by $P(6, 3)$. ■

Permutations

Let S be a set containing n distinguishable objects. Any *ordered arrangement* of r of these objects, where $0 \le r \le n$, obtained by successive selections from S *without replacement* is called a **permutation**. In such a case, we say that we have a **permutation of n things taken r at a time.** The total number of such permutations is denoted by $P(n, r)$. (Calculators frequently use the notation $_nP_r$ or nPr.

Graphing calculators have factorial and permutation commands. On the TI-83 Plus, the factorial command "!" and permutation command nPr are found under the PRB submenu of the MATH menu. As for spreadsheets, Excel has the factorial function FACT in the Math & Trig function category and the permutation function PERMUT under the Statistical function category.

EXAMPLE 4

Computing the Number of Permutations

Compute $P(6, 4)$.

SOLUTION We think of successively selecting (without replacement) four elements from a set of six elements. There are six choices for the first selection, five choices for the second, four for the third, and three for the fourth. That is,

$$6 \cdot 5 \cdot 4 \cdot 3 = 360 = P(6, 4). \quad \blacksquare$$

EXAMPLE 5

Computing the Number of Permutations

The order of administering five drugs to a recuperating patient is important. In how many ways can all five of the drugs be administered?

SOLUTION Think of a set of five drugs from which we are to select five successively without replacement. The total number of ways this may be done is

$$P(5, 5) = 5 \cdot 4 \cdot 3 \cdot 2 \cdot 1 = 5! = 120.$$

This is also the number of possible ordered arrangements of a set of five drugs. $\blacksquare$

Even though we have seen that the multiplication principle will solve any permutation problem, there is a special "permutations formula" that will do the same thing. This formula may be obtained by generalizing the observation that

$$P(7, 3) = 7 \cdot 6 \cdot 5$$
$$= 7 \cdot 6 \cdot 5 \cdot \frac{4 \cdot 3 \cdot 2 \cdot 1}{4 \cdot 3 \cdot 2 \cdot 1} = \frac{7!}{4!} = \frac{7!}{(7 - 3)!}.$$

In general, the multiplication principle states that

$$P(n, r) = n \cdot (n - 1) \cdot (n - 2) \cdot \cdots \cdot (n - r + 1),$$

which may be rewritten as

$$P(n, r) = n \cdot (n - 1) \cdot (n - 2) \cdot \cdots \cdot (n - r + 1) \cdot \frac{(n - r) \cdot \cdots \cdot 2 \cdot 1}{(n - r) \cdot \cdots \cdot 2 \cdot 1}$$
$$= \frac{n!}{(n - r)!}.$$

The Number of Permutations of _n_ Distinguishable Objects Taken _r_ at a Time

$$P(n, r) = n \cdot (n - 1) \cdot (n - 2) \cdot \cdots \cdot (n - r + 1)$$
$$= \frac{n!}{(n - r)!}.$$

Because 0! is defined to be 1, this formula is also correct when $r = 0$ or $r = n$.

EXAMPLE 6

The Number of
Permutations (by the
Formula)

Calculate $P(9, 4)$ by using the permutations formula.

SOLUTION $P(9, 4) = \dfrac{9!}{(9-4)!} = \dfrac{9!}{5!} = \dfrac{9 \cdot 8 \cdot 7 \cdot 6 \cdot 5 \cdot 4 \cdot 3 \cdot 2 \cdot 1}{5 \cdot 4 \cdot 3 \cdot 2 \cdot 1}$

$\qquad\qquad = 9 \cdot 8 \cdot 7 \cdot 6$

$\qquad\qquad = 3024.$ ∎

We summarize the link between the multiplication principle and permutations by recognizing that counting permutations of distinguishable objects may be done by a straightforward use of the multiplication principle, as well as with the formula we have developed. However, not every multiplication principle problem is a permutation problem. (See Example 3 of Section 7.5, for instance.)

Mixture of Distinguishable and Indistinguishable Objects

Suppose we now consider a set of n elements, not all of which are distinguishable, and ask about the number of permutations possible when the elements are selected n at a time.

EXAMPLE 7

Not All Objects to Be
Selected Are
Distinguishable

How many distinguishable permutations of the letters $\{a, a, a, b,$ and $c\}$ are there when taken five at a time?

SOLUTION Consider any particular arrangement of these five letters, such as a, b, a, a, c. Now, for the moment, consider the letters a to be distinguishable so that they can be labeled a_1, a_2, and a_3. Then there are $3 \cdot 2 \cdot 1 = 3! = 6$ arrangements in the distinguishable case that are not in the indistinguishable case:

Indistinguishable	Distinguishable
a, b, a, a, c	a_1, b, a_2, a_3, c
	a_1, b, a_3, a_2, c
	a_2, b, a_1, a_3, c
	a_2, b, a_3, a_1, c
	a_3, b, a_1, a_2, c
	a_3, b, a_2, a_1, c

Let x be the total number of permutations that may be obtained when the a's are considered indistinguishable. We have just discovered that, for each indistinguishable arrangement, there are $3!$ distinguishable arrangements. The multiplication principle now tells us that there must be a total of $3! \cdot x$ permutations when the a's are considered distinguishable. But we also know that there are $5 \cdot 4 \cdot 3 \cdot 2 \cdot 1 = 5!$ arrangements of $\{a_1, a_2, a_3, b, c\}$ in the distinguishable case. Therefore,

$$3! \cdot x = 5!$$

$$x = \frac{5!}{3!} = 20.$$ ∎

After one more example, the general pattern for permutations with indistinguishable objects will emerge.

EXAMPLE 8

Not All Objects to Be
Selected Are
Distinguishable

Find the number of distinguishable permutations of the letters $\{a, a, a, b, b, b, b, c, c, d\}$, when taken 10 at a time.

SOLUTION Again, let x be the number of such permutations. Reasoning as we did in Example 7, there are $3! \cdot x$ permutations if we replace the a's by a_1, a_2, and a_3. Likewise, there are $(4!)(3!)x$ permutations if we then replace the b's by b_1, b_2, b_3, and b_4, corresponding to the number of ways in which b_1, b_2, b_3, b_4, can be arranged. Continuing, we find that there are $(2!)(4!)(3!)x$ permutations if we also replace the c's by c_1 and c_2. But we know that there are 10! permutations of $\{a_1, a_2, a_3, b_1, b_2, b_3, b_4, c_1, c_2, d\}$. Therefore,

$$(2!)(4!)(3!)x = 10!$$

or

$$x = \frac{10!}{2!\ 4!\ 3!}. \quad \blacksquare$$

The pattern emerging from these example leads to the conclusion that whenever there are k indistinguishable objects in the set from which selections are to be made, a division by $k!$ needs to be made to count only the different permutations. We state this result in more specific terms next.

Permutations with Indistinguishable Objects or Permutations with Repetitions

Let S be a set with n elements (objects).

Suppose that a subset containing k_1 of these elements is indistinguishable, a different subset containing k_2 elements is also indistinguishable, a different subset containing k_3 elements is also indistinguishable, and so on, until, finally, a different subset containing k_m elements is also indistinguishable. Then the number of distinguishable permutations of these n elements taken n at a time is

$$\frac{n!}{(k_1!)(k_2!)(k_3!)\cdots(k_m!)}.$$

EXAMPLE 9

Permutations with
Indistinguishable
Objects

Find the number of permutations (ordered linear arrangements) of the letters in the word "TALLAHASSEE."

SOLUTION There are 11 letters in this word, and we consider arrangements of them 11 at a time. Following the patterns just discussed, we get

$$\frac{11!}{3!\ 2!\ 2!\ 2!\ 1!\ 1!} = 831,600. \quad \blacksquare$$

EXAMPLE 10

Permutations with
Indistinguishable
Objects

A box contains five red indistinguishable marbles and three green indistinguishable marbles. The marbles are to be successively selected and arranged in a line. How many such distinguishable linear arrangements are there?

SOLUTION The answer is

$$\frac{8!}{5!\,3!} = 56. \; \blacksquare$$

As we have done previously, we will end this section with an example to demonstrate how counting what we do not want can lead us to faster solutions to some counting problems. Again, we use the complement principle—the technique of computing the size of the complement of a set—and subtract this from the size of the universal set to obtain the size of the set in question.

EXAMPLE 11

Counting What We Do
Not Want

The Oxford Accounting Firm employs five female and three male CPA's. Three of the CPA's are to be named director, associate director, and assistant director for a new project. How many ways can this be done if there must be at least one male CPA assigned to one of these positions?

SOLUTION We begin by defining the following outcome sets:

U: The set of all possible assignments.

M: The set of all assignments that have at least one male.

M': The set of all assignments that have no males.

Next, we count the size of U, the set of all possible outcomes in which these three positions can be assigned from the CPA's in the firm. Since there are eight CPA's, there are

$$n(U) = P(8,3) = \frac{8!}{5!} = 336$$

different possible assignments with no restrictions. Next, we count the number of assignments in M', the set of assignments that have no males. This is the same as counting the number of assignments that can be made from the subset of female CPA's. Since there are five female CPA's, there are

$$n(M') = P(5,3) = \frac{5!}{2!} = 60$$

ways to assign these three positions to only female CPA's. Using the two numbers obtained, we can count the number of assignments that have some male CPA's included:

$$n(M) = n(U) - n(M') = 336 - 60 = 276.$$

Thus, there are 276 different ways to assign the three positions so that at least one male CPA has been assigned. $\blacksquare$

Exercises 7.6

Do the calculations required in Exercises 1 through 24, and simplify your answer to a counting number.

1. $6!$

2. $4!$

3. $8!$

4. $\dfrac{6!}{3!}$

5. $\dfrac{8!}{4!}$

6. $\dfrac{6!}{4!\,2!}$

7. $\dfrac{12!}{7!\,5!}$

8. $\dfrac{51!}{48!\,3!}$

9. $\dfrac{52!}{47!\,5!}$

10. $(5!)^2$

11. $(2!)!$

12. $(3!)!$

13. $P(8, 4)$

14. $P(12, 3)$

15. $P(6, 1)$

16. $P(10, 5)$

17. $P(7, 7)$

18. $P(5, 0)$

19. $P(10, 0)$

20. $P(12, 12)$

21. $P(4, 3)$

22. $P(4, 1)$

23. $P(0, 0)$

24. $(P(6, 2))^2$

In Exercises 25 through 28,
(a) Restate each in permutation notation.
(b) Solve the problem.

25. In how many ways can four of eight books be arranged on a shelf?

26. **Electing Officers** In how many ways can 3 of 10 people be elected to the positions of president, secretary, and treasurer?

27. In how many ways can seven monkeys be arranged in a row for a poster picture?

28. How many arrangements are there of the letters in the word *computer?*

29. **Ballot Construction** In how many ways can five candidates for an office be listed on a ballot?

30. **Election Outcome** A state political convention is selecting nominees for the office of governor and lieutenant governor from among six candidates. A publicity firm has been hired to design campaign buttons for each possible outcome before the convention. How many different buttons must be designed?

31. **Electing Officers** The eight-person board of directors of the Triangle Corporation is to elect a president, a vice president, and a treasurer. In how many ways can these officers be elected?

32. **Committee Assignments** The 12-person executive committee of the Flying Ace Corporation is to select an environmental compliance officer, a federal governmental relations officer, a state governmental relations officer, and a county governmental relations officer. In how many ways can these officers be selected?

33. **Concert Program** The Music Department at State University will give a concert consisting of five pieces of music. In how many ways can the selections be listed in the program?

34. Six members of the Bryan Fire Department are to receive awards for outstanding performance of duties during the past year. In how many ways can the awards ceremony be arranged?

35. How many linear arrangements are there of the letters in the word *page?*

36. How many linear arrangements are there of the letters in the word *beautiful?*

37. How many permutations are there of the letters in the word *college?*

38. How many permutations are there of the letters in the word *intelligible?*

39. For an experiment, 12 sociology students are to be divided into two groups, one containing 7 students and the other containing 5 students. In how many ways can this grouping be done?

40. Fifteen students are going hiking on their spring break. They plan to travel in three vehicles—one seating 7, one seating 5, and one seating 3 students. In how many ways can the students group themselves for their trip?

41. **Concert Performers** A star musical performer is giving a concert at State University. At the concert, she will present 13 songs, 4 of which are country and western, 3 of which are rock 'n' roll, and 6 of which are pop numbers. In how many ways can the star arrange her concert if songs within a category are considered indistinguishable?

42. Six indistinguishable red balls and four indistinguishable green balls are arranged in a line. How many distinguishable arrangements of these balls are possible?

43. Five indistinguishable red balls, three indistinguishable green balls, and two indistinguishable blue balls are arranged in a line. How many distinguishable arrangements of these balls are possible?

44. Ten books are to be arranged on a shelf. Four of the books have red covers, 3 have green covers, and another 3 have gray covers. In how many ways can the books be arranged on a shelf if books of the same color must be arranged together?

45. How many two-digit or three-digit I.D. numbers can be constructed from the digits 1, 3, 4, 5, 6, 8, and 9 if no repetition of digits is allowed?

46. How many three-letter or four-letter code words can be made if no repetition of letters is allowed?

47. A firm has 750 employees. Explain why at least 2 of the employees would have the same pair of initials for their first and last names.

Section 7.7   Combinations

Another important counting problem arises when we ask, "How many 5-card hands can be dealt from a standard deck of 52 cards?" (See accompanying table.) The cards are selected in succession, without replacement. However, once the 5 cards have been dealt (selected), any reordering of those cards is *not* considered a new hand. In the final analysis, the order of selection is unimportant; we are interested only in finding out how many *different subsets* of 5 cards can be created from the cards in the deck. In this sense, we might rephrase the original question as "How many subsets of 5 cards each are possible in a deck of 52 cards?" The next two paragraphs present two other examples of this type of counting.

Standard Deck of 52 Cards				
four suits →	Clubs (♣)	Diamonds (♦)	Hearts (♥)	Spades (♠)
13 denominations ↓	Ace of ♣	Ace of ♦	Ace of ♥	Ace of ♠
	King of ♣	King of ♦	King of ♥	King of ♠
	Queen of ♣	Queen of ♦	Queen of ♥	Queen of ♠
	Jack of ♣	Jack of ♦	Jack of ♥	Jack of ♠
	10 of ♣	10 of ♦	10 of ♥	10 of ♠
	9 of ♣	9 of ♦	9 of ♥	9 of ♠
	8 of ♣	8 of ♦	8 of ♥	8 of ♠
	. . .	. . .	. . .	. . .
	2 of ♣	2 of ♦	2 of ♥	2 of ♠
	↑	↑	↑	↑
	All are black	All are red	All are red	All are black

(The kings, queens, and jacks are face cards. Depending on the game, aces may be the highest or the lowest card in the deck.)

Suppose the Internal Revenue Service has 50 personal income tax forms from which it is to select a sample of 6 to check for accuracy. In how many ways can such a sample be obtained? The forms will be randomly selected, in succession, without replacement, from the 50 distinguishable forms. Once the forms are selected, our interest is not in the order of their selection, but rather their accuracy. In other words, any reordering of the 6 forms does not give a new sample of 6 to be checked for accuracy. Once again, the question might have been rephrased as "How many sets of 6 tax forms can be created from among 50 forms?"

In how many ways can a committee of 3 be selected from among 12 people? Once more, a person selected is not returned to the pool of people from which the remaining committee members are chosen. Furthermore, once a committee is chosen, any rearrangement of its members *does not* result in a new committee.

Whenever objects are successively selected from a set of distinguishable objects and any rearrangement of that outcome does not give a new outcome, we say that we have a **combination** of those objects. Stated differently, a combination of objects is a set of objects; the order is not important. In general, counting problems having outcomes such as these are called **combination problems.** There are three central themes running through these examples and, hence, through all combination problems:

Selection Rules for Combinations

1. Objects are successively selected from a single set *without* replacement.
2. The objects in the set are considered *distinguishable.*
3. Any permutation (interchange, rearrangement) within a given outcome *does not* give a new outcome to be counted.

Combinations

Let S be a set of n distinguishable elements. The successive selection of r elements, where $0 \leq r \leq n$, *without regard to order,* is called a **combination of n things taken r at a time.** The total number of combinations of n things taken r at a time is denoted by $C(n, r)$.

NOTE Other notations for the total number of combinations of n things taken r at a time are $_nC_r$ and $\binom{n}{r}$. The first is a common notation for calculators. On the TI-83 Plus, the combination command nCr is found under the PRB submenu of the MATH menu. As for spreadsheets, Excel has the combination function COMBIN under the Math & Trig function category.

NOTE Again, we assume that sets contain only distinguishable objects, unless specifically noted. In particular, if a sample is to be taken from 100 lightbulbs, think of each bulb as having a different number written on it. If a sample of balls is to be taken from a box containing 20 red balls, 8 green balls, and 3 purple balls, think of each of the 31 balls as having a different number written on it.

EXAMPLE 1

Comparing
Combinations with
Permutations

For selections from the set $S = \{a, b, c, d\}$, compare $C(4, 3)$ with $P(4, 3)$.

SOLUTION The set of possible outcomes for $C(4, 3)$ and $P(4, 3)$ can be listed in a table:

Outcomes in $C(4, 3)$	Outcomes in $P(4, 3)$
abc	abc, acb, bac, bca, cab, cba
abd	abd, adb, bad, bda, dab, dba
acd	acd, adc, cad, cda, dac, dca
bcd	bcd, bdc, cbd, cdb, dbc, dcb
Total: 4	Total: 24

In Example 1, $C(4, 3)$ is smaller than $P(4, 3)$. In fact, for each combination, there are $3 \cdot 2 \cdot 1 = 3!$ permutations. Hence, $C(4, 3) \cdot 3! = P(4, 3)$. In general, $C(n, r)$ is smaller than $P(n, r)$ because, within each selection of r elements, there are $r!$ permutations that are counted in $P(n, r)$, but are not counted in $C(n, r)$. Therefore,

$$C(n, r) \cdot r! = P(n, r)$$

or

$$C(n, r) = \frac{P(n, r)}{r!} = \frac{1}{r!} \cdot \frac{n!}{(n - r)!} = \frac{n!}{r! \, (n - r)!}. \quad \blacksquare$$

The Number of Combinations

$$C(n, r) = \frac{n!}{r! \, (n - r)!}.$$

EXAMPLE 2

Calculating with the Combination Formula

(a) $C(8, 5) = \dfrac{8!}{5! \cdot 3!} = \dfrac{8 \cdot 7 \cdot 6 \cdot 5 \cdot 4 \cdot 3 \cdot 2 \cdot 1}{5 \cdot 4 \cdot 3 \cdot 2 \cdot 1 \cdot 3 \cdot 2 \cdot 1} = 56.$

(b) $C(7, 2) = \dfrac{7!}{2! \cdot 5!} = \dfrac{7 \cdot 6 \cdot 5 \cdot 4 \cdot 3 \cdot 2 \cdot 1}{2 \cdot 1 \cdot 5 \cdot 4 \cdot 3 \cdot 2 \cdot 1} = 21.$

(c) $C(5, 5) = \dfrac{5!}{5! \cdot 0!} = \dfrac{5 \cdot 4 \cdot 3 \cdot 2 \cdot 1}{5 \cdot 4 \cdot 3 \cdot 2 \cdot 1 \cdot 1} = 1.$

(d) $C(4, 0) = \dfrac{4!}{0! \cdot 4!} = \dfrac{4 \cdot 3 \cdot 2 \cdot 1}{1 \cdot 4 \cdot 3 \cdot 2 \cdot 1} = 1. \quad \blacksquare$

EXAMPLE 3

Computing the Number of Combinations

How many different 5-card hands are there in a deck of 52 cards?

SOLUTION A hand is considered to be the successive selection of 5 cards drawn without replacement from a standard deck of 52 cards in which any reordering of the 5 cards drawn does not give a new hand. Therefore, the answer is

$$C(52, 5) = \frac{52!}{5! \cdot 47!} = 2{,}598{,}960. \quad \blacksquare$$

EXAMPLE 4

Quality Control: A Sample of Lightbulbs

The quality control unit in a lightbulb factory is to test a sample of 5 lightbulbs selected from among 60 bulbs. How many different samples, selected without replacement, are possible?

SOLUTION We assume that, once five bulbs have been selected, any rearrangement of those bulbs does not give a new sample to be tested. Therefore, the answer is

$$C(60, 5) = \frac{60!}{5! \cdot 55!} = 5,461,512. \quad \blacksquare$$

EXAMPLE 5

Linking Combinations
with the Multiplication
Principle

Among 20 calculators, 8 are defective. In how many ways can 6 calculators be selected from among the 20, without replacement and without regard to order, in which

(a) All are defective?

SOLUTION We are to select only from the eight defective calculators. Since calculators are selected without replacement, and since the order of selection is immaterial, this question could be rephrased as "Find the number of combinations of the eight defective calculators taken six at at time." The answer is

$$C(8, 6) = \frac{8!}{6! \, 2!} = 28.$$

(b) Exactly four in the sample are defective?

SOLUTION The preceding sentence means that four of the calculators selected are to be defective and two are to have no defect, making a total of six selected. Think of the calculators as being divided into two subsets, the first with the eight defective ones and the second containing the 12 with no defect. Selecting from the first set, in which order is unimportant, we find that there are $C(8, 4)$ ways of selecting four defective calculators, and for each of these ways, there are $C(12, 2)$ ways of selecting the two remaining calculators from the second set. Since we are selecting from different sets, the multiplication principle tells us that the resulting two numbers should be multiplied together to give the total number of ways of selecting four defective calculators, followed by two with no defect. However, once the calculators are selected, any reordering is not considered to be a new selection of calculators. Therefore, the answer is given by the ordering in which the initial selections were made:

(Choose 4 Lightbulbs from Set of 8 Defectives)

$\times$ (Choose 2 Lightbulbs from Set of 12 Good Ones)

$$= C(8, 4) \cdot C(12, 2) = \frac{8!}{4! \, 4!} \cdot \frac{12!}{2! \, 10!} = 4,620.$$

(c) At least one in the sample is defective?

SOLUTION This is a perfect time to apply the complement principle. The total number of different outcomes is given by the combination of selecting 6 calculators out of 20, or

$$C(20, 6) = 38,760.$$

Now we count all of the possible samples that have no defective calculators:

$$C(12, 6) = 924.$$

Next, by the complement principle, we can count the number of samples that have at least one defective calculator. We obtain

$$n(\text{Samples with at least one defective}) = n(\text{Samples}) - n(\text{Samples with no defectives})$$
$$= 38{,}760 - 924 = 37{,}836. \quad \blacksquare$$

EXAMPLE 6

An Application of Combinations

State University, State Tech, and Fields College send four representatives each to a conference on student government. At the conference, a committee of 5 is to be selected from among the 12 representatives. In how many ways can the committee be selected if

(a) There are no restrictions?

SOLUTION In this case, selection occurs without replacement: Once selected, any rearrangement of a committee does not give a new committee. (Contrast this case with the selection of people for specific titles.) Therefore, all 12 representatives are eligible for selection, and the question asks, "How many combinations of 12 things taken 5 at a time are there?" The answer is

$$C(12, 5) = 792.$$

(b) Exactly two of the committee members must come from Fields College.

SOLUTION Think of the 12 representatives divided into two subsets, one containing the four Fields College students and the other containing the eight students from State University and State Tech. We are to select two students from the first-mentioned set and the remaining three students from the second-mentioned set; in each case, the order of selection is not important. The multiplication principle tells us that we should multiply the number of possible selections in each case together to get the total number of committees in which Fields students are selected first and the others second. But once a committee is selected, all reorderings give the same committee. The answer is

$$(\text{Choose 2 from 4 Fields Students}) \times (\text{Choose 3 from 8 State Students})$$
$$= C(4, 2) \cdot C(8, 3) = 6 \cdot 56 = 336.$$

(c) Two students must come from State University, two from State Tech, and one from Fields College.

SOLUTION The number of possible committees under these conditions is

$$C(4, 2) \cdot C(4, 2) \cdot C(4, 1) = 6 \cdot 6 \cdot 4 = 144. \quad \blacksquare$$

EXAMPLE 7

An Application of Combinations

Seven cards are drawn without replacement from a deck of 52 cards. In how many ways can this be done if

(a) All of them must be clubs?

SOLUTION There are 13 clubs, so there are $C(13, 7) = 1716$ different seven-card hands consisting of clubs only.

(b) All of them must be of the same suit?

SOLUTION First, select a suit. There are $C(4, 1) = 4$ ways to do this. Now, for each selection of a suit, there are $C(13, 7)$ ways of selecting the seven cards. The multiplication principle then gives a final answer of

$$\text{(Choose 1 from 4 Suits)} \times \text{(Choose 7 from 13 Face Values)}$$
$$= C(4, 1) \cdot C(13, 7) = 4 \cdot 1716 = 6864. \ \blacksquare$$

Reflect on Examples 5, 6, and 7 to see common threads. All may be thought of as a deck of cards, for instance, in which there may be more or fewer than four suit and each suit still need not have the same number of cards. In Example 5, think of a deck of cards wth two "suits": *defective,* with eight cards; and *no defect,* with 12 cards. In Example 6, think of a deck of cards with three "suits"—State University, State Tech, and Fields College, each having 4 cards. So, in a sense, any problem asking for the number of possible outcomes, selected without replacement and in which order is not important, can be thought of as a hand of cards if you wish.

EXAMPLE 8

Combinations and Inclusion–Exclusion

A box contains five red, three green, and five blue balls.

(a) In how many ways is it possible to select a sample of three balls, without replacement, in which all of the balls are red or all of the balls are green?

SOLUTION By the inclusion–exclusion principle,

$$n(\text{Three Red Balls or Three Green Balls})$$
$$= n(\text{Three Red Balls}) + n(\text{Three Green Balls})$$
$$- n(\text{Three Balls That are Both Red and Green})$$
$$= C(5, 3) + C(3, 3) - 0 = 10 + 1 = 11$$

(b) In how many ways is it possible to select a sample of four balls, without replacement, in which at least three are red?

SOLUTION We translate "at least three are red" into the logically equivalent ("three are red and one is not red") or ("all four are red"). Because it is impossible for the latter two statements to happen simultaneously, the inclusion–exclusion principle states that our answer is found with the formula

$$n(\text{Three Red Balls and One Non-red Ball}) + n(\text{Four Red Balls})$$
$$= \text{(Choose 3 from 5 Red Balls)} \times \text{(Choose 1 from 8 Non-red Balls)}$$
$$+ \ \text{(Choose 4 from 5 Red Balls)} \times \text{(Choose 0 from 8 Non-red Balls)}$$
$$= C(5, 3) \cdot C(8, 1) + C(5, 4) \cdot C(8, 0) = 10 \cdot 8 + 5 \cdot 1 = 85. \ \blacksquare$$

EXAMPLE 9

An Application of Combinations

How many different arrangements in a row can be made from three S's and five F's in which all of the letters are used each time?

SOLUTION Note that, for any arrangement of these eight letters, such as

$$S \ F \ S \ F \ F \ F \ S \ F,$$

any interchange of elements among the S's (or the F's) does not give a different arrangement to be counted, because the S's (or the F's) are indistinguishable. First, find the number of ways to place three S's among the eight slots so that any interchange of any of the S's does not give a new arrangement. There are $C(8, 3)$ ways. Now, how many ways are there of placing F's in the remaining slots so that any interchange does not give a new arrangement? The number is 1. The multiplication principle then gives $C(8, 3) \cdot 1 = 56 \cdot 1 = 56$ as the answer. (We could have selected slots for the five F's first, then the S's, and then obtained the answer of $C(8, 5)$. However $C(8, 5) = 56$ also.) ■

We also could have gotten the answer to Example 9 by using permutations of indistinguishable objects; consider both the S's and the F's to be indistinguishable:

$$\frac{8!}{3! \, 5!} = 56.$$

But this is precisely the meaning of combinations: Objects are considered distinguishable during the *selection* process, but once selected, they are considered indistinguishable!

Example 10 solves the problem posed at the outset of this chapter by utilizing the complement principle.

EXAMPLE 10

Counting What We Do Not Want

The Galax city council is composed of 10 members, of which 6 are Democrats and 4 are Republicans. A committee of 4 council members is to be formed so that there is at least one representative of each political affiliation. How many committees can be formed under this restriction?

SOLUTION We begin by calculating the total number of committees of 4 that can be formed from the 10 council members, without any restrictions:

$$C(10, 4) = \frac{10!}{4! \, 6!} = 210.$$

Now, let us determine what we do not want. We do not want a committee consisting of only Democrats and there are

$$C(6, 4) = \frac{6!}{4! \, 2!} = 15$$

such committees. We also do not want a committee consisting of only Republicans, and there is

$$C(4, 4) = \frac{4!}{4! \, 0!} = 1$$

such committee. Thus, of the 210 possible committees of 4, there are

$$210 - 15 - 1 = 194$$

that contain at least one member of each political party represented on the council. ■

While working through the solution to Example 10, you may have wondered whether using the complement principle was really the most efficient approach to solving the problem. Let us consider the problem again, taking the more direct approach of counting only what we *do* want. Since there are six Democrats and four Republicans, the committees of four that we do wish to count are summarized in the following table:

Number of Democrats	Number of Republicans	Count of Possible Number of Committees
3	1	$C(6,3) \cdot C(4,1) = 20 \cdot 4 = 80$
2	2	$C(6,2) \cdot C(4,2) = 15 \cdot 6 = 90$
1	3	$C(6,1) \cdot C(4,3) = 6 \cdot 4 = 24$
	Total:	$= 80 + 90 + 24 = 194$

As you can see, in our solution to Example 10, we needed to calculate only three combinations and then perform some subtraction. If we had chosen instead to count all the possible committees we *wanted*, we would have needed to calculate six combinations, perform some multiplication, and then perform some addition. Thus, the "count what we do not want" strategy can save time by reducing the number of calculations needed.

Exercises 7.7

In Exercises 1 through 10, calculate each combination.

1. $C(6,2)$
2. $C(8,5)$
3. $C(9,3)$
4. $C(12,4)$
5. $C(15,2)$
6. $C(52,3)$
7. $C(48,4)$
8. $C(12,6)$
9. $C(14,7)$
10. $C(12,0)$

11. (a) Calculate $C(5,2)$ and $C(5,3)$.
    (b) Calculate $C(9,7)$ and $C(9,2)$.
    (c) Calculate $C(8,5)$ and $C(8,3)$.
    (d) State a rule you discovered about combinations while working Exercise 11(a) through (c).

12. Calculate $C(5,r)$ for all possible values of r.

13. Calculate $C(6,r)$ for all possible values of r.

In the remaining exercises, assume that sets contain distinguishable elements (objects). Imagine that they are numbered if you wish. Use technology where appropriate.

14. **Tennis Teams** How many tennis doubles teams can be formed from 12 players?

15. **Geometry** How many straight lines are determined by five points in the plane, no three of which are on the same line?

16. How many 4-card hands are possible from a standard deck of 52 cards?

17. How many 3-card hands are possible from a standard deck of 52 cards?

18. Among 18 computers, 12 are in working order and six are defective. How many samples of 4 are possible, selected without replacement and without regard to order, wherein all are defective?

19. Among 23 computer modems, 15 are in working order and 8 are defective. How many samples of 6 are possible, selected without replacement and without regard to order, if all of the modems are in working order?

20. A warehouse contains 12 refrigerators, 2 of which are not in working order. In how many ways can a sample of 5 of these refrigerators be selected, without replacement and without regard to order, if

    (a) All are to be in working order?
    (b) Exactly 2 are not in working order?
    (c) All 5 are not in working order?

21. A box contains 20 computer disks, 5 of which are known to have bad sectors. In how many ways can 3 of these disks be selected, without replacement and without regard to order, so that

(a) None have bad sectors?

(b) All have bad sectors?

(c) Exactly 2 do not have bad sectors?

22. Committee Assignments A committee of four is to be selected from among eight graduate students and a professor. The committee is to meet with the dean about new classroom equipment. In how many ways can the committee be selected if

(a) The professor cannot be on the committee?

(b) The professor must be on the committee?

(c) There are no restrictions?

23. Committee Assignments Five people from among the mayor and seven other city officials are to be selected to make an inspection of solid-waste facilities. In how many ways can these five people be selected if

(a) The mayor must be included in the five?

(b) The mayor is not to be among the five selected?

(c) There are no restrictions?

24. Committee Assignments Companies Archer *and* Sisko send three representatives each to a conference. At the conference, a committee of four is to be selected from among these representatives. In how many ways can this be done if

(a) All of the committee members are to come from Archer's?

(b) Exactly two are to come from Archer's?

(c) There are no restrictions?

25. Committee Assignments Data Fix, Data Mix, and Data Six each send five representatives to a conference. At the conference, a committee of four from among the company representatives is selected. In how many ways can this be done if

(a) All of the committee members are to come from Data Fix?

(b) None are to come from Data Fix?

(c) All are to come from Data Mix or Data Six?

(d) All are to come from Data Mix or all are to come from Data Six?

26. In how many ways can 5 cards be drawn, without replacement, from a deck of 52 cards if

(a) All of the cards drawn are to be clubs?

(b) Exactly two are clubs, two are diamonds, and one is a spade?

(c) All are to be clubs or diamonds?

(d) All are to be hearts or all are to be kings?

27. Eight cards are to be drawn, without replacement, from a deck of 52 cards. In how many ways can this be done if

(a) All of the cards drawn are to be spades?

(b) All are to be from the same suit?

(c) Exactly five are to be clubs?

(d) All are to be hearts or all are to be spades?

28. There are five red and seven green balls in a box, and a sample of three is to be chosen without replacement and without regard to order. In how many ways can this be done if

(a) None of the balls selected are to be green?

(b) At least two are to be green?

29. Government Committee From among three conservatives and five liberals, a committee of three is to be selected. In how many ways can this be done if

(a) None of the committee members are to be conservatives?

(b) Exactly one is to be a conservative?

(c) At least two are liberals?

30. In how many ways can a 3-card hand be dealt from a standard deck of 52 cards if

(a) All of the cards in the hand are to be red cards?

(b) All are to be face cards?

(c) At least one is to be a face card?

31. In how many ways can a 4-card hand be dealt from a standard deck of 52 cards if

(a) All of the cards in the hand are to be black cards?

(b) All are to be eights?

(c) At least two are to be hearts?

32. A box contains four red, three green, and five blue marbles. In how many ways can three marbles be selected, without replacement and without regard to order, if

(a) All of the marbles selected are to be blue?

(b) All are to be red or all are to be blue?

(c) Exactly one is red?

33. A box contains five red, six blue, and seven purple balls. In how many ways can four balls be selected, without replacement and without regard to order, if

(a) All of the balls selected are to be blue or all are to be purple?

(b) Exactly two are to be blue?

(c) Exactly one is to be purple?

34. **Job Assignments** From a company of 15 soldiers, a squad of 4 is selected for patrol duty each night.

    (a) For how many nights in a row could a squad go on duty without two of the squads being identical?

    (b) In how many ways could the squad be chosen if Joe, a soldier in the company, must always be on the squad?

    (c) In how many could the squad be chosen if Joe, a soldier in the company, can never be on the squad?

35. How many different arrangements are there of four F's and seven S's in a row?

36. **Committee Assignments** In how many ways can a committee of nine be divided into two subcommittees, one having five members and the other having four?

37. In a lot of 100 lightbulbs, 6 are defective. In how many ways can a sample of 3 be drawn from the lot, without replacements and without regard to order, if

    (a) All of the bulbs drawn are good?

    (b) All are defective?

    (c) Exactly two are defective?

38. Six cards are to be drawn (or dealt), without replacement, from a deck of 52 cards. In how many ways can this be done if

    (a) All of the cards drawn are to be spades?

    (b) All are to be kings?

    (c) None are to be diamonds?

    (d) All are to be clubs or all are to be kings?

39. A sample of five marbles is to be selected, without replacement and without regard to order, from a box that contains three red, four green, and six purple marbles. In how many ways can this be done if

    (a) Exactly two of the marbles selected are to be red, two are to be green, and one is to be purple?

    (b) All are red or purple?

    (c) All are to be red or all are to be purple?

    (d) At least four are to be purple?

40. A box contains two red, five green, six black, and four blue jelly beans. In how many ways can a sample of four jelly beans be selected from the box, without replacement, and without regard to order, if

    (a) All of the jelly beans are to be the same color?

    (b) Exactly three are to be the same color?

41. A shipment of eight men's suits has three of size 38, three of size 40, and two of size 42. In how many ways can a sample of three suits be selected for a style show if

    (a) All are the same size?

    (b) Exactly two are the same size?

42. **Pizza Toppings** A pizza chain advertises that it has 10 toppings to choose from. How many ways are there of ordering a pizza from this firm with up to 4 toppings? (Assume that a pizza has at least 1 topping)

43. **Vaults** Why should *combination locks* actually be called *permutation locks?*

Chapter 7 Summary

CHAPTER HIGHLIGHTS

Problems that ask "How many?" play an important role in computer science, statistics, and probability. These *counting problems* involve special key words of logic—such as *and*, *or*, and *not*—and their *set operation* counterparts of intersection, union, and complementation. *Venn diagrams* are a useful tool for exploring set relationships and for solving some types of counting problems. The *multiplication principle* is a basic computational procedure for counting the number of different sequences formed from successive choices of objects from the same set (called per mutations) or of objects from different sets. Counting problems involving the number of sets that can be selected from a collection of objects are called *combinations*. In some situations, counting can be done most efficiently by counting what we do not want.

GUIDE TO COUNTING TECHNIQUES

• *Problems solved by Venn diagrams:*
  The distinctive wording of Venn diagram problems points toward the use of such a diagram. Careful labeling of the separate regions eliminates double-counting.

- *Problems that involve the connective "or":*
  When *or* connects two sentences, use the inclusion–exclusion principle or a Venn diagram.

- *Problems that involve "at least" or "at most":*
  Translate these problems into logically equivalent ones by using the *or* connective, or use a count of what you do not want.

- *The multiplication principle:*
  In making successive selections from the same set of distinct elements, with or without replacement, when the order of selection is important, apply this principle. Also, apply it when making successive selections from different sets.

- *The complement principle:*
  May be used when counting what you *do not want* is easier.

- *Permutations:*
  Use permutations when selecting distinguishable objects without replacement and the order of selection *is* important. There is a special formula to use when there is a mixture of distinguishable and indistinguishable objects.

- *Combinations:*
  Use combinations when selecting distinguishable objects without replacement and the order of selection *is not* important.

EXERCISES

1. Use truth tables to determine whether the statements $p \vee (q \wedge p)$ and $(p \vee q) \wedge q$ are equivalent.

2. Use truth tables to determine whether the statements $p \Rightarrow \sim q$ and $\sim p \vee \sim q$ are equivalent.

3. Write the negation of "If the current machine breaks down, then we will buy a new machine."

4. Write the negation of "All of the lights are burned out."

In Exercises 5 through 13, for the sets $U = \{1, 2, 3, 4, 5, 6\}$, $A = \{1, 2\}$, $B = \{3, 5, 6\}$, and $C = \{2, 4, 5, 6\}$, compute the results of the given set operation(s).

5. $B \cap C$ 6. $A - C$ 7. B'
8. $A \cup C'$ 9. $(C')'$ 10. $(A \cup C)'$
11. $A \cap B$ 12. $C \cap (A \cup B)$ 13. $(C \cap A) \cup B$

14. Use set operations to describe the shaded set in the following Venn diagram:

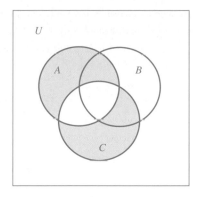

15. How many possible outcomes are there when a coin is tossed seven times?

16. How many five-digit identification numbers can be formed if at least one of the digits must be a zero?

In Exercises 17 through 21, calculate and simplify to a counting number.

17. $P(96, 5)$ 18. $C(96, 5)$ 19. $C(96, 91)$
20. $P(23, 7)$ 21. $C(47, 38)$

22. What logical connective(s) should be used between the constraints of a linear programming problem to completely determine the feasible region? (For example, $x + 2y \leq 8$, $4x + y \leq 20$, $x \geq 0$, $y \geq 0$.)

23. Write the negation of "All Fords are red or blue."

24. Write the negation of "Some college students drink tea and coffee."

25. The Food-4-Less grocery store questioned 75 of its shoppers about their bread purchases during the past week. The results indicated that 43 bought white, 39 bought whole wheat, 16 bought rye, 19 bought both white and whole wheat, 6 bought both whole wheat and rye, 5 bought both white and rye, and 3 bought no bread. How many bought all three kinds of bread?

26. Eight students arrived at Cinema Ten to see a newly released movie. In how many ways could four of them line up to buy tickets?

27. An astronomy club has 20 members, including 4 officers. How many possible delegations of 5 persons could be selected to visit Chicago's Museum of Science and Industry if the president must go, but none of the other officers may go?

The following problem appeared on an actuarial examination:

28. Which of the following events is identical to $(B \cap C) \cup (A' \cap B \cap C')$?

   I. $B \cap (A' \cup C)$

  II. $(A' \cap B) \cup (B \cap C)$

  III. $(A' \cap C') \cup (B \cap C)$

  (a) I and II only

  (b) I and III only

  (c) II and III only

  (d) I, II, and III

  (e) The correct answer is not given by (a), (b), (c), or (d).

29. Calculate $C(83, 61)$.

CUMULATIVE REVIEW

30. Find the least-squares regression line for the points (10, 21), (20, 27), (30, 43), (40, 56), and (50, 58).

31. Write a linear cost function for traveling x miles in a taxicab, if 3 miles cost \$3.70 and 10 miles cost \$8.25.

32. Write the general solution of the system

$$5x + 2y - z = 6$$
$$-3x + y = 2$$

in parametric form.

33. Calculate AB and BA for the matrices

$$A = \begin{bmatrix} 2 & 0 & -1 \\ 1 & -3 & 0 \end{bmatrix} \quad \text{and} \quad B = \begin{bmatrix} 0 & 2 \\ 5 & -3 \\ 1 & -1 \end{bmatrix}.$$

34. Solve this linear programming problem:

$$\text{Maximize } f = 3x + 4y$$
$$\text{subject to } 5x + 3y \leq 30$$
$$x + 4y \geq 8$$
$$-x + y \leq 6$$
$$x \geq 0, y \geq 0.$$

Chapter 7 Sample Test Items

Consider the statements

       p: The first ball drawn is red.

       q: The second ball drawn is not blue.

In Questions 1 through 4, rewrite the indicated sentences in English. Make use of logically equivalent sentences where appropriate.

1. $\sim p \vee \sim q$ **2.** $\sim p \wedge q$

3. $\sim(\sim p \vee q)$ **4.** $p \Rightarrow \sim q$

In Questions 5 through 11, let $U = \{\varnothing, 2, \{2\}, 3, 4, 5\}$, $A = \{\varnothing, 2, 3, 5\}$, $B = \{2, 4, 5\}$, and $C = \{\varnothing, \{2\}, 3, 5\}$. Decide whether each statement is true (T) or false (F), and circle your choice.

T F **5.** $2 \in C$ T F **6.** $\{2\} \subseteq C$

T F **7.** $\varnothing \subseteq B$ T F **8.** $2 \subseteq A$

T F **9.** $\varnothing \in C$ T F **10.** $A = C$

T F **11.** All students like mathematics, or some students drink tea.

Using sets U, A, B, and C from Questions 5 through 10, list the elements of each indicated set in Questions 12 through 14.

12. $(A \cup C)'$ **13.** $A \cap C'$ **14.** $A - B$

15. Shade the accompanying Venn Diagram to represent $A \cup (B \cap C')$.

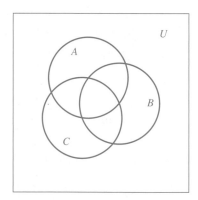

In Questions 16 and 17, 50 students were randomly surveyed from among those who ate regularly at a university dining hall. The survey revealed that 38 ate breakfast, 25 ate lunch, 18 ate breakfast and lunch, 35 ate breakfast and dinner, 24 ate lunch and dinner, and 17 ate all three meals.

16. How many of these students ate lunch only?

17. How many of these students ate dinner?

18. How many three-digit numbers are there among the integers if zero cannot be the first digit?

19. In how many ways can a 5-card hand be drawn, without replacement, from a standard deck of 52 cards if exactly 3 of the cards drawn are to be hearts?

20. In how many ways can the letters in the word *banana* be arranged?

In Questions 21 and 22, nine people are to travel to dinner in a five-passenger van and a four-passenger sports car.

21. How many different groups of five and four are possible for the trip?

22. How many seating arrangements are possible if two persons—Juanita and Kareem—are designated drivers and the others can sit in any of the remaining seats?

23. You have two different pairs of tennis shoes, three different pairs of shorts, and six different T-shirts. How many different outfits do you have?

24. In how many ways can four couples be seated in a row of eight seats at a theater if each couple is seated together?

25. How many ways can a 10-question true–false test be answered so that at least one answer is correct?

In-Depth Application

Arrangements of Base Pairs in DNA

Objectives

Information about all the human body's functions is stored in permutations of nucleotides in strands of DNA. You will use the skills demonstrated in this chapter to count the number of possible permutations under different restrictions.

Introduction

Gigantic molecules called **deoxyribonucleic acid (DNA)** are found in every cell of the human body. Each DNA molecule is a superlong double helix, composed of two outer backbone strands of alternating deoxyribose and phosphate groups twirling around each other. The two strands are held together by **nucleotides** (or **bases**) run-

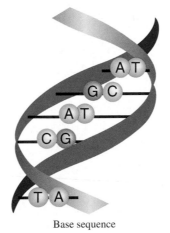

Base sequence

(continued)

ning along in pairs down the middle of the double helix. The nucleotides—**adenine (A), thymine (T), cytosine (C),** and **guanine (G)**—always come in pairs, adenine with thymine and cytosine with guanine. When one speaks of a **base sequence,** he or she means listing off the letters, in order, of the bases along one side of some portion of the DNA. For example, in the figure, the base sequence illustrated is AGACA. The activities of each cell are determined by base sequences.

Base-Sequence Permutations

Suppose we are interested in a sequence of 10 bases (although the length of such a sequence is miniscule in comparison to the lengths of base sequences found in human cells).

1. What is the number of possible base sequences of length 10, if any number of each base is allowed in the sequence?

2. Human DNA tends to be composed of about 60% adenine–thymine pairs. How many sequences (remember, of length 10) have *at least* six bases that are adenine or thymine? How many sequences have *exactly* six bases of adenine or thymine? How many sequences are possible with exactly three adenine bases, three thymine bases, two cytosine bases, and two guanine bases?

3. Sometimes certain minisequences appear almost exclusively within other sequences. Suppose, in our sequence, every adenine base is followed by a guanine base. How many sequences are there with this restriction? How many sequences are there in which adenine and thymine appear only in triples (AAA or TTT)?

8

Basic Concepts of Probability

PROBLEM

A company sells one-year term policies designed for 40-year-old women. The face value of the policy is $10,000, and each policy sells for $300. If the company's mortality tables show that a 40-year-old woman will live for another year with probability 0.98, what is the average earnings per policy the company can expect from the sale of such policies? (See Example 2, Section 8.3.)

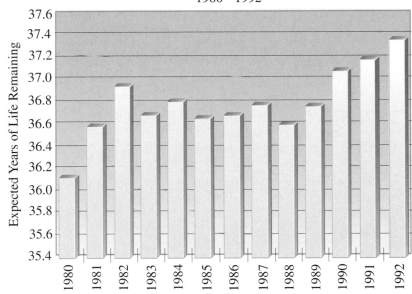

Mortality Figures for 40-year-old Women in the U.S. 1980 - 1992

Expected Years of Life Remaining (y-axis: 35.4, 35.6, 35.8, 36.0, 36.2, 36.4, 36.6, 36.8, 37.0, 37.2, 37.4, 37.6)

(x-axis: 1980, 1981, 1982, 1983, 1984, 1985, 1986, 1987, 1988, 1989, 1990, 1991, 1992)

SOURCE: Berkeley Mortality Database, University of California-Berkeley (http://demog.berkeley.edu/wilmoth/mortality), based on data from the Office of the Chief Actuary, U.S. Social Security Administration.

CHAPTER PREVIEW

Suppose an action is to be taken that will produce one of several possibilities. (For instance, a state income tax form is to be checked and classified as correct, overpaid, or underpaid.) Before the action takes place, the concept of probability can be used to measure, usually with a fraction or a decimal number, the likelihood that each of these possibilities will indeed happen. Sections 1 and 2 of this chapter investigate how these numbers are assigned to each of these several possibilities, as well as to certain subsets of them. In Section 3, the concept of "expected value" is explored as a useful application of probability to long-term trends. Liberal use will be made of the counting techniques discussed in Chapter 7.

Technology Opportunities

In this chapter, we will be using many of the counting techniques discussed in Chapter 7. For this reason, a calculator with permutation, combination, and factorial commands will be useful for performing calculations throughout the chapter. Spreadsheets also have commands for performing these types of calculations. In addition, we will often be organizing information in tables, so a spreadsheet package or a calculator with the ability to store lists will be helpful as well.

Section 8.1 Sample Spaces with Equally Likely Outcomes

The subject of probability can be traced back to the 17th century, when it arose out of the study of gambling problems. The range of applications today goes far beyond gambling problems, reaching into such areas as medical science, business decisions, botanical experiments, and weather forecasting, to mention just a few. In fact, even though the term *probability* is not commonly used in everyday publications, we still come into contact with it in disguised ways. For example, a weather forecast telling of "an 80% chance of rain tomorrow" or a sportscaster mentioning that "the odds for the Eagles winning the Super Bowl are six to five" are actually ways of stating probabilities.

Generally speaking, questions of probability arise when we see that a number of possibilities can occur and we do not know beforehand just which one will actually happen. We then attempt to assign a number to each possibility that reflects the proportional share of the time that possibility is expected to happen. If one outcome is more likely than another, then the more likely one should be assigned a larger fractional share than the less likely one. If, by contrast, all possibilities are **equally likely,** then each one should be assigned identical fractional shares of time. Since the equally likely case occurs so often in the natural settings, we take it as our starting point in the study of probability.

HISTORICAL NOTE

BLAISE PASCAL (1623–1662) AND PIERRE DE FERMAT (1601–1665)

The rigorous mathematical foundations of modern probability theory were jointly laid by two French mathematicians: Blaise Pascal and Pierre de Fermat (Fer-mä). Probability arose from twin roots: games of chance and statistical data for such matters as mortality tables and insurance rates.

EXAMPLE 1
Tossing a Fair Coin

Suppose a fair or balanced coin is to be tossed. What is the probability that it will fall with heads up?

SOLUTION We reason this way: Upon being tossed, the coin can fall in only two ways—heads or tails—and since the coin is balanced, it is equally likely to fall either way. Since there is a 1 in 2 chance heads will be showing, the number $\frac{1}{2}$ is used to predict the likelihood that this possibility will occur. In the language of probability, we say that the **probability** that the coin will fall with heads up is $\frac{1}{2}$. (The probability of the coin falling with tails up is also $\frac{1}{2}$.) ■

Even though the outcome of any single trial of tossing a balanced coin is uncertain, the importance of the probabilities found in Example 1 is that they are *reliable predictors* of the *long-term* behavior of repeatedly tossing a single balanced coin many times: About half of the time a head will turn up, and about half the time a tail will turn up. To demonstrate our reasoning, three records of 100 tosses of a balanced coin, simulated with a spreadsheet, are shown next.

First Trial	Second Trial	Third Trial
49 heads	56 heads	50 heads
51 tails	44 tails	50 tails

It is the long-term predictability feature that makes probability so useful. In the science of genetics, for example, it is uncertain whether an offspring will be male or female, but accurate long-term predictions can be made as to the percentage of males and females. An insurance company cannot predict which persons in the United States will die at the age of 60, but it can predict with great accuracy *how many* people will die at the age of 60.

Example 1 provides a framework for terminology used in the study of probability. An **experiment** (tossing a coin) is to be performed, and there are two possible **outcomes:** heads (H) or tails (T). The set of all possible outcomes from an experiment gives a universal set within which all probability activity takes place and is given the special name **sample space.** In Example 1, the sample space is $S = \{H, T\}$.

Experiments

An **experiment** is an activity or a procedure that produces distinct, well-defined possibilities called **outcomes** which can be observed or measured, but which cannot be predicted with certainty. Each outcome is an element in the universal set of all possible outcomes for the experiment. The universal set is called the **sample space.**

EXAMPLE 2

Tossing a Fair Coin Twice

A balanced coin is to be tossed twice and the result of each toss recorded. Find the probability of each outcome in the sample space.

SOLUTION The experiment is the act of tossing a fair coin twice and recording the results of the first toss and then the second toss. The sample space, composed of the possible outcomes, can be written as $\{HH, HT, TH, TT\}$, where H represents *heads* and T represents *tails*. Since the coin is fair, there is reason to believe that the outcomes are equally likely. Because there are four possible outcomes, it seems natural to assign each outcome a probability of $\frac{1}{4}$. ■

Outcome	Assigned Probability
HH	$\frac{1}{4}$
HT	$\frac{1}{4}$
TH	$\frac{1}{4}$
TT	$\frac{1}{4}$
Sum	1

The probabilities in Example 2 do two things. First, they show the relative strength of each outcome, given as a fractional part of 1; thus, the sum of the probabilities of all outcomes in the sample space should always be 1. Second, the predicted long-term effects of repeating this experiment over and over are for two heads to turn up about $\frac{1}{4}$ of the time, a head first and a tail second about $\frac{1}{4}$ of the time, a tail first and a head second about $\frac{1}{4}$ of the time, and two tails about $\frac{1}{4}$ of the time.

NOTE Unless stated to the contrary, we will henceforth always assume that coins are balanced, or fair.

EXAMPLE 3
Rolling a Single Die

A single, fair, six-sided die is to be rolled and the number of dots or pips showing on the top face recorded. Find the probability of each outcome in the sample space.

SOLUTION Figure 1 shows the possible outcomes for this experiment. The sample space may then be written as $S = \{1, 2, 3, 4, 5, 6\}$. Since the die is balanced, there is no reason to believe that any one of these outcomes would occur more frequently than any other. Hence, it is reasonable to assume that each outcome is equally likely, with probability $\frac{1}{6}$.

FIGURE 1
Sample Space for a Single Six-Sided Die

Outcome	Assigned Probability
1	$\frac{1}{6}$
2	$\frac{1}{6}$
3	$\frac{1}{6}$
4	$\frac{1}{6}$
5	$\frac{1}{6}$
6	$\frac{1}{6}$
Sum	1

Using Excel, we can simulate rolling a balanced, six-sided die with the function = RANDBETWEEN(1, 6), which returns an integer between 1 and 6, inclusive, and which assumes that each of the integers is equally likely. The following table shows three different simulations of repeating this experiment 500 times and recording the outcomes:

Pips Showing	First Repetition of 500 Rolls		Second Repetition of 500 Rolls		Third Repetition of 500 Rolls		Combined Repetition of 1500 Rolls	
	Number	Relative Frequency	Number	Relative Frequency	Number	Relative Frequency	Number	Relative Frequency
1	85	0.170	86	0.172	87	0.174	258	0.172
2	84	0.168	78	0.156	77	0.154	239	0.159
3	76	0.152	80	0.160	88	0.176	244	0.163
4	86	0.172	85	0.170	88	0.176	259	0.173
5	80	0.160	88	0.176	82	0.164	250	0.167
6	89	0.178	83	0.166	78	0.156	250	0.167

Note that the term **relative frequency** in the table refers to the proportion of the 500 rolls in a simulation that a particular outcome was obtained. We had assigned, to each outcome in the sample space, the probability $\frac{1}{6}$, which is approximately 0.167. Notice that the spreadsheet simulations appear to agree with these assignments. ∎

NOTE Henceforth, all dice will be considered fair or balanced unless otherwise noted.

EXAMPLE 4
Rolling a Pair of Dice

A pair of dice is to be rolled and the number of pips on the upper face of each die recorded. To aid in the thought process, think of one die as red and one as green. An outcome then consists of two numbers—one denoting the number of pips on the upper face of the red die and one denoting the number of pips on the upper face of the green die. Find the number of elements in the sample space and the probability of each outcome in that sample space.

SOLUTION There are six possible ways to record the pips on the top side of the red die, and for each of these, there are six possible ways to record the pips on the top side of the green die. Applying the multiplication principle then gives $6 \cdot 6 = 36$ possible outcomes for this experiment. Figure 2 gives a visual picture of the sample space. Note that a red 2

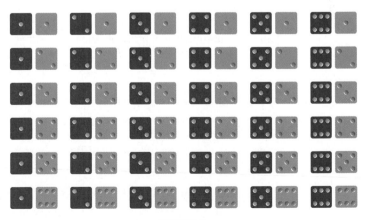

FIGURE 2
Sample Space for Two Dice

and a green 3 is a different outcome from a red 3 and a green 2. A "double," such as a red 4 and a green 4, however, would be listed only once as an outcome.

If we agree that each outcome is equally likely, then each should be assigned a probability of $\frac{1}{36}$. ∎

As it was with counting problems, the simplest way to view any probability problem is to write out the list of outcomes, as was done in Examples 1 through 4. After studying those examples, you have probably sensed that, for equally likely outcomes, the *number* of outcomes in the sample space is the key to assigning probability; if the number of outcomes is k, then each outcome is assigned the probability $\frac{1}{k}$. Knowing this, we can utilize the counting techniques of Chapter 7 to find the number of outcomes when it is not practical to write them all down. Example 5 illustrates this idea.

EXAMPLE 5

Dealing a Hand from
a Deck of Cards

A 5-card hand is dealt from a well-shuffled deck of 52 cards. (See Section 7.6 for a review of a standard deck of 52 cards.) How many outcomes are there from such an activity, and what is the probability of each outcome?

SOLUTION The experiment is that of selecting or dealing five cards from the deck, without replacement and without regard to order. The number of such hands, and hence the number of outcomes from the experiment, is found by computing the number of combinations of 52 things taken 5 at a time: $C(52, 5) = 2,598,960$. Since the cards were dealt from a well-shuffled deck, no single five-card hand can be assumed more likely to be dealt than any other. This leads to the conclusion that each five-card hand (outcome) is equally likely and therefore has probability $\frac{1}{2,598,960}$. ∎

EXAMPLE 6

Computer Passwords

To access the security system in the ABC Company's computer network, a password consisting of three digits followed by two letters must be correctly entered. If an unauthorized person knows how the password is constructed, what is the probability that that person will guess the password correctly on the first try?

SOLUTION The experiment is the act of selecting and entering three digits followed by two letters into the computer. Each *ordered* sequence of three digits followed by two letters is an outcome of the experiment. The selections in the sequence are made with replacement (repetitions are allowed), and the order in which they are selected is important. The multiplication principle then tell us that there are

$$10 \cdot 10 \cdot 10 \cdot 26 \cdot 26 = 676,000$$

possible outcomes (passwords) of the experiment. Since the unauthorized person is guessing at the correct makeup of the password, we assume that each password is equally likely to be guessed. Therefore, the probability $\frac{1}{676,000}$ is assigned to each outcome. This means that the probablility of correctly guessing the password with one try is $\frac{1}{676,000}$. ∎

Suppose we now return to Example 2, in which a coin was to be tossed twice with a sample space of $S = \{HH, HT, TH, TT\}$ and each outcome has probability $\frac{1}{4}$. The question "What is the probability that at least one head will turn up?" does not concern itself with just one outcome, but rather with a subset of *three* outcomes. In the language of probability, such a subset is called an **event.** In fact, the term "event" is used to describe *any* subset of a sample space, including subsets consisting of a single outcome, the empty set, or the entire sample space. An event consisting of a single outcome is called a **simple event,** while an event containing more than one outcome is called a **compound event.** If E denotes an event in a sample space S, the notation $P(E)$ is used to denote the probability of E. For the particular question at hand, $E = \{HH, HT, TH\}$. Figure 3 shows a Venn diagram of this event. The event E is said to **occur** if a trial of the experiment produces an outcome that belongs to E.

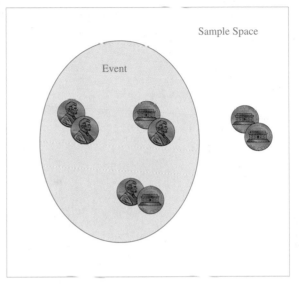

FIGURE 3
The Event "At Least One Head"

Event in a Sample Space

Given an experiment with sample space S, a set E is called an **event** in S if E is a subset of S ($E \subseteq S$).

Now, how likely is a trial of this experiment to produce one of the outcomes HH, HT, or TH? Since each outcome in the sample space is equally likely, three times out of four would sound reasonable. Accordingly, we write $P(E) = \frac{3}{4}$. Stated differently, the probability is the ratio of the number of outcomes in E, denoted by $n(E)$, to the total number of outcomes in the sample space S, denoted by $n(S)$.

NOTE This construction of probability is proper only when we have equally likely outcomes.

EXAMPLE 7

Rolling a Pair of Dice

A pair of dice is to be rolled. What is the probability that

(a) A sum of 5 is rolled?

SOLUTION Consider one of the pair to be red and one to be green, as in Example 4. There are 36 outcomes in the sample space. Among these, "roll a sum of 5" describes an event E that, as a subset, can be listed with the number of pips on one die, say, red, first, and the number of pips on the other (green) die second, as follows:

$$E = \{(1, 4), (2, 3), (3, 2), (4, 1)\}.$$

Since all outcomes are equally likely and $n(E) = 4$, we reason that there are 4 chances out of 36 of getting one of the outcomes in E on a single trial of the experiment "roll a pair of dice." The probability of rolling a sum of 5 is then given by ratio

$$P(E) = \frac{n(E)}{n(S)} = \frac{4}{36}.$$

(b) A sum of 8 is rolled?

SOLUTION This time the event described may be listed as

$$E = \{(2, 6), (3, 5), (4, 4), (5, 3), (6, 2)\}.$$

Because $n(E) = 5$, there are 5 chances out of 36 of getting one of the outcomes in E. Therefore, the probability of rolling a sum of 8 is given by

$$P(E) = \frac{5}{36}. \quad \blacksquare$$

In view of the development of events presented here, any experiment that can be realistically modeled by assuming equally likely outcomes is, from a probability viewpoint, a counting problem: Count the number of elements (outcomes) in E and the number of elements (outcomes) in S, and then divide the one by the other.

The Probability of an Event

Let S be a sample space with $n(S)$ equally likely outcomes. Then the probability of each outcome is

$$\frac{1}{n(S)}.$$

If E is an event in this sample space with $n(E)$ outcomes, then the probability of E is

$$P(E) = \frac{n(E)}{n(S)}.$$

On the one hand, the smallest number of elements that E can have is 0, which occurs when E is the empty set. On the other hand, the largest number of elements E can have is $n(S)$, which occurs when $E = S$. This means that the smallest that $P(E)$ could be is $0 \div n(S) = 0$, while the largest that $P(E)$ could be is $n(s) \div n(s) - 1$. Therefore, the probability of any event is always between 0 and 1, inclusive.

EXAMPLE 8

Selecting a Card

One card is to be drawn from a deck of 52 cards and the denomination and suit recorded.

(a) Find the probability that a spade is drawn.

SOLUTION There are 52 possible cards that may be drawn and hence 52 outcomes in the sample space. Since each card is just as likely to be drawn as another, we assign each outcome (card) a probability of $\frac{1}{52}$. The event E described by "the card is a spade" has 13 outcomes. Therefore, $P(\text{spade}) = \frac{13}{52} = \frac{1}{4}$.

(b) Find the probability that a queen is drawn.

SOLUTION The event "the card is a queen" contains 4 outcomes. It follows that $P(\text{queen}) = \frac{4}{52} = \frac{1}{13}$. ■

EXAMPLE 9

Checking for Flaws

A maker of sweaters selects a typical rack of 50 sweaters for inspection by a buyer for a large discount chain. Assume that 2 of the 50 sweaters have a flaw. Suppose the buyer randomly selects a sample of 6 sweaters from the rack, without replacement, and checks them for a flaw. (Random selection means that no one sweater is more likely to be selected than another.)

(a) What is the probability that none of the 6 sweaters will have a flaw?

SOLUTION Because they are selected without replacement and the order in which they are inspected for a flaw is not important, the experiment may be construed as selecting a *set* of 6 sweaters from 50. The number of possible outcomes (sets of 6 sweaters) from such an experiment is $C(50, 6) = 15{,}890{,}700$.

Now, among all possible outcomes, how many meet the criteria of the event E described by "none of the 6 sweaters has a flaw"? The answer is only sets of 6, selected from the 48 that have no flaw. The number of such sets is $C(48, 6) = 12{,}271{,}512$. Therefore, the probability of selecting 6 sweaters in which none has a flaw is

$$P(E) = \frac{C(48, 6)}{C(50, 6)} = \frac{12{,}271{,}512}{15{,}890{,}700} \approx 0.7722.$$

(b) What is the probability that exactly 1 of the 6 sweaters has a flaw?

SOLUTION Among all possible sets of 6, this event asks about the number having 1 sweater with a flaw and 5 without a flaw. Since there are 2 sweaters with a flaw and 48 without a flaw, that number is $C(2, 1) \cdot C(48, 5)$. Therefore, the probability we seek is

$$\frac{C(2, 1) \cdot C(48, 5)}{C(50, 6)} = \frac{2 \cdot (1{,}712{,}304)}{15{,}890{,}700} \approx 0.2155. \quad ■$$

EXAMPLE 10

Downsizing

A company has 13 manufacturing plants throughout the country. It plans to close 3 of these plants, and the order in which they will be closed is based on what least disrupts supply schedules. Assume that economic conditions are such that any one plant is just as likely to be closed as any other. If the plants are numbered 1 through 13, what is the probability that plants 2, 6, and 7 will be closed in the order stated?

SOLUTION The experiment is the act of selecting 3 plants for closing from a total of 13, without replacement, in which order is important. The number of outcomes from this experiment is the number of permutations of 13 plants taken 3 at a time: $P(13, 3) = 1716$. From among these outcomes, the event E of interest to us is the single ordered selection 2, 6, and 7. It follows that

$$P(E) = \frac{1}{P(13, 3)} = \frac{1}{1716} \approx 0.0006. \quad \blacksquare$$

In our last example in this section, we want to reemphasize the importance of listing the outcomes in a sample space whenever possible. Doing so will greatly aid in the counting process that yields probabilities.

EXAMPLE 11

Selecting Slips of Paper

Three slips of paper, each with just the number 1, 2, or 3 written on it, are placed in a box. A slip is randomly selected, the number on it is recorded, and the slip is then returned to the box. Finally, a second slip is randomly selected and the number on it recorded.

(a) Write the sample space for this experiment.

SOLUTION Each outcome from the experiment may be viewed as an ordered pair of numbers, selected with replacement from set $\{1, 2, 3\}$, where no pair is more likely to be selected than another. The sample space consists of the following nine ordered pairs:

$$(1, 1) \quad (1, 2) \quad (1, 3)$$
$$(2, 1) \quad (2, 2) \quad (2, 3)$$
$$(3, 1) \quad (3, 2) \quad (3, 3).$$

(b) What is the probability that the sum of the two numbers drawn is 4?

SOLUTION Of the nine possibilities, those in the event "the sum of the two numbers is 4" are (1, 3), (2, 2), and (3, 1). Therefore the probability we seek is $\frac{3}{9} = \frac{1}{3}$.

(c) What is the probability that the first number drawn is a 2 *or* that the sum of the two numbers drawn is 3?

SOLUTION The outcomes in the event described are (2, 1), (2, 2), (2, 3), and (1, 2). As a consequence, the probability of the event described is $\frac{4}{9}$.

(d) What is the probability that the sum of the two numbers drawn is 8?

SOLUTION The event described is empty: There is no sum that equals 8 in the sample space. Since $n(\varnothing) = 0$, the probability of this event is $\frac{0}{9} = 0$. An impossible event always has probability 0. $\blacksquare$

Exercises 8.1

1. **Spinners** The arrow in the following figure is spun, and the letter representing the area in which the arrow points is then recorded:

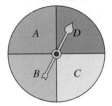

   (a) Write the outcomes in the sample space for this experiment.

   (b) Assign a probability to each outcome in the sample space.

   (c) Find the probability that the outcome is not a vowel.

2. **Spinners** The arrow in the following figure is spun, and the number representing the area in which the arrow points is then recorded:

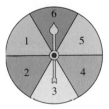

   (a) Write the outcomes in the sample space for this experiment.

   (b) Assign a probability to each outcome in the sample space.

   (c) What is the probability that the arrow points to an even number?

   (d) What is the probability that the arrow points to a number greater than 4?

3. **Family Composition** A sociologist is studying two-child families. For probability considerations, she records the sex of each child in a sample space and assumes that each outcome is as likely as the other.

   (a) Write the possible outcomes of the sociologist's sample space.

   (b) Assign a probability to each outcome in the sample space.

   (c) What is the probability that such families will have at least one girl?

   (d) What is the probability that such families will have a child of each sex?

4. **Tossing a Coin** A coin is tossed three times. On each toss, a record is made of whether the coins lands with heads or tails turned up.

   (a) Write the outcomes of the sample space for this experiment.

   (b) Assign a probability to each outcome in the sample space.

   (c) What is the probability that at least one head turns up?

   (d) What is the probability that exactly two heads turn up?

5. **Family Composition** Terry Smith, a sociologist, is studying three-child families. For probability considerations, she records the sex of each child and assumes that each possible outcome is just as likely as any other.

   (a) Write the outcomes of the sociologist's sample space.

   (b) Assign a probability to each outcome in that sample space.

   (c) What is the probability that such families will have at least one girl?

   (d) What is the probability that such families will have exactly two girls?

6. **Committee Selection** A committee of three is selected from among Alice, Bob, Carl, and Donna. Assume that the committee is selected in such a way that each person is just as likely as any other to be a member of the committee.

   (a) Write the outcomes of the sample space for the experiment.

   (b) Assign a probability to each outcome.

   (c) What is the probability that Alice will be on the committee?

   (d) What is the probability that Bob or Donna will be on the committee?

7. **Selecting Officers** A president and a secretary are selected from among Tigist, Noora, and Ran. Assume that the officers are selected in such a way that each slate of officers is equally likely to occur.

   (a) Write the outcomes of the sample space for the experiment.

   (b) Assign a probability to each outcome.

   (c) What is the probability that Tigist will hold one of the offices?

(d) What is the probability that Noora will be elected president?

8. **Seating Arrangements** Jacob, Lina, and Shawna are randomly seated in a row. (Each arrangement is just as likely as any other to occur.)

   (a) Write the outcomes of the sample space for the experiment of seating three people.

   (b) Assign the probability outcome.

   (c) What is the probability that Shawna will be seated on the left end of the row?

   (d) What is the probability that Jacob or Lina will be seated in the middle?

9. **A Peculiar Die** A four-sided pyramidal die (a tetrahedron) is rolled, and the number of pips on the bottom side is counted. (Assume that one, two, three, and four pips, respectively, appear on each of the sides.)

   (a) Write the outcomes of the sample space for the experiment.

   (b) Assign a probability to each outcome.

   (c) What is the probability of rolling an even number?

   (d) What is the probability of rolling a number greater than 3?

10. **Peculiar Dice** A pair of four-sided pyramidal dice (tetrahedrons), one red and one green, is rolled, and the number of pips on the bottom side of each die is recorded. (Assume that one, two, three, and four pips, respectively, appear on each of the sides.)

    (a) Write the outcomes of the sample space for the experiment.

    (b) Assign a probability to each outcome.

    (c) What is the probability of rolling a sum of 6?

    (d) What is the probability of rolling a sum greater than 5?

11. An experiment consists of tossing a coin and recording whether it lands with heads or tails turned up; then a single four-sided pyramidal die is rolled, and the number of pips on the bottom face is recorded. (See Exercise 9.)

    (a) Write the outcomes in the sample space for the experiment.

    (b) Assign a probability to each outcome.

    (c) What is the probability that the coin turns up heads and an even number of pips shows on the bottom face of the die?

    (d) What is the probability that the die shows a number greater than three on the bottom?

12. An experiment consists of tossing a coin and recording whether it lands heads or tails turned up; then a single standard die is rolled and the number of pips on the upper face is recorded.

    (a) Write the outcomes in the sample space for the experiment.

    (b) Assign a probability to each outcome.

    (c) What is the probability that the coin lands heads up and an even number of pips shows on the upper face of the die?

    (d) What is the probability that the die shows a number greater than 4?

13. **Sampling** Nine balls numbered from 1 through 9, respectively, are in a box. One ball is to be randomly selected from the box and the number on it recorded.

    (a) Write the outcomes of the sample space for this experiment.

    (b) Assign a probability to each outcome.

    (c) What is the probability that the number on the ball selected is greater than seven?

    (d) What is the probability that the number on the ball selected is odd?

14. **Testing** Dr. Sun, a mathematics professor, gives a two-question, multiple-choice quiz in which each question has four possible answers: *A, B, C,* and *D.* Assume that a student guesses at the answers on this quiz.

    (a) Write the outcomes of the sample space for the experiment.

    (b) Assign a probability to each outcome.

    (c) What is the probability that the student will answer both questions correctly?

    (d) What is the probability that the student will answer one question correctly?

15. **Sampling** Three slips of paper, each with just the number 1, 2, or 3 written on it, are placed in a box. A slip is to be drawn and the number on it recorded. That slip is laid aside, and a second slip drawn and the number on it recorded.

    (a) Write the outcomes of the sample space for this experiment.

    (b) Assign a probability to each outcome.

    (c) Find the probability that the sum of the two numbers drawn is 5.

16. **Sampling** Four slips of paper, each with just the number 1, 2, 3, or 4 written on it, are placed in a box. A slip is to be drawn and the number on it recorded. That slip is then

returned to the box and a second slip drawn and the number on it recorded.

(a) Write the outcomes of the sample space for this experiment.

(b) Assign a probability to each outcome.

(c) Find the probability that the sum of the two numbers drawn is 6.

17. Rolling Dice A pair of standard dice is rolled. Find the probability that

(a) A sum of 7 is rolled.

(b) A sum of 11 is rolled.

(c) A sum greater than 8 is rolled.

18. Rolling Dice A pair of standard dice, one red and one green, is to be rolled. Find the probability that

(a) A sum of 10 is rolled.

(b) The red die shows five pips on the top side.

(c) The red die does not show five pips on the top side.

19. Rolling Dice A pair of standard dice, one red and one green, is to be rolled. Find the probability that

(a) A sum of 1 is rolled.

(b) A sum of 2 is rolled.

(c) The red die shows two pips on the top side.

20. Rolling Dice A pair of standard dice is to be rolled. Find the probability that

(a) A sum greater than 1 is rolled.

(b) A sum less than 5 is rolled.

(c) A sum less than 3 or greater than 8 is rolled.

(d) A sum less than 3 and greater than 8 is rolled.

21. Rolling Dice Three standard dice, one red, one green, and one blue, are to be rolled.

(a) What is the probability of rolling a sum of 2?

(b) What is the probability of rolling a sum of 4?

(c) What is the probability of rolling a sum of 7?

22. Rolling Dice Three standard dice, one red, one green, and one blue, are to be rolled.

(a) How many outcomes are in the sample space?

(b) What probability is assigned to each outcome?

(c) What is the probability of rolling a sum of 8?

23. Drawing a Card A single card is to be drawn from a standard deck of 52 cards.

(a) How many outcomes are in the sample space for this experiment?

(b) What is the probability of drawing a club?

(c) What is the probability of drawing a king?

24. Drawing a Card A single card is to be drawn from a standard deck of 52 cards.

(a) What is the probability that the card will have a 4 on it?

(b) What is the probability that the card is red?

(c) What is the probability that the card is a face card (a jack, queen, or king)?

25. Card Hands A 5-card hand is to be dealt from a standard deck of 52 cards.

(a) What is the probability that all five of the cards are clubs?

(b) What is the probability that all five of the cards are from the same suit?

(c) What is the probability that none of the five cards are hearts?

26. Card Hands A 3-card hand is to be dealt from a standard deck of 52 cards.

(a) What is the probability that all 3 cards are spades?

(b) What is the probability that all 3 cards are from the same suit?

(c) What is the probability that none of the 3 cards are hearts?

27. Card Hands A 6-card hand is to be dealt from a standard deck of 52 cards.

(a) What is the probability that all 6 of the cards are face cards?

(b) What is the probability that all 6 of the cards are kings?

(c) What is the probability that none of the 6 cards is a king?

28. Card Hands A 2-card hand is to be dealt from a standard deck of 52 cards.

(a) What is the probability that both of the cards are jacks?

(b) What is the probability that both of the cards are clubs?

(c) What is the probability that neither of the cards is a heart?

29. Card Hands A 5-card hand is to be dealt from a standard deck of 52 cards.

(a) What is the probability that exactly 2 of the cards are spades?

(b) What is the probability that exactly 2 of the cards are queens?

(c) What is the probability that exactly 3 of the 5 cards are not hearts?

30. Card Hands An 8-card hand is to be dealt from a standard deck of 52 cards.

   (a) What is the probability that exactly 5 of the cards are clubs?

   (b) What is the probability that exactly 4 of the cards are face cards?

   (c) What is the probability that exactly 4 of the cards are aces?

31. Quality Control A buyer for a small chain store randomly selects 10 shirts, without replacement, from a lot of 200 shirts to inspect them for a flaw. It is known that 10 of the shirts among the 200 in fact have a flaw. What is the probability that the sample selected by the buyer

   (a) Contains no shirt with a flaw?

   (b) Contains exactly 2 shirts with a flaw?

   (c) Contains exactly 5 shirts with a flaw?

32. Quality Control A campus bookstore buys 100 calculators. Assume that 2 have a defect of some kind. The mathematics department buys 8 of these calculators from the bookstore. What is the probability that, of those calculators bought by the mathematics department,

   (a) All are free of defects?

   (b) Exactly 1 will have a defect?

   (c) Exactly 2 will have a defect?

33. Three boys and two girls are to be randomly seated in a row. What is the probability that

   (a) A boy will be seated in the middle seat?

   (b) Girls will be seated on both ends?

   (c) Boys and girls will be alternately seated?

34. Three boys and three girls are to be randomly seated in a row. What is the probability that

   (a) Boys will be seated on both ends?

   (b) Boys and girls will be alternately seated?

   (c) A particular girl, Tonya, will be seated on the left end?

35. License Plates Suppose the license plates in your state are made with three letters followed by three digits. If you obtain a new license plate, what is the probability that it will

   (a) Start with the letter A and end with the digit 8?

   (b) Have no letter or digit repeated?

   (c) Not contain the letter Z or the digit 0?

36. Identification Numbers Suppose a bank is to issue you an ATM card with a personal identification number (PIN)

consisting of four digits randomly selected by a computer. What is the probability that your PIN will

   (a) Begin with an 8?

   (b) Have no digit repeated?

   (c) Not contain the digit 7?

NOTE Recall that we consider all objects, such as lightbulbs and colored balls, to be distinguishable unless otherwise noted.

37. Quality Control Among 80 lightbulbs, 20 are known to have defects. A sample of 6 bulbs is to be selected, without replacement and without regard to order. What is the probability that the sample will

   (a) Contain no defective bulbs?

   (b) Contain exactly 3 defective bulbs?

   (c) Contain all defective bulbs?

38. Quality Control Among 30 microwave ovens, 10 are known to have defects. A sample of 5 ovens is to be selected, without replacement and without regard to order. What is the probability that the sample will

   (a) Contain no defective ovens?

   (b) Contain exactly 2 defective ovens?

   (c) Contain all defective ovens?

39. Quality Control From among 10 spark plugs, 3 are known to be defective. An ordered sample of 4 spark plugs is to be selected at random, with replacement, from the 10 plugs and tested for the defect. What is the probability that

   (a) None of the plugs will be defective?

   (b) The first 2 plugs tested will be defective?

   (c) All of the plugs tested will be defective?

40. Sampling From among four red and eight green balls, an ordered sample of five balls is to be randomly selected, with replacement. What is the probability that

   (a) All of the balls will be green?

   (b) The first two balls selected will be red?

   (c) All of the balls will be red?

41. Sampling From among three red and five green balls, an ordered sample of three balls is to be randomly selected, without replacement. What is the probability that

   (a) All of the balls will be green?

   (b) All of the balls will be red?

   (c) The first ball will be red and the remaining two will be green?

42. Quality Control From a lot of 10 swimsuits, 4 are known to have a flaw. An ordered sample of 3 swimsuits is to be randomly selected, without replacement, from the 10 swimsuits. What is the probability that

(a) None will have a flaw?

(b) The first two will have a flaw and the third will not have a flaw?

(c) The first one selected will have a flaw and the remaining two will not have a flaw?

43. **Sampling** From among four red and six green balls, a sample of four balls is to be randomly drawn, without replacement and without regard to order. What is the probability that

(a) All of the balls will be green?

(b) Exactly two of the balls will be red?

(c) All of the balls will be red?

44. **Sampling** From among three red, four green, and three purple balls, a sample of three balls is to be randomly drawn, without replacement and without regard to order. What is the probability that

(a) All of the balls will be green?

(b) Exactly two of the balls will be red?

(c) Exactly one ball will be purple?

45. **Committee Selection** An executive committee of 5 people is to be selected from among 10 managers, 4 of whom are men and 6 of whom are women. If the committee is selected by drawing lots (meaning that each committee is just as likely as any other) what is the probability that

(a) All members of the committee will be women?

(b) Exactly 2 of the committee members will be men?

(c) Exactly 3 of the committee members will be women?

46. **Committee Selection** From among the mayor and five other city council members, a committee of three is to be selected for inspection of city waste management facilities. If this committee is selected by drawing lots (meaning that each committee is just as likely as any other), what is the probability that

(a) The mayor will not be on the committee?

(b) The mayor will be on the committee?

Section 8.2 Outcomes with Unequal Probability; Odds

Our investigation of sample spaces will be complete when we consider experiments that do not have equally likely outcomes. Sometimes examples of such experiments arise in a basic sense, such as the spinner shown in Figure 4.

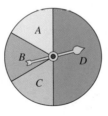

Outcome	Probability
A	$\frac{1}{6}$
B	$\frac{1}{6}$
C	$\frac{1}{6}$
D	$\frac{1}{2}$
	Sum: 1

FIGURE 4

If the arrow is spun and the letter representing the area to which the arrow points is recorded, then the possible outcomes are A, B, C, and D and the sample space is $S = \{A, B, C, D\}$. From the appearance of the spinner, we cannot assign equal probabilities to each of these four outcomes; outcome D deserves a larger number than the other three. In fact, since the area labeled D is one-half of the total area of the circle, we would expect the arrow to point to D about half of the time, and hence D should be assigned a probability of $\frac{1}{2}$. Similarly, the areas labeled A, B, and C each occupy $\frac{1}{3}$ of $\frac{1}{2}$, or $\frac{1}{6}$, of the total area of the circle, which means that they should be assigned a probability of $\frac{1}{6}$ each.

These probability assignments tell us that if the experiment is repeated over and over, then the long-term effects are that about $\frac{1}{6}$ of the time the arrow will point to one of the regions A, B, and C, respectively, and about half of the time it will point to region D. Notice that the probabilities are assigned in such a way that their sum is 1, just as in the equally likely case of Section 8.1.

How would we find the probability of the event $E = \{C, D\}$, for example? Stated differently, "What proportionate share of the total should the region $C \cup D$ be assigned?" Would you agree that $\frac{1}{6} + \frac{1}{2} = \frac{4}{6} = \frac{2}{3}$ should be the answer? In other words, just add the probabilities of the various outcomes in the event E:

$$P(E) = P(\{C, D\}) = P(C) + P(D) = \frac{1}{6} + \frac{1}{2} = \frac{2}{3}.$$

Following this line of reasoning, the probabilities of some other events in this sample space are listed next:

$$P(\{A, B, D\}) = P(A) + P(B) + P(D) = \frac{1}{6} + \frac{1}{6} + \frac{1}{2} = \frac{5}{6}.$$

$$P(\{A, B\}) = P(A) + P(B) = \frac{1}{6} + \frac{1}{6} = \frac{2}{6} = \frac{1}{3}.$$

It seems reasonable to define the probability of an event to be the sum of the probabilities of all outcomes that make up the event. This definition also encompasses the way events were handled in the equally likely case.

Thus far in this chapter, we have observed several properties of probabilities. Since an outcome is assigned a probability based on the proportion of the time we expect the outcome to occur, that probability must be a number between 0 and 1. The sum of the probabilities of all the outcomes in a sample space must always be 1. Finally, the probability that no outcome in the sample space occurs is 0. These properties, together with our observations regarding the probability of an event, are summarized next.

General Rules for Probability

If a sample space S has a finite number of outcomes $O_1, O_2, \ldots O_n$, then the assignment of probabilities to that space, denoted by

$$P(O_1) = p_1, P(O_2) = p_2, \ldots, P(O_n) = p_n,$$

must obey the following three rules:

1. Each probability $p_1, p_2, \ldots, p_n$ is a number between 0 and 1, inclusive.
2. $p_1 + p_2 + \cdots + p_n = 1$.
3. $p(\emptyset) = 0$.

The **probability of an event E** in S is found by adding the probabilities of all outcomes making up E.

EXAMPLE 1

Applying the General
Rules of Probability

Suppose a particular experiment results in the sample space $S = \{a, b, c, d\}$.

(a) If all outcomes of S are equally likely, what is the probability of a single outcome?

SOLUTION Let p represent the equal probability of each outcome; that is, let $P(a) = P(b) = P(c) = P(d) = p$. Then, since the sum of the probabilities of all the outcomes in S must equal 1, we have

$$P(a) + P(b) + P(c) + P(d) = 1$$
$$p + p + p + p = 1 \qquad \text{Substitute } p.$$
$$4p = 1 \qquad \text{Simplify.}$$
$$p = \frac{1}{4}. \qquad \text{Divide both sides by 4.}$$

(b) If not all outcomes in S are equally likely—for instance, $P(a) = 0.3$, $P(b) = 0.2$, and $P(c) = 0.1$—what is $P(d)$?

SOLUTION Again, since the sum of the probabilities of all the outcomes in S must be equal to 1, we have

$$P(a) + P(b) + P(c) + P(d) = 1$$
$$0.3 + 0.2 + 0.1 + P(d) = 1 \qquad \text{Substitute known probabilities.}$$
$$0.6 + P(d) = 1 \qquad \text{Simplify.}$$
$$P(d) = 0.4. \qquad \text{Solve for } P(d).$$

(c) If $E = \{a, c, d\}$ is an event in S with probability $P(E) = 0.8$, what is $P(b)$?

SOLUTION Since the probability of an event in S is found by adding the probabilities of all outcomes making up the event, we have

$$P(a) + P(c) + P(d) = P(E) = 0.8.$$

Therefore,

$$P(a) + P(b) + P(c) + P(d) = 1$$
$$0.8 + P(b) = 1 \qquad \text{Substitute known value for } P(E).$$
$$P(b) = 0.2. \qquad \text{Solve for } P(b).$$

(d) Explain why the probability assignment $P(a) = 0.4$, $P(b) = 0.4$, $P(c) = 0.4$, $P(d) = -0.2$ is not valid.

SOLUTION Even though the sum of the probabilities is equal to 1, outcome d has been assigned a *negative* probability. Since each outcome must be assigned a probability between 0 and 1, inclusive, this probability assignment is not valid. ■

Another way in which a sample space having outcomes that are not equally likely can arise occurs when there are two components to an experiment: (1) an activity and (2) what is recorded from that activity. In essence, then, there can be many sample spaces arising from a given activity, depending upon what is recorded. Example 2 shows how a familiar activity can result in a sample space that is different from our previous considerations.

EXAMPLE 2

Another Look at Tossing Coins

A coin is to be tossed twice and the number of times it lands heads up is recorded. Find the sample space and assign a probability to each outcome of that space.

SOLUTION We begin with the fundamental premise that tossing a coin twice will result in four outcomes, each of which is an ordered pair $\{HH, HT, TH, TT\}$, and where each of these outcomes has probability $\frac{1}{4}$. This creates an **underlying sample space** with equally likely outcomes. From this sample space, a new sample space is created whose outcomes reflect the number of heads that could turn up: 0 (TT), 1 (HT or TH), or 2 (HH). Thus, the outcomes in the new sample space—0, 1, and 2—define *events* in the underlying space. For that reason, the term **event space** is sometimes applied to the new sample space. The probability assignments for the outcomes in the new sample space are those of corresponding events in the underlying sample space of equally likely outcomes. A list of those assignments is as follows:

Outcome (number of heads)	Event in Underlying Space	Probability
0	$\{TT\}$	$\frac{1}{4}$
1	$\{HT, TH\}$	$\frac{1}{4} + \frac{1}{4} = \frac{1}{2}$
2	$\{HH\}$	$\frac{1}{4}$
		Sum: 1

The first and last columns of this list show how the probability is distributed over the outcomes 0, 1, and 2 and is called the **probability distribution** for the sample space. (The middle column merely clarifies how the assignments are made.) This distribution can be pictured in a type of bar graph called a **probability histogram,** as shown in Figure 5. ▄

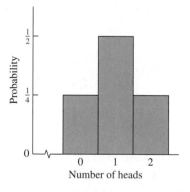

FIGURE 5
Probability Distribution Histogram
for the Number of Heads on
Two Tosses of a Coin

EXAMPLE 3

Another Look at
Tossing a Die

A standard die is rolled, and the number of pips on the top side is observed. A "success" is recorded if four or fewer pips are observed; otherwise, a "failure" is recorded. Find the sample space for this experiment and the probability of each outcome of the sample space.

SOLUTION The underlying sample space is the familiar one that records the number of pips on the top side of the die—$\{1, 2, 3, 4, 5, 6\}$—where each outcome has probability $\frac{1}{6}$. From this sample space, a new one is constructed that uses the desired outcomes "success" and "failure," whose probabilities are obtained from the events they define in the underlying sample space. The probability distribution of the new sample space is as follows: ▄▄

Outcome	Event in Underlying Space	Probability
Success	$\{1, 2, 3, 4\}$	$\frac{4}{6} = \frac{2}{3}$
Failure	$\{5, 6\}$	$\frac{2}{6} = \frac{1}{3}$
		Sum: 1

NOTE Recall that all objects, such as colored balls and lightbulbs, are considered distinguishable unless otherwise noted.

EXAMPLE 4

Selecting a Ball

A box contains five red and three green balls. One ball is to be randomly selected from the box and its color recorded. Find the sample space for this experiment and the probability of each outcome of this sample space.

SOLUTION Since the balls are distinguishable, visualize the red ones to be labeled with the numerals 1 through 5, respectively, and the green ones to be labeled with the numerals 6 through 8, respectively. We consider the selection of any one ball to be as likely as that of any other ball. The outcome "red" then defines the event $\{1, 2, 3, 4, 5\}$ with probability $\frac{5}{8}$, and the outcome "green" defines the even $\{6, 7, 8\}$ with probability $\frac{3}{8}$. The probability assignments for these two outcomes are as follows: ▄▄

Outcome (color of ball)	Event in Underlying Space	Probability
Red	$\{1, 2, 3, 4, 5\}$	$\frac{5}{8}$
Green	$\{6, 7, 8\}$	$\frac{3}{8}$
		Sum: 1

EXAMPLE 5

Selecting a Card

Suppose that, in a standard deck of cards, each ace is assigned the value 0, each face card is assigned the value 1, and the remaining cards are assigned the number on the card. One card is to be selected and its value noted.

(a) Find the sample space for this experiment and the probability of each outcome of that sample space.

SOLUTION The underlying sample space consists of 52 outcomes (cards), each with probability $\frac{1}{52}$. The new sample space has 11 outcomes—0, 1, 2, 3, ... , 10. The number 0 is assigned probability $\frac{4}{52} = \frac{1}{13}$ because there are 4 aces out of 52 cards in the underlying sample space; the outcome 1 is assigned $\frac{12}{52} = \frac{3}{13}$ because there are 12 face cards out of 52 cards in the underlying sample space; that outcome 2 is assigned probability $\frac{4}{52} = \frac{1}{13}$ because there are four 2's out of 52 cards in the underlying sample space, ... , the outcome 10 is assigned probability $\frac{4}{52} = \frac{1}{13}$ because there are four 10's out of 52 cards in the underlying space. Hence, we have the following table:

Outcome	0	1	2	3	4	5	6	7	8	9	10
Probability	$\frac{1}{13}$	$\frac{3}{13}$	$\frac{1}{13}$	$\frac{1}{13}$	$\frac{1}{13}$	$\frac{1}{13}$	$\frac{1}{13}$	$\frac{1}{13}$	$\frac{1}{13}$	$\frac{1}{13}$	$\frac{1}{13}$

Note that the sum of these probabilities is 1.

(b) Find the probability that a card with value of at least 7 will be drawn.

SOLUTION The event described consists of the outcomes $E = \{7, 8, 9, 10\}$ and has probability

$$P(E) = P(7) + P(8) + P(9) + P(10)$$
$$= \frac{1}{13} + \frac{1}{13} + \frac{1}{13} + \frac{1}{13} = \frac{4}{13}.$$

(c) Find the probability that a card with value less than 3 will be drawn.

SOLUTION The event described consists of the outcomes $E = \{0, 1, 2\}$ and has probability

$$P(E) = P(0) + P(1) + P(2)$$
$$= \frac{1}{13} + \frac{3}{13} + \frac{1}{13} = \frac{5}{13}. \quad \blacksquare$$

The probability assignments made throughout Section 8.1 and thus far in this section were **theoretical** in nature. That is, they were made by deductive reasoning alone. For example, when we said that the probability of a coin falling heads up was $\frac{1}{2}$, we did not require that a coin be tossed or even that a coin be at hand; we had a conceptual experiment and an idealized outcome. Nonetheless, we feel comfortable with such theoretical concepts because they seem logical and can be experimentally verified. In contrast, suppose we had a certain coin that was slightly bent in such a way that, upon 1000 tosses, it turned up heads 750 times. In this case, we would be much more inclined to rely on experimental evidence, rather than an idealized concept, and assign heads a probability of $\frac{750}{1000} = 0.75$ based on the **relative frequency** with which heads turned up. Similarly, the probability or relative frequency that a tail would turn up would be $\frac{250}{1000} = 0.25$. Example 6 explores this concept further.

EXAMPLE 6

A Botany Experiment

From a large volume of pea pods, a botanist broke open 20 pods and counted the number of peas in each pod. The results were as follow:

5 pods contained 4 peas each

9 pods contained 5 peas each

4 pods contained 6 peas each

2 pods contained 7 peas each

(a) Use the preceding data to assign a probability to the number of peas that will be found in a typical pea pod.

SOLUTION The experiment consists of breaking open pea pods and recording the number of peas in each pod. The outcomes are 4, 5, 6, or 7 peas. On the basis of these data, 5 of the 20 pods contained 4 peas each. Therefore, it seems reasonable to assign the outcome 4 a probability of $\frac{5}{20}$. Stated differently, the **frequency** with which a pod contained 4 peas was 5; therefore, the **relative frequency** with which a pod contained 4 peas was $\frac{5}{20}$. Similarly, the relative frequency with which a pod contained 5 peas—and hence the probability—was $\frac{9}{20}$. The probability (relative frequency) distribution for the given pea pods is shown in the following table:

Outcome of Experiment (number of peas in pod)	Probability Assignment (relative frequency)
4	$\frac{5}{20}$
5	$\frac{9}{20}$
6	$\frac{4}{20}$
7	$\frac{2}{20}$
	Sum:1

A visual picture of this probability distribution is shown in the relative frequency histogram of Figure 6.

(b) Based on these data, what is the probability that a pea pod will contain fewer than 6 peas?

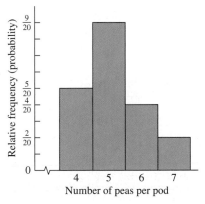

FIGURE 6

Relative-Frequency Histogram

SOLUTION Stated differently, this question asks for the probability of the event $E = \{4, 5\}$. That probability is $\frac{5}{20} + \frac{9}{20} = \frac{14}{20}$.

(c) On the basis of the data shown, what percentage of a large volume of pea pods will contain 6 or more peas? This question can be restated by asking for the probability of the event $E = \{6, 7\}$ and then converting the probability to a percentage. The probability of E is $\frac{4}{20} + \frac{2}{20} = \frac{6}{20} = 0.30$, or about 30%. ▪

The assignment of probabilities in Example 6 is unlike that employed in the theoretical models in two ways. First, the sample size of 20 leads only to an approximation of the true probabilities for number of peas per pod. However, as the sample size increases, it portrays the true proportion of pods containing 4, 5, 6, or 7 peas more accurately. Second, there is no realistic way to perform this experiment repeatedly under exactly the same conditions—as is possible with tossing a coin. So the best we can do is use the relative frequencies from a collected sample to assign probabilities and assume that they are valid over the entire population, even though *they are approximate* in that sense. Relative frequency assignments are also called **empirical probabilities,** because they are based on empirical evidence.

EXAMPLE 7

Flu Shots: An Application of Empirical Probability

Medical authorities in a particular state examined randomly selected records of 250 residents who had been given flu shots in the fall and found that 6 of the residents had contracted the flu sometime during the next winter. The authorities then released the following figures showing the chances of getting the flu even though a shot was given:

Outcome	Probability (relative frequency)
Had shot, will get flu	$\frac{6}{250} = 0.024$
Had shot, will not get flu	$\frac{244}{250} = 0.976$
	Sum: 1

Another way of expressing this probability would be to say that, among the residents who had been given flu shots, 2.4% would get the flu, and 97.6% would not. ▪

EXAMPLE 8

Auditing: An Application of Empirical Probability

An auditor of personal income tax forms submitted to a certain state gathered the following information from a randomly selected sample of 300 forms:

Income Level	Correct	Understated Tax	Overstated Tax
$30,000–$50,000	125	20	10
$50,000–$100,000	110	30	5

Find the following empirical probabilities for a randomly selected individual income tax form submitted to the state in question.

(a) Find the probability that a person will submit a correct form.

SOLUTION We assume that any of the 300 forms are equally likely to be selected. According to the data collected, $125 + 110 = 235$ of the forms were correct. Therefore, the probability assigned to this event is $\frac{235}{300}$.

(b) Find the probability that a person will be in the \$30,000–\$50,000 income level *and* will submit a tax form that understates his or her tax.

SOLUTION There are 20 tax forms meeting this criterion, implying that the probability assigned to this event is $\frac{20}{300}$.

(c) Find the probability that a person will be in the \$30,000–\$50,000 income level *or* will submit a tax form that understates his or her tax.

SOLUTION There are $125 + 20 + 10 + 30 = 185$ tax forms meeting this criterion. Consequently, the probability assignment to this event is $\frac{185}{300}$. ■

Odds

Another form of probability is in the statements of "odds" for or against some event. For example, "the odds are 7 to 5 that the Eagles will beat the Giants" or "the odds are 3 to 2 against having a recession next year" are actually statements about the likelihood of an event occurring and may therefore be put in the context of probability. Specifically, the statement "the odds are 7 to 5 that the Eagles will beat the Giants" means that if a 12-game series $(7 + 5 = 12)$ is played over and over between these two teams under identical conditions, the tendency would be for the Eagles to win 7 of the games and lose 5. Accordingly, the odds statement implies that the probability of the Eagles winning against the Giants is $\frac{7}{12}$, while the probability that the Eagles will lose is $\frac{5}{12}$. Following this pattern, if the odds *for* an event E are stated as a to b, then

$$P(E) = \frac{a}{a + b} \quad \text{and} \quad P(E') = \frac{b}{a + b}.$$

The preceding discussion shows how an odds statement can be converted to a probability statement. It also shows that the odds statement "a to b" may be thought of as a comparison of the probability that E will occur to the probability that E will not occur.

EXAMPLE 9
Computing Probability
When the Odds
Are Known

If the odds for rain today are estimated to be 1 to 3, what is the probability of rain today?

SOLUTION From the meaning of odds, $P(\text{rain}) = \frac{1}{1 + 3} = \frac{1}{4}$. ■

Now supposed we are told that the probability is $\frac{7}{12}$ that the Eagles will beat the Giants. How do we find the *odds* of the Eagles beating the Giants? Note that the probability that the Eagles will lose is $\frac{(12 - 7)}{12} = \frac{5}{12}$. The numerators provide the comparison of winning to losing: 7 to 5. This ratio can also be obtained by dividing the probability of the Eagles winning by the probability of the Eagles losing: $(\frac{7}{12})/(\frac{5}{12}) = \frac{7}{5}$. In general, suppose we know that $P(E) = m/n$. How can we find the odds for E? First, note that $P(E') = 1 - m/n = (n - m)/n$. Then think of odds as being a comparison of the probability of E occurring to the probability of E not occurring: m to $n - m$. This ratio can be obtained by computing

$$\frac{P(E)}{P(E')} = \frac{\dfrac{m}{n}}{\dfrac{n-m}{n}} = \frac{m}{n-m}$$

Odds

Let E be an event in a sample space S where E is not all of S. Then the **odds for E** are found by reducing $P(E)/P(E')$ to lowest terms a/b and writing "a to b" or "$a : b$". Similarly, the **odds against E** are found by reducing $P(E')/P(E)$ to lowest terms.

EXAMPLE 10

Computing Odds

A single card is drawn from a deck of 52 cards, and the card is noted. What are the odds that the card will be an ace? What are the odds against the card being an ace?

SOLUTION Let E be the event "an ace is drawn." Then $P(E) = \frac{4}{52}$ and $P(E') = \frac{48}{52}$. Therefor, the odds for E are

$$\frac{P(E)}{P(E')} = \frac{\dfrac{4}{52}}{\dfrac{48}{52}} = \frac{4}{48} = \frac{1}{12},$$

or "1 to 12," or 1:12. The odds *against* an ace being drawn are 12:1. ∎

Exercises 8.2

In Exercises 1 through 4, consider the sample space $S = \{a, b, c, d\}$. Use the information given to find the desired probabilities.

1. $P(a) = \frac{1}{3}$, $P(b) = \frac{1}{3}$, $P(c) = \frac{1}{6}$; find $P(d)$.

2. $P(a) = P(b) = P(d) = \frac{1}{9}$; find $P(c)$.

3. $E = \{b, c, d\}$, $P(E) = \frac{1}{2}$; find $P(a)$.

4. $E = \{a, c\}$, $P(E) = \frac{1}{6}$, $P(b) = P(d)$; find $P(b)$.

In Exercises 5 through 8, consider the sample space $S = \{a, b, c, d\}$. Use the general rules for probability to explain why the given probability assignment is not valid.

5. $P(a) = \frac{1}{3}$, $P(b) = \frac{1}{3}$, $P(c) = \frac{1}{3}$; $P(d) = \frac{1}{3}$.

6. $P(a) = \frac{1}{2}$, $P(b) = \frac{1}{4}$, $P(c) = \frac{1}{4}$; $P(d) = \frac{1}{4}$.

7. $P(a) = \frac{1}{6}$, $P(b) = \frac{1}{6}$, $P(c) = \frac{1}{12}$; $P(d) = \frac{1}{12}$.

8. $P(a) = \frac{1}{3}$, $P(b) = \frac{1}{3}$, $P(c) = \frac{2}{3}$; $P(d) = -\frac{1}{3}$.

9. **Spinners** The arrow on the spinner in the accompanying figure is spun, and the letter representing the area the arrow points to is then recorded.

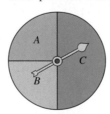

(a) Find the sample space for this experiment and assign probabilities to each outcome of the sample space.

(b) What is the probability that the arrow will point to the area labeled A or the area labeled C?

(c) What is the probability that the arrow will not point to the area labeled A?

10. **Spinners** The arrow on the spinner in the following figure is spun, and the letter representing the area the arrow points to is then recorded.

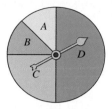

(a) Find the sample space for this experiment, and assign probabilities to each outcome of the sample space.

(b) What is the probability that the arrow will point to the area labeled *A* or the area labeled *D*?

(c) What is the probability that the arrow will point to the area labeled *B* or the area labeled *C*?

11. **Rolling a Die** A die is to be rolled. If three pips show on the upper side, a "success" is recorded. Otherwise, a "failure" is recorded. Find the sample space for this experiment, and assign probabilities to each outcome of the sample space.

12. **Colored Balls** Five red and nine green balls are in a box. One ball is to be randomly drawn from the box and the color recorded.

(a) Find the sample space for this experiment, and assign probabilities to each outcome of the sample space.

(b) What is the probability that the ball will not be red?

13. **Colored Balls** Six red, two green, and seven blue balls are in a box. One ball is to be randomly drawn from the box and the color recorded.

(a) Find the sample space for this experiment, and assign probabilities to each outcome of the sample space.

(b) What is the probability that the ball is red or green?

(c) What is the probability that the ball will not be red?

14. **Colored Balls** Eight red, five green, seven blue, and ten purple balls are in a box. One ball is to be randomly drawn from the box and the color recorded.

(a) Find the sample space for this experiment, and assign probabilities to each outcome of the sample space.

(b) What is the probability that the ball is red or green?

(c) What is the probability that the ball will not be red?

15. **Dice** A pair of dice is to be rolled. If a sum of 7 or 8 is rolled, a "success" is recorded. Otherwise a "failure" is recorded. Find the sample space for this experiment, and assign probabilities to each outcome of the sample space.

16. **Dice** A pair of dice is to be rolled. If a sum of 2 or 10 is rolled, a "success" is recorded. Otherwise a "failure" is recorded. Find the sample space for this experiment, and assign probabilities to each outcome of the sample space.

17. **Cards** One card is to be drawn from a deck of 52 cards. If the card is a heart, a "success" is recorded. Otherwise a "failure" is recorded. Find the sample space for this experiment, and assign probabilities to each outcome of the sample space.

18. **Cards** One card is to be drawn from a deck of 52 cards. If the card is a jack, a "success" is recorded. Otherwise a "failure" is recorded. Find the sample space for this experiment, and assign probabilities to each outcome of the sample space.

19. **Tossing Coins** A coin is to be tossed twice. If two heads turn up, a "success" is recorded. Otherwise a "failure" is recorded. Find the sample space for this experiment, and assign probabilities to each outcome of the sample space.

20. **Tossing Coins** A coin is to be tossed twice. If one head and one tail turn up, a "success" is recorded. Otherwise, a "failure" is recorded. Find the sample space for this experiment, and assign probabilities to each outcome of the sample space.

21. **Tossing Coins** A coin is to be tossed three times and the number of heads recorded.

(a) Find the sample space for this experiment and the probability of each outcome of that sample space.

(b) What is the probability that at least two heads will turn up?

(c) What is the probability that two or fewer heads will turn up?

(d) Construct the probability distribution histogram for this experiment.

22. **Tossing Coins** A coin is to be tossed three times, and the number of tails on the last two tosses is recorded.

(a) Find the sample space for this experiment and the probability of each outcome of that sample space.

(b) What is the probability that at least one tail will turn up on the last two tosses?

(c) What is the probability that fewer than two tails will turn up on the last two tosses?

(d) Construct the probability distribution histogram for this experiment.

New Car Sales In Exercises 23 through 26, refer to the following relative-frequency histogram for new car sales in the United States during 1999, organized by type of vehicle:

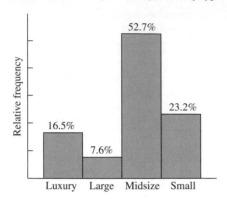

SOURCE: World Almanac and Book of Facts 2001.

A customer who purchased a new car in 1999 is randomly selected, and the type of car purchased is recorded.

23. What is the probability that the selected customer purchased a large car?

24. What is the probability that the selected customer purchased a luxury car?

25. What is the probability that the selected customer purchased a midsize or a small car?

26. What is the probability that the selected customer did not purchase a small car?

27. **Dice** A pair of dice, one red and one green, is to be rolled. The number of pips on the top face of the red die is recorded.

   (a) Find the sample space for this experiment and the probability of each outcome of that sample space.

   (b) What is the probability that the red die shows at least four pips on the top face?

   (c) What is the probability that the red die shows at most two pips on the top face?

   (d) Construct the probability distribution histogram for this experiment.

28. **Seating Arrangements** Three boys and two girls are to be seated randomly in a row, and the number of boys who sit in the last two chairs of the row is recorded.

   (a) Find the sample space for this experiment and the probability of each outcome of that sample space.

   (b) Find the probability that at least one boy sits in the last two chairs.

   (c) Construct the probability distribution histogram for this experiment.

29. **Seating Arrangements** Three boys and two girls are to be seated randomly in a row, and the number of girls who sit in the last two chairs of the row is recorded.

   (a) Find the sample space for this experiment and the probability of each outcome of that sample space.

   (b) Find the probability that at least one boy sits in the last two chairs.

   (c) Construct the probability distribution histogram for this experiment.

30. **I. D. Numbers** A four-digit identification number is to be selected at random, and the number of zeros in the last two digit positions is recorded.

   (a) Find the sample space for this experiment and the probability of each outcome of that sample space.

   (b) Find the probability that at most one zero appears in the last two digit positions.

   (c) Construct the probability distribution histogram for this experiment.

31. **I. D. Numbers** A three-digit identification number is to be selected at random, and the number of twos in the last two digit positions is recorded.

   (a) Find the sample space for this experiment and the probability of each outcome of that sample space.

   (b) Find the probability that at most one of the last two digit positions contains the number 2.

   (c) Construct the probability distribution histogram for this experiment.

32. **Biased Die** A die is biased in such a way that the probability of a particular side turning up is proportional to the number of pips on that side. Such a die is to be rolled and the number of pips on the top side recorded.

   (a) Find the sample space for this experiment and the probability of each outcome of that sample space.

   (b) Find the probability that at least a four will be rolled.

   (c) Construct the probability distribution histogram for this experiment.

33. **Cards** Two cards are to be dealt, without replacement, from a deck of 52 cards, and the number of kings is to be recorded.

   (a) Find the sample space for this experiment and the probability of each outcome of that sample space.

   (b) Find the probability that at least one king will be dealt.

   (c) Construct the probability distribution histogram for this experiment.

34. Cards Two cards are to be dealt, without replacement, from a deck of 52 cards, and the number of clubs is to be recorded.

(a) Find the sample space for this experiment and the probability of each outcome of that sample space.

(b) Find the probability that at least one club will be dealt.

(c) Construct the probability distribution histogram for this experiment.

35. Quality Control Among 10 computers, 3 are known to have a defect. A sample of 2 computers is to be selected, without replacement and without regard to order, and the number of defective computers recorded.

(a) Find the sample space for this experiment and the probability of each outcome of that sample space.

(b) Find the probability that at least 1 of the computers will have a defect.

(c) Construct the probability distribution histogram for this experiment.

36. Quality Control Among 12 microwave ovens, 4 are known to have a defect. A sample of 2 ovens is to be selected, without replacement and without regard to order, and the number of defective ovens recorded.

(a) Find the sample space for this experiment and the probability of each outcome of that sample space.

(b) Find the probability that at least 1 of the ovens will have a defect.

(c) Construct the probability distribution histogram for this experiment.

37. Marketing Survey A survey of 20 households asked how many radios they owned. The results were as follows:

5 households answered 3

8 households answered 4

7 households answered 5

(a) Construct the relative-frequency (probability) distribution for this experiment.

(b) Construct the relative-frequency histogram.

(c) Find the probability that a household will own at least four radios.

38. Marketing Survey A survey of 25 households asked how many television sets they owned. The results were as follows:

2 households answered 0

8 households answered 1

12 households answered 2

3 households answered 3

(a) Construct the relative-frequency (probability) distribution for this experiment.

(b) Construct the relative-frequency histogram.

(c) Find the probability that a household will own at least two television sets.

39. Agriculture Survey A farmer broke open 18 soybean pods and counted the number of beans in each pod. The results were as follow:

2 pods contained 2 beans

5 pods contained 3 beans

8 pods contained 4 beans

3 pods contained 5 beans

(a) Construct the relative-frequency (probability) distribution for this experiment.

(b) Construct the relative-frequency histogram.

(c) Find the probability that a bean pod will contain three or more beans.

40. Agriculture Survey On a certain day early in the growing season, a commercial tomato grower checked the number of blooms on each of 26 plants. The results were as follows:

8 plants had 5 blooms

12 plants had 8 blooms

6 plants had 10 blooms

(a) Construct the relative-frequency (probability) distribution for this experiment.

(b) Construct the relative-frequency histogram.

(c) Find the probability that a plant will have fewer than 8 blooms.

41. **Biased Coin** A coin is tossed twice and the number of heads recorded. The experiment was repeated 1000 times, with the following results:

Outcome Recorded	Frequency of Outcome
0 heads	290
1 head	510
2 heads	200

(a) Make an empirical probability distribution for this information.

(b) Find the probability that at least one head will occur in two tosses of the coin.

42. **Biased Die** A die was tossed 600 times and the number of pips on the top face recorded. The results were as follows:

Number of Pips	Frequency of Outcome
1	100
2	90
3	140
4	80
5	105
6	85

(a) Make an empirical probability distribution for this information. If the die is rolled, find the empirical probability that

(b) A 4 or greater turns up.

(c) An even number turns up.

(d) An odd number turns up.

43. **Economic Trends** A survey of 120 economists shows that 64 predict a recession, 42 predict that the economy will show slow growth, and 14 predict sharply higher growth within the next year.

(a) Make an empirical probability distribution for this information.

(b) Based on the given information, find the probability that there will be no recession within the next year.

44. **Political Polls** Jones, Brown, and Smith are in a three-way race for the Democratic party's nomination for governor. A week before the election, 80 potential Democratic voters were surveyed, with 32 of them favoring Jones, 18 favoring Brown, 15 favoring Smith, and 15 undecided.

(a) Make an empirical probability distribution for this information.

(b) Assume that the undecided vote is split equally among the three candidates. Find the probability that a voter will favor Jones in the election.

45. **Product Reliability** A large number of computer chips was tested, and it was found that 92% were usable.

(a) Make an empirical probability distribution for this information.

(b) If one of these chips is randomly selected, what is the probability that it will be usable?

46. **Product Reliability** An inspection of 54 children's coats checked for accuracy of sizing revealed that 6 were too small, 3 were too large, and 45 were within accepted tolerances.

(a) Make an empirical probability distribution for this information.

(b) Find the probability that a randomly selected coat is not correctly sized.

(c) Find the probability that a randomly selected coat is not too small.

In Exercises 47 and 48, the maker of a certain brand of automobile gathered the following information on 450 persons who leased their automobiles for two-year periods:

Age	Number of Times Lease Renewed			
	1	2	3	4
20–29	10	8	0	0
30–39	12	20	16	3
40–49	6	28	50	31
50–59	20	46	29	35
60 or over	8	42	36	50

47. **Marketing Analysis** Find the empirical probabilities for each of the following events:

(a) Being in the 30–39 age range and renewed a lease three times.

(b) Being in the 50–59 age range and renewed a lease at least twice.

(c) Being in the 50–59 age range or renewed a lease one time.

(d) Renewed a lease twice.

48. **Market Analysis** Find the empirical probabilities for each of the following events:

   (a) Being in the 20–29 age range and renewed a lease three times.

   (b) Being in the 60 or over age range and renewed a lease at most three times.

   (c) Being in the 60 or over age range or renewed a lease at most three times.

   (d) Being in the 40–49 age range and renewed a least at least once.

In Exercises 49 and 50, an audit of 250 billing statements from a large trucking firm for both short-haul and long-haul routes revealed the following information:

	Short-haul Route	Long-haul Route
Correct	100	82
Overbilled	30	20
Underbilled	10	8

49. **Market Analysis** If a billing statement from the firm is randomly selected, find the empirical probability that

   (a) The statement is correct.

   (b) The statement is for a long-haul route.

   (c) The statement overbills and is for a short-haul route.

   (d) The statement underbills or is for a long-haul route.

50. **Market Analysis** If a billing statement from the firm is randomly selected, find the empirical probability that

   (a) The statement underbills.

   (b) The statement is for a short-haul route.

   (c) The statement is correct and is for a long-haul route.

   (d) The statement overbills or is for a long-haul route.

Exercises 51 and 52 refer to the following: In order to learn whether there was a relationship between the number of visits a student made to the main campus library and the student's class standing, the library at a large state university asked 200 randomly selected students how many times they visited the main campus library each week. The number of students, by classification and number of library visits, is shown in the following table:

Class Standing	Number of Visits to the Main Library Each Week				
	0	1	2	3	Over 3
Freshman	3	4	5	3	1
Sophomore	8	7	8	4	3
Junior	4	12	20	18	12
Senior	1	25	30	22	10

51. **Library Habits** If an undergraduate student attending the state university is chosen at random, find the empirical probability that the student is

   (a) A freshman who made three or more visits to the library in a week.

   (b) A sophomore or a junior and did not visit the library in a week.

   (c) A junior and visited the library twice a week.

   (d) A senior or visited the library three times in a week.

52. **Library Habits** If an undergraduate student attending the state university is chosen at random, find the empirical probability that the student is

   (a) A sophomore and visited the library more than three times a week.

   (b) A junior or senior who visited the library once in a week.

   (c) A senior and visited the library once in a week.

   (d) A senior or visited the library once in a week.

53. **Odds** A single card is to be selected from a deck of 52 cards. Find the odds

   (a) For the card being a king.

   (b) For the card being a spade.

   (c) For the card being red.

   (d) Against the card being a club.

54. **Odds** A pair of dice, one red and one green, is to be rolled. Find the odds

   (a) For the sum of the pips on the top sides to be 6.

   (b) For the sum of the pips on the top sides to be 11.

   (c) For the red die to show three pips on the top side.

   (d) Against the pips' sum on the top side being 3.

55. **Odds** A three-card hand is to be dealt. Find the odds

   (a) For the hand to contain all spades.

   (b) For the hand to contain all kings.

   (c) Against the hand containing all jacks.

56. **Odds** Three boys and four girls are to be seated in a row. Find the odds

   (a) For a boy to be seated on either end.

   (b) For a girl to be seated in the middle seat.

   (c) Against a boy to be seated in the middle seat.

57. **Odds** A survey of 20 households revealed that 6 households owned no chain saws, 10 households owned one chain saw, and 4 households owned two chain saws. Find the odds

(a) For a household owning two chain saws.

(b) For a household owning at least one chain saw.

(c) Against a household owning two chain saws.

58. Odds The owner of an apple orchard took a sample of 30 apples. The check revealed that 20 apples contained no worms, 6 contained one, and 4 contained two. Find the odds

(a) For an apple containing no worm.

(b) For an apple containing at least one worm.

(c) Against an apple containing two worms.

59. Winning a Promotional Game A fast-food chain is conducting a game in which the odds of winning a double cheeseburger are reported to be 1:100. Find the probability of winning a double cheeseburger.

60. Asteroids The April 5, 2002, issue of the journal *Science* published an article indicating that the probability of a recently discovered asteroid (referred to as 1950 DA)

striking the Earth is $\frac{1}{300}$. [*Source:* http://impact.arc.nasa.gov]. A local news station reported the story, but announced that the *odds* of the asteroid hitting Earth were 1 in 300. Find the difference between the probability reported by the *Science* article and the probability implied by the local news report.

Section 8.3 Discrete Random Variables and Expected Value

The concept of probability allows us to compute what is "expected" in terms of an average for probability distributions that have numerical outcomes. Applications include games of chance and decision making in the business world.

Suppose you consider playing this game with a friend: A coin is tossed. If it turns up heads, you give your friend $1; if it turns up tails, your friend gives you $1. If the game is repeated 10 times, you might record the following results:

$$\{H, T, H, H, T, T, H, H, T, H\}.$$

Each time the coin is tossed, money changes hands. If we assign the values $+1$ to tails (because you gain $1) and -1 to heads (because you lose $1), the preceding results could be rewritten as

$$\{-1, +1, -1, -1, +1, +1, -1, -1, +1, -1\}.$$

At the end of the 10 trials, you have gained a total of $4 and lost a total of $6, for a net earnings of $4 - $6 = -$2. Since the game was repeated 10 times, you have averaged $-$2/10 = -$0.20, or -20¢, per game. (Note that the results may be different each time the game is repeated 10 times.)

Now suppose that you and your friend continued playing the game for a very long time. What would you expect to be your average earnings per game in the long run? Since the probability of the coin turning up heads is $\frac{1}{2}$, you expect to lose $1 about half the time. Since the probability of the coin turning up tails is $\frac{1}{2}$, you also expect to gain $1 about half the time. So, in the long run, if we were to list the monetary results of playing the game as before, we would expect the list to contain -1 half the time and $+1$ half the time. For example, if the game was repeated 1000 times, you would expect to lose \$1, or the coin to turn up heads, $\frac{1}{2}(1000) = 500$ times, and you would expect to gain \$1, or the coin to turn up tails, $\frac{1}{2}(1000) = 500$ times. So, your *average earnings per game played* are expected to be

$$\frac{(-1)(\text{no. of heads}) + (+1)(\text{no. of tails})}{\text{number of repetitions}}$$

$$= \frac{(-1)\left(\frac{1}{2}\right)(1000) + (+1)\left(\frac{1}{2}\right)(1000)}{1000}$$

$$= (-1)\left(\frac{1}{2}\right) + (+1)\left(\frac{1}{2}\right) \qquad \text{\textit{upon cancelling the 1000's}}$$

$$= -\frac{1}{2} + \frac{1}{2}$$

$$= 0.$$

Thus, in the long run, you expect to have earned nothing, \$0, from playing the game. The number 0 in this demonstration is called the **expected value** of this game; **expected value** means *long-term average.*

Because we assigned values to the outcomes in our sample space (-1 to heads, $+1$ to tails), we have associated a function with the sample space. This function **varies** in a **random** fashion, depending upon the chance outcome of tossing the coin. For that reason, we refer to such a function as a **random variable.** The **values** of the random variable are, in this case, -1 and $+1$. Since each outcome in the sample space is associated with a probability, the corresponding value of the random variable is also associated with this probability. The following table summarizes these associations for the coin-tossing game:

Outcome in Sample Space	Random Variable Value	Assigned Probability
H	-1	$\frac{1}{2}$
T	$+1$	$\frac{1}{2}$

The preceding table is called the **probability distribution** for the random variable. Referring back to our calculation of the expected value of the coin-tossing game, notice that the number of times the game was played "cancelled out" of our calculation, and our final result depended only upon the random variable values and the respective probabilities:

$$(-1)\left(\frac{1}{2}\right) + (+1)\left(\frac{1}{2}\right) = 0.$$

This is the key to finding expected values.

We give another example of a random variable and its expected value before formalizing these concepts.

EXAMPLE 1

Calculating the Expected Value of a Random Variable

To raise money for a charity, a local organization plans a carnival game. Each player will pay $2 to roll a die. If a 6 is rolled, the player wins $5; if any other number is rolled, the player wins nothing. What is a player's expected earning from this game?

SOLUTION The underlying sample space contains six equally likely outcomes. Because the result of rolling the die dictates how much money, if any, changes hands, we can create a new sample space with two outcomes, *win* and *lose*, with probabilities $\frac{1}{6}$ and $\frac{5}{6}$, respectively. As a player plays this game repeatedly, his or her "earnings" from each game will vary among the numbers $3 (paid $2, won $5) and -2 (paid $2, won $0), depending on the chance occurrence of the respective outcomes *win* or *lose* in the sample space. The numbers $+3$ and -2 then become the values of the random variable associated with the game. A table showing the probability distribution for this random variable may be constructed as follows:

Outcome of Sample Space	Value of Random Variable	Assigned Probability
Win	$+3$	$\frac{1}{6}$
Lose	-2	$\frac{5}{6}$

The table shows that, as the game is played over and over for a long period of time, a player can *expect* the value $+3$ to appear $\frac{1}{6}$ of the time and the value -2 to appear $\frac{5}{6}$ of the time. As we noted in our introductory discussion, these values and not the actual (large) number of times the game is played, are the key ingredients to finding the expected value.

$$\text{Expected Value} = (+3)\left(\frac{1}{6}\right) + (-2)\left(\frac{5}{6}\right) = \frac{3}{6} - \frac{10}{6} = -\frac{7}{6} \approx -1.17.$$

As we learn from this calculation, the average of a player's earnings, in the long run, should be about $-$1.17$ per game. If, for example, the game was played 3000 times, a player could expect to lose a total of $\left(-\$\frac{7}{6}\right)(3000) = -\$3500.$ ∎

Our examples show that the values of random variables are numbers that are decided by the various outcomes in a sample space. In the introductory example of tossing a coin, the outcome of heads was assigned to $-$1$ while the outcome of tails was assigned to $+$1$. In Example 1, the assignments were "win" to $+$3$ and "lose" to $-$2$. Such assignments may be thought of as a function whose domain is a sample space and whose functional values are particular real numbers. The function is called a **random variable function** and is usually denoted by X rather than f or g, while the functional values, or random-variable values, are denoted by $x_1, x_2, \ldots, x_n$ in the finite case. In the final analysis, it is the probability distribution of the random variables that allows us to find the expected value.

Random Variables and Expected Value

A **random variable function** X is a rule that assigns a numerical value to each outcome in a sample space. Each numerical value is known as a **random variable value.** If the random-variable values are $x_1, x_2, x_3, \ldots, x_n$, with respective probabilities $p_1, p_2, p_3, \ldots, p_n$, then the **expected value of the random variable** is

$$E(X) = x_1 p_1 + x_2 p_2 + x_3 p_3 + \cdots + x_n p_n.$$

Insurance companies rely on mortality tables prepared by actuaries to tell them the probability that persons of a certain age will live through a specified interval of time. Example 2 solves the problem that was presented at the outset of this chapter.

EXAMPLE 2

Expected Value of a
Life Insurance Policy

An insurance company sells one-year term life insurance policies designed for 40-year-old women. The face value of the policy is $10,000, and each policy sells for $300. Suppose that the company's mortality tables show that the probability that a 40-year-old woman will live for another year is 0.98. What are the company's expected earnings from this type of policy?

SOLUTION The sale of these policies is a game of chance; we think of a sample space with two outcomes: "live," with probability 0.98, and "die," with probability 0.02. The company's "earnings" on each such policy depend on whether the buyer lives or dies during the year after purchasing the policy. If the buyer lives, then the company earns $300, whereas if the buyer dies, the company "earns" $300 − $10,000 = −$9700. It follows that company earnings per policy can be thought of as a random variable X with values $300 and −$9700. Assuming that a large number of these policies are sold, the company would expect that 0.98, or 98%, of the values would be $300 (the policyholder lives) and 0.02, or 2%, of the values would be −$9700 (the policyholder dies). The probability distribution for these random-variable values is, accordingly, as follows:

Outcome of Sample Space	Value of Random Variable	Assigned Probability
Live	$300	0.98
Die	−$9700	0.02

The expected value is

$$E(X) = \$300(0.98) + (-\$9700)(0.02) = \$100.$$

Knowing that expected value means "long-term average," we conclude that the company should profit an average of $100 per policy sold. ■

EXAMPLE 3

Insuring Your Computer

Suppose you plan to insure your new laptop computer, which you will be taking to campus, against theft for the amount of $3000. An insurance company claims that its records indicate that 3% of such computers on college campuses are stolen within one year. The company offers to insure your computer for an annual premium of $100. What is *your* expected return per year from the insurance company on this policy?

SOLUTION You may think of insuring the computer as a game of chance for which you pay $100 (the premium) to play. If the computer is stolen during the year, you will gain $3000 − $100 = $2900 with empirical probability 0.03. If the computer is not stolen during the year, you will lose the premium of $100 and gain nothing from the insurance company, with empirical probability 0.97. Thus, $2900 and −$100 are values of a random variable X, and we organize your earnings in the following table:

Outcome of Sample Space	Value of Random Variable	Assigned Probability
Stolen	$2900	0.03
Not stolen	−$100	0.97

Your expected earnings from this policy are then

$$E(X) = (\$2900)(0.03) + (-\$100)(0.97) = \$87 - \$97 = -\$10.$$

This means that if you insure your computer with this company over many years under the same circumstances, you will average a loss of $10 each year to the company. ▪

The next example shows that expected value need not always involve money.

EXAMPLE 4

Dealing Cards

A 2-card hand is to be dealt from a deck of 52 cards. Find the expected number of aces in the hand.

SOLUTION The number of aces in a 2-card hand is a random variable X whose values are 0 (no aces), 1 (1 ace, 1 not an ace), and 2 (both cards are aces). Using the fact that there are 4 aces and 48 cards that are not aces, we have, for the probability of such hands,

$$P(\text{no aces}) = \frac{C(4, 0) \cdot C(48, 2)}{C(52, 2)} = \frac{1128}{1326}$$

$$P(\text{1 ace}) = \frac{C(4, 1) \cdot C(48, 1)}{C(52, 2)} = \frac{192}{1326}$$

$$P(\text{2 aces}) = \frac{C(4, 2)}{C(52, 2)} = \frac{6}{1326}.$$

The expected value of the random variable X, or the expected number of aces, is then

$$E(X) = 0\left(\frac{1128}{1326}\right) + 1\left(\frac{192}{1326}\right) + 2\left(\frac{6}{1326}\right) = \frac{204}{1326} \approx 0.1538.$$

The number 0.1538 is interpreted as follows: If two-card hands are dealt over and over for a long period of time and the number of aces are counted in each hand, then the average of those numbers would be approximately 0.1538. ▪

EXAMPLE 5

Outcomes May Be the Random Variable Values

Suppose a coin is tossed twice and the number of heads that turn up each time is recorded. Find the expected number of heads that turn up if this experiment is repeated over a long period of time.

SOLUTION The sample space for the experiment has outcomes 0 (no heads turned up), 1 (1 head, 1 tail turned up), or 2 (both coins turned up heads). The probability distribution for the experiment is shown in the following table:

Number of Heads That Turn Up in Two Tosses of a Coin	Assigned Probability
0	$\frac{1}{4}$
1	$\frac{1}{2}$
2	$\frac{1}{4}$

This time, the random variable values are precisely those of the outcomes in the experiment: 0, 1, and 2. The expected number of heads is then

$$0\left(\frac{1}{4}\right) + 1\left(\frac{1}{2}\right) + 2\left(\frac{1}{4}\right) = 0 + \frac{1}{2} + \frac{1}{2} = 1.$$

If this experiment is repeated over a long period of time, then we would expect to average 1 head per trial. ■

EXAMPLE 6

A Botany Experiment

A botanist broke open 20 pea pods and counted the number of peas in each pod. The results were as follows: 2, 4, 3, 2, 6, 4, 5, 4, 3, 3, 4, 6, 5, 2, 5, 4, 3, 3, 4, and 4. Find the expected number of peas per pod.

SOLUTION The number of peas per pod can be considered a random variable X with values 2, 3, 4, 5, and 6. These values and their associated probabilities can be organized in a table such as the following one:

Number of Peas in Pod	Frequency with which That Number Appears	Assigned Probability
2	3	$\frac{3}{20}$
3	5	$\frac{5}{20}$
4	7	$\frac{7}{20}$
5	3	$\frac{3}{20}$
6	2	$\frac{2}{20}$

Note that a probability is assigned to each random-variable value in accordance with the relative frequency with which it occurs, as is shown in the last column of the table. The expected number of peas per pod is now calculated as follows:

$$E(X) = 2\left(\frac{3}{20}\right) + 3\left(\frac{5}{20}\right) + 4\left(\frac{7}{20}\right) + 5\left(\frac{3}{20}\right) + 6\left(\frac{2}{20}\right) = \frac{76}{20} = 3.8 \text{ peas.}$$

We interpret 3.8 as the average number of peas per pod. In reality, it means that if this experiment could be repeated over and over, then, in the long run, we would expect to find an average of 3.8 peas per pod. The probability distribution histogram of this random variable is shown in Figure 7.

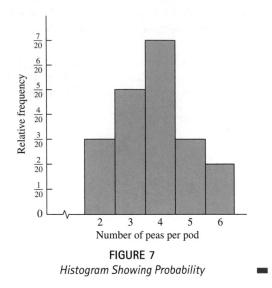

FIGURE 7
Histogram Showing Probability

EXAMPLE 7

A Business Decision

Two locations, uptown and downtown, are under consideration for a new fast-food franchise. Traffic counts reveal that uptown can expect 50% of its customers in the morning, 20% in the afternoon, and 30% during the evening hours. Downtown can expect 25% of its customers in the morning, 20% in the afternoon, and 55% during the evening hours. Statistics from the parent company show that a typical customer will spend $2.50 in the morning, $2 in the afternoon, and $4.50 in the evening. Assuming that each location is projected to have about 1500 customers per day, which location will have the larger revenue?

SOLUTION We view this problem as one of finding the expected value of two experiments, one occurring uptown and the other downtown. The random-variable values are the same in each location—$2.50, $2, and $4.50—but the respective probabilities are different. The following table shows the pertinent information:

Period of day	Value of Random Variable (revenue per customer)	Percentage of Volume Stated as a Probability	
		Uptown	Downtown
Morning	2.50	0.50	0.25
Afternoon	2.00	0.20	0.20
Evening	4.50	0.30	0.55

Expected value for uptown: $2.50(0.50) + $2(0.20) + $4.50(0.30) = $3.00.

Expected value for downtown: $2.50(0.25) + $2(0.20) + $4.50(0.55) = $3.50.

The calculations show that downtown has the better expected value: Each customer will spend an average of $3.50 there, while each will spend only $3 uptown. For the 1500 projected customers, the expected revenue per day from downtown will be $1500(\$3.50) = \$5250$, whereas only $1500(\$3) = \$4500$ is expected at uptown. ■

EXAMPLE 8
A Game of Chance

A typical roulette wheel has 38 equally spaced slots numbered 00, 0, 1, 2, 3, . . . , 36. Half of the slots numbered 1 through 36 are red, the other half are black, and the slots numbered 00 and 0 are green. Among the various types of betting on a roulette wheel is that of betting on a color. For instance, if a player places a sum of money on black and the winning number is black, then the player wins a sum of money equal to the bet placed. Otherwise, the bet is lost. Find the expected value of the winnings if a $20 bet is placed on black.

SOLUTION The winnings from such betting may be considered to be a random variable X with values 20 and -20. These values, along with their probabilities, are organized in the following table:

Color	Values of Random Variable	Assigned Probability
Black	20	$\frac{18}{38}$
Other	-20	$\frac{20}{38}$

The expected value of the winnings is then

$$E(X) = \$20\left(\frac{18}{38}\right) - \$20\left(\frac{20}{38}\right) = \$\frac{360}{38} - \$\frac{400}{38} = -\$\frac{40}{38} \approx -\$1.05.$$

This calculation shows that if a player places a $20 bet on black over and over, then the player will average a loss of $1.05 *per game.* ■

Suppose an unbiased coin is tossed. If it comes up heads, you give your friend $1; if it comes up tails, your friend gives you $1. Your expected winnings from such a game are $\$1\left(\frac{1}{2}\right) - \$1\left(\frac{1}{2}\right) = \$0$. When the expected value of the winnings from a game are zero, the game is called **fair.** In a fair game, neither player has an advantage.

A Fair Game

A game is called **fair** if the expected value of a player's winnings are zero. That is, if X is a random variable whose values represent a player's winnings from a game, then that game is fair if

$$E(X) = 0.$$

Notice that the game described in Example 8 is *not* fair.

In general, random variables fall into on of two classification: **discrete** and **continuous**. A discrete random variable is a random variable that has a finite number of values or whose values can be arranged in a sequence. All of the examples in this section have had a finite number of random-variable values; hence, all fall into the discrete classification. Continuous random variables will be introduced in Chapter 10.

Exercises 8.3

To find the requested expected values of the exercises in this section, you may find it helpful to construct a table that will clearly identify the random-variable values and their respective probabilities, as was done in several of the examples.

1. Suppose a coin is to be tossed. If the coin turns up heads, you win $1; if the coin turns up tails, you lose $2.

    (a) What are your expected earnings from such a game?

    (b) If you played this game 1500 times, how much money would you expect to gain or lose?

2. Suppose a die is to be tossed. If the number of pips on the top face is at least five, you win $10. Otherwise you lose $5.

    (a) What are your expected earnings from this game?

    (b) If you played the game 3000 times, how much money would you expect to gain or lose?

3. To move from your current position on a game board, you draw one ball, with replacement, from a container that has five red and six green balls. If the ball you draw is red, you advance four places; if the ball is green, you move back three places.

    (a) What is your expected movement in this game?

    (b) In 44 turns at drawing a ball, where would you expect to be located on the game board relative to your present position?

4. **Advertising** A firm allocates one-sixth of its advertising dollar to newspaper ads, one-third to radio ads, and one-half to television ads. It is estimated that the return on that portion of a dollar spent on each of these three media is, respectively, $4, $3, and $6.

    (a) What is the firm's expected value from the given advertising sources?

    (b) If $5000 is spent for ads in the proportion indicated, what is the firm's expected dollar return?

5. Assume that a boy is just as likely as a girl at each birth. In a two-child family, what is the expected number of boys?

6. Assume that a boy is just as likely as a girl at each birth. In a three-child family, what is the expected number of girls?

7. A box contains four slips of paper, on each of which is written just the number 4, 6, 9, or 18. One slip of paper is randomly drawn, and the number on the slip is considered to be a value of the random variable. Find the expected value of this random variable.

8. A box contains five slips of paper, on each of which is written just the number 3, 4, 8, 9, or 20. One slip of paper is randomly drawn, and the number on the slip is considered to be a value of the random variable. Find the expected value of this random variable.

9. Suppose that you pay $5 to play this game: From a box containing six red and eight green marbles, you are allowed to select one marble after being blindfolded. If the marble is red, you win $10, but if the marble is green, you win nothing. What are your expected earnings from the game?

10. Suppose that you pay $25 to play this game: After being blindfolded, you draw one ball from a box containing five red, three green, and six white balls. If the ball is red, you win $5; if the ball is green, you win $100; and if the ball is white, you win nothing. What are your expected earnings from the game?

11. Suppose you draw 1 card from a deck of 52 cards. If the card is a spade, you win $5; if the card is not a spade, you lose $2. What are your expected earnings from this game?

12. Suppose you draw 1 card from a deck of 52 cards. If the card is a king or a queen, you win $50; otherwise you win nothing. What are your expected earnings from playing this game?

13. Suppose that you pay $2 to play this game: A pair of dice is rolled. If the sum of the pips showing on top is 6, 7, or 8, you win $5. Otherwise you win nothing. What are your expected earnings from this game?

14. Consider the following game: A pair of dice is rolled. If the sum of the pips on top is 6 or less, your friend gives you $10. For any other sum rolled, you give your friend $7. What are your expected earnings from this game?

15. Suppose you match pennies with a friend. If both coins show heads or both show tails, you win $1. Otherwise you give your friend $1. What are your expected earnings from this game?

16. If you bet $1 on any three-digit number from 000 to 999 and your number is drawn, you win $500. What are your expected earnings from this game?

17. Carnival Games Suppose that you pay $1 to play this carnival game: The operator has a box containing 80 balls, each marked with a different number from 1 to 80, respectively. You select a number from 1 through 80. Then the game operator randomly selects 20 balls. If your number is among the 20 on the balls selected, you get $3. What are your expected earnings?

18. Gambling A single die is rolled, and the amount of money you lose or win is the number of pips on the top side minus four. (A negative number means a loss; a positive number means a gain.) What are your expected earnings?

19. Insurance Avery is considering insuring against theft the new $300 CD-player he just installed in his automobile. His insurance agent tells him that such an option could be added to his present policy for $12 per year. The agent further states that the probability of theft is 0.2 in a given year. If he takes this insurance, what is his expected return per year?

20. Insurance Mesha is considering an insurance policy for the new automobile she just purchased. One policy option is for towing insurance which costs $5 per year. The insurance agent tells her that company statistics show that she will need towing during the year with probability 0.1 and that an average towing charge within her city is $60. If she takes this insurance, what is her expected return per year?

21. Insurance An insurance company is going to sell one-year life insurance policies with a face value of $100,000 to 25-year-old women for $1100. The company's mortality tables show that these women will live for one year with probability 0.99. Consider the company's earnings from such policies to be values of a random variable.

   (a) Find the company's expected earnings per policy.

   (b) If the company hopes to sell 2000 of these policies, how much income could it expect from this source?

22. Insurance An insurance company is going to sell one-year life insurance policies with a face value of $50,000 to 25-year-old men for $5500. The company's mortality tables show that these men will live for one year with probability 0.9. Consider the company's earnings from such policies to be values of a random variable.

   (a) Find the company's expected earnings per policy.

   (b) If the company hopes to sell 3000 of these policies, how much income could it expect form this source?

23. Insurance A life insurance company plans to sell one-year term life insurance policies with a face value of $50,000 to 50-year-old men. The company's mortality tables show that these men will live for one year with probability 0.92. How should the company price such policies so that its average earnings per policy will be $500?

24. Insurance A company plans to sell one-year term life insurance policies with a face value of $25,000 to 35-year-old women. The company's mortality tables show that these women will live for one year with a probability of 0.97. How should the company price such policies to have average earnings of $50 per policy?

25. Testing An upcoming test in a music theory class will have four true–false questions and eight multiple-choice questions. The latter will have five answer choices. If a student guesses at the answer on every question on the test, what is the expected number of questions he or she will get correct?

26. Testing An upcoming test in an art history class will have four true–false questions, five standard multiple-choice questions with four choices each, and 12 ten-response questions that require the selection of 1 of 10 possible answers. If a student guesses at the answer on every question on the test, what is the expected number of questions he or she will get correct?

27. Roulette Roulette wheels have 38 equally spaced slots numbered 00, 0, 1, 2, 3, . . . , 36. To play roulette, a player bets $1 on one of the 38 numbers. If the ball comes to rest on that number, the player wins $35, plus the $1 that was bet. Otherwise, the $1 bet is lost. What is the expected value for a player of this game?

28. Business Decisions A chain of video rental stores is considering Alma and Boone for the location of a new store. Alma has 30% of its population 20 years of age or under, 30% from 21 through 30, and 40% 31 or older. Boone has 25% of its population 20 years of age or under, 20% from 21 through 30, and 55% 31 or older. The company estimates that the 20-and-under age group rents an average of five videos per week, the 21-through-30 age group rents two videos per week, and the 31-and-over age group rents one video per week. In which city should the chain locate its new store to rent the most videos per week?

29. Business Decisions Antonio is comparing how a copy shop would fare at a campus location versus a location in the heart of the business district. Estimates are that the campus location would get 20% of its business in the morning, 30% of its business in the afternoon, and 50% of its business during the evening hours. The business district location would get 40% of its business in the morning, 50% of its business in the afternoon, and 10% of its business during the evening hours. If it is projected that copy jobs average $5 in the morning, $10 in the afternoon, and $3 in the evening and that each location will get about 2000 customers each day, which location has the largest expected earnings?

30. **Agriculture Survey** Twenty bean pods were broken open and the number of beans in each pod counted. The record looked like this:

Number of Beans in Pod	Frequency
2	5
3	8
4	4
5	2
6	1

Consider the number of beans in each pod to be a random variable. Find the expected number of beans per pod.

31. **Demographic Survey** Sixty people were asked about the number of years they had been in college. The responses are as follows:

Number of Years in College	Frequency
0	5
1	7
2	8
3	10
4	20
5	6
6	2
7	2

Find the expected number of years of college attended.

32. If Zal bet his best friend $25 that Zal's favorite team would win the Super Bowl, and the announced odds for his team winning are 9 to 7, what were his expected earnings?

33. If Garrett bet his sister Jerrica $5 that his favorite baseball team would win the World Series, and the announced odds for his team winning were 89 to 10, what were Garrett's expected earnings?

34. **Lottery Earnings** A local civic organization sells 10,000 lottery tickets for $1 each. Prizes are awarded by a random drawing of three ticket numbers. The holder of the first number drawn wins a new portable TV set worth $300, the holder of the second number drawn wins a $200 shopping spree at a local supermarket, and the holder of the third number drawn wins $50 in cash. If you buy one ticket, what are your expected earnings?

35. If the odds for winning a $500 lottery prize are 1 to 62,586, and you spend $2 for the ticket and $1 for

gasoline to go purchase the ticket, what are your expected earnings for buying one ticket?

36. **Quality Control** Among a rack of 30 college sweatshirts, four have a flaw of some kind. Suppose you buy two of these sweatshirts and consider the number having a flaw to be a random variable.

   (a) Make a probability distribution table for the random variable.

   (b) Construct a probability histogram for the distribution.

   (c) Of the two sweatshirts you bought, what is the expected number with a flaw?

37. **Sampling** Two balls are drawn, without replacement and without regard to order, from a box containing three red and five black balls. The number of red balls drawn is considered to be a random variable.

   (a) Make a probability distribution table for the random variable.

   (b) Construct a probability histogram for the distribution.

   (c) Find the expected number of red balls in the sample drawn.

38. **Committee Selection** A committee of three is selected from among five men and four women, and the number of women on the committee is considered to be a random variable.

   (a) Make a probability distribution table for the random variable.

   (b) Construct a probability histogram for the distribution.

   (c) Find the expected number of women on the committee.

39. Two cards are drawn, without replacement and without regard to order, from a standard deck of 52 cards. What is the expected number of kings?

40. Three cards are drawn, without replacement and without regard to order, from a standard deck of 52 cards. What is the expected number of kings?

41. A box contains five red and three green marbles. Two marbles are randomly selected from this box, without replacement and without regard to order. What is the expected number of red marbles?

42. A bag contains three red, two green and five black jelly beans. If you reach into the bag and randomly select three jelly beans, without replacement and without regard to order, what is the expected number of red jelly beans? Of green jelly beans?

43. A box contains one red, two green, four blue, and five black balls. If a sample of two balls is drawn from this box, without replacement and without regard to order, what is the expected number of red balls? Of green balls?

44. A 2-card hand is dealt from a deck of 52 cards. Define a random variable X to have values that are the number of eights in the hand. Find the expected number of eights.

45. A 2-card hand is dealt from a deck of 52 cards. Define a random variable X to have values that are the number of hearts in the hand. Find the expected number of hearts in the hand.

46. On each of three slips of paper in a box is written exactly one of the numbers 5, 8, and 11, without duplication. Two slips of paper are drawn, in succession, with replacement. Define the random variable X to have values that are one-half the sum of the two numbers drawn. Find the expected value of this random variable.

47. On each of three slips of paper in a box is written exactly one of the numbers 2, 4, and 9, without duplication. Two slips of papers are drawn, without replacement. Define

the random variable X to have values that are three times the sum of the two numbers drawn. Find the expected value of this random variable.

48. A pair of dice is rolled. Let X be the random variable defined to have the value 8 if a sum of 2, 3, or 4 is rolled and -5 otherwise. Find the expected value of this random variable.

49. A pair of dice is rolled. Let X be the random variable defined to have values equal to the sum of the pips showing on the top faces. Find the expected value of this random variable.

50. One card is drawn from a deck of 52 cards. Define the random variable X to have a value equal to the number on the card if the card has a number on it and equal to 15 if the card does not have a number on it. Find the expected value of this random variable.

Chapter 8 Summary

CHAPTER HIGHLIGHTS

Probability numerically expresses the likelihood of particular outcomes occurring relative to the other possibilities in an *experiment*. *Events* are subsets of the sample space and have probabilities of at least 0, but never more than 1. *Simple events* consist of a single outcome, whereas *compound events* contain more than one outcome. Probabilities can be assigned *theoretically*, by deductive reasoning, or *empirically*, on the basis of relative frequencies. The *odds in favor* of an event is the ratio of the number of elements in the sample space that are in the event to those which are not; the *odds against* an event is the reverse ratio. A *random variable* assigns a number to each outcome in a sample space. The long-term average of these numerical values is called the *expected value* of the random variable.

EXERCISES

In problems 1 and 2, find the number of outcomes in the sample space for the given experiments.

1. Draw a card from a standard deck of 52 cards and then roll a die.

2. Construct a license plate having any three letters, followed by any four nonzero digits.

3. What are the odds in favor of getting a queen or a king when drawing one card from a standard deck?

4. What are the odds against an event that has a probability of $\frac{23}{79}$ of occurring?

A finite mathematics class has 17 women and 12 men. A group of 5 is randomly selected to work on a project.

5. Find the probability that the group has exactly 2 women.

6. Find the probability that 3 or more women are in the group of 5.

In problems 7 and 8, a pair of dice is rolled. If a sum of 6 or more turns up, a "success" is recorded. Otherwise a "failure" is recorded.

7. Find the sample space for this experiment.

8. Determine the probability of a success.

9. Consider the number of girls in a three-child family to be a random variable. Find the expected value.

For problems 10 and 11, 50 people were randomly surveyed to determine the number of DVD's they rented each month, as shown in the following table:

Number of DVD's	Frequency
0	5
1	7
2	6
3	8
4	11
5 or more	13

10. Find the empirical probability that a person rents three or more DVD's per month.

11. Determine the expected number of DVD's rented per person each month.

12. A lottery commission sells 1,000,000 tickets for $1 each. There are 10 third prizes of $500 each, 5 second prizes of $5000 each, and 1 grand prize of $50,000. What is Corey's expected value if he buys one ticket?

13. An insurance company sells one-year term life insurance policies with a face value of $50,000 to 40-year-old women for $1800. The company's mortality tables show that such women will live for one year with probability 0.98. Find the company's expected earnings per policy.

14. Describe the characteristics of an event E if $P(E) = 0$. Give an example of such an event.

15. Describe the characteristics of an event E if $P(E) = 1$. Give an example of such an event.

16. Define *random variable* in your own words.

17. Define *expected value* in your own words.

18. ⬚ Use a computer or a calculator program to simulate the roll of a single die 120 times. Read the number of 1s, 2s, . . . , 6s obtained, and compute the relative frequency of each of the outcomes. Compare these numbers with the *theoretical relative* frequency of each outcome and the *theoretical* number of 1s, 2s, . . . , 6s that would be obtained. (*Note*: Use the function RANDBETWEEN(1,6) on Excel for this simulation.)

The following problem appeared on an actuarial examination:

19. What is the probability that a hand of 5 cards, chosen randomly and without replacement from a standard deck of 52 cards, contains the king of spades, exactly one other king, and exactly two queens?

(a) $\dfrac{C(4, 2) \cdot C(4, 2) \cdot C(44, 1)}{C(52, 5)}$

(b) $\dfrac{C(3, 1) \cdot C(4, 2) \cdot C(44, 1)}{C(52, 5)}$

(c) $\dfrac{C(7, 3) \cdot C(44, 1)}{C(52, 5)}$

(d) $\dfrac{C(3, 1) \cdot C(4, 2) \cdot C(11, 1)}{C(52, 5)}$

(e) $\dfrac{C(3, 1) \cdot C(4, 2)}{C(52, 5)}$

The following exercise is similar to one that appeared in the Sample Test Questions for a PRAXIS examination:

20. The numbers 1 through 25, inclusive, are individually written on 25 cards, and the cards are placed into a box. One card is drawn at random. What is the probability that the number on the card chosen is not a multiple of 4?

The following exercise is similar to one that appeared in Consumer Duffy CPA Review: Accounting and Reporting, 1998:

21. A distributor decides to increase its daily bagel purchases by 200 boxes. A box of bagels costs $5 and sells for $9 through supermarkets. Any boxes not sold through supermarkets are sold through the distributor's outlet store for $4. The following probabilities have been assigned to selling additional boxes:

Additional Boxes	Probability of Selling
150	0.7
200	0.3

What is the expected value of the distributor's decision to buy 200 additional boxes of bagels?

(a) $485   (b) $200

(c) $625   (d) $450

CUMULATIVE REVIEW

Problems 22 and 23 refer to the following matrix:

$$A = \begin{bmatrix} 1 & 0 & -3 \\ 0 & 2 & 4 \\ 1 & 0 & -4 \end{bmatrix}.$$

22. Calculate the inverse of A.

23. Compute and simplify $A^2 - 3A$ for the given matrix A.

24. Solve this linear programming problem:

$$\text{Minimize } f = 2x + 5y$$
$$\text{subject to } 2x + 5y \geq 20$$
$$3x + 4y \geq 24$$
$$6x + y \geq 12$$
$$x \geq 0, y \geq 0.$$

25. In a family of four children, which outcome is most likely, that four are of the same sex, three are of one sex, or two are of each sex? Explain your answer.

Chapter 8 Sample Test Items

1. Find the number of outcomes in the sample space for the experiment of selecting a sample of three marbles from a bag containing eight marbles—five blue and three orange.

2. Two boys and two girls randomly sit in a row of four seats. Find the probability that both boys and both girls sit next to each other.

3. What are the *odds* in favor of two heads and two tails coming up when a coin is tossed four times?

4. If the odds against an event E occurring are 5 to 9, then find the probability that E occurs.

In Problems 5 through 7, two standard dice, one red and one green, are to be rolled. Find the probability that

5. A sum of six will be rolled.

6. Both dice will show four pips on the top side.

7. The green die will not show two pips on the top side.

In Problems 8 through 10, a 3-card hand is to be dealt from a standard deck of 52 cards. Find the probability that

8. None of the cards will be aces.

9. Exactly two of the cards will be diamonds.

10. All three cards will be from the same suit.

Among eight graphing calculators, three are know to have a defect. A sample of three calculators is to be selected, without replacement and without regard to order, and the number of defective calculators will be recorded.

11. Find the probability that exactly two calculators will be defective.

12. Find the probability that at least one calculator will be defective.

13. Construct the probability distribution histogram for this experiment.

In a five-child family, excluding multiple births and assuming that girls and boys are equally likely at birth, what is the

14. Probability of having two girls?

15. Probability of having at most two boys?

16. Expected number of girls?

17. Two cards are drawn, without replacement, from a standard deck of 52 cards. What is the expected number of jacks?

18. An upcoming sociology test will have 60 true–false questions and 40 multiple-choice questions with five options per multiple-choice item. What is the expected number of questions that can be answered correctly by a student who guesses on every question?

In-Depth Application

The Expected Value of Powerball®

Objectives

In this application, you will apply your knowledge of the general rules of probability, odds, and expected value to the multistate lottery game of Powerball®.

Introduction

Powerball® is a lottery game played in many states. Customers purchase a lottery ticket for $1 and choose a series of six numbers. The first five numbers are chosen from the integers 1 through 49, inclusive, and are associated with numbered white balls. The last number is chosen from the integers 1 through 42, inclusive, and is associated with numbered red balls. (*Note*: Throughout the remainder of this application, we will refer to the first five numbers as the *white* numbers and the sixth number as the *red* number.) The lottery ticket is valid for one drawing. When the drawing occurs, five white balls are drawn from five drums, corresponding to the white numbers selected on the lottery ticket. One red ball is drawn from a sixth drum, corresponding only to the red number on the lottery ticket. Several monetary prizes are awarded for matching the white and red numbers; white numbers can be matched with the white balls drawn in any order.

(continued)

The Data

According to the Multi-State Lottery Association, the odds in favor of winning monetary prizes were as follows on August 31, 2001:

Matching Numbers	Amount of Prize	Odds
5 white, 1 red	Jackpot	1:80,089,128
5 white only	$100,000	1:1,953,393
4 white, 1 red	$5000	1:364,042
4 white only	$100	1:8,879
3 white, 1 red	$100	1:8,466
3 white only	$7	1:207
2 white, 1 red	$7	1:605
1 white, 1 red	$4	1:118
red only	$3	1:74

The "jackpot" is a large sum of money that builds from drawing to drawing if no jackpot winner is determined. For the purposes of this application, let us assume that the jackpot stands at $20,000,000.

1. Recall that the odds in favor of an event E are given by

$$\frac{P(E)}{P(E')} = \frac{P(E)}{1 - P(E)}.$$

   Using this formula, convert the odds in the given table to probabilities. (Round your results to four significant figures.)

2. Notice that the table does not provide a probability for the outcome of winning *no* monetary prize. Under what conditions would a single lottery ticket result in that outcome? Apply the general rules of probability to determine the probability of such an outcome.

3. What is the probability of winning at least $100 in this lottery game? What is the probability of winning no more than $5000 in this game?

4. Assuming that a single lottery ticket is purchased for $1, consider the random variable that assigns the net earnings to each outcome of the sample space. For example, such a random variable would assign the value $2, calculated from $3 − $1, to the outcome "red only." Construct the probability distribution table for this random variable.

5. Find the expected value of the random variable in part 3. Explain carefully what this value represents.

6. According to the results of part 4, if you purchased 1000 Powerball® tickets at a price of $1 each, and each with a *different* set of numbers chosen, what would you expect your net earnings to be?

7. How many *different* lottery tickets would you have to purchase to guarantee that you would win *some* money? Discuss your findings.

9

Additional Topics in Probability

PROBLEM

A particular rocket assembly depends on two key components. If either component is working properly, then the assembly will function properly, but if both fail, the assembly will not function properly. The probability of failure for one component is 0.002 and the other is 0.01. If the ability or the failure to function of either component has no effect on the other, what is the probability that both components will fail?
(See Example 9, Section 9.3.)

CHAPTER PREVIEW

The set operations of union, intersection, and complement can be used to create new events from given events in a sample space. How to find the probability of a union and an intersection will be among the highlights of this chapter. In particular, an addition formula will be given for the probability of the union of two or more events. The formula reduces to an even simpler addition formula for events that are mutually exclusive. Similarly, a multiplication formula will be given for finding the probability of the intersection of two or more events and this formula becomes simpler when the events are independent. The idea of conditional probability is introduced, not only for its own value and natural display on probability tree diagrams, but as a transitional vehicle to calculating the probability of the intersection of events. Bayes's theorem is shown to be an important application of conditional probability. Finally, the useful concept of binomial experiments is explored.

36. A pair of dice is rolled. What is the probability of rolling at least a sum of 4?

37.  Three cards are dealt from a deck of 52 cards. What is the probability that at least 1 of the cards that is dealt is a king?

38. Four cards are dealt from a deck of 52 cards. What is the probability that at least 1 of the cards that is dealt is a spade?

39. A box contains six slips of paper, on each of which is written just the number 1, 2, 3, 4, 5, or 6, respectively. One slip is randomly drawn. What is the probability of drawing a number greater than 2?

40. Rolling Dice A pair of four-sided pyramidal dice, one red and one green, is rolled, and the pips on the downward-facing sides are noted. What is the probability that the sum of the pips will be

(a) A number less than 4 and greater than 6?

(b) A number less than 4 or greater than 6?

(c) What is the probability that the red die show a 2?

41. Rolling Dice A pair of fair dice is rolled, and the sum of the pips showing on the upper faces is noted. What is the probability that

(a) A sum less than 5 is rolled?

(b) A sum less than 5 or greater than 10 is rolled?

(c) A sum greater than 5 or less than 8 is rolled?

(d) A sum greater than 2 is rolled?

(e) A sum greater than 3 is rolled?

(f) The sum of 1 is rolled?

42. Rolling Dice A pair of standard dice, one red and one green, is rolled. What is the probability that

(a) The sum of the pips on the top faces is 6 or 8?

(b) The sum of the pips on the top faces is 4 or a "double" is rolled?

(c) The red die shows 4 pips on the top face or the green die shows a 5 on the top face?

43. Rolling Dice A pair of fair dice, one red and one green, is rolled. What is the probability that

(a) The sum of the pips on the top faces is 2 or 9?

(b) The sum of the pips on the top faces is 7 or the sum of the pips on the top faces is greater than 5?

(c) The red die shows 4 pips on the top face or the sum of the pips on the top faces is greater than 6?

44. Cards Five cards are drawn from a standard deck of 52 cards. What is the probability that

(a) All of the cards drawn are hearts or all are spades?

(b) At least two of the cards drawn are spades?

(c) All of the cards drawn are kings or all are queens?

45. Cards Three cards are dealt from a standard deck of 52 cards. What is the probability that

(a) At least one of the cards that is dealt is a club?

(b) All of the cards that are dealt are kings or all are queens?

(c) All of the cards that are dealt are kings or all are face cards?

46. Cards Seven cards are dealt from a standard deck of 52 cards. What is the probability that

(a) All of the cards that are dealt are hearts or all are diamonds?

(b) All of the cards that are dealt are hearts or all are face cards?

(c) All of the cards that are dealt are hearts or each of the cards has a number on it?

(d) At least one of the cards that are dealt is a heart?

47. Committee Selection The sophomore, junior, and senior classes each send three representatives to a meeting about campus recycling. At the meeting, a committee of four is selected from among the representatives. What is the probability that

(a) The sophomores will have at least two representatives on the committee?

(b) The sophomores will have at least one representative on the committee?

(c) The sophomores will have no representative on the committee?

48. Quality Control A box contains 20 lightbulbs, 8 of which are 60-watt bulbs, 5 of which are 75-watt bulbs, and 7 of which are 100-watt bulbs. A sample of three bulbs is randomly selected from the box, without replacement and without regard to order. What is the probability that

(a) At least one of the bulbs that are selected is a 60-watt bulb?

(b) All of the bulbs selected are 60-watt bulbs or all are 100-watt bulbs?

(c) Exactly one of the bulbs that are selected is a 75-watt bulb?

49. Education Survey Tests revealed that 30 of the 150 children in a particular school have a vision problem of some type. If five children are selected from this school, without replacement and without regard to order, what is the probability that at least one child will have a vision problem?

50. Selecting a Committee Each of the 50 states has 2 senators. A committee of 20 senators is selected. What is the probability that Texas will have

(a) One or more senators on the committee?

(b) No senators on the committee?

51. Family Composition In families with three children, what is the probability that

(a) The first child is a boy?

(b) The first child is a boy or the third child is a boy?

(c) The first child is a boy and the third child is a boy?

(d) At least one of the children is a girl?

52. Drawing Slips of Paper A box contains four slips of paper, on each of which is written just the number 1, 2, 3, or 4, without duplication. A slip is drawn and the number noted. The slip is not returned to the box. Then a second slip is drawn and the number noted. What is the probability that

(a) The first number drawn is a 2 or the second is a 4?

(b) The first number drawn is a 2 or the sum of the two numbers drawn is six?

(c) The first number drawn is greater than two, or the second is less than three?

(d) The first number drawn is at least two?

53. Matching Socks A drawer contains six blue socks and four white socks. Three of the socks are chosen at random without replacement. What is the probability of getting a pair of the same color?

54. Let A and B be any two events in a sample space. Write the event $A \cup B$ as the union of two mutually exclusive events.

55. Let A, B, and C be any three events in a sample space. Find a formula similar to the addition theorem for two events that will give $P(A \cup B \cup C)$.

56. A coin is tossed until, for the first time, either two heads or two tails appear in succession. What is the probability that the experiment will end before the sixth toss?

Section 9.2 Conditional Probability

Suppose that someone rolls a single die, out of your sight, and tells you it came up an even number. You are then asked, "What is the probability that a 2 has been rolled?" The answer to the question is certainly affected by the information that is known to you, namely, that the die is known to have come up a 2, a 4, or a 6. This, in effect, has reduced the sample space to $S^* = \{2, 4, 6\}$, from which the question is now easily answered as "$\frac{1}{3}$." The mathematical notation for this question is

$$P(\text{a two comes up} \mid \text{an even number has been rolled}) = \frac{1}{3},$$

where the *vertical bar* is read "given that" and the event to the right of the bar is the condition under which the question on the left of the bar is to be answered—hence the name **conditional probability.** In generic terms, $P(A \mid B)$ is read "the probability of A, given that B has already occurred."

In many cases, conditional probability can be computed in a straightforward way by considering a **reduced sample space.** The next few examples will give specifics.

EXAMPLE 1

Computing Conditional Probability

If 2 cards are randomly drawn, in succession, without replacement, from a deck of 52 cards, what is the probability that the second card drawn is a spade, given that the first card was a spade?

SOLUTION In conditional probability language, the problem becomes P (second card is a spade | first card was a spade). Because a spade has been removed from the deck on the first draw, the reduced sample space S^* from which the second card is to be drawn contains 51 cards: 12 spades and 39 nonspades. The probability of drawing a spade from S^* is then $\frac{12}{51}$. ∎

EXAMPLE 2

Computing Conditional Probability

If 2 cards are randomly drawn, in succession, with replacement, from a deck of 52 cards, what is the probability that the second card drawn is a king, given that the first card was a king?

SOLUTION In symbols, the problem may be stated as P(second card is a king | first card was a king). We are given the first card drawn, a king, but this time it was replaced. After the first draw, the sample space S^* looks just like the original sample space, still containing all 52 cards. The probability that the second card is a king is then $\frac{4}{52} = \frac{1}{13}$. ∎

EXAMPLE 3

Computing Conditional Probability

If three jelly beans are randomly drawn, in succession, and without replacement, from a bowl containing six red and four green jelly beans, what is the probability that the third jelly bean drawn is red, given that the first two jelly beans drawn were green?

SOLUTION Expressed in symbols, what we are able to find is P(third jelly bean is red | first two jelly beans drawn were green). We know that two green jelly beans have already been drawn from the bowl, so that the sample space for the third draw has been reduced to S^*, consisting of six red and two green jelly beans. That is, $S^* = \{R, R, R, R, R, R, G, G\}$. Since there are now only 8 jelly beans, 6 of which are red, the answer to our question is $\frac{6}{8} = \frac{3}{4}$. (See Figure 8.)

Original sample   Reduced sample
space S space S^*

FIGURE 8
Computing Conditional Probability   ∎

EXAMPLE 4

Drawing a Card

One card is drawn from a deck of 52 cards. What is the probability that the card is

(a) A king (K), given that the card drawn was a heart (H)?

SOLUTION Answering $P(K|H)$ means considering the reduced sample space S^* consisting of 13 hearts. The probability that a king is among them is $\frac{1}{13}$.

(b) A diamond, given that the card drawn is a diamond (D) or a heart (H)?

SOLUTION In symbols, the question asks for $P(D|D \cup H)$. The given condition implies that we may think of the reduced sample space S^* of 26 cards, of which 13 are diamonds and 13 are hearts. Under these circumstances, the chance of the drawn card being a diamond is $\frac{13}{26} = \frac{1}{2}$.

(c) A diamond, given that the card drawn is a diamond (D) and a jack (J)?

SOLUTION We are to find $P(D|D \cap J)$. Note that the given conditions tells us that the card drawn is the jack of diamonds, so the probability that a diamond has been drawn is 1.

(d) A heart, given that the card drawn was a 10 or a diamond (D)?

SOLUTION $P(H|10 \cup D)$ is found by considering the reduced sample space S^* representing the given information: four 10's along with the 12 other diamonds, making a total of 16 cards. Among these, there is only one heart, namely, the 10 of hearts. Our answer is therefore $\frac{1}{16}$. ∎

EXAMPLE 5 Computing Conditional Probability	A coin is tossed three times. Find the probability that the first two tosses resulted in heads, given that at least one tail was tossed.

SOLUTION The visual effect of writing out the sample space for tossing a coin three times will be helpful:

$$S = \{\text{HHH, HHT, HTH, THH, TTH, THT, HTT, TTT}\}.$$

We are given that at least one tail was tossed. This means that one of the following seven outcomes making up the reduced sample space occurred:

$$S^* = \{\text{HHT, HTH, THH, TTH, THT, HTT, TTT}\}.$$

Among these outcomes, only the first one listed has the first two tosses coming up heads. Consequently, the answer is $\frac{1}{7}$. ∎

EXAMPLE 6 Rolling Dice	A pair of dice, one red and one green, is rolled. Find the probability that a sum of 5 was rolled, given the green die has two pips on the upper side.

SOLUTION There are six possible outcomes that have the green die showing two pips on the upper side. They are, upon listing the red die first,

$$S^* = \{(1, 2), (2, 2), (3, 2), (4, 2), (5, 2), (6, 2)\}.$$

Among these outcomes, $(3, 2)$ is the only one whose pips sum to 5. It follows that the requested probability is $\frac{1}{6}$. ∎

EXAMPLE 7 Employment Data: Pointing toward the Conditional Probability Formula	Employment data regarding adults in the workforce in Midway revealed the following information:

	Employed	Unemployed
Male	500	50
Female	200	60

What is the probability that a person selected from this workforce is employed, given that the person is a male?

SOLUTION The condition "given . . . a male" reduces the sample space to the information in the top row of the chart, of which there are a total of $500 + 50 = 550$ persons. The question asks what proportion of *these* are employed. So we focus on the employed column, but consider only the 500 (the intersection of male and employed), to get

$$P(\text{employed} \mid \text{male}) = \frac{500}{550}.$$

This answer may be viewed in terms of sets as

$$P(\text{employed} \mid \text{male}) = \frac{500}{550} = \frac{n(\text{employed and male})}{n(\text{male})}.$$

Now, if the numerator and denominator of the expression on the right are both divided by the number in the sample space, $n(S)$, an expression for the conditional probability in terms of probability in the entire sample space of 810 is obtained:

$$P(\text{employed} \mid \text{male}) = \frac{\dfrac{n(\text{employed and male})}{n(S)}}{\dfrac{n(\text{male})}{n(S)}} = \frac{P(\text{employed and male})}{P(\text{male})}$$

$$= \frac{\dfrac{500}{810}}{\dfrac{550}{810}} = \frac{500}{550}. \ \blacksquare$$

To replay the process of Example 7 in a general setting, let A and B be two events in a sample space S, and consider $P(A \mid B)$. Then the given event B becomes the reduced sample space S^*. (See Figure 9.) Using the Venn diagram in shown in the figure, we obtain

$$P(A \mid B) = \frac{n(A \cap B)}{n(B)} = \frac{\dfrac{n(A \cap B)}{n(S)}}{\dfrac{n(B)}{n(S)}} = \frac{P(A \cap B)}{P(B)},$$

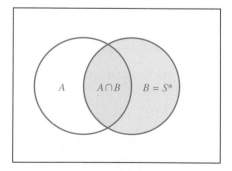

FIGURE 9
Diagram for Conditional Probability

provided that $P(B) \neq 0$. This justifies the **conditional probability formula** for sample spaces of equally likely outcomes, but the formula is also valid for all sample spaces.

Conditional Probability Formula

For any two events in a sample space, $P(A|B) = \dfrac{P(A \cap B)}{P(B)}$, provided that $P(B) \neq 0$.

This formula may be used to solve *any* conditional probability problem. Note that the probabilities involved are computed from the *entire original sample space,* not from the reduced sample space. Note further that the "given" event is the one that appears in the denominator of the quotient on the right.

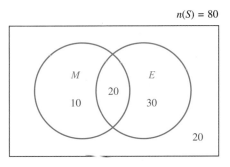

$n(S) = 80$

FIGURE 10
Class Selections at Washburn High School

EXAMPLE 8

Using the Conditional Probability Formula

The senior class of Washburn High School has 80 students, 50 of whom are taking English, 30 mathematics, and 20 both subjects. (See Figure 10.)

(a) What is the probability that a senior is taking English, given that the senior is taking mathematics?

SOLUTION Figure 10 will help organize the given data. Applying the conditional probability formula to this question gives $P(\text{taking English} \mid \text{taking mathematics}) =$

$$\frac{P(\text{taking English and mathematics})}{P(\text{taking mathematics})} = \frac{\dfrac{20}{80}}{\dfrac{30}{80}} = \frac{20}{30} = \frac{2}{3}.$$

(b) What is the probability that a senior is taking mathematics, given that she is taking English?

SOLUTION Applying the conditional probability formula to this question gives $P(\text{taking mathematics} \mid \textbf{taking English}) =$

$$\frac{P(\text{taking mathematics and English})}{P(\textbf{taking English})} = \frac{\dfrac{20}{80}}{\dfrac{50}{80}} = \frac{20}{50} = \frac{2}{5}. \ \blacksquare$$

The conditional probability formula *always* gives an expression that will solve a conditional probability problem. We can use the formula as a primary approach to such problems or as a foolproof option in case the reduced-sample-space approach becomes too complicated.

EXAMPLE 9
Drawing Slips of Paper

Three slips of paper, each with exactly one of the numbers 1, 2, and 3 written on it, without duplication, are placed in a box. The sample space for successively drawing two slips of paper, with replacement, and recording the numbers on them is as follows:

$$S = \{(1, 1), (1, 2), (1, 3), (2, 1), (2, 2), (2, 3), (3, 1), (3, 2), (3, 3)\}.$$

Use this sample space of 9 outcomes to find the probability that the sum of the numbers on the slips is 5, given that the first number drawn was a 2.

SOLUTION The conditional probability formula is used first to solve this problem:

$$P(\text{sum is 5} \mid \textbf{first number is 2}) = \frac{P(\text{sum is 5 and first number is 2})}{P(\textbf{first number is 2})}$$

$$= \frac{\dfrac{1}{9}}{\dfrac{3}{9}} = \frac{1}{3}.$$

The reduced-sample-space approach to this question, of course, gives the same answer: The given information leads to the reduced sample space $S^* = \{(2, 1), (2, 2), (2, 3)\}$, and among these three outcomes, only one has a sum of 5. Therefore, the probability that the sum is 5, given that the first number drawn was a 2, is $\frac{1}{3}$. $\blacksquare$

The conditional probability formula

$$P(A|B) = \frac{P(A \text{ and } B)}{P(B)}$$

involves three probabilities: $P(A|B)$, $P(A \text{ and } B)$, and $P(B)$. Anytime we know two of these probabilities, we can algebraically solve for the third. Example 10 shows how $P(A \text{ and } B)$ may be found, provided that we know the other two probabilities. In the next section, we will further explore this useful way to find the probability of an intersection of two events.

EXAMPLE 10
Using the Conditional Probability Formula

If $P(A|B) = \frac{2}{5}$ and $P(B) = \frac{7}{8}$, find $P(A \cap B)$.

SOLUTION The conditional probability formula

$$P(A|B) = \frac{P(A \cap B)}{P(B)}$$

may be solved for $P(A \cap B)$, to give

$$P(A \cap B) = P(B) \cdot P(A|B).$$

The given data may now be substituted to produce

$$P(A \cap B) = \left(\frac{7}{8}\right)\left(\frac{2}{5}\right) = \frac{14}{40} = \frac{7}{20}. \quad \blacksquare$$

A natural way to visually display probabilities of events that occur in an order, such as a first draw, a second draw, and a third draw, is a **probability tree diagram.** Example 11 shows how such diagrams are constructed. Even more generally, probability tree diagrams can be constructed for *any* conditional probability problem, regardless of whether there is a natural order. The main value of a probability tree diagram lies in its visual organization of conditional probability problems and its linkage to the probability of an intersection of events via the technique demonstrated in Example 10.

EXAMPLE 11

Constructing a Probability Tree Diagram

Suppose that two balls are randomly drawn in succession, without replacement, from a box containing five red and eight green balls. Draw and label a tree diagram that will describe the probabilities of the various outcomes.

SOLUTION There are two generations of branches, the first representing the possible outcomes on the first draw, the second the possible outcomes on the second draw. Each branch in the "second-draw" generation represents a possible outcome after the outcome of the branch to which it is attached has *already occurred from the first draw.* For instance, in Figure 11, arriving at point A from the starting point means that a red ball has been drawn on the first draw and has not been replaced; the branches attached to A on the right are then possibilities for the second draw *after the removal of a red ball on the first draw.* Similarly, the branches to the right of point B reflect possibilities for the second draw after the removal of a green ball on the first draw. Probabilities are now assigned to each branch, reflecting the outcome of the preceding branch to which it is attached, as shown in Figure 12.

The symbolism of conditional probability may now be used to relabel the tree diagram of Figure 12 into a final product as shown in Figure 13.

In view of Example 10, the conditional probability problem

$$P(2\text{nd } R \mid 1\text{st } R) = \frac{P(1\text{st } R \text{ and } 2\text{nd } R)}{P(1\text{st } R)}$$

may be algebraically solved for $P(1\text{st } R \text{ and } 2\text{nd } R)$, giving

$$P(1\text{st } R \text{ and } 2\text{nd } R) = P(1\text{st } R) \cdot P(2\text{nd } R \mid 1\text{st } R).$$

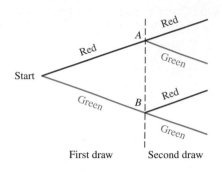

FIGURE 11
Construction of a Probability Tree Diagram

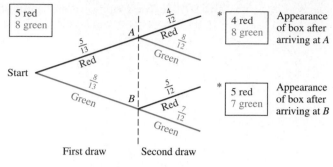

FIGURE 12
Drawing Balls without Replacement

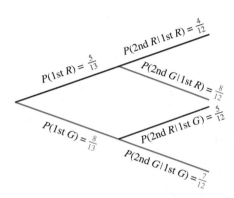

FIGURE 13
Final Appearance of Tree Diagram

Now observe that the product on the right side is precisely the product of the two probabilities along the top two branches of the tree! Thus,

$$P(\text{1st } R \text{ and 2nd } R) = \left(\frac{5}{13}\right)\left(\frac{4}{12}\right) = \frac{5}{39}. \blacksquare$$

Exercises 9.2

In each of Exercises 1 through 8, (a) state the sample space, (b) find the requested probability by using a reduced sample space, and (c) find the requested probability by using the conditional probability formula.

1. A single die has been rolled. What is the probability that three pips show on the top side, given that an odd number was rolled?

2. A single die has been rolled. What is the probability that four pips show on the top side, given that at least a two has been rolled?

3. A single die has been rolled. What is the probability that two pips show on the top side, given that an odd number was rolled?

4. A single die has been rolled. What is the probability that at least a four has been rolled, given that an odd number was rolled?

5. A coin is tossed three times. What is the probability that exactly two tails appeared, given that at least one head appeared?

6. A coin is tossed three times. What is the probability that exactly two tails appeared, given that exactly one head appeared?

7. A coin is tossed three times. What is the probability that the second toss resulted in heads, given that exactly one head appeared?

8. A coin is tossed three times. What is the probability that heads appeared on the last two tosses, given that exactly one head appeared?

Selections *Exercises 9 through 13 refer to the following experiment: Two cards are drawn in succession, without replacement, from a deck of 52 cards. What is the probability that*

9. The second card drawn was a jack, given that the first card was not a jack?

10. The second card drawn was a spade, given that the first card was a spade?

11. The second card drawn is an ace, given that the first card is a face card?

12. The second card drawn was a club, given that the first card was a heart or a spade?

13. The second card drawn was a nine, given that the first card was a diamond and a nine?

Selections *Exercises 14 through 16 refer to the following experiment: Three cards are drawn in succession, with replacement, from a deck of 52 cards. What is the probability that*

14. The third card is a spade, given that the first two cards drawn were not spades?

15. The second card is a heart, given that the first and third cards drawn were hearts?

16. The third card is a queen, given that the first two cards drawn were queens?

17. Selections From a box containing five red and nine green marbles, two marbles are drawn in succession, with replacement. What is the probability that the second marble drawn is

   (a) Green, given that the first was green?

   (b) Red, given that the first was red?

   (c) Red, given that the first was green?

18. Selections A box contains three red and seven green marbles. If three marbles are drawn in succession, without replacement, what is the probability that

   (a) The third marble is red, given that the first two marbles drawn were red?

   (b) The third marble is red, given that the first two marbles drawn were green?

   (c) The third marble is green, given that one red and one green marble have been drawn?

   (d) All three marbles are red?

Retail Sales *Exercises 19 through 24 refer to the following data:*

During November, a store selling TV sets recorded these brands in inventory:

	28-inch Screen	36-inch Screen
NuView	30	8
VuTech	40	12

According to the data, what is the probability that a TV set in inventory in November

19. Was a NuView brand, given that it had a 28-inch screen?

20. Had a 28-inch screen, given that it was a NuView brand?

21. Had a 36-inch screen, given that it was a VuTech brand?

22. Was a VuTech brand, given that it had a 36-inch screen?

23. Was a NuView brand?

24. Had a 36-inch screen?

Grade Distribution *Exercises 25 through 29 refer to the grade distribution in a particular finite mathematics class populated by freshmen and sophomores only:*

	A	*B*	*C*	*D*	*F*
Freshmen	2	5	8	3	2
Sophomores	3	4	8	2	3

If a person is randomly selected from this class, what is the probability that the person

25. Is a freshman?

26. Is a freshman and made an *A*?

27. Is a freshman or made an *A*?

28. Is a freshman, given that the student made an *A*?

29. Made a *C*, given that the student was a sophomore?

Personal Ads *Exercises 30 through 34 refer to the following data: A journalism major made a study of personal ads in the classified section of newspapers from cities of 50,000 people. She found that, on the average, there were 25 such ads in each paper, and among them were the following classifications:*

	Communication with Lover	Looking for Mate	Other Matters
Placed by male	3	6	4
Placed by female	4	2	6

Based on these findings, what percentage of such ads

30. Are placed by a male, given that the ads communicate with a lover?

31. Communicated with a lover or seek a mate, given that the ads are placed by a female?

32. Are looking for a mate or other matters, given that the ads are placed by a male?

33. Communicate with a lover or seek a mate?

34. Are placed by a male?

Political Survey *Exercises 35 through 39 refer to the following data: One hundred adults were asked their political party affiliation and the question "Will the European Common Market benefit the United States?" The results were as follows:*

	Yes	No	No Opinion
Democrat	20	12	12
Republican	25	8	5
Libertarian	6	5	7

Based on this survey, what is the probability that a person

35. Will answer "yes," given that he or she is a Democrat?

36. Will be a Republican or a Libertarian, given that he or she had no opinion?

37. Will not be a Republican, given an answer of "no"?

38. Had no opinion?

39. Was a Libertarian?

Exercises 40 through 45 refer to the following experiment: A pair of dice (one red and one green) is thrown. What is the probability that

40. A sum of 10 is rolled, given that at least a sum of 9 has been rolled?

41. A sum of 3 or 9 is rolled?

42. A sum of 7 is rolled, given that a sum of 3 or 9 has been rolled?

43. A sum of 3 or 4 is thrown, given that the sum of the pips showing is less than 5?

44. The red die shows a 5, given that a sum of 10 has been rolled?

45. A sum of 7 is rolled, given that the green die shows 4 pips?

46. Four slips of paper, on each of which is written just the number 1, 2, 3, or 4, without duplication, are put into a box. One of the slips is drawn, and the number on it is

recorded; then the slip is returned to the box. Now a second slip is drawn and the number on it noted. What is the probability that

(a) The sum of the two numbers drawn is five?

(b) The sum is six, given that the first number drawn is a three?

(c) The first number drawn is a two, given that the sum of the two numbers is four?

47. Four slips of paper, on each of which is written just the number 1, 2, 3, or 4, without duplication, are put into a box. Two slips are drawn in succession without replacement. What is the probability that

(a) The sum of the numbers drawn is five?

(b) The second number drawn was a three, given that the sum of the two numbers was four?

(c) The sum of the two numbers is nine, given that the first number drawn is a two?

(d) The sum of the numbers is five, given that the first number drawn is a four or a three?

Family Composition *Exercises 48 through 50 refer to three-children families. What is the probability that*

48. The first child is a boy, given that the last two are girls?

49. Two of the children are boys, given that at least one is a girl?

50. At least one is a boy, given that at least one is a girl?

Agricultural Spending *Exercises 51 through 53 refer to the following data: Fifteen percent of the farmers in an agricultural state bought a new tractor last year, 21% bought a new car, and 8% bought both a tractor and a new car. What is the probability that a farmer in this state will buy*

51. A new tractor, given that he or she bought a new car?

52. A new tractor or a new car?

53. Exactly one of these two items, given that he or she bought at least one?

Drug Testing *Exercises 54 through 56 refer to the following experiment: A new drug is administered to 100 people. Six of the people reported a rise in blood pressure, 5 reported loss of sleep, and 2 reported both of these side effects. According to these statistics, what is the probability that a person taking the drug*

54. Will experience a rise in blood pressure, given a report of lost sleep?

55. Will experience loss of sleep, given a reported rise in blood pressure?

56. Will experience neither of the two side effects?

Use the conditional probability formula

$$P(A|B) = \frac{P(A \cap B)}{P(B)}$$

to solve each of Exercises 57 to 62.

57. Find $P(A|B)$ if $P(A \cap B) = 0.3$ and $P(B) = 0.7$.

58. Find $P(B)$ if $P(A|B) = 0.8$ and $P(A \cap B) = \frac{2}{3}$.

59. Find $P(A \cap B)$ if $P(A|B) = \frac{4}{5}$ and $P(B) = \frac{5}{7}$.

60. Find $P(A|B)$ if $P(A \cap B) = 0.3$ and $P(B') = 0.2$.

61. Find $P(A|B)$ if $P(A) = 0.4$, $P(B) = 0.6$, and $P(A \cup B) = 0.8$.

62. Find $P(A|B)$ if $P(A) = 0.5$, $P(B') = 0.7$, and $P(A \cup B) = 0.7$.

In Exercises 63 through 68, draw a tree diagram, and label each branch with the correct probability.

63. A sample of two refrigerators is selected in succession, without replacement, from among six good ones and four defective ones.

64. A sample of two marbles is selected in succession, without replacement, from a box containing three red, four white, and two green marbles.

65. A sample of three marbles is selected in succession, without replacement, from a box containing eight red and three blue marbles.

66. A sample of 2 cards is drawn in succession, without replacement, from a deck of 52 cards; predict whether each card is a club.

67. A sample of 3 cards out of 52 is selected in succession, with replacement; predict whether each card is a spade.

68. A sample of three balls is selected, with replacement, from a box containing four red, five green, and seven white balls.

Codes *Exercises 69 through 72 refer to the following experiment: Four-letter code words are to be made from among the upper-case letters of the alphabet, with repetition of letters allowed. What is the probability that such a code*

69. Has exactly one A in it?

70. Has exactly two A's, given that it has at least one A in it?

71. Has exactly one A, given that it has at least one A in it?

72. Has at least two A's, given that it has at least one A in it?

Section 9.3   Multiplication Rules for Probability: Independent Events

The focus of this section is on finding the probability of the intersection of events. Examples 10 and 11 of the previous section set the stage for our work; the conditional probability formula is used to show that the probability of an intersection (the connective "and") of events may be found by multiplying probabilities. Events that occur in a natural order offer a natural entry point into these ideas.

EXAMPLE 1

The Intersection of Two Successive Events

Suppose two calculators are to be randomly selected, in succession, without replacement, from a box that contains four defective and nine good calculators; after each selection, the calculator is checked to see whether it is good or defective. What is the probability that the first calculator selected is good *and* the second calculator selected is defective?

SOLUTION We are to find

$$P(\text{1st good } and \text{ second defective}) = P(\text{1st good} \cap \text{2nd defective}).$$

We reason like this: First, we can find the probability that the first selection will be good because there are *four defective and nine good calculators in the box*. The answer is

$$P(\text{1st } G) = \frac{9}{13}.$$

Next, we can find the probability that the second calculator is defective, given that the first calculator selected was good because the box would have *four defective and eight good* calculators in it. The answer is

$$P(\text{2nd } D \mid \text{1st } G) = \frac{4}{12}.$$

The conditional probability formula may now be used to rewrite this last expression as

$$P(\text{2nd } D \mid \text{1st } G) = \frac{P(\text{1st } G \text{ and 2nd } D)}{P(\text{1st } G)}.$$

The probability in the numerator, $P(\text{1st } G \text{ and 2nd } D)$, is precisely the one we seek! Therefore, upon solving algebraically for that probability, we get the product

$$P(\text{1st } G \text{ and 2nd } D) = P(\text{1st } G) \cdot P(\text{2nd } D \mid \text{1st } G).$$

Since we know the two probabilities on the right, our problem is solved:

$$P(\text{1st } G \text{ and 2nd } D) = \left(\frac{9}{13}\right) \cdot \left(\frac{4}{12}\right) = \frac{3}{13}. \quad \blacksquare$$

Our solution to Example 1 shows us that "and" (intersection) means that we should multiply: Multiply the probability of the first selection by the probability of the second selection, *conditioned by what happened on the first selection.*

A tree diagram such as that shown in Figure 14 is a natural way to view events like those in Example 1. Notice how the result obtained implies that *"and' means to multiply the probabilities along the specified branches of a tree.*

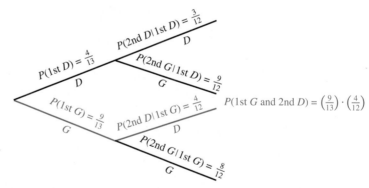

FIGURE 14
The Probability of Select-
ing Defective (D) versus
Good (G) Calculators

The conditional probability formula guides us to the probability of any intersection, regardless of whether the events occur in a natural order. The general statement of the formula,

$$P(A \mid B) = \frac{(P \cap B)}{P(B)}, \quad P(B) \neq 0,$$

concerns *any* two events A and B wherein information about B is given and the probability of A is to be found. Upon solving algebraically for $P(A \cap B)$, we have the multiplication formula

$$P(A \cap B) = P(B) \cdot P(A|B).$$

If the roles of A and B are reversed—that is, if information about A is given and the probability of B is to be found—the formula would read

$$P(B \cap A) = P(A) \cdot P(B|A).$$

But because $A \cap B = B \cap A$, the left sides of these two expressions are equal. It follows that

$$P(A \cap B) = P(B) \cdot P(A|B) = P(A) \cdot P(B|A).$$

The preceding formula is a general **multiplication rule for the intersection of two events** that is used with either of the two events in the "given" slot. Notice the ordering of A and B in this rule:

Same event (either A or B)

$$P(A \cap B) = P(\ \ )P(\ \ |\ \ ).$$

Other event (either B or A)

Multiplication Rule for the Intersection of Two Events

Let A and B be *any* events in a sample space for which $P(A) \neq 0$ and $P(B) \neq 0$. Then

$$P(A \text{ and } B) = P(A \cap B) = P(A) \cdot P(B|A) = P(B) \cdot P(A|B).$$

Here is an example of how this rule may be applied to Example 1:

$$P(\text{1st good and 2nd defective}) = P(\text{1st good}) \cdot P(\text{2nd defective} \mid \text{1st good})$$
$$= P(\text{2nd defective}) \cdot P(\text{1st good} \mid \text{2nd defective}).$$

Both expressions are equally valid for finding the requested intersection. Of course, the first expression would usually be more natural and easier to work with than the second, as we saw in Example 1. Another example of how the rule may be applied, where no order is implied, is

$$P(\text{male and diabetic}) = P(\text{male}) \cdot P(\text{diabetic} \mid \text{male})$$
$$= P(\text{diabetic}) \cdot P(\text{male} \mid \text{diabetic}).$$

Again, both expressions are valid and the choice of which to use would depend upon what data were available to calculate the required probabilities on the right sides.

The general multiplication rule can be extended to more than two events:

Multiplication Rules for the Intersection of Several Events

The multiplication rule can be extended to several events as follows:

$$P(A \cap B \cap C \cap D \ldots)$$
$$= P(A) \cdot P(B|A) \cdot P(C|A \cap B) \cdot P(D|A \cap B \cap C) \ldots.$$

Applied to events that occur in a natural order, 1st, 2nd, 3rd, and so on, the rule says

$$P(\text{1st and 2nd and 3rd} \ldots)$$

$$= P(\text{1st}) \cdot P(\text{2nd}|\text{1st has occurred}) \cdot P(\text{3rd}|\text{1st and 2nd have occurred}) \ldots$$

We show three more examples of events that occur in a natural order.

EXAMPLE 2
Drawing Cards

Two cards are to be randomly selected, in succession, without replacement, from a deck of 52 cards. What is the probability that the first card drawn will be a heart *and* the second card drawn will be a spade?

SOLUTION We need to find

$$P(\text{1st heart and 2nd spade}).$$

The product rule implies that

$$P(\text{1st heart and 2nd spade}) = P(\text{1st heart}) \cdot P(\text{2nd spade}|\text{1st heart}).$$

The probability that the first card drawn will be a heart is $\frac{13}{52}$. the probability that the second card will be a spade, given that the first card was a heart, is $\frac{13}{51}$. (There are now 51 cards in the deck and all 13 spades are still present.) Therefore,

$$P(\text{1st heart and 2nd spade}) = \left(\frac{13}{52}\right)\left(\frac{13}{51}\right) = \frac{13}{204} \approx 0.0637. \ \blacksquare$$

EXAMPLE 3
Selecting Balls
from a Box

From a box containing five red balls and three green balls, four balls are to be randomly selected, in succession, without replacement. What is the probability that the first ball drawn will be red, the second will be green, and the last two will be red?

SOLUTION In symbols, the question asks

$$P(\text{1st } R \text{ and 2nd } G \text{ and 3rd } R \text{ and 4th } R).$$

The connective "and" implies that the multiplication rule should be used to answer the question. So we write the probability of each selection, taking into account what has happened on the previous selections, and then multiply all the probabilities together:

$$P(1\text{st } R) \cdot P(2\text{nd } G \,|\, 1\text{st } R) \cdot P(3\text{rd } R \,|\, 1\text{st } R \text{ and } 2\text{nd } G) \cdot P(4\text{th } R \,|\, 1\text{st } R \text{ and } 2\text{nd } G \text{ and } 3\text{rd } R).$$

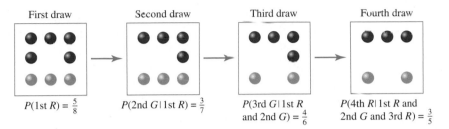

FIGURE 15
Successive Selections without Replacement.

Figure 15 makes use of successive pictorial representations of these conditionals to aid in finding the probability of each selection. The answer is then the product

$$\left(\frac{5}{8}\right)\left(\frac{3}{7}\right)\left(\frac{4}{6}\right)\left(\frac{3}{5}\right) = \frac{3}{28}. \quad \blacksquare$$

EXAMPLE 4

Selecting Cards

Three cards are to be randomly selected, in succession, *with replacement,* from a deck of 52 cards. What is the probability that the first card drawn will be a heart, the second a heart, and the third a spade?

SOLUTION The probability that the first card drawn will be a heart is $\frac{13}{52}$. The probability that the second card will be a heart, given that the first card was a heart, is again $\frac{13}{52}$, because the first card drawn was replaced in the deck. The probability that the third card will be a spade, given that the first was a heart and the second was a heart, is also $\frac{13}{52}$, because, again, the second card has been returned to the deck. This means that

$$P(1\text{st heart and 2nd heart and 3rd spade})$$

$$= \left(\frac{13}{52}\right)\left(\frac{13}{52}\right)\left(\frac{13}{52}\right) = \frac{1}{64} \approx 0.0156. \quad \blacksquare$$

We now turn to an example in which there is no "natural order" to the two events involved.

EXAMPLE 5

Consumer Credit

Research by a department store chain revealed that 80% of the people who make purchases are women and that 75% of those women's purchases are charged on the chain's credit card. What is the probability that a person making a purchase from this chain will be a woman *and* charge the purchase on her credit card?

SOLUTION The data show that the probability that a purchase is made with a credit card, *given* that purchase was made by a woman, is 0.75. That is, in abbreviated form, the purchasing information tells us that $P(\text{use credit card}|\text{woman}) = 0.75$. This conditional information tells us how the product should be written in order to take advantage of the given probabilities:

$$P(\text{woman and use credit card}) = P(\text{woman}) \cdot P(\text{use credit card}|\text{woman})$$
$$= (0.80)(0.75) = 0.6. \quad \blacksquare$$

Example 6 shows that three diagrams can also be used to display information about events that have no "natural" ordering. It is the conditional information that allows this to be done. Example 6 also shows that a tree diagram can be used to answer many other types of probability problems, in addition to those involving the connective *and*.

EXAMPLE 6

Using a Tree Diagram

A contractor buys bags of cement from two suppliers—Rodrigues's Lumber and Builder's Mart—and stores them in a warehouse. She buys 60% from Rodrigues's and 40% from Builder's Mart and, upon delivery, discovers that 2% of the bags from Rodrigues's are damaged and 4% of the bags from Builder's Mart are damaged.

(a) Draw a tree diagram for these data.

SOLUTION The tree diagram, with abbreviated notation, appears in Figure 16.

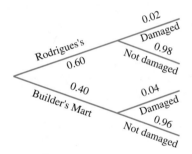

FIGURE 16
*The Probability of Damage
to Bags of Cement*

(b) For a randomly selected bag, what is the probability that it came from Rodrigues's and was damaged?

SOLUTION $P(\text{from Rodrigues's and damaged}) = (0.60)(0.02) = 0.012$, from the tree diagram.

(c) For a randomly selected bag, what is the probability that it was damaged, given that it came from Rodrigues's?

SOLUTION The answer is read directly from the tree diagram:

$$P(\text{damaged}|\text{Rodrigues's}) = 0.02.$$

(d) For a randomly selected bag, what is the probability that it was damaged?

SOLUTION This would be easy if we only knew whether the bag came from Rodrigues's or from Builder's Mart. Because it must come from one or the other, we interpret the problem as the union of two mutually exclusive events: $P(\text{damaged}) = P(\text{"from Rodrigues's and damaged" or "from Builder's Mart and damaged"}) = P(\text{from Rodrigues's and damaged}) + P(\text{from Builder's Mart and damaged}) = P(\text{from Rodrigues's}) \cdot P(\text{damaged}|\text{from Rodrigues's}) + P(\text{from Builder's Mart}) \cdot P(\text{damaged}|\text{from Builder's Mart}) = (0.60)(0.02) + (0.40)(0.04) = 0.028$. In other words, 2.8% of the bags in the warehouse are damaged. ∎

We now turn our attention to the concepts of dependent and independent events. A few examples will illustrate the difference between these concepts. For instance, consider the successive selection of 2 cards from a deck of 52 cards. If the cards are drawn *without replacement,* what is the probability that the first card drawn is a king and the second card is a queen? The answer is $\left(\frac{4}{52}\right)\left(\frac{4}{51}\right)$. That is, when selecting *without replacement,* the probability of the second selection definitely **depends,** or is **dependent,** upon what happened on the first selection. By contrast, if the cards are drawn *with replacement,* this same probability question would be answered by $\left(\frac{4}{52}\right)\left(\frac{4}{52}\right)$. In this case, the probability of the second selection is unaffected by what happened on the first selection. Stated differently, we say that the probability of the second selection is **independent** of what happened on the first selection.

Independent Events

Two events A and B are **independent,** provided that the occurrence of one has no effect on any probability question asked about the other. That is, $P(A|B) = P(A)$ and $P(B|A) = P(B)$. Events that are not independent are called **dependent.**

Here are some examples to help clarify these concepts:

Independent Events	Dependent Events
Outcome of successive draws of cards, with replacement	Outcomes of successive draws of cards, without replacement
Outcomes of repeated rolls of a die	Outcomes of successive draws of marbles, without replacement
The main computer fails; the backup computer fails	Type of bait in mousetrap; mouse being trapped
Outcomes of successive tosses of a coin	The weather today; the weather tomorrow

Although the phrases *independent events* and *mutually exclusive events* may sound similar, they are different. Suppose we toss two coins and consider the event A that the first coin comes up heads and the event B that the second coin comes up tails. Events A and B are independent, since the first toss of the coin has no impact on the probability that the second toss comes up tails; in other words, $P(B|A) = P(B)$. However, events A and B are *not* mutually exclusive, since the outcomes in A are $\{HH, HT\}$ and the outcomes in B are

$\{HT, TT\}$; that is, $A \cap B = \{HT\}$. In contrast, suppose we toss two coins and consider the event C that both coins come up heads and the event D that both coins come up tails. Events C and D *are* mutually exclusive, since $C = \{HH\}$ and $D = \{TT\}$. But, events C and D are *not* independent, since $P(D|C) = 0$ and $P(C|D) = 0$; that is, once we know that both coins have come up heads (or tails), there is no chance that they have also both come up tails (or heads).

The convenience of independent events lies in the fact that the multiplication rule for two independent events A and B now becomes very simple, because

$$P(A \cap B) = P(A) \cdot P(B|A) = P(A) \cdot P(B).$$

The Multiplication Rule for Independent Events

If A and B are independent events in a sample space, then

$$P(A \cap B) = P(A) \cdot P(B).$$

The extension of the multiplication rule for independent events is particularly easy: Just multiply the probabilities together!

The Extended Multiplication Rule for Independent Events

If $A_1, A_2, \ldots, A_n$ are independent events in a sample space, then

$$P(A_1 \cap A_2 \cap \cdots \cap A_n) = P(A_1) \cdot P(A_2) \cdot \cdots \cdot P(A_n).$$

EXAMPLE 7

Probability of Independent Events

A single die is rolled twice. What is the probability that the first roll is a four and the second roll is not a four?

SOLUTION The independence of the two events gives

$$P(\text{1st roll a four and 2nd roll not a four})$$
$$= P(\text{1st roll a four}) \cdot P(\text{2nd roll not a four})$$
$$= \left(\frac{1}{6}\right)\left(\frac{5}{6}\right) = \frac{5}{36}. \quad \blacksquare$$

EXAMPLE 8

Effectiveness of Surgery: Probability of Independent Events

Suppose that cataract surgery is successful 95% of the time. Of the next four people who have such surgery, what is the probability that

(a) The first three will be successful and the fourth unsuccessful?

SOLUTION The success of the surgery on one person is assumed to have no bearing on the success or failure of the surgery on another person; hence, independence is assumed. Thus, $P(\text{1st successful and 2nd successful and 3rd successful and 4th unsuc-}$

cessful) $= (0.95)(0.95)(0.95)(0.05) \approx 0.0429$. In other words, this particular sequence of events will happen about 4.29% of the time.

(b) All four will be successful.

SOLUTION Translating P(all four successful) into P(1st successful and 2nd successful and 3rd successful and 4th successful) means that, due to independence, the answer is found by $(0.95)(0.95)(0.95)(0.95) \approx 0.8145 \approx 81.45\%$.

(c) Exactly two will be successful.

SOLUTION Among the four surgeries, the exact order of the two successful and two unsuccessful surgeries is not specified. Therefore, all possibilities of order must be considered. Using S to stand for "successful" and S' to stand for "unsuccessful," we note that the surgeries might have occurred as "S and S and S' and S'," "S and S' and S' and S," and so on. Writing these possibilities in abbreviated form as follows, we find that there are six mutually exclusive orderings in which the surgeries could have taken place:

P(exactly 2 successful)

$= P(SSS'S' \text{ or } SS'S'S \text{ or } S'S'SS \text{ or } SS'SS' \text{ or } S'SS'S \text{ or } S'SSS')$

$= (0.95)(0.95)(0.05)(0.05) + (0.95)(0.05)(0.05)(0.95) \ldots + (0.05)(0.95)(0.95)(0.05).$

Observing that in each of the six products there are two 0.95's and two 0.05's, we find the answer to be

$$6(0.95)^2(0.05)^2 \approx 0.0135. \ \blacksquare$$

Example 9 answers the question posed at the outset of this chapter.

EXAMPLE 9

Computing Probability for Independent Events

A particular rocket assembly depends on two components, A and B. If either is working properly, the assembly will function, whereas if both fail, the assembly will not function. Assume that the functioning of either component is independent of the other, that the probability of failure of A is 0.002, and that the probability of failure of B is 0.01.

(a) What is the probability that both components will fail simultaneously?

SOLUTION We first find $P(A \text{ fails and } B \text{ fails})$. Because the functioning of these components is independent, the answer is $P(A \text{ fails}) \cdot P(B \text{ fails}) = (0.002)(0.01) = 0.00002$.

(b) What is the probability that at least one of the components is working properly?

SOLUTION The answer is $1 - P(\text{both fail}) = 1 - 0.00002 = 0.99998$.

(c) What is the probability that one component will be functioning properly and one will not?

SOLUTION This question may be restated as P("A is functioning and B failed" or "A failed and B is functioning"). The *or* separates two mutually exclusive events, so that a further restatement becomes $P(A \text{ is functioning and } B \text{ failed}) + P(A \text{ failed and } B \text{ is functioning}) = (0.998)(0.01) + (0.002)(0.99) \approx 0.012. \ \blacksquare$

<u>EXAMPLE 10</u>

Selecting Marbles

A box contains three red and five green marbles. Two marbles are to be selected, without replacement and without regard to order. What is the probability that one of the marbles will be red and the other green?

SOLUTION

(a) Since marbles are not replaced and order is disregarded, the use of combinations will give the answer:

$$P(\text{one } R, \text{one } G) = \frac{C(3, 1) \cdot C(5, 1)}{C(8, 2)} \approx 0.5357.$$

SOLUTION

(b) The same answer may be obtained by thinking in terms of ordered selections:

$$P(\text{one } R, \text{one } G) = P((\text{1st } R \text{ and 2nd } G) \text{ or } (\text{1st } G \text{ and 2nd } R))$$
$$= P(\text{1st } R \text{ and 2nd } G) + P(\text{1st } G \text{ and 2nd } R)$$
$$= \left(\frac{3}{8}\right)\left(\frac{5}{7}\right) + \left(\frac{5}{8}\right)\left(\frac{3}{7}\right) = \frac{30}{56} \approx 0.5357 \quad \blacksquare$$

<u>EXAMPLE 11</u>

Sampling

Among 20 computer chips, 5 are known to be defective. If a sample of three of the chips is to be selected, without replacement and without regard to order, what is the probability that all of the chips will be defective?

SOLUTION

(a) The answer is thought of as selecting a set:

$$P(\text{all 3 defective}) = \frac{C(5, 3)}{C(20, 3)} \approx 0.0088.$$

SOLUTION

(b) The solution is thought of in terms of ordered selections:

$$P(\text{all 3 defective}) = P(\text{1st } D \text{ and 2nd } D \text{ and 3rd } D)$$
$$= \left(\frac{5}{20}\right)\left(\frac{4}{19}\right)\left(\frac{3}{18}\right) = \frac{1}{114} \approx 0.0088. \quad \blacksquare$$

Exercises 9.3

The events in Exercises 1 through 9 are dependent.

Exercises 1 through 5 refer to the following experiment: Two cards are drawn in succession without replacement. What is the probability that

1. The second card is a heart, given that the first is a heart?

2. The first is an ace and the second is not an ace?

3. The first is a spade and the second is not a spade?

4. The first is a club and the second is a club?

5. Both cards are aces?

Quality Control *Exercises 6 through 9 refer to the following experiment: A sample of 3 lightbulbs is drawn in succession, without replacement, from a lot of 20 bulbs, 4 of which are known to be defective. What is the probability that*

6. The first two bulbs are good and the last one is defective?

7. The first one is good and the last two are defective?

8. All three bulbs are good?

9. All three bulbs are defective?

The events in Exercises 10 through 18 are independent.

Exercises 10 through 12 refer to the following experiment: A coin is tossed five times. What is the probability that

10. All tosses come up heads?

11. The first three tosses come up heads and the last two tails?

12. The last three tosses come up heads, given that the first two were tails?

Test Construction *Exercises 13 through 15 refer to the following experiment: A multiple-choice test has five questions, and each question has four choices. If you guess at the answers, what is the probability that you will*

13. Get all of them right?

14. Get the first three right and the last two wrong?

15. Get the first right, the second wrong, the third right, the fourth wrong, and the last one right?

Batting Averages *Exercises 16 through 18 refer to the following data: A baseball player has a batting average of 0.245. Assume that each turn at bat is not affected by the previous turns and that no walks occur on the times at bat considered. What is the probability that, for the next three times at bat, the player will*

16. Get a hit all three times?

17. Get exactly two hits?

18. Get at least two hits?

In Exercises 19 through 22, assume that A and B are independent events.

19. (a) Find $P(A)$ if $P(B) = 0.6$ and $P(A \cap B) = 0.4$.

(b) Find $P(A)$ if $P(B') = 0.7$ and $P(A \cap B) = 0.1$.

20. (a) Find $P(B)$ if $P(A) = 0.8$ and $P(A \cap B) = \frac{7}{16}$.

(b) Find $P(A \cup B)$ if $P(A) = 0.7$ and $P(B) = 0.5$.

21. (a) Find $P(B)$ if $P(A') = 0.5$ and $P((A \cap B)') = 0.6$.

(b) Find $P(A \cup B)$ if $P(A') = 0.4$ and $P(B) = 0.5$.

22. (a) Find $P(A \cap B)$ if $P(A) = \frac{1}{4}$ and $P(B) = \frac{2}{7}$.

(b) Find $P(A' \cap B')$ if $P(A) = \frac{1}{3}$ and $P(B) = \frac{4}{5}$.

In Exercises 23 and 24, find the requested probabilities associated with the tree diagrams shown.

23. (a) $P(R|N)$
(b) $P(M \cap S)$
(c) $P(N \cap R)$
(d) $P(R)$
(e) $P(S)$

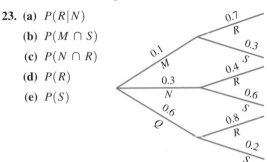

24. (a) $P(X|A)$
(b) $P(Y|B)$
(c) $P(B \text{ and } Y)$
(d) $P(X)$
(e) $P(Y)$

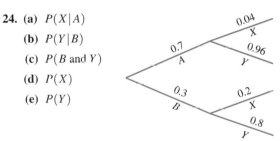

25. Three marbles are drawn, in succession and without replacement, from a box containing four red and five green marbles. Find the probability that

(a) The first marble drawn will be red and the last two green.

(b) The first marble drawn will be red, the second green, and the third red.

(c) All three marbles will be red.

26. Four cards are drawn, in succession and without replacement, from a standard deck of 52 cards. Find the probability that

(a) The first card drawn will be a spade, the second a club, the third a spade, and the fourth a club.

(b) The first two cards drawn will be kings and the last two will be queens.

(c) All of the cards will be hearts.

27. A box contains five red, three green, and four purple balls. Three balls are drawn from the box, in succession and with replacement. Find the probability that

    (a) The first ball will be red and the next two purple.

    (b) The first ball will be red, the second green, and the third purple.

    (c) All of the balls will be green.

28. Quality Control A shipment of computer chips contains 12 that are good and 3 that are defective. Three chips are selected, in succession, with replacement, and checked for a defect. Find the probability that

    (a) The first chip will be defective, the second defective, and the third good.

    (b) None of the chips will be defective.

    (c) All of the chips will be defective.

29. A single die is rolled four times. Find the probability that

    (a) The successive rolls will result in the following sequence of pips showing on the top side: 2, 2, not a 2, not a 2.

    (b) The successive rolls will result in the following sequence of pips showing on the top side: 3, 5, not a 5, 2.

    (c) The successive rolls will result in the following sequence of events: a number greater than 4, a 3, a number less than 5, a 4.

30. A single die is rolled five times. Find the probability that

    (a) The successive rolls will result in the following sequence of pips showing on the top side: not a 3, 4, 4, not a 5, not a 5.

    (b) A 3 turns up every time.

    (c) A 3 never turns up.

31. A pair of dice is rolled twice. Find the probability that

    (a) A sum of six will be rolled first, and a sum of nine will be rolled second.

    (b) A sum of seven will be rolled both times.

    (c) A sum of five will be rolled first, and a sum that is not five will be rolled second.

32. A pair of dice is rolled three times. Find the probability that

    (a) A sum of six will be rolled first, a sum of nine second, and a sum of three third.

(b) A sum of four will be rolled all three times.

(c) A sum of five will be rolled the first time, and a sum of three will be rolled the second and third times.

Effectiveness of a Vaccine *Exercises 33 through 35 refer to the following data: Suppose that it is known that flu shots are 97% effective. Suppose also that two people who live in different parts of the country take flu shots. What is the probability that*

33. Both will get the flu?

34. Neither will get the flu?

35. At least one will get the flu?

Committee Selection *Exercises 36 through 38 refer to the following experiment: A committee of three is selected at random from four juniors and five seniors. What is the probability that*

36. The first two persons selected are juniors and the third is a senior?

37. All three persons selected are juniors?

38. The last person selected is a junior, given that the first two selected are seniors?

Mortality Tables *Exercises 39 through 41 refer to the following data: According to insurance mortality tables, the probability that a 50-year-old woman will be alive 10 years from now is 0.91. College roommates Maria and Magda have mothers who are 50 years old and live in different states. What is the probability that*

39. Both of their mothers will be alive 10 years from now?

40. At least one of their mothers will be alive 10 years from now?

41. Neither of their mothers will be alive 10 years from now?

Product Reliability *Exercises 42 through 45 refer to the fol-*

lowing data: A rocket scientist estimates that a particular component in a rocket is 99.8% reliable. Three such components are installed so that if one fails, the task of the component is automatically passed to the next component. Assume that each component works properly or fails independently of the others. What is the probability that

42. All three components will fail?

43. The first two components will fail, but the third will not?

44. At least one component will function properly?

45. How many components should be installed so that at least one will function properly, with a probability of 0.99999?

Product Reliability *Exercises 46 through 48 refer to the following data: A bank has a main computer and a backup computer. The main computer has a reliability of 0.96, and the backup computer has a reliability of 0.90. What is the probability that*

46. Both computers will fail at the same time?

47. At least one will function properly?

48. One will be working properly and one will be down?

Tossing Coins *Exercises 49 through 51 refer to the following experiment: A coin is tossed seven times. What is the probability that*

49. All tosses turn up heads?

50. The last toss turns up heads, given that the first six turned up heads?

51. The last three tosses turn up heads, given that the first four turned up tails?

Product Durability *Exercises 52 through 54 refer to the following data: A particular type of lightbulb has a probability of 0.01 of burning out in less than 800 hours.*

52. If four of the bulbs are put into four different lamps, what is the probability that all four will burn out in less than 800 hours?

53. If four of the bulbs are put into four different lamps, what is the probability that at least one will still be burning after 800 hours?

54. Now suppose that the bulbs are used in a particular security application in which it is required that at least one bulb remain lighted for 800 consecutive hours. How many bulbs are required in order to assure that the probability of success is at least 0.999?

Quality Control *Exercises 55 through 57 refer to the following data: A sample of 3 TV sets is selected, in succession and without replacement, from among 12 sets, 3 of which are defective. What is the probability that*

55. All three of the TV sets chosen are defective?

56. The first two sets chosen are not defective, but the third one is defective?

57. At least one set is defective?

Demographic Survey *Exercises 58 through 60 are about families who plan to have three children. What is the probability that*

58. The first two children will be boys and the third a girl?

59. All three children will be girls?

60. There will be more boys than girls?

Voter Survey *Exercises 61 through 63 refer to the following data: Suppose that 28% of the voting-age adults consider themselves Democrats, 25% consider themselves Republican, and the rest consider themselves Independents. If 10 voters are randomly contacted, what is the probability that*

61. All are Republicans?

62. None are Republicans?

63. At least one is a Republican?

Safety Survey *Exercises 64 through 66 refer to the following data: A survey in Star City revealed that 40% of the drivers do not wear a seat belt. If three cars are to be stopped during a routine traffic check in the town, what is the probability that*

64. None of the three drivers will be wearing a seat belt?

65. At least one will be wearing a seat belt?

66. One will be wearing a seat belt and two will not?

Security *Exercises 67 and 68 refer to the following data: Nuclear power plants have a threefold security system, each of which is 98% reliable and independent of the others, to prevent unauthorized persons from entering the premises. What is the probability that an unauthorized person will*

67. Get through all three security systems?

68. Get through the first two systems, but not the third?

Fruit Flies *Exercises 69 through 71 refer to the following data: Examination of a large number of fruit flies reveals that 10% have red eyes. If three of the flies are selected at random, in succession and without replacement, what is the probability that*

69. All will have red eyes?

70. None will have red eyes?

71. Exactly one will have red eyes?

Section 9.4 Bayes's Theorem

Bayes's theorem is a special application of conditional probability. An example leads to the theorem.

EXAMPLE 1

Pointing toward Bayes's Theorem

Surf Mart, which sells shirts under its own private label, buys 40% of its shirts from supplier A, 50% from supplier B, and 10% from supplier C. It is found that 2% of the shirts from A have flaws, 3% from B have flaws, and 5% from C have flaws. A probability tree diagram representing these purchases and flaw rates, is shown in Figure 17. The first-generation branches are labeled with probabilities representing the random selection of a supplier. The second-generation of branches are labeled with conditional probabilities that a randomly selected shirt will have a flaw or no flaw, depending on which supplier produced the shirt. If one of these shirts is bought from Surf Mart, what is the probability that

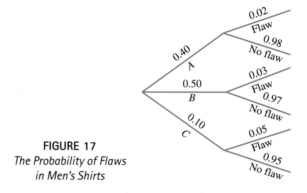

FIGURE 17
The Probability of Flaws in Men's Shirts

(a) The shirt has a flaw, given that it came from A?

SOLUTION Knowing that the shirt came from A means that we have already traversed the A branch and now look for the probability, which is .02, along the attached *flaw* branch. Therefore, $P(\text{flaw}|\text{from } A) = .02$. (Note that the answer was found by reading the tree diagram from left to right.)

(b) The shirt has a flaw?

SOLUTION The solution is

$$P(\text{"from } A \text{ and a flaw" or "from } B \text{ and a flaw" or "from } C \text{ and a flaw"})$$
$$= P(\text{from } A) \cdot P(\text{a flaw}|\text{from } A) + P(\text{from } B) \cdot P(\text{a flaw}|\text{from } B)$$
$$+ P(\text{from } C) \cdot P(\text{a flaw}|\text{from } C)$$
$$= (0.40)(0.02) + (0.50)(0.03) + (0.10)(0.05) = 0.028.$$

(c) The shirt came from A, given that it has a flaw?

SOLUTION This time, the given event *flaw* is on the right, and we are confronted with somehow trying to read back to the left to reach the event A. This is the *reverse* probability to which Bayes referred and which, from the viewpoint of a tree diagram, will always be the clue that the problem is a Bayes's theorem problem. How do we find the solution? Just use the conditional probability formula and read from the tree:

$$P(A|\text{flaw}) = \frac{P(A \text{ and flaw})}{P(\text{flaw})}$$

$$= \frac{(0.40)(0.02)}{(0.40)(0.02) + (0.50)(0.03) + (0.10)(0.05)} \approx 0.286. \quad \blacksquare$$

Figure 18 illustrates the Venn diagram for Example 1. Notice how the whole sample space is the union of the three mutually exclusive events, *A, B,* and *C.* Thus, we can view the event *F* as the union of three mutually exclusive events: $A \cap F$, $B \cap F$, and $C \cap F$. The general principle is that when our whole sample space can be written as the union of mutually exclusive events, every possible event in the space can be viewed as being somehow split or partitioned by those events. Thus, in Example 1, when we used algebra to write $P(A|F) = \frac{P(A \cap F)}{P(F)}$, we were able to break $P(F)$ down further into $P(A \cap F) + P(B \cap F) + P(C \cap F)$. **Bayes's theorem** pulls all of these steps together.

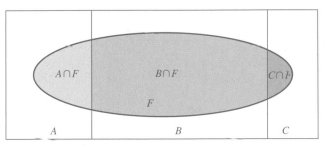

FIGURE 18
The Diagram for Bayes's Theorem

Bayes's Theorem

Let *A* and *B* be mutually exclusive events, the union of which is the entire sample space *S.* Let *F* be any event with a probability that is not zero. Then

$$P(A|F) = \frac{P(A \text{ and } F)}{P(A \text{ and } F) + P(B \text{ and } F)}$$

$$= \frac{P(A) \cdot P(F|A)}{P(A) \cdot P(F|A) + P(B) \cdot P(F|B)}.$$

A similar formula for $P(B|F)$ can be found by interchanging *A* and *B.* This formula can be extended to any finite number of mutually exclusive events, $A_1, A_2, \ldots, A_n$ whose union is *S,* as follows:

$$P(A_i|F) = \frac{P(A_i) \cdot P(F|A_i)}{P(A_1) \cdot P(F|A_1) + P(A_2) \cdot P(F|A_2) + \cdots + P(A_n) \cdot P(F|A_n)}.$$

HISTORICAL NOTE

THOMAS BAYES (1702–1761)

The Reverend Thomas Bayes was an English clergyman who discovered the formula for what he called "reverse" probabilities in a sample space. His mathematical work was a hobby, and his famous formula was not published until after his death. Even so, a correct proof was not given (by R. A. Fisher) until the 1930s.

As our development has suggested, a tree diagram is useful when using Bayes's theorem.

EXAMPLE 2

Tree Diagrams and Bayes's Theorem

Two marbles are randomly drawn, in succession and without replacement, from a box containing two red and five white marbles. The tree diagram looks like Figure 19.

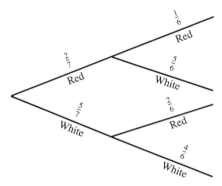

FIGURE 19
Drawing Marbles without Replacement

What is the probability that

(a) The second marble drawn is red?

SOLUTION The answer is $\left(\frac{2}{7}\right)\left(\frac{1}{6}\right) + \left(\frac{5}{7}\right)\left(\frac{2}{6}\right) = \frac{2}{42} + \frac{10}{42} = \frac{12}{42} = \frac{2}{7}$.

(b) The second marble drawn is red, given that the first marble is white?

SOLUTION After locating the given event "first marble white," we continue reading naturally from *left to right* to arrive at the event "second marble red" and the answer, $\frac{2}{6}$.

(c) The first marble drawn is white, given that the second marble is red?

SOLUTION This time the given event "second marble red" is located on the right (in two different places) of the tree diagram, and we are being asked to read from *right to left,* to locate the event "first marble white." This is a Bayes's theorem problem, and we write

$$P(\text{1st white}\,|\,\text{2nd red}) = \frac{P(\text{1st white and 2nd red})}{P(\text{2nd red})}$$

$$= \frac{\left(\dfrac{5}{7}\right)\left(\dfrac{2}{6}\right)}{\left(\dfrac{2}{7}\right)\left(\dfrac{1}{6}\right) + \left(\dfrac{5}{7}\right)\left(\dfrac{2}{6}\right)} = \frac{\dfrac{10}{42}}{\dfrac{12}{42}} = \frac{10}{12} = \frac{5}{6}. \quad \blacksquare$$

EXAMPLE 3

Tree Diagrams and
Bayes's Theorem

A department store made a study of the personal checks it received for payment of goods. It discovered that 40% of the checks with insufficient funds had the wrong date on them, while only 2% of all good checks had the wrong date on them. It also found that 0.5% of all checks received had insufficient funds to cover them. The tree diagram for this situation is shown in Figure 20. If a clerk in this store receives a personal check from a customer, what is the probability that

(a) The check has insufficient funds and a wrong date?

SOLUTION The answer is $(0.005)(0.40) = 0.002$.

(b) The check has the wrong date, given that it has insufficient funds?

SOLUTION $P(\text{wrong date}\,|\,\text{insufficient funds}) = 0.40$.

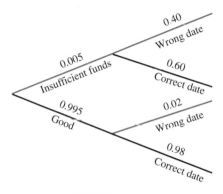

FIGURE 20
Personal Checks

(c) The check has insufficient funds, given that it has the wrong date?

SOLUTION This is a Bayes's theorem problem. We have

$$P(\text{insufficient funds}\,|\,\text{wrong date})$$
$$= \frac{P(\text{insufficient funds and wrong date})}{P(\text{wrong date})}$$
$$= \frac{(0.005)(0.40)}{(0.005)(0.40) + (0.995)(0.02)} \approx 0.0913.\ \blacksquare$$

The examples in this section show that a tree diagram is helpful in identifying a Bayes's theorem problem and in finding its solution. We suggest the use of such a diagram whenever practical.

Exercises 9.4

In Exercises 1 through 5, use the accompanying tree diagram to determine each probability.

1. $P(A \text{ and } D)$
2. $P(D)$
3. $P(E|B)$
4. $P(B|E)$
5. $P(A|D)$

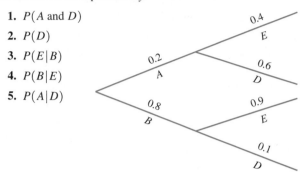

In Exercises 6 through 11, use the accompanying tree diagram to determine each probability.

6. $P(E|A)$
7. $P(A|E)$
8. $P(E)$
9. $P(C|D)$
10. $P(D|C)$
11. $P(B|E)$

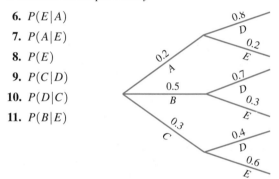

Every remaining exercise can be modeled with a tree diagram. Draw one for each set of exercises, and use it to answer the questions. Some of the exercises involve Bayes's theorem.

Sampling *Exercises 12 through 14 refer to the following experiment: Two balls are drawn, in succession and with replacement, from a box containing two red, five blue, and seven purple balls. What is the probability that*

12. The first ball drawn is blue and the second ball is purple?

13. The second ball is drawn is blue, given that the first is purple?

14. The first ball drawn is blue, given that the second ball is purple?

Sampling *Exercises 15 through 17 refer to the following experiment: Three balls are drawn in succession, without replacement, from a box containing three red and six green balls. What is the probability that*

15. All three balls drawn are red?

16. The first two balls drawn are red and the last one is green?

17. The third ball drawn is green?

Sampling *Exercises 18 through 20 refer to the following experiment: A box contains three slips of paper, on each of which is written just the number 1, 2, or 3, without duplication. Two slips of paper are drawn, in succession and with replacement, from the box. What is the probability that*

18. The second number drawn is a 3, given that the first number drawn is a 2?

19. The first number drawn is a 3, and the second number is a 1?

20. The first number is a 2, given that the second number is a 3?

Sampling *Exercises 21 through 24 refer to the following experiment: One box, A, contains three red and four green balls, while a second box, B, contains eight red and two green balls. A box is selected at random and a ball is drawn. What is the probability that*

21. The ball is red, given that it came from box A?

22. The ball is red and came from box B?

23. The ball is red?

24. The ball was drawn from box B, given that it is red?

Demographic Survey *Exercises 25 through 27 refer to the following data: Thirty-five percent of the students at State Tech are female, while 55% of the students at State University are female. If one of these universities is selected at random (assume a 50–50 chance for each), and a student from that university is selected at random, what is the probability that the student will be*

25. Female?

26. Female, given that the student comes from State Tech?

27. From State Tech, given that the student is female?

Color Blindness *Exercises 28 through 31 refer to the following data: It is known that about 5% of all men and 0.25% of all women are color blind. Assume that half of the population are men and half women. If a person is selected at random, what is the probability that the person is*

28. Female and color blind?

29. Color blind, given that the person is male?

30. Color blind?

31. Male, given that the person is color blind?

Tossing a Biased Coin *Exercises 32 through 34 refer to the following experiment: A coin is biased to show 60% heads and 40% tails. The coin is tossed twice. What is the probability that*

32. The first toss lands heads up and the second toss lands tails up?

33. The second toss comes up heads?

34. The first toss comes up heads, given that the second toss landed tails up?

Applicant Screening *Exercises 35 through 37 refer to the following data: The human resources director of a certain organization estimates from experience that, for a particular job opening, 80% of the applicants are qualified. To help in the selection process, a test has been designed so that a qualified applicant has a probability of 0.85 of passing the test, while an unqualified applicant has a probability of only 0.05 of passing the test. If an applicant is selected at random, what is the probability that the person*

35. Is qualified and will pass the test? What are the odds for this event?

36. Can pass the test, given that the applicant is unqualified for the job?

37. Is unqualified for the job, given a passing score on the test?

Insurance Risk *Exercises 38 through 40 refer to the following data: An auto insurance company classifies its drivers as good, medium, or bad risks. Of its policyholders, 50% are considered good risks, 35% medium risks, and 15% bad risks. Company records indicate that the probability of a driver from these three risk categories having an accident during a given year is 0.02, 0.05, and 0.10, respectively. What is the probability that one of the company's insured drivers*

38. Is a good risk and will have an accident during the next year?

39. Will have an accident during the next year?

40. Will be rated a bad risk, given that during the next year the driver has an accident?

Vaccine Effectiveness *Exercises 41 through 43 refer to the following data: During a given winter, 23% of the population received a flu shot, and, of those, 1% got the flu. For those who did not receive the flu shot, 12% got the flu. What is the probability that a person*

41. Did not get the flu, given that he or she had received the flu shot?

42. Did not receive the flu shot and got the flu?

43. Did not receive the shot, given that he or she had the flu?

Medical Test Reliability *Exercises 44 through 46 refer to the following data: Records indicate that 2% of the population has a certain kind of cancer. A medical test has been devised to help detect this kind of cancer. If a person does have the cancer, the test will detect it 98% of the time. However, 3% of the time the test will indicate that a person has the cancer when, in fact, he or she does not. For persons using this test, what is the probability that*

44. The person has this type of cancer, and the test indicates he or she has it?

45. The person has this type of cancer, given that the test indicates that he or she has it?

46. The person does not have this type of cancer, given a positive test result for it?

Product Reliability *Exercises 47 through 50 refer to the following data: A manufacturer makes two models of personal digital assistants (PDAs): a standard model and a deluxe model, with 60% of production allotted to the standard model. About 2% of the standard models and 1% of the deluxe models have some kind of defect. If one of the PDAs is selected at random, what is the probability that*

47. It is a deluxe model and has a defect?

48. It is defective, given that it is a deluxe model?

49. It is a deluxe model, given that it is defective?

50. It is defective? What percentage of the PDAs have a defect?

Product Reliability *Exercises 51 through 54 refer to the following data: An auto mechanic knows that 1% of the new fuel pumps installed fail within one year, but that about 5% of the rebuilt ones fail within one year. A mechanic selects a fuel pump from an inventory of 20, not realizing that 5 of the 20 are rebuilt fuel pumps. What is the probability that*

51. The pump selected was rebuilt and will fail within a year?

52. The pump selected will fail within a year, given that it was rebuilt?

53. The pump selected was rebuilt, given that it failed during the year?

54. Any fuel pump in the inventory will fail within a year? What percentage of the inventory will fail within a year?

Product Reliability *Exercises 55 through 59 refer to the following data: Downhill Express buys ski poles in the following percentages from four suppliers: 30% from A1 Poles, 20% from Stick-Em, 40% from Down Force, and 10% from Pole Pro. Downhill Express finds that the defective rates in shipments received are 4%, 5%, 3%, and 6%, respectively. What percentage of the ski poles*

55. Have a defect?

56. Have a defect, given that the pole came from A1 Poles or Stick-Em?

57. Came from Down Force and does not have a defect?

58. Came from Pole Pro, given that the pole has a defect?

59. Came from Stick-Em, given that the pole has a defect?

Salary Structure *Exercises 60 through 62 refer to the following data: In a particular firm, 5% of the men and 3% of the women have salaries that exceed $50,000. Sixty percent of the firm's employees are men, and 40% are women. For persons working for this firm, what is the probability that*

60. The employee is a woman and has a salary that exceeds $50,000?

61. The employee is a man, given that the salary exceeds $50,000?

62. The employee's salary exceeds $50,000? What percentage of the firm's employees earn more than $50,000?

63. Sampling A box contains three balls numbered 1 through 3. One ball is drawn from the box and laid aside. Then a second ball is drawn. What is the probability that the second ball is numbered 2?

Section 9.5   Binomial Experiment; A Guide to Probability

Experiments classified as **binomial** were first studied by James Bernoulli in the late 1600s. The term *binomial* comes from the fact that these experiments can be classified as having exactly two outcomes. For an experiment to qualify as binomial, several conditions, already familiar to us, must be met.

HISTORICAL NOTE

JAMES BERNOULLI (1654–1705)

James (Jacob, Jacques) Bernoulli was a Swiss mathematician who taught at the University of Basel. The Bernoulli distribution and the Bernoulli theorem of probability and statistics are among his many notable contributions to mathematics. The mathematical careers of James and his brother John form the connecting link between the mathematics of the 17th and 18th centuries.

Conditions for a Binomial Experiment

1. *Experiment repeated n times.* The experiment is to be repeated a finite number of times, say, n, under the same conditions. Each individual repetition is referred to as a *Bernoulli trial.*
2. *Independence.* Each time the experiment is performed, the outcome has no effect on the outcome of any other trial of the experiment.
3. *There are exactly two possible outcomes: success and failure.* Each time the experiment is performed, the outcome can be classified in exactly one of two ways; (1) success (S), with a probability of p; or (2) failure (F), with a probability of $q = 1 - p$. The probability of success never changes from trial to trial.

The concepts of repeated trials and independence are already familiar to us, so the only point of concern might be regarding the third condition, wherein the outcomes must be reduced to a success (S) or a failure (F).

The central question in a binomial experiment is to find the probability of r successes from among n trials; the outcome labeled "success" will be determined by the question asked, and all other outcomes will be lumped together and labeled "failure." Some examples clarify the terminology. Solutions follow later.

EXAMPLE 1

Determining Success and Failure in a Binomial Experiment

A single die is rolled 10 times, and we want to know the probability that exactly three fives will be rolled. The experiment is repeated $n = 10$ times, there are to be $r = 3$ successes, and the outcome of each trial is independent of the others.

Because the question asks about rolling fives, we call "roll a 5" a success S and "roll a 1, 2, 3, 4, or 6" a failure F. It follows that $p = P(S) = \frac{1}{6}$ and $q = P(F) = \frac{5}{6}$. ▪

EXAMPLE 2

Determining Success and Failure in a Binomial Experiment

A coin is tossed 20 times, and we are interested in knowing the probability that exactly 12 heads will turn up. The experiment is repeated $n = 20$ times (for 20 trials) and the outcomes of the trials are independent. Because the questions ask about heads, success will denote the event "a head lands up" and a failure will denote the event "a tail lands up." In this case, we have

$$p = P(S) = \frac{1}{2}, \quad q = P(F) = \frac{1}{2}, \quad \text{and} \quad r = 12. \ \blacksquare$$

EXAMPLE 3

Effectiveness of Surgery: Determining Success and Failure in a Binomial Experiment

Suppose that it is known that kidney transplants are successful 80% of the time. Of the next 12 people who receive such transplants, we are interested in knowing the probability that at least 10 transplants will be successful. The experiment is repeated $n = 12$ times, and we consider the success or failure of the transplant for any one person to be independent of the results of the others. Here, S represents the event "the transplant will be successful" and F represents the event "the transplant will not be successful." It follows that $p = P(S) = 0.80$, $q = P(F) = 0.20$, and r is 10, 11, or 12. ▪

EXAMPLE 4

Product Reliability: Determining Success and Failure in a Binomial Experiment

Tests show that 4% of the hair dryers made by the Hot-Dri Company have a defect. Suppose that 50 of the company's hair dryers are ordered by a discount store, and we are interested in the probability that exactly one of them will be defective. In this case, $n = 50$, and because we are interested in *defects,* success will denote the event "the dryer is defective" and failure will denote the event "the dryer is not defective." We therefore have

$$p = P(S) = 0.04, \quad q = P(F) = 0.96, \quad \text{and} \quad r = 1. \ \blacksquare$$

In the first three of these examples, the labeling of success and failure was rather natural, but in Example 4, it seemed a bit unnatural. If desired, such questions can be restated so that they seem more comfortable. For instance, in Example 4, we could restate the question as "What is the probability that exactly 49 of the dryers are not defective?" This will reverse the designations of success and failure (and their probabilities). Either interpretation will lead to the same answer.

Just how do we go about answering probability questions regarding binomial experiments such as those carried out in Examples 1 through 4? Two examples will lead us to a useful formula that will give us the probabilities in any binomial experiment. ▪

EXAMPLE 5

Political Poll: How Probability in a Binomial Experiment Is Calculated

In a poll of probable voters in an upcoming gubernatorial election, candidate A gets 30% of the votes. Assume that this percentage holds for the actual voting population. If four probable voters are asked about their choice of candidates, what is the probability that exactly three of them will favor candidate A?

SOLUTION We are interested in candidate A. Therefore, let S (success) denote "vote for A" and F (failure) denote "vote for another candidate." Then, according to the data given, $P(S) = p = 0.30$ and $P(F) = q = 0.70$; the number of trials, n, is 4; and the number of successes we seek is $r = 3$.

We assume that how any one person answers the poll is unaffected by how another person answers the poll; hence, the trials are independent. The importance of independence is that the probability of any sequence of S's and F's joined by the connective "and" is found by *multiplying the individual probabilities in the sequence together*. Thus, to get the probability that exactly three will vote for A, we have these mutually exclusive *ordered possibilities* of "vote for A" (S) and "will not vote for A" (F) among the four persons polled:

$P(\text{exactly three will vote for } A)$

$\quad = P(SSSF \text{ or } SSFS \text{ or } SFSS \text{ or } FSSS)$

$\quad = (0.30)(0.30)(0.30)(0.70) + (0.30)(0.30)(0.70)(0.30) + (0.30)(0.70)(0.30)(0.30)$
$$\qquad\qquad + (0.70)(0.30)(0.30)(0.30)$$

$\quad = 4(0.30)^3(0.70) = 0.0756.$

Notice that the product in each term is the same, $(0.30)^3(0.70)$, because there are always 3 S's and 1 F. Therefore, if the number of terms is known, then the desired probability can be found by multiplying that number, 4, by $(0.30)^3(0.70)$, as we have just done. ■

EXAMPLE 6

How the Probability of a Binomial Experiment Is Calculated

A single die is rolled five times. What is the probability that exactly three of the rolls show two pips?

SOLUTION Here, the number of trials is $n = 5$, S represents "roll a two" with $P(S) = p = \frac{1}{6}$, F represents "roll something not a two" with $P(F) = q = \frac{5}{6}$, and the number of desired successes is $r = 3$. There are several sequences of the five rolls that meet the condition of three successes, and because the outcome on each roll of the die is independent of the others, the probability of each sequence is found by multiplying the probabilities of the various outcomes together:

Sequence satisfying three successes	Probability of sequence occurring
$SFFSS$	$\left(\frac{1}{6}\right)\left(\frac{5}{6}\right)\left(\frac{5}{6}\right)\left(\frac{1}{6}\right)\left(\frac{1}{6}\right) = \left(\frac{1}{6}\right)^3\left(\frac{5}{6}\right)^2$
$FSSFS$	$\left(\frac{5}{6}\right)\left(\frac{1}{6}\right)\left(\frac{1}{6}\right)\left(\frac{5}{6}\right)\left(\frac{1}{6}\right) = \left(\frac{1}{6}\right)^3\left(\frac{5}{6}\right)^2$
$SFSSF$	$\left(\frac{1}{6}\right)\left(\frac{5}{6}\right)\left(\frac{1}{6}\right)\left(\frac{1}{6}\right)\left(\frac{5}{6}\right) = \left(\frac{1}{6}\right)^3\left(\frac{5}{6}\right)^2$
. . . .	. . . .

Notice that the probability is the same—namely, $\left(\frac{1}{6}\right)^3\left(\frac{5}{6}\right)^2$—every time we have a sequence of throws involving three S's and two F's. Notice also that the sequences are *mutually exclusive* events. That is, no two of them can happen simultaneously. It follows that we can just add all of the probabilities in the right-hand column to answer one question. Better yet, if we only knew how many such sequences were on the list, we could just multiply that number by $\left(\frac{1}{6}\right)^3\left(\frac{5}{6}\right)^2$ to answer our question. The number of such sequences is the number of different-appearing arrangements in a row of the five letters—three S's and two F's—where the S's are considered indistinguishable and the F's are considered indistinguishable. That number is

$$\frac{5!}{3! \cdot 2!} = C(5, 3) = 10.$$

Therefore, the answer to our question is

$$10\left(\frac{1}{6}\right)^3\left(\frac{5}{6}\right)^2 \approx 0.0322,$$

or, symbolically, $C(5, 3)(\frac{1}{6})^3(\frac{5}{6})^2$. ■

In terms of n, p, q, and r, the final calculation for Example 6 becomes

$$C(n, r) \cdot p^r \cdot q^{n-r}.$$

Other examples show that this general pattern of thought prevails every time, leading to the following formula:

The General Formula for Binomial Experiments

For a binomial experiment with n trials, where $P(S) = p$ and $P(F) = q = 1 - p$, the probability of exactly r successes is $C(n, r)p^r q^{n-r}$.

EXAMPLE 7

Drug Effectiveness: Using the General Formula for Binomial Experiments

Tests show that about 4% of the people who take a particular drug are subject to side effects. Of 20 people taking the drug, what is the probability that exactly 5 of them will experience side effects?

SOLUTION We assume that one person's having side effects will have no bearing on another person's having side effects from the drug. With independence thus assumed, we may take the outcomes to be S: "will have side effects," with a probability of 0.04, and F: "will not have side effects," with a probability of 0.96. With $n = 20$ and $r = 5$, the binomial-experiment formula gives $C(20, 5)(0.04)^5(0.96)^{15} \approx 0.00086$. ■

EXAMPLE 8

Market Analysis: Using the General Formula for Binomial Experiments

A particular company polled a large number of potential customers and found that 25% of them would buy the company's product. If a salesperson calls on 10 potential customers, what is the probability that

(a) Exactly 8 will buy the product?

SOLUTION Assuming that the customers are independent in their buying decisions, we have $n = 10$, S represents "will buy the product," F represents "will not buy the product," $p = 0.25$, $q = 0.75$, and $r = 8$. Therefore, our answer is

$$C(10, 8)(0.25)^8(0.75)^2 \approx 0.00039.$$

(b) Exactly 3 will buy the product?

SOLUTION The answer is $C(10, 3)(0.25)^3(0.75)^7 \approx 0.250$.

(c) At least 8 will buy the product?

SOLUTION The event "at least 8 will buy the product" is, as usual, rewritten as an equivalent statement using mutually exclusive events:

$$P(\text{at least 8 will buy}) = P(\text{exactly 8 will buy}) + P(\text{exactly 9 will buy})$$
$$+ P(\text{exactly 10 will buy}) = C(10, 8)(0.25)^8(0.75)^2 + C(10, 9)(0.25)^9(0.75)^1$$
$$+ C(10, 10)(0.25)^{10}(0.75)^0 \approx 0.00039 + 0.00003 + 0.00000 = 0.00042.$$

(d) At least 1 will buy the product?

SOLUTION The complement theorem will give the easiest translation:

$$P(\text{at least 1 will buy}) = 1 - P(\text{none will buy})$$
$$= 1 - (0.75)^{10} \approx 0.944. \quad \blacksquare$$

Finally, we explore the connection between binomial experiments and expected value. Suppose that the probability that a driver in a certain city will be wearing a seat belt is 0.70. Taking success to be "wearing a seat belt" and failure to be "not wearing a seat belt," and assuming that drivers are independent in their wearing of a belt, we find the probability that r drivers among n will be wearing seat belts by the binomial formula $C(n, r)p^r q^{n-r}$. If 4 drivers in this city are checked, then the number of those wearing seat belts, 0, 1, 2, 3, or 4, may be considered to be values of a random variable X, and a probability distribution for this random variable may be constructed as follows:

Number Wearing Seat Belts	Probability
0	$C(4, 0)(0.70)^0(0.30)^4 = 0.0081$
1	$C(4, 1)(0.70)^1(0.30)^3 = 0.0756$
2	$C(4, 2)(0.70)^2(0.30)^2 = 0.2646$
3	$C(4, 3)(0.70)^3(0.30)^1 = 0.4116$
4	$C(4, 4)(0.70)^4(0.30)^0 = \underline{0.2401}$
	Sum: 1

Since the conditions for a binomial experiment are met, the probability distribution is called a **binomial probability distribution,** or a **binomial distribution.**

The expected value of this random variable may be found using the probability distribution table just constructed:

$$E(X) = 0(0.0081) + 1(0.0756) + 2(0.2646) + 3(0.4116) + 4(0.2401) = 2.8.$$

This means that if the experiment of recording the number of drivers wearing seat belts among 4 drivers is repeated over and over for a long period of time, the average number of those wearing seat belts would be 2.8. But for binomial probability distributions, the expected value can be found much easier than going through the preceding calculation. We reason this way: If the probability of wearing a seat belt is 0.70 and 4 drivers are checked, then we would *expect* that $4(0.70) = 2.8$ drivers would be wearing seat belts. If 800 drivers are checked, we would *expect* that $800(0.70) = 560$ drivers would be wearing seat belts, and so on. In the case of binomial distributions, to yield these expectations can be mathematically proven

$$E(X) = np,$$

where n is the number of trials in the experiment and p is the probability of success.

Guide to Probability

Near the end of the exercises for this section, you will find probability problems of all types that we have studied in Chapters 8 and 9. The following guide to the various types and how their solutions are obtained is presented as a helpful review:

Outcomes that are equally likely. We write out a sample space and use counting techniques to determine the ratio of the number of outcomes in the event to those in the entire sample space.

Outcomes with unequal probability. We write out a sample space and add the probabilities of the outcomes in an event to find the probability of that event.

Venn diagrams. A Venn diagram may be useful when we have several events and are given information regarding the probabilities of their intersections, unions, or complements.

The union of events. Union, or "or," means to add the probabilities of each event, but don't double-count the probability of any outcome. If the events are mutually exclusive, just add their probabilities.

Intersection of events. Intersection, or "and," means to multiply the probabilities of each event, but always take into account the events that already have occurred. If the events are independent, just multiply their probabilities.

Conditional probability. Reduce the sample space dictated by the "given" information, use the conditional probability formula.

Bayes' theorem. This is a special application of conditional probability. Use the conditional probability formula and a tree diagram to aid in finding the probabilities sought.

Binomial probability. This is an application of independent events. Determine the events "success" and "failure" and then multiply the number of sequences consisting of S's and the probability of each sequence.

Exercises 9.5

The first 55 exercises in this section involve binomial experiments. The remainder involve a random selection of all types of probability problems that we have studied.

Technology will be helpful in obtaining a final numerical answer in exercises involving binomial experiments. Setting them up for calculation will be a good learning experience, even if you do not calculate any particular number.

Tossing a Biased Coin *Exercises 1 through 3 refer to the following experiment: A coin found to be biased 55% for heads and 45% for tails is tossed six times. What is the probability of getting*

1. Exactly four heads?

2. Exactly two heads?

3. At least five heads?

Batting Averages *Exercises 4 through 6 refer to the following data: A baseball player is batting 0.335. During his next five times at bat, what is the probability that he will*

4. Get exactly three hits?

5. Get no hits?

6. Get at least one hit?

Product Reliability *Exercises 7 through 9 refer to the following data: A popular brand of lightbulbs has a 1% defect rate. If Dr. Harrell buys a six-pack of such bulbs, what is the probability that*

7. Exactly one will be defective?

8. Four of the bulbs will be good and two will be defective?

9. All six of the bulbs will be good?

Political Survey *Exercises 10 through 12 refer to the following data: In a particular city, 38% of the voters consider themselves Republican. If 10 voters are randomly called during an upcoming election campaign, what is the probability that*

10. Exactly 6 will be Republican?

11. Exactly 8 will be of other political persuasions?

12. At least 8 will be Republican?

Effectiveness of Surgery *Exercises 13 through 15 refer to the following data: In a particular medical facility, cataract surgery is considered to be 98% successful. Of the next eight patients having such surgery, what percentage of the time*

13. Will exactly six surgeries be successful?

14. Will at least six be successful?

15. Will all eight be successful?

Drug Effectiveness *Exercises 16 through 18 refer to the following data: The probability that women using a particular type of fertility drug will have multiple births is 0.20. Among 12 pregnant women who have taken the drug, what percentage of the time*

16. Will exactly 10 have multiple births?

17. Will none have multiple births?

18. Will at least 1 have a multiple birth?

Political Polls *Exercises 19 through 21 refer to the following data: A city bond issue is favored by 56% of the voters polled. Assuming a representative sample, what is the probability that, among five voters,*

19. Exactly four will favor the bond issue?

20. A majority of the five will favor the bond issue?

21. All five will favor the bond issue?

Family Planning *Exercises 22 through 24 refer to the following experiment: What is the probability that, in a family that wishes to have eight children,*

22. All will be girls?

23. Exactly four will be boys?

24. At least one will be a boy?

Rolling Dice *Exercises 25 through 28 refer to the following experiment: A pair of dice is rolled four times. What is the probability that*

25. A sum of 7 is rolled every time?

26. At least one sum of 7 is rolled?

27. No sums of 7 are rolled?

28. No sum of 3 or 4 is rolled?

Market Survey *Exercises 29 through 31 refer to the following data: In taste tests, 62% of the participants said they preferred Diet Raspberry drink over Diet Banana. If this taste test is given to 10 people, what is the probability*

29. All 10 will prefer Diet Raspberry?

30. None will prefer Diet Raspberry?

31. At least 1 will prefer Diet Banana?

Testing *Exercises 32 through 34 refer to the following experiment: A student guesses on every question of a multiple-choice test that has 10 questions, each with four possible answers. What is the probability that the student will*

32. Get exactly 8 of the questions right?

33. Get at least 8 of the questions right?

34. Get none of the questions right?

Quiz Taking *Exercises 35 through 37 refer to the following experiment: A true–false quiz in English has 10 questions. If a student in the class guesses at the answer to every question, what is the probability that*

35. Exactly 5 are answered correctly?

36. Exactly 2 are answered correctly?

37. At least 2 are answered correctly?

Sampling *Exercises 38 through 40 refer to the following experiment: Suppose that 5 cards are drawn in succession, with replacement, from a deck of 52 cards. What is the probability that*

38. Exactly 4 of the cards drawn are clubs?

39. All 5 are clubs?

40. At least 1 is a club?

Product Reliability *Exercises 41 and 42 refer to the following data: Cost-Mart buys handheld mirrors in packages of 100 from a supplier. On the average, one broken mirror is found in each package. If an employee randomly selects 12 mirrors from one of these packages, what is the probability that*

41. None of the mirrors will be broken?

42. At least one mirror will be broken?

Product Reliability *Exercises 43 through 45 refer to the following data: A seed company claims that 90% of its petunia seeds will germinate. If 10 seeds are planted in warm, moist soil, what is the probability that*

43. Exactly 9 of them will germinate?

44. At least 9 of them will germinate?

45. At least 7 of them will germinate?

Safety Survey *Exercises 46 through 48 refer to the following data: A newspaper reporter surveyed drivers in Burlington and found that 48% of the drivers actually wore seat belts. Based on this information, if 10 motorists in the city were stopped by the police and checked for seat-belt violations, what is the probability that*

46. All 10 will be wearing seat belts?

47. Exactly 8 of the 10 will be wearing seat belts?

48. None of the 10 will be wearing seat belts?

Service Claims *Exercises 49 and 50 refer to the following data: An overnight delivery service claims that it delivers messages on time 92% of the time. A firm has a message that absolutely, positively, must be delivered the next morning. To make sure that the message will be delivered on time, the firm sends the same message in four separate envelopes through the delivery service. Assuming that the messages travel independently, what is the probability that*

49. None of the four envelopes will be delivered on time the next morning?

50. At least one of the envelopes will be delivered on time the next morning?

51. Successful Sales Calls A salesperson has a success rate of 60% with the customers he calls on.

   (a) Make a binomial distribution table for the possible number of successful calls on 5 customers.

   (b) Calculate, in two ways, the expected number of successes for the 5 calls made.

   (c) What is the expected number of successes if the salesperson calls on 200 customers? On 800 customers?

52. Product Reliability A seed company claims that 90% of its marigold seeds will germinate.

   (a) Make a binomial distribution table for the possible number of germinations when 4 seeds are planted.

   (b) Calculate, in two ways, the expected number of germinations for the 4 seeds.

   (c) What is the expected number of germinations if 300 seeds are planted? If 1500 seeds are planted?

53. Field-Goal Percentage A field-goal kicker is 95% accurate within the 30-yard range.

   (a) Make a binomial distribution table for the possible number of field goals made in 3 attempts within the 30-yard range.

   (b) Calculate, in two ways, the expected number of field goals made in 3 attempts within the 30-yard range.

   (c) What is the expected number of field goals made if 50 are attempted? If 200 are attempted?

54. Kidney Transplants Medical record indicate that kidney transplants are 92% successful.

   (a) Make a binomial distribution table for the number of possible successful transplants among the next 4 patients.

   (b) Calculate, in two ways, the expected number of successful transplants in the next 4 patients.

   (c) Find the expected number of successful transplants in the next 300 and 600 patients.

55. Caller ID Kwasi checks the caller ID box on his telephone and notes that 65% of the calls coming into his home are for his 14-year-old daughter, Amma. Assuming that this trend continues,

   (a) Calculate the expected number of phone calls out of the next 10 that will be for Amma.

   (b) Find the probability that at least 8 of the next 10 phone calls will be for Amma.

(c) How many of the next 100 phone calls should Kwasi expect to be for someone in the home other than Amma?

56. Find the probability that a 5-card hand dealt without replacement will contain all spades. The deck contains 52 cards.

57. Quality Control A sample of 3 computers is selected, without replacement and without regard to order, from among 50, 2 of which are known to be defective. What is the probability that 2 of the computers are defective and 1 is not?

58. Two marbles are selected in succession and without replacement from a box containing five red and three green marbles. What is the probability that the second marble is red, given that the first marble was green?

59. Quiz Taking A student takes a German quiz and guesses on every question. The quiz has five questions, and each question has three possible answers. What is the probability that he will get at least three of the five answers correct?

60. One card is drawn for a deck of 52 cards. What is the probability that the card will be a queen or a club?

61. Diabetes Screening About 6% of the general public has diabetes. A particular test for the condition shows positive results 97% of the time when a person has diabetes and also indicates a positive reading 2% of the time when a person does not have the condition. What is the probability that a person would actually have diabetes if she tested positive?

62. If 5 cards are drawn in succession and without replacement from a deck of 52 cards, what is the probability that all will be hearts or all will be spades?

63. Mortality Tables Mortality tables show a probability of $\frac{1}{5}$ that Juan will live at least 20 more years and a probability of $\frac{4}{9}$ that Graciela will live at least 20 more years. What is the probability that both will live at least 20 more years? That at least one of the two will be alive 20 years from now?

64. Two dice are rolled. What is the probability that one shows an even number and one an odd number?

65. Two cards are drawn in succession and without replacement from a deck of 52 cards. What is the probability that one card is an ace and one is a queen?

66. Two marbles are selected, without replacement, from a box containing three red and two green marbles. What is the probability that one marble is red and one is green?

67. Three boys and three girls are to be seated in a row. What is the probability that boys and girls will be in alternate seats?

68. Product Reliability Sixty percent of the toasters in a warehouse come from the Hot-Slice Co., and of those toasters, 3% are defective. Forty percent come from the Warm Morning Co., and of those, 5% are defective. What percentage of the toasters are defective?

69. If 3 cards are selected in succession, with replacement, from a deck of 52 cards, what is the probability that all of the cards selected are clubs?

70. Product Safety A safety engineer claims that 15% of all automobile accidents are due to mechanical failure. Assuming this to be correct, what is the probability that, of the next eight automobile accidents, at least one will be due to mechanical failure?

71. If $P(A) = 0.6$, $P(B) = 0.5$, and $P(A \cup B) = 0.8$, what is $P(A \cap B)$?

72. A hand selected from a deck of 52 cards contains the king of clubs, the queen of hearts, the eight of hearts, and the three of spades. Two more cards are drawn in succession, without replacement. What is the probability that at least one of these cards will be a diamond?

73. Course Failure Rates At State University, 35% of the freshmen failed mathematics, 20% failed English, and 10% failed both mathematics and English. What is the probability that a freshman failed exactly one of the two courses?

74. If 5 cards are dealt, without replacement, from a well-shuffled deck of 52 cards, what is the probability that exactly 2 of the cards dealt will be hearts?

75. What is the probability that the last two digits of a four-digit telephone number are both even?

Chapter 9 Summary

CHAPTER HIGHLIGHTS

Probabilities involving combinations of events can be determined by using the set operations of *union, intersection,* and *complement.* Pairs of events that *cannot occur simultaneously* are called *mutually exclusive. Conditional probability* refers to the likelihood of an event happening, *given* that *another event has occurred;* if the occurrence of this event has no effect on the probability of the other, the events are said to be *independent. Tree diagrams* and *Venn diagrams* can be helpful visual tools for *comprehensively* and *logically* displaying pertinent probabilities in an experiment. The *addition theorem*, the *complement theorem*, the *conditional probability formula*, *Bayes's theorem*, and the *general formula for binomial experiments* are all important in calculating particular types of probabilistic outcomes.

EXERCISES

For Exercises 1 through 6, use the accompanying tree diagram to determine the requested probabilities.

1. $P(W|B)$

2. $P(C \text{ and } Y)$

3. $P(W)$

4. $P(A|W)$

5. $P(B \text{ or } Y)$

6. $P(C|Y)$

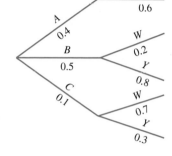

For events A and B, $P(A) = 0.5$ and $P(A \cup B) = 0.8$.

7. Find (B') if A and B are mutually exclusive.

8. Find $P(B)$ if $P(A \cap B) = 0.3$.

9. Find $P(B)$ if A and B are independent.

10. Find $P(A \cap B)$ if $P(B) = 0.7$.

11. Find $P(A|B)$ if $P(B) = 0.7$.

12. Find $P(B|A)$ if $P(A' \cup B') = 0.9$.

For mutually exclusive events C and D, $P(C) = \frac{3}{7}$ and $P(D) = \frac{4}{7}$. For event E, $P(E|C) = \frac{2}{3}$ and $P(E|D) = \frac{3}{4}$.

13. Find $P(E)$.

14. Find $P(C|E)$.

15. Find $P(D|E.)$

A tennis player has a 60% success rate on her first serves. When she does miss her first serve, 90% of her second serves succeed.

16. Find the probability that she double-faults (misses both serves).

17. If a serve is good, find the probability that it was a first serve.

A hazard forecaster predicts that, for a particular ski slope in Utah, the probability of an avalanche triggering naturally during the next 10 hours is $\frac{1}{100}$, but if a skier is on it, the probability is $\frac{1}{40}$. If the slope releases an avalanche, the probability that the road below it has traffic and an accident occurs on it is $\frac{1}{25}$.

18. What is the probability of a traffic accident caused by a naturally triggered avalanche?

19. What is the probability of traffic being on the road below the slope, given that an avalanche was triggered naturally?

20. If a skier is on the slope, what is the probability that an avalanche occurs and that there is traffic on the road below the slope?

21. If a skier is on the slope, what is the probability of an avalanche occurring?

22. What is the probability of a traffic accident being caused by an avalanche if the odds that a skier is on the slope are 3 to 2?

A professional football player has a 58% pass-completion rate. For his next 10 passes, what is the probability that he

23. Completes exactly 5 passes?

24. Completes the first 5 passes and misses the next 5 passes?

25. Has more completions than incompletions or interceptions?

26. Analyze the situation described next, and decide who is correct. Then write a paragraph, perhaps using empirical or theoretical data, that supports your solution.

In the September 9, 1990, "Ask Marilyn" column of *Parade Magazine,* Marilyn vos Savant was asked this question: "Suppose you're on a game show, and you're given a choice of three doors. Behind one door is a car; behind the others, goats. You pick a door—say, No. 1—and the host, who knows what's behind the doors, opens another door—say, No. 3—which has a goat. He then says to you, 'Do you want to pick door No. 2?' Is it to your advantage to switch your choice?"

Marilyn said that you should switch, because the first door has a $\frac{1}{3}$ chance of winning, but the second door has a $\frac{2}{3}$ chance. Numerous people wrote her, including many Ph.D. mathematicians, saying she was wrong. One wrote,

"You blew it, and you blew it big! I'll explain: After the host reveals a goat, you now have a one-in-two chance of being correct. Whether you change your answer or not, the odds are the same. There is enough mathematical illiteracy in this country, and we don't need the world's highest IQ propagating more. Shame!"

Who is right? Why? (*Hint*: Remember that the host knows what is behind each door.)

Exercise 27 appeared on an actuarial exam.

27. Let S and T be independent events. Choose (a), (b), (c), (d), or (e) to complete this equation: $P(S \cap T) = \frac{1}{10}$ and $P(S \cap T') = \frac{1}{5}$. $P[(S \cup T)'] =$

(a) $\dfrac{3}{10}$ (b) $\dfrac{11}{30}$ (c) $\dfrac{7}{15}$

(d) $\dfrac{8}{15}$ (e) $\dfrac{9}{10}$

Exercise 28 appeared on an actuarial exam.

28. A card hand selected from a standard card deck consists of two kings, a queen, and jack, and a 10. Three additional cards are selected at random and without replacement from the remaining cards in the deck. What is the probability that the enlarged hand contains at least three kings?

(a) $\dfrac{3}{1081}$ (b) $\dfrac{132}{1081}$ (c) $\dfrac{135}{1081}$

(d) $\dfrac{264}{1081}$ (e) $\dfrac{267}{1081}$

29. Use a computer or a calculator to simulate the roll of a pair of dice 360 times. Read the number of sums of 2, 3, 4, . . . , 12 obtained, and compute the relative frequency of each of these outcomes. Compare these numbers with the theoretical probabilities of each sum.

CUMULATIVE REVIEW

30. Write an equation for the line that contains the points $(5, -2)$ and $(-1, -10)$.

31. Solve the given system of equations, using any appropriate method:

$$3x - y + 2z = 19$$
$$-x + 3y + 10z = -1$$
$$2x + 4y + 5z = 7.$$

32. Sketch and shade a Venn diagram, using three overlapping sets, A, B, and C, that represents the region given by

$$(B' \cap C' \cap A) \cup (B \cap C \cap A').$$

33. Determine the matrix product AB, given that

$$A = \begin{bmatrix} 1 & 0 & 3 & 2 \\ 2 & -5 & 1 & 0 \end{bmatrix} \text{ and } B = \begin{bmatrix} 3 & -1 & 0 \\ -2 & 1 & 2 \\ 1 & 0 & -1 \\ 2 & -1 & 0 \end{bmatrix}.$$

34. Write the negation of the sentence "Some cats are not quick."

Chapter 9 Sample Test Items

In Exercise 1 through 4, three boys and four girls are randomly seated in a row. What is the probability that

1. A girl will be seated in the middle seat and a boy will be seated in the right end seat?

2. A girl will be seated in the middle seat or a boy will be seated in the right end seat?

3. Girls and boys are seated alternately or a girl is seated in the right end seat?

4. A girl will not be seated in the middle seat or the girls will not be seated all together?

In Exercises 5 through 11, the finite mathematics students at State University are categorized as follows:

	Freshman	Sophomore	Junior	Senior
Male	64	37	11	8
Female	86	43	9	2

Find the probability that a finite mathematics student at State University is

5. Female.

6. A freshman.

7. Female and a freshman.

8. Female or a freshman.

9. Female, given that the student is a freshman.

10. A senior, given that the student is male.

11. Not a male or not a freshman.

12. Find the probability that a 3-card hand dealt from a standard deck of 52 cards has exactly two queens.

In Exercises 13 and 14, 1 card is drawn from a standard deck of 52 cards:

13. Find the probability that the card is a 9, given that it is a 10.

14. Find the probability that the card is a 4 or a club.

In Exercises 15 and 16, the probability that an earthquake occurs is 0.002. The probability that you get an A in finite math is 0.2.

15. What is the probability that an earthquake occurs and you get an *A* in finite math?

16. What is the probability that an earthquake occurs or you get an *A* in finite math?

17. Three six-sided dice are rolled. Find the probability that at least one die shows six pips.

A survey of supermarket shoppers revealed that 47% bought bread, 22% bought crackers, and 61% bought bread or crackers. Find the probability that a person from the survey bought

18. Bread only.

19. Bread, given that the person did not buy crackers.

Consider families of three children. What is the probability that

20. Exactly two are girls?

21. Exactly two are girls, given that the third child is a girl?

A chef's school is 60% male and 40% female. Seventy percent of the males and 90% of the females like eating crab's legs for dinner. What is the probability that a member of this chef's school

22. Is male or likes eating crab's legs for dinner?

23. If female, given that the member likes eating crab's legs for dinner?

A special variety of beans sprout 90% of the time. If 20 beans are planted, find the probability that

24. Exactly 18 sprout.

25. At least 18 sprout.

In-Depth Application

Applications of Probability: Age, Education, and Smoking

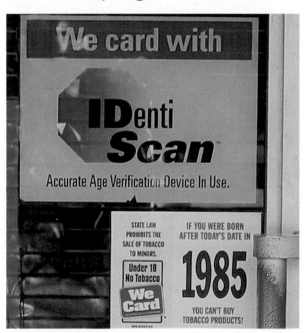

Objectives

In this application, you will apply your knowledge of conditional probability to analyze data regarding the age distribution of people in the United States, their attained level of education, and smoking.

(continued)

Introduction

Health-care professionals involved in community education often target specific demographic groups. For instance, in a community education campaign designed to help people quit smoking, officials will often choose a target group for whom smoking is particularly prevalent. This target group may be determined from demographic data, such as that gathered by the U.S. Bureau of the Census, regarding age, race, gender, income, and education level, as well as from information gathered by the Centers for Disease Control regarding the group's smoking habits. A particular campaign might be more effective with smokers over the age of 35 who have no college education, while another may be more effective with smokers who have attained a college degree. Thus, knowing the target population will help professionals design and implement programs with a higher success rate.

The Data

The following table contains data extrapolated from the 2000 *Statistical Abstract of the United States* and from the Centers for Disease Control:

Characteristic	Not a High School Graduate	High School Graduate	Some College	College Degree	Smokes	Population (1000s)
Age:						
25 to 34 years old	4732	11,812	10,888	11,042	10,200	38,474
35 to 44 years old	5369	15,168	12,126	12,081	12,300	44,744
45 to 54 years old	4122	11,169	9442	10,464	8500	35,197
55 to 64 years old	4307	8453	4948	5200	4600	22,909
65 years old or over	10,352	11,292	5778	4972	3100	32,395
Smokes	9200	15,000	10,100	4400		
Population (1000s)	28,883	57,894	43,182	43,759		

Note that the education attainment levels indicated refer to the highest level attained. The category "high school graduate" also includes those who have received the equivalent of a high school diploma, such as a GED. The category "some college" also includes those persons with two-year degrees.

Using the information from the table, and assuming that all persons mentioned live in the United States, answer the following questions:

1. For each of the five age groups presented in the table, what is the probability of being a smoker in that age group?

2. For each of the four education-level groups presented in the table, what is the probability of being a smoker in that group?

3. What is the probability that a person is 45 years of age or older and smokes?

4. What is the probability that a person has at least graduated from high school and smokes?

5. Given that a person is at least 35 years of age and has a college degree, what is the probability that the person also smokes?

6. Given that a person is at least 35 years of age and did not graduate from high school, what is the probability that the person also smokes?

7. Given that a person smokes, what is the probability that the person has a college degree?

8. Given that a person smokes, what is the probability that the person is under 45 years of age?

CHAPTER 10

Statistics

PROBLEM

An automaker guarantees its particular type of automatic transmission for 60,000 miles. Tests have shown that such transmissions have an average life of 90,000 miles, with a standard deviation of 15,000 miles. If the lives of these transmissions are normally distributed, what percentage of cars will be returned to the company for transmission work while the cars are still under warranty? (See Example 6, Section 10.4.)

CHAPTER PREVIEW

Statistics is the science of collecting, organizing, and analyzing data. Many of the basic techniques used in such analyses, including grouping and displaying data by histograms, finding various kinds of averages, calculating the standard deviation, and working with the "bell-shaped" normal curve, are discussed in this chapter. Applications range from the analysis of surveys taken to determine how many respondents own a certain product to the analysis of failure to satisfy a guarantee.

Technology Opportunities

Calculators with statistical capabilities will be invaluable throughout this chapter. Many modern calculators, such as the TI-83 Plus, have the ability to create histograms, as well as other statistical graphs. Many calculators also have special distribution functions, such as the normal distribution, built into them. Spreadsheet packages have built-in functions for calculating statistical values, utilizing statistical distributions, and organizing data into charts and graphs.

Section   10.1   Organizing Data: Frequency Distributions

Statistics is concerned with the collection and analysis of data and with making predictions from the data. Data are collected in many ways, often in the form of numerical quantities that measure specific characteristics. Each of these numerical quantities is called an **observation.** For example, data collected about a retail business might include observations of gross sales, net profit, inventory, earnings per share, and so forth. Data collected for medical studies might include observations of blood pressure, height, weight, and so on.

HISTORICAL NOTE

FLORENCE NIGHTINGALE (1820–1910)

During the Crimean War, Florence Nightingale served as a nursing administrator at military hospitals. After noting the harsh and unsanitary conditions in which injured soldiers were being treated, she began a serious undertaking to reform the military hospital system. By collecting data on mortality rates, Nightingale was able to use statistical analysis to verify that improving the sanitary conditions of the military hospitals would result in a lower mortality rate for injured soldiers. Her work eventually earned her the distinction of being the first woman elected to the Royal Statistical Society in 1858.

In statistical terminology, the set of things from which data are collected is called the **population.** Sometimes, we collect data from an entire population, in which case we simply summarize the data as factual matter. You might think of the grade distribution on a recent test or the "stats" on a recent basketball game as examples of this type of statistics, known as **descriptive statistics.**

Another major activity of statistics is known as **inferential statistics,** a subject that deals with the methods of collecting data from a subset of the population, called a **sample.** Based on the information from the sample, we draw conclusions about the entire population. Sampling is used when it is impossible, impractical, or too expensive to collect data on the entire population. Inferential statistics then determines how to select the samples and, with the use of probability, how accurately the sample portrays the entire population. A prime example of how inferential statistics can be used is the prediction of winners during an election. This is also a prime example of how inferences made from a sample of data may *not* be correct. In the 2000 presidential election in the United States, many networks and newspapers declared that candidate Al Gore "won" the state of Florida, based on an exit poll of a sample of voters in that state. Not all voters in Florida were polled, of course, but the sample was assumed to be large enough and the statistics calculated strong enough to infer that Gore would receive more votes than his opponent, George W. Bush. However, later in the evening, the statistical evidence was refuted, and Bush was declared the winner of Florida's electoral votes. It took weeks to learn the actual outcome of that election. Finally, it was determined that Bush would be the 43rd president of the United States. During the 2002 election, because the flaws in the gathering and analysis of voting data from 2000 were being reviewed, exit polls were not relied upon quite as heavily to predict the outcome of the 2002 senatorial elections.

When collecting data for the purpose of making inferences, it is important that the collection process adhere to certain rules, to prevent biased samples and to increase the chances of making a correct inference. For instance, in the case of exit polls during an election, you would not want to poll only the voters in a district that contains mostly registered Democrats, since you would not be obtaining a sample that reflected the entire population of voters. Designing a good collection process and the probabilities associated with making inferences from the collected data are topics covered in more in-depth courses in statistics.

For whatever purpose data are collected, their usefulness may be enhanced by some specialized arrangements and organization. One way this may be done is by sorting the data into **classes** and then displaying the classes in a **frequency distribution table.** Sometimes, the data will fall naturally into classifications with little or no effort, such as the data in Figure 1 and the data in Example 1.

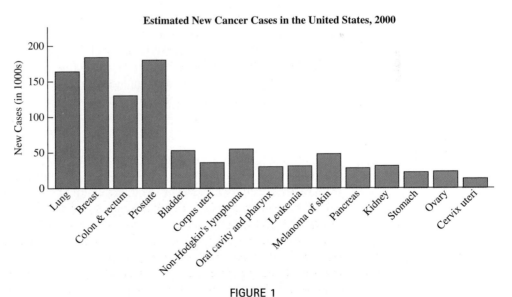

FIGURE 1

Estimated New Cancer Cases in the United States (2000)
SOURCE: *Statistical Abstract of the United States 2000.*

<u>**EXAMPLE 1**</u>

A Typical Survey

Thirty people in a shopping mall were asked how many telephones were in their home. The responses were as follows:

$$
\begin{array}{ccccc}
2 & 1 & 2 & 2 & 4 \\
3 & 2 & 1 & 2 & 1 \\
3 & 3 & 2 & 3 & 3 \\
1 & 1 & 0 & 2 & 2 \\
2 & 1 & 2 & 2 & 4 \\
4 & 3 & 4 & 2 & 2 \\
\end{array}
$$

Because the subject of this survey is the number of telephones in each household, those numbers—ranging from 0 to 4—form the classes of interest. The **frequency distribution table** (see Table 1) for this data set is completed by tallying the number of 0's, 1's, 2's, 3's, and 4's and then counting the total. Now we understand the data on phone ownership more clearly.

Class: Number of Telephones	Tally	Frequency
0	I	1
1	⦀⦀ I	6
2	⦀⦀ ⦀⦀ III	13
3	⦀⦀ I	6
4	IIII	4
		Total: 30

TABLE 1. FREQUENCY DISTRIBUTION TABLE   ■

We can get other statistical information from Example 1 by computing the **relative frequency** for each class. Recall that in Chapter 8 we calculated **empirical probabilities** from frequency distributions related to probability experiments. The empirical probabilities were calculated by dividing each frequency by the total of all frequencies. In general, we were making a statistical inference in those problems when we used experimental data to estimate the probabilities of individual outcomes or events. The idea of a **relative frequency** is the same and is calculated in a similar manner.

Relative Frequency

The **relative frequency** of a class is found by dividing the frequency of that class by the total of all frequencies. The relative frequency is the same as the empirical probability that an individual data value will fall within that class.

Two other types of frequency can be used to describe data sets. The **cumulative frequency** is the sum of the frequencies for a specific class and all preceding classes. The **cumulative relative frequency** is the sum of the relative frequencies for a specific class and all preceding classes. These cumulative values have importance when we are dealing with data sets that seem to "fit" a certain pattern. In a later section, we will study the **normal probability distribution,** which is used often in statistics and probability and for which many tables exist containing cumulative probabilities, much like the cumulative relative frequencies discussed here.

Cumulative Frequency and Cumulative Relative Frequency

The **cumulative frequency** of a class is the sum of the frequency of that class and the frequencies of all the preceding classes when classes are listed in some sensible order (numerical order or, in the case of nonquantitative data, alphabetical order or some other reasonable order).

The **cumulative relative frequency** of a class is the sum of the frequency of that class and the frequencies of all the preceding classes when classes are listed in some sensible order.

Table 2 shows the relative, cumulative, and cumulative relative frequencies for the telephone ownership data from Example 1. Note that sums are not given at the bottom of the table for the cumulative or cumulative relative frequencies, since the totals for each of those columns are contained in the last entry in the column.

Number of Telephones	Frequency	Relative Frequency	Cumulative Frequency	Cumulative Relative Frequency
0	1	$\dfrac{1}{30}$	1	$\dfrac{1}{30}$
1	6	$\dfrac{6}{30}$	$1 + 6 = 7$	$\dfrac{7}{30}$
2	13	$\dfrac{13}{30}$	$1 + 6 + 13 = 20$	$\dfrac{20}{30}$
3	6	$\dfrac{6}{30}$	$1 + 6 + 13 + 6 = 26$	$\dfrac{26}{30}$
4	4	$\dfrac{4}{30}$	$1 + 6 + 13 + 6 + 4 = 30$	$\dfrac{30}{30} = 1$
	Total = 30	Sum = 1		

TABLE 2. FREQUENCY, RELATIVE FREQUENCY, CUMULATIVE FREQUENCY, AND CUMULATIVE RELATIVE FREQUENCY TABLES

In Chapter 8, we used **probability histograms** to visually represent both theoretical and empirical probability assignments. Any column of information in Table 2 can also be visually displayed in a **histogram,** using the actual classes (in this case, the number of telephones) as labels for the horizontal axis.

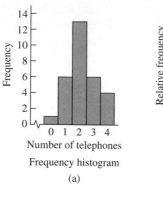

Frequency histogram

(a)

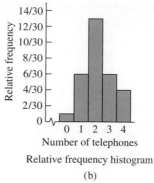

Relative frequency histogram

(b)

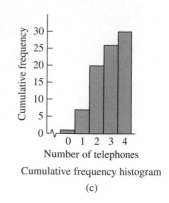

Cumulative frequency histogram

(c)

FIGURE 2
Histograms

One obvious use of histograms is to show the nature of particular data in an immediately visual way. (See Figure 2.) Note that in constructing each of the histograms, a break was indicated in the horizontal axis so that the first bar does not overlap the vertical axis. Similar breaks can be used on either axis to show expansion or contraction of that axis for a better visual picture. Note also that the frequency and the relative frequency histograms have the same shape. A moment's reflection will convince us that this will always be the case: The *frequency* histogram shows how frequently the collected data fall within a given class, while the *relative frequency* tells us the percentage of our sample that falls into that class. One item of interest later in the chapter is that if each bar is constructed to have a *width of one,* then the *area* of each bar is precisely the same as the estimated probability associated with that bar in a relative frequency histogram. In addition, the cumulative and cumulative relative frequency histograms share a common shape.

Sometimes, collected data may not fall so quickly into nice, neat classes, as did the data from the telephone ownership survey in Example 1. The amount of time you might wait at a checkout counter for service in a department store, the life span of a flashlight battery, and the price of a given stock over a one-year period are examples of data that might contain decimals or fractions—and many of them! In fact, no two data values may be equal, so a frequency distribution would not be a good way to describe the data if we simply listed each different data value as we did in Example 1. When confronted with such data, we find it useful to group the data into **class intervals.** While no standard method exists for deciding how many class intervals to use, or even what size of class intervals to use, we will adopt the following rule of thumb: Group the data into 5 to 15 well-chosen class intervals, each containing at least one data value. We demonstrate how to do this in Example 2.

EXAMPLE 2

The Use of Class
Intervals

The person in charge of a college computer room observed a random sample of 25 students who came to the room and used the computers. She recorded the time (in hours) that each student spent on a computer at one sitting during a particular day. The results were as follows:

2.0	3.25	1.53	1.56	0.3
1.0	0.2	1.3	2.58	1.0

0.6	1.1	0.8	1.2	1.5
1.5	1.58	3.52	0.75	0.75
2.56	2.52	1.25	2.2	1.4

Group the data into class intervals of equal width, compute the frequency for each class interval, and construct a frequency histogram for the grouped data.

SOLUTION How do we decide how many class intervals to use? We will begin by choosing a width for each interval. We want to choose class intervals of a reasonable width and to utilize units that will be easily understood by anyone who might look at our grouped data in the future. To choose the width of our class intervals, we first identify the smallest and largest values in the data set, which are 0.2 hour and 3.52 hours, respectively. Since $3.52 - 0.2 = 3.32$ is not a particularly large number, intervals of width 0.5 hour (or 30 minutes) seem appropriate.

Next, we choose a starting point for our intervals. We do not have to begin at 0.2, the smallest number in our data set, but we do want to make sure that the first class interval includes that value. We also should make sure that the starting point is clearly understood by subsequent viewers of our table or histograms. A reasonable choice here would be to start our intervals at 0. Thus, our class intervals would be 0–0.5 hour, 0.5–1.0 hour, 1.0–1.5 hours, 1.5–2.0 hours, and so on.

Since we have chosen to use the number 1.5 as a "dividing point" between two intervals, and since 1.5 is a value in our data set, we have one last decision to make: In which class interval do we want to place the value 1.5? We certainly do not want to list this observation twice, in two different intervals, because we would be misrepresenting the data we actually collected. One option would be to adjust the class intervals to be 0–0.49 hour, 0.5–0.99 hour, 1.0–1.49 hours, and so on. Another, which would allow us to keep our nice dividing points between intervals, would be to choose to include the dividing point only when it is the smallest value in that interval; that is, we include the observation 1.5 in the interval 1.5–2.0, but we do *not* include it in the interval 1.0–1.5. If you choose the latter option, which is that chosen in the construction of Table 3, you should make sure that your readers are aware that you have made this choice. The frequency and relative frequency histograms for these grouped data are shown in Figure 3, with the class interval endpoints interpreted as in Table 3.

Class (time in hours)	Tally	Frequency	Relative Frequency	Cumulative Frequency
0 up to 0.5	‖	2	$\frac{2}{25}$	2
0.5 up to 1.0	‖‖	4	$\frac{4}{25}$	6
1.0 up to 1.5	‖‖‖ ‖	7	$\frac{7}{25}$	13
1.5 up to 2.0	‖‖‖	5	$\frac{5}{25}$	18
2.0 up to 2.5	‖	2	$\frac{2}{25}$	20
2.5 up to 3.0	‖‖	3	$\frac{3}{25}$	23
3.0 up to 3.5	‖	1	$\frac{1}{25}$	24
3.5 up to 4.0	‖	1	$\frac{1}{25}$	25
		Total: 25	Sum: 1	

TABLE 3.

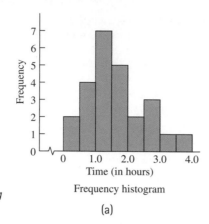

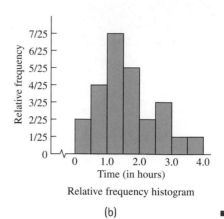

FIGURE 3
Histograms Showing
Computer Usage

Frequency histogram

(a)

Relative frequency histogram

(b)

Sometimes, class intervals are used to get a clearer picture of integer data values that are widely distributed. Example 3 is such a case.

EXAMPLE 3

Using Class Intervals
with Integer Data

Thirty contributions to a charitable cause were recorded as follows:

$50	$65	$50	$35	$75	$40
$25	$75	$50	$60	$80	$60
$60	$90	$100	$65	$100	$50
$40	$65	$25	$50	$30	$75
$50	$100	$50	$125	$100	$25

Organize the data into class intervals, construct the frequency, relative frequency, and cumulative frequency tables, and construct the frequency and cumulative frequency histograms for the data.

SOLUTION Since the smallest value in the data is $25 and the largest is $125, the range of the data is $125 − $25 = $100. A reasonable class interval length for this data would be $25. Thus, adopting the same convention as in Example 2, where each interval contains its left endpoint, but not its right endpoint, the required tables can be constructed as in Table 4. The requested histograms are shown in Figure 4.

Class (dollars)	Frequency	Relative Frequency	Cumulative Frequency
$25 up to $50	7	$\frac{7}{30}$	7
$50 up to $75	13	$\frac{13}{30}$	20
$75 up to $100	5	$\frac{5}{30}$	25
$100 up to $125	4	$\frac{4}{30}$	29
$125 up to $150	1	$\frac{1}{30}$	30
	Total: 30	Sum: 1	

TABLE 4.

EXAMPLE 4

Creating Histograms
with Excel

Use the information in Table 4 to create a frequency histogram with Excel.

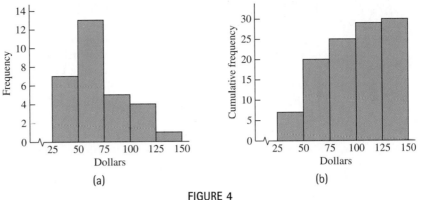

FIGURE 4

Histograms Showing Charitable Giving

	A	B
1	Charitable Contribution	Frequency
2	$25 up to $50	7
3	$50 up to $75	13
4	$75 up to $100	5
5	$100 up to $125	4
6	$125 up to $150	1

FIGURE 5

Excel Worksheet for Table 4

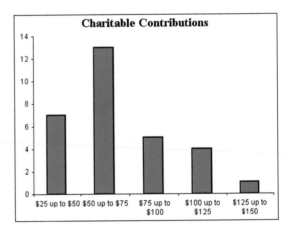

FIGURE 6

Frequency Histogram Created with Excel

SOLUTION We begin by putting the information from the table into a worksheet, as shown in Figure 5.

To create a frequency histogram from these data, highlight the entire table and choose Insert and then Chart. The current Excel chart settings are appropriate for what we want, so click Next to see what the chart will look like. Click Finish to draw the chart. The result should appear similar to Figure 6. ▄

NOTE Many instructors will prefer that successive rectangles in a histogram actually touch, rather than having a large gap between rectangles as in Figure 6. To make this adjustment on Excel, right-click on a rectangle in the chart and select Format data series, then click on the Options tab. Reduce the Gap width to 0 to force successive rectangles to touch.

Graphing calculators, such as the TI-83 Plus, have the ability to create histograms from stored data. (See the Graphing Calculator Manual that accompanies this text for instructions.)

Let's look back at Example 3. Can you look at Figure 4(a) and tell the size of most gifts? How are the sizes of the gifts distributed? Do you see how Figure 4(a) relates to

Figure 4(b), as far as the growth of larger gifts is concerned? These questions again point out one reason for using histograms: to give an instant visual analysis of statistical data. Another very important reason for using histograms, in the statistical sense, is to learn the shape of the probability distribution of the data collected. As noted after Figure 2, if the bars have width one, then area can be equated to probability. (Otherwise, area is proportional to probability.) This suggests that we might be able to replace the histogram with a curve fitted by a mathematical formula and then interpret probability as an area between the curve and the *x*-axis. That is indeed the case for data whose histograms have some specific shapes and characteristics. The *bell-shaped curve,* or **normal probability distribution,** is undoubtedly the most famous of these special probability distributions, and it is examined in detail in Section 10.4. Figure 7 shows collected data that are approximately normally distributed. Some statistical terminology for various distributions is shown in Figure 8.

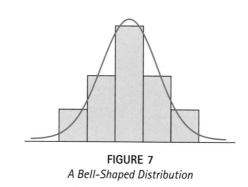

FIGURE 7
A Bell-Shaped Distribution

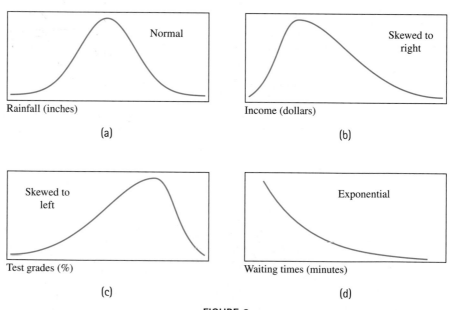

FIGURE 8
Some Types of Distributions

The use of frequency tables and frequency histograms gives us a better picture of the data we are working with, but we lose some of the identity of the original data. It is possible, however, to determine the shape of the probability distribution and still retain all of the data collected. Figure 9 shows one way of doing this, using the data and class intervals of Example 3. (This technique applies to noninteger data also.) The **histogram table** presented in the figure shows the data skewed to the right, as did the original frequency and relative frequency histograms.

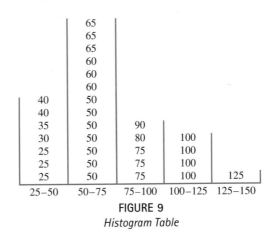

FIGURE 9

Histogram Table

Abuses of Statistics

Telling the Story the Way You *Want* it Told Two very different conclusions may often be obtained from the same data. Figure 10 shows two different frequency histograms, which at first appear to represent different data. However, both histograms represent the data set {6, 7, 9, 9, 9, 10, 11, 13, 15, 16}. Figure 10(a) displays the data in groups in which the data at the right end of each class interval are included; Figure 10(b) displays the data in groups in which the data at the *left* end of each class interval are included. Each of the histograms is technically correct. The first shows a steady decline and might be used by someone who wishes to make the point that higher values in the data set are less likely than lower values. The second shows a normally distributed situation. The moral of the story is that one must be very careful not to draw conclusions from graphically displayed data without more information about the original set of data.

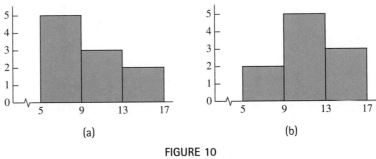

FIGURE 10

Different Ways of Displaying the Same Data

Figure 11 represents the same set of data as does Figure 10, with data at the left end of each class interval included. But, again, we see that very different stories are being told! The difference this time is in the scaling on the vertical axis. In Figure 11(a), the vertical axis is scaled evenly with each tick mark representing one unit. In Figure 11(b), uneven scaling is used, with each tick mark representing different units. Notice that Figure 11(a) would lead one to believe that there is a large jump in the frequencies from the first interval to the second. Figure 11(b), by contrast, would lead one to believe that the differences between the frequencies are not as dramatic. Again, you should be very careful not to jump to any quick conclusions on the basis of a graphical representation of data, without more information; here, the fact that no values are given for the locations of the tick marks on the vertical axis should leave you suspicious of the story being told by the frequency histogram.

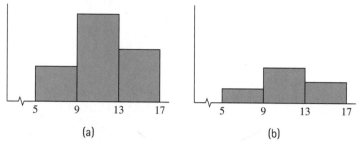

(a)   (b)

FIGURE 11
Different Scales on the Vertical Axis

Exercises 10.1

In Exercises 1 through 8, expand the given table to include the relative frequency and the cumulative frequency. Then draw frequency, relative frequency, and cumulative frequency histograms, and tell whether the data appear to be normally distributed, skewed, or neither.

1.

Class	Frequency
0 up to 3	4
3 up to 6	8
6 up to 9	12
9 up to 12	7
12 up to 15	4

2.

Class	Frequency
0	3
1	5
2	9
3	12
4	8
5	2

3.

Class	Frequency
3	9
4	12
5	4
6	3
7	2

4.

Class	Frequency
3 up to 5	2
5 up to 7	5
7 up to 9	9
9 up to 11	4
11 up to 13	7
13 up to 15	4
15 up to 17	2

5.

Class	Frequency
150 up to 180	30
180 up to 210	50
210 up to 240	80
240 up to 270	60
270 up to 300	40

6.

Class	Frequency
200 up to 250	6
250 up to 300	20
300 up to 350	50
350 up to 400	45
400 up to 450	40
450 up to 500	25
500 up to 550	10

7.

Class	Frequency
0	2
1	3
2	8
3	5
4	2

8.

Class	Frequency
0	2
1	3
2	3
3	3
4	5
5	9

Waiting for Services *Exercises 9 through 12 refer to the following frequency histogram, which depicts the time (in minutes) people spent waiting in line at a teller's window in a bank.*

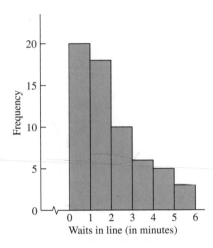

Interpret the time intervals at the bottom as including the time at the left endpoint and going up to, but not including, the right endpoint. Then answer the following questions.

9. How many people spent three or more minutes waiting in line?

10. What is the probability that a person will spend at least two minutes waiting in line?

11. What percentage of those observed spent less than three minutes waiting in line?

12. What are the odds that a person will wait less than two minutes in line?

Battery Length of Life *In Exercises 13 through 15, 50 flashlight batteries were tested for length of life, with the following results (in hours):*

13.5   14.6   13.2   15.0   14.5   12.4   12.0
12.6   12.8   11.1   11.6   11.5   10.6   10.2
13.0   13.5   13.2   13.8   13.6   14.0   12.1
14.2   14.5   15.2   11.6   12.8   12.5   14.2
13.9   13.5   12.6   12.3   12.9   13.6   12.2

13.5   13.7   13.3   15.4   14.6   14.8   13.4
15.4   14.8   14.6   13.0   13.5   15.8   15.6
14.4

13. Construct the frequency, relative frequency, and cumulative frequency distribution tables by organizing the data into six classes of equal width.

14. Construct the frequency, relative frequency, and cumulative frequency histograms. Do these data appear to be normally distributed or skewed?

15. About what percentage of the batteries have a life of 12 or more hours?

Botany Experiment *Exercises 16 through 21 refer to an experiment in which a botanist broke open a sample of 50 bean pods and counted the number of beans in each pod. The results obtained were as follows:*

5	4	1	5	2	0	4	5	2	4
2	3	4	4	4	6	4	3	5	4
0	2	3	2	3	5	4	4	4	6
1	3	2	4	4	3	1	5	3	5
3	3	3	1	3	4	2	3	4	3

16. Make a frequency and relative frequency table for these data.

17. Construct a frequency and relative frequency histogram for the data.

18. Arrange the data into a histogram table similar to that of Figure 9. In this arrangement, as well as in the frequency distribution histogram, do the data appear to be normally distributed?

19. Use the relative frequency histogram to compute the percentage of pods that have four or more beans in them.

20. What is the probability that a pod will have two or fewer beans in it?

21. What are the odds that a pod will have three or more beans in it?

Market Survey *In Exercises 22 through 27, a random sample of 28 households was questioned about the number of can openers each household possessed. The responses were as follows:*

2	2	2	4	2	3	0
1	3	1	1	1	3	3
3	1	2	2	1	2	2
2	3	3	1	2	2	1

22. Make a frequency and relative frequency table for these data.

23. Construct a frequency and relative frequency histogram for the data. Do you think the data are normally distributed or skewed?

24. Arrange the data in a histogram table similar to the one in Figure 9.

25. What percentage of homes has exactly two can openers, according to this survey?

26. What is the probability that a home will have at least two can openers?

27. What are the odds that a home will have three or more can openers?

Take-Home Pay *In Exercises 28 through 30, the weekly take-home pay (in dollars) for 20 employees of Z-Mart was as follows:*

172.63	187.60	87.42	187.21	207.35
148.23	129.36	133.71	152.37	165.70
173.19	153.60	105.51	165.70	158.26
161.87	190.35	201.40	162.50	148.23

28. Organize these data into a frequency distribution table that will adequately show the take-home pay for the employees.

29. Construct the frequency and relative frequency histograms for the data.

30. Using the same class intervals as in Exercise 28, display the salaries in a histogram table similar to that in Figure 9.

Mail Delivery Times *In Exercises 31 through 34, the time (in minutes) it took a particular mail carrier to make each of 30 deliveries was recorded as follows:*

1.3	5.1	7.2	4.2	7.0	7.1
2.6	1.4	2.5	3.8	2.4	3.7
3.1	3.1	3.6	6.5	4.2	4.2
1.6	6.2	6.9	4.5	6.3	3.1
2.0	1.1	5.1	2.6	5.0	4.6

31. Make a frequency and a relative frequency table for these data.

32. Display your distribution of the data in a frequency and a relative frequency histogram.

33. Construct a cumulative frequency histogram for the data.

34. Using the class intervals of Exercise 31, display the data in a histogram table similar to the one in Figure 9. Do these delivery times appear to be normally distributed or skewed?

Fleet Purchases *In Exercises 35 through 37, the Speedy Cab Co. purchased 24 new cars and recorded the number of months that each car was driven before being replaced. The recorded data were as follows:*

8	10	12	9	14	3
5	12	14	15	3	8
9	10	11	4	13	10
6	9	14	12	9	6

35. Make a frequency and relative frequency table for these data.

36. Using the class intervals of Exercise 35, make a histogram table for the data similar to the one in Figure 9. Do the data appear to be normally distributed? Skewed? Neither?

37. Approximate the probability that the cab company will keep a car 11 months or longer.

Weekly Salaries *In Exercises 38 through 40, the weekly salaries of 20 persons in a particular department of a manufacturing plant are recorded as follows:*

$300	$500	$475	$325	$225
$175	$275	$375	$425	$425
$305	$180	$400	$525	$385
$500	$292	$305	$390	$475

38. Construct a frequency distribution table and an accompanying frequency histogram to portray the preceding data as accurately as you can.

39. Construct a frequency distribution table and an accompanying frequency histogram to make the salaries appear as low as possible.

40. Construct a frequency distribution table and an accompanying frequency histogram to make the salaries appear as high as possible.

41. Men's Heights The cumulative relative frequencies for the heights of men aged 20 to 29 in the United States are given in the following table. (*Source*: U.S. Bureau of the Census, *Statistical Abstract of the United States 2000*):

Height (in feet and inches)	Cumulative Relative Frequency
5'1"	0.001
5'2"	0.005
5'3"	0.013
5'4"	0.034
5'5"	0.069
5'6"	0.117
5'7"	0.208
5'8"	0.320
5'9"	0.463
5'10"	0.587
5'11"	0.701
6'0"	0.812
6'1"	0.874
6'2"	0.947
6'3"	0.979

Age (in years)	Relative Frequency
Under 5	0.069
5–9	0.073
10–14	0.072
15–19	0.072
20–24	0.066
25–29	0.067
30–34	0.072
35–39	0.083
40–44	0.082
45–49	0.071
50–54	0.060
55–59	0.047
60–64	0.039
65–74	0.067
75–84	0.045
85 and older	0.015

(a) Construct the cumulative relative frequency histogram for the heights shown in the table.

(b) Calculate the *relative* frequency for each height.

(c) Construct the relative frequency histogram for the heights shown in the table.

42. Charitable Contributions The relative frequencies for the annual amount of household contributions to charities in 1998 are given in the following table. (*Source*: U.S. Bureau of the Census, *Statistical Abstract of the United States 2000*):

Contribution (in dollars)	Relative Frequency
$1 to $100	0.222
$101 to $200	0.125
$201 to $300	0.099
$301 to $400	0.080
$401 to $500	0.061
$501 to $600	0.047
$601 to $700	0.037
$701 to $999	0.061
$1,000 or more	0.268

(a) Construct the relative frequency histogram for this grouped data set.

(b) Compute the cumulative relative frequency for each class interval, and construct the cumulative relative frequency histogram.

43. Demographics The relative frequencies for the ages of U.S. residents in 1999 are given in the following table. (*Source*: U.S. Bureau of the Census, *Statistical Abstract of the United States 2000*):

(a) Construct the relative frequency histogram for the data.

(b) Compute the cumulative relative frequency for each class interval, and construct the cumulative relative frequency distribution.

Despite a lack of training in how and where to take a sample that would best represent an entire population, do your best to select your own random sample of 15 students on your campus, and collect and analyze data, using the ideas of this section that would give a picture of the information asked for in each of Exercises 44 through 47.

44. The heights of women on your campus

45. The number of hours students sleep at night

46. How much money each student is carrying

47. How many books each student carries on campus

Select your own random sample of 35 students on your campus, and perform an analysis, using the ideas of this section that would give a picture of the information asked for in each of the Exercises 48 through 52.

48. The heights of men on your campus

49. Grade point averages for students

50. How many hours a student listens to the radio each week

51. How much money students earn on summer jobs

52. Randomly select a sample of 20 freshmen and 20 seniors from your college. Ask them how many hours they watched television the previous week, and analyze the information gathered.

Section 10.2 Measures of Central Tendency

We often hear statements such as "The **average** weekly salary is $254" or "The **median** price of a house in the United States is $98,750." These statements use a single number that is considered to be the best representative of all observations under consideration. Such numbers, in some sense, measure the center of the observations; hence, the term **measure of central tendency** is often applied to them.

Several specific measures of central tendency and the merits of each are discussed in this section. Note that a particular measure may *not* be appropriate for a specific application, and at other times, it might be a matter of preference as to which measure is used. With this in mind, never accept one of these measures unless you know exactly what kind it is, and if you are giving a measure, use one that is appropriate for the application.

The easiest measure of central tendency to compute is the **mode.** The mode is the observation that is recorded most often in the set of data.

The Mode

The **mode** of a set of observations is the observation that occurs most frequently.

EXAMPLE 1

Calculating the Mode

A set of observations may have one or more modes. In the set

$$30, 70, 80, 30, 70, 30,$$

the mode is 30, because 30 is the most frequently recorded observation. However, for the data

$$8, 6, 8, 10, 26, 17, 26,$$

there are two modes, namely, 8 and 26. These data are called **bimodal.** Alternatively, a set of observations may have no mode at all. The data set

$$6, 9, 11, 14, 81$$

has no mode, because no observation occurs more frequently than any other. ■

The mode is very easy to find and represents the data in its own special way. However, it is not very stable: In the first set of observations in Example 1, if one of the 30s is replaced by a 70, the mode suddenly becomes 70 rather than 30. Nonetheless, if you are interested in the **most common value** among the observations, then the mode is that value.

Another measure of central tendency is the **median.** The use of the median as a measure of central tendency is widespread in statistical analysis. (See Figure 12.)

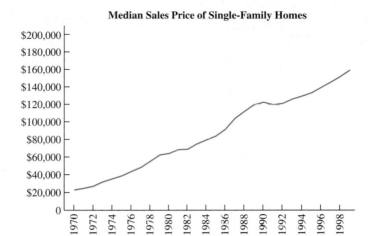

Median Sales Price of Single-Family Homes

FIGURE 12
Use of the Median to Display Sale Prices of Single-Family Homes
Source: Statistical Abstract of the United States 2000.

The Median

The **median** of a data set is the numerical value which has the property that, when the data are listed from left to right in numerical order, the number of data values to the left of the median is equal to the number of data values to the right of the median. The median need *not* be an element of the data set.

To find the median of a given set of data, use the following steps:

1. Arrange the observations in numerical order.
2. If there is an odd number of observations, the median is the middle observation.
3. If there is an even number of observations, the median is one-half the sum of the two middle observations.

EXAMPLE 2

Finding the Median

(a) Find the median of these observations: 8, 3, 7, 12, 3.

SOLUTION The first step is to arrange all of the observations, including *both* 3s, from smallest to largest in numerical value:

$$3, 3, 7, 8, 12.$$

Because there is an odd number ($n = 5$) of observations, there is a middle observation, and that observation is 7. Therefore, the median is 7.

Notice that if the observation 12 is replaced with 5280 in these data, the median would still be 7:

$$3, 3, 7, 8, 5280.$$

The median is not affected by extremely large or extremely small observations. This is one reason the median is a favored measure of central tendency.

(b) Find the median of these observations: 0, 5, 0, 9, 6, 2.

SOLUTION Again, we begin by arranging the data from smallest to largest in numerical value:

$$0, 0, 2, 5, 6, 9.$$

This time there are six observations $(n = 6)$. Hence, there is no one middle observation, but rather two middle observations: 2 and 5. According to the definition, the median is

$$\frac{2 + 5}{2} = \frac{7}{2} = 3.5.$$

This example shows that the median need not be one of the given observations. ■

Once the median is found, there will *always* be an equal number of observations greater than it and less than it. This is the idea of "center" or "measure of central tendency" for the median. The median is *not* affected by *extreme* observations, but rather is determined by its *position*. The *median* is an appropriate measure of central tendency to use *if the data are spread over a large range.*

EXAMPLE 3

The Mode and Median of Data

A Girl Scout cookie sales record for a particular troop had these entries in the "boxes sold" column:

$$2, 4, 4, 2, 8, 6, 12, 2, 2, 2, 6.$$

When arranged from lowest to highest, the number of boxes sold appears as

$$2, 2, 2, 2, 2, 4, 4, 6, 6, 8, 12.$$

The mode is 2, while the median is 4. Even though the most common order size (the mode) is two boxes, the median number of boxes sold is four. Some might view the median as the better single number to represent the "central measure" of the number of boxes sold per customer. ■

Perhaps the most common measure of central tendency used in statistics is the **arithmetic mean,** sometimes just called the **mean.** This measure is the one most of us think of when we use the term "average." For example, if you wanted to calculate the average of your exam scores, you would calculate the mean by finding the sum of all your scores then dividing by the number of exams. Since you would probably use all of your exam scores in a particular course to find this average, you would be calculating the **population mean.** In other words, all data values in the population are used to calculate the mean.

The Population Mean

Given a population with n elements, $\{x_1, x_2, \ldots, x_n\}$, the **population mean,** denoted by μ (the Greek letter *mu*), is the sum of all values in the population, divided by n:

$$\mu = \frac{(x_1 + x_2 + \cdots + x_n)}{n} = \frac{\sum x}{n}.$$

EXAMPLE 4

Calculating the
Population Mean

Alyssa scored 93, 72, 84, and 86 on the four papers in her French literature class. Her final grade in the course is the mean of these four scores. What is Alyssa's final grade?

SOLUTION Since the four scores represent the entire population (i.e., all of Alyssa's scores), we calculate the population mean to determine Alyssa's final grade:

$$\mu = \frac{93 + 72 + 84 + 86}{4} = \frac{335}{4} = 83.75.$$

Thus, Alyssa's final grade in this class is 83.75. ∎

If we have collected only a *sample* from the entire population, the calculation of the mean of the values collected is likely to be only an approximation of the true mean of the population. In statistics, we refer to such calculations from samples as **estimators**. Thus, the mean of a sample, denoted by $\overline{x}$, is an **estimator** of the population mean μ. By calculating $\overline{x}$, we are making an inference that the true mean is approximated by the sample mean. If we collect a different sample, we would likely calculate a different sample mean. However, we expect that our sample means do not vary widely from the population mean. To use terminology from Chapter 8, we say that the **expected value** of $\overline{x}$ is precisely equal to μ; that is $E(\overline{x}) = \mu$. Thus, in the long run, if we calculate the sample mean for many different samples, we expect the average of those sample means to be the population mean. In this case, we call $\overline{x}$ an **unbiased estimator** of μ.

The Sample Mean

Given a sample of n observations, $\{x_1, x_2, \ldots, x_n\}$, the **sample mean,** denoted by $\overline{x}$, (read "x-bar"), is the sum of all observations, divided by n:

$$\overline{x} = \frac{x_1 + x_2 + \cdots + x_n}{n} = \frac{\sum x}{n}.$$

EXAMPLE 5

Calculating the Sample
Mean

Ten customers in a bookstore on a Friday afternoon were asked how many books they planned to purchase from the store that day. The results were 0, 2, 2, 3, 1, 0, 4, 2, 4, and 1. What is the mean of this sample?

SOLUTION Since the observations listed represent the responses of only a sample of 10 shoppers, and not the responses of *all* shoppers in the bookstore, we calculate the sample mean:

$$\overline{x} = \frac{0 + 2 + 2 + 3 + 1 + 0 + 4 + 2 + 4 + 1}{10} = \frac{19}{10} = 1.9.$$

Therefore, we estimate that shoppers in this store on Friday afternoon planned to purchase, on average, 1.9 books that day. ∎

In Examples 4 and 5, the means that were calculated were not actual values from the respective data sets. In Example 4, even though Alyssa's final grade was 83.75, she never *actually* scored an 83.75 on one of her four papers. In Example 5, the sample mean was 1.9 books, and it does not seem possible to purchase a fraction of a book. Of the three measures of central tendency we have discussed thus far, only the *mode* must be an actual value in the data set. The median and the mean may or may not correspond to actual values in the data set.

EXAMPLE 6

Comparing the Mean with the Median

A small firm makes hammers for the army. The firm's employees are the owner, a plant superintendent, and four workers. Their respective annual salaries are $60,000, $40,000, $15,000, $15,000, $15,000, and $15,000. The owner boasts that the average salary at the plant is over $26,000 per year. Is the owner referring to the mean salary when she uses the word "average"?

SOLUTION The mean salary is

$$\mu = \frac{60{,}000 + 40{,}000 + 15{,}000 + 15{,}000 + 15{,}000 + 15{,}000}{6} = \$26{,}666.67,$$

so that must be the one the owner used. If you were considering a job with this firm, would that average give a realistic salary picture? Probably not. The median, $15,000, best represents the company's salaries, and this just happens to be the mode for the data as well. ■

As shown in Example 6, the mean has the disadvantage of being affected by extreme values. Therefore, it may not be too meaningful to represent the data by the mean when the data set has extreme values.

In Example 7, we introduce the concept of a **weighted mean.**

EXAMPLE 7

Comparing the Weighted Mean with Other Measures

Suppose a product is bought at three different times, at the prices and quantities shown in the following table:

Purchase Price per Unit	Number of Units Purchased	Relative Frequency
$2.00	500	$\frac{500}{800}$
$3.00	200	$\frac{200}{800}$
$5.00	100	$\frac{100}{800}$

Questions with very different answers come to mind about such purchases.

(a) What is the median price per unit?

SOLUTION Think of arranging the observations from smallest to largest:

$$\underbrace{2, 2, 2, \ldots \qquad , 2}_{500 \text{ observations}} \quad \underbrace{3, 3, 3, \ldots \qquad , 3}_{200 \text{ observations}} \quad \underbrace{5, 5, 5, \ldots \qquad , 5}_{100 \text{ observations}}.$$

There will be 500 2s, followed by 200 3s, followed by 100 5s. There are $n = 800$ observations, so the middle two observations are in the 400th and 401st positions. This is really midway between two 2s, so the median price per unit is $2.00.

(b) What is the average (mean) purchase price per unit?

SOLUTION This question is answered by finding the total dollars spent for the purchase and then dividing by the number of units purchased. The dollars spent amount to

$$\$2(500) + \$3(200) + \$5(100) = \$2100,$$

while the total number of units purchased is

$$500 + 200 + 100 = 800.$$

Therefore, the mean price paid per unit is

$$\frac{\$2(500) + \$3(200) + \$5(100)}{800} = \frac{\$2100}{800} \approx \$2.63.$$

The left side of this calculation may be rewritten as

$$\$2\left[\frac{500}{800}\right] + \$3\left[\frac{200}{800}\right] + \$5\left[\frac{100}{800}\right] \approx \$2.63,$$

to show each of the dollar amounts "weighted" with its proportional share (relative frequency or probability) of the total number of purchases. For this reason, such an average is sometimes given the name **weighted mean** or **weighted average.**

(c) What is the average (mean) *of the purchase prices?*

SOLUTION The answer is $\dfrac{\$2 + \$3 + \$5}{3} = \$3.33.$ ■

Example 7 serves as a guidepost to finding an average of grouped data. The basic lesson is this: *When in doubt, write it out.* By this, we mean make a list of the data such as was done in part (a). First, a list will give visual help in locating the middle or two middle numbers when finding the median. Second, a list shows how the mean, as calculated in part (b), may be viewed either as a traditional mean or as a weighted mean. Once these concepts are fully understood, such averages can be found by direct computations from the table displaying the groups.

The Weighted Mean

If $x_1, x_2, \ldots, x_n$ are n observations with respective relative frequencies $p_1, p_2, \ldots, p_n$, then the **weighted mean** is given by $x_1 p_1 + x_2 p_2 + \cdots + x_n p_n = \sum xp.$

NOTE A **mean** of n observations may be thought of as a **weighted mean:** Each observation is assigned the weight (relative frequency) $\frac{1}{n}$.

The weighted mean (and hence the mean) may be interpreted geometrically as the balance point or the center of gravity of the system, using the observations as points on a number line and the relative frequencies as the respective weights at each point. Such a system is shown in Figure 13 for the data in Example 7(b). The figure visually depicts the measure of central tendency for the weighted mean to be the center of gravity.

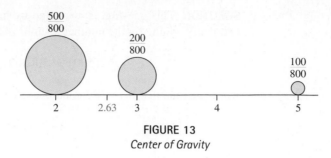

FIGURE 13
Center of Gravity

   Suppose one is presented with a frequency histogram whose original observations were no longer available. Is it possible to compute the mean of the data? The answer is "no," but we can calculate a reasonably good estimate by *assuming* that each observation lies at the midpoint of the interval in which it is located.

EXAMPLE 8

Estimating the Mean of Grouped Data

Consider these test scores on a recent calculus test:

Scores	Frequency
91 to 99	3
82 to 90	7
73 to 81	12
64 to 72	5
55 to 63	2
Total:	29

Estimate the mean test score.

SOLUTION The accepted method for estimating the mean is to assume, for lack of more accurate information, that each test score lies at the midpoint of its respective class interval. Therefore, we consider the test scores to be

3 scores of **95**   (95 is midway between 91 and 99)

7 scores of **86**   (86 is midway between 82 and 90)

12 scores of **77**   (77 is midway between 73 and 81)

5 scores of **68**   (68 is midway between 64 and 72)

2 scores of **59**   (59 is midway between 55 and 63)

Because there are 29 test scores, the mean (actually, the weighted mean) of the scores is

$$\frac{3(95) + 7(86) + 12(77) + 5(68) + 2(59)}{29} \approx 78.24. \quad \blacksquare$$

Estimating the Mean of Grouped Data

If data are grouped within intervals having midpoints $m_1, m_2, \ldots, m_n$, with corresponding frequencies $f_1, f_2, \ldots, f_n$, then the **estimated mean of the data** is given by

$$\frac{m_1 f_1 + m_2 f_2 + \cdots m_n f_n}{f_1 + f_2 + \cdots + f_n} = \frac{\Sigma m f}{\Sigma f}$$

The median for the test scores in Example 8 can also be estimated by considering each score to lie at the midpoint of its respective interval (the same way as is done in estimating the mean). In this case, the estimate of the median test score is 77.

EXAMPLE 9

Calculating the Mean, Median, and Mode with Excel

With the Girl Scout cookie sales data from Example 3, use Excel to calculate the mean, median, and mode of the number of boxes sold by the troop.

SOLUTION Begin with a new worksheet, and enter the data, {2, 4, 4, 2, 8, 6, 12, 2, 2, 2, 6}, in column A, beginning in cell A1. To calculate the mean of the data, in any cell, type =AVERAGE(A1:A11) and press ENTER. The mean number of boxes sold by members of the troop is calculated to be 4.5454. In another empty cell, type =MEDIAN(A1:A11) and press ENTER to see that the median number of boxes sold is 4. Finally, type =MODE(A1:A11) in any empty cell to see that the mode for the data set is 2. ▬

EXAMPLE 10

Calculating the Mean and Median with the TI-83 Plus

With the Girl Scout cookie sales data from Example 3, use the TI-83 Plus to calculate the mean, median, and mode of the number of boxes sold by the troop.

SOLUTION Enter the data, {2, 4, 4, 2, 8, 6, 12, 2, 2, 2, 6}, into list L₁. From the home screen, access the LIST MATH menu and select mean. Enter the name of the list under which the data are stored—in this case, L₁; then press ENTER. The mean number of boxes of cookies sold by members of the troop is found to be 4.5454. Access the LIST MATH menu again and select median. Enter the name of the list under which the data are stored and press ENTER. The median number of boxes of cookies sold is 4. ▬

Abuses of Statistics

One should be very careful when reading statements involving statistics. Without access to the original data set, the information provided may be misleading.

What Does an Average Really Tell You? A growth company has 30 million shares of stock outstanding. Two people own 20 million of these shares, and the other 10 million shares are owned by another 50,000 people. The company states that its average stockholder owns 600 shares of company stock. This is correct (using the arithmetic mean as the method of averaging), but it portrays an inflated voting power among the stockholders. Excluding the two major stockholders, the average number of shares per stockholder is 200. The former average thus overstates the voting power of most stockholders by a factor of three!

If you were asked to find the average earnings of those in your high school graduating class, you would probably do so by sending them a survey asking about their earnings. Then, on the basis of the surveys returned, an average could be computed. However, did you have the addresses of *all* classmates, or just those who were prominent, highly successful classmates? Who responded? Surely not everyone! Which method of averaging did you use? What does your average really mean?

CAUTION Do not accept an average without asking questions!

Exercises 10.2

In Exercises 1 through 6, compute the following measures of central tendency:

(a) *The mode (if any)*

(b) *The mean*

(c) *The median*

1. The population is 6, 9, 26, 18, and 43.

2. The population is 17, 12, 29, 17, 20, 16, 20, and 20.

3. The sample is 225, 102, 96, 173, 102, and 112.

4. The sample is 101, 86, 137, 96, and 120.

5. The population is -5, -6, 12, 8, -8, 6, 0, -12, 3, 5, and -3.

6. The sample is -22, 3, 7, 8, -15, 0, 2, -10, 5, and 0.

In Exercises 7 through 14, decide which measure of central tendency—the mean, the mode, or the median—best describes the data. Explain your reasoning.

7. Salaries of $20,000, $23,000, $21,000, and $60,000

8. Salaries of $30,000, $60,000, $65,000, $63,000, and $80,000

9. The number of boxes of Girl Scout cookies sold per scout: 23, 18, 25, 15, 80, 20, and 19

10. Test scores of 76, 83, 92, 75, 83, 98, 65, 73, 82, 56, 43, 78, and 80

11. Ladies' shoe sizes of 5, 7, 7, 6, 7, 8, 7, 7, 6, 7, and 7

12. Grade point averages of 3.51, 2.78, 3.01, 2.60, 2.95, and 3.26 for six students

13. Ages of students: 18, 19, 18, 20, 22, 19, 30, 20, 42, 19, and 20

14. Men's suit sizes of 40, 38, 40, 42, 44, 40, 40, 46, 40, and 38

In Exercises 15 through 19, find the center of gravity (also interpreted as the weighted mean).

15.
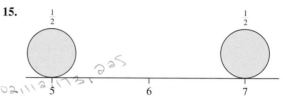

16.
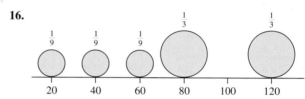

17.
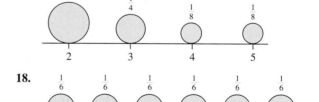

18.
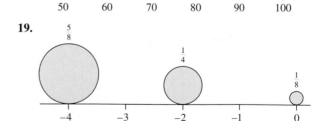

19.

20. Truth in Advertising An apartment owner has four two-bedroom units that rent for $700 per month and two efficiency units that rent for $400 per month. He advertises that his average rent is $600 per month. Which average is he using? Is this truthful advertising? Does it present an accurate picture of the apartments to potential renters?

21. Charitable Donations The director of a charitable cause reports that gifts received on a particular day were $20, $50, $30, $250, $50, $25, $500, $50, $25, $25, $30, $50, $25, $50, $25, and $25. The director claims that the average gift exceeds $75. What average is the director using? What average do you think best describes the giving for that day?

Math Anxiety Study *In Exercises 22 through 24, a psychology department developed a math-anxiety test totaling 1000 points. The department gave the test to a sample of 10 students, with the following results: 360, 240, 250, 300, 400, 990, 370, 210, 570, and 380.* 210, 240, 250, 300, 360, 370, 380, 400,

22. Compute the mean for the entire sample.

23. Compute the median for the entire sample. Which best describes the scores for the entire sample, the mean or the median?

24. Delete the 2 exceptionally high scores in the sample, and compute the mean and median of the remaining 8 scores. Compare these with the mean and median found using all 10 test scores.

Faculty Survey *In Exercises 25 and 26, a college journalism department has professors with the following number of years of teaching experience: 4, 12, 6, 8, 12, and 9.*

25. Find the mean and median for this population.

26. Suppose that the department hires a new professor with 23 years of teaching experience. Now find the mean and median for the years of teaching experience possessed by the faculty, and compare them with the calculations of Exercise 25.

Fund Raising *In Exercises 27 through 30, a student organization bought 1000 replicas of the school mascot for $4.00 each*

to sell at an upcoming football game. The organization sold 700 for $6.00 each at the game and discounted the rest to a wholesaler for $3.50 each.

27. What was the mean price for which all 1000 mascots were sold?

28. What was the median selling price for all 1000?

29. What was the mean of the two selling prices?

30. What was the mean profit per mascot?

Retail Prices *In Exercises 31 through 34, two people bought soft drinks for a party. Maria bought 20 six-packs for $1.79 each, and Bradley bought 15 six-packs for $1.59 each. They sold the drinks for 40 cents per can.*

31. What was the mean purchase price per six-pack?

32. What was the median purchase price per six-pack?

33. What was the mean of the two purchase prices?

34. What was the mean profit per six-pack?

In Exercises 35 and 36, the Pump-Up Fitness Company recently bought three treadmill models from a wholesaler. The prices and number of each model purchased are displayed in the following table:

Purchase Price per Unit	Number of Units Purchased
$150	50
$260	30
$400	10

35. Find the median purchase price per unit.

36. Find the mean purchase price per unit.

Sales *In Exercises 37 through 39, a manufacturer's 2001 sales record for handheld calculators was as follows:*

Model	Unit Price	Number of Units Sold
KU-1	$25	2000
KU-2	$35	1500
KU-20	$80	1000
KU-102	$95	800
KU-150	$120	500

37. Find the mean price per unit at which the calculators were sold.

38. Find the median selling price per unit.

39. Find the mean of the unit prices.

Quiz Scores *In Exercises 40 through 42, scores on a 10-point quiz in a history course were as follows:*

Score	Frequency
10	5
9	7
8	12
7	9
6	2
5	2

40. Find the mean of the scores 5, 6, 7, 8, 9, and 10.

41. Find the mean quiz score.

42. Find the median quiz score.

Market Survey *In Exercises 43 through 45, a survey of 27 households asked how many automobiles were owned by the household. The results were as follows:*

Number of Automobiles	Frequency
0	1
1	3
2	12
3	9
4	2

43. Find the mean number of automobiles per household, as represented by this survey.

44. Find the mean of the numbers 0, 1, 2, 3, and 4.

45. Find the median number of automobiles per household.

Agriculture Survey *In Exercises 46 through 49, a seed company is testing a new variety of garden peas. A worker broke open 28 pea pods and counted the number of peas in each pod. The worker recorded the following data:*

Number of Peas in Pod	Frequency
1	3
2	4
3	5
4	8
5	6
6	2

46. Find the mean number of peas per pod.

47. Find the median number of peas per pod.

48. What is the mean of the numbers 1, 2, 3, 4, 5, and 6?

49. **Financial Survey** Thirty students were asked how much money was in their bank accounts at the moment. Their responses were recorded as follows:

Amount in Account	Frequency
$0 to $50	5
$50 to $100	16
$100 to $200	5
$200 to $300	4

Estimate the mean amount in the students' accounts.

50. **Salary Ranges** A company has 20 people on its payroll in the following salary ranges:

Salary	Frequency
$15,000 to $20,000	2
$20,000 to $25,000	6
$25,000 to $30,000	8
$30,000 to $50,000	4

Estimate the mean salary of a person employed by the company.

51. Estimate the mean and median for the data displayed in the following histogram:

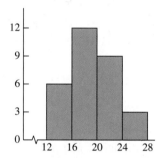

52. Estimate the mean and median for the data displayed in the following histogram:

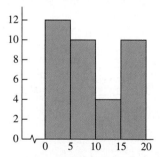

53. **Demographics** A group of senior citizens consisting of eight people in their 60s, five in their 70s, and eight in their 80s went on a sightseeing trip. Estimate the mean age of the people.

54. **Test Scores** On a recent journalism test, 5 students scored in the 90's, 8 in the 80's, 15 in the 70's, 4 in the 60's, and 3 in the 50's. Estimate the mean test score.

55. Course Grades Suppose that a student has one-hour test scores of 78, 92, and 84 and a final exam score of 85. The professor announces that the final will be weighted as two one-hour exams. What is the student's class average?

56. Course Grades Latonya has a homework average of 92; one-hour test scores of 86, 90, and 89; and a final exam score of 86. The professor announces that the homework will be weighted as one-fourth of a one-hour test score and the final will be weighted as three one-hour test scores. What is Latonya's average?

57. Course Grades Suppose that you have one-hour test scores of 88, 79, and 91 thus far in a course, with only the final exam left. The final is weighted as two one-hour exams, and you need a 90 average (mean) for an *A* in the course. What is the minimum score you can get on the final and still make an *A*?

58. Course Grades Suppose that you have one-hour test scores of 92, 90, and 84, a homework average of 90%, and a quiz average of 86%. If the homework is weighted as one-third of a one-hour test, the quizzes as one-half of a one-hour test, and the final as two one-hour tests, what score would you need on the final to have a 92% average (weighted mean)?

Charitable Contributions *In Exercises 59–61, a sample of 2719 households in the United States were polled regarding their charitable contributions in 1998. The relative frequencies for the annual charitable contributions are given in the following table. (Source:* U.S. Bureau of the Census, *Statistical Abstract of the United States 2000):*

Contribution (in dollars)	Relative Frequency
$1 to $100	0.222
$101 to $200	0.125
$201 to $300	0.099
$301 to $400	0.080
$401 to $500	0.061
$501 to $600	0.047
$601 to $700	0.037
$701 to $999	0.061
$1000 or more	0.268

59. Use the fact that the sample size is $n = 2719$ to estimate the frequency for each class interval.

60. Estimate the mean U.S. household charitable contribution in 1998.

61. Estimate the median U.S. household charitable contribution in 1998.

Demographics *In Exercises 62–64, the relative frequencies for the ages of U.S. residents in 1999 are given in the following table. (Source:* U.S. Bureau of the Census, *Statistical Abstract of the United States 2000):*

Age (in years)	Relative Frequency
Under 5	0.069
5–9	0.073
10–14	0.072
15–19	0.072
20–24	0.066
25–29	0.067
30–34	0.072
35–39	0.083
40–44	0.082
45–49	0.071
50–54	0.060
55–59	0.047
60–64	0.039
65–74	0.067
75–84	0.045
85 and older	0.015

62. Use the relative frequencies to estimate the mean age of U.S. residents in 1999. (*Note:* This mean is actually a weighted mean.)

63. Estimate the median age of U.S. residents in 1999.

64. Estimate the mode of the ages of U.S. residents in 1999, and use this statistic to estimate which year persons of that age were born.

65. Study Time Ask 20 students how much time, to the nearest hour, they studied last week. Analyze the data you collect by finding the arithmetic mean, the median, and the mode and then constructing a histogram to give a visual picture of the data.

66. Data in a **box-and-whisker** plot are displayed as follows: Arrange the data from the smallest observation *s* to the largest observation ℓ along a line. Next, find the median *m*. Then find the median m_1 of the data from *s* to *m* and the median m_2 of the data from *m* to ℓ. Draw a box like the following, and "whisker out" to the extremes of the data:

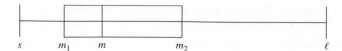

$$s \qquad m_1 \qquad m \qquad m_2 \qquad\qquad\qquad \ell$$

Such a display, like the histogram, shows how the data are shaped. The five numbers s, m_1, m, m_2, and ℓ are called **Tukey's five-number summary.** For each of the following sets of data, find Tukey's five-number summary, and construct the box-and-whisker plot:

(a) 15, 3, 22, 18, 5, 12, 21, 18, 19, 9, 20, 10, 16, 21, 17

(b) The consecutive positive integers from 1 through 100

(c) The integers 1, 2, 3, followed by 30 ninety-seven times

Section 10.3 Measures of Dispersion

Another quantity that aids in analyzing data is a number that will measure how widely spread out or dispersed the observations are, either as a set or from a central point such as the mean. One of the simplest ways to get such a number is to compute the **range** of the observations. The range is the difference between the largest observation and the smallest observation. Even though it is easy to calculate, the range is not a very good measurement of spread, because it tells us nothing about the other observations. Another method uses the **deviation** (difference) between the mean and each of the observations in the list of data. This method leads to a measurement called the **standard deviation.** The more concentrated the observations are around the mean, the smaller the standard deviation is, while the further the observations are from the mean, the larger the standard deviation is. Although not as easy to calculate as the range, the standard deviation is considered to be a better measurement, in the sense that it involves every observation in the set of data. Example 1 shows how the standard deviation is calculated.

EXAMPLE 1

How the Standard Deviation Is Calculated

A campus group recorded the following samples of 12 delivery times of pizza from each of two local pizza parlors. (See Table 5.) Even though the average (mean) delivery times are the same, which pizza parlor would you prefer to order from and why?

Anton's Pizza Delivery Time (in minutes)	Barry's Pizza Delivery Time (in minutes)
35	36
38	50
31	20
34	44
30	18
32	36
36	65
37	30
32	20
33	25
34	34
36	30
Mean delivery time: $\bar{x} = 34$ minutes	Mean delivery time: $\bar{x} = 34$ minutes

TABLE 5.

SOLUTION The **range** of a data set is the difference between the highest and lowest values in the set. The range for Anton's pizza delivery times is $38 - 30 = 8$ minutes. The range for Barry's pizza delivery times is $65 - 18 = 47$ minutes. These values indicate that Anton's delivery times are less dispersed than Barry's. On the basis of this information, most of us would probably choose Anton's. In many cases, however, deciding solely on the basis of the range of the data set may cause us to make a bad decision. Thus, we look more closely at how the observations deviate from the respective sample means.

Table 5 can now be expanded (see Table 6) to calculate how the delivery times in each case deviate from the respective means. At this point, some kind of "average" of the deviations in each case is needed, to measure how the delivery times vary or deviate from the mean. Notice, however, that the sum of the deviations in each case is zero. This is no accident: The sum of the deviations from the mean of *any* set of observations is always zero. The reason is that the mean is the balance point, or the center of gravity, of the data. This implies that the sum of the distances (deviations) of the various observations to the left of the mean is precisely the same as the sum of the distances (deviations) of the various observations to the right of the mean.

Deviations from the Sample Mean, Anton's	Deviations from the Sample Mean, Barry's
$x - \bar{x}$	$x - \bar{x}$
$35 - 34 = \quad 1$	$36 - 34 = \quad 2$
$38 - 34 = \quad 4$	$50 - 34 = \quad 16$
$31 - 34 = -3$	$20 - 34 = -14$
$34 - 34 = \quad 0$	$44 - 34 = \quad 10$
$30 - 34 = -4$	$18 - 34 = -16$
$32 - 34 = -2$	$36 - 34 = \quad 2$
$36 - 34 = \quad 2$	$65 - 34 = \quad 31$
$37 - 34 = \quad 3$	$30 - 34 = -4$
$32 - 34 = -2$	$20 - 34 = -14$
$33 - 34 = -1$	$25 - 34 = -9$
$34 - 34 = \quad 0$	$34 - 34 = \quad 0$
$36 - 34 = \quad 2$	$30 - 34 = -4$
Sum $= \quad 0$	Sum $= \quad 0$

TABLE 6.

Deviations Always Sum to Zero

For any set of observations, the sum of the deviations from the mean of those observations is *always* zero.

Given that the average of the deviations from the mean will *always* be zero, what do statisticians do? They *square* each of the deviations and then compute an average. (The mathematical theory is easier to use with squaring rather than with absolute values.) Table 7 shows the result of squaring each deviation from the mean for both firms. Now we have numbers whose average (mean) will have significance. However, there is yet another twist to the averaging process: If we have collected *every* observation in the entire population of n observations, then we know precisely the mean for the population and all possible deviations from the mean can be calculated, so the average of the squares of the deviations

Deviations Squared, Anton's	Deviations Squared, Barry's
$(x - \overline{x})^2$	$(x - \overline{x})^2$
1	4
16	256
9	196
0	100
16	256
4	4
4	961
9	16
4	196
1	81
0	0
4	16
Sum: 68	Sum: 2086

TABLE 7.

can be found as usual by adding and then dividing by n. This value is called the **population variance** and is denoted by σ^2 (sigma squared).

Recall from Section 9.2 that the sample mean $\overline{x}$ is an **unbiased estimator** of the true population mean μ. Since we have collected only a *sample* from Anton's and Barry's pizza delivery times, the calculation of the sample mean of these numbers is likely to be only an approximation to the true mean of the pizza delivery times. Thus, the deviations from this sample mean, and hence the square of those deviations from the sample mean, may not be reflective of the variance of the entire population of pizza delivery times. When we are working with a sample, we want to make sure that the estimated variance we calculate from our data is an **unbiased estimator** of the true population variance σ^2. In more advanced books on statistics, we learn that this is achieved by dividing the sum of the squares of the deviations from the sample mean by $n - 1$, rather than n. We call this estimator the **sample variance** and denote it by s^2. Since this estimator is unbiased, we say that $E(s^2) = \sigma^2$; that is, in the long run, if we calculate the sample variance for many different samples, we expect the average of those sample variances to be the population variance. The formulas for the population variance and the sample variance are summarized next.

The Variance

For n observations, the **sample variance** is

$$s^2 = \frac{[\sum(x - \overline{x})^2]}{(n - 1)}$$

and the **population variance** is

$$\sigma^2 = \frac{[\sum(x - \mu)^2]}{n},$$

where x is any observation, $\overline{x}$ is the sample mean, and μ is the population mean.

For our present *sample* of pizza delivery times, the deviations squared will be averaged using $n - 1 = 12 - 1 = 11$, rather than 12. The sample variance for Anton's is

$$s^2 = \frac{68}{n - 1} = \frac{68}{11} \approx 6.18 \text{ (minutes)}^2,$$

and the sample variance for Barry's is

$$s^2 = \frac{2086}{n - 1} = \frac{2086}{11} \approx 189.64 \text{ (minutes)}^2.$$

Notice that each variance is in units of $(\text{minutes})^2$. We might wonder about the utility of such a calculation, but there is a solution to this problem: Take the square root of the variance to bring the units back to the original units of time. The result of taking the square root of the variance is known as the **standard deviation.** The sample standard deviation is then $\sqrt{s^2} = s = \sqrt{6.18} \approx 2.49$ minutes for Anton's and $\sqrt{s^2} = s = \sqrt{189.64} \approx 13.77$ minutes for Barry's. Notice how these two numbers reflect the deviations of delivery times from the mean (small deviations for Anton's and large deviations for Barry's). Since Anton's pizza delivery times are less dispersed (i.e., Anton's delivery times have a smaller sample standard deviation), we have more statistical evidence to justify choosing Anton's pizza over Barry's when having a reliable delivery time is of primary concern. ■

The Standard Deviation

The **standard deviation** is the square root of the variance. For n observations, the **sample standard deviation** is

$$s = \sqrt{\frac{\Sigma(x - \overline{x})^2}{(n - 1)}}$$

and the **population standard deviation** is

$$\sigma = \sqrt{\frac{\Sigma(x - \mu)^2}{n}},$$

where x is any observation, $\overline{x}$ is the sample mean, and μ is the population mean.

We want to emphasize that, even though we are using different notations to distinguish between a sample mean and a population mean, there is only one way to find the mean of *any* set of n observations: Add the observations and divide by n. Contrast this with the averaging process used to find the variance.

EXAMPLE 2

Finding the Population Mean and Standard Deviation

The weights of all 10 people who are about to undergo a controlled diet program are as follows: 237, 304, 296, 345, 322, 315, 295, 317, 352, and 365 pounds. Compute the mean and standard deviation for these weights.

SOLUTION The population mean is the sum of the weights divided by 10:

$$\mu = \frac{3148}{10} = 314.8 \text{ pounds.}$$

Table 8 summarizes the calculations that lead to the variance and then to the standard deviation.

x	$x - \mu$	$(x - \mu)^2$
237	−77.8	6,052.84
304	−10.8	116.64
296	−18.8	353.44
345	30.2	912.04
322	7.2	51.84
315	0.2	0.04
295	−19.8	392.04
317	2.2	4.84
352	37.2	1,383.84
365	50.2	2,520.04
		Sum: 11,787.60

TABLE 8.

Therefore, the variance is

$$\sigma^2 = \frac{11{,}787.60}{10} = 1178.76 \ (\text{lb})^2$$

and the standard deviation is

$$\sigma = \sqrt{1178.76} \approx 34.33 \ \text{lb.} \ \blacksquare$$

Knowing how many of the observations lie within one standard deviation of the mean gives another indication as to how the data points are spread out from the mean. For the data in Example 2, the mean is $\mu = 314.8$ lb, and the standard deviation is $\sigma = 34.33$ lb. Thus, *one* standard deviation is 34.33 lb. Weights that lie within one standard deviation of the mean are at least as large as $\mu - \sigma = 314.8 - 34.33 = 280.47$ and not larger than $\mu + \sigma = 314.8 + 34.33 = 349.13$. As shown in Figure 14, seven of the weights lie within one standard deviation of the mean.

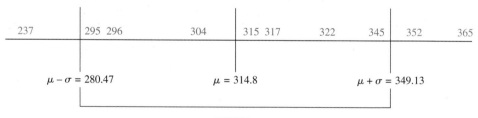

FIGURE 14

Chebychev's Theorem

Even if nothing is known about the distribution of the observations, the famous Chebychev's theorem tells us something about the *probability* that an observation lies within k units of the mean.

Chebychev's Theorem

For any collection of numerical observations with mean μ and standard deviation σ, the probability that an observation x lies within k standard deviations of the mean μ is at least as large as $1 - (1/k)^2$. In symbols, the theorem states that $P[(\mu - \sigma k) \leq x \leq (\mu + \sigma k)] \leq 1 - (1/k)^2$.

EXAMPLE 3

Using Chebychev's Theorem

Estimate the probability that a car battery will last between 57 and 63 months if it is known that such batteries have a mean life of 60 months and a standard deviation of 2 months.

SOLUTION According to Chebychev's theorem, the probability that the battery will last between 57 and 63 months is *at least* $1 - \left(\frac{2}{3}\right)^2 = 1 - \frac{4}{9} = \frac{5}{9}$. ∎

Estimating Standard Deviation from Grouped Data

If observations have been grouped in a frequency distribution table and the observations are not available, we can estimate both the variance and the standard deviation by assuming that each observation lies at the midpoint of its respective class interval.

EXAMPLE 4

Estimating the Standard Deviation from a Frequency Distribution

A sample of 20 flashlight batteries was tested, and the lifetimes of the batteries were recorded as follows:

Life in Hours	Frequency
8 to 10	4
10 to 12	6
12 to 14	8
14 to 16	2

Estimate the mean and standard deviation for these batteries.

SOLUTION We assume that the lifetime of each battery is located at the midpoint of the interval in which it lies. This implies that four of the batteries last for 9 hours, six of the batteries last for 11 hours, eight of the batteries last for 13 hours, and two of the batteries last for 15 hours. Therefore, the sample mean is

$$\overline{x} = \frac{4(9) + 6(11) + 8(13) + 2(15)}{20} = \frac{236}{20} = 11.8 \text{ hours.}$$

The sample standard deviation is

$$s = \sqrt{\frac{(9 - 11.8)^2 4 + (11 - 11.8)^2 6 + (13 - 11.8)^2 8 + (15 - 11.8)^2 2}{19}}$$

≈ 1.88 hours. ∎

Estimating the Standard Deviation from a Frequency Distribution

An estimate of the sample deviation is

$$s = \sqrt{\frac{\Sigma[(x - \bar{x})^2 f]}{(n - 1)}}$$

and an estimate of the population standard deviation is

$$\sigma = \sqrt{\frac{\Sigma[(x - \mu)^2 f]}{n}},$$

where x represents the midpoint of any class interval, f is the frequency of that class interval, n is the number of observations, $\bar{x}$ is the sample mean, and μ is the population mean.

Finally, we can apply the concept of the standard deviation of a population to discrete random variables, such as those we discussed in Chapter 8. Example 5 demonstrates this application.

EXAMPLE 5

Find the Standard Deviation of a Random Variable

Suppose that you pay $10 to play a game in which you draw one ball from a box containing eight red and three green balls. If the ball is red, you win $5, whereas if the ball is green, you win $22. Consider your earnings to be values of a random variable X. Find your expected earnings and the standard deviation of these values from the expected value.

SOLUTION If the ball that is drawn is red, your earnings are $5 − $10 = −$5. That is, you actually lose $5. If, however, the ball is green, your earnings are $22 − $10 = $12. Therefore, the numbers of interest are −$5 and $12, so they will become the values of our random variable. Accordingly, we construct Table 9.

Value of Random Variable x	Assigned Probability $P(x)$
−$5	$\frac{8}{11}$
+$12	$\frac{3}{11}$

TABLE 9. PROBABILITY DISTRIBUTION TABLE FOR THE GAME IN EXAMPLE 5

The expected value is then

$$E(X) = -\$5\left(\frac{8}{11}\right) + \$12\left(\frac{3}{11}\right) = \$\frac{-40 + 36}{11} = -\$\frac{4}{11} \approx -\$0.36.$$

Table 10 sets out the steps required to find the standard deviation.

Value of Random Variable x	Probability $P(x)$	Deviation $x - E(X)$	Deviation^2 $(x - E(x))^2$	$(x - E(x))^2 P(x)$
−5	8/11	−5 − (−0.36) = −4.64	21.53	$(21.53)(8/11) \approx 15.66$
12	3/11	12 − (−0.36) = 12.36	152.77	$(152.77)(3/11) \approx 41.67$
				Variance: 57.33

TABLE 10. CALCULATING THE VARIANCE OF A RANDOM VARIABLE

Therefore, the standard deviation is $\sigma(X) = \sqrt{57.33} \approx \$7.57$. ∎

Abuses of Statistics

Consider High and Low Data Points When thinking about accepting a job offer with a firm in Hartsfield, you might be impressed with the fact that the city claims that its unemployment rate over the past 10 years has averaged only 4%. You might conclude that a very stable job climate exists there. However, the city managers did not tell you that these rates ranged from a low 2% (for several earlier years) to a high of 14% (in recent years). Extremes do matter, and the timing of the extremes might also matter!

Suppose that you take a standardized college entrance exam and score 146. The average score for students entering Tech University is 135, while students entering State University have an average of 140. Should you conclude that you have a better chance of getting into Tech University? No! No one told you that the standard deviation for such scores at Tech University is 12 and at State University is 2.5. Your score is exceptionally good compared with the scores of other students at State University, but it is within just one standard deviation of the mean (average) compared with the scores of other students at Tech University.

CAUTION The standard deviation and extremes in the data do matter. Ask questions!

Exercises 10.3

In Exercises 1 through 6,
(a) Calculate the appropriate mean, $\bar{x}$ or μ
(b) Calculate the appropriate deviation, $x - \bar{x}$ or $x - \mu$, for each observation
(c) Sum the deviations
(d) Sum the deviations squared
(e) Find the appropriate variance (sample or population)
(f) Find the appropriate standard deviation (sample or population)

1. The sample: 2, 3, 4, 5, 6

2. The population: 2, 3, 4, 5, 6

3. The population: 12, 18, 20, 30, 32, 33, 80, 82, 84, 86

4. The sample: 12, 18, 20, 30, 32, 33, 80, 82, 84, 86

5. The sample: 50, 51, 52, 52.5, 53, 53.5, 54, 54.5, 55, 55.5

6. The population: 50, 51, 52, 52.5, 53, 53.5, 54, 54.5, 55, 55.5

In Exercises 7 through 10, decide, by inspection, which set of data has the largest standard deviation.

7. (a) 5, 7, 9, 10, 11, 12
   (b) 5, 7, 15, 18, 18, 20

8. (a) 50, 52, 48, 50, 51, 50, 53, 50, 49, 49
   (b) 80, 81, 80, 86, 80, 87, 80, 90, 79, 90

9. (a) 23, 25, 50, 25, 23, 25, 23, 25

   (b) 46, 46.5, 46, 46.5, 46, 45, 46, 46

10. (a) 101, 120, 102, 130, 102, 140, 102
   (b) 800, 801, 802, 803, 804, 800, 801, 802, 803, 804

11. Find the mean and standard deviation of these hourly wages, considered as the population:
   (a) 8.60, 12.30, 10.42, 9.60, 7.08
   (b) 4.30, 5.20, 9.60, 12.10, 16.80

12. Find the mean and standard deviation of these annual salaries, considered as the population:
   (a) 20,000, 20,000, 25,000, 25,000
   (b) 20,000, 25,000, 12,500, 32,000

13. **Insurance Claims** The Somers Insurance Company took a random sample of 10 auto claims in Washington, DC. The amounts were $1690.00, $823.68, $1261.18, $687.25, $972.40, $1391.52, $1560.20, $2580.12, $876.12, and $920.63.
   (a) Find the mean and standard deviation of the claims.
   (b) How many claims are within one standard deviation of the mean?

14. **Production** A machine produces what is called a "3-inch" bolt. From a day's production run, five bolts are selected and measured for length. The results, in inches, are 3.01, 2.97, 3.10, 3.11, and 2.98.

(a) Find the mean and standard deviation of these lengths.

(b) How many lengths are within one standard deviation of the mean?

15. **Exam Scores** Eight graduate students took their written examination for their master's degree, with these composite scores: 88, 72, 91, 65, 73, 79, 82, 85.

(a) Find the population mean and standard deviation of the scores.

(b) How many scores lie within one standard deviation of the mean?

16. **Demographics** Seven students were awarded out-of-state scholarships by the History Department of State University. The distance, in miles, from the hometowns of the students to the university were 250, 875, 1060, 780, 300, 1200, and 920.

(a) Find the population mean and standard deviation of these mileages.

(b) How many mileages lie within two standard deviations of the mean?

17. **Heart Rate** The heart rates of eight people undertaking a physical fitness program were recorded as 72, 81, 83, 64, 72, 78, 85, and 75, respectively.

(a) Find the population mean and the standard deviation of the heart rates.

(b) How many observations lie within two standard deviations of the mean?

18. **Blood Pressure Survey** The diastolic blood pressures of 10 people about to undertake a supervised diet program are 136, 150, 129, 145, 152, 140, 160, 129, 136, and 142.

(a) Find the population mean and standard deviation of these data.

(b) How many observations lie within three standard deviations of the mean?

19. **Climate** A random sample of five January noontime temperatures in Albany was 42 degrees, 46 degrees, 61 degrees, 37 degrees, and 40 degrees Fahrenheit.

(a) Find the sample mean.

(b) Find the variance. What units are expressed in the variance?

(c) Find the standard deviation. What units are expressed in the standard deviation?

20. **Charitable Donations** A wealthy woman donated a bank account to each of four charities. The amount in each account, to the nearest dollar, was $28,625, $46,228, $37,258, and $63,520.

(a) Find the population mean.

(b) Find the variance. What units are expressed in the variance?

(c) Find the standard deviation. What units are expressed in the standard deviation?

21. **Tourist Visits** The number of tourists visiting the Gold Nugget Saloon during the first five days in July was 236, 345, 362, 425, and 210.

(a) Find the population mean and median of these data.

(b) Find the variance. What units are expressed in the variance?

(c) Find the standard deviation. What units are expressed in the standard deviations?

22. **Driving Speed** A radar check of the speed of a random sample of 15 cars during rush-hour traffic gave these readings: 48, 56, 50, 63, 58, 56, 60, 52, 65, 53, 60, 58, 58, 51, 57.

(a) Find the sample mean and the median.

(b) Find the variance. What units are expressed in the variance?

(c) Find the standard deviation. What units are expressed in the standard deviation?

23. **Plant Sample** A botanist, Leigh Camp, counted the number of blooms on 10 blueberry plants. The results were 86, 70, 65, 81, 72, 88, 80, 82, 81, and 80.

(a) Find the sample mean and median.

(b) Find the variance. What units are expressed in the variance?

(c) Find the standard deviation. What units are expressed in the standard deviation?

24. **Highway Beautification Survey** A group dedicated to beautifying its state's highways counted the number of billboards per mile on 12 stretches of highway. The data looked like this: 4, 8, 3, 4, 3, 5, 20, 21, 6, 3, 5, 5.

(a) Find the sample mean and median.

(b) Find the variance. What units are expressed in the variance?

(c) Find the standard deviation. What units are expressed in the standard deviation?

In Exercises 25 through 30, use Chebychev's theorem to estimate the probability of the outcome described.

25. **Product Life Expectancy** Estimate the probability that a particular brand of dishwasher will last between 6 and 10 years if the average life of such dishwashers is 8 years, with a standard deviation of 1 year.

26. **Product Life Expectancy** Estimate the probability that a particular brand of hair dryer will last between three and four years if the average life of such dryers is 3.5 years, with a standard deviation of three months.

27. **Tire Life Expectancy** Estimate the probability that a tire will last less than 30,000 miles or more than 40,000 miles if the average life span of such tires is 35,000 miles, with a standard deviation of 2500 miles.

28. **Product Labeling** Estimate the probability that a can of beans will weigh less than 15 ounces (oz) or more than 17 oz if such cans have an average weight of 16 oz, with a standard deviation of 1 oz.

29. Estimate that probability that an observation will always lie within two standard deviations of the mean.

30. Estimate the probability that an observation will always lie within three standard deviations of the mean.

31. **Test Scores** Midterm test scores for a class of 100 accounting students were distributed as follows:

Test Scores	Frequency
50 to 60	5
60 to 70	15
70 to 80	50
80 to 90	26
90 to 100, inclusive	4

(a) Estimate the population mean.

(b) Estimate the population standard deviation.

32. **Product Life Expectancy** A sample of 20 lightbulbs was tested for length of life. The results were as follows:

Length of Life (hours)	Frequency
700 to 725	5
725 to 750	6
750 to 775	6
775 to 800	3

(a) Estimate the sample mean.

(b) Estimate the sample standard deviation.

33. **Gas Mileage** Fifty cars of a particular make were checked for gasoline mileage, with these results:

Miles per Gallon	Frequency
22 to 24	6
24 to 26	12
26 to 28	20
28 to 30	10
30 to 32	2

(a) Estimate the sample mean.

(b) Estimate the sample standard deviation.

34. **Height Survey** Twenty people recently were admitted to a new class in the police academy in a large city. Their heights are recorded as shown in the following table:

Heights	Frequency
5'4" to 5'6"	4
5'6" to 5'8"	3
5'8" to 5'10"	6
5'10" to 6'0"	4
6'0" to 6'2"	3

(a) Estimate the population mean.

(b) Estimate the population standard deviation.

35. **Charitable Contributions** The following table contains the frequencies for the amount of charitable contributions per U.S. household in 1998, based on a sample of 2719 households:

Contribution (in dollars)	Frequency
$1 to $100	604
$101 to $200	340
$201 to $300	269
$301 to $400	218
$401 to $500	166
$501 to $600	128
$601 to $700	101
$701 to $999	166
$1000 or more	729

Estimate the sample variance and standard deviation for charitable contributions. (*Source*: U.S. Bureau of the Census, *Statistical Abstract of the United States 2000.*)

36. Fertility Rates The fertility rate is measured as the expected number of children born to a woman of child-bearing age. Fertility rates for 32 African countries in the year 2000 are summarized in the following table:

Fertility Rate (in children per woman)	Number of Countries
1–2	1
2–3	7
3–4	5
4–5	7
5–6	6
6–7	6

Estimate the mean, variance, and standard deviation for fertility rates in these countries. (*Source:* U.S. Bureau of the Census, *Statistical Abstract of the United States 2000.*)

37. Game of Chance A box contains three red and five green balls. Suppose you randomly draw one ball from the box. If the ball is red, you win $5; if the ball is green, you lose $2. Consider your earnings to be values of a random variable. Find your expected earnings and the standard deviation of those values from the expected value.

38. Game of Chance A box contains four red and two green marbles. Suppose you pay $10 to draw one marble from the box. If the marble is red, you win $7; if the marble is green, you win $12. Consider your earnings to be values of a random variable. Find your expected earnings and the standard deviation of those values from the expected value.

39. Insurance A company sells one-year term life insurance policies with a face value of $50,000 to 20-year-olds for $1000. The company mortality tables show that people of that age will live for another year with probability 0.98. Consider the company's earnings per policy to be values of a random variable. Find the company's expected value from the sale of the policies and the standard deviation of the company's earnings from the expected value.

40. Insurance A company offers promotional one-year term life insurance policies with a face value of $100,000 to 40-year-olds for $4000. The company mortality tables show that people of this age will live for another year with probability 0.97. Consider the company's earnings per policy to be values of a random variable. Find the company's expected value from the sale of the policies and the standard deviation of the company's earnings from the expected value.

Section 10.4 Continuous Random Variables and the Normal Distribution

In Chapter 7, we studied random variables that had a finite number of values. These fell under the classification of **discrete random variables,** which may loosely be described as random variables having the property that, between any two consecutive ones, there is an interval of numbers, none of which are themselves random variables. **Continuous random variables,** by contrast, are random variables whose values can be *any* number within a specified interval. The fuel mileage of an automobile, the weight of a newborn infant, and the actual time of arrival of a particular airplane flight are some examples of continuous random variables. Here is how such data are handled: First, the sample mean and sample standard deviation are calculated, to obtain estimates for the population mean and population standard deviation. Then a relative frequency histogram is constructed, and from that, a **probability distribution curve** is drawn to represent the underlying statistical population. (See Figure 15.) Finally, we try to match the shape of the sketched curve with the graph of one of several known **probability density functions,** which are described precisely by mathematical equations. Undoubtedly, the most famous of all probability distributions is the one known as the **normal probability distribution,** or the **bell-shaped curve,** which we discussed earlier in the chapter. This distribution has a symmetrical shape, as shown in Figure 16. The curve used to represent such a probability distribution is called a **normal probability density function,** or just the **normal curve** or the **bell curve.** There is not just one normal curve, but a different one for each value of the mean μ and standard deviation σ of the underlying normally distributed population. The **standard normal distribution,** which has a mean of zero and standard deviation of 1, is used to make inferences about normally distributed populations with other means and standard deviations. The graph of the standard normal distribution curve is shown in Figure 16.

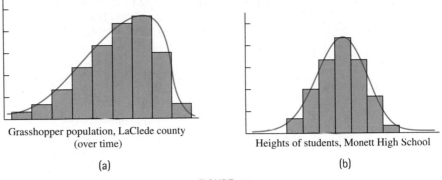

Grasshopper population, LaClede county
(over time)

(a)

Heights of students, Monett High School

(b)

FIGURE 15
Probability Distribution Curves

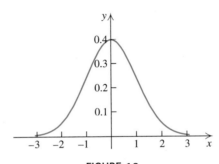

FIGURE 16
The Standard Normal Curve

Even though different normal curves have different graphs, depending on the values of μ and σ, they have many properties in common with the standard normal curve. While methods from calculus are needed to verify them, we outline some of these properties as follows:

1. The highest point on the graph is attained at the mean of the population.
2. The graph is always symmetric about the vertical line through the mean, as shown in Figure 17(a). Furthermore, the graph never touches the x-axis, although it gets infinitesimally close on both ends.

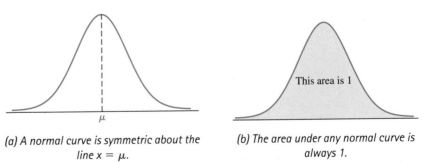

(a) A normal curve is symmetric about the
line $x = \mu$.

(b) The area under any normal curve is
always 1.

FIGURE 17
Symmetry and Area Properties of a Normal Curve

3. The area below the curve and above the x-axis is always equal to 1, as shown in Figure 17(b). This is because the area is associated with probability.

4. The area below the curve and *to the left* of the mean is 0.5; the area below the curve and *to the right* of the mean is also 0.5. This is because the graph is symmetric about the mean and the total area below the curve is equal to 1.

5. As shown in Figure 18, the concavity of the graph of a normal curve changes from *concave up* to *concave down* at the point that is one standard deviation to the left of the mean. The concavity of the graph changes from *concave down* to *concave up* at the point that is one standard deviation to the right of the mean.

For instance, if the annual rainfall in the town of Campbellsville is normally distributed with $\mu = 42$ inches and $\sigma = 3$ inches, then the graph of the normal curve representing the annual rainfalls in this town will look like Figure 19(a). However, if the rainfall in Cheyenne is also normally distributed, but with mean $\mu = 14$ inches and $\sigma = 1$ inch, then the normal curve representing the annual rainfalls in that town would be similar to Figure 19(b).

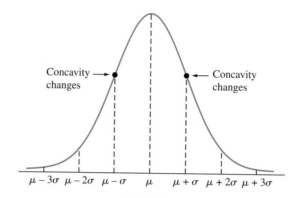

FIGURE 18
Concavity Properties of a Normal Curve

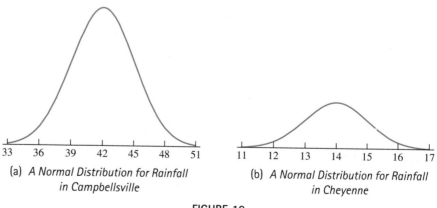

(a) *A Normal Distribution for Rainfall in Campbellsville*

(b) *A Normal Distribution for Rainfall in Cheyenne*

FIGURE 19
Normal Curves

All normally distributed random variables obey the **empirical rule,** which provides good estimates for the area below the curve that falls between certain values. We summarize the **empirical rule** next and represent it graphically in Figure 20.

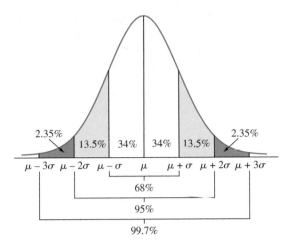

FIGURE 20
Approximate Areas under the Normal Curve

The Empirical Rule

For a normally distributed random variable with mean μ and standard deviation σ,

1. Approximately 68% of the area under the graph of the normal curve lies within *one* standard deviation from the mean.
2. Approximately 95% of the area under the graph of the normal curve lies within *two* standard deviations from the mean.
3. Approximately 99.7% of the area lies within *three* standard deviations from the mean.

For example, if the I.Q. scores for a particular group of students were normally distributed with mean $\mu = 100$ and standard deviation $\sigma = 12$, then about 68% of the students will have I.Q. scores between 88 and 112 ($\mu - \sigma = 88$, $\mu + \sigma = 112$), about 34% of the students will have I.Q. scores between 100 and 112 ($\mu = 100, \mu + \sigma = 112$), and about $\frac{1}{2}(0.3\%) = 0.15\%$ of the students will have I.Q. scores higher than 136 ($\mu + 3\sigma = 136$).

EXAMPLE 1

Counting Standard Deviations

A particular pizza chain requires that, on a large pepperoni pizza, the weight of the pepperoni used be a normally distributed random variable with a mean of 12 oz and a standard deviation of 0.5 oz. A franchise can be revoked if a check reveals that the amount of pepperoni used is below the mean by more than three standard deviations. What is the minimum amount of pepperoni that can be put on such a pizza without violating franchise standards?

SOLUTION Each standard deviation is 0.5 oz, so that three standard deviations will be $3(0.5) = 1.5$ oz. Therefore, $12 - 1.5 = 10.5$ oz is the minimum amount of pepperoni that should be put on such pizzas. Notice that it is *possible* that a "good" store might accidentally use less than 10.5 oz of pepperoni, but this will happen only 0.15% of the time. ∎

EXAMPLE 2

Counting Standard
Deviations

If the gasoline mileage on a particular make and model of automobile is normally distrib-uted with a mean $\mu = 32$ mpg and a standard deviation $\sigma = 2$ mpg, what percentage of these cars gets between 30 and 36 mpg?

SOLUTION Because *each* standard deviation is 2 mpg, it follows that 30 lies one standard devi-ation to the left of the mean $\mu = 32$ and 36 lies two standard deviations to the right of the mean. Using Figure 20, we find that the approximate percentage of area under the normal curve from one standard deviation to the left of the mean to two standard deviations to the right of the mean is $34\% + 34\% + 13.5\% = 81.5\%$. This tells us that approximately 81.5% of such cars will attain a gasoline mileage between 30 and 36 mpg. Stated differently, the approximate proba-bility that such a car will get a gasoline mileage between 30 and 36 mpg is 0.815. ■

The connection between Example 2, Figure 20, and the empirical rule gives us the secret to finding probabilities (percentages) for normally distributed random variables: *A proba-bility will be associated with the area under the normal curve, and the area will depend upon how far away the value of the random variable is from the mean.* Example 2 also points the way toward calculating the number of standard deviations of any observation X from the mean. As Figure 21 indicates, for $x = 30$, $30 = 32 + (-1) \times 2$, which may be written symbolically as $x = \mu + (-1)\sigma$; for $x = 36$, $36 = 32 + (2) \times 2$, which, again, may be expressed as $x = \mu + 2\sigma$.

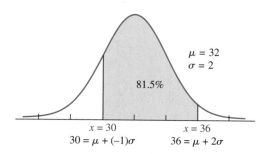

FIGURE 21
Gasoline Mileages

In general, as shown in Figure 22, for any x, $x = \mu + z\sigma$, where z is negative if x lies to the left of the mean, z is positive if x lies to the right of the mean, and $z = 0$ if x is the

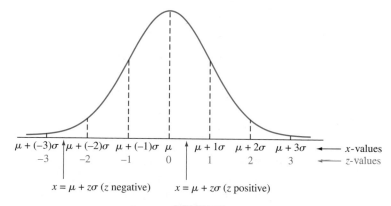

FIGURE 22
The Relationship between x and z

mean. For a given value x, how do we find the corresponding value of z? We write $x = \mu + z\sigma$ and solve for z:

$$z = \frac{x - \mu}{\sigma}.$$

This means that we can convert an area under any normal curve to an equal area under *one* normal curve, called the **standard normal curve,** as shown in Figure 23. This particular random variable is known by the distinguishing letter Z, and its values are called **z-scores.**

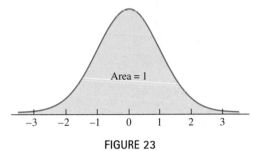

Area = 1

FIGURE 23
The Standard Normal Curve

Transforming to *z*-scores

Let X be a normally distributed random variable with mean μ and standard deviation σ. Let X be one of the values of x. Then the **z-score** for x is given by

$$z = \frac{x - \mu}{\sigma}.$$

A table showing the standard normal probability to four-decimal accuracy (the percentage of area to the *left* of a given z-score) is provided in an appendix of this book.

Spreadsheet packages such as Excel have built-in capabilities for calculating these areas. For instance, to find the area to the left of the z-score 1.35, type $=$ Normsdist(1.35) in any empty cell and press ENTER. The TI-83 Plus also has this capability. The normalcdf function, in the DISTR menu, calculates the area *between* two values of a normally distributed random variable. Figure 24 shows how the calculator displays the area between $z = 1.20$ and $z = 1.45$. (See the *Graphing Calculator Manual* that accompanies this text for more information on how to use this function.)

```
normalcdf(1.20,1
.45,0,1)
             .0415404312
```

FIGURE 24
TI-83 Plus Screen Displaying Probabilities
Associated with a Normal Distribution

EXAMPLE 3

Using the Standard
Normal Curve

The yearly snowfall in Pinnacle is normally distributed with a mean of 53 inches and a standard deviation of 4 inches. What is the probability that snowfall in a given year will be less than 58 inches?

SOLUTION For this example $x = 58$, $\mu = 53$, and $\sigma = 4$. Therefore $z = \frac{58-53}{4} = 1.25$. Thus, the shaded area in Figure 25(a) is the same as that in Figure 25(b). In symbols, $P(x < 58) = P(z < 1.25)$. A table of z-scores gives the shaded area as 0.8944 (to four-decimal accuracy), which is also the answer to our probability question.

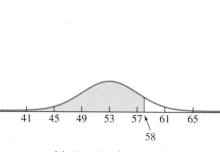

(a) Snowfall (in inches)

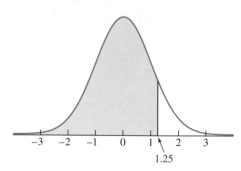

(b) *Standard Normal Curve*

FIGURE 25
Using the Standard Normal Curve ▬

EXAMPLE 4

Using the Standard
Normal Curve

State University found that the GPAs of its freshman students at the end of the fall semester were normally distributed with mean 2.74 and standard deviation 0.3. What percentage of these students had a GPA of 2.5 or more?

SOLUTION For the given data, $x = 2.5$, $\mu = 2.74$, and $\sigma = 0.3$. The z-score is

$$z = \frac{2.5 - 2.74}{0.3} = -0.80.$$

The equal shaded areas shown in Figure 26 give our answer. According to a table of z-scores, the area *to the left of* -0.80 is 0.2119, which means that the desired area *to the right of* -0.80 is $1 - 0.2119 = 0.7881$. It follows that 78.81% of the students have a GPA of 2.5 or more.

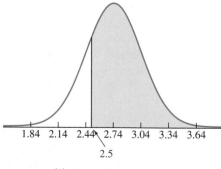

(a) *Grade Point Averages*

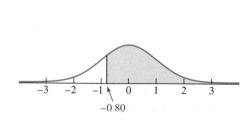

(b) *Standard Normal Curve*

FIGURE 26
Using the Standard Normal Curve ▬

EXAMPLE 5

Using the Standard
Normal Curve

The weights of newborn infants in a particular hospital are normally distributed with a mean of 7.8 lb and a standard deviation of 8 oz. What is the probability that a baby born in this hospital will weigh between 8.0 and 8.5 lb?

SOLUTION As shown in Figure 27, the desired area will be found by subtracting the area to the left of the *leftmost* z-score from the area to the left of the *rightmost* z-score. Noting that 8 ounces is 0.5 lb, or $\sigma = 0.5$, we find the leftmost z-score to be

$$\frac{8.0 - 7.8}{0.5} = 0.40,$$

and, from the table of z-scores, the area to the left of 0.40 is 0.6554. The rightmost z-score is

$$\frac{8.5 - 7.8}{0.5} = 1.40,$$

and the area to its left is 0.9192. The desired area is then $0.9192 - 0.6554 = 0.2638$.

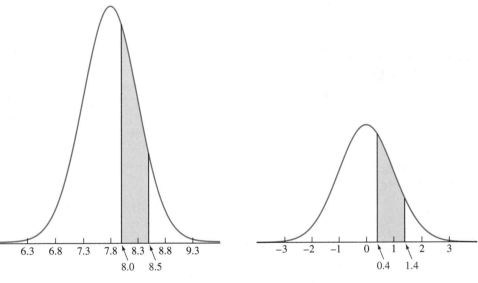

(a) *Weights of Newborn Infants*   (b) *Standard Normal Curve*

FIGURE 27
Using the Standard Normal Curve

The question posed at the outset of this chapter is now answered in Example 6.

EXAMPLE 6

Using the Standard
Normal Curve

An automaker guarantees its particular type of automatic transmission for 60,000 miles. Tests have shown that such transmissions have an average life of 90,000 miles, with a standard deviation of 15,000 miles, as shown in Figure 28. What percentage of cars having these transmissions will be returned to the company for transmission work while the cars are still under warranty?

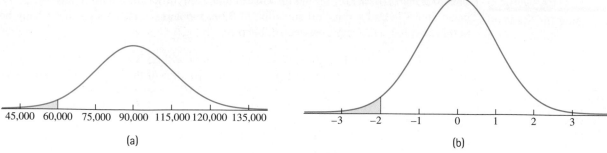

FIGURE 28
Using the Standard Normal Curve

SOLUTION The problem may be restated as asking for the probability that a transmission will last less than $x = 60{,}000$ miles. The z-score for $x = 60{,}000$ is

$$z = \frac{60{,}000 - 90{,}000}{15{,}000} = -2,$$

Thus, the percentage of the area to the left of $x = 60{,}000$ is the same as the area to the left of $z = -2$ under their respective normal curves. The area is 0.0227, so the percentage that will fail to meet the guarantee is 2.27%. ▬

Occasionally, we are faced with the task of finding the z-score such that a given percentage of the area under the standard normal curve is to the right (or left) of that z-score. In such cases, we locate, as closely as possible, the required percentage in a standard normal distribution table and record the corresponding z-score. The remaining examples in this section demonstrate how this is accomplished.

EXAMPLE 7

Locating a Particular z-score

Using the standard normal distribution table in the appendix, find the z-scores so that the following statements hold:

(a) The area to the left of z is 0.8508.

(b) The area to the left of z is 0.90.

(c) The area to the left of z is 0.01.

(d) The area to the right of z is 0.01.

SOLUTION

(a) Searching through the standard normal distribution table, we find that the z-score with an area of 0.8508 to its left is $z = 1.04$.

(b) Our search for an area of 0.90 in our standard normal distribution table fails, but we do find that $z = 1.28$ has an area of 0.8997 to its left and that $z = 1.29$ has an area of 0.9015 to its left. So, the z-score with an area of 0.90 to its left is somewhere between 1.28 and 1.29. We then approximate the desired value with the average of these two z-scores: $\frac{1.28 + 1.29}{2} = 1.285$. We say that $z = 1.285$ has *approximately* an area of 0.90 to its left.

(c) Again, our search for an area of 0.01 fails. We do, however, find that $z = -2.33$ has an area of 0.0099 to its left and $z = -2.32$ has an area of 0.0102 to its left. Thus, we say that $z = -2.325$ the average of these two z-scores, has an area of approximately 0.01 to its left.

(d) Since we now want the area to the *right* of our z-score to be 0.01, we search for a z-score with an area of $1 - 0.01 = 0.99$ to its *left*. Since no z-score in our table has such an area, we approximate the desired z-score by the average of $z = 2.32$ and $z = 2.33$. In other words, $z = 2.325$ has an area of approximately 0.99 to its left, so it also has an area of approximately 0.01 to its *right*. ■

EXAMPLE 8

Finding a Particular
z-score

The automaker in Example 6 decides to change the time limit on its transmission warranty. The company wants to make sure that no more than 1% of all transmissions are returned to the company for work under the warranty. For how many miles should the automaker guarantee its transmissions?

SOLUTION We begin by locating a z-score with 1% of the area to its left. In Example 7(c), we approximated this z-score by $z = -2.325$. However, that value of z does not answer the question posed in this example. We must now convert our z-score back into the proper units for the transmission life. In Example 6, we were told that the transmissions have an average life of 90,000 miles with a standard deviation of 15,000 miles. Since the formula $z = \frac{x - \mu}{\sigma}$ is used to transform a normally distributed random variable X into a standard normal random variable, we rewrite the formula so that we can transform from the standard normal random variable back to the original random variable:

$$z = \frac{x - \mu}{\sigma}$$

$\sigma z = x - \mu$ Multiply both sides by σ.

$\sigma z + \mu = x$ Add μ to both sides.

$x = \sigma z + \mu$ Rearrange terms.

Therefore, if $z = -2.325$, then $x = 15,000(-2.325) + 90,000 = 55,125$. So, if the automaker warrantees its transmissions for 55,125 miles, then approximately 1% of the transmissions will be returned while they are under warranty. ■

Abuses of Statistics

How Normal Are Normal Distributions? As the manager of a Z-Mart store, you purchase a large quantity of children's play watches from a firm that certifies the watches to have a mean life of one year with a standard deviation of one month. After consulting your z-score tables, you decide to guarantee the watches for nine months. A few months later, many irate mothers with children in tow come to your store and demand their money back because the watches have failed. Should you sue the firm you bought the watches from? No. You have *assumed* a normal distribution of the life of the watches when none exists, because a few long-lived watches distorted the mean.

Does your professor grade on a "normal curve"? Should you question whether the class grades are normally distributed? Often, they are not.

CAUTION Do not apply the normal distribution to data, unless there is clear evidence that the data are in fact normally distributed. Ask questions!

Exercises 10.4

In Exercises 1 and 2, determine which normal probability curve has the larger standard deviation.

1.

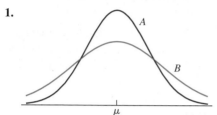

2.

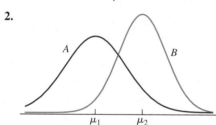

3. At which points on the x-axis does the normal curve with a mean of 20 and a standard deviation of 2 change concavity?

4. At which points on the x-axis does the normal curve with a mean of 60 and a standard deviation of 8 change concavity?

5. Suppose that X is a normally distributed random variable with a mean of 20 and a standard deviation of 1.5. Find the coordinate of the point on the x-axis that lies

  (a) 2 standard deviations above the mean.

  (b) 3 standard deviations below the mean.

  (c) 1.6 standard deviations below the mean.

6. Suppose that X is a normally distributed random variable with a mean of 206.3 and a standard deviation of 16.8. Find the coordinate of the point on the x-axis that lies

  (a) 2 standard deviations below the mean.

  (b) 1.6 standard deviations above the mean.

  (c) 5 standard deviations above the mean.

Exercises 7 through 11 refer to a normally distributed random variable X with a mean of 20 and a standard deviation of 3. Use Figure 20 to approximate the percentage of area under the curve.

7. Between $x = 20$ and $x = 23$

8. Between $x = 17$ and $x = 23$

9. To the left of $x = 20$

10. To the right of $x = 23$

11. To the left of $x = 11$

Exercises 12 through 15 refer to a normally distributed random variable X with a mean of 100 and a standard deviation of 6. Use Figure 20 to find the approximate percentage of area under the normal curve.

12. Between $x = 94$ and $x = 106$

13. To the left of $x = 88$

14. To the right of $x = 112$

15. To the left of $x = 106$

Exercises 16 through 20 refer to the normally distributed random variable Z with a mean of 0 and a standard deviation of 1. Use Figure 20 to find the approximate percentage of area under the standard normal curve.

16. Between $z = 0$ and $z = 1$

17. Between $z = -1$ and $z = 1$

18. To the left of $z = -1$

19. To the right of $z = -1$

20. To the right of $z = 2$

In the remainder of the exercises, use the appendix or appropriate calculator or computer technology to find the required area for a given z-score.

In Exercises 21 through 30, find the requested area under the standard normal curve.

21. The area to the left of $z = 2$

22. The area to the left of $z = 1.6$

23. The area to the left of $z = -2.3$

24. The area to the right of $z = 2.07$

25. The area to the right of $z = 0.03$

26. The area between $z = 1.85$ and $z = 2.05$

27. The area between $z = -2.1$ and $z = 0.67$

28. The area between $z = -2.15$ and $z = -1.79$

29. The area between $z = -6$ and $z = 0$

30. The area between $z = -7$ and $z = 9.3$

In Exercises 31 through 38, X is a continuous normally distributed random variable with mean $\mu = 80$ and standard deviation $\sigma = 5$.

(a) Convert each probability problem involving x into a probability problem involving z.

(b) Find the indicated probability.

31. $P(x > 90)$ **32.** $P(70 < x < 85)$

33. $P(x < 62.8)$ **34.** $P(68 < x < 72)$

35. $P(x < 93.67)$ **36.** $P(83 < x < 93)$

37. $P(x > 76.5)$ **38.** $P(80 < x < 88)$

39. Product Labeling If boxes of cereal have weights that are normally distributed with a mean of 283 grams and a standard deviation of 4 grams, what should be the minimum amount in each box so that the weight will not be below the mean by more than 2 standard deviations?

40. Production The length of what are considered "1-inch" bolts is found to be normally distributed random variable with a mean of 1.001 inches and a standard deviation of 0.002 inch. If a bolt measures more than 2 standard deviations from the mean, it is rejected as not meeting factory tolerances.

  (a) What is the minimum and maximum length a bolt must have to meet factory tolerances?

  (b) What percentage of bolts will be rejected?

41. Exam Scores Crystal scored 193 points on Professor Schein's final exam, for which the scores were normally distributed with a mean of 180 points and a standard deviation of 20 points. Meanwhile, Xiou scored 182 points on the same exam in Professor Feldman's section of the same class, for which the scores were normally distributed with a mean of 165 points and a standard deviation of 10 points. If both professors grade on a normal curve, who gets the higher grade, Crystal or Xiou? (Those within one standard deviation of the mean get C's, then the B's and D's, and then the A's and F's.)

42. Exam Scores Jorge scored 630 points on a particular section of a CPA examination for which the scores were normally distributed with a mean of 600 points and a standard deviation of 70 points. At a different time, Guy scored 530 points on the same section, for which the scores were normally distributed with a mean of 500 points and a standard deviation of 25 points. If both apply to the Weick Accounting Firm for a job, who has the better chance of being hired on the basis of his excellence in relation to the norms?

43. Car Battery Life Expectancy Tests of a particular brand of car battery reveal that the length of its life is a normally distributed random variable X with a mean of 52 months and a standard deviation of 3 months.

  (a) What percentage of the batteries can be expected to last more than 55 months?

  (b) What percentage of the batteries can be expected to last less than 46 months?

  (c) If a store sells 200 batteries, how many can be expected to last less than 46 months?

  (d) If the batteries are guaranteed for 48 months, what percentage of them can the company expect to have returned under warranty?

44. Product Labeling The weights of cans of corn were found to be a normally distributed random variable X with a mean of 14 oz and a standard deviation of 1 oz.

  (a) What percentage of the cans weigh less than 13 oz?

  (b) What percentage of the cans weigh more than 16 oz?

  (c) What percentage of the cans weight more than 20 oz?

  (d) What percentage of the cans weigh between 12 oz and 16 oz?

  (e) If the store has 680 cans of corn, how many will weigh more than 16 oz?

45. Product Life Expectancy Lightbulbs were tested for length of life, which was found to be a normally distributed random variable X with a mean of 750 hours and a standard deviation of 25 hours. A sample of 10,000 such lightbulbs is selected.

  (a) How many of the bulbs can be expected to last at least 775 hours?

  (b) How many of the bulbs can be expected to last less than 700 hours?

  (c) How many of the bulbs can be expected to last between 700 and 800 hours?

46. Production Production samples from a bolt factory indicate that a certain type of bolt has lengths that are a normally distributed random variable with a mean of 8 cm and a standard deviation of 0.06 cm. From a day's production of 2000 such bolts, how many would have lengths

(a) Greater than 8.1 cm?

(b) Less than 7.5 cm?

(c) Between 7.9 and 8.2 cm?

47. **Complaints** Bernice Foust, a mathematics department secretary, has a large number of students come by her office each semester to complain about their professors. She decided to document the time she spent listening to such complaints. After recording the times she spent with several students, she found them to be normally distributed with a mean of 4 minutes, 30 seconds, and a standard deviation of 1 minutes, 45 seconds. Based on the analysis of her sample,

(a) What percentage of the students spend more than 5 minutes complaining?

(b) What is the probability that a student will spend less than 2 minutes complaining?

48. **Carrot Lengths** Ruby measured the lengths of several carrots from her garden and decided, after analysis, that the lengths of her carrots were normally distributed with a mean length of 15.3 cm and a standard deviation of 1.2 cm. What percentage of her carrot crop will

(a) Have lengths between 15.5 and 17 cm?

(b) Have lengths more than 18 cm (to be eligible for the county fair)?

49. **Commuting Time** At the beginning of the semester, Sharokh clocks the time it takes to leave his apartment and arrive in the classroom for his first class. He found the times to be normally distributed with a mean of 18 minutes, 30 seconds, and a standard deviation of 2 minutes, 12 seconds.

(a) What is the probability that Sharokh will require more than 20 minutes to get to his first class?

(b) As the semester goes on, Sharokh has the habit of leaving at 7:15 A.M. for his first class, which is at 7:30 A.M. What percentage of time does Sharokh get to class on time?

50. **Gas Mileage** A government agency checked the gasoline mileage of a particular make of automobile and found the mileages it got to be normally distributed, with a mean of 28.6 mpg and a standard deviation of 2.3 mpg. For one of these automobiles, what is the probability that the mileage will be

(a) At least 30 mpg?

(b) Between 28 and 32 mpg?

51. **Product Labeling** A government agency checked the weights of bags of peanuts, on which were stamped "net weight 14 oz." The agency found that the weights on the bags that were checked were normally distributed, with a mean of 14.1 oz and a standard deviation of 0.2 oz. Based on this information, what is the probability that a bag of these peanuts

(a) Will weigh at least 14 oz?

(b) Will weigh between 13.8 and 14.5 oz?

52. **Billings** Hunt's department store has 50,000 credit card holders. During the month of January, the average billing to credit card holders (rounded to dollars) was $85, with a standard deviation of $12. Assuming that these billings are normally distributed,

(a) How many billings were for less than $75?

(b) How many billing were for more than $100?

53. **Account Balances** A particular bank has 100,000 checking accounts, which collectively have an average balance of $10,000, with a standard deviation of $2500. Assuming that these account balances are normally distributed,

(a) How many accounts have balances of more than $12,000?

(b) What is the percentage of accounts that have balances of more than $5000?

54. **Product Survey** Benjamin is a connoisseur of chocolate chip cookies. He decided to check the "chocolateness" of his favorite brand by buying 10 packages and counting the number of chocolate chips in every cookie. After analyzing his count, he decided that the number of chocolate chips in each cookie was a normally distributed random variable, with a mean of 6.8 chips and a standard deviation of 1.7 chips. Based on these data, what is the probability that one of the cookies will

(a) Have at least eight chocolate chips in it?

(b) Have fewer than four chocolate chips in it?

55. **Demographic Survey** The ages of faculty members at State University are found to be approximately normally distributed, with a mean of 48 years, 6 months, and a standard deviation of 7 years, 2 months.

(a) The new president institutes an early retirement incentive program for all faculty over 65 years of age. What percentage of the faculty is eligible for this program?

(b) What is the probability that a faculty member is under 40 years of age?

56. **SAT Scores** State University claims that its students' scores on the mathematics portion of the SAT are normally distributed, with a mean of 600 points and a standard deviation of 80 points. Melanie scores 700 on the test, and Brad scores 675.

(a) What is the probability that a student at State University has a score below that of Melanie's?

(b) What is the probability that a student at State University has a score below that of Brad's?

(c) Why are the probabilities in parts (a) and (b) not so radically different?

57. Tire Life Expectancy A tire manufacturer claims that the life of its tires, calculated in miles, is a normally distributed random variable, with a mean of 25,000 miles and a standard deviation of 1600 miles.

(a) What percentage of these tires will last between 23,000 and 28,000 miles?

(b) If the tires are guaranteed for 22,500 miles, what percentage will be returned under warranty?

58. Product Life Expectancy The length of life of trouble-free Bright Screen TV sets is found to be a normally distributed random variable, with a mean of six years, four months, and a standard deviation of six months.

(a) What is the probability that one of these TV sets will give at least eight years of trouble-free service?

(b) If the sets are guaranteed to be trouble free for five years and a store sells 150 of them, how many of the sets do they expect to need warranty work?

59. Cat Weights Suppose the weights of domestic short hair cats is normally distributed with a mean of 10 lbs and a standard deviation of 2 lbs.

(a) What is the probability that a domestic short hair cat weighs between 7 and 13 lbs?

(b) Ender, a domestic short hair cat, weighs 19 lbs. What percentage of all domestic short hair cats weigh as much or more than Ender?

60. Disease Recovery Assume that the number of days required to completely recover from a particular virus is normally distributed with a mean of 4 days and a standard deviation of 1.2 days. Sanjay just became infected with the virus, but only has 3 days of sick leave left. What is the probability that Sanjay will completely recover within 3 days?

In Exercises 61–66, find the requested z-score.

61. The are to the left of z is 0.9370.

62. The area to the left of z is 0.0080.

63. The area to the left of z is 0.1.

64. The area to the left of z is 0.95.

65. The area to the right of z is 0.001.

66. The area to the right of z is 0.05.

67. Product Warranties If the life, in years, of a refrigerator is normally distributed with a mean of 15 years and a standard deviation of 3 years, what should be the warranty period if the company wants less than 4% of the refrigerators to fail while under warranty?

68. Product Warranties If the life span of a tire, calculated in miles, is normally distributed with a mean of 30,000 miles and a standard deviation of 2000 miles, what should be the warranty period if the company wants less than 2% of the tires to fail to last through the warranty period?

69. Product Warranties Residential air conditioners of a particular make have a life span, in years, that is normally distributed with a mean of 10 years and a standard deviation of 2.5 years. For what period of time should the air conditioners be under warranty if the company wants to be sure that less than 5% of them will fail while under warranty?

70. Product Warranties A particular brand of roof shingles has a life span, in years, that is normally distributed with a mean of 20 years and a standard deviation of 2 years. For how long should the company guarantee such shingles if they want to be sure that less than 3% will fail while under warranty?

Chapter 10 Summary

CHAPTER HIGHLIGHTS

The realm of *statistics* concerns the collection, analysis, and presentation of data, usually in numerical form. The set of things from which the data are collected is called the *population*. When it is impractical or impossible to use the entire population, subsets called *samples* are collected and conclusions are drawn about the entire population. Sorting, arranging, or organizing data can be accomplished by using *class intervals*, *frequency-distribution tables*, or *histograms*. Some distributions have shapes that are *normal, skewed to the right,* or *skewed to the left*. *Averages*, which are single numbers representing an entire set, include the *mode*, the *median*, the *arithmetic mean*, and the *weighted mean*. The *standard deviation* indicates the variability of the data around the mean.

Random variables assign numbers to each outcome of an experiment and to their related probability distributions. Normal random variables are analyzed with the use of the *standard normal curve*.

EXERCISES

A sample of five students on campus were asked how much money (to the nearest dollar) they were carrying. The responses were $2, $28, $13, $2, and $57. For this sample, find the

1. Mode.
2. Median.
3. Mean.
4. Standard deviation.

For the standard normal curve, find the area

5. To the left of $z = -1.3$.
6. To the right of $z = -1.9$.
7. Between $z = -2.4$ and $z = 0.6$.

Exercises 8–9: A sample of 20 calculator batteries were tested for length of life. The results were as shown in the following table:

Length of Life (months)	Frequency
6 to 9	2
9 to 12	6
12 to 15	8
15 to 18	3
18 to 21	1

8. Estimate the mean.
9. Estimate the standard deviation.

Exercises 10–12: The weights of bags of Kay's Potato Chips were found to be a normally distributed random variable with a mean of 6 oz and a standard deviation of $\frac{1}{2}$ oz.

10. What percentage of the bags weighs less than $5\frac{1}{2}$ oz?
11. What is the probability that a randomly selected bag weighs between $5\frac{3}{4}$ oz and $6\frac{1}{4}$ oz?
12. If a store has 500 bags of Kay's chips in stock, about how many will weigh more than 7 oz?

Exercises 13–17: The 20 families in the Somerset neighborhood were surveyed regarding how many times they called a professional to make household repairs in the past year. The results were as follows:

3	2	1	0	1	4	1	0	2	0
2	3	1	0	3	0	1	2	1	1

13. Organize the data into a frequency table and a relative frequency table.
14. Construct the frequency and relative frequency histograms for these data.
15. What percentage of the families surveyed called a professional for household repairs more than two times in the past year?
16. Find the mean, median, and mode of this population. Which measure of central tendency best represents this data set? Why?
17. Find the variance and standard deviation of the population.
18. A collection of observations has a range from 10 to 125, with a median of 80.
    (a) Make a sketch of a frequency distribution that could fit this collection of data.
    (b) Construct two data sets, one with $n = 8$ and one with $n = 13$, that fit the given properties.
19. Organize frequency tables using 5, 6, 7, 8, 9, and 10 class intervals for these data:

    3.7, 4.5, 6.7, 8.2, 4.6, 4.5, 5.8, 6.2, 8.1, 6.8, 6.8, 7.3, 4.7, 3.6, 7.7, 6.8, 7.1, 7.1, 6.9, 6.5, 6.4, 5.5, 3.1, 4.5

    From each of the groupings of data, approximate the mean and the standard deviation. Compare your approximations with the actual mean and standard deviation.

CUMULATIVE REVIEW

20. Write the equation of a vertical line that contains the point $(-5, 7)$.

21. Solve the following matrix equation:

$$\begin{bmatrix} 1 & 5 & -1 \\ 1 & 0 & 2 \\ 1 & 1 & 1 \end{bmatrix} \begin{bmatrix} x \\ y \\ z \end{bmatrix} = \begin{bmatrix} 4 \\ 1 \\ 0 \end{bmatrix}.$$

Problems 22 and 23 refer to the following system of inequalities:

$$\begin{aligned} x + \ y &\le 10 \\ -2x + \ y &\le 1 \\ 7x + 3y &\ge 21 \\ -x + 2y &\ge 0 \end{aligned}$$

22. Graph the feasible region determined by the given constraints.

23. Find all the corner points of the feasible region determined by the given constraints.

24. Rosanna plans to retire in 10 years. If she invests $250 per quarter in an annuity that earns 4.25% interest compounded quarterly, how much will the annuity be worth upon her retirement?

Chapter 10   Sample Test Items

Problems 1–4: A survey of the firms in an industrial park revealed the following information about the number of secretaries working there:

Number of Secretaries	Frequency	Relative Frequency
1	2	
3	3	
5	5	

1. Complete the relative frequency (probability) column.

2. What is the mean number of secretaries per firm?

3. What is the median number of secretaries per firm?

4. If a secretary from these firms is selected at random, what is the probability that he or she works for a firm with three or fewer secretaries?

Problems 5–8: A random sample of five students revealed the following number of books they were carrying: 1, 8, 2, 1, 3. For these data, calculate the indicated statistic for Questions 5 through 8:

5. The median

6. The mode

7. The mean

8. The standard deviation

9. The daily salaries of the Lhila Corporation employees are $90, $80, $110, $90, $100, $97, and $70. These salaries have a mean of $91 and a standard deviation of $13.2.

How many of the salaries lie within one standard deviation of the mean?

10. Complete the probability distribution table for this experiment: Three cards are drawn (without replacement) from a standard deck of 52 cards, and the number of hearts is counted. Use the following table:

X	p
0	
1	
2	
3	

11. Summarize these data, using appropriate statistics:

76, 82, 61, 54, 28, 98, 97, 78, 70, 82

12. A business manager is faced with a decision to expand her business or leave things as they are. She feels that, if the business expands, there is a 0.6 probability of making an additional $25,000 profit, while there is a 0.4 probability of losing $12,000, through market saturation. What is the expected profit (or loss)?

13. Suppose that the lifetimes of a particular type of lightbulb are normally distributed with $\mu = 1200$ hours and $\sigma = 160$ hours. Find the probability that a bulb will burn out in less than 1100 hours.

14. Assume that SAT verbal scores for a freshman class at a university are normally distributed with a mean of 520 and a standard deviation of 75. The top 10% of the students are placed in the honors program for English. What is the lowest score for admittance into the honors program?

Internet Access and Annual Income *In the year 2000, U.S. households with Internet access were distributed among various annual household income ranges as shown in the following table (Source: U.S. Bureau of the Census,* Statistical Abstract of the United States 2000):

Annual Household Income	Relative Frequency
Less than $10,000	0.0230
$10,000 to $19,999	0.0382
$20,000 to $29,999	0.0685
$30,000 to $39,999	0.0997
$40,000 to $49,999	0.1063
$50,000 or more	0.6644

Households with Internet Access

Use the table to answer Exercises 15 through 19.

15. Construct the relative frequency histogram for the data shown.

16. Estimate the mean annual income of U.S. households with Internet access in the year 2000.

17. Estimate the median annual income of U.S. households with Internet access in the year 2000.

18. Estimate the standard deviation of the annual income of U.S. households with Internet access in the year 2000.

19. If a household is selected at random from those U.S. households with Internet access in the year 2000, what is the probability that the household had an annual income of at least $20,000?

In-Depth Application

The "Top 25" CEO's

Objectives

In this application, you will apply your knowledge of measures of central tendency, measures of dispersion, and graphical representation of data to the annual salaries, years with company, and ages of the "Top 25" chief executive officers (CEO's) in 2001, according to *Forbes* magazine.

(continued)

Introduction

It seems that people are always interested in ranking individuals or groups on the basis of various factors. During the NCAA basketball season, the "Top 25" ranking of men's college basketball teams is released every Sunday. In the late summer or early fall of each year, various magazines release the "Top 10" new-car models. Even professions are ranked by the *Jobs Rated Almanac* each year. (It is interesting to note that jobs which rely heavily on mathematics, including "statistician," are always rated among the "Top 10.")

All of these "top" rankings are based on sets of numerical data that are somehow averaged and combined into a single number. Each year, *Forbes* magazine ranks the CEO's of major companies in the United States on the basis of such factors as their annual compensation, five-year compensation, years with the company, and age, as well as the value of company stock they own.

The Data

Following are data taken from the *Forbes* website (www.forbes.com, 2001) for the Top 25 CEO's at large U.S. companies in the year 2001.

CEO Rank	Total Compensation (in $1000)	Years with Company	Age	CEO Rank	Total Compensation (in $1000)	Years with Company	Age
1	235,912	17	36	14	72,848	17	59
2	216,183	15	68	15	61,412	18	60
3	164,388	29	62	16	59,246	40	76
4	157,305	10	51	17	59,130	21	51
5	137,447	11	60	18	56,179	38	59
6	103,410	8	59	19	54,130	13	53
7	97,387	4	51	20	53,977	32	55
8	92,246	32	64	21	53,674	8	48
9	90,000	26	47	22	51,208	7	46
10	84,449	11	47	23	51,135	20	63
11	77,062	23	57	24	49,039	21	61
12	76,425	41	65	25	48,194	25	62
13	75,232	24	56				

1. Organize the data regarding "years with company" into class intervals, and then create a frequency histogram for the data.

2. Do the data represented by the histogram appear to be *normal?* Explain.

3. Find the mean and median of "years with company" for the "Top 25" CEO's. Which of these measures of central tendency seems to describe the set of data best? Explain.

4. Find the variance and standard deviation for "years with company."

5. What percentage of the data values fall within one standard deviation from the mean? What percentage fall within two standard deviations from the mean?

6. If a CEO from among the group of 25 is selected at random, what is the probability that the person had been with his or her company for no more than 20 years in the year 2001?

7. Organize the data regarding "total compensation" into class intervals, and then create a frequency histogram for the data.

8. Do the data represented by the histogram appear to be *normal?* Explain.

9. Find the mean and median of "total compensation" for the "Top 25" CEO's. Which of these measures of central tendency seems to describe the set of data best? Explain.

(continued)

10. Find the variance and standard deviation for "total compensation."

11. What percentage of the data values fall within one standard deviation from the mean? What percentage fall within two standard deviations from the mean?

12. If a CEO from among the group of 25 is selected at random, what is the probability that the person has a total compensation in excess of $100 million dollars in the year 2001?

13. Organize the data regarding "age" into class intervals, and then create a frequency histogram for the data.

14. Do the data represented by the histogram appear to be *normal?* Explain.

15. Find the mean and median of the "age" for the "Top 25" CEO's. Which of these measures of central tendency seems to describe the set of data best? Explain.

16. Find the variance and standard deviation for "age."

17. What percentage of the data values fall within one standard deviation from the mean? What percentage fall within two standard deviations from the mean?

18. If a CEO from among the group of 25 is selected at random, what is the probability that the person was over 60 years of age in the year 2001?

11

Markov Chains

A private boarding school for girls has 50 students, one of whom, Svetlana, is infected with a cold virus. Assume that no student is initially immune to the virus. Let S denote the students who are susceptible to the virus and who can catch it from Svetlana; let I represent the students who are infected with the virus; and let R designate the students who have had the virus and are then immune to it. Assume that, after one week, 90% of those students currently susceptible to the virus will still be susceptible and 10% will be infected; all of the students currently infected will be immune; and any student who is immune will stay immune. How many students will be in each state one week later? (See Example 3, Section 11.2.)

CHAPTER PREVIEW

Markov chains provide a connection between the matrix algebra studied in Chapter 3 and the probability theory studied in Chapters 8 and 9 with the goal of answering questions about repeated trials of some types of experiments. Markov chains are used in many types of applications, including concepts from economics, genetics, population biology, and epidemiology. At the beginning of the chapter, the types of experiments that can be modeled with Markov chains will be studied, as will be ways to represent those experiments with matrices and transition diagrams. Finally, the long-term effects of repeated trials will be explored.

Technology Opportunities

This chapter relies heavily upon matrix multiplication, which can be expedited with a graphing calculator or a spreadsheet program. You may wish to review the matrix multiplication capabilities of calculators and spreadsheets discussed in Chapter 3 before beginning the current chapter.

Section   11.1   Markov Chains as Mathematical Models

We have already studied repeated trials of experiments that lead to binomial distributions. In these cases, each trial had only two outcomes and each trial was *independent* of all other trials. We now want to generalize these conditions in order to solve problems involving more than two outcomes, wherein each trial (except for the first) is *dependent* on the outcome of the previous trial. The fact that an outcome depends on the previous outcome suggests that repeated trials can be thought of as a *chain* of events, in the sense that each trial is linked to the one preceding it. The Russian mathematician Andrei Andreyevich Markov first studied such chains; hence, the name "Markov chain" is applied to a problem meeting these conditions.

Experiments such as those described in the previous paragraph have wide-ranging applications. For instance, if a flu epidemic infects a small community, the members of the community can be divided into three mutually exclusive events or **states**: susceptible to the virus (S), infected with the virus (I), and immune to the virus and so removed from the process (R). On any given day, we can assume that a person's state depends only upon his or her state on the previous day. One possible chain of events might be $S \rightarrow S \rightarrow I \rightarrow R$, indicating that, in a sequence of four days, a person was susceptible to being infected on the first two days, was infected on the third day, and was removed (perhaps via quarantine) on the fourth day. Among the many questions that can arise from such a situation are "What percentage of the population can be expected to avoid the virus?" and "How long can we expect the epidemic to last?" These are the types of questions that can be answered relatively quickly with the methods we will study in this chapter.

Conditions for a Markov Chain

1. The experiment is repeated a finite number of times. (We do not consider cases of infinite repetitions in this text.)
2. Each trial has the *same sample space.* The sample space consists of a finite number of outcomes, called **states,** and the probabilities assigned to them are called **transition probabilities.**
3. On any trial except the first, the probability that a given state is entered depends only on the state previously occupied. This condition is known as the **Markov property.**

Other applications of Markov chains arise in the study of economics, financial planning, genetics, the weather, and population dynamics. In this section, we will demonstrate ways of organizing and visualizing the information involved in a situation that can be modeled with a Markov chain.

HISTORICAL NOTE

ANDREI ANDREYEVICH MARKOV (1856–1922)

Andrei Andreyevich Markov (mär-käf) was a Russian mathematician whose initial work led to the development of a field that is today called "stochastic processes." One type of stochastic process (a probabilistic process that takes place in steps) is known as a *Markov chain* in honor of his early work.

EXAMPLE 1

Tree Diagrams for Markov Chains

The Aurora Planning Commission found that, each year, 92% of those commuting to work by mass transportation continued to do so the next year, while 8% switched to private auto. Of those commuting by private auto, 95% continued to do so the next year, while 5% switched to mass transportation. Construct a tree diagram that represents these transitions.

SOLUTION The sample space, which contains all the people in Aurora who commute to work, can be divided into two mutually exclusive events or **states:** commuting by mass transportation (M) and commuting by private auto (A). In a given year, a commuter who was in state M the previous year may move to state A or remain in state M. Similarly, a commuter who was in state A the previous year may move to state M or remain in state A. Thus, the probability of entering a given state in a particular year depends only upon the state occupied the previous year. Table 1 describes the fixed **transition probabilities** for these states.

As with the conditional probability problems we discussed in Chapter 9, a tree diagram will visually portray these probabilities and emphasize that the probability of being in one state next year is conditional upon being in the current state, as shown in Figure 1.

Event	Probability
M to A	0.08
M to M	0.92
A to M	0.05
A to A	0.95

TABLE 1. PROBABILITY
DISTRIBUTION TABLE FOR
EXAMPLE 1

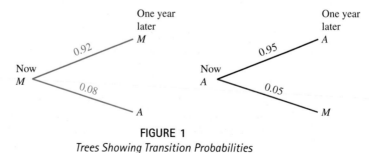

FIGURE 1
Trees Showing Transition Probabilities

Another way to visualize a Markov chain is via a **transition diagram,** which shows the transition probabilities from one outcome (the **current state**) to the following outcome (the **next state**). Transition diagrams are also helpful for displaying the "movement" from one state to the next. Example 2 demonstrates how to construct a transition diagram.

EXAMPLE 2

Transition Diagrams for Markov Chains

Construct a transition diagram for the commuters in Example 1.

SOLUTION We begin by drawing circles, much like a Venn diagram, to represent the mutually exclusive states (events); in this case, we draw one circle to represent being in state M and another to represent being in state A. Since commuters who travel by mass transit this year may continue to do so next year, we draw an arrow from M back to M. Because these same commuters may switch to commuting by private auto next year, we draw another arrow from M to A. Similar arrows are drawn for the transitions from A to M and from A to A. Finally, we label each arrow with its corresponding transition probability, as determined in Example 1. The final result is shown in Figure 2.

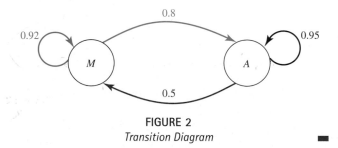

FIGURE 2
Transition Diagram   ■

The information contained in a tree diagram or transition diagram for a Markov chain can also be organized into a matrix, called a **transition matrix.** The rows and columns of a transition matrix represent each of the mutually exclusive states, and the entries in the matrix represent the corresponding transition probabilities. For instance, the entry in the first row and second column of the transition matrix

$$T = \begin{array}{c} \\ A \\ B \end{array} \begin{array}{cc} A & B \\ \begin{bmatrix} 0.2 & 0.8 \\ 0.7 & 0.3 \end{bmatrix} \end{array}$$

represents the probability of moving from state A to state B. In general, an entry in a transition matrix occupies the row corresponding to its **current state** and the column corresponding to its **next state.** This configuration is demonstrated further in Example 3.

EXAMPLE 3

Constructing a Transition Matrix for a Markov Chain

Construct the transition matrix for the Markov chain described in Example 1.

SOLUTION Using the probabilities summarized in Table 1, we construct the transition matrix T for the commuters in Aurora:

$$T = \begin{array}{c} \\ M \\ A \end{array} \begin{array}{cc} M & A \\ \begin{bmatrix} 0.92 & 0.08 \\ 0.05 & 0.95 \end{bmatrix} \end{array}.$$ ■

As we will see in the next several examples, this transition matrix and matrix multiplication can be used to provide a great deal of insight into problems that can be modeled with Markov chains.

EXAMPLE 4
Calculating Future
Probabilities

Consider the commuters studied by the Aurora Planning Commission, as described in Example 1.

(a) Find the probability that a person who uses mass transportation now will still be using mass transportation two years from now.

SOLUTION A tree diagram such as that shown in Figure 3 will help us analyze this problem.

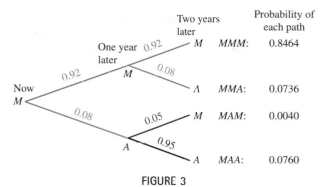

FIGURE 3
Two-Stage Tree Diagram Beginning with State M

From the tree diagram, we can see that there are two mutually exclusive paths in the tree that begin with state *M* and end with state *M* two stages (years) later: *MMM* and *MAM*. Recall from Chapter 9 that the probabilities of these events are calculated by using the multiplication rules for probability. In particular,

$$P(MMM) = P(\text{2nd is } M \,|\, \text{1st is } M) \times P(\text{3rd is } M \,|\, \text{2nd is } M)$$
$$= 0.92 \times 0.92$$
$$= 0.8464.$$

Similarly,

$$P(MAM) = P(\text{2nd is } A \,|\, \text{1st is } M) \times P(\text{3rd is } M \,|\, \text{2nd is } A)$$
$$= 0.08 \times 0.05$$
$$= 0.0040.$$

Therefore, the probability that we start with *M* and end with *M* two stages later is

$$P(MMM \text{ or } MAM) = P(MMM) + P(MAM) = 0.8464 + 0.0040 = 0.8504.$$

(b) What is the probability that a person who now uses mass transportation will be commuting by private auto in two years?

SOLUTION The tree diagram in Figure 3 again gives us the answer:

$$P(MMA \text{ or } MAA) = P(MMA) + P(MAA) = 0.0736 + 0.0760 = 0.1496.$$

(c) What is the probability that a person who now commutes by private auto will be using mass transportation in two years?

SOLUTION A tree diagram beginning with state A as the current state, as shown in Figure 4, will help solve this problem.

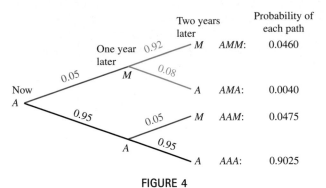

FIGURE 4
Two-Stage Tree Diagram Beginning with State A

Two paths begin with A and end with M two stages later: AMM and AAM. Consequently,

$$P(AMM \text{ or } AAM) = P(AMM) + P(AAM) = 0.0460 + 0.0475 = 0.0935.$$

(d) What is the probability that a person using a private auto now will still be using a private auto in two years?

SOLUTION Again, from Figure 4, we see that two paths begin with A and end with A. Thus, the probability that a person using a private auto now will still be using a private auto in two years is

$$P(AMA \text{ or } AAA) = P(AMA) + P(AAA) = 0.0040 + 0.9025 = 0.9065. \quad \blacksquare$$

Example 4 demonstrates that a tree diagram can be quite helpful in calculating future probabilities for a Markov chain. However, using a tree diagram to calculate probabilities for more than two stages in the future can be quite cumbersome. Let us now turn our attention back to the transition matrix and see how it can aid us in calculating the same probabilities.

Recall from Example 3 that the transition matrix for the Aurora commuters is

$$T = \begin{matrix} & \begin{matrix} M & A \end{matrix} \\ \begin{matrix} M \\ A \end{matrix} & \begin{bmatrix} 0.92 & 0.08 \\ 0.05 & 0.95 \end{bmatrix} \end{matrix}.$$

Let us calculate T^2 and see what connections that matrix will have with the probabilities calculated in Example 4:

$$T^2 = \begin{bmatrix} 0.92 & 0.08 \\ 0.05 & 0.95 \end{bmatrix}\begin{bmatrix} 0.92 & 0.08 \\ 0.05 & 0.95 \end{bmatrix}$$

$$= \begin{bmatrix} 0.92 \times 0.92 + 0.08 \times 0.05 & 0.92 \times 0.08 + 0.08 \times 0.95 \\ 0.05 \times 0.92 + 0.95 \times 0.05 & 0.05 \times 0.08 + 0.95 \times 0.95 \end{bmatrix}$$

$$= \begin{bmatrix} 0.8504 & 0.1496 \\ 0.0935 & 0.9065 \end{bmatrix}.$$

Notice that the entry in the first row and first column of the matrix is precisely the probability we calculated in Example 4(a): $P(MMM$ or $MAM)$. Also, the entry in the first row, second column, is precisely $P(MMA$ or $MAA)$. In fact, each entry in the matrix resulted from the same calculations used in Example 4. In general, T^2 describes the probabilities two stages later:

Two Stages Later

$$T^2 = \textit{Current Stage} \begin{array}{c} \\ M \\ A \end{array} \begin{array}{cc} M & A \\ \begin{bmatrix} 0.8504 & 0.1496 \\ 0.0935 & 0.9065 \end{bmatrix} \end{array}.$$

If T is cubed, we get

Three Stages Later

$$T^3 = \textit{Current Stage} \begin{array}{c} \\ M \\ A \end{array} \begin{array}{cc} M & A \\ \begin{bmatrix} 0.7900 & 0.2100 \\ 0.1310 & 0.8690 \end{bmatrix} \end{array}.$$

In general, suppose there are k states: $S_1, S_2, \ldots, S_k$. If we start in one of these states, say, S_i, then, after n transitions, we will be in state S_j with a probability equal to the number that lies in the ith row and jth column of T^n.

The Powers of a Transition Matrix

Let T^n be the nth power of a transition matrix T. If the process (or chain) starts in state S_i, then, after n repetitions of the experiment (transitions), it will be in state S_j with a probability equal to the number in the ith row and jth column of T^n.

For example, the entry 0.2100 in the first-row, second-column position $(i = 1, j = 2)$ of T^3 tells us that if we start in state $S_i = S_1 = M$, then, after three transitions, we will be in a state $S_j = S_2 = A$ with a probability 0.2100.

EXAMPLE 5

Brand Loyalty

A marketing researcher studied people's loyalty to three brands of popular laundry detergents. She found that 97% of the people who brought brand B still bought that brand a year later, while 2% had switched to brand C and 1% had switched to brand D. Of those who bought brand C, 95% still bought that brand after a year, while 2% had switched to B and 3% had switched to D. Of those who bought D, 93% still bought that brand a year later, while 4% had switched to B and 3% had switched to C.

(a)   Find the transition matrix for this survey.

SOLUTION We have

$$
T = \begin{array}{c} \\ B \\ C \\ D \end{array}
\begin{array}{ccc} B & C & D \\ \left[\begin{array}{ccc} 0.97 & 0.02 & 0.01 \\ 0.02 & 0.95 & 0.03 \\ 0.04 & 0.03 & 0.93 \end{array}\right]. \end{array}
$$

(b)   What percentage of those buying brand B now will be buying brand D two years from now?

SOLUTION We first find T^2, keeping the labels as before:

$$
T^2 = \begin{array}{c} \\ B \\ C \\ D \end{array}
\begin{array}{ccc} B & C & D \\ \left[\begin{array}{ccc} 0.917 & 0.0387 & 0.0196 \\ 0.0396 & 0.9038 & 0.0566 \\ 0.0766 & 0.0572 & 0.8662 \end{array}\right]. \end{array}
$$

The answer to the question is located in the first-row, third-column position of T^2: 0.0196, or about 2%. ∎

With some Markov chains, it is impossible to pass from one particular state to another or even from a given state back to that state. The transition diagram shown in Figure 5 indicates that it is impossible to move directly from state S_2 back to itself or from S_3 directly into S_1 or S_2. The probabilities assigned to these events would, of course, be 0, as demonstrated in the next example.

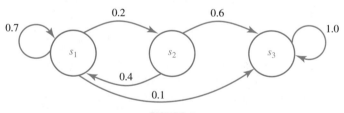

FIGURE 5
Transition Diagram

EXAMPLE 6
More on the
Transition Matrix

Construct a transition matrix for the transition diagram in Figure 5.

SOLUTION The transition matrix for such a Markov chain is

$$
T = \begin{array}{c} \\ S_1 \\ S_2 \\ S_3 \end{array}
\begin{array}{ccc} S_1 & S_2 & S_3 \\ \left[\begin{array}{ccc} 0.7 & 0.2 & 0.1 \\ 0.4 & 0 & 0.6 \\ 0 & 0 & 1 \end{array}\right]. \end{array} \quad ∎
$$

EXAMPLE 7
More about the
Transition Matrix

Construct the transition diagram for the transition matrix

$$T = \begin{array}{c} \\ S_1 \\ S_2 \end{array} \begin{array}{cc} S_1 & S_2 \\ \begin{bmatrix} 0.3 & 0.7 \\ 0 & 1 \end{bmatrix} \end{array}.$$

SOLUTION The transition matrix T shows that it is possible to pass from state S_1 to state S_2 and from state S_2 to state S_2, but, once in state S_2, it is impossible to leave that state. The transition diagram is shown in Figure 6.

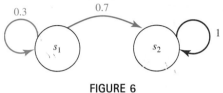

FIGURE 6
The Transition Diagram   ▬

States such as S_3 in Example 6 or S_2 in Example 7 are called **absorbing states.**

Exercises 11.1

In Exercises 1 through 6, find the missing probabilities between the various states shown, and construct the transition matrix.

1.

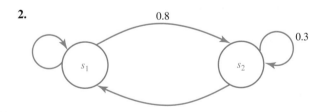

2.

3.

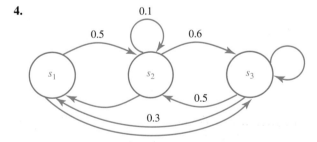

4.

5.

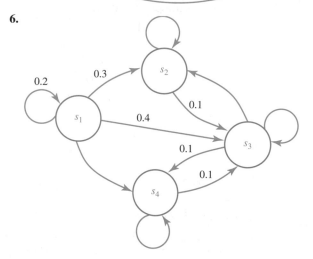

6.

In Exercises 7 through 10, construct the transition diagram for each of the given matrices.

7. $\begin{bmatrix} 0.2 & 0.8 \\ 0.6 & 0.4 \end{bmatrix}$ **8.** $\begin{bmatrix} 0 & 1 \\ 0.4 & 0.6 \end{bmatrix}$

9. $\begin{bmatrix} 0.2 & 0.4 & 0.4 \\ 0 & 0.3 & 0.7 \\ 0.9 & 0 & 0.1 \end{bmatrix}$ **10.** $\begin{bmatrix} 1 & 0 & 0 \\ 0.4 & 0.6 & 0 \\ 0.02 & 0.8 & 0.18 \end{bmatrix}$

In Exercises 11 through 15, construct the transition diagram and the transition matrix that represent the given data.

11. Customer Loyalty Eighty percent of the people who have bought a foreign make of automobile buy one on their next purchase of an automobile, while 20% switch to an American make. Ninety percent of the people who have bought an American make buy one on their next purchase of an automobile, while 10% switch to a foreign make.

12. Weather Forecasting If it rains today, there is a 60% chance of rain tomorrow; however, if it does not rain today, there is an 80% chance that it will not rain tomorrow.

13. Genetic Traits Sixty percent of the people in one generation who have a particular psychological characteristic will pass that characteristic on to the next generation. Fifty percent of the people in one generation who do not have the characteristic will pass it on to the next generation.

14. Voter Trends Eighty-five percent of those who voted Democratic in the last election will vote Democratic in the next election, 10% will vote Republican, and 5% will vote Independent. Ninety percent of those who voted Republican in the last election will do so in the next election, while 7% will vote Democratic, and 3% will vote Independent. Fifty-eight percent of those who voted Independent in the last election will do so in the next election. 26% will vote Democratic, and 16% will vote Republican.

15. Newspaper Subscriptions Ninety percent of the subscribers to the *Evening Star* renew their subscriptions each year, while 6% change to the *Morning Tribune* and 4% change to the *Gazette.* Seventy percent of the subscribers to the *Morning Tribune* renew their subscription each year, while 20% change to the *Evening Star* and 10% change to the *Gazette.* Seventy-five percent of the subscribers to the *Gazette* renew their subscriptions each year, while 16% change to the *Evening Star* and 9% change to the *Morning Tribune.*

In Exercises 16 through 20, decide which of the matrices are transition matrices. For those which are not, tell why they are not.

16. $\begin{bmatrix} 0.6 & 0.4 \\ 0.3 & 0.7 \end{bmatrix}$ **17.** $\begin{bmatrix} 0.1 & 0.9 \\ 0.3 & 0.8 \end{bmatrix}$

18. $\begin{bmatrix} 0.05 & 0.95 \\ 0.85 & 0.15 \end{bmatrix}$ **19.** $\begin{bmatrix} 0.2 & 0.4 & 0.4 \\ 0 & 0.8 & 0.2 \\ 0.5 & 0.3 & 0.2 \end{bmatrix}$

20. $\begin{bmatrix} 0.4 & 0 & 0.6 \\ 0 & 0.9 & 0.1 \\ 0.5 & 0.3 & 0.4 \end{bmatrix}$

21. Customer Preferences Suppose that people who own NumberKrunch computers for home use will purchase another NumberKrunch with a probability of 0.80 and will switch to a QuickDigit computer with a probability of 0.20. Those who own a QuickDigit will purchase another with a probability of 0.60 and will switch to a NumberKrunch with a probability of 0.40.

(a) Construct the transition matrix for this two-state Markov chain.

(b) Use a tree diagram to find the probability that a person now using a NumberKrunch computer will have switched to a QuickDigit computer two purchases later.

(c) Find the probability asked for in Exercise 21(b) by using matrix multiplication.

(d) Find the probability that if a person now has a NumberKrunch computer, two computer purchases later he or she will also buy a NumberKrunch computer.

22. Disease Modeling For simple diseases, people can be categorized as being in three states, *Susceptible (S)*, *Infected (I)*, and *Removed (R)*. Those people classified as susceptible to the disease do not have it and are not immune to it. Those who are classified as infected have the disease and can transmit it to others. Those people classified as removed have had the disease, cannot transmit it any longer, and are now immune to it. A Markov chain model, called an SIR model, can be used to describe the effect a disease has on a population. Assume that the transition matrix

$$T = \begin{matrix} & \begin{matrix} S & I & R \end{matrix} \\ \begin{matrix} S \\ I \\ R \end{matrix} & \begin{bmatrix} 0.75 & 0.25 & 0 \\ 0 & 0 & 1 \\ 0 & 0 & 1 \end{bmatrix} \end{matrix}$$

represents the movement from state to state, in terms of days, in a small family exposed to a cold virus.

(a) Draw the transition diagram for this transition matrix.

(b) Which states in the SIR model can be classified as *absorbing* states?

23. Product Reliability A space probe contains equipment that will diagnose a misalignment of the probe and correct it as the probe travels through space. If the probe is aligned one minute, then 60% of the time it is still aligned one minute later. On the other hand, if it is misaligned one minute, then 40% of the time it is misaligned the next minute.

(a) Construct the transition diagram and the transition matrix for these data.

(b) Use a tree diagram to find the probability that if the probe is aligned now, it will still be aligned two minutes from now.

(c) Use matrix multiplication to find the probability asked for in part (b).

(d) Find the probability that if the probe is not aligned now, then it still will not be aligned three minutes from now.

24. College Majors A study of freshmen, sophomores, and juniors at State University revealed that, each year, 10% of the engineering students change to liberal arts, 5% change to business, and the rest remain in engineering. Three percent of the liberal arts students change to engineering, 15% change to business, and the rest stay in liberal arts. Five percent of the business students change to engineering, 8% change to liberal arts, and the rest remain in business.

(a) Construct the transition diagram and transition matrix for these data.

(b) What percentage of the students now in engineering will change to business three years from now?

(c) What percentage of the students now in liberal arts will still be in liberal arts three years from now?

25. Customer Preferences The records of an automobile-leasing company give the following information about preferences for two-door sedans (2-d), four-door sedans (4-d), and minivans (MV):

		Next Choice		
		2-d	4-d	MV
Choice	2-d	70%	20%	10%
Now	4-d	5%	90%	5%
	SW	10%	30%	60%

(a) Construct the transition diagram and transition matrix for these data.

(b) Construct a tree diagram to find the probability that a person leasing a two-door sedan now ill be leasing a station wagon three choices later.

(c) Find the probability asked for in part (b) by matrix multiplication.

26. Tennis Matches Gordon and Dennis have a standing tennis match every Saturday. A record of their matches shows that if Gordon wins on any given Saturday, he wins the next Saturday with a probability of 0.60, whereas if Dennis wins, he wins the match the following Saturday with a probability of 0.45. If Gordon wins the match this

Saturday, what is the probability that he will win the match three Saturdays from now?

27.  **Free-throw Percentages** Records indicate that Denita's free-throw shooting can be represented by a Markov chain: If she makes a free throw, she makes the next one 80% of the time and misses 20% of the time. If she misses a free throw, she makes the next one 60% of the time and misses 40% of the time. If Denita makes her first free throw of a game, what is the probability that she will miss the third free throw of that game?

28.  **Cognition Experiment** A mouse is put into a maze that looks like the following diagram:

1	2	3	4

There are two types of transition: moving to an adjacent room and staying in the same room for a given length of time, say, one minute. Assume that the mouse is twice as likely to remain in the same room as to leave, but that if it does leave, it is equally likely to go into any adjacent room. Find the transition matrix for this experiment, and find the probability that if the mouse starts in room 2, it will be in room 4 three transitions from now.

29. For a transition matrix

$$P = \begin{bmatrix} 1 & 0 \\ 0.3 & 0.7 \end{bmatrix}$$

for a Markov chain, explain what the first row means in terms of being able to go from one state to another. Does this property apply to all powers of P?

30. For a transition matrix

$$P = \begin{bmatrix} 1 & 0 & 0 \\ 0 & 1 & 0 \\ 0.3 & 0.4 & 0.3 \end{bmatrix}$$

for a Markov chain, explain what the first two rows mean in terms of being able to go from one state to another. Does the property apply to all powers of P?

Section 11.2 State Vectors

In the last section, we focused our attention on transition probabilities and calculated future probabilities for each state in a Markov chain. However, we never considered how the current sample space might be distributed over the mutually exclusive states. For instance, if the Aurora Planning Commission determined that this year 25% of the commuters in Aurora commute by mass transportation and 75% commute by private auto, it would be interesting to know what percentage of the commuters will commute by mass transportation next year and in subsequent years. To answer such questions, we will utilize an **initial state vector,** which will contain the percentage of the sample space that is currently in each state.

EXAMPLE 1

The State Vector

Recall from Example 1 of Section 11.1 that commuters in Aurora travel either by mass transportation (state M) or by private auto (state A). Suppose the Aurora Planning Commission determines that, currently, 25% of the commuters commute by mass transportation and 75% travel by private auto. Construct the initial state vector P_0.

SOLUTION The initial state vector, representing how the initial sample space is distributed over the mutually exclusive states M and A is given by

$$\begin{array}{cc} M & A \end{array}$$
$$P_0 = [0.75 \quad 0.25]. \quad \blacksquare$$

For a given Markov chain with k states, $S_1, S_2, \ldots, S_k$, where the percentage of the sample space in each state is given by $p_1, p_2, \ldots, p_k$, respectively, the **initial state vector** is

$$P_0 = [p_1 \quad p_2 \quad \cdots \quad p_k],$$

where $p_1 + p_2 + \cdots + p_k = 1$.

Recall from Example 3 of Section 11.1 that the transition matrix for the Aurora commuters is

$$T = \begin{array}{c} M \\ A \end{array}\begin{bmatrix} \begin{array}{cc} M & A \end{array} \\ 0.92 & 0.08 \\ 0.05 & 0.95 \end{bmatrix}.$$

Using P_0 from the previous example, let us now calculate P_0T and determine what the entries in the resulting product represent. First, from Chapter 3, since P_0 has dimension 1×2 and T has dimension 2×2, the product, P_0T, will have dimension 1×2 as well. Now,

$$P_0T = \begin{array}{cc} M & A \end{array} \\ [0.25 \quad 0.75] \begin{array}{c} M \\ A \end{array}\begin{bmatrix} \begin{array}{cc} M & A \end{array} \\ 0.92 & 0.08 \\ 0.05 & 0.95 \end{bmatrix}$$

$$\begin{array}{cc} M\ to\ M\ +\ A\ to\ M & M\ to\ A\ +\ A\ to\ A \end{array}$$
$$= [(0.25)(0.92) + (0.75)(0.05) \quad (0.25)(0.08) + (0.75)(0.95)]$$

$$\begin{array}{cc} Total\ to\ M & Total\ to\ A \end{array}$$
$$= [0.27 \quad 0.73].$$

That is, one year later, 27% of the commuters will use mass transportation and 73% will use private auto. The vector we just obtained is another **state vector,** representing the state after one transition; thus, we will call it P_1.

If we now take the state vector P_1 and multiply on the right by T, the result should show the distribution for the commuters' choice of transportation after two years:

$$P_2 = P_1T = (P_0T)T = P_0T^2 = [0.25 \quad 0.75]\begin{bmatrix} 0.8504 & 0.1496 \\ 0.0935 & 0.9065 \end{bmatrix} \approx [0.2827 \quad 0.7173].$$

In other words, approximately 28.3% will be commuting by mass transportation and 71.7% by private auto.

The preceding calculations show that there are two ways in which the distribution for the choice of transportation of the commuters may be found through n stages with matrix algebra. One way is to multiply the initial state vector by T to create a new state vector for the end of the first year, then multiply this state vector by T to create a new state vector for the end of the second year, etc. After n transitions, the result will be

$$P_0, P_1 = P_0 T, P_2 = P_1 T, P_3 = P_2 T, \ldots,$$
$$P_n = P_{n-1} T. \tag{1}$$

The second, equivalent, way is to use powers of the transition matrix:

$$P_0, P_1 = P_0 T, P_2 = P_1 T = (P_0 T)T = P_0 T^2,$$
$$P_3 = P_2 T = (P_0 T^2)T = P_0 T^3, \ldots, P_n = P_0 T^n. \tag{2}$$

Each of these methods has its own advantages and disadvantages. For instance, with a spreadsheet, the first method is more efficient; with a graphing calculator, the second method is more efficient.

EXAMPLE 2

Brand Loyalty

Errol, a marketing researcher, studied customers' loyalty to three brands of popular laundry detergents: Brite, Calcite, and Delite. The following transition matrix represents the transition probabilities for switching from one brand to another after one year:

$$T = \begin{array}{c} \\ B \\ C \\ D \end{array} \begin{array}{ccc} B & C & D \\ \left[\begin{array}{ccc} 0.97 & 0.02 & 0.01 \\ 0.02 & 0.95 & 0.03 \\ 0.04 & 0.03 & 0.93 \end{array} \right] \end{array}.$$

Errol determined that the market shares among the three brands this year were 38%, 42%, and 20%, respectively.

(a) Construct the initial state vector for the three brands of laundry detergent.

SOLUTION The initial state vector is $P_0 = \begin{bmatrix} 0.38 & 0.42 & 0.20 \end{bmatrix}$.

(b) What will the market shares be one year from now?

SOLUTION To determine the market shares for the three brands after one year, we calculate

$$P_1 = P_0 T = \begin{bmatrix} 0.38 & 0.42 & 0.20 \end{bmatrix} \begin{bmatrix} 0.97 & 0.02 & 0.01 \\ 0.02 & 0.95 & 0.03 \\ 0.04 & 0.03 & 0.93 \end{bmatrix} = \begin{bmatrix} 0.385 & 0.413 & 0.202 \end{bmatrix}.$$

Hence, 38.5% of the households will be buying Brite, 41.3% will be buying Calcite, and 20.2% will be buying Delite in one year.

(c) Calculate T^2 and use it to determine the market shares in two years.

SOLUTION Keeping the same labels as before, we calculate

$$
T^2 = \begin{array}{c} \\ B \\ C \\ D \end{array}
\begin{array}{ccc} B & C & D \\ \left[\begin{array}{ccc} 0.942 & 0.039 & 0.020 \\ 0.040 & 0.904 & 0.057 \\ 0.077 & 0.057 & 0.866 \end{array}\right]. \end{array}
$$

To find the market shares in two years, we can calculate

$$
P_2 = P_0 T^2 = \begin{bmatrix} 0.38 & 0.42 & 0.20 \end{bmatrix} \begin{bmatrix} 0.942 & 0.039 & 0.020 \\ 0.040 & 0.904 & 0.057 \\ 0.077 & 0.057 & 0.866 \end{bmatrix}
$$

$$
= \begin{bmatrix} 0.390 & 0.406 & 0.204 \end{bmatrix}.
$$

Thus, in two years, 39% of the households will be buying Brite, 40.6% will be buying Calcite, and 20.4% will be buying Delite. ∎

Example 3 answers the question posed at the outset of this chapter.

EXAMPLE 3

Model of an Epidemic

A private boarding school for girls has 50 students, one of whom, Svetlana, is infected with a cold virus. Assume that no student is initially immune to the virus. Let S denote the students who are susceptible to the virus and who can catch it from Svetlana; let I represent the students who are infected with the virus; and let R designate the students who have had the virus and are then immune to it. Assume that the transition matrix, from one week to the next, for this cold virus is

$$
T = \begin{array}{c} \\ S \\ I \\ R \end{array}
\begin{array}{ccc} S & I & R \\ \left[\begin{array}{ccc} 0.9 & 0.1 & 0 \\ 0 & 0 & 1 \\ 0 & 0 & 1 \end{array}\right]. \end{array}
$$

How many students will be in each state one week later? Two weeks later?

SOLUTION The initial state vector is $P_0 = \begin{bmatrix} 0.98 & 0.02 & 0 \end{bmatrix}$, since $\frac{49}{50} = 98\%$ of the students are initially susceptible to the virus and $\frac{1}{50} = 2\%$ are initially infected with the virus. To find the state vector after one week, we compute

$$
P_1 = P_0 T = \begin{bmatrix} 0.98 & 0.02 & 0 \end{bmatrix} \begin{bmatrix} 0.9 & 0.1 & 0 \\ 0 & 0 & 1 \\ 0 & 0 & 1 \end{bmatrix} = \begin{bmatrix} 0.882 & 0.098 & 0.02 \end{bmatrix}.
$$

This result indicates that 88.2%, or $0.882 \times 50 = 44.1$ students, will still be susceptible to the disease next week; 9.8%, or $0.098 \times 50 = 4.9$ students, will be infected with the cold virus next week; and 2%, or $0.02 \times 50 = 1$ student, will be immune to the virus next week.

To find the number of students in each state two weeks later, we calculate

$$P_2 = P_0 T^2 = \begin{bmatrix} 0.98 & 0.02 & 0 \end{bmatrix} \begin{bmatrix} 0.81 & 0.09 & 0.1 \\ 0 & 0 & 1 \\ 0 & 0 & 1 \end{bmatrix} = \begin{bmatrix} 0.7938 & 0.0882 & 0.1180 \end{bmatrix}.$$

Thus, in the second week, we anticipate that 79.38%, or $0.7938 \times 50 = 39.69$ students, will still be susceptible to the virus; 8.82%, or $0.0882 \times 50 = 4.41$ students, will be infected with the cold; and 11.8%, or $0.1180 \times 50 = 5.9$ students, will be immune to the virus. ■

Exercises 11.2

 In Exercises 1–10, use the given initial state vector P_0 and transition matrix T to find the desired state vector.

1. Find P_1 if $P_0 = \begin{bmatrix} 0.3 & 0.7 \end{bmatrix}$ and $T = \begin{bmatrix} 1 & 0 \\ 0.2 & 0.8 \end{bmatrix}$.

2. Find P_1 if $P_0 = \begin{bmatrix} 0.4 & 0.6 \end{bmatrix}$ and $T = \begin{bmatrix} 0.3 & 0.7 \\ 0.5 & 0.5 \end{bmatrix}$.

3. Find P_2 if $P_0 = \begin{bmatrix} 0.25 & 0.75 \end{bmatrix}$ and $T = \begin{bmatrix} 1 & 0 \\ 0.2 & 0.8 \end{bmatrix}$.

4. Find P_2 if $P_0 = \begin{bmatrix} 0.65 & 0.35 \end{bmatrix}$ and $T = \begin{bmatrix} 0.3 & 0.7 \\ 0.5 & 0.5 \end{bmatrix}$.

5. Find P_1 if $P_0 = \begin{bmatrix} 0.60 & 0.20 & 0.20 \end{bmatrix}$ and

$$T = \begin{bmatrix} 0.25 & 0.50 & 0.25 \\ 0.30 & 0.30 & 0.40 \\ 0.50 & 0 & 0.50 \end{bmatrix}.$$

6. Find P_1 if $P_0 = \begin{bmatrix} 0.2 & 0.5 & 0.3 \end{bmatrix}$ and

$$T = \begin{bmatrix} 0.6 & 0.2 & 0.2 \\ 0.1 & 0.8 & 0.1 \\ 0.4 & 0.2 & 0.4 \end{bmatrix}.$$

7. Find P_2 if $P_0 = \begin{bmatrix} 0.70 & 0.15 & 0.15 \end{bmatrix}$ and

$$T = \begin{bmatrix} 0.25 & 0.50 & 0.25 \\ 0.30 & 0.30 & 0.40 \\ 0.50 & 0 & 0.50 \end{bmatrix}.$$

8. Find P_2 if $P_0 = \begin{bmatrix} 0.1 & 0.2 & 0.7 \end{bmatrix}$ and

$$T = \begin{bmatrix} 0.6 & 0.2 & 0.2 \\ 0.1 & 0.8 & 0.1 \\ 0.4 & 0.2 & 0.4 \end{bmatrix}.$$

9. Find P_3 if $P_0 = \begin{bmatrix} 0.5 & 0.5 \end{bmatrix}$ and $T = \begin{bmatrix} 1 & 0 \\ 0.2 & 0.8 \end{bmatrix}$.

10. Find P_3 if $P_0 = \begin{bmatrix} 0.4 & 0.4 & 0.2 \end{bmatrix}$ and

$$T = \begin{bmatrix} 0.6 & 0.2 & 0.2 \\ 0.1 & 0.8 & 0.1 \\ 0.4 & 0.2 & 0.4 \end{bmatrix}.$$

11. For the transition matrix $T = \begin{bmatrix} 0.6 & 0.4 \\ 0.2 & 0.8 \end{bmatrix}$ and the state vector $\begin{bmatrix} 0.7 & 0.3 \end{bmatrix}$, find the state vector

   (a) One transition later.

   (b) Two transitions later.

12. For the transition matrix $T = \begin{bmatrix} 0.02 & 0.98 \\ 0.3 & 0.7 \end{bmatrix}$ and the state vector $\begin{bmatrix} 0.40 & 0.60 \end{bmatrix}$, find the state vector

   (a) Two transitions later.

   (b) Three transitions later.

13. For the transition matrix $T = \begin{bmatrix} 0.3 & 0.2 & 0.5 \\ 0.4 & 0 & 0.6 \\ 0.1 & 0.8 & 0.1 \end{bmatrix}$ and the state vector $\begin{bmatrix} 0.1 & 0.4 & 0.5 \end{bmatrix}$, find the state vector

   (a) One transition later.

   (b) Two transitions later.

14. For the transition matrix $T = \begin{bmatrix} 0.4 & 0.1 & 0.5 \\ 0.2 & 0.6 & 0.2 \\ 0.9 & 0.1 & 0 \end{bmatrix}$ and the state vector $\begin{bmatrix} 0.9 & 0 & 0.1 \end{bmatrix}$, find the state vector

   (a) Two transitions later.

   (b) Three transitions later.

15. **Customer Loyalty** Eighty percent of the people who buy a foreign make of automobile buy one on their next purchase of automobile, while 20% switch to an American make. Ninety percent of the people who buy an American make buy one on their next purchase, while 10% switch to a foreign make. Of the people making their first automobile purchase, 40% bought American makes and 60% bought foreign makes. Find the percentage of these people who will buy an American make and the percentage who will buy a foreign make on their third automobile purchase.

16. **Newspaper Subscriptions** In Campbellsville, subscribers to three local newspapers—the *Evening Star* (*E*), the *Morning Tribune* (*T*), and the *Gazette* (*G*)—have the following transition matrix for subscription renewals at the end of a given year:

$$T = \begin{array}{c} \\ E \\ T \\ G \end{array} \begin{array}{ccc} E & T & G \\ \left[\begin{array}{ccc} 0.90 & 0.06 & 0.04 \\ 0.20 & 0.70 & 0.10 \\ 0.16 & 0.09 & 0.75 \end{array} \right] \end{array}.$$

Of the current newspaper subscribers, 30% subscribe to the *Evening Star,* 45% subscribe to the *Morning Tribune,* and 25% subscribe to the *Gazette.* Assuming that no one in Campbellsville ever subscribes to more than one paper, find the percentage that will be subscribing to each newspaper in three years.

17. **College Majors** A study of freshmen and sophomores at State University revealed that, each year, students change their majors from among majors in business (*B*), engineering (*E*), and mathematics (*M*) according to the transition matrix

$$T = \begin{array}{c} \\ B \\ E \\ M \end{array} \begin{array}{ccc} B & E & M \\ \left[\begin{array}{ccc} 0.87 & 0.05 & 0.08 \\ 0.05 & 0.85 & 0.10 \\ 0.15 & 0.03 & 0.82 \end{array} \right] \end{array}.$$

If, currently, 50% of the students in the study are business majors, 40% are engineering majors, and 10% are mathematics majors, how will the percentages of each major be distributed in two years?

18. **Genetic Traits** A particular genetic trait is passed from mothers to daughters. Suppose 75% of the mothers with the trait pass it on to their daughters and 80% of the mothers without the trait do not pass it on to their daughters.

(a) Construct the transition diagram for this genetic trait.

(b) Construct the transition matrix for the genetic trait.

(c) If, currently, the trait is present in 45% of all mothers, what percentage of the population will have the trait three generations later?

19. **Resistance to Antibiotics** A particular strain of bacteria carries a gene that can cause immunity to one of two antibiotics. If a single bacterium is immune to sulfasalazine, then 40% of its offspring will also be immune to sulfasalazine, 20% will be immune to Cipro®, and the rest will not be immune to either antibiotic. If a single bacterium is immune to Cipro®, then 45% of its offspring will be immune to Cipro®, 30% will be immune to sulfasalazine, and the rest will not be immune to either antibiotic. If a single bacterium is not immune to either antibiotic, 5% of its offspring will be immune to sulfasalazine, 8% will be immune to Cipro®, and the rest will not be immune to either antibiotic.

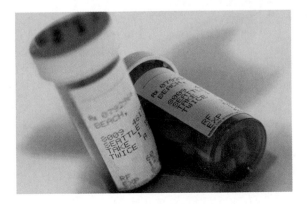

(a) Construct the transition diagram for the resistance of this strain of bacteria to the two antibiotics.

(b) Construct the transition matrix for the resistance of the strain of bacteria to the two antibiotics.

(c) Currently, 20% of this strain of bacteria are immune to sulfasalazine, 30% are immune to Cipro®, and the rest are not immune to either antibiotic. Find the percentage of the bacteria that will be immune to sulfasalazine, the percentage that will be immune to Cipro®, and the percentage that will be immune to neither antibiotic after three generations.

20. **Social Strata** Sociological studies indicate that, from generation to generation, there is some linkage between the social stratum of a parent and that of his or her offspring. Assume that the population is divided into three classes—low (L), middle (M), and high (H)—with the probability of moving from one class to another in the next generation given by the transition matrix

$$T = \begin{array}{c} L \\ M \\ H \end{array} \begin{array}{ccc} L & M & H \\ \begin{bmatrix} 0.65 & 0.23 & 0.12 \\ 0.08 & 0.80 & 0.12 \\ 0.03 & 0.37 & 0.60 \end{bmatrix} \end{array}.$$

(a) Construct the transition diagram for this transition matrix.

(b) If the population is currently divided so that 18% are classified as low class, 70% are classified as middle class, and 12% are classified as high class, how will the population be divided among the classes after four generations?

Since $P_1 = P_0 T$, if T has an inverse, then $P_1 T^{-1} = P_0$. In Exercises 21–26, find T^{-1} of each transition matrix given, and then use the inverse and P_1 to find the initial state vector.

21. $T = \begin{bmatrix} 0.5 & 0.5 \\ 0.3 & 0.7 \end{bmatrix}$; $P_1 = \begin{bmatrix} 0.4 & 0.6 \end{bmatrix}$.

22. $T = \begin{bmatrix} 0.5 & 0.5 \\ 0.3 & 0.7 \end{bmatrix}$; $P_1 = \begin{bmatrix} 0.25 & 0.75 \end{bmatrix}$.

23. $T = \begin{bmatrix} 0.24 & 0.76 \\ 0.91 & 0.09 \end{bmatrix}$; $P_1 = \begin{bmatrix} 0.03 & 0.97 \end{bmatrix}$.

24. $T = \begin{bmatrix} 0.24 & 0.76 \\ 0.91 & 0.09 \end{bmatrix}$; $P_1 = \begin{bmatrix} 0.97 & 0.03 \end{bmatrix}$.

25. $T = \begin{bmatrix} 0.2 & 0.4 & 0.4 \\ 0.9 & 0.1 & 0 \\ 0 & 1 & 0 \end{bmatrix}$; $P_1 = \begin{bmatrix} 0.25 & 0.45 & 0.30 \end{bmatrix}$.

26. $T = \begin{bmatrix} 0.2 & 0.4 & 0.4 \\ 0.9 & 0.1 & 0 \\ 0 & 1 & 0 \end{bmatrix}$; $P_1 = \begin{bmatrix} 0.44 & 0.39 & 0.17 \end{bmatrix}$.

In Exercises 27–30, consider the transition matrix $T = \begin{bmatrix} 0.2 & 0.8 \\ 0.5 & 0.5 \end{bmatrix}$, and find $P_1, P_2, P_3, \ldots, P_{15}$ for each initial state vector.

27. $P_0 = \begin{bmatrix} 1 & 0 \end{bmatrix}$ **28.** $P_0 = \begin{bmatrix} 0 & 1 \end{bmatrix}$

29. $P_0 = \begin{bmatrix} 0.4 & 0.6 \end{bmatrix}$ **30.** $P_0 = \begin{bmatrix} 0.7 & 0.3 \end{bmatrix}$

In Exercises 31–35, consider the transition matrix $T = \begin{bmatrix} 1 & 0 & 0 \\ 0.5 & 0 & 0.5 \\ 0 & 1 & 0 \end{bmatrix}$, and find $P_1, P_2, P_3, \ldots, P_{15}$ for each initial state vector.

31. $P_0 = \begin{bmatrix} 1 & 0 & 0 \end{bmatrix}$ **32.** $P_0 = \begin{bmatrix} 0.1 & 0.3 & 0.6 \end{bmatrix}$

33. $P_0 = \begin{bmatrix} 0.25 & 0.5 & 0.25 \end{bmatrix}$ **34.** $P_0 = \begin{bmatrix} 0 & 1 & 0 \end{bmatrix}$

35. $P_0 = \begin{bmatrix} 0 & 0 & 1 \end{bmatrix}$

Section 11.3 Regular Markov Chains

Some interesting and useful results are found in the long-term behavior of a Markov chain. In some cases the powers of the transition matrix tend to get closer and closer to one fixed matrix as the number of steps in the chain increases, while in other cases those powers continue to exhibit noticeable change. As an example of the latter, the transition matrix

$$T = \begin{bmatrix} 0 & 1 \\ 1 & 0 \end{bmatrix} \text{ has } T^2 = \begin{bmatrix} 1 & 0 \\ 0 & 1 \end{bmatrix}, T^3 = \begin{bmatrix} 0 & 1 \\ 1 & 0 \end{bmatrix} = T, T^4 = \begin{bmatrix} 1 & 0 \\ 0 & 1 \end{bmatrix} = T^2,$$

and so on, showing that the powers of T change back and forth between T and T^2 as the power increases. By contrast, consider the transition matrix

$$T = \begin{bmatrix} 0.1 & 0.9 \\ 0.7 & 0.3 \end{bmatrix}.$$

Notice the pattern developing in each entry of the various powers of T (for convenience, entries of T^3, and so on, are rounded to two decimals):

$$T^2 = \begin{bmatrix} 0.64 & 0.36 \\ 0.28 & 0.72 \end{bmatrix}; \quad T^3 = \begin{bmatrix} 0.32 & 0.68 \\ 0.53 & 0.47 \end{bmatrix}; \quad T^4 = \begin{bmatrix} 0.51 & 0.49 \\ 0.38 & 0.63 \end{bmatrix};$$

$$T^5 = \begin{bmatrix} 0.39 & 0.61 \\ 0.47 & 0.53 \end{bmatrix}; \quad T^6 = \begin{bmatrix} 0.46 & 0.54 \\ 0.42 & 0.58 \end{bmatrix}; \quad T^7 = \begin{bmatrix} 0.42 & 0.58 \\ 0.45 & 0.55 \end{bmatrix};$$

$$T^8 = \begin{bmatrix} 0.45 & 0.55 \\ 0.43 & 0.57 \end{bmatrix}; \quad T^9 = \begin{bmatrix} 0.43 & 0.57 \\ 0.44 & 0.56 \end{bmatrix}; \quad T^{10} = \begin{bmatrix} 0.44 & 0.56 \\ 0.44 & 0.57 \end{bmatrix};$$

$$T^{11} = \begin{bmatrix} 0.44 & 0.57 \\ 0.44 & 0.56 \end{bmatrix}; \quad T^{12} = \begin{bmatrix} 0.44 & 0.56 \\ 0.44 & 0.56 \end{bmatrix}; \quad T^{13} = \begin{bmatrix} 0.44 & 0.56 \\ 0.44 & 0.56 \end{bmatrix};$$

Our calculations show that powers of 13 and above applied to T will give the same matrix as T^{12} upon rounding to two decimals. This fact is described by saying that, in the long run, the powers of T tend to approach the **stable** or **limiting matrix**

$$L = \begin{bmatrix} 0.44 & 0.56 \\ 0.44 & 0.56 \end{bmatrix}.$$

A transition matrix having such a property is called **regular.** Is there a way to tell when such a phenomenon will occur? Was it just luck causing the rows in the limiting matrix to be the same? The answers to these questions are revealed next.

Regular Transition Matrices

A transition matrix T is **regular** if all entries of T^n are *positive* for some positive integer n. The Markov chain associated with T is called a **regular Markov chain.** The importance of a regular matrix T lies in the fact that, as n increases, T^n always approaches some stable or limiting matrix L having identical rows.

A word of caution about round-off errors is in order in raising a matrix to higher and higher powers with a calculator or computer. Often, when a matrix has small decimal entries, increasingly higher powers of that matrix can result in errors of such magnitude that they actually exceed the original entries themselves. Without going into error analysis, a subject in its own right, be cautious of the accuracy of the last few digits on a calculator that supposedly displays several-digit accuracy.

EXAMPLE 1

Deciding when a Matrix
is Regular

(a) The matrix $T = \begin{bmatrix} 0.5 & 0.5 \\ 0.2 & 0.8 \end{bmatrix}$ is regular, because the entries in T^1 are all positive.

(b) The matrix $T = \begin{bmatrix} 1 & 0 \\ 0 & 1 \end{bmatrix}$ is not regular, because T^n has a 0 entry for every positive integer n.

(c) The matrix $T = \begin{bmatrix} 0.1 & 0.9 \\ 0 & 1 \end{bmatrix}$ is not regular, because every positive integer power of T has a 0 in the second-row, first-column position.

(d) The matrix $T = \begin{bmatrix} 0.1 & 0.2 & 0.7 \\ 0.5 & 0 & 0.5 \\ 0.4 & 0.4 & 0.2 \end{bmatrix}$ is regular, because $T^2 = \begin{bmatrix} 0.39 & 0.3 & 0.31 \\ 0.25 & 0.3 & 0.45 \\ 0.32 & 0.16 & 0.52 \end{bmatrix}$.

∎

Regular Markov chains have another important and almost amazing property: Regardless of what the initial distribution in a state vector may be, as the number of steps in the chain increases, the distribution tends to approach a fixed **stable, or limiting, distribution.** To see why this is so, let T be a regular matrix and let L be the limiting matrix. This means that we may select a positive integer n large enough so that, after rounding the decimal approximations for both T^n and L to a reasonable number of places, $T^n = L$. It follows that $T^n = T^{n+1} = T^{n+2} = \cdots = L$. Now, if P_0 is a state vector representing some initial distribution, then, for the nth stage and beyond, the following equalities hold:

$$P_0 T^n = P_0 T^{n+1} = P_0 T^{n+2} = \cdots = P_0 L. \tag{1}$$

In other words, from the nth stage and thereafter, the state vector does not change, but instead remains stable at the constant limiting value of $P_0 L$.

The stable or long-term state vector in a regular Markov chain is of great interest in applications. However, as you can imagine, it will usually require considerable calculational effort to find the limiting matrix L required for computing $P_0 L$. Fortunately, there is an easier way: For simplicity, rename the stable state vector S. That is, let $P_0 L = S$. Then the first two and the last terms of Equation (1) allow us to write

$$P_0 T^n = P_0 T^{n+1}$$
$$P_0 T^n = (P_0 T^n)T$$
$$P_0 L = (P_0 L)T$$
$$S = ST.$$

Furthermore, we know that the entries in S are probabilities, so their sum must be 1. We now have the defining properties of S.

Finding the Stable State Vector

In any regular Markov chain with transition matrix T, any initial distribution given in an initial state vector P_0 eventually reaches the stable state vector $S = [p_1, p_2, \ldots, p_n]$, which has these properties:

$$S = ST \quad \text{and} \quad p_1 + p_2 + \cdots + p_n = 1.$$

EXAMPLE 2

Finding a Stable State Vector

The Aurora planning commission (Section 11.1) found that, each year, 92% of those commuting by mass transportation (M) continued to commute by mass transportation the next year, while 8% changed to private automobile (A). Of those commuting by private automobile, 95% continued to do so the next year, while 5% changed to mass transportation. What is the eventual choice of transportation of commuters among the two modes?

SOLUTION The transition matrix for this Markov chain was found (in Section 11.1) to be

$$T = \begin{array}{c} \\ M \\ A \end{array} \begin{array}{c} \begin{array}{cc} M & A \end{array} \\ \begin{bmatrix} 0.92 & 0.08 \\ 0.05 & 0.95 \end{bmatrix} \end{array}.$$

We seek a stable vector $S = \begin{bmatrix} u & v \end{bmatrix}$ such that $S = ST$ and $u + v = 1$. Specifically, we are to find u and v such that

$$\begin{bmatrix} u & v \end{bmatrix} = \begin{bmatrix} u & v \end{bmatrix} \begin{bmatrix} 0.92 & 0.08 \\ 0.05 & 0.95 \end{bmatrix} \quad \text{and} \quad u + v = 1.$$

Performing the indicated matrix multiplication, we find that this system becomes

$$\begin{bmatrix} u & v \end{bmatrix} = \begin{bmatrix} 0.92u + 0.05v & 0.08u + 0.95v \end{bmatrix} \quad \text{and} \quad u + v = 1.$$

Equating the entries in like addresses of the matrices gives

$$u = 0.92u + 0.05v$$
$$v = 0.08u + 0.95v$$
$$u + v = 1,$$

or, upon combining like terms in the first two equations,

$$0.08u - 0.05v = 0$$
$$-0.08u + 0.05v = 0$$
$$u + \quad v = 1.$$

Because the first two equations are the equivalent, the system to be solved is

$$0.08u - 0.05v = 0$$
$$u + \quad v = 1.$$

The solution is $u = 0.38$ and $v = 0.62$. In other words, regardless of what the choice of transportation may have been at any step of the Markov process, eventually about 38% of the commuters will use mass transportation and about 62% will use private automobile. ∎

EXAMPLE 3

Finding a Stable
State Vector

The regular transition matrix

$$\text{This year} \quad T = \begin{matrix} B \\ C \\ D \end{matrix} \begin{bmatrix} 0.97 & 0.02 & 0.01 \\ 0.02 & 0.95 & 0.03 \\ 0.04 & 0.03 & 0.93 \end{bmatrix}$$

Next year
$\quad B \quad C \quad D$

describes the changes in brand loyalty each year by shoppers who bought brands B, C, and D. (See Example 5, Section 11.1.) What is the eventual market share for each of the three brands?

SOLUTION We are to find the state vector $S = \begin{bmatrix} u & v & w \end{bmatrix}$ for which $S = ST$ and $u + v + w = 1$. Specifically,

$$\begin{bmatrix} u & v & w \end{bmatrix} = \begin{bmatrix} u & v & w \end{bmatrix} \begin{bmatrix} 0.97 & 0.02 & 0.01 \\ 0.02 & 0.95 & 0.03 \\ 0.04 & 0.03 & 0.93 \end{bmatrix} \quad \text{and} \quad u + v + w = 1$$

are to be satisfied. The indicated matrix multiplication gives

$$\begin{bmatrix} u & v & w \end{bmatrix} = \begin{bmatrix} 0.97u + 0.02v + 0.04w & 0.02u + 0.95v + 0.03w & 0.01u + 0.03v + 0.93w \end{bmatrix},$$

so that

$$u = 0.97u + 0.02v + 0.04w$$
$$v = 0.02u + 0.95v + 0.03w$$
$$w = 0.01u + 0.03v + 0.93w.$$

Combining like terms yields

$$-0.03u + 0.02v + 0.04w = 0$$
$$0.02u - 0.05v + 0.03w = 0$$
$$0.01u + 0.03v - 0.07w = 0.$$

Since u, v, and w must satisfy these three equations, as well as $u + v + w = 1$, it follows that our stable state vector $S = \begin{bmatrix} u & v & w \end{bmatrix}$ may be derived from the solution of the following system of equations:

$$-0.03u + 0.02v + 0.04w = 0$$
$$0.02u - 0.05v + 0.03w = 0$$
$$0.01u + 0.03v - 0.07w = 0$$
$$u + \quad v + \quad w = 1.$$

Upon inspection, we see that the third equation is equivalent to the sum of the first two equations. Thus, we can reduce the problem to finding the solution of the system

$$-0.03u + 0.02v + 0.04w = 0$$
$$0.02u - 0.05v + 0.03w = 0$$
$$u + \quad v + \quad w = 1.$$

The solution (to two-decimal accuracy) is $u \approx 0.48$, $v \approx 0.31$, and $w \approx 0.20$, which means that, among the three brands, in the long run, brand B will have about 48% of the market, brand C will have about 31% of the market, and brand D will have about 20% of the market. ▬

Earlier in the section, we stated that the limiting matrix L obtained when a regular transition matrix is raised to successively higher powers has identical rows. We raised

$$T = \begin{bmatrix} 0.1 & 0.9 \\ 0.7 & 0.3 \end{bmatrix}$$

to the 13th power, whereupon we saw that

$$L = \begin{bmatrix} 0.44 & 0.56 \\ 0.44 & 0.56 \end{bmatrix}.$$

For the matrix

$$T = \begin{bmatrix} 0.92 & 0.08 \\ 0.05 & 0.95 \end{bmatrix}$$

of Example 2, which shows how commuters changed their mode of commuting, it takes well over 100 powers of T to arrive at

$$L = \begin{bmatrix} 0.38 & 0.62 \\ 0.38 & 0.62 \end{bmatrix}$$

Looking back at Example 2, do you see a relationship between the rows of L and the stable state vector S found there? What do you think the stable state vector for the matrix

$$T = \begin{bmatrix} 0.1 & 0.9 \\ 0.7 & 0.3 \end{bmatrix}$$

will be, given that its limit matrix is

$$L = \begin{bmatrix} 0.44 & 0.56 \\ 0.44 & 0.56 \end{bmatrix}?$$

If you answered $S = [0.44 \quad 0.56]$, you are correct! Mathematically, we can prove that, for a regular transition matrix T, each row of the limiting matrix L is precisely the stable state vector S. So, if we can find S, we can find L upon deciding what decimal accuracy we desire. When constructing L from S, we do not know how many steps it takes to reach the stable condition.

Limiting Matrix for Regular Markov Chains

Let T be the transition matrix for a regular Markov chain. As the positive integer n increases, the matrix T^n approaches a **limiting matrix** L, each of whose rows is the stable state vector for the chain.

EXAMPLE 4

Constructing the
Limiting Matrix from
the Stable State Vector

Find the stable state vector S and the limiting matrix L for the regular matrix

$$T = \begin{bmatrix} 0.2 & 0.8 \\ 0.6 & 0.4 \end{bmatrix}.$$

SOLUTION We find the stable state vector $S = [u \quad v]$ by solving the system of equations;

$$u = 0.2u + 0.8v$$
$$v = 0.6u + 0.4v$$
$$u + \quad v = \quad 1,$$

which simplifies to

$$-0.8u + 0.8v = 0$$
$$0.6u - 0.6v = 0$$
$$u + \quad v = 1.$$

The solution of this system of equations is $u = v = 0.5$. Thus, the stable state vector is $S = [0.5 \quad 0.5]$, and the limiting matrix is

$$L = \begin{bmatrix} 0.5 & 0.5 \\ 0.5 & 0.5 \end{bmatrix}. \quad \blacksquare$$

Exercises 11.3

In Exercises 1 through 6, decide which of the given transition matrices are regular.

1. $\begin{bmatrix} 0.5 & 0.5 \\ 0.9 & 0.1 \end{bmatrix}$ **2.** $\begin{bmatrix} 0 & 1 \\ 0.4 & 0.6 \end{bmatrix}$

3. $\begin{bmatrix} 0 & 0.5 & 0.5 \\ 0.2 & 0.4 & 0.4 \\ 0.8 & 0.1 & 0.1 \end{bmatrix}$ **4.** $\begin{bmatrix} 0 & 0.1 & 0.9 \\ 0.2 & 0 & 0.8 \\ 0.4 & 0.6 & 0 \end{bmatrix}$

5. $\begin{bmatrix} 1 & 0 & 0 \\ 0.5 & 0.5 & 0 \\ 0.2 & 0.3 & 0.5 \end{bmatrix}$ **6.** $\begin{bmatrix} 0 & 0.4 & 0.6 \\ 1 & 0 & 0 \\ 0 & 0 & 1 \end{bmatrix}$

For each of the regular matrices in Exercises 7 through 12, find the limiting matrix L by calculating the powers of the matrix.

7. $\begin{bmatrix} 0.3 & 0.7 \\ 0.5 & 0.5 \end{bmatrix}$ **8.** $\begin{bmatrix} 0.6 & 0.4 \\ 0.3 & 0.7 \end{bmatrix}$

9. $\begin{bmatrix} 0.9 & 0.1 \\ 0.2 & 0.8 \end{bmatrix}$ **10.** $\begin{bmatrix} 0 & 1 \\ 0.9 & 0.1 \end{bmatrix}$

11. $\begin{bmatrix} 0.6 & 0.4 \\ 1 & 0 \end{bmatrix}$ **12.** $\begin{bmatrix} 0.3 & 0.4 & 0.3 \\ 0.4 & 0.4 & 0.2 \\ 0.1 & 0.6 & 0.3 \end{bmatrix}$

For each of the regular matrices in Exercises 13 through 21, find the stable state vector S, and, from that, find the limiting matrix L.

13. $\begin{bmatrix} 0.4 & 0.6 \\ 0.2 & 0.8 \end{bmatrix}$ **14.** $\begin{bmatrix} \frac{2}{5} & \frac{3}{5} \\ \frac{2}{3} & \frac{1}{3} \end{bmatrix}$

15. $\begin{bmatrix} \frac{7}{8} & \frac{1}{8} \\ \frac{1}{6} & \frac{5}{6} \end{bmatrix}$ **16.** $\begin{bmatrix} 0 & 1 \\ 0.8 & 0.2 \end{bmatrix}$

17. $\begin{bmatrix} \frac{1}{7} & \frac{6}{7} \\ \frac{9}{10} & \frac{1}{10} \end{bmatrix}$ **18.** $\begin{bmatrix} 0.33 & 0.67 \\ 0.08 & 0.92 \end{bmatrix}$

19. $\begin{bmatrix} \frac{1}{5} & \frac{2}{5} & \frac{2}{5} \\ \frac{1}{2} & \frac{1}{4} & \frac{1}{4} \\ \frac{1}{3} & \frac{1}{6} & \frac{1}{2} \end{bmatrix}$ **20.** $\begin{bmatrix} 0 & 0.4 & 0.6 \\ 0.2 & 0.6 & 0.2 \\ 0.8 & 0.1 & 0.1 \end{bmatrix}$

21. $\begin{bmatrix} 0 & 0.9 & 0.1 \\ 0.8 & 0 & 0.2 \\ 0.3 & 0.3 & 0.4 \end{bmatrix}$

22. Newspaper Subscriptions Circulation statistics in Charlotte show that the *Evening Star* keeps 65% of its subscribers and loses 35% to the *Tribune* each year. On the other hand, the *Tribune* keeps 80% of its subscribers and loses 20% to the *Evening Star* each year.

(a) Construct the transition diagram and transition matrix representing these statistics.

(b) In the long run, what percentage of the subscribers will each paper have?

23. Brand Loyalty Surveys indicate that, among people who buy an American-built car, 85% will buy American on their next purchase and, among those who buy a foreign make, 80% will buy a foreign make on their next purchase. Assuming that this trend holds, what will the distribution of buyers between American and foreign cars eventually become?

24. Advertising A retail store with one major competitor is considering an intensive advertising campaign. The designers of the ad campaign claim that each week 20% of the competitor's customers will switch to their store, while about 5% of their own customers will be annoyed by the ad campaign and switch to the competitor.

(a) If the store now has 70% of the market and the competitor has 30%, what will be the long-term gain in market share if the ad campaign continues for several weeks?

(b) How many weeks of the ad campaign will be needed before the market share stabilizes?

25. Voter Trends In Mt. Vernon, the voting population remains rather constant at 12,000. From election to election, the changes among Democrats, Republicans, and Independents are as shown in the following transition matrix:

		Next election		
		D	*R*	*I*
This election	*D*	0.80	0.15	0.05
	R	0.08	0.90	0.02
	I	0.20	0.15	0.65

In the long run, how many voters will each party have?

26. Social Strata Sociological studies indicate that, from generation to generation, there is some linkage between the social stratum of the parent and that of the offspring. Assume that the population is divided into three classes—low, middle, and high—with the probability of moving from one class to another in the next generation given by the following transition matrix:

Next generation

$$\begin{array}{c} \text{This} \\ \text{generation} \end{array} \begin{array}{c} \\ L \\ M \\ H \end{array} \begin{array}{ccc} L & M & H \\ \left[\begin{array}{ccc} 0.65 & 0.23 & 0.12 \\ 0.08 & 0.80 & 0.12 \\ 0.03 & 0.37 & 0.60 \end{array} \right]. \end{array}$$

(a) If these probabilities hold true generation after generation, what is the long-term percentage of people in each class?

(b) What is the long-term probability of a person going from L in one generation to H in the next, as opposed to those staying in L?

27. Genetic Traits Suppose that the transition matrix for the inheritance of a particular trait—being right handed (RH), left handed (LH), or ambidextrous (AM)—is given by the following transition matrix.

Offspring

$$\begin{array}{c} \\ \text{Parent} \end{array} \begin{array}{c} \\ RH \\ LH \\ AM \end{array} \begin{array}{ccc} RH & LH & AM \\ \left[\begin{array}{ccc} 0.70 & 0.30 & 0 \\ 0.25 & 0.50 & 0.25 \\ 0.40 & 0.40 & 0.20 \end{array} \right]. \end{array}$$

(a) Find the stable state vector S.

(b) Find the limiting matrix L, and interpret the results.

28. Cognition Experiment A rat placed in a small area is faced with three choices of holes to go into. In one hole (C), the rat is rewarded with a piece of cheese; in another (M), a toy mouse awaits; in the third (S), a shock awaits. Repeated trials of this experiment reveal that if the rat goes into C on one trial, then, on the next trial, it will go into C 80% of the time. into M 15% of the time, and into S 5% of the time. If the rat goes into M on one trial, then, on the next trial, it will go into C 40% of the time, into M 50% of the time, and into S 10% of the time. If the rat goes into S on one trial, then, on the next trial, it will go into C 40% of the time, into M 40% of the time, and back into S 20% of the time.

(a) Construct the transition diagram and the transition matrix for this experiment.

(b) Find the stable state vector S, and interpret the entries.

(c) Find the limiting matrix L, and interpret the entries.

29. Insurance An auto insurance company rates the drivers it insures as low risk (L), medium risk (M), and high risk (H). Records indicate that, each year, in the low-risk category, about 95% of those drivers remain in the category, while 4% move down to M, and 1% fall to H. Of those in the medium-risk category, 8% rise to L, 90% remain in M, and 2% fall to H. Of those in the high-risk category, 1% rise to L, 15% rise to M, and 84% remain in H.

(a) Construct the transition diagram and the transition matrix for these data.

(b) In the long run, what will be the company's distribution of drivers in each category?

(c) Find the limiting matrix L, and interpret the entries.

Chapter 11 Summary

CHAPTER HIGHLIGHTS

Markov chains combine matrix algebra and probability in order to answer questions about repeated trials of an experiment. *Transition diagrams* and *transition matrices* are both convenient vehicles for summarizing the pertinent data relating to such situations. If some power of a transition matrix has all positive entries, the matrix is called *regular,* and successive powers always approach a *limiting matrix* having stable entries. Such *regular Markov chains* eventually reach the same *stable state vector*, regardless of the initial distribution in a state vector.

EXERCISES

In Questions 1 and 2, calculate the missing probabilities in each transition diagram, and write the transition matrix.

1.

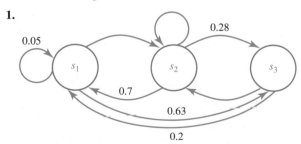

2.

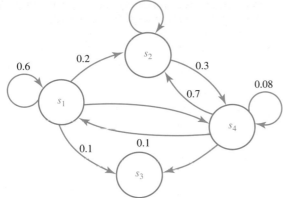

For Questions 3 and 4, given the transition matrix

$$T = \begin{bmatrix} 0.2 & 0.5 & 0.3 \\ 0.1 & 0.7 & 0.2 \\ 0.6 & 0 & 0.4 \end{bmatrix}$$

and the state vector $P_0 \begin{bmatrix} 0.1 & 0.6 & 0.3 \end{bmatrix}$, find the state vector

3. P_1 one transition later.

4. P_2 two transitions later.

For Questions 5 through 7, find the limiting matrix L for each regular matrix.

5. $\begin{bmatrix} 0.1 & 0.9 \\ 0.7 & 0.3 \end{bmatrix}$ **6.** $\begin{bmatrix} 0 & 1 \\ 0.8 & 0.2 \end{bmatrix}$

7. $\begin{bmatrix} 0.5 & 0.5 \\ 0.6 & 0.4 \end{bmatrix}$

For the matrices in Questions 8 through 10, explain why each matrix cannot be a transition matrix.

8. $\begin{bmatrix} 1 & 0 \\ 0 & 0 \end{bmatrix}$ **9.** $\begin{bmatrix} 0.2 & 0.8 \\ 1.2 & -0.2 \end{bmatrix}$

10. $\begin{bmatrix} 0.3 & 0.3 & 0.4 \\ 0.2 & 0.9 & 0.2 \\ 0.5 & 0 & 0.5 \end{bmatrix}$

 An archer's record indicates that her performance can be represented by a Markov chain. If she hits the bull's-eye, she hits the next one 82% of the time and misses 18% of the time. If she misses the bull's eye, she hits the next one 73% of the time and misses 27% of the time. Use this information to solve Questions 11 and 12.

11. Given that the archer misses the bull's-eye on her first shot, find the probability that she hits the bull's-eye on her third shot.

12. Determine the archer's limiting matrix and the number of steps it takes her to reach this stable condition.

13. To four-decimal accuracy, determine the limiting matrix L for the transition matrix

$$T = \begin{bmatrix} 0.5 & 0.2 & 0.3 \\ 0.1 & 0.6 & 0.3 \\ 0.3 & 0.5 & 0.2 \end{bmatrix}$$

by raising T to consecutive positive integer powers, using a graphics calculator or computer. What is the smallest power of T that gives the desired limiting matrix?

CUMULATIVE REVIEW

14. Write the equation of a line having x-intercept 7.2 and that is parallel to the line $4x + 2y = 23$.

15. Solve this system:

$$\begin{aligned} x + 2y + z &= 4 \\ 2x + y + 3z &= 11 \\ 3x + 5y - 2z &= -2. \end{aligned}$$

16. Calculate $(A + B)^2$ for the matrices

$$A = \begin{bmatrix} 1 & 2 \\ 3 & 4 \end{bmatrix} \quad \text{and} \quad B = \begin{bmatrix} 0 & 5 \\ -2 & 1 \end{bmatrix}.$$

17. Shade the following Venn diagram to represent the set $A' \cap (B' - C')$:

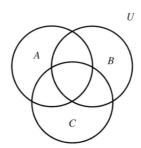

18. When a pair of dice is rolled, find the probability that the sum is not six or the product is not six.

19. Solve the following linear programming problem.

$$\begin{aligned} \text{Maximize } f &= 3x + 5y \\ \text{subject to} \quad x + y &\leq 20 \\ 2x + 3y &\leq 51 \\ x, y &\geq 0. \end{aligned}$$

Chapter 11   Sample Test Items

Problems 1 and 2 refer to the following transition diagram:

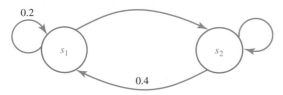

1. Complete the transition diagram.

2. Find the transition matrix.

Problems 3 and 4 refer to the following transition diagram

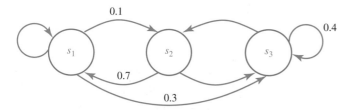

3. Complete the transition diagram.

4. Find the transition matrix.

For Problems 5 and 6, use each transition matrix to construct the transition diagram.

5. $\begin{bmatrix} 0.3 & 0.7 \\ 0.8 & 0.2 \end{bmatrix}$

6. $\begin{bmatrix} 0.5 & 0.2 & 0.3 \\ 0.4 & 0 & 0.6 \\ 0 & 0.1 & 0.9 \end{bmatrix}$

The following statement refers to Problems 7 and 8: If it snows today, there is a 30% chance that it will snow tomor-row. If it does not snow today, there is a 90% chance that it will not snow tomorrow.

7. Construct the transition diagram.

8. Write the transition matrix.

In Problems 9 and 10, use the transition matrix

$$\begin{bmatrix} 0.4 & 0.5 & 0.1 \\ 0 & 0.7 & 0.3 \\ 0.6 & 0.4 & 0 \end{bmatrix}$$

and the state vector $P_0 = \begin{bmatrix} 0.2 & 0.5 & 0.3 \end{bmatrix}$, to find the state vector

9. P_1 one transition later.

10. P_2 two transitions later.

In Problems 11 and 12, decide whether each transition matrix is regular.

11. $\begin{bmatrix} 1 & 0 \\ 0.8 & 0.2 \end{bmatrix}$

12. $\begin{bmatrix} 0.1 & 0 & 0.9 \\ 0.5 & 0.3 & 0.2 \\ 0.6 & 0.4 & 0 \end{bmatrix}$

For Problems 13 and 14, find the limiting matrix L by calculating the powers of each regular matrix.

13. $\begin{bmatrix} 0.1 & 0.9 \\ 1 & 0 \end{bmatrix}$

14. $\begin{bmatrix} 0.7 & 0.3 \\ 0.4 & 0.6 \end{bmatrix}$

In Problems 15 and 16, for each regular matrix, find the stable state vector S and the limiting matrix L.

15. $\begin{bmatrix} 0.47 & 0.53 \\ 0.26 & 0.74 \end{bmatrix}$

16. $\begin{bmatrix} 0 & 1 \\ 0.7 & 0.3 \end{bmatrix}$

In-Depth Application

The Billboard® Top 100 Albums

Objectives

In this application, you will apply your knowledge of probability and Markov chains to predict the position of an album on the *Billboard*® "Top 100" in a given week.

Introduction

Each week, *Billboard*® ranks albums from data compiled by sales reports from retail stores in the United States. The ranking can affect the radio airplay of singles from an album, as well as future sales. People may be more likely to buy an album that makes it into the "Top 100," and some devoted fans may be more likely to buy an album in order to keep it there.

(continued)

The Data

In this application, we will use the following categories for albums ranked by *Billboard*®:

S_1: ranked 1–25;
S_2: ranked 26–50;
S_3: ranked 51–75;
S_4: ranked 76–100;
S_5: not in the "Top 100."

These categories will be considered the *states*. Note that S_5 includes those albums which have just been released for the first time in the initial week, as well as those which are no longer in the "Top 100." Probabilities were generated from data compiled from the *Billboard*® "Top 100" for the week ending February 24, 2002, and were adjusted slightly to compensate for new releases. The estimated probabilities of this set of albums moving from one state to another after one week are summarized in the following table:

State	S_1	S_2	S_3	S_4	S_5
S_1	0.88	0.12	0	0	0
S_2	0.08	0.64	0.28	0	0
S_3	0	0.16	0.56	0.28	0
S_4	0	0.08	0.16	0.60	0.16
S_5	0.005	0	0	0.15	0.845

According to the table, for example, an album that is in state S_1 one week will move to state S_2 the next week with probability 0.12.

1. Draw the transition diagram for this system.

2. What is the probability that an album will stay in the "Top 25" for 2 weeks in a row? For 3 weeks in a row?

3. Twenty-five albums, of course, are in the "Top 25" in a given week. How many of those will still be in the "Top 25" 5 weeks later?

4. What percentage of the albums ranked 76–100 will move into the "Top 25" after 3 weeks?

5. What percentage of the albums in the "Top 25" will fall out of the "Top 100" after 10 weeks?

CHAPTER

12

Game Theory: Two-Player, Zero-Sum Games

PROBLEM

A campaign debate between Senator Smith and her opponent will focus on foreign policy and defense. The senator's managers feel that her opponent is strong on the defense issue, but not on foreign policy. Senator Smith would likely gain 3 points when both discuss foreign policy; her opponent, 5 points, when both discuss defense; the senator, 10 points, if she discusses foreign policy while her opponent discusses defense; and the senator, 6 points, if she discusses defense while her opponent discusses foreign policy. What proportion of the debate should the senator spend on each issue to gain the best advantage from the debate, regardless of how her opponent divides his time? (See Example 2, Section 12.3.)

CHAPTER PREVIEW

Our introduction to game theory encompasses "two-person" and "zero-sum" games. The mathematical analysis involves setting up the payoff matrix and deciding from that matrix whether the game is classified as "pure strategy" or "mixed strategy." The solution to a pure strategy game may be found by inspection, while games of mixed strategy are solved by linear programming. The solution gives the best strategy for each player and the expected value of the game (in the sense of probability), indicating which player the game favors.

Technology Opportunities

As seen in Chapter 3, both graphing calculators and spreadsheet packages have matrix manipulation capabilities. In the current chapter, we will explore how games, strategies, and values can be viewed as matrices and matrix multiplication leads to the solution to game-theoretic problems. In the last section, the strong link between dual linear programming problems and zero-sum games is demonstrated. In this regard, Excel's `Solver` function will prove quite helpful.

Section 12.1 Strictly Determined Games

To most of us, the word *game* suggests that competitive forces are at work, each trying to devise some strategy that will allow one of the forces to win the game. Some games—such as a tennis match, a chess game, a basketball game, and the "game" between negotiating teams for labor and management—are played by only two competing forces. These are known as **two-person games** (regardless of the actual number of players). Other games have several competing forces. An example of the latter might be the fast-food establishments in a city competing for business. In any event, many situations in the business world, psychology, economics, the military, and so forth may be considered "games" in the sense that we wish to consider in this chapter. This approach to conflict, competition, bargaining, and even cooperation, was first introduced in the late 1920's and early 1930's by one of the greatest mathematicians of the 20th century: John Von Neumann. In the 1950's, the importance of his work was finally recognized and built upon by the science and mathematics community in a flurry of activity that continues today.

For mathematical simplicity, we restrict our study to games involving only two competing forces—in other words, **two-person games.** In addition, and again for simplicity, we consider only games in which the amount gained by one of the players is equal to the amount lost by the opposing player. That is, at the end of the game, the sum of the earnings of two players is zero. (A loss is denoted by a negative number.) These games are known as **zero-sum games.** Thus, our games are limited to *two-person, zero-sum games.*

HISTORICAL NOTE

John Von Neumann (1903–1957)

John Von Neumann (Fun nŏi-mən) was a brilliant mathematician and one of the six original professors in the School of Mathematics at the Princeton Institute for Advanced Study founded in 1933. He is the founder of game theory. Many of his original applications were in economics. His other important mathematical contributions were in the theory of maxima and minima of nonlinear problems and in theories of computer architecture.

EXAMPLE 1
A Two-Person, Zero-Sum Game

Two friends, Match and Mismatch, play a game of "matching pennies." Upon a signal, each shows a penny. If the pennies turn up both heads or both tails, Match wins $1 from Mismatch; however, if only one of the coins turns up heads, Mismatch wins $1 from

Match. This is a two-person game. Because any amount won by one player precisely equals the amount lost by the other, the game is also a zero-sum game. ▬

Both Match and Mismatch have a choice to make each time they play the preceding game: whether to show heads or to show tails. If Match, for example, decides to show heads each time, it will not take Mismatch long to notice this and to begin to show tails each time, thereby winning $1 on every game. Thus, always showing heads would not be a good strategy for Match. So, just what *is* the best course of action for Match? Before answering this question, let us consider one more example of a two-player, zero-sum game.

EXAMPLE 2

A Two-Person, Zero-Sum Game

The city of Graham has two appliance stores: Appliance City and Appliance World. The two stores offer discounts of 20% or 30%, depending on how aggressively they want business. If both stores offer the same discount, Appliance City gets 60% of the business. If Appliance City discounts more than Appliance World, then Appliance City will get 70% of the business. If Appliance World discounts more than Appliance City, the latter will get 55% of the business. This is evidently a two-person (actually two-business game), but it is not clear that it is a zero-sum game. Notice that there is a fixed total amount of market share to be had. We can see that any business one store gains the other store loses. If we consider the different market shares in terms of a baseline share of 50%, then we can rewrite the game as a zero-sum game. For example, we said that when both stores offer the same discount, Appliance City gets a 60% share of the business. Instead, think of this as Appliance City getting an additional 10% of business (above and beyond 50%) and Appliance World losing 10%. ▬

In this game, each business has a choice of how much to discount, and each choice leads to a different configuration of market share. For instance, if Appliance City's strategy is to discount 20% and Appliance World's strategy is to discount 30%, then Appliance City would gain 5% of the market and Appliance World's share would drop by 5%. From among the available courses of action, which is the best for each business?

The games of Example 1 and Example 2 demonstrate what this chapter covers: Each player has a course of action to decide, and the decisions affect the outcome of the game. Any course of action taken by a player is known as a **strategy** for that player. How, then, does each player decide on what is the best strategy after taking into account all of the options the opposing player has available? When we find the best or **optimal strategy** for each player, we say we have "solved the game." Solving the game is our ultimate objective.

Games such as these may be mathematically analyzed if we express the **payoff** to one of the players in matrix form. Such a matrix is called a **payoff matrix.** To obtain the entries in a payoff matrix, we arbitrarily choose one of the players (called the **row player**) and use the rows in the matrix to represent that player's gains as positive numbers and losses as negative numbers. Of course, if no gain occurs, a zero is recorded. In the game of Example 1, suppose we call Match the row player. Then the resulting payoff matrix (in dollars) is

$$\begin{array}{c} \\ \text{Match} \begin{array}{c} H \\ T \end{array} \end{array} \overset{\overset{\displaystyle \text{Mismatch}}{\begin{array}{cc} H & T \end{array}}}{\left[\begin{array}{cc} 1 & -1 \\ -1 & 1 \end{array} \right]}.$$

For the players in Example 2, if we let Appliance City be the row player, then the pay-off matrix (in percentage of market share) becomes

$$
\begin{array}{c}
 & \text{Appliance World} \\
 & \begin{array}{cc} 20\% & 30\% \end{array} \\
\begin{array}{c} \text{Appliance} \\ \text{City} \end{array}
\begin{array}{c} 20\% \\ 30\% \end{array}
\left[\begin{array}{cc} 0.10 & 0.05 \\ 0.20 & 0.10 \end{array} \right]
\end{array}
$$

Each entry represents Appliance City's gain over a half share of the market.

The payoff matrix for a game helps us classify the game into one of two classifications: a **strictly determined** game and a game of **mixed strategy.** As suggested by the name, a game is strictly determined if it has a solution consisting of a *single course of action* that never varies from game to game. This type of solution is called a **pure strategy.** In contrast, a game of mixed strategy has a solution involving more than one course of action, each of which is played a certain percentage of the time.

The payoff matrix for Appliance City and Appliance World shows us that a single course of action by each (a pure strategy) is the best. Obviously, Appliance City will want to choose a 30% discount, because both market share gains shown in the second row are larger than those in the first row. However, Appliance World is going to counter with a 30% discount to minimize that impact. This is the best strategy for both stores in the sense that Appliance City can do no worse than garner an additional 10% of the market, regardless of what Appliance World does. This is also the best Appliance World can do under the circumstances. So the term *best strategy* is based, not necessarily on obtaining the maximum possible payoff, but rather on doing the best a player can, given the available options. Notice that the entry signaling this strategy is the 0.10 in the second-row, second-column position. It is the minimum entry in the second row (signaling "no worse than" for the row player) and the maximum in the second column (signaling the best the column player can do under the circumstances).

In terms of the payoff matrix, it is easy to tell when a game is strictly determined.

Strictly Determined Games

A game is called **strictly determined** if its payoff matrix has an entry that is the minimum element in its row and is also the maximum element in its column. Such an entry is called the **saddle point,** or **value,** of the game. The row in which the saddle point lies is the best strategy for the row player, and the column in which the saddle point lies is the best strategy for the column player. Just like the point in the middle of a saddle, the value is the peak when viewed in one direction and the valley when viewed in another (Figure 1). If a game has positive value, it favors the row player; if it has a negative value, it favors the column player. If a game has zero value, it is called a **fair game.**

According to these definitions, the game in Example 1 is not strictly determined, because there is no entry in the payoff matrix that could be classified as a saddle point. This means that no single course of action, such as always showing heads, is the best strategy for either player.

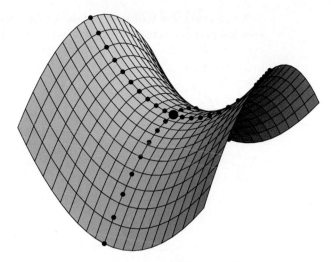

FIGURE 1
A Saddle Point is a Peak or a Valley, Relative to Direction Traveled Along the Saddle.

The game of Example 2 is, however, a strictly determined game, because the entry 0.10 in the second-row, second-column position is the minimum value in the second row and the maximum value in the second column. Therefore, 0.10 is a saddle point or value for the game. This means that the row player, Appliance City in this case, will get an additional 10% of the business when both businesses discount by 30%. The payoff matrix, from the location of the saddle point, also tells us that the best strategy for the row player is to pursue the single course of action (pure strategy) of discounting 30% all of the time and for the column player to do the same. The game is now solved.

EXAMPLE 3

A Strictly Determined, Two-Person Game

The following payoff matrix comes from a two-person, zero-sum game:

$$\begin{bmatrix} 1 & -1 & -3 \\ -1 & 1 & -2 \\ 2 & 5 & -4 \end{bmatrix} \rightarrow \text{saddle point value} = -2$$

Determine whether this game is strictly determined. If it is, what is the saddle point and best strategy for the row and column players?

SOLUTION In the first row, the entry with the minimum value is -3, located in the third column, but it is not the maximum value of entries in the third column. So -3 is not a saddle point. In the second row, the entry -2 in the third column is the entry with the minimum value, and it is also the largest value in the third column. Therefore, it is a saddle point. The third row does not have a candidate for a saddle point. It follows that the game is strictly determined and has a value of -2. The row player's best strategy is row 2, while the column player's best strategy is column 3. This game favors the column player because the value of the game is -2; the game is not fair.

A visual aid that will help decide on the location of any saddle points is circling the smallest number in each row and then boxing the largest number in each column. A number that is both circled and boxed will then be a saddle point. This scheme is used on the original matrix:

$$\begin{bmatrix} 1 & -1 & \boxed{\text{\Large\textcircled{-3}}} \\ -1 & 1 & \boxed{\text{\Large\textcircled{-2}}} \\ \boxed{2} & \boxed{5} & \text{\Large\textcircled{-4}} \end{bmatrix}$$

Smallest number in row is circled.

Largest number in column is boxed. ■

EXAMPLE 4

A Strictly Determined,
Two-Person Game

Graph-It has decided to bring out a new graphics calculator to compete with Tech Image's products. Graph-It is to aim its product at either the public school market only or the business market only. The company will make a decision as to which market to pursue on the basis of a financial analysis of the markets. Tech Image has the choice of putting its resources into the public school market only, putting them into the business market only, or splitting its resources 50:50 between the two markets. Market analysts have determined that there is $15 million of profit to be had in the total market. Estimates are that if Graph-It invests its money in the public school market, then its profit will be $1 million, $8 million, or $6 million, respectively, depending on Tech Image's investments in the public school market only, the business market only, or both markets. If, however, Graph-It invests its money in the business market only and Tech Image invests in public schools only, business only, or both, then Graph-It's profit will be, respectively, $10 million, $2 million, or $5 million. Under these assumptions, what is each company's best strategy for monetary gain?

SOLUTION Suppose that we consider Graph-It as the row player and abbreviate public schools with P, business with B, and a split with S. We must rewrite the payoffs as a zero-sum game by assuming an even baseline profit of $7.5 million (half of $15 million) for each company. Subtracting this amount from each of the possible profits gives our payoff matrix:

Tech Image

Graph-It:
$$\begin{array}{c c} & \begin{array}{ccc} P & B & S \end{array} \\ \begin{array}{c} P \\ B \end{array} & \begin{bmatrix} 1.5 & \boxed{0.5} & \boxed{\text{\Large\textcircled{-1.5}}} \\ \boxed{2.5} & \text{\Large\textcircled{-5.5}} & -2.5 \end{bmatrix} \end{array}.$$

The -1.5 in the first-row, third-column position is the only saddle point in the matrix. Consequently, the game is strictly determined, with a profit of $6 million for Graph-It and a million dollars for Tech Image. The best strategy for Graph-It is to invest its money in the public school market, and that for Tech Image is to split its resources 50:50 between the public school and the business sectors. Notice that Graph-It could make $10 million by investing in the business market if it could be guaranteed that Tech Image will invest only in the public school market. However, Graph-It cannot be sure that Tech Image will do that. Should Tech Image also decide to put its resources into the business sector, Graph-It would make only $2 million. The saddle-point option tells us that, regardless of what Tech Image does, if Graph-It puts its resources in the public school market, Graph-It can be assured of a $6 million profit. This is not a fair game: If Tech Image plays it "smart," Graph-It cannot earn half the available profits. But Graph-It can also play "smart" and minimize its disadvantage. This again points out the true meaning of **best strategy:** The best strategy is based, not necessarily on obtaining the maximum possible payoff, but rather on *doing the best* a player can *under the available options.* ■

Is it possible for there to be more than one solution to a strictly determined game? Indeed, it is, and as far as the mathematics is concerned, it does not matter which of these solution strategies the players choose. Is it possible that the "maximin" payoff entry for the row player will not agree with the "minimax" payoff entry for the column player? Unfortunately, yes, and in these cases the choice of one strategy by one player would result in a change in strategy by the other player, neither ever settling down to a single strategy (unless one of the players became lazy or charitable.) The next section will take a look at such games and provide a more satisfactory solution.

Exercises 12.1

In Exercises 1 through 10, decide whether the game represented by the given payoff matrix is strictly determined. If so, give the value of the game and whom it favors (the row or the column player), and suggest the best strategy for the row and column players.

1. $\begin{bmatrix} 3 & -1 \\ 5 & 2 \end{bmatrix}$

2. $\begin{bmatrix} 4 & 3 \\ 2 & 1 \end{bmatrix}$

3. $\begin{bmatrix} 5 & 3 \\ 2 & 7 \end{bmatrix}$

4. $\begin{bmatrix} 0 & 4 \\ -2 & 3 \end{bmatrix}$

5. $\begin{bmatrix} -3 & 3 & -5 \\ 2 & 4 & 6 \\ 1 & 2 & -2 \end{bmatrix}$

6. $\begin{bmatrix} 2 & 0 & -1 \\ 4 & -3 & 2 \\ 1 & 2 & 3 \end{bmatrix}$

7. $\begin{bmatrix} 1 & 4 & 7 \\ 0 & 2 & 4 \\ -1 & 3 & 5 \end{bmatrix}$

8. $\begin{bmatrix} 4 & -4 & -3 \\ 2 & -3 & 4 \\ 1 & -2 & 3 \end{bmatrix}$

9. $\begin{bmatrix} 2 & 1 & -2 & 4 \\ 1 & -3 & -5 & 1 \\ 3 & 2 & 4 & -2 \end{bmatrix}$

10. $\begin{bmatrix} 3 & -2 & 5 & -3 \\ 2 & 1 & 4 & -2 \\ 2 & 3 & 1 & -1 \end{bmatrix}$

11. For what values of x is the following payoff matrix strictly determined?

$$\begin{bmatrix} x & 9 & 3 \\ 0 & x & -8 \\ -4 & 5 & x \end{bmatrix}.$$

12. For what values of x is the following payoff matrix strictly determined?

$$\begin{bmatrix} x & 4 & 5 \\ 3 & -2 & x \\ -3 & 1 & -4 \end{bmatrix}.$$

13. **Business Decisions** A retail store is exploring the possibility of remodeling to give a more modern look to its physical facilities. If the store remodels and the economic situation is favorable, the business will earn $100,000. However, if the store remodels and the economy goes into a recession, the business will lose $40,000. If the store is not remodeled and the economic situation is favorable, the business will lose $50,000 due to the loss of customers to other stores. If the store does not remodel and there is a recession, the business will lose $45,000. Assume this is to be a two-person game, with the store playing against the economy. Set up the payoff matrix and decide which is the best strategy for the store: remodel or leave the facilities as they are.

14. **Political Campaign** The managers of a political campaign are trying to decide whether to reveal a skeleton in their candidate's closet. On the one hand, they figure that if the skeleton is revealed and the news media are critical, the candidate will lose 20,000 votes; but if the media are not critical, she will gain 3000 sympathy votes. On the other hand, if the media learn of the skeleton first and are critical, the candidate will lose 120,000 votes; but if the

media do not learn of the skeleton and it is not revealed, she will neither gain nor lose votes. Consider this a game between the candidate and the media. In the interest of getting the most votes possible in the upcoming election, should the candidate confess or try to keep the skeleton secret? Construct a payoff matrix to help you decide.

15. **Health Care Providers** Two health clinics, Nutone and Flabless, are considering offering expanded services in their town. At present, their clientele is rather stable, but they expect to divide 600 new customers with city growth and the expanded services. Nutone is considering offering a sports-medicine program, a senior-citizen program, and a drug-therapy program, while Flabless is considering a sports-medicine and a senior-citizen program. Nutone figures that if it starts a sports-medicine program, it will get 250 new accounts if Flabless starts a sports-medicine program and 400 new accounts if Flabless starts a senior-citizen program. If Nutone starts a senior-citizen program, it will get 350 new accounts if Flabless starts a sports-medicine program and 300 new accounts if Flabless starts a senior-citizen program. If Nutone starts a drug-therapy program, it will get 375 new accounts if Flabless starts a sports-medicine program and 400 new accounts if Flabless starts a senior-citizen program.

    (a) Construct the payoff matrix for this two-person game.

    (b) Find the saddle point.

    (c) What is the best strategy for each of the two clinics?

16. **Market Analysis** Harley and Honda are considering offering new changes in their bikes. Harley is contemplating offering either a new body style or a new transmission, while Honda is considering changing the suspension, the engine, or the body style of its bikes. Market studies show that, on the one hand, if Harley offers a new body style, it will get 60% of the market if Honda offers a new suspension, 55% if Honda offers a new engine, and 50% if Honda offers a new body style. On the other hand, if Harley offers a new transmission, it will get 40% of the market if Honda offers a new suspension, 35% if Honda offers a new engine, and 30% if Honda offers a new body style.

(a) Construct the payoff matrix for this two-person game.

(b) What is the value of the game?

(c) What are the best strategies for the two companies?

17. **Cards** Juan has two cards in his hand—a red 4 and a black 9—and Hector has three cards in his hand—a red 5, a black 7, and a black 8. Each simultaneously places one card on the table. If the cards are the same color, Juan gets the difference between the numbers from Hector; but if the cards are of different colors, Hector gets the minimum of the two numbers on the cards from Juan.

    (a) Construct a payoff matrix for this game.

    (b) Decide whether this game is strictly determined. If it is, what are the best strategies for both players?

18. **Cards** Sue and Jill each have three cards in their hands. Sue has an ace, a red 3, and a black 8. Jill has a red 5, a black 7, and an ace. Each simultaneously lays a card on the table. If the cards are both aces, Sue gives Jill 1 cent; but if one plays an ace and the other does not, then the person playing the ace receives pennies equal to the number on the opponent's card. Otherwise, if both cards are the same color, Jill gives Sue pennies equal to the sum of the numbers on the cards; if the cards are not the same color, Sue gives Jill pennies equal to twice the difference of the numbers on the cards.

    (a) Construct a payoff matrix for this game.

    (b) Is the game strictly determined? If so, give the saddle point and the best strategy for both players.

19. **Locating Retail Outlets** Two companies, Computer Software and Computer Hardware, are each studying whether to locate a new store in Tallahassee or in Austin. Studies of the competitive nature of the business indicate that if Computer Software locates in Tallahassee and Computer Hardware locates in Austin, then Computer Software's annual profit will be $500,000 more than Computer Hardware's profit. If Computer Software locates in Austin and Computer Hardware locates in Tallahassee, then Computer Hardware's profit will be $250,000 greater than Computer Software's profit. If they both locate in Tallahassee, their annual profits will be equal; but if both locate in Austin, Computer Software's annual profit will be $50,000 greater than Computer Hardware's profit.

    (a) Construct the payoff matrix for this two-company game.

    (b) Decide whether the game is strictly determined. If it is, find the saddle point.

(c) If the game is strictly determined, what is the best strategy for both companies?

20. Locating Retail Outlets Competing dry cleaners Peter Pan and Cleanrite are considering putting in a new branch, either in a shopping center or in the suburbs. If both select the shopping center site or the suburban site, then each will get 50% of the business. If, however, one selects the shopping center and the other selects the sub-

urban site, then the business is split 60:40 in favor of the shopping center site.

(a) Construct the payoff matrix for this two-business game.

(b) Find the saddle point, and determine the best strategy for each dry-cleaning firm.

Section 12.2 The Expected Value of Games with Mixed Strategies

In Section 12.1, we saw a way to find the optimal strategy for two-person, strictly determined games. The optimal strategy was found by locating a saddle point, which, in turn, meant that the row player always selected the strategy of the row in which the point resided and the column player always selected the strategy of the column in which that point resided. The term *strictly determined,* or *pure strategy,* meant that every time the game was played, the optimal strategy never deviated from the row–column selections determined by the saddle point. If both players select their optimal strategies, the outcome of the game is fixed and is called the *value of the game.* Even if one player discovers the strategy of the other, the first player will not change strategy, because the selection of a different strategy will not improve her or his standing in the game. A little reflection will convince us that most games are *not* strictly determined, but rather call for different strategies if the opposing player changes strategies. Such games are called games of **mixed strategy,** compared with the **pure strategy** of a strictly determined game.

A simple example of such a game is the game in Example 1 of Section 12.1, in which Match and Mismatch each select heads or tails on a penny. Upon revealing their choices, Match gets $1 from Mismatch if the coins match; if they do not match, Mismatch gets $1 from Match. Consider Match as the row player, the payoff matrix is as before:

$$
\begin{array}{c}
 & \text{Mismatch} \\
 & \begin{array}{cc} H & T \end{array} \\
\text{Match} \begin{array}{c} H \\ T \end{array} & \left[\begin{array}{cc} 1 & -1 \\ -1 & 1 \end{array} \right].
\end{array}
$$

The payoff matrix reveals that this is not a strictly determined game. Further verification comes from the fact that if one of the players adopts a pure strategy, the other will soon discover it, counter, and win every time. If there is a best strategy for the game, it must involve random choices among the available options of showing heads or tails. It turns out that if any such strategy is determined, the expected value (expected earnings) to the row player may be computed. (See Section 8.3 for a review of expected values.) To see how this is done, suppose that Match decides to reveal heads $\frac{3}{5}$ of the time and tails $\frac{2}{5}$ of the time, while Mismatch decides to reveal heads $\frac{1}{6}$ of the time and tails $\frac{5}{6}$ of the time, each in some random order. Because these selections are independent in the probability sense, the probability of each head–tail combination is a product of the probabilities shown in the following table:

Match–Mismatch	Probability	Payoff matrix entry
Head–head	$\left(\frac{3}{5}\right)\left(\frac{1}{6}\right) = \frac{3}{30}$	1
Head–tail	$\left(\frac{3}{5}\right)\left(\frac{5}{6}\right) = \frac{15}{30}$	−1
Tail–head	$\left(\frac{2}{5}\right)\left(\frac{1}{6}\right) = \frac{2}{30}$	−1
Tail–tail	$\left(\frac{2}{5}\right)\left(\frac{5}{6}\right) = \frac{10}{30}$	1

TABLE 1.   PAYOFF PROBABILITIES, GIVEN A MIXED STRATEGY OF $\left(\frac{3}{5}, \frac{2}{5}\right)$ FOR MATCH AND $\left(\frac{1}{6}, \frac{5}{6}\right)$ FOR MISMATCH

Because the payoff matrix expresses gains and losses in terms of the row player—in this case, Match—the expected value of Match's earnings is

$$E(\text{Match}) = \left(\frac{3}{30}\right)(1) + \left(\frac{15}{30}\right)(-1) + \left(\frac{2}{30}\right)(-1) + \left(\frac{10}{30}\right)(1) = -\frac{4}{30}$$

or a loss of $\$\frac{4}{30} \approx 13$ cents per game. Thus, the game favors Mismatch, so if Mismatch stays with this strategy, Match should change in hope of gaining a better advantage.

The expected value just calculated can be obtained by putting the proportion of times (probabilities) for selecting heads and tails of the row player (Match) into a row matrix

$$P = \begin{bmatrix} \frac{3}{5} & \frac{2}{5} \end{bmatrix},$$

putting the probabilities of the two choices for the column player (Mismatch) into a column matrix

$$Q = \begin{bmatrix} \frac{1}{6} \\ \frac{5}{6} \end{bmatrix},$$

and letting the payoff matrix be represented by A. Then the row player's expected value is

$$E(\text{Row}) = PAQ = \begin{bmatrix} \frac{3}{5} & \frac{2}{5} \end{bmatrix} \begin{bmatrix} 1 & -1 \\ -1 & 1 \end{bmatrix} \begin{bmatrix} \frac{1}{6} \\ \frac{5}{6} \end{bmatrix} = -\frac{4}{30}.$$

The preceding example may be generalized by denoting the payoff and mixed-strategies matrices as

$$A = \begin{bmatrix} a_{11} & a_{12} \\ a_{21} & a_{22} \end{bmatrix}, \quad P = \begin{bmatrix} p_1 & p_2 \end{bmatrix}, \quad \text{and} \quad Q = \begin{bmatrix} q_1 \\ q_2 \end{bmatrix},$$

where p_1, p_2, and q_1, q_2 are the respective proportions of time (probabilities) that the row and column players choose the first and second options. Note that $p_1 + p_2 = 1$ and $q_1 + q_2 = 1$. If the row player and the column player both choose the first option, with payoff a_{11} to the row player, the expected gain for the row player is $(p_1 q_1)a_{11}$; if the row player chooses option 1 and the column player chooses option 2, then the expected gain for the row player is $(p_1 q_2)a_{12}$; and so forth. The sum of such products gives the expected value for the row player. In matrix form, the expected value can be expressed as

$$E(\text{Row}) = PAQ = \begin{bmatrix} p_1 & p_2 \end{bmatrix} \begin{bmatrix} a_{11} & a_{12} \\ a_{21} & a_{22} \end{bmatrix} \begin{bmatrix} q_1 \\ q_2 \end{bmatrix}$$

$$= \begin{bmatrix} p_1 a_{11} + p_2 a_{21} & p_1 a_{12} + p_2 a_{22} \end{bmatrix} \begin{bmatrix} q_1 \\ q_2 \end{bmatrix}$$

$$= (p_1 q_1)a_{11} + (p_2 q_1)a_{21} + (p_1 q_2)a_{12} + (p_2 q_2)a_{22}.$$

Since this is a zero-sum game, the expected value for the column player will be

$$E(\text{Column}) = -E(\text{Row}).$$

We define the **expected value** E of a set of strategies as precisely the row player's expected value.

The Expected Value of a Two-Person, Zero-Sum Game

If the row player chooses the ith option with probability p_i and the column player chooses the jth option with probability q_j, then the two together occur with probability $p_i q_j$. If the entry in the ith row and jth column of the payoff matrix is a_{ij}, then the product $(p_i q_j)a_{ij}$ is the expected return to the row player for this combination of options. The **expected value** for the game is $\sum (p_i q_j)a_{ij}$, which may be written in matrix form as PAQ, where P is the row matrix giving the proportion of time the row player chooses the individual strategies available to him or her, A is the payoff matrix, and Q is the column matrix giving the proportion of time the column player chooses the individual strategies available to him or her.

EXAMPLE 1

The Expected Value of a
Two-Person Game

Suppose that Match and Mismatch were simply tossing their coins before revealing whether heads or tails appeared. In this case, the strategies of both Match and Mismatch are randomly determined, with heads and tails appearing equally likely with probability $\frac{1}{2}$. Thus,

$$P = \begin{bmatrix} \frac{1}{2} & \frac{1}{2} \end{bmatrix} \quad \text{and} \quad Q = \begin{bmatrix} \frac{1}{2} \\ \frac{1}{2} \end{bmatrix} \quad \text{so that } PAQ = \begin{bmatrix} \frac{1}{2} & \frac{1}{2} \end{bmatrix} \begin{bmatrix} 1 & -1 \\ -1 & 1 \end{bmatrix} \begin{bmatrix} \frac{1}{2} \\ \frac{1}{2} \end{bmatrix} = 0.$$

This result is not unexpected: Even without the computation, we would expect the game to be fair. ∎

EXAMPLE 2

The Expected Value of a
Two-Person Game

Return to the game Match and Mismatch played in Section 12.1, and suppose that, because an earlier decision to show heads $\frac{3}{5}$ of the time resulted in a negative expected payoff, Match decides to change strategy and show heads $\frac{1}{5}$ of the time and tails $\frac{4}{5}$ of the time. In this case, the expected value of the game (and, more specifically, for the row player, Match) is

$$PAQ = \begin{bmatrix} \frac{1}{5} & \frac{4}{5} \end{bmatrix} \begin{bmatrix} 1 & -1 \\ -1 & 1 \end{bmatrix} \begin{bmatrix} \frac{1}{6} \\ \frac{5}{6} \end{bmatrix} = \frac{12}{30} = 0.4,$$

a much more desirable payoff. ■

Just guessing a proportion of time to reveal heads and tails for either Match or Mismatch is not a very logical method of proceeding if either truly wants to maximize the payoff. We have seen that the expected values can vary significantly upon changing the proportions of time that heads and tails are revealed by Match, and the same would be true for Mismatch. Is there a best way to play this game, as well as other games of a similar nature? The fundamental theorem of game theory tells us that *every* game has a solution—that is, a best strategy.

The Fundamental Theorem of Game Theory

Using the notation of this section, there are optimal sets of probabilities (in matrix form) P_{opt} and Q_{opt} for the row and column players, respectively, and a number v such that the matrix product

$$P_{opt}AQ \geq v \text{ for every strategy } Q \text{ available to the column player,}$$

$$PAQ_{opt} \leq v \text{ for every strategy } P \text{ available to the row player,}$$

and

$$P_{opt}AQ_{opt} = v.$$

The number v is called the **value** of the game, and P_{opt} and Q_{opt} are called the **optimal strategies** for the row and column players, respectively. Every game has such a solution.

The guarantee that such an optimal solution always exists for two-player, zero-sum games is also known in the literature as the maximin theorem. Although Von Neumann gave the first proof of this theorem, George Dantzig later gave a proof using linear programming and the simplex method, while the recently celebrated John Nash gave what is considered the most elegant proof.

HISTORICAL NOTE
JOHN F. NASH (1928–PRESENT)

John Nash has made many major contributions in multiple areas of mathematics, including ground-breaking work in game theory. It is for the latter and associated results that Nash was honored with the Nobel prize in economics in 1994. (Interestingly, Nash is credited with saying that the prize cited his most trivial work!) Nash also has the distinction of being one of the most well-known mathematicians to the general public; his life is the subject of the acclaimed book and movie *A Beautiful Mind*.

For strictly determined games, it is easy to find P_{opt} and Q_{opt}. For instance, in Example 2 of Section 12.1, the optimal strategy for Appliance City was the second option—discounting 30%—and the same was true for the column player, Appliance World. Therefore,

$$P_{opt} = \begin{bmatrix} 0 & 1 \end{bmatrix} \quad \text{and} \quad Q_{opt} \begin{bmatrix} 0 \\ 1 \end{bmatrix}.$$

We can establish that this is indeed the case by noting the following:

$$\begin{bmatrix} 0 & 1 \end{bmatrix} \begin{bmatrix} 0.10 & 0.05 \\ 0.20 & 0.10 \end{bmatrix} Q = \begin{bmatrix} 0.20 & 0.10 \end{bmatrix} Q = \begin{bmatrix} 0.20 & 0.10 \end{bmatrix} \begin{bmatrix} q_1 \\ q_2 \end{bmatrix}$$
$$= 0.20 q_1 + 0.10 q_2 \geq 0.10.$$

The inequality follows from the fact that, since $q_1 + q_2 = 1$, it follows that $q_2 = 1 - q_1$, and substitution into $0.20 q_1 + 0.10 q_2$ gives

$$0.20 q_1 + 0.10(1 - q_1) = 0.20 q_1 + 0.10 - 0.10 q_1 = 0.10 q_1 + 0.10.$$

Thus, $0.20 q_1 + 0.10 q_2 = 0.10 q_1 + 0.10 \geq 0.10$, because the probability q_1 must be between 0 and 1, inclusive.
    Likewise,

$$P \begin{bmatrix} 0.10 & 0.05 \\ 0.20 & 0.10 \end{bmatrix} \begin{bmatrix} 0 \\ 1 \end{bmatrix} = P \begin{bmatrix} 0.05 \\ 0.10 \end{bmatrix} = \begin{bmatrix} p_1 & p_2 \end{bmatrix} \begin{bmatrix} 0.05 \\ 0.10 \end{bmatrix} = 0.05 p_1 + 0.10 p_2 \leq 0.10,$$

because $p_1 + p_2 = 1$. Therefore, the fundamental theorem tells us that 0.10 is the expected value of the game, as we had discovered before. As was the case here, any pure-strategy row or column matrix will consist of one 1 and all other entries 0.

EXAMPLE 3

Showing That a Strategy is Optimal

Consider the game with payoff matrix

$$\begin{bmatrix} 3 & 1 \\ 2 & 4 \end{bmatrix}.$$

Show that the strategies

$$P = \begin{bmatrix} \frac{1}{2} & \frac{1}{2} \end{bmatrix} \quad \text{and} \quad Q = \begin{bmatrix} \frac{3}{4} \\ \frac{1}{4} \end{bmatrix}$$

are optimal.

SOLUTION First, the expected value is computed, with the column player's strategy left unknown. From this, we can find a lower bound on the expected value:

$$E = \begin{bmatrix} 0.5 & 0.5 \end{bmatrix} \begin{bmatrix} 3 & 1 \\ 2 & 4 \end{bmatrix} Q = \begin{bmatrix} 2.5 & 2.5 \end{bmatrix} \begin{bmatrix} q_1 \\ q_2 \end{bmatrix} = 2.5 q_1 + 2.5 q_2 \geq 2.5.$$

Regardless of what strategy the column player chooses, the expected value of the game will be greater than 2.5. We then do a similar computation to find an upper bound on the expected value:

$$E = P \begin{bmatrix} 3 & 1 \\ 2 & 4 \end{bmatrix} \begin{bmatrix} 0.75 \\ 0.25 \end{bmatrix} = \begin{bmatrix} p_1 & p_2 \end{bmatrix} \begin{bmatrix} 2.5 \\ 2.5 \end{bmatrix} = 2.5p_1 + 2.5p_2 \leq 2.5.$$

Putting the two inequalities together indicates that the row player's strategy guarantees at least an expected value of 2.5, while the column player's strategy guarantees an expected value of no more than 2.5. Thus, by our fundamental theorem, these strategies are optimal, and 2.5 is the value for the game. As a check, we have

$$v = \begin{bmatrix} 0.5 & 0.5 \end{bmatrix} \begin{bmatrix} 3 & 1 \\ 2 & 4 \end{bmatrix} \begin{bmatrix} 0.75 \\ 0.25 \end{bmatrix} = \begin{bmatrix} 2.5 & 2.5 \end{bmatrix} \begin{bmatrix} 0.75 \\ 0.25 \end{bmatrix} = 2.5. \quad \blacksquare$$

Even though finding P_{opt} and Q_{opt} is rather easy for pure-strategy games, it is not quite as easy for games of mixed strategy. The next section shows how that is done and completes our solution-finding efforts for game theory. As we shall see, the value of a game, v, may be determined with the aid of linear programming.

Exercises 12.2

In Exercises 1 through 8, find the expected value of the game whose payoff matrix is given in which the row player and the column player each choose strategies that have equal probabilities.

1. $\begin{bmatrix} 2 & -1 \\ 0 & 3 \end{bmatrix}$

2. $\begin{bmatrix} 4 & -5 \\ -6 & 8 \end{bmatrix}$

3. $\begin{bmatrix} 2 & -1 & 4 \\ 0 & 3 & -2 \\ 1 & 2 & -5 \end{bmatrix}$

4. $\begin{bmatrix} 0 & 1 & -1 \\ 1 & 1 & 1 \\ -3 & 2 & 1 \end{bmatrix}$

5. $\begin{bmatrix} -1 & 5 & 4 \\ 2 & 3 & -6 \end{bmatrix}$

6. $\begin{bmatrix} -3 & 4 \\ 2 & -3 \\ 1 & -1 \end{bmatrix}$

7. $\begin{bmatrix} -3 & 2 & 5 \\ 1 & 4 & -3 \\ -5 & 1 & 3 \end{bmatrix}$

8. $\begin{bmatrix} 2 & -3 & -5 \\ -1 & 4 & -3 \end{bmatrix}$

9. Suppose that the row player chooses a strategy

$$P = \begin{bmatrix} \frac{3}{8} & \frac{5}{8} \end{bmatrix}$$

and the column player chooses a strategy

$$Q = \begin{bmatrix} \frac{4}{11} \\ \frac{5}{11} \\ \frac{2}{11} \end{bmatrix}, \quad \text{where } A = \begin{bmatrix} -3 & 5 & 7 \\ 0 & -2 & -4 \end{bmatrix}.$$

Find the expected value of this game.

10. Suppose that the row player chooses a strategy

$$P = \begin{bmatrix} \frac{1}{6} & \frac{2}{6} & \frac{3}{6} \end{bmatrix}$$

and the column player chooses a strategy

$$Q = \begin{bmatrix} \frac{5}{8} \\ \frac{3}{8} \end{bmatrix}, \quad \text{where } A = \begin{bmatrix} 4 & -2 \\ 0 & 5 \\ -3 & 6 \end{bmatrix}.$$

Find the expected value of this game.

In Exercises 11 through 16, suppose that the row player has chosen a strategy

$$P = \begin{bmatrix} \frac{3}{5} & \frac{2}{5} \end{bmatrix}$$

and that the column player decides to choose either

$$Q_1 = \begin{bmatrix} \frac{2}{3} \\ \frac{1}{3} \end{bmatrix} \quad or \quad Q_2 = \begin{bmatrix} \frac{7}{11} \\ \frac{4}{11} \end{bmatrix}.$$

Use the expected value to decide which of the column player's strategies is better for each of the payoff matrices listed.

11. $\begin{bmatrix} 0 & 2 \\ 1 & -3 \end{bmatrix}$ **12.** $\begin{bmatrix} -2 & 3 \\ 4 & -1 \end{bmatrix}$

13. $\begin{bmatrix} \frac{3}{4} & \frac{1}{2} \\ \frac{1}{3} & -\frac{5}{4} \end{bmatrix}$ **14.** $\begin{bmatrix} 0.2 & -0.5 \\ -0.8 & 0.7 \end{bmatrix}$

15. $\begin{bmatrix} -3 & 4 \\ 2 & -1 \end{bmatrix}$ **16.** $\begin{bmatrix} \frac{1}{2} & \frac{5}{6} \\ \frac{3}{5} & -\frac{2}{3} \end{bmatrix}$

In Exercises 17 through 20, each payoff matrix is for a game of pure strategy. Write the row and column matrices that represent the optimal strategy, and show that those matrices fulfil the conditions of the fundamental theorem of game theory.

17. $\begin{bmatrix} 3 & -5 \\ 4 & 6 \end{bmatrix}$ **18.** $\begin{bmatrix} -2 & -3 \\ 4 & 3 \end{bmatrix}$

19. $\begin{bmatrix} 4 & -7 \\ -3 & -8 \end{bmatrix}$ **20.** $\begin{bmatrix} -4 & 3 \\ -2 & 5 \end{bmatrix}$

21. Digit Game Tanya and Ilke play a game in which they simultaneously hold out either one or two fingers. If the sum of the fingers held out is even, Tanya wins 5 cents from Ilke; if the sum of the fingers held out is odd, Ilke wins 5 cents from Tanya.

(a) Find the payoff matrix for this game.

(b) Find the expected value for the game if each player has a strategy of holding out one finger twice as often as two fingers. Whom does this strategy favor?

22. Raffle Tickets Suppose that 1000 raffle tickets are sold for $1 each and two prizes are to be awarded: one of $400 and one of $100. You are to be the row player, and you have two choices: to buy a ticket or not to buy a ticket. The luck of the draw is the column player, and there are three choices: Win a net of $399, win a net of $99, or lose $1. Write the payoff matrix for this game, and determine your optimum strategy.

23. A Coin Game Suppose that Jack and Jill both have a penny and a dime in their hand. They simultaneously show a coin on the table. If the sum of the amounts is even, Jack wins an amount equal to the coin he has shown. If the sum is odd, Jill wins an amount equal to the coin she has shown.

(a) Write the payoff matrix for this game, considering Jack as the row player.

(b) If Jack always plays his penny and Jill always plays her dime, what is the expected value of the game?

(c) If Jack always plays his penny and Jill plays her coins in a 50–50 split, what is the value of the game?

(d) Which of the two strategies, the one from part (b) or the one from part (c), is best for Jill?

24. A Card Game Terrence hold three cards in his hand—a 6, an 8, and a 10. He puts one of the cards face down on the table. Jenny then guesses which of the cards is on the table. If Jenny guesses correctly, she wins the value of the card, in cents, from Terrence. If Jenny does not guess correctly, she gives Terrence the value, in cents, of the card on the table.

(a) Construct the payoff matrix for this game, considering Terrence as the row player.

(b) Suppose that Terrence selects his card in an equally likely manner, but Jenny always guesses that the card is an 8. What is the expected value of this game with these strategies?

(c) Suppose that Terrence selects his card so that the probability of selection is proportional to the value of the card and Jenny guesses the 6 three-fifths of the time, the 8 two-fifths of the time, and the 10 never. Find the expected value of the game for these strategies.

(d) Which strategy, the one in part (b) or the one in part (c), favors Terrence the most?

Section 12.3 Solving Mixed-Strategy Games

The fundamental theorem of game theory tells us how to identify the solution to a mixed-strategy game, but not how to find that solution. For games that are not strictly determined, we show in this section how the optimal strategy for both the row and the column player may be found by solving dual linear programming problems.

The fundamental theorem of game theory leads us to a result that can be used to search for optimal solutions.

**Solution Theorem for Mixed-Strategy Games
That Are Not Strictly Determined**

Let A be the payoff matrix for a game that is not strictly determined, and let R' and C' be strategies for the row and the column players, respectively. Let v be a number such that

1. $R'AC_i \geq v$ for all *pure* strategies C_i of the column player and
2. $R_jAC' \leq v$ for all *pure* strategies R_j of the row player.

Then the solution to the game is $R' = R_{\text{opt}}$ and $C' = C_{\text{opt}}$; v is the **value of the game.**

An example of how the optimal strategies of the row player can be cast in the light of a linear programming problem is given next.

EXAMPLE 1

The Row Player's
Optimal Strategy

Construct a linear programming problem that will determine the row player's optimal strategy for the payoff matrix.

$$A = \begin{bmatrix} 3 & 1 \\ 2 & 4 \end{bmatrix}.$$

SOLUTION Let

$$P = \begin{bmatrix} p_1 & p_2 \end{bmatrix} \quad \text{and} \quad Q = \begin{bmatrix} q_1 \\ q_2 \end{bmatrix}$$

represent any strategies for the row and column players, respectively. Then the expected value of the game is

$$E = PAQ = \begin{bmatrix} p_1 & p_2 \end{bmatrix} \begin{bmatrix} 3 & 1 \\ 2 & 4 \end{bmatrix} \begin{bmatrix} q_1 \\ q_2 \end{bmatrix}$$

$$= \begin{bmatrix} 3p_1 + 2p_2 & p_1 + 4p_2 \end{bmatrix} \begin{bmatrix} q_1 \\ q_2 \end{bmatrix}$$

$$= (3p_1 + 2p_2)q_1 + (p_1 + 4p_2)q_2.$$

The column player must choose either Column 1 or Column 2. If Column 1 is the choice, then $q_1 = 1$, $q_2 = 0$, and the expected value becomes

$$E = 3p_1 + 2p_2. \tag{1}$$

If Column 2 is the column player's choice, then $q_1 = 0$, $q_2 = 1$, and the expected value becomes

$$E = p_1 + 4p_2. \tag{2}$$

The goal of the row player is to find values of p_1 and p_2 such that $p_1 + p_2 = 1$, which will maximize the expected value of the game. Since the column player will select the lesser of the two expected values in Equations (1) and (2), let v be the smaller of those values. Then, in either case, $E \geq v$, which means that if v is maximized, so will E be. We can also assert that $v > 0$, because all entries in the matrix A are positive. Equations (1) and (2) may now be written as

$$3p_1 + 2p_2 \geq v$$
$$p_1 + 4p_2 \geq v.$$

Next, we divide each of these inequalities by v to obtain

$$3\left(\frac{p_1}{v}\right) + 2\left(\frac{p_2}{v}\right) \geq 1$$
$$\left(\frac{p_1}{v}\right) + 4\left(\frac{p_2}{v}\right) \geq 1.$$

For simplicity, substitute $X = p_1/v$ and $Y = p_2/v$ to make these preceding two equations read

$$3X + 2Y \geq 1 \tag{3}$$
$$X + 4Y \geq 1. \tag{4}$$

Since $p_1 \geq 0$ and $p_2 \geq 0$, we have

$$X \geq 0 \quad \text{and} \quad Y \geq 0. \tag{5}$$

Equations (3), (4), and (5) are now recognizable as the constraint system for a linear programming problem. The objective function for this problem comes from the fact that $p_1 + p_2 = 1$. Divide both sides by v to obtain

$$\frac{p_1}{v} + \frac{p_2}{v} = \frac{1}{v} \quad \text{or} \quad X + Y = \frac{1}{v}. \tag{6}$$

Keep in mind that we are to maximize v. This will be done if $1/v$ is *minimized* or, in view of line (6), if $X + Y$ is minimized. Finally, a linear programming problem to do just that can be stated from lines (6), (3), (4), and (5):

$$\text{Minimize } g = X + Y$$
$$\text{subject to } 3X + 2Y \geq 1$$
$$X + 4Y \geq 1$$
$$X \geq 0, Y \geq 0.$$

Notice how the coefficients of X and Y in the main constraints are the *transpose* of the original payoff matrix A. Once the values of X and Y have been determined, line **(6)** allows us to find $v = 1/(X + Y)$. The substitutions $X = p_1/v$, and $Y = p_2/v$ finally allow us to write the optimal row strategy as $p_1 = vX$ and $p_2 = vY$. ■

Let us summarize the procedure for finding the optimal row strategy for 2×2 payoff matrices. This statement can be generalized in an evident way to larger payoff matrices.

How to Determine the Optimal Row Strategy

Let the payoff matrix be $\begin{bmatrix} a & b \\ c & d \end{bmatrix}$, where a, b, c, and d are positive numbers. Solve the linear programming problem

$$\text{Minimize } g = X + Y$$
$$\text{subject to } aX + cY \geq 1$$
$$bX + dY \geq 1$$
$$X \geq 0, \ y \geq 0.$$

The expected value of the game is $v = \dfrac{1}{X + Y}$.

The optimal row strategy is $[p_1 \quad p_2]$, where $p_1 = vX$ and $p_2 = vY$.

Going through a similar analysis, we can arrive at the following method for finding the optimal column strategy:

How to Determine the Optimal Column Strategy

Let the payoff matrix be $\begin{bmatrix} a & b \\ c & d \end{bmatrix}$, where a, b, c, and d are positive numbers. Solve the linear programming problem

$$\text{Maximize } f = U + V$$
$$\text{subject to } aU + bV \leq 1$$
$$cU + dV \leq 1$$
$$U \geq 0, V \geq 0.$$

The expected value of the game is $v = \dfrac{1}{U + V}$.

The optimal column strategy is $\begin{bmatrix} q_1 \\ q_2 \end{bmatrix}$, where $q_1 = vU$ and $q_2 = vV$.

Notice that this time the coefficient matrix of the structural constraints is precisely the payoff matrix. Upon comparing the linear programming problems for the row and column strategies, we find that they are *dual problems*. This means that both problems can efficiently be solved simultaneously by using the standard maximization problem for the column player's optimal strategy and then using the concept of duality to read the row player's optimal strategy from the bottom of the slack-variables column in the final tableau. To demonstrate these solutions, suppose we solve the linear programming problem that will determine the column player's optimal strategy, using the payoff matrix of Example 1, namely,

$$A = \begin{bmatrix} 3 & 1 \\ 2 & 4 \end{bmatrix}.$$

The entries in A are all positive. From the column player's point of view, the linear programming problem is the following standard maximization problem:

$$\text{Maximize } f = U + V$$
$$\text{subject to } 3U + V \le 1$$
$$2U + 4V \le 1$$
$$U \ge 0, V \ge 0.$$

The initial tableau with the first pivot element circled is as follows:

$$\begin{bmatrix} 3 & 1 & 1 & 0 & 0 & 1 \\ 2 & ④ & 0 & 1 & 0 & 1 \\ -1 & -1 & 0 & 0 & 1 & 0 \end{bmatrix}.$$

Completion of the first pivot yields this tableau:

$$\begin{bmatrix} ⑤/② & 0 & 1 & -\frac{1}{4} & 0 & \frac{3}{4} \\ \frac{1}{2} & 1 & 0 & \frac{1}{4} & 0 & \frac{1}{4} \\ -\frac{1}{2} & 0 & 0 & \frac{1}{4} & 1 & \frac{1}{4} \end{bmatrix}.$$

Pivoting on the circled element in the preceding matrix gives the final tableau:

$$\begin{bmatrix} ① & 0 & \frac{2}{5} & -\frac{1}{10} & 0 & \frac{3}{10} \\ 0 & 1 & -\frac{1}{5} & \frac{3}{10} & 0 & \frac{1}{10} \\ 0 & 0 & \frac{1}{5} & \frac{1}{5} & 1 & \frac{2}{5} \end{bmatrix}.$$

From this matrix, we find that $U = \frac{3}{10}$ and $V = \frac{1}{10}$, so that $U + V = \frac{4}{10} = \frac{2}{5}$. This means that the expected value for this game is $v = \frac{5}{2}$. Therefore, $q_1 = vU = \left(\frac{5}{2}\right)\left(\frac{3}{10}\right) = \frac{3}{4}$ and $q_2 = vV = \left(\frac{5}{2}\right)\left(\frac{1}{10}\right) = \frac{1}{4}$. The column player's optimal strategy is then given by $\begin{bmatrix} \frac{3}{4} \\ \frac{1}{4} \end{bmatrix}$.

The final tableau in the simplex algorithm calculations shows that the dual problem has solutions $X = \frac{1}{5}$ and $Y = \frac{1}{5}$, which we read from the bottom of the slack-variable columns. Thus, $X + Y = \frac{2}{5}$, from which we find that, again, $v = \frac{5}{2}$, as we expect. In this case, $p_1 = vX = (\frac{5}{2})(\frac{1}{5}) = \frac{1}{2}$ and $p_2 = vY = (\frac{5}{2})(\frac{1}{5}) = \frac{1}{2}$. The row player's optimal strategy is then $[\frac{1}{2} \quad \frac{1}{2}]$.

If we want further justification of how this solution relates to the previous section, the maximum expected value, or value of the game, $v = \frac{5}{2}$, should be the result of the matrix multiplication PAQ. For our present example,

$$PAQ = \begin{bmatrix} \frac{1}{2} & \frac{1}{2} \end{bmatrix} \begin{bmatrix} 3 & 1 \\ 2 & 4 \end{bmatrix} \begin{bmatrix} \frac{3}{4} \\ \frac{1}{4} \end{bmatrix} = \begin{bmatrix} \frac{5}{2} & \frac{5}{2} \end{bmatrix} \begin{bmatrix} \frac{3}{4} \\ \frac{1}{4} \end{bmatrix} = \frac{5}{2}.$$

The calculations we have just made with the simplex algorithm show that both the row and column player's optimal strategies can be computed with one linear program. However, we should note that the row player's optimal strategy could have been computed independently of that of the column player, and it could be done in the case of a 2×2 payoff matrix either by the graphical method or by Crown's method applied to a minimization problem. Similarly, the column player's optimal strategy could have been found independently of the row player's, and, in the 2×2 payoff matrix, either by the graphical method or the simplex algorithm applied to a standard maximization problem.

Finally, let us comment on the need for all entries in the payoff matrix A to be positive. This assumption was adopted so that v would be positive and hence we could be assured that the inequalities just above Equations (3) and (4) in Example 1 would be preserved. However, we can relax that restriction now. If some entries in the payoff matrix are not positive, we find some positive number n such that, when n is added to each entry in the payoff matrix A, a new matrix B will result in which all entries are positive. The use of B will produce precisely the same optimal strategies as would be found by using A. The value of the game, however, will be that found by using matrix B and then subtracting the number n. We demonstrate this concept in the next example, which also answers the question posed at the outset of this chapter.

EXAMPLE 2
Solving a Game That Is Not Strictly Determined

A debate between Senator Smith and her opponent for reelection is almost certain to center on two issues: foreign policy (FP) and defense (D). The senator's managers feel that her opponent is strong on the defense issue, but not as knowledgeable as the senator is about foreign policy, and they estimate the points earned in the debate in various combinations of these two issues as follows:

A gain of 3 points for Senator Smith when both discuss foreign policy,

A gain of 5 points for her opponent when both discuss the defense issue,

A gain of 10 points for Senator Smith if she discusses foreign policy while her opponent discusses defense, and

A gain of 6 points for Senator Smith if she discusses defense while her opponent discusses foreign policy.

What fraction of the debate should Senator Smith spend on each issue to take best advantage of the point estimates? (*Note:* One point represents 1% of the voting public.)

SOLUTION Suppose that we consider Senator Smith the row player. The payoff matrix then has this appearance:

$$
\begin{array}{cc}
 & \begin{array}{cc} \text{Opponent} & \\ \text{FP} & \text{D} \end{array} \\
\text{Senator Smith} \begin{array}{c} \text{FP} \\ \text{D} \end{array} & \begin{bmatrix} 3 & 10 \\ 6 & -5 \end{bmatrix}.
\end{array}
$$

This is evidently not a strictly determined game, which means that we may apply the linear programming techniques just discussed. We begin by adding 6 to each entry in the payoff matrix to produce the matrix

$$
B = \begin{bmatrix} 9 & 16 \\ 12 & 1 \end{bmatrix}
$$

with all positive entries. From the column player's viewpoint, the linear programming problem would appear as the standard maximization problem

$$
\text{Maximize } f = U + V
$$
$$
\text{subject to} \quad 9U + 16V \le 1
$$
$$
12U + \quad V \le 1
$$
$$
U \ge 0, V \ge 0.
$$

The tableaus needed to solve this problem are as follows:

$$
\begin{bmatrix}
9 & 16 & 1 & 0 & 0 & 1 \\
\boxed{12} & 1 & 0 & 1 & 0 & 1 \\
\hline
-1 & -1 & 0 & 0 & 1 & 0
\end{bmatrix};
\begin{bmatrix}
0 & \boxed{\frac{61}{4}} & 1 & -\frac{3}{4} & 0 & \frac{1}{4} \\
1 & \frac{1}{12} & 0 & \frac{1}{12} & 0 & \frac{1}{12} \\
\hline
0 & -\frac{11}{12} & 0 & \frac{1}{12} & 1 & \frac{1}{12}
\end{bmatrix};
$$

$$
\begin{bmatrix}
0 & 1 & \frac{4}{61} & -\frac{3}{61} & 0 & \frac{1}{61} \\
1 & 0 & -\frac{1}{183} & \frac{16}{183} & 0 & \frac{5}{61} \\
\hline
0 & 0 & \frac{11}{183} & \frac{7}{183} & 1 & \frac{6}{61}
\end{bmatrix}.
$$

Since we are interested in Senator Smith's optimal strategy, and since she is considered the row player, we use the duality principle and read the values of X and Y from the bottom of the slack-variable columns:

$$
X = \frac{11}{183} \text{ and } Y = \frac{7}{183} \text{ so that } X + Y = \frac{18}{183} \text{ and hence } v = \frac{183}{18}.
$$

Therefore, $p_1 = vX = \left(\frac{183}{18}\right)\left(\frac{11}{183}\right) = \frac{11}{18}$ and $p_2 = vY = \left(\frac{183}{18}\right)\left(\frac{7}{183}\right) = \frac{7}{18}$, which implies that Senator Smith's optimal strategy is $\begin{bmatrix} \frac{11}{18} & \frac{7}{18} \end{bmatrix}$. That is, the senator should spend $\frac{11}{18}$ of her debate time talking about foreign policy and $\frac{7}{18}$ of the time talking about defense to gain the maximum number of points. The number of points gained by doing this is the value of the game. That value is $v = \frac{183}{18}$ minus the number 6, which was added to the entries in the original payoff matrix at the outset. Since $\frac{183}{18} \approx 10.167$, the value of the original game is $10.167 - 6 = 4.167$, or about 4.2 points. ∎

The next example shows that the linear programming techniques discussed for finding optimal strategies apply to matrices of any size.

EXAMPLE 3

Larger Payoff Matrices

Find the optimal strategies for both the row and column players in a game with the payoff matrix

$$A = \begin{bmatrix} 2 & -2 & 3 \\ -4 & 0 & -1 \end{bmatrix}.$$

SOLUTION We begin by adding the number 5 to each entry in A, resulting in the matrix

$$B = \begin{bmatrix} 7 & 3 & 8 \\ 1 & 5 & 4 \end{bmatrix}.$$

The standard maximization problem for the column player's optimal strategy becomes

$$\text{Maximize } f = U + V + W$$
$$\text{subject to } 7U + 3V + 8W \leq 1$$
$$U + 5V + 4W \leq 1$$
$$U \geq 0, V \geq 0, W \geq 0.$$

The tableaus leading to the solution are as follows:

$$\begin{bmatrix} \boxed{7} & 3 & 8 & 1 & 0 & 0 & 1 \\ 1 & 5 & 4 & 0 & 1 & 0 & 1 \\ \hline -1 & -1 & -1 & 0 & 0 & 1 & 0 \end{bmatrix}; \begin{bmatrix} 1 & \frac{3}{7} & \frac{8}{7} & \frac{1}{7} & 0 & 0 & \frac{1}{7} \\ 0 & \boxed{\frac{32}{7}} & \frac{20}{7} & -\frac{1}{7} & 1 & 0 & \frac{6}{7} \\ \hline 0 & -\frac{4}{7} & \frac{1}{7} & \frac{1}{7} & 0 & 1 & \frac{1}{7} \end{bmatrix};$$

$$\begin{bmatrix} 1 & 0 & \frac{7}{8} & \frac{5}{32} & -\frac{3}{32} & 0 & \frac{1}{16} \\ 0 & 1 & \frac{5}{8} & -\frac{1}{32} & \frac{7}{32} & 0 & \frac{3}{16} \\ \hline 0 & 0 & \frac{1}{2} & \frac{1}{8} & \frac{1}{8} & 1 & \frac{1}{4} \end{bmatrix}.$$

The last tableau shows that $U = \frac{1}{16}$, $V = \frac{3}{16}$, and $W = 0$. Thus, $U + V + W = \frac{4}{16} = \frac{1}{4}$, from which it follows that $v = 4$. This means that the column player's optimal strategy is

$$q_1 = vU = (4)\left(\frac{1}{16}\right) = \frac{1}{4}, \qquad q_2 = vV = (4)\left(\frac{3}{16}\right) = \frac{3}{4}, \qquad \text{and}$$
$$q_3 = vU = (4)(0) = 0.$$

Using the duality principle, we find that $X = \frac{1}{8}$ and $Y = \frac{1}{8}$. It follows that the row player's optimal strategy is then

$$p_1 = vX = (4)\left(\frac{1}{8}\right) = \frac{1}{2} \qquad \text{and} \qquad p_2 = vY = (4)\left(\frac{1}{8}\right) = \frac{1}{2}.$$

The value of this game is $4 - 5 = -1$. ■

Our last example introduces the use of Excel's `Solver` to find the optimal solution to a two-person, zero-sum game.

EXAMPLE 4

Using Excel to Find Optimal Strategies

Consider again the game given in Example 3, and use Excel to find the optimal strategy for the row player.

SOLUTION We will approach this problem similarly to the way we did in Section 5.2. First, a spreadsheet should be labeled and filled with pertinent information, as in Figure 2. All of the cell entries up to this point are text, not formulas. It is important to recognize the four main parts of our spreadsheet: the range in which we are going to load the values of the variables (B2 : B3), the cell that will contain the objective function's value (B5), the ranges containing the constraints (B7 : B9 and D7 : D9), and the ranges containing the value and optimal strategies of the game (B11 and B14 : B15). This last set of ranges helps us immediately interpret the results for our variables. Notice that we have already added the constant 5 to the payoff entries, so we need to take care to readjust our game value at the end.

	A	B	C	D
1	**Variables**			
2	x			
3	y			
4	**Objective Function**			
5	Min x+y			
6	**Structural Constraints**			
7	7x+y >= 1		>=	
8	3x+5y >= 1		>=	
9	8x+4y >= 1		>=	
10	**Game Value**			
11	v = 1/(x+y)			
12	**Optimal Strategy**			
13	p1 = v x			
14	p2 = v y			

FIGURE 2
Initial Labeling of Spreadsheet

As in Section 5.2, all of the formulas are entered as follows:

Cell	Content	Cell	Content
B2	1		
B3	2		
B5	=B2+B3		
B7	=7*B2+B3	D7	1
B8	=3*B2+5*B3	D8	1
B9	=8*B2+4*B3	D9	1
B11	=1/(B2+B3)		
B13	=B11*B2		
B14	=B11*B3		

After this material is entered, the spreadsheet will look like Figure 3.

	B9	▼	f_x =8*B2+4*B3		
	A		B	C	D
1	**Variables**				
2	x		1.000		
3	y		2.000		
4	**Objective Function**				
5	Min x+y		3.000		
6	**Structural Constraints**				
7	7x+y >= 1		9.000	>=	1
8	3x+5y >= 1		13.000	>=	1
9	8x+4y >= 1		16.000	>=	1
10	**Game Value**				
11	v = 1/(x+y)		0.333		
12	**Optimal Strategy**				
13	p1 = v x		0.333		
14	p2 = v y		0.667		

FIGURE 3
The Initial Values at the Origin of the Feasible Region.

Notice that the initial values of the variables are not zeros. This is due to the fact that we are actually solving a minimization problem directly; thus, our feasible region does not contain the origin. We must select initial values in the feasible region. In general, we pick values that satisfy the constraints.

Now we run the Solver function, which can be found under the Tools menu. The safest first step is to click and confirm the Reset All button, which clears all previous entries. Next, enter the cell ranges exactly as seen in Figure 4. Notice how the addresses

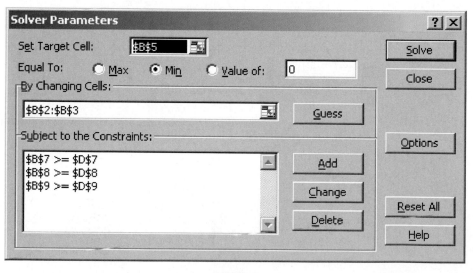

FIGURE 4
The Solver Ready to Run.

needed by `Solver` are divided into precisely the first three areas we set up in the spreadsheet. Since this is a minimization problem, we make sure the correct radio button (`Min`) is selected in the `Equal To` area.

Before clicking the `Solve` button, there is one last step: We select `Options` and make sure that both `Assume Linear Model` and `Assume Non-Negative` are checked. The first option allows the program to be more efficient in its computations; the second is the reason there was no need to include the non-negativity constraints in the spreadsheet.

After `Solver` runs, a pop-up window asks whether the results should be accepted, which, in this case, they should. Our final solution can be read off of the screen, as seen in Figure 5. The optimal strategy for the row player is $P = \lfloor 0.5 \quad 0.5 \rfloor$ (as we already saw in Example 3). Finally, we need to subtract 5 from the stated game value to get the original value of -1.

	A	B	C	D
1	**Variables**			
2	x	0.125		
3	y	0.125		
4	**Objective Function**			
5	Min x+y	0.250		
6	**Structural Constraints**			
7	7x+y >= 1	1.000	>=	1
8	3x+5y >= 1	1.000	>=	1
9	8x+4y >= 1	1.500	>=	1
10	**Game Value**			
11	v = 1/(x+y)	4.000		
12	**Optimal Strategy**			
13	p1 = v x	0.500		
14	p2 = v y	0.500		

FIGURE 5

The Optimal Strategy for the Row Player.   ■

Sometimes, it is possible to reduce the size of the payoff matrix if there is a row that *dominates* another row or column that *dominates* another column. By the term **dominate,** we mean that each entry in one of the rows (columns) is greater than or equal to the corresponding entry in the other row (column). For instance, in the matrix

$$\begin{bmatrix} 4 & -3 \\ -2 & 5 \\ 3 & -6 \end{bmatrix},$$

the first row dominates the third row, because each element of the first row is *greater than* the corresponding element of the third row. When this happens, the row player need never play the third row in any optimal strategy, because he or she could always do at least as well by playing the first row. Consequently, the game may be considered to have the payoff matrix

$$\begin{bmatrix} 4 & -3 \\ -2 & 5 \end{bmatrix}.$$

Thus, even though we are restricting our games that are not strictly determined to ones having 2×2 payoff matrices, it may still be possible to analyze games with larger matrices if dominance is present.

Exercises 12.3

In Exercises 1 through 10, each matrix is the payoff matrix for a two-person, zero-sum game, some of which are strictly determined and some of which are not.

(a) Solve each game. For those that are not strictly determined, solve by using a linear program.

(b) Find the value of each game.

1. $\begin{bmatrix} 2 & -3 \\ 1 & 1 \end{bmatrix}$
2. $\begin{bmatrix} 4 & -2 \\ -3 & 1 \end{bmatrix}$

3. $\begin{bmatrix} -2 & 3 \\ 4 & -5 \end{bmatrix}$
4. $\begin{bmatrix} 2 & 5 \\ 1 & -3 \end{bmatrix}$

5. $\begin{bmatrix} 6 & 4 \\ 3 & -2 \end{bmatrix}$
6. $\begin{bmatrix} 4 & -2 \\ 3 & 4 \end{bmatrix}$

7. $\begin{bmatrix} -1 & 6 \\ 4 & -2 \end{bmatrix}$
8. $\begin{bmatrix} 0 & -2 \\ 3 & 4 \end{bmatrix}$

9. $\begin{bmatrix} 5 & -2 \\ 1 & -3 \\ -2 & 7 \end{bmatrix}$
10. $\begin{bmatrix} 8 & -1 & 6 \\ -3 & 0 & 2 \end{bmatrix}$

In Exercises 11 through 22, use the duality of a linear program to find the row player's and column player's optimal strategy and the value of the game.

11. $\begin{bmatrix} -2 & 4 \\ 3 & -1 \end{bmatrix}$
12. $\begin{bmatrix} -5 & 8 \\ 6 & -2 \end{bmatrix}$

13. $\begin{bmatrix} 1 & 1 \\ 4 & -3 \end{bmatrix}$
14. $\begin{bmatrix} \frac{7}{8} & \frac{1}{4} \\ -3 & \frac{5}{6} \end{bmatrix}$

15. $\begin{bmatrix} -2 & 6 \\ 8 & 0 \\ 6 & -1 \end{bmatrix}$
16. $\begin{bmatrix} 2 & 1 & -4 \\ -3 & -5 & 6 \end{bmatrix}$

17. $\begin{bmatrix} 1 & 3 & 5 \\ 6 & 2 & 1 \\ 4 & 1 & 4 \end{bmatrix}$
18. $\begin{bmatrix} 2 & 5 & 7 \\ 1 & 2 & 4 \\ 6 & 1 & 3 \end{bmatrix}$

19. $\begin{bmatrix} 1 & 0 & 1 \\ -1 & 1 & -1 \\ 0 & -1 & 2 \end{bmatrix}$
20. $\begin{bmatrix} 3 & -1 & 9 \\ 1 & 2 & -3 \\ -2 & 3 & 4 \end{bmatrix}$

21. $\begin{bmatrix} 5 & 0 & 4 \\ -4 & 1 & 2 \end{bmatrix}$
22. $\begin{bmatrix} 6 & 1 \\ -2 & 3 \\ 1 & 7 \end{bmatrix}$

23. **Coin Guessing** Match and Mismatch play this game: Match conceals a coin in one of her hands. If Mismatch guesses which hand the coin is in, then Match gives Mismatch \$3. If Mismatch does not guess correctly, then Mismatch gives Match \$2 if the coin is in the right hand and \$5 if the coin is in the left hand.

(a) What is the optimal strategy for Match?

(b) What is the optimal strategy for Mismatch?

(c) Which player does the game favor?

24. **Business Decisions** A polling firm believes that its livelihood will be affected by the political party in power. The firm must decide on one of two courses of action before the election. The relative strengths to the company, depending on which party wins the election, are shown in the following payoff matrix:

$$\begin{array}{c} \\ \text{Action 1} \\ \text{Action 2} \end{array} \begin{array}{cc} \text{Dem} & \text{Rep} \\ \begin{bmatrix} 4 & -3 \\ -5 & 2 \end{bmatrix} \end{array}.$$

Determine the optimal row strategy for the polling firm, and find the value of this game.

25. **Market Analysis** M-Mart and Q-Mart have stores in a local mall. Each company has daily specials to attract customers. M-Mart features clothing or sporting goods, while Q-Mart features electronics or kitchenware. The swing customers are affected by the specials as follows, according to surveys run by the mall management:

M-Mart gains 200 such customers if it features clothing and Q-Mart features electronics, but loses 100 of them if Q-Mart features kitchenware.

M-Mart loses 300 such customers if it features sporting goods and Q-Mart features electronics, but gains 175 if Q-Mart features kitchenware.

(a) Determine the optimum strategy for each store.

(b) Which store does the game favor?

26. Tomato Farming Tomato growers have two varieties of plants available, one of which does very well if the growing season is dry. The other variety does very well if the season is wet. The payoff matrix showing the relative performances under the two weather scenarios is

$$\begin{array}{cc} & \begin{array}{cc} \text{Dry} & \text{Wet} \end{array} \\ \begin{array}{c} \text{Dry Variety} \\ \text{Wet Variety} \end{array} & \begin{bmatrix} 20 & 8 \\ 10 & 20 \end{bmatrix}. \end{array}$$

Long-range weather bureau predictions indicate that the probability of a dry season is 60%; hence, Mother Nature's strategy is fixed at

$$\begin{bmatrix} 0.60 \\ 0.40 \end{bmatrix}.$$

What is the tomato growers' best strategy? What is the value of this game?

27. A Coin Game Bob and John each have a number of nickels and dimes. They choose one coin each and simultaneously put their coins on the table. If the sum of the amounts is even, Bob takes both coins. If the sum of the amounts is odd, John takes both coins. Determine the optimal strategy for both players.

28. A European Game Hans and Wolfgang play a game in which the objective is to accumulate as many stones from a previously collected pile as they can. They make a fist and then, upon a signal, hold forth either one or two finger. If the number of fingers held forth is even, Hans gets that many stones, but if the number of fingers held forth is odd, Wolfgang gets that many stones. Find the optimal strategy for both players.

Chapter 12 Summary

CHAPTER HIGHLIGHTS

Two-person, zero-sum games involve two competing forces, each trying to devise a winning strategy. When the optimal strategy for each player is found, the game is considered solved. A *strictly determined,* or *pure strategy, game* has a single, invariant course of action as its solution, whereas a *mixed strategy game* involves more than one course of action for its solution. In strictly determined games, the *payoff matrix* has a single saddle point or value that indicates which player is favored or that the game is fair. In mixed-strategy games, different strategies may produce different *expected values,* but optimal strategies still can always be determined.

EXERCISES

In Exercises 1 through 4, calculate the value of each pure-strategy game.

1. $\begin{bmatrix} 10 & 9 \\ 8 & 7 \end{bmatrix}$ 　　 **2.** $\begin{bmatrix} 2 & 0 \\ 4 & 1 \end{bmatrix}$

3. $\begin{bmatrix} -4 & 5 & -6 \\ 2 & 3 & 5 \\ 0 & 1 & -3 \end{bmatrix}$ 　 **4.** $\begin{bmatrix} 8 & 3 & 10 & 2 \\ 7 & 6 & 9 & 3 \\ 7 & 4 & 6 & 4 \end{bmatrix}$

In Exercises 5 through 10, for each payoff matrix A, calculate the expected value, given that the players choose strategies having equal probability.

5. $\begin{bmatrix} 0 & -3 \\ -2 & 1 \end{bmatrix}$ 　　 **6.** $\begin{bmatrix} 11 & 2 \\ 1 & 15 \end{bmatrix}$

7. $\begin{bmatrix} 1 & 2 & 0 \\ 2 & 2 & 2 \\ -2 & 3 & 2 \end{bmatrix}$ 　 **8.** $\begin{bmatrix} -3 & 3 & 2 \\ 0 & 1 & -8 \end{bmatrix}$

9. $\begin{bmatrix} -9 & 12 \\ 6 & -9 \\ 3 & -3 \end{bmatrix}$ **10.** $\begin{bmatrix} -4 & 1 & 4 \\ 0 & 3 & -4 \\ -6 & 0 & 2 \end{bmatrix}$

In Exercises 11 through 14, solve each two-person, zero-sum game.

11. $\begin{bmatrix} 1 & -1 \\ 4 & 5 \end{bmatrix}$ **12.** $\begin{bmatrix} 9 & 0 & 7 \\ -2 & 1 & 3 \end{bmatrix}$

13. $\begin{bmatrix} 2 & 2 \\ 8 & -6 \end{bmatrix}$ **14.** $\begin{bmatrix} -1 & 3 \\ 4 & 0 \\ 3 & -1 \end{bmatrix}$

15. Jamie and Twyla are playing a professional clay-court tennis match. The commentators estimate that Jamie will win 60% of the points when both stay at the baseline and 65% when both are at the net. When Jamie is at the baseline and Twyla at the net, Twyla wins 70% of the points. When Twyla is at the baseline and Jamie at the net, Twyla wins 55% of the points. What percentage of the time should Jamie come to the net, regardless of where her opponent plays?

16. Describe the distinction between mixed-strategy games and pure-strategy games.

17. Indicate the significance of a saddle point for a game.

18. Explain why the term *saddle point* is appropriate for the value of a strictly determined game.

19. Describe the meaning of *expected value* of a two-person, mixed-strategy game.

20. Explain the meaning and significance of the term *dominate* with respect to a payoff matrix.

CUMULATIVE REVIEW

21. Write the equation of a linear cost function if 30 items cost $400 and 55 items cost $600.

22. Complete the pivot operation on the circled element of the augmented matrix

$$\begin{bmatrix} 0 & -7 & 8 & 14 & 3 \\ 5 & 2 & 0 & 4 & 12 \\ 6 & 3 & -2 & ① & 5 \end{bmatrix}.$$

23. Bus routes are available through the cities showing in the following diagram, in the direction of the arrows only:

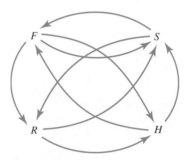

Write a matrix that displays this transportation network.

24. Construct a truth table for $\sim p \vee (p \wedge \sim q)$.

25. Calculate the mean, median, mode, and range for the data 1, 3, 7, 8, 10, 10, and 10.

Chapter 12 Sample Test Items

In Problems 1 through 3, use the given payoff matrix to decide whether the game is strictly determined. If so, find the saddle point for the game, and find whom the game favors.

1. $\begin{bmatrix} 4 & 1 \\ 6 & 3 \end{bmatrix}$ **2.** $\begin{bmatrix} 8 & 5 \\ -2 & 14 \end{bmatrix}$

3. $\begin{bmatrix} 3 & 1 & -4 \\ 0 & 5 & -2 \\ -3 & 4 & -6 \end{bmatrix}$

In Problems 4 and 5, for what values of x is the given payoff matrix strictly determined?

4. $\begin{bmatrix} x & -2 \\ 0 & 3 \end{bmatrix}$ **5.** $\begin{bmatrix} x & -2 \\ 0 & x \end{bmatrix}$

The following description refers to Problems 6 and 7: The First and Second National Banks are each planning to open a branch on campus or at the mall. If they both select the campus or they both select the mall, they will share the business equally. However, if one selects the campus and the other the mall, the bank selecting the campus will get a 55% share of the business versus a 45% share for the other.

6. Construct the payoff matrix for this two-bank game.

7. Find the value of the game, and determine the best strategy for each bank.

In Problems 8 through 10, for each given payoff matrix, find the expected value of the game if both players choose strategies that have equal probabilities.

8. $\begin{bmatrix} 3 & -5 \\ 0 & 8 \end{bmatrix}$ **9.** $\begin{bmatrix} -2 & 6 & 4 \\ 3 & 5 & -7 \end{bmatrix}$

10. $\begin{bmatrix} -1 & 3 & 7 \\ 2 & 3 & -2 \\ -4 & 1 & 5 \end{bmatrix}$

In Problems 11 through 13, the row player has settled on strategy

$$P = \begin{bmatrix} \frac{1}{3} & \frac{2}{3} \end{bmatrix}$$

and the column player decides to choose either

$$Q_1 = \begin{bmatrix} \frac{1}{4} \\ \frac{3}{4} \end{bmatrix} \quad \text{or} \quad Q_2 = \begin{bmatrix} \frac{4}{13} \\ \frac{9}{13} \end{bmatrix}.$$

Use the expected value to decide whether Q_1 or Q_2 is the best strategy for the column player, given each payoff matrix.

11. $\begin{bmatrix} -1 & 2 \\ 3 & 0 \end{bmatrix}$ **12.** $\begin{bmatrix} 0.3 & -0.6 \\ -0.9 & 0.8 \end{bmatrix}$

13. $\begin{bmatrix} \frac{1}{3} & \frac{2}{5} \\ \frac{7}{8} & -\frac{1}{2} \end{bmatrix}$

In Problems 14 through 19, solve each two-person, zero-sum game, using the given payoff matrix.

14. $\begin{bmatrix} 5 & -1 \\ 4 & 5 \end{bmatrix}$ **15.** $\begin{bmatrix} 4 & -1 \\ 3 & 3 \end{bmatrix}$

16. $\begin{bmatrix} 0 & -4 \\ 6 & 8 \end{bmatrix}$ **17.** $\begin{bmatrix} -10 & 16 \\ 12 & -4 \end{bmatrix}$

18. $\begin{bmatrix} 10 & -4 \\ 2 & -6 \\ -4 & 14 \end{bmatrix}$ **19.** $\begin{bmatrix} 9 & 0 & 7 \\ -2 & 1 & 3 \end{bmatrix}$

Mia and Elena have a dime and a quarter each. They each choose one coin and simultaneously put their selected coins on a table. If the coins match, Mia gets the product (in cents) of the two coins. If the coins do not match, Elena gets 10 times the sum (in cents) of the two coins. Use this game to answer 20 and 21.

20. Determine the optimum strategy for each player.

21. Whom does this game favor?

In-Depth Application

Routing Packets Around a Message Interceptor

Objective

In this project, we explore the game theory behind choosing paths in a network to minimize time. Our antagonist is a message interceptor, but the idea can be extended to multiple players, each seeking the fastest route by avoiding congestion.

Introduction

There are many competitive games that are quite enjoyable to play, but there also are a few that are not so enjoyable and that are extremely serious. One such game is avoiding messages being intercepted between two sites on the Internet. The two players are the message sender and the interceptor. The message is divided into multiple packets and sent across a network of paths, each packet passing though several computers (routers) and all eventually to be reassembled at the message's destination computer. Physical eavesdropping along the route lines is virtually useless, as the messages sent between routers will be encrypted. This leaves the starting computer, the routers, and the ending computer as the primary weak links in the system. Assuming that the sender and receiver of the message are extremely careful about their own security, the primary focus of the interceptor will be the routers. The messenger wants a strategy for picking a path through the routers that will minimize both time and interceptions.

SOURCE: http://mathlab.usc.edu/~bohacek/AMICT/Overlays.htm, April 2002, Stephen Bohacek, University of Southern California.

The Network

Figure 6 shows a simple network with five routers between the sender and the receiver, with a total of five paths along which an individual packet can be sent:

(continued)

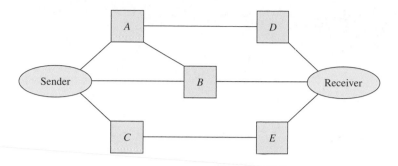

FIGURE 6
The Network Between the Sender and the Receiver

1. Sender–*A*–*D*–Receiver (*AD*)

2. Sender–*A*–*B*–Receiver (*AB*)

3. Sender–*B*–Receiver (*B*)

4. Sender–*B*–*A*–*D*–Receiver (*BAD*)

5. Sender–*C*–*E*–Receiver (*CE*).

Each router has a level of security: Router *A* is security level 4, router *D* is security level 3, routers *B* and *E* are security level 2, and router *C* is security level 1. These levels are important in two ways: They describe how long it takes for a packet to pass through the router, and they indicate the level of resources an interceptor needs to infiltrate the router. For example, it will take a nonintercepted packet $4 + 3 = 7$ time units to travel the path *AD*, and it will take an interceptor 7 resource units to tap into both routers *A* and *D*.

Assume that the infiltrator has four units of resources, giving her several options as to which router(s) she can infiltrate. A final, simplifying assumption is that this is a zero-sum game in which the messenger wants to minimize time while the interceptor wants to maximize time by blocking as many packets as possible. If a packet is intercepted, it will automatically have a time cost of 20 assigned to it.

1. List the time cost for each of the five paths, assuming that the packet is not intercepted.

2. The strategies of the messenger are the five possible paths from which to choose. What are all of the possible strategies for the interceptor? (*Hint:* There are nine.)

3. Make a payoff matrix for this game, with the interceptor as the row player. The matrix should be 9×5.

4. Reflect upon the payoff matrix, network, and dominating strategies in order to reduce the number of useful strategies to two each for the messenger and the interceptor. Record the new 2×2 payoff matrix.

5. From part 4, why did the network structure fail to prevent the interceptor from being a terrible nuisance? If you could add links to the network, where would you put them to foil the interceptor? What principle does this illustrate in regard to the general construction of networks?

6. Find the optimal strategies for the messenger and interceptor. What is the game value?

7. How do the optimal strategies and game values change when the penalty is set at 10, 30, and 40? What happens when the interception penalty becomes exceedingly large compared with the time costs of the paths? How are these results mitigated by your network constructions discussed in part 5?

Appendix I

Algebra Review

Some of the basic algebra skills needed to study finite mathematics are reviewed in this appendix. A list of the topics included and the chapters in which these skills are utilized follows for your convenience:

The real numbers: All chapters

Properties of arithmetic: All chapters

Solving linear equations: Chapters 1, 2, 3, 4, 8, 9, 11, 12

Properties of exponents: Chapters 6, 7, 8, 9, 10, 11

Properties of logarithms: Chapter 6

Solving equations involving variable exponents: Chapter 6

Sums of geometric progressions: Chapter 6

We have not attempted to include all prerequisite material from algebra in this appendix; instead, we have included just the highlights of those areas which our own students ask questions about and that are necessary for an understanding of various chapters in the text.

The Real Numbers

The set of **real numbers,** denoted by R, consists of all numbers that can be located on the **real number line.** (See Figure 1.) The numbers used for counting, namely $\{1, 2, 3, 4, \ldots\}$, are called the **natural numbers** and are denoted by N. The natural

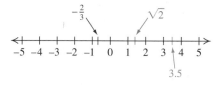

FIGURE 1
The Real Number Line with
Various Real Numbers Labeled

numbers are a **subset** of the real numbers; that is, $N \subseteq R$. (See Section 7.3 for an introduction to **sets.**) The **integers,** denoted by Z, consist of all the natural numbers, the number 0, and the negatives of all the natural numbers; that is, $Z = \{ \ldots, -4, -3, -2, -1, 0, 1, 2, 3, 4, \ldots \}$.

The **rational numbers,** denoted by Q (for *quotient*), make up the set of all quotients of integers. Since any integer z can be written in the form $z = \frac{z}{1}$, we can think of all integers themselves as quotients of integers. Thus, we have the following relationship between the various sets of numbers mentioned so far: $N \subseteq Z \subseteq Q \subseteq R$. The rational numbers are often defined to be those real numbers whose decimal form is either a **terminating decimal** or a **repeating decimal.** For instance, the rational number $\frac{3}{4}$ can be written in decimal form as 0.75, which is a **terminating decimal.** The rational number $\frac{1}{3} = 0.333333 \ldots$, sometimes abbreviated as $0.\overline{3}$ (the bar indicates that the number 3 repeats indefinitely in the decimal representation), is a **repeating decimal.** Some other examples of repeating decimals are $\frac{5}{6} = 0.83333 \ldots = 0.8\overline{3}$ and $\frac{1}{7} = 0.142857142857 \ldots = 0.\overline{142857}$.

Some real numbers cannot be expressed as quotients of integers; that is, they cannot be written as terminating or as repeating decimals. The famous number π is such a number; as a decimal, we often write $\pi \approx 3.1417$, but this is merely an approximation or abbreviation of the number, since the decimal continues indefinitely without repeating any pattern. Other such numbers include $\sqrt{2}$ and the number e, which is the base for the natural logarithm. All of the real numbers whose decimal representation is a **non-terminating, non-repeating decimal** are referred to as the **irrational numbers.** The use of the name "irrational" is not due to the fact that the numbers somehow do not make sense; it refers to the fact that these numbers *cannot* be written as a quotient, or *ratio,* of two integers. We will use I to represent the set of irrational numbers.

It is important to realize that when we round off an irrational number to a fixed number of decimal places, we are merely approximating the value of the number and consequently lose a certain amount of accuracy. For instance, when we write $\sqrt{3}$, we are referring to the exact value of a particular irrational number. If, however, we write $\sqrt{3} \approx 1.732$, we are *approximating* the value of the number $\sqrt{3}$. **It is never correct to write $\sqrt{3} = 1.732$,** which erroneously implies that the number $\sqrt{3}$ is a terminating decimal.

We are now in a position to define the set of all real numbers.

The Real Numbers

The set of **real numbers,** denoted by R, consists of all rational numbers Q, together with all irrational numbers I. Using set notation, we write $R = Q \cup I$. Every real number can be represented by a point of the **real number line.**

Properties of Arithmetic

Addition and multiplication of real numbers are governed by various properties. We divide the basic properties into two groups: properties of addition and properties of multiplication.

Properties of Addition

If a, b, and c are real numbers, then

1. $a + b = b + a$ (***commutative property of addition***).
2. $a + (b + c) = (a + b) + c$ (***associative property of addition***).
3. $a + 0 = 0 + a = a$ (The number 0 is the ***additive identity***).
4. $a + (-a) = (-a) + a = 0$ (Every real number has an ***additive inverse***).

In light of these properties, subtraction can be thought of as a special case of addition. We might write $4 + (-3) = 1$, but we usually write the simpler, but equivalent, expression, $4 - 3 = 1$.

Properties of Multiplication

If a, b, and c are real numbers, then

1. $a \cdot b = b \cdot a$ (***commutative property of multiplication***).
2. $a \cdot (b \cdot c) = (a \cdot b) \cdot c$ (***associative property of multiplication***).
3. $a \cdot 1 = 1 \cdot a = a$ (The number 1 is the ***multiplicative identity***).
4. If $a \neq 0$, then $a \cdot \dfrac{1}{a} = \dfrac{1}{a} \cdot a = 1$ (Every real number ***except*** 0 has a ***multiplicative inverse***).
5. $a \cdot (b + c) = a \cdot b + a \cdot c$ and $(a + b) \cdot c = a \cdot c + b \cdot c$ (***distributive property of multiplication over addition***).

Just as subtraction can be considered a special case of addition, division can be thought of as a special case of multiplication. The expression $10 \div 5$ can be written as $\frac{10}{5}$, which is equivalent to both $10 \cdot \frac{1}{5}$ and $\frac{1}{5} \cdot 10$. All of these expressions are equal to 2. The fact that all of them are equivalent is the reason we were taught as children to "invert and multiply" when dividing fractions.

You have likely heard the statement "Every real number except 0 has a multiplicative inverse" as "Division by 0 is undefined," which is essentially an equivalent statement. We can see why these statements are true by considering what we would need to happen in order for 0 to have a multiplicative inverse. Suppose there were a real number z that served as the multiplicative inverse of 0; then it would have to be true that $z \cdot 0 = 0 \cdot z = 1$, by the preceding property 4. However, we know, and state formally later, that 0 multiplied by any number always results in a product of 0. Thus, it must also be true that, for any number, including our hypothetical z, that $z \cdot 0 = 0 \cdot z = 0$. So we arrive at a contradiction, namely, $1 = 0$, which leads us to the logical conclusion that our assumption was wrong; there *is no such* z that serves as the multiplicative inverse of 0.

The **distributive property of multiplication over addition** is often used to simplify algebraic expressions. For instance, we can use the distributive property to simplify the following expression:

$$2 \cdot (x + 3)$$
$$2 \cdot x + 2 \cdot 3 \qquad \text{Apply the Distributive Property}$$
$$2x + 6 \qquad \text{Multiply } 2 \cdot 3$$

The following box lists some additional properties of the real numbers that are quite useful:

Further Properties of Real Numbers

If a, b, and c are real numbers, then

1. $a \cdot 0 = 0 \cdot a = 0$.
2. $-(-a) = a$.
3. $a \cdot (-b) = (-a) \cdot b = -(a \cdot b)$.
4. $(-a) \cdot (-b) = a \cdot b$.
5. If $c \neq 0$ and $a \cdot c = b \cdot c$, then $a = b$. If $c \neq 0$ and $c \cdot a = c \cdot b$, then $a = b$ (*cancellation laws*).
6. If $a \neq 0$, then $\dfrac{1}{\left(\dfrac{1}{a}\right)} = a$.

Occasionally in mathematics, we are faced with expressions involving more than one arithmetic operation. Without specific rules for everyone to follow, several different people might arrive at several different answers to the same expression. To preserve the properties of addition and multiplication just outlined, we use the following **order of operations:**

The Order of Operations

If an expression involves more than one arithmetic operation, proceed as follows:

1. Start with the innermost set of **parentheses** (or other grouping symbols, such as [ ] or { }).
2. Within the parentheses, perform all **multiplication** and **division,** *in the order that those operations appear from left to right.*
3. Within the parentheses, perform all **addition** and **subtraction,** *in the order that those operations appear from left to right.*
4. Repeat steps 1 through 3 until all parentheses are removed. (It may be necessary to apply the distributive property in order to remove all parentheses.)

The mnemonic device "Please, My Dear Aunt Sally" is a well-known way to remember "Parentheses, multiplication and division, addition and subtraction."

To see how to apply the **order of operations,** let us consider the expression $3(4 + 20 \div 5 - 7) + 1$. We simplify this expression as follows:

$3(4 + 20 \div 5 - 7)$	Work inside the parentheses first.
$= 3(4 + 20 \div 5 - 7) = 3(4 + 4 - 7)$	Perform all multiplication and division, from left to right.
$= 3(4 + 4 - 7) = 3(8 - 7)$	Perform all addition and subtraction, from left to right.
$= 3(8 - 7) = 3(1)$	
$= 3$	Complete the last operation.

Let us now take a moment to review some notation. There are two traditional symbols for multiplication: a dot "·" and a cross "×". Either one is appropriate to use when you write expressions involving products of real numbers. Parentheses are also used to represent multiplication; we often refer to this usage as **implied multiplication.** For instance, instead of writing 2×7 or $2 \cdot 7$, we might write $2(7)$. The latter expression is called **implied multiplication** because no symbol for the operation of multiplication is used in the expression.

Another form of implied multiplication is common when algebraic expressions involving variables are written out. We often write $2x$ rather than $2 \cdot x$ to represent "2 multiplied by x." This convention is certainly less confusing than $2 \times x$, which can easily be misread. Parentheses used for implied multiplication can also be confusing when more than one variable is involved. For example, a reader might confuse the expression $x(y)$ with **function notation,** as utilized in Chapter 1, so it is better to write xy or $x \cdot y$ to represent multiplication in this case.

There are other ways to denote multiplication, which you should understand if you are using any sort of technological tools. Computer programs typically require the use of an asterisk $(*)$ to represent multiplication. While many calculators understand the use of parentheses for implied multiplication, most computer programs do not. For example, the spreadsheet program Excel understands the command $=2 * 3$, but it does *not* understand the command $=2(3)$. Most calculators and computers also have notational limitations when multiplication involves more than one variable. For instance, Excel will regard the expression xy as a strange word, not as "x times y." In general, regardless of the technological tool you may be using, *always* use the symbol "$*$" when multiplication is required.

Solving Linear Equations

In several chapters of this text, it is necessary to be able to **solve an equation** for a single variable. In general, a mathematical equation involving a single variable can be classified into one of three categories:

Category I: Conditional Equations.

> **Conditional equations** are equations that are true only for certain values of the variable. A value of the variable that makes the equation true is called a **solution** of the equation.

Category II: Consistent Equations.

> **Consistent equations,** also called **identities,** are equations that are *always* true, regardless of the value of the variable.

Category III: Inconsistent Equations.

Inconsistent equations are *never* true, regardless of the value of the variable.

The process of **solving an equation** is the process of determining which of the three categories the equation belongs to and, if it belongs to the first category, finding those values of the variable which make the equation true. The steps involved in this process depend on many factors. Here, we discuss only the steps required to solve a **linear equation** involving one variable.

While there are several formal ways to define a **linear equation in a single variable,** the linear equations encountered in this text usually have the form $ax + b = cx + d$, where a, b, c and d are real numbers. To **solve** such an equation, we utilize the following steps:

Steps Involved in Solving a Linear Equation in a Single Variable

To solve the equation $ax + b = cx + d$,

1. Subtract b from both sides of the equation and simplify.
2. Subtract cx from both sides of the equation and simplify.
3. If a term involving the variable x remains on the left-hand side of the equation, divide both sides by the coefficient of x and simplify.

Once the steps just outlined are performed, we should know not only which of the three categories our equation belongs to, but also the solution if the equation is conditional. We demonstrate with three examples.

EXAMPLE 1

Solving a Conditional
Linear Equation

Solve the equation $3x + 7 = -x + 5$.

SOLUTION Following the steps outlined, we obtain the following chain of reasoning:

$$3x + 7 = -x + 5$$
$$3x + 7 - 7 = -x + 5 - 7 \qquad \text{Subtract } 7 \text{ from both sides of the equation.}$$
$$3x + 0 = -x - 2 \qquad \text{Simplify.}$$
$$3x = -x - 2$$
$$3x - (-x) = -x - (-x) - 2 \qquad \text{Subtract } -x \text{ from both sides of the equation.}$$
$$3x + x = -x + x - 2 \qquad \text{Simplify.}$$
$$4x = 0 - 2$$
$$4x = -2$$
$$\frac{1}{4} \cdot 4x = \frac{1}{4} \cdot (-2) \qquad \text{Multiply both sides of the equation by } \frac{1}{4}.$$
$$1x = -\frac{2}{4} \qquad \text{Simplify.}$$
$$x = -\frac{1}{2}.$$

The equation is a conditional equation that is true when x is equal to $-\frac{1}{2}$. ∎

EXAMPLE 2

Solving a Consistent
Equation

Solve the equation $2x + 4 = 2(x + 2)$.

SOLUTION Before proceeding with the steps outlined, we should rewrite the right-hand side of the equation, using the distributive property, as follows:

$$2x + 4 = 2(x + 2)$$
$$2x + 4 = 2 \cdot x + 2 \cdot 2$$
$$2x + 4 = 2x + 4.$$

Notice that the left- and right-hand sides of the equation are now identical! No matter what value we choose for the variable x, the equation will be true. Thus, the equation is a consistent equation and is true for all values of x. ▬

EXAMPLE 3

Solving an Inconsistent
Equation

Solve the equation $3x + 9 = 3x + 8$.

SOLUTION Again, we proceed with the steps outlined:

$$3x + 9 = 3x + 8$$

$3x + 9 - 9 = 3x + 8 - 9$ Subtract 9 from both sides of the equation.

$3x + 0 = 3x - 1$ Simplify.

$3x = 3x - 1$

$3x - 3x = 3x - 3x - 1$ Subtract $3x$ from both sides of the equation.

$0 = 0 - 1$ Simplify.

$0 = -1$ This equation is never true; 0 is not equal to -1.

The equation is inconsistent and hence has no solution. ▬

We end this discussion with a note about checking solutions. If you solve an equation by finding a solution, you should *always* check to make sure that your solution is correct by substituting the value into the *original* equation. In Example 1, we found $x = -\frac{1}{2}$ to be the only solution of the equation. We check this answer as follows:

Left-Hand Side	Right-Hand Side	
$3x + 7$	$-x + 5$	
$3 \cdot \left(-\dfrac{1}{2}\right) + 7$	$-\left(-\dfrac{1}{2}\right) + 5$	Substitute $-\dfrac{1}{2}$ for x.
$-\dfrac{3}{2} + 7$	$\dfrac{1}{2} + 5$	Simplify.
$-\dfrac{3}{2} + \dfrac{14}{2}$	$\dfrac{1}{2} + \dfrac{10}{2}$	Obtain common denominators.
$\dfrac{11}{2}$	$\dfrac{11}{2}$	Add.

Since the computations resulted in the same value for the left- and right-hand sides of the equation, we see that $-\frac{1}{2}$ is indeed a solution of the equation.

Properties of Exponents

The expression 2^3 is a convenient way to write the expression $2 \cdot 2 \cdot 2$. Thus, $2^3 = 2 \cdot 2 \cdot 2 = 8$. The number 2 in the expression 2^3 is called the **base** and the number 3 in the expression 2^3 is called the **exponent.** When a natural number is used as an exponent, it represents the number of times the base appears as a factor in a string of multiplications. Because of the various properties of multiplication of real numbers, we can use *any rational number* as an exponent. In calculus, we verify that, in reality, *any real number* can be used as an exponent, although we encounter this fact only in Chapter 6 of the text. Following are some of the important properties of exponents that hold when the base is a real number:

Properties of Exponents

Let x be any real number, and let m and n be positive integers.

1. *The multiplication rule:* $x^m x^n = x^{m+n}$.
2. *The power rule:* $(x^m)^n = x^{m \cdot n}$.
3. *The quotient rule:* $\dfrac{x^m}{x^n} = x^{m-n}$ if $x \neq 0$.
4. *The zero-exponent rule:* $x^0 = 1$ if $x \neq 0$.
5. *The negative exponent rule:* $x^{-n} = \dfrac{1}{x^n}$ if $x \neq 0$.

The requirement in the quotient rule that x cannot be 0 should be clear: Division by zero is undefined. But why would we require that x be nonzero in the zero-exponent rule? Let's consider an example. Suppose we have the expression $2^3/2^3$. Since we can perform the arithmetic rather quickly, we see that $2^3/2^3 = \frac{8}{8} = 1$. If we apply the quotient rule for exponents, we could also rewrite this expression as $2^3/2^3 = 2^{3-3} = 2^0$. Since both results come from the same initial expression, it certainly must be true that $2^0 = 1$. However, this line of reasoning would not be valid if we used 0 rather than 2 as the base for our exponential expression, because division by zero is undefined.

We next present properties of rational exponents.

Properties of Rational Exponents

Let x be any *positive* real number and let m and n be integers. Then

1. $x^{1/n} = \sqrt[n]{x}$.

2. $x^{m/n} = \sqrt[n]{x^m} = (\sqrt[n]{x})^m$.

In this text, the only time we encounter rational exponents is in Chapter 6, in working with problems having to do with the mathematics of finance. In such cases, you will likely be using calculators or computers to perform the arithmetic operations, and you may not be asked to work with such exponents by hand. However, it is still important to understand the connection between rational exponents and radical expressions in order to understand many of the formulas derived in that chapter.

To see how the first property works, consider the expression $9^{1/2}$. If we multiply $(9^{1/2})^2 = 9^{1/2} \cdot 9^{1/2}$, we can apply the multiplication rule for exponents and obtain

$$(9^{1/2})^2 = 9^{1/2} \cdot 9^{1/2} = 9^{1/2 + 1/2} = 9^1 = 9.$$

Is there another positive number that we can square (i.e., raise to an exponent of 2) and obtain an answer of 9? This is just another way of asking the question, "What is $\sqrt{9}$?" Thus, $\sqrt{9} = 9^{1/2} = 3$.

The properties of exponents listed here are also valid for any real number exponents when the underlying base makes sense with that exponent. However, in this text, all bases for exponents are positive real numbers, so we will not discuss what happens when the base of the exponential expression is negative.

In working with long or complicated expressions involving exponents, it is important to remember that *an exponent applies only to its base,* which is always to its immediate left. Thus, the expression $(-2)^2 = (-2) \cdot (-2) = 4$, since the exponent has the number -2 as its base. However, the expression -2^2 means take the negative of 2^2, which is -4.

With exponents reviewed, we now complete the list of the order of operations for real numbers.

The Order of Operations

If an expression involves more than one arithmetic operation, proceed as follows:

1. Start with the innermost set of **parentheses** (or other grouping symbols, such as [ ] or { }).
2. Apply all **exponents.**
3. Within the parentheses, perform all **multiplication** and **division,** *in the order that those operations appear from left to right.*
4. Within the parentheses, perform all **addition** and **subtraction,** *in the order that those operations appear from left to right.*
5. Repeat until all parentheses are removed. (It may be necessary to apply the distributive property in order to remove all parentheses.)

Properties of Logarithms

Historically, logarithms were extremely important in performing arithmetic operations involving very large or very small numbers, such as those often encountered in physics and chemistry. With the rise of computers and inexpensive calculators, the use of logarithms as computational tools has diminished, but they are still necessary for solving equations when the variable appears in the exponent of an expression in the equation.

Definition of a Logarithm

Let a be a positive real number. Then the expression $\log_a x$ (read "the log, to the base a, of x), represents the number y such that $a^y = x$.

To see how the definition works, consider $\log_2 8$. This expression represents the exponent that, when applied to a base of 2, gives us 8 as the answer. In other words, $\log_2 8 = 3$, since $2^3 = 8$. With this definition in hand, we have the following properties:

Properties of Logarithms

Let a, x, and y be positive real numbers. Then we have

1. *The multiplication rule:* $\log_a(xy) = \log_a x + \log_a y$.
2. *The quotient rule:* $\log_a \dfrac{x}{y} = \log_a x - \log_a y$.
3. *The power rule:* $\log_a x^r = r \log_a x$.
4. $\log_a 1 = 0$.

The most useful base for a logarithm is the number e, which is an irrational number that is approximately equal to 2.7182. It may not seem natural to use such a base, but e does arise quite naturally in problems involving population growth, the decay of a radioactive chemical, and even compound interest. (For the latter, see Chapter 6.) For this reason, the logarithm with a base of e is referred to as the **natural logarithm** and is denoted by $\ln x$, read "el-en of x." Modern calculators usually have a key, or at least a function, that calculates the natural logarithm of a number. The TI-83 Plus, for example, has a key labeled LN for that purpose. We conclude this section with a restatement of the properties of logarithms for the case of the natural logarithm.

Properties of the Natural Logarithm

Let x and y be positive real numbers. Then we have

1. *The multiplication rule:* $\ln(xy) = \ln x + \ln y$.
2. *The quotient rule:* $\ln \dfrac{x}{y} = \ln x - \ln y$.
3. *The power rule:* $\ln x^r = r \ln x$.
4. $\ln 1 = 0$.

Solving Equations Involving Variable Exponents

In Chapter 6, there are several formulas whose derivation requires solving an equation for a variable that appears in an exponent. When the equation is not too complicated, we can sometimes solve the equation by inspection. For instance, if we are asked to solve the equation $2^x = 8$, we just need to think, "What exponent would we apply to a base of 2 in order to get

8 as our answer?" After a moment's reflection, we arrive at the solution: $x = 3$. The question just posed in order to solve the equation is merely a restatement of the definition of a logarithm. This observation will be useful to us as we try to solve more complicated equations.

EXAMPLE 4

Using Logarithms to Solve Equations

Solve the equation $2^x = 5$.

SOLUTION Unfortunately, it is not as clear as before what exponent we should apply to the base of 2 in order to obtain 5 as our answer. We can see that the number lies somewhere between 2 and 3, since $2^2 = 4$, which is smaller than 5, and $2^3 = 8$, which is larger than 5. So, we can reason that $2 < x < 3$, but this does not provide us with a reasonable answer. Using the properties of logarithms—particularly the power rule—we can find the value of x as follows:

$2^x = 5$

$\ln(2^x) = \ln(5)$ Take the natural logarithm of both sides of the equation.

$x\ln(2) = \ln(5)$ Simplify, using the power rule for logarithms.

$x = \dfrac{\ln(5)}{\ln(2)} \approx 2.3219$ Divide both sides by the coefficient of x, and compute the value.

Our assessment was correct: The solution does lie between 2 and 3; it is approximately 2.3219. ■

While we could use *any* logarithm with *any* positive real number as its base to solve the equation in Example 2, the natural logarithm is convenient because we have quick access to its values via calculators and computers.

Geometric Progressions

A **geometric progression** is a sequence of numbers with a **coefficient** a and a **ratio** r such that:

1. The first term is a.
2. Each successive term is obtained by multiplying the previous term by r.
3. The progression has n terms, where n is a natural number.

For example, the geometric progression $\{2, 6, 18, 54\}$ represents the progression obtained when the coefficient $a = 2$, the ratio $r = 3$, and the progression has $n = 4$ terms. To see that these values are correct, we follow the pattern described:

1. The first term is $a = 2$.
2. Each successive term is obtained by multiplying the previous term by $r = 3$.
3. The progression has $n = 4$ terms.

Thus, we can think of the progression as having the form

$$\{2, 6, 18, 54\} = \{2, 2 \cdot 3, (2 \cdot 3) \cdot 3, ((2 \cdot 3) \cdot 3) \cdot 3\}$$
$$= \{2, 2 \cdot 3, 2 \cdot 3^2, 2 \cdot 3^3\}.$$

This suggests the following definition:

Geometric Progressions

If a and r are real numbers and n is a natural number, then the **geometric progression** with n **terms, coefficient** a, and **ratio** r is given by $\{a, ar, ar^2, \ldots, ar^{n-1}\}$.

In Chapter 6, in discussing long-term investments such as annuities, the sums required for calculating the future value of an investment can be considered geometric progressions. It is therefore convenient to have a "closed form" (a nice formula) to represent the value of the sum of a geometric progression.

Consider the geometric progression with 7 terms, coefficient 3, and ratio $\frac{1}{2}$. Let S_7 represent the sum of these 7 terms:

$$S_7 = 3 + 3\left(\frac{1}{2}\right) + 3\left(\frac{1}{2}\right)^2 + 3\left(\frac{1}{2}\right)^3 + 3\left(\frac{1}{2}\right)^4 + 3\left(\frac{1}{2}\right)^5 + 3\left(\frac{1}{2}\right)^6.$$

If we multiply both sides of this equation by the ratio $\frac{1}{2}$, we obtain

$$\frac{1}{2}S_7 = 3\left(\frac{1}{2}\right) + 3\left(\frac{1}{2}\right)^2 + 3\left(\frac{1}{2}\right)^3 + 3\left(\frac{1}{2}\right)^4 + 3\left(\frac{1}{2}\right)^5 + 3\left(\frac{1}{2}\right)^6 + 3\left(\frac{1}{2}\right)^7.$$

Now, let us subtract the second equation from the first:

$$S_7 - \frac{1}{2}S_7 = \left[3 + 3\left(\frac{1}{2}\right) + 3\left(\frac{1}{2}\right)^2 + 3\left(\frac{1}{2}\right)^3 + 3\left(\frac{1}{2}\right)^4 + 3\left(\frac{1}{2}\right)^5 + 3\left(\frac{1}{2}\right)^6\right]$$

$$- \left[3\left(\frac{1}{2}\right) + 3\left(\frac{1}{2}\right)^2 + 3\left(\frac{1}{2}\right)^3 + 3\left(\frac{1}{2}\right)^4 + 3\left(\frac{1}{2}\right)^5 + 3\left(\frac{1}{2}\right)^6 + 3\left(\frac{1}{2}\right)^7\right]$$

$$\left(1 - \frac{1}{2}\right)S_7 = 3 - 3\left(\frac{1}{2}\right)^7$$

$$S_7 = \frac{3 - 3\left(\frac{1}{2}\right)^7}{1 - \frac{1}{2}} = \frac{3\left[1 - \left(\frac{1}{2}\right)^7\right]}{1 - \frac{1}{2}}.$$

The sum of this geometric progression can then be calculated as $\frac{381}{64}$.

We could have rewritten the rightmost expression of the final equation in a more computationally convenient form, but the format leads us to the following generalization:

Sum of a Geometric Progression

For a geometric progression with n terms, coefficient a, and ratio r, the sum of the terms of the progression is given by $S_n = \dfrac{a[1 - r^n]}{1 - r}$.

We conclude this appendix with an example.

EXAMPLE 5

The Sum of a Geometric Progression

Find the coefficient and ratio for the geometric progression $\{1, \frac{2}{3}, \frac{4}{9}, \frac{8}{27}, \frac{16}{81}, \frac{32}{243}\}$. Then find the sum of the terms of the progression.

SOLUTION The coefficient of the geometric progression is $a = 1$, since 1 is the first term in the progression. The ratio is $r = \frac{2}{3}$, since each term can be found by multiplying the previous term by $\frac{2}{3}$. There are $n = 6$ terms; therefore, the sum of the terms of the given geometric progression is

$$S_6 = \frac{1\left[1 - \left(\frac{2}{3}\right)^6\right]}{1 - \frac{2}{3}} = \frac{1 - \frac{2^6}{3^6}}{\frac{1}{3}} = 3 - \frac{2^6}{3^5} = \frac{665}{243} \approx 2.7366.\ \blacksquare$$

Exercises*

In Exercises 1–5, determine whether each real number is (a) a natural number, (b) an integer, (c) a rational number, or (d) an irrational number.

1. 0

2. -2

3. $\frac{2}{5}$

4. $\sqrt{2}$

5. $\sqrt{\frac{16}{4}}$

In Exercises 6–10, draw a real number line and estimate the location of each real number on the line.

6. -2 and 2.1

7. $\frac{-3}{2}$ and 4.25

8. $1.\overline{1}$ and 1.2

9. $-\sqrt{3}$ and $\frac{10}{3}$

10. π and 3

In Exercises 11–16, use the properties of real numbers to simplify each arithmetic expression.

11. $2(3 - 7)$

12. $2 \div [1 + 1(3 \cdot 2 - 6)]$

13. $-3(2 - x)$

14. $-4(x - 8) - 30$

15. $-2(2 - 3^2)$

16. $-2(2 - 3)^2$

In Exercises 17–25, solve each equation for the variable x. Identify each equation as conditional, consistent, or inconsistent.

17. $3x + 2 = 11$

18. $2 - 2x = 4$

19. $3x + 7 = x - 5$

20. $0.25x + 200 = 0.3x + 320$

21. $0.25x + 100 = 0.25(x + 400)$

22. $0.25x + 100 = 0.25(x + 100)$

23. $0.25x + 100 = 0.25(400 - x)$

24. $2(x - 3) + 4 = 2x + 1$

25. $2(x - 3) + 4 = 2x - 2$

Use the properties of exponents and of the real numbers to simplify the expressions in Exercises 26–30 as much as possible.

26. $(3^2)(3^{-1})$

27. $\frac{2^4}{2^5}$

28. $\frac{4 \cdot 3 \cdot 2 \cdot 1}{(2 \cdot 1)(2 \cdot 1)}\left(\frac{1}{3}\right)^2\left(\frac{2}{3}\right)^2$

19. $\frac{3 \cdot 2 \cdot 1}{(2 \cdot 1)(1)}\left(\frac{1}{4}\right)^2\left(\frac{3}{4}\right)^1$

30. $4\left(\frac{1}{3}\right)^0\left(\frac{2}{3}\right)^4$

In Exercises 31–35, solve each equation, using the natural logarithm.

31. $2^x = 3$

32. $3^x = 2$

---

*The solutions to all the exercises in this appendix are included in Answers to Selected Exercises.

33. $12 \cdot 4^x = 36$

34. $100(1.01)^x = 500$

35. $\dfrac{100[1 - (0.01)^x]}{1 - 0.01} = 101$

In Exercises 36–40, find the sum of the geometric progression with the given coefficient, ratio, and number of terms.

36. $a = 2, r = 3, n = 5$

37. $a = 3, r = \dfrac{1}{2}, n = 10$

38. $a = 1, r = 0.1, n = 20$

39. $a = 200, r = 0.01, n = 25$

40. $a = 50, r = 1.01, n = 5$

Appendix II

One could argue that the advent of spreadsheet and word-processing software ushered in the personal computer as a primary tool in every workplace. In today's world, wherever there is a table of data, it is stored in a spreadsheet. Following are some of the things one can do with data via spreadsheets:

- Data Arrangement
- Calculation
- Visualization
- Database Management

- Statistical Analysis
- Predictions
- Optimization

Spreadsheet packages also come with powerful built-in programming languages, making their versatility almost limitless. The best part about spreadsheets is that the initial learning curve is very short! We use Microsoft Excel 2000 for all demonstrations in this appendix, but due to the standardization in today's spreadsheets, one should be able to apply the material presented to almost any spreadsheet package. Throughout this appendix, spreadsheet and computer terminology will be used. In Figure 1, the typical window one sees is displayed with many of the names we use.

Files

Collections of sheets are stored in files called **workbooks.** When you open Excel, it starts with a brand new workbook called "Book1.xls". Figure 1 is very close to what you will see. Notice the title of the workbook in the **title bar** at the top of the window. If you want to open a previously saved workbook, you can do so with `File arrow Open` on the **menu bar,** or open the workbook directly by finding the file in which it is saved and double-clicking on the icon for the file.

Each workbook consists of sheets, accessed via the **sheet tabs** at the bottom of the screen. (See Figure 1.) Each sheet can be either a **chart** or a **worksheet.** Charts are any kind of visualization of data, such as scatter plots or bar graphs. Worksheets are the actual spreadsheets, the place where we (and the program) do all of the work.

Cells

Everything begins with the **cell.** If you can understand exactly what a cell is and what properties it has, the rest should be relatively easy. *It is crucial that you understand this section!*

A cell is one of the many rectangles you see on a worksheet. One or more cells are **selected** when they are surrounded by a bold outline with a **fill handle** in the lower right-hand corner of the outline. (We will discuss the purpose of the fill handle later.) You can select a single cell by clicking on it, or select a rectangular set of cells by click-dragging. (See Figure 2.)

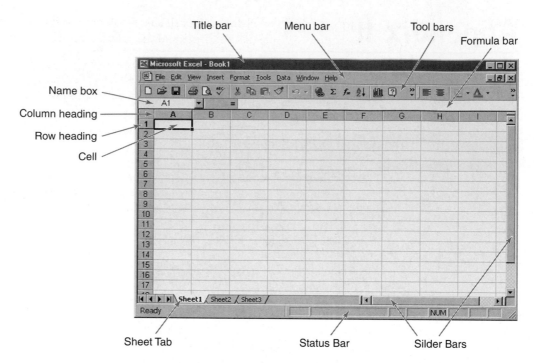

FIGURE 1
The Basic Components of a Spreadsheet

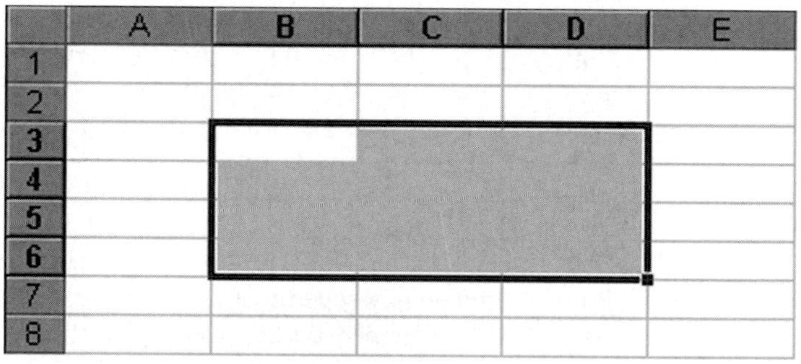

FIGURE 2
The Range B3:D6 is Selected.

The cell's primary function is to hold and display data of some kind. *The cell has four major attributes, which you must learn:*

- Address
- Format
- Content
- Value

Cell Address The **address** of a cell is indicated by the row and column in which the cell sits. (If you have ever played Battleship®, you know exactly how this works!) The **column heading** is always one or two letters (ranging from A to IV), and the **row heading** is always a number (from 1 to 65536). A cell's address consists of the column letter(s) and the row number. For example, if a cell sits in column B and row 3, then its address is B3. (See Figure 3.). Note that the address of the selected cell is given in the **name box.**

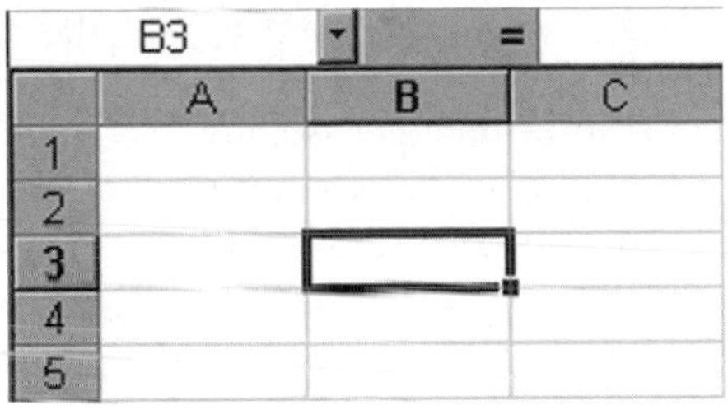

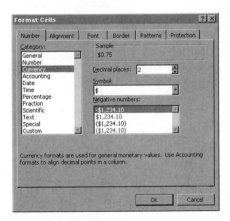

FIGURE 3
The Address of the Selected Cell is B3.

When a rectangular **range** of cells is selected, the address of the range is given by the address of the upper left cell and the lower right cell, separated by a colon. In Figure 2, the upper left cell of the selected range is B3 and the lower right cell is D6; thus, the range is denoted by B3:D6.

NOTE When a range of cells forms a single column or row, the two end cells are used to denote the range.

Cell Format A cell can contain many things, such as numbers and text, but how the cell displays those things can vary greatly. Suppose you type the number 0.75 into every cell of a range (say, B1:B5). Without any formatting, the cells all display 0.75. By right-clicking a cell, you bring up a plethora of options, including **Format Cells. . . .** (Like most everything else, this option can also be obtained through the menu bar). The first tab you get in the Format Cells pop-up window is Number. (See Figure 4.) Here is where you can format the cell to display its value as anything from dates to currency. You can even dictate how many decimal places the cell will display (which can cause Excel to round the displayed value, but *not* the actual value). In Figure 5, you can see some of the possibilities; keep in mind that all the cells contain the same value: 0.75.

FIGURE 4
The Format Cells Window

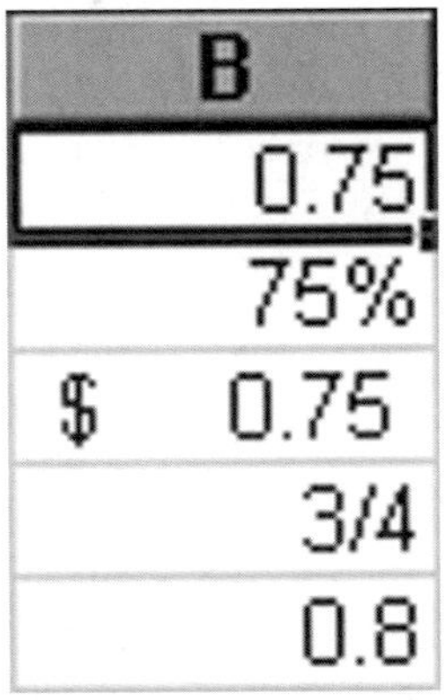

FIGURE 5
Assorted Formatting of 0.75

Cell Content The **content** of a cell is whatever you type into the cell, which is usually a number, text, or a formula. The content of the cell is not necessarily what the cell displays! In other words, you may type a formula into a cell and press `Enter`. Then, what the cell *displays* is the value of the formula, while the *content* of the cell is the formula. Whenever you select a cell, the content of the cell is shown in the **formula bar.** For example, suppose you select a cell, type the formula =3/4, and press `Enter`. If you formatted the cell as a percentage, then the cell will display the value 75%. Now reselect the cell, look in the formula bar, and you should see the formula =3/4. (See Figure 6.)

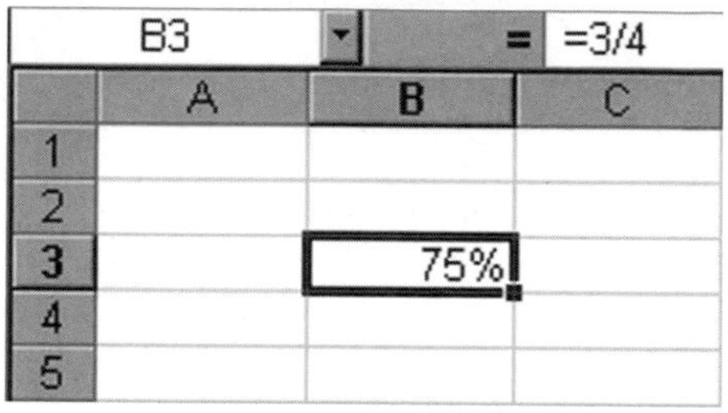

FIGURE 6
Cell B3 Displays 75%, Has a Value of 0.75,
and Contains the Formula =3/4.

This is what makes spreadsheets so powerful: You will be able to enter formulas into cells that use values from other cells. In turn, these other cells may contain formulas themselves!

There is something very important to remember when you use longer, more complicated formulas in Excel: *Excel does not correctly obey the standard order of operations!* Recall from basic algebra that $-3^2 = -9$ and $(-3)^2 = 9$, because an exponent applies *only* to whatever is to its immediate left. Unfortunately, Excel is programmed to apply negation (the negative sign) *before* applying exponents, whether parentheses are present or not. So the formula =-A1^2 will return the value of -A1 squared, not the negative of A1 squared. Therefore, when using Excel, you should adopt the convention of enclosing the base *and* the exponent in parentheses. Table 1 shows the result of three different formulas. If you wish to calculate -3^2, then you should use the formula shown in the last column of the table.

A2	= A2^2	= -A2^2	= -(A2^2)
3	9	9	-9

TABLE 1.

Cell Value Every cell has a numeric **value**—even an empty one (which has a default value of 0). If you type a number into a cell, then that number will be the cell's value. If you type a numeric formula into a cell, then that cell's value will be the outcome of the formula. When a cell contains text or has a formula that returns text, the value of that cell will default to 0. Remember, the value of a cell, what the cell displays, and the cell's content are three different things. As seen in Figure 6, the value of the cell is 0.75, while the cell's content is the formula =3/4 and the cell's display is 75%.

Formulas

As we saw earlier, you can type formulas into a cell. All formulas start with "=". A formula can have numbers, algebraic operations, **functions,** and addresses for cells and ranges. Notice that we do not have variables! Anytime we want to use a value from another cell in a formula, we can use the cell's address instead. Let us consider an example.

Suppose you wanted to store, in cell D2, the average of three numbers that you have stored in cells B1, B2, and B3. This can be done with a formula. In cell D2, type =(B1+B2+B3)/3 and press Enter. Cell D2 displays the average. If you reselect cell D2, the formula bar will display the formula that the cell contains, while the cell displays its value. (See Figure 7.)

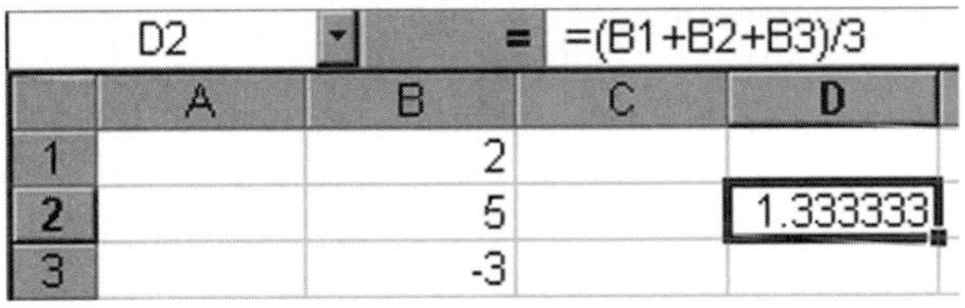

FIGURE 7
Average of Cells via a Simple Formula

For many common calculations (such as averaging) there are built-in functions. To view a list of these functions, click on a cell and type "=". The name box changes and becomes a drop-down list of functions you have used recently, but right now the most interesting option is More Functions... (See Figure 8.) Through this option, you can explore and

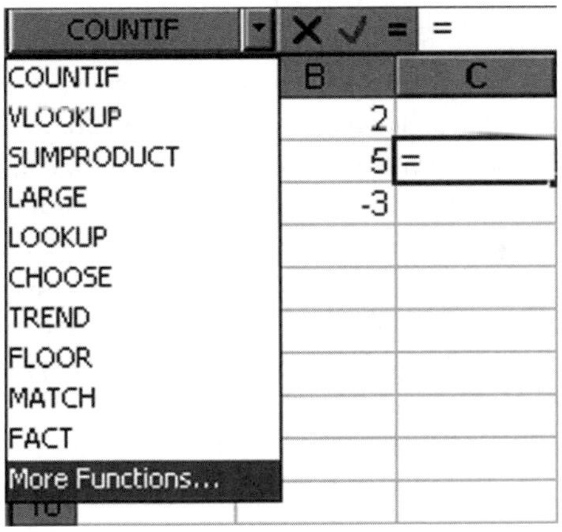

FIGURE 8
Average of Cells via a Simple
Formula

use a whole host of functions; it is certainly worth your while to check this out! Another way of obtaining a function is through the menu bar Insert → Function. You can also type in functions directly if you know the correct spelling and appropriate arguments. In Figure 9, the function =AVERAGE(B1:B3) gives the same result as seen in Figure 7.

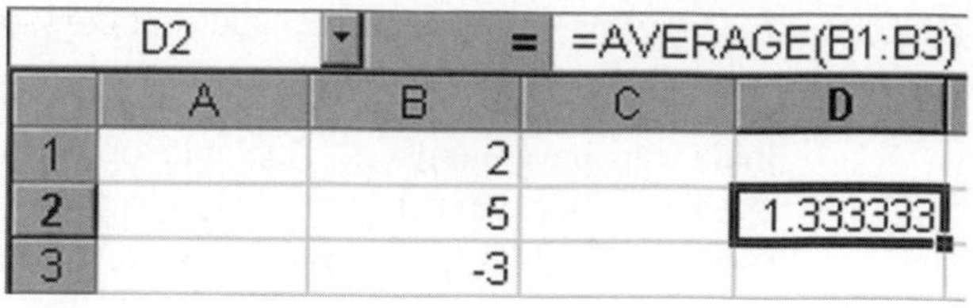

FIGURE 9
Average of Cells via a Simple Formula

Copy, Paste, and Fill

Anyone who has used a personal computer is probably quite familiar with cut, copy, and paste. In copying cells, if the cell's content is anything other than a formula, the contents are transcribed identically. However, when the content of a copied cell is a function, the addresses within the function are adjusted by the relative distance from the copied cell to the pasted cell.

As an example, consider Figure 10, which shows a table of values of which you want the average of each column. Type in the appropriate formula for averaging the first column, copy that cell, and paste its contents into the cell under the next column. (See Figure 11.)

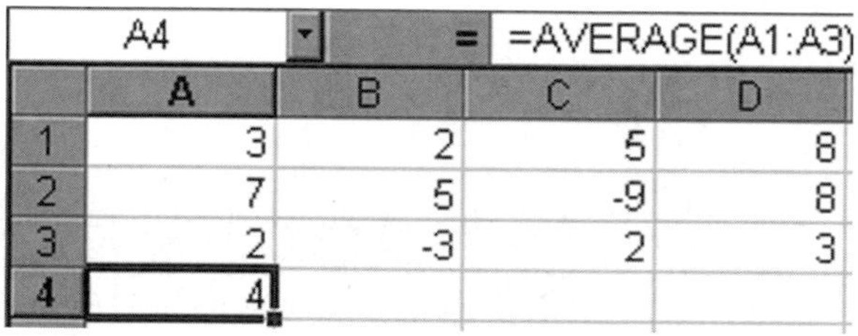

FIGURE 10
The First Average Function Is Typed In.

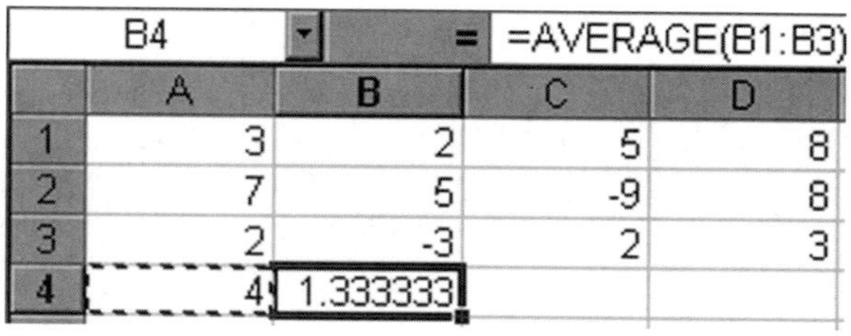

FIGURE 11
The Next Average Is Copied from the First.

Notice how the address range A1:A3 is shifted by exactly one column value, to B1:B3. This is precisely the relative change from the copied cell (A4) to the pasted cell (B4). If you pasted from the cell A4 to the cell C9, the relative change would be two to the left and five down; thus, the pasted formula would be =AVERAGE(C6:C8).

There is a much quicker way of copying one cell to multiple cells. The technique is called **filling,** and as with most things, there are several ways of accomplishing it. As noted earlier, every highlighted cell has a fill handle in the lower right-hand corner. If you hold-click this handle, grab it and drag in any direction, an automatic copy–paste is performed on all covered cells. In Figure 12, the handle on cell B4 is shown grabbed and dragged across cells C4 and D4. When the mouse is released, the average function is pasted into both C4 and D4, with the appropriate adjustment to the address ranges within the function. (See Figure 13.)

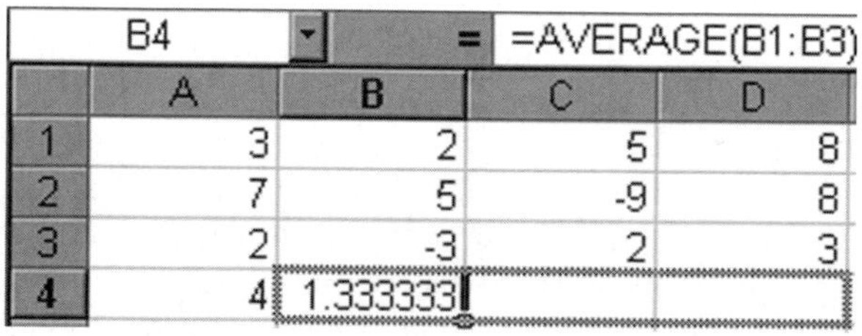

FIGURE 12
The Fill Handle Is Grabbed and Dragged.

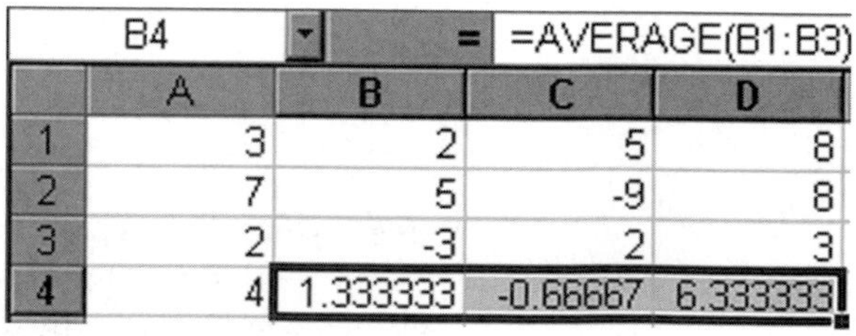

FIGURE 13
All of the Averages Are Correctly Filled In.

There may be times when you will copy–paste a cell and you want one or more parts of the address ranges in the function's arguments *not* to change. The trick is to use a $ in front of every part of the addresses you want to freeze. For instance, suppose you are filling from a cell with the formula =A1+B3-C7. If you do not want the column of the address A1 to change, you would instead use the formula =$A1+B3-C7. If you do not want the row address of B3 to change, you would use the formula =A1+B$3-C7. If you do not want the address C7 to change at all, you would use the formula =A1+B3-C7. When an address is completely frozen, (e.g., C7), this is called an **absolute reference.** When no part of the cell reference is frozen, (e.g. C7), this is called a **relative reference.** When part of a cell reference is frozen (e.g., $C7 or C$7), this is called a **mixed reference.**

Charts

A picture is worth a thousand words. Spreadsheets are loaded with graphical tools for displaying sets of data. The best part is how easy it is to create a **chart** of data. Consider the data in Figure 14 of homework and test grades over the first four chapters of a class. Select the data to be visualized (range B1:C5) and engage the **chart wizard** via the tool bar or the menu under Insert. As seen in Figure 15, the first screen of the chart wizard gives many choices of possible charts, including bar charts, pie charts, and scatter plots. The default choice is the standard grouped bar graph.

	A	B	C
1	Chapter	HW	Test
2	1	75	73
3	2	82	78
4	3	78	80
5	4	91	88

FIGURE 14
These Are Class Grades to Be Visualized.

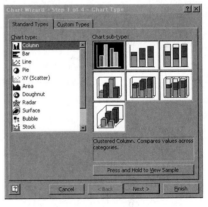

FIGURE 15
The Chart Wizard

As you move through the wizard, you are given a vast number of choices to fine-tune and annotate your graph. (See Figure 16.) The last choice you need to make is whether the chart will be displayed on the data spreadsheet or given a sheet of its own. (See Figure 17.)

After the chart is created, changes can be made by simply right-clicking on the part of the chart in which you are interested. For example, if you want to change the format of the title, right-click on it and select Format Chart Title from the resulting menu. You can make major changes to the whole chart by right-clicking in the white area of the chart, outside of any parts. The resulting menu allows you to change everything from the chart type to grid lines. Chapter 1 of the text shows how, by right-clicking the data graphics, one can even include trend lines!

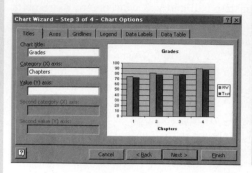

FIGURE 16

The Chart Wizard Gives Lots of Control over the Chart.

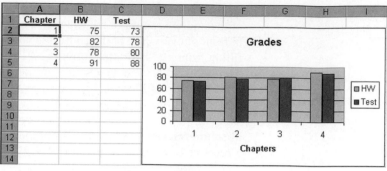

FIGURE 17

The Chart Can be Added to the Data Spreadsheet or Placed on a Separate Sheet.

Conclusion

Once you become comfortable with the spreadsheet environment, you should experiment with all the available charts, formulas, and built-in capabilities the spreadsheet has. The possibilities for effectively and efficiently using Excel, or any other spreadsheet program, are practically endless. If you are using Microsoft Word or Microsoft Power-Point, all tables and charts from Excel can be copied and pasted into a Word document or PowerPoint presentation so that you can create attractive papers and presentations. You can even import data from other programs and databases, as well as from the World Wide Web.

You may also find spreadsheets useful for keeping track of your college credits, for handling your personal finances, and even as an address book for e-mail and mailing addresses and phone numbers. You are limited only by your imagination and patience!

Appendix III

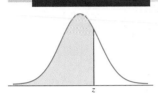

The following table gives values for the area $A(z)$ under the graph of the standard normal curve and to the left of the given z-value:

z	A(z)	z	A(z)	z	A(z)	z	A(z)
−3.00	0.0013	−2.66	0.0039	−2.32	0.0102	−1.98	0.0239
−2.99	0.0014	−2.65	0.0040	−2.31	0.0104	−1.97	0.0244
−2.98	0.0014	−2.64	0.0041	−2.30	0.0107	−1.96	0.0250
−2.97	0.0015	−2.63	0.0043	−2.29	0.0110	−1.95	0.0256
−2.96	0.0015	−2.62	0.0044	−2.28	0.0113	−1.94	0.0262
−2.95	0.0016	−2.61	0.0045	−2.27	0.0116	−1.93	0.0268
−2.94	0.0016	−2.60	0.0047	−2.26	0.0119	−1.92	0.0274
−2.93	0.0017	−2.59	0.0048	−2.25	0.0122	−1.91	0.0281
−2.92	0.0018	−2.58	0.0049	−2.24	0.0125	−1.90	0.0287
−2.91	0.0018	−2.57	0.0051	−2.23	0.0129	−1.89	0.0294
−2.90	0.0019	−2.56	0.0052	−2.22	0.0132	−1.88	0.0301
−2.89	0.0019	−2.55	0.0054	−2.21	0.0136	−1.87	0.0307
−2.88	0.0020	−2.54	0.0055	−2.20	0.0139	−1.86	0.0314
−2.87	0.0021	−2.53	0.0057	−2.19	0.0143	−1.85	0.0322
−2.86	0.0021	−2.52	0.0059	−2.18	0.0146	−1.84	0.0329
−2.85	0.0022	−2.51	0.0060	−2.17	0.0150	−1.83	0.0336
−2.84	0.0023	−2.50	0.0062	−2.16	0.0154	−1.82	0.0344
−2.83	0.0023	−2.49	0.0064	−2.15	0.0158	−1.81	0.0351
−2.82	0.0024	−2.48	0.0066	−2.14	0.0162	−1.80	0.0359
−2.81	0.0025	−2.47	0.0068	−2.13	0.0166	−1.79	0.0367
−2.80	0.0026	−2.46	0.0069	−2.12	0.0170	−1.78	0.0375
−2.79	0.0026	−2.45	0.0071	−2.11	0.0174	−1.77	0.0384
−2.78	0.0027	−2.44	0.0073	−2.10	0.0179	−1.76	0.0392
−2.77	0.0028	−2.43	0.0075	−2.09	0.0183	−1.75	0.0401
−2.76	0.0029	−2.42	0.0078	−2.08	0.0188	−1.74	0.0409
−2.75	0.0030	−2.41	0.0080	−2.07	0.0192	−1.73	0.0418
−2.74	0.0031	−2.40	0.0082	−2.06	0.0197	−1.72	0.0427
−2.73	0.0032	−2.39	0.0084	−2.05	0.0202	−1.71	0.0436
−2.72	0.0033	−2.38	0.0087	−2.04	0.0207	−1.70	0.0446
−2.71	0.0034	−2.37	0.0089	−2.03	0.0212	−1.69	0.0455
−2.70	0.0035	−2.36	0.0091	−2.02	0.0217	−1.68	0.0465
−2.69	0.0036	−2.35	0.0094	−2.01	0.0222	−1.67	0.0475
−2.68	0.0037	−2.34	0.0096	−2.00	0.0228	−1.66	0.0485
−2.67	0.0038	−2.33	0.0099	−1.99	0.0233	−1.65	0.0495

z	A(z)	z	A(z)	z	A(z)	z	A(z)
−1.64	0.0505	−1.13	0.1292	−0.62	0.2676	−0.11	0.4562
−1.63	0.0516	−1.12	0.1314	−0.61	0.2709	−0.10	0.4602
−1.62	0.0526	−1.11	0.1335	−0.60	0.2743	−0.09	0.4641
−1.61	0.0537	−1.10	0.1357	−0.59	0.2776	−0.08	0.4681
−1.60	0.0548	−1.09	0.1379	−0.58	0.2810	−0.07	0.4721
−1.59	0.0559	−1.08	0.1401	−0.57	0.2843	−0.06	0.4761
−1.58	0.0571	−1.07	0.1423	−0.56	0.2877	−0.05	0.4801
−1.57	0.0582	−1.06	0.1446	−0.55	0.2912	−0.04	0.4840
−1.56	0.0594	−1.05	0.1469	−0.54	0.2946	−0.03	0.4880
−1.55	0.0606	−1.04	0.1492	−0.53	0.2981	−0.02	0.4920
−1.54	0.0618	−1.03	0.1515	−0.52	0.3015	−0.01	0.4960
−1.53	0.0630	−1.02	0.1539	−0.51	0.3050	0.00	0.5000
−1.52	0.0643	−1.01	0.1562	−0.50	0.3085	0.01	0.5040
−1.51	0.0655	−1.00	0.1587	−0.49	0.3121	0.02	0.5080
−1.50	0.0668	−0.99	0.1611	−0.48	0.3156	0.03	0.5120
−1.49	0.0681	−0.98	0.1635	−0.47	0.3192	0.04	0.5160
−1.48	0.0694	−0.97	0.1660	−0.46	0.3228	0.05	0.5199
−1.47	0.0708	−0.96	0.1685	−0.45	0.3264	0.06	0.5239
−1.46	0.0721	−0.95	0.1711	−0.44	0.3300	0.07	0.5279
−1.45	0.0735	−0.94	0.1736	−0.43	0.3336	0.08	0.5319
−1.44	0.0749	−0.93	0.1762	−0.42	0.3372	0.09	0.5359
−1.43	0.0764	−0.92	0.1788	−0.41	0.3409	0.10	0.5398
−1.42	0.0778	−0.91	0.1814	−0.40	0.3446	0.11	0.5438
−1.41	0.0793	−0.90	0.1841	−0.39	0.3483	0.12	0.5478
−1.40	0.0808	−0.89	0.1867	−0.38	0.3520	0.13	0.5517
−1.39	0.0823	−0.88	0.1894	−0.37	0.3557	0.14	0.5557
−1.38	0.0838	−0.87	0.1922	−0.36	0.3594	0.15	0.5596
−1.37	0.0853	−0.86	0.1949	−0.35	0.3632	0.16	0.5636
−1.36	0.0869	−0.85	0.1977	−0.34	0.3669	0.17	0.5675
−1.35	0.0885	−0.84	0.2005	−0.33	0.3707	0.18	0.5714
−1.34	0.0901	−0.83	0.2033	−0.32	0.3745	0.19	0.5753
−1.33	0.0918	−0.82	0.2061	−0.31	0.3783	0.20	0.5793
−1.32	0.0934	−0.81	0.2090	−0.30	0.3821	0.21	0.5832
−1.31	0.0951	−0.80	0.2119	−0.29	0.3859	0.22	0.5871
−1.30	0.0968	−0.79	0.2148	−0.28	0.3897	0.23	0.5910
−1.29	0.0985	−0.78	0.2177	−0.27	0.3936	0.24	0.5948
−1.28	0.1003	−0.77	0.2206	−0.26	0.3974	0.25	0.5987
−1.27	0.1020	−0.76	0.2236	−0.25	0.4013	0.26	0.6026
−1.26	0.1038	−0.75	0.2266	−0.24	0.4052	0.27	0.6064
−1.25	0.1056	−0.74	0.2296	−0.23	0.4090	0.28	0.6103
−1.24	0.1075	−0.73	0.2327	−0.22	0.4129	0.29	0.6141
−1.23	0.1093	−0.72	0.2358	−0.21	0.4168	0.30	0.6179
−1.22	0.1112	−0.71	0.2389	−0.20	0.4207	0.31	0.6217
−1.21	0.1131	−0.70	0.2420	−0.19	0.4247	0.32	0.6255
−1.20	0.1151	−0.69	0.2451	−0.18	0.4286	0.33	0.6293
−1.19	0.1170	−0.68	0.2483	−0.17	0.4325	0.34	0.6331
−1.18	0.1190	−0.67	0.2514	−0.16	0.4364	0.35	0.6368
−1.17	0.1210	−0.66	0.2546	−0.15	0.4404	0.36	0.6406
−1.16	0.1230	−0.65	0.2578	−0.14	0.4443	0.37	0.6443
−1.15	0.1251	−0.64	0.2611	−0.13	0.4483	0.38	0.6480
−1.14	0.1271	−0.63	0.2643	−0.12	0.4522	0.39	0.6517

z	A(z)	z	A(z)	z	A(z)	z	A(z)
0.40	0.6554	0.91	0.8186	1.42	0.9222	1.93	0.9732
0.41	0.6591	0.92	0.8212	1.43	0.9236	1.94	0.9738
0.42	0.6628	0.93	0.8238	1.44	0.9251	1.95	0.9744
0.43	0.6664	0.94	0.8264	1.45	0.9265	1.96	0.9750
0.44	0.6700	0.95	0.8289	1.46	0.9279	1.97	0.9756
0.45	0.6736	0.96	0.8315	1.47	0.9292	1.98	0.9761
0.46	0.6772	0.97	0.8340	1.48	0.9306	1.99	0.9767
0.47	0.6808	0.98	0.8365	1.49	0.9319	2.00	0.9772
0.48	0.6844	0.99	0.8389	1.50	0.9332	2.01	0.9778
0.49	0.6879	1.00	0.8413	1.51	0.9345	2.02	0.9783
0.50	0.6915	1.01	0.8438	1.52	0.9357	2.03	0.9788
0.51	0.6950	1.02	0.8461	1.53	0.9370	2.04	0.9793
0.52	0.6985	1.03	0.8485	1.54	0.9382	2.05	0.9798
0.53	0.7019	1.04	0.8508	1.55	0.9394	2.06	0.9803
0.54	0.7054	1.05	0.8531	1.56	0.9406	2.07	0.9808
0.55	0.7088	1.06	0.8554	1.57	0.9418	2.08	0.9812
0.56	0.7123	1.07	0.8577	1.58	0.9429	2.09	0.9817
0.57	0.7157	1.08	0.8599	1.59	0.9441	2.10	0.9821
0.58	0.7190	1.09	0.8621	1.60	0.9452	2.11	0.9826
0.59	0.7224	1.10	0.8643	1.61	0.9463	2.12	0.9830
0.60	0.7257	1.11	0.8665	1.62	0.9474	2.13	0.9834
0.61	0.7291	1.12	0.8686	1.63	0.9484	2.14	0.9838
0.62	0.7324	1.13	0.8708	1.64	0.9495	2.15	0.9842
0.63	0.7357	1.14	0.8729	1.65	0.9505	2.16	0.9846
0.64	0.7389	1.15	0.8749	1.66	0.9515	2.17	0.9850
0.65	0.7422	1.16	0.8770	1.67	0.9525	2.18	0.9854
0.66	0.7454	1.17	0.8790	1.68	0.9535	2.19	0.9857
0.67	0.7486	1.18	0.8810	1.69	0.9545	2.20	0.9861
0.68	0.7517	1.19	0.8830	1.70	0.9554	2.21	0.9864
0.69	0.7549	1.20	0.8849	1.71	0.9564	2.22	0.9868
0.70	0.7580	1.21	0.8869	1.72	0.9573	2.23	0.9871
0.71	0.7611	1.22	0.8888	1.73	0.9582	2.24	0.9875
0.72	0.7642	1.23	0.8907	1.74	0.9591	2.25	0.9878
0.73	0.7673	1.24	0.8925	1.75	0.9599	2.26	0.9881
0.74	0.7704	1.25	0.8944	1.76	0.9608	2.27	0.9884
0.75	0.7734	1.26	0.8962	1.77	0.9616	2.28	0.9887
0.76	0.7764	1.27	0.8980	1.78	0.9625	2.29	0.9890
0.77	0.7794	1.28	0.8997	1.79	0.9633	2.30	0.9893
0.78	0.7823	1.29	0.9015	1.80	0.9641	2.31	0.9896
0.79	0.7852	1.30	0.9032	1.81	0.9649	2.32	0.9898
0.80	0.7881	1.31	0.9049	1.82	0.9656	2.33	0.9901
0.81	0.7910	1.32	0.9066	1.83	0.9664	2.34	0.9904
0.82	0.7939	1.33	0.9082	1.84	0.9671	2.35	0.9906
0.83	0.7967	1.34	0.9099	1.85	0.9678	2.35	0.9906
0.84	0.7995	1.35	0.9115	1.86	0.9686	2.36	0.9909
0.85	0.8023	1.36	0.9131	1.87	0.9693	2.37	0.9911
0.86	0.8051	1.37	0.9147	1.88	0.9699	2.38	0.9913
0.87	0.8078	1.38	0.9162	1.89	0.9706	2.39	0.9916
0.88	0.8106	1.39	0.9177	1.90	0.9713	2.40	0.9918
0.89	0.8133	1.40	0.9192	1.91	0.9719	2.41	0.9920
0.90	0.8159	1.41	0.9207	1.92	0.9726	2.42	0.9922

z	A(z)	z	A(z)	z	A(z)	z	A(z)
2.43	0.9925	2.57	0.9949	2.72	0.9967	2.87	0.9979
2.44	0.9927	2.58	0.9951	2.73	0.9968	2.88	0.9980
2.45	0.9929	2.59	0.9952	2.74	0.9969	2.89	0.9981
2.46	0.9931	2.60	0.9953	2.75	0.9970	2.90	0.9981
2.47	0.9932	2.61	0.9955	2.76	0.9971	2.91	0.9982
2.47	0.9932	2.62	0.9956	2.77	0.9972	2.92	0.9982
2.48	0.9934	2.63	0.9957	2.78	0.9973	2.93	0.9983
2.49	0.9936	2.64	0.9959	2.79	0.9974	2.94	0.9984
2.50	0.9938	2.65	0.9960	2.80	0.9974	2.95	0.9984
2.51	0.9940	2.66	0.9961	2.81	0.9975	2.96	0.9985
2.52	0.9941	2.67	0.9962	2.82	0.9976	2.97	0.9985
2.53	0.9943	2.68	0.9963	2.83	0.9977	2.98	0.9986
2.54	0.9945	2.69	0.9964	2.84	0.9977	2.99	0.9986
2.55	0.9946	2.70	0.9965	2.85	0.9978	3.00	0.9987
2.56	0.9948	2.71	0.9966	2.86	0.9979		

Answers to Selected Exercises

CHAPTER 1 APPLICATIONS OF LINEAR EQUATIONS

1.1 Exercises

1.

3.

5.
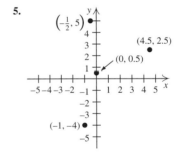

7.
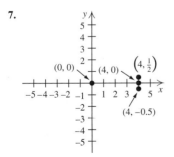

9.
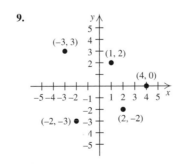

11.

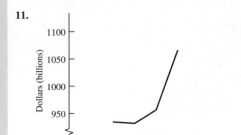

13.

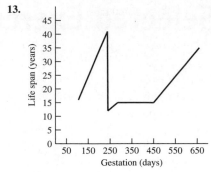

15.

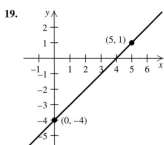

17.

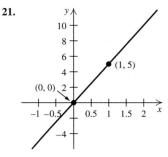

19.

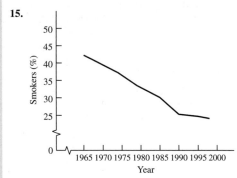

21.

23.

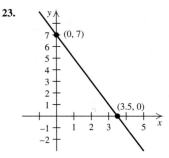

25.

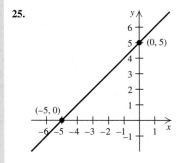

27.

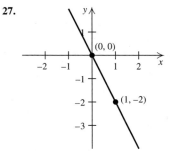

29.

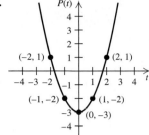

31.

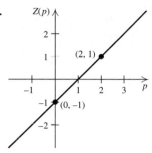

33.

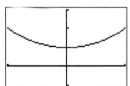

35.

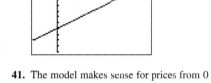

37.

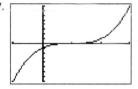

39. The model makes sense for mileage from 0 to 40 thousand miles.

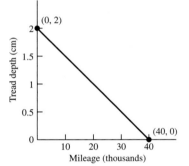

41. The model makes sense for prices from 0 to 8 hundred dollars.

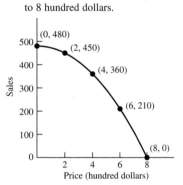

43. The store owner may use from 0 to 40 work shifts per week.

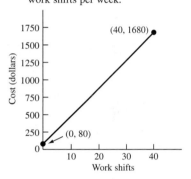

1.2 Exercises

1. a. Slope $= -3$
   b.

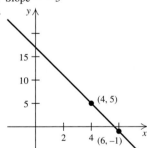

3. a. Slope $= 0$
   b.

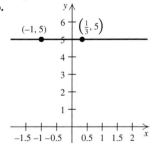

5. a. Slope $= 16.8$
   b.

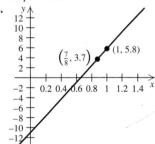

7. a. Slope is undefined.
b.

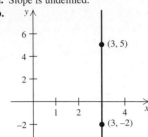

9. a. Slope = 1
b.

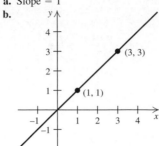

11. a. Slope–Intercept Form: $y = -2x + 9$
General Form: $2x + y = 9$
b.

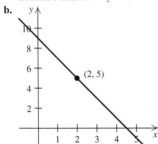

13. a. Slope–Intercept Form: $y = (2/3)x + 5.7$
General Form: $-2x + 3y = 17.1$
b.

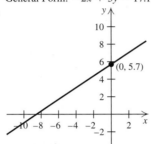

15. a. Slope–Intercept Form: $y = (-5/3)x + 9$
General Form: $5x + 3y = 27$
b.

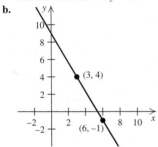

17. a. Slope–Intercept Form: $y = (-20/23)x + (29/23)$
General Form: $20x + 23y = 29$
b.

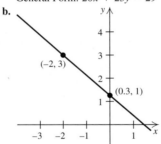

19. a. Slope–Intercept Form: $y = -25x + 725$
General Form: $25x + y = 725$
b.

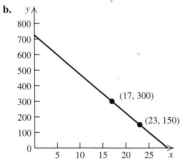

21. a. Slope–Intercept Form: $y = (-43/16)x + 4.3$
General Form: $43x + 16y = 68.8$
b.

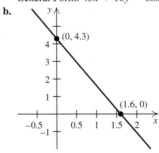

23. $x = -1/2$
$y = 4$

25. $x = 1.2$
$y = -8.6$

27. $y = (4/7)x$
$y = (4/7)x + 2$
$y = (4/7)x - 2$

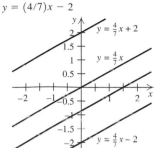

29. $x = 1$
$x = 3$
$x = -2$

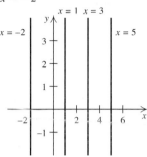

31. $y = (1/2)x$
$y = (1/2)x + 2$
$y = (1/2)x - 3$

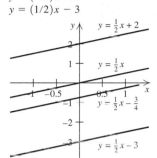

33. $y = -3x + 5.9$

35. $y = (-2/3)x - 3.5$

37. a. y-intercept $= (0, -4)$
$y = x - 4$
$y = 5x - 4$
$y = -4$
b. x-intercept $= (-4, 0)$
$y = x + 4$
$y = -3(x + 4)$
$y = 5(x + 4)$

39. a. y-intercept $= (0, -5/2)$
$y = x - 5/2$
$y = 11x - 5/2$
$y = -10x - 5/2$
b. x-intercept $= (5, 0)$
$y = x - 5$
$y = 10(x - 5)$
$y = -(x - 5)$

41. a. y-intercept $= (0, 0)$
$y = 2x$
$y = -7x$
$y = 11x$
b. x-intercept $= (0, 0)$
$y = 2x$
$y = -7x$
$y = 11x$

43. a. $\Delta y = 3$
b. $\Delta y = 6$
c. $\Delta y = 15$

45. a. $\Delta y = -2$
b. $\Delta y = -4$
c. $\Delta y = -10$

47. a. $\Delta y = -2/5$
b. $\Delta y = -4/5$
c. $\Delta y = -2$

49. a. After 5 weeks, there were 10,000 subscribers; after 20 weeks, there were 22,000 subscribers.
b. The subscribers are increasing at a rate of 800 per week.
c. The slope 800 gives the rate at which the subscribers are changing; the y-intercept $(0, 6000)$ gives the number of subscribers with which the paper starts.
d.

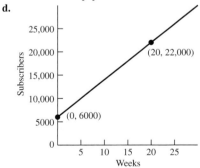

e. The model gives 18,000 subscribers in 15 weeks .

51. a. On September 15, the population is predicted to be approximately 70 thousand grasshoppers per acre; on September 20, it will be 60 thousand grasshoppers per acre.
b. The grasshopper population per acre is predicted to be decreasing at a rate of 2 thousand grasshoppers per day.
c. The slope -2 gives the rate at which the grasshopper population per acre is changing, while the y-intercept $(0, 100)$ approximates the grasshopper population per acre at the end of August.
d.

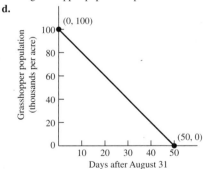

e. In 40 days (on October 10), the population is predicted to reach 20,000 grasshoppers per acre.

53. a. The temperature increases by $5/9°$ Celsius for each $1°$ Fahrenheit increase.
b. A temperature of $80°$ Fahrenheit is approximately $26.67°$ Celsius.
c. $F = (9/5)C + 32$
d. The temperature increases by $9/5°$ Fahrenheit for each $1°$ Celsius increase.

55. a.

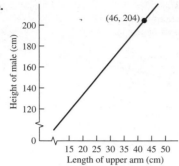

b. If the length of the bone from the elbow to the shoulder of a male is 46 cm, then he has a predicted height of 204 cm.
c. A female of height 180 cm should measure 38.75 cm from elbow to shoulder.

57. a. Sales in 2006 should be approximately 38.97 thousand cases of soft drinks.
b. Sales should break 50 thousand cases in 2019.
c.

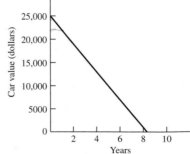

1.3 Exercises

1. This is not modeled by a linear function.   **3.** This is modeled by the linear function $Y(t) = 20t + 500$.   **5.** This is modeled by the linear function $Y(t) = 60$.   **7. a.** The marginal cost is $3 per item.   **b.** The fixed cost is $20.   **c.** The total cost of 20 items is $80.   **d.** The average cost of producing the first 20 items is $4 per item, of producing 100 items is $3.20 per item, and of producing 200 items is $3.10 per item.   **9. a.** The marginal cost is $3.20 per item.   **b.** The fixed cost is $1680.   **c.** The total cost of 20 items is $1744.   **d.** The average cost of producing the first 20 items is $87.20 per item, of producing 100 items is $20 per item, and of producing 200 items is $11.60 per item.
11. a. The marginal cost is $1.60 per item.   **b.** The fixed cost is $5000.   **c.** The total cost of 20 items is $5032.   **d.** The average cost of producing the first 20 items is $251.60 per item, of producing 100 items is $51.60 per item, and of producing 200 items is $26.60 per item.

13.

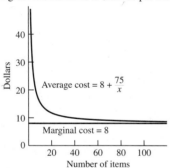

15. $C(x) = 300x + 4000$
17. $C(x) = 25x + 3000$
19. $C(x) = 140x + 600$
21. $C(x) = 20x$
23. $C(x) = 44x + 240$
25. $y = 0.25x$

27. a. $y = 600x + 12{,}000,\ 0 \le x$
b.

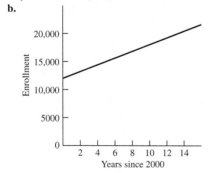

c. The slope, 600, is the number of increases in student enrollment at State University, and the y-intercept, 12,000, is the initial enrollment in the year 2000.
d. The enrollment will pass 20,000 students in 14 years, or in the year 2004.

29. a. $y = -3000x + 25{,}000,\ 0 \le x \le 8.33$
b. The slope, -3000, is the amount the value of the car decreases each year, and the y-intercept, 25,000, is the initial value of the car.
c. The value of the car drops to $5000 in approximately 6.67 years, and it is at $1500 in approximately 7.83 years.
d.

[graph showing Car value (dollars) vs Years, line from (0, 25,000) decreasing to about 8 on x-axis]

31. a. $y = 20x + 2000, 0 \le x$

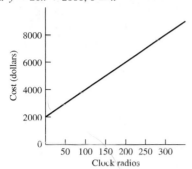

b. One hundred fifty clock radios cost $5000 to produce.
c. The additional cost of making the 201st clock radio is $20.
d. The average cost of the first 200 clock radios is $30 per radio.
e. The average cost of the first 500 clock radios is $24 per radio, of 1000 clock radios is $22 per radio, and of 2000 clock radios is $21 per radio.

33. a. $y = 1500x + 25,000, 0 \le x$
b. 50 on-site inspections require a total cost of $100,000.
c. The additional cost of making the 51st on-site inspection is $1,500.
d. The average cost of making the first 50 on-site inspections is $2,000 per inspection, of making 100 on-site inspections is $1750 per inspection, and of making 200 on-site inspections is $1625 per inspection.
e.

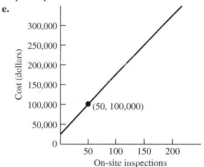

35. a. $y = 0.04x + 1500, 0 \le x$
b.

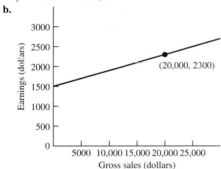

c. At gross monthly sales of $20,000, you would earn $2300.
d. You would earn 4¢ for the 10,001st dollar earned in gross sales.
e. To earn $3000 per month, you need $37,500 in gross monthly sales.

37. a. $y = -0.2x + 220, 700 \le x \le 1100$

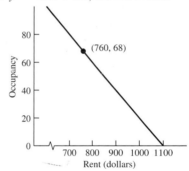

b. $-0.2(700) + 220 = 80$
$-0.2(800) + 220 = 60$
c. At a rent of $760 per month, the predicted occupancy is 68 apartments.

39. a. $y = 0.45x + 3.6$, x is the number of years since 2000. **b.** In the year 2005, the predicted sales are $5.85 million. **c.** In the year 2015, sales will top $10 million. **d.** Financial markets are too complicated and the data are too scant to accurately predict sales many years into the future. **41. a.** $y = 0.895x + 5.4, 0 \le x$, where x represents the number of years after 1960. **b.** The slope 0.895 is the predicted amount of increase we see in yearly production of paper waste. **c.** The predicted waste generation for 1996 is 37.6 million tons, which is under the actual amount by 0.4 million ton. **d.** The predicted waste generation for 1998 is 19.4 million tons, which is over the actual amount by 1.2 million tons.
43. a. The average growth of waste from 1980 to 1998 is approximately 3.817 million tons per year. **b.** $y = 3.817x + 151.5, 0 \le x$, where x represents the number of years since 1980. The predicted waste for 1995 is approximately 208.8 million tons, which is under the actual amount by 2.6 million tons. **c.** In 2010, the predicted waste is 266.0 million tons. **d.** The model predicts that 2019 will be the first year when wastes generated will be over 300 million tons.

45. a. The building is to be depreciated by $5000 per year.
   b. $y = -5000x + 100,000, 0 \leq x \leq 20$
   c.

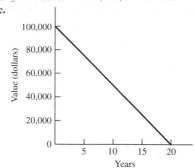

47. a. The machine is to be depreciated by $3500 per year.
   b. $y = -3500x + 80,000, 0 \leq x \leq 20$
   c.

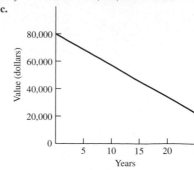

49. a. The building is to be depreciated by $20,000 per year.
   b. $y = -20,000x + 550,000, 0 \leq x \leq 25$
   c.

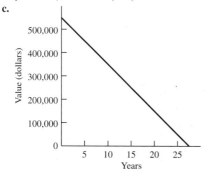

51. a. At a price of $20 per unit, we have a demand of 40 units, which is below the supply of 120 units. Thus, we have a surplus.
   b. A demand of 100 units comes at a price of $10 per unit. At this price, the supply is 60 units, which gives us a shortage.
   c.

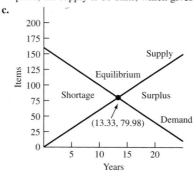

53. a. $D(p) = -1.5p + 18$ **b.** At a price of $5.25 per bushel, the demand is 10.125 million bushels.   **c.** A price of $7.33 per bushel generates a demand of 7 million bushels.   **d.** The model predicts a demand of 18 million bushels if the soybeans are given away free. This is highly unrealistic, however, since free soybeans would fill the short-term needs of every country with food shortages or high food costs.

1.4 Exercises

1. one point   **3.** no point   **5.** one point   **7. a.** one point **b.** $(5/3, -1/3)$ **9. a.** all points **b.** $y = -2x + 3$ **11. a.** one point
b. $(0, 0)$ **13. a.** all points **b.** $y = (-1/2)x$ **15. a.** no points   **17. a.** one point **b.** $(12, -8/3)$ **19. a.** one point **b.** $(2, 1)$
21. a. one point **b.** $(0, 0)$ **23. a.** one point **b.** $(3, 0)$ **25.** $(15, 40)$ **27.** $y = 0.75x - 2$
29. a. We break even at 30 units.
   b. $(30, 1500)$
   c.

31. a. We break even at 250 units.
   b. $(250, 5000)$
   c.

33. a. We break even at 70 units.
  b. $(70, 1050)$
  c.

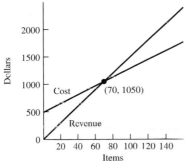

35. a. We break even at 12 units.
  b. $(12, 36)$
  c.

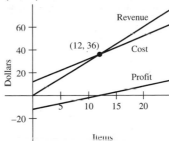

37. a. We break even at 61.25 units.
  b. $(61.25, 1225)$
  c.

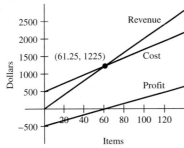

39. a. $C(x) = 8.36x + 8000$
    $R(x) = 10.80x$
    $P(x) = 2.44x - 8000$
  b. We break even at approximately 3279 CD's sold monthly.
  c.

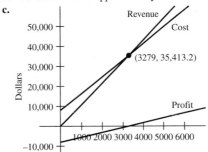

  d. A loss will occur at sales of both 800 and 2000 CD's.

41. a. We reach equilibrium at a price of $2 per item.
  b. At equilibrium, the supply and demand are both at 14 units.
  c. $D(p) = 15 - 0.5p$
    $S(p) = p + 12$

43. a. We reach equilibrium at a price of $20 per item.
  b. At equilibrium, the supply and demand are both 200 items.
  c. $D(p) = 400 - 10p$
    $S(p) = p + 180$

45. a. We reach equilibrium at a price of approximately $41.33 per hundred software packages.
  b. At equilibrium, the supply and demand are both approximately 29 software packages.

47. a. $C(x) = 0.40x + 20,000$
  b. $R(x) = 0.50x + 8000$
  c. The car must be driven for 120,000 miles in the first year in order for the owner to break even.
  d.

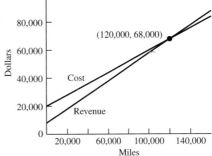

49. a. $C(x) = 1.5x + 50,000$
$R(x) = 2x$
b. $P(x) = 0.5x - 50,000$

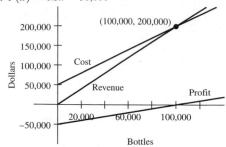

c. If sales are at 60,000 bottles, the company will lose $20,000.
d. 100,000 bottles must be sold in order for the company to break even.
e. The cost of producing 20,000 bottles is $80,000, with an average cost of $4 per bottle. The cost of producing 40,000 bottles is $110,000, with an average cost of $2.75 per bottle.
f. The additional cost of producing the 20,001st bottle is $1.50.

53. a. Brand A: $C(x) = 500$
Brand B: $C(x) = 35x + 350$
b.

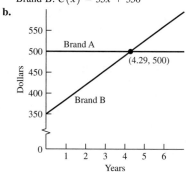

c. The two brands have equal costs at 4.29 years.
d. Brand A costs the least at the 5-year mark.

51. a. Compute method: $C(x) = 25x + 300$
Offset method: $C(x) = 20x + 500$
b.

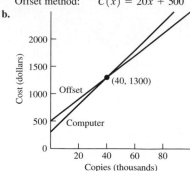

c. The methods have equal costs at 40,000 copies.
d. The cost for 30,000 copies is less with the computer method, while the offset method is the cheaper for the higher volume of 70,000 copies.

55. B-Mart will over take A-Mart in 17 years.
57. The two cities will be recycling the same amount in 2042.
59. Boundary equations: $y = 6$, $y = 2$, $x = 8$, $y = (-4/3)x + 8$
Corner points: $(8, 6)$, $(8, 2)$, $(4.5, 2)$, $(1.5, 6)$
61. Boundary equations: $x = 0$, $y = (-1/2)x + 6$,
$y = (2/3)x + 4/3$
Corner points: $(0, 4/3)$, $(0, 6)$, $(4, 4)$
63. Boundary equations: $x = 0$, $y = (1/2)x + 1$, $y = (-3/5)x + 6$,
$y = (-8/5)x + 8$
Corner points: $(0, 1)$, $(0, 6)$, $(2, 4.8)$, $(10/3, 8/3)$

1.5 Exercises

1. a. $y = 2.5x - 0.5$
b.

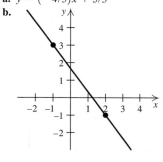

3. a. $y = (-4/3)x + 5/3$
b.

5. a. $y = 3.5x - 2.667$
b.

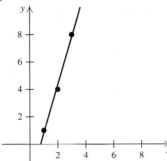

7. a. $y - 1.5x$
b.

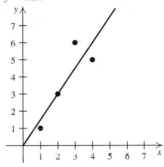

9. a.

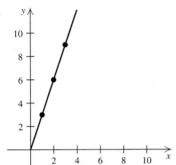

11. a.

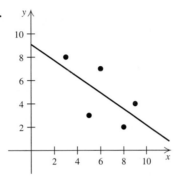

13. a.

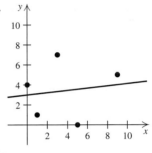

b. $y = 3x$
c. $r = 1$

b. $y = -0.693x + 9.097$
c. $r = -0.64$

b. $y = 0.113x + 2.992$
c. $r = 0.14$

15. Very high positive correlation, perhaps above 0.9.   **17.** Fairly low positive correlation, perhaps between 0.1 and 0.2.
19. Very high negative correlation, perhaps below -0.9.

21. a.

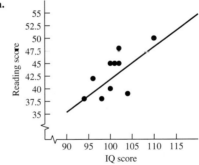

$y = -0.466x + 48.457$
b. $r = -0.99$
c. For each ton of pollutant, the population is dropping by 0.466 thousand, or 466 black bass.
d. At 65 tons of pollutants, the regression model predicts a population of 18,200 fish.
e. 28.8 tons of pollutant will result in a population of 35 thousand fish.

23. a.

b. $y = 0.647x - 22.891$
c. $r = 0.67$

25. a. $y = 11.801x - 209.382$
b. For each inch increase in the diameter, the regression model predicts an increase of 11.801 hundred feet of boards.
c. The regression model predicts that a tree with a diameter of 45 inches will produce 32,168 feet of boards.
d. The regression model predicts that a tree with a diameter of 38.93 inches will produce 25,000 feet of boards.

27. a. Current dollars: $y = 0.210x + 4.807, r = 0.99$
  Constant dollars: $y = -0.023x + 5.528, r = -0.60$
b.

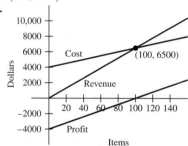

c. The regression model predicts that in 2005 the average current-dollar earnings will be $10.06 per hour, while the average constant-dollar earnings will be $4.96.
29. a. $y = 12.916x + 65.20, r = 0.84$ **b.** The regression model predicts that in 2010 $452.68 million will be spent on drug advertisements in magazines.   **31. a.** Waste generated: $y = 3.574x + 84.925, r = 0.984$; Waste recovered: $y = 1.417x - 5.741, r = 0.853$ **b.** No, the models predict a maximum recovery rate of about 40% in the long run.   **c.** The regression models predict that in the year 2010 approximately 263.60 million tons of waste will be produced and approximately 65.104 million tons will be recovered.

Chapter 1 Summary

1. x-intercepts: $(10, 0)$ **3.** Hor. Line: $y = 8$ **5.** $y = 3x + 10$ **7.** $y = (5/2)x$ **9.** $(3, 2/3)$
  y-intercepts: $(0, -6)$ Vert. Line: $x = -3$
11. The lines are equivalent; thus, the common points are all points that satisfy the equation $5x + 3y = 9$.   **13. a.** The total cost of 60 items is $2994.   **b.** The average cost of the first 60 items is $49.90 per item.   **c.** The additional cost of producing the 61st item is $7.90.
15. a. $P(x) = 40x - 4000$
  b. $(100, 6500)$
  c.

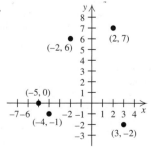

17. choice (b)

Sample Test Items for Chapter 1

1.

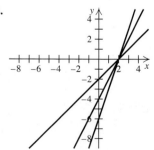

2.

3. $y = 3.7x - 2.8$ **4.** $C(x) = 3.2x + 490$ **5.** $y = 250x + 2300$ **6.** $x = -2$
7. Slope $= 0.5$
    x-intercept $= (9, 0)$
    y-intercept $= (0, -4.5)$
8. $y = 3.2x + 6.4$ **9.** $y = -20x + 173$ **10.** $y = -350x + 5600$ **11.** They do not have equal slopes; thus, they cross at exactly one point.   **12.** $(17/3, -2/3)$ **13.** all points given by $y = -1.4x + 15/4$
14. $P(x) = 125x - 40,000$
    Break-even: $(1600, 156,800)$
15. Equilibrium is reached at a price of $588.89 per quilt. At this price, both supply and demand will be 8.67 quilts.
16. a.

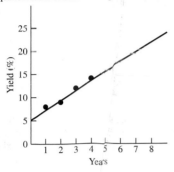

b. $y = 2.1x + 5.5$
c. The regression model predicts a return of 16% in the fifth year.
d. The regression model predicts that one must wait until the seventh year to see a return of 20%.

CHAPTER 2

2.1 Exercises

1. $\begin{bmatrix} 1 & 3 \\ 2 & -1 \end{bmatrix}; \begin{bmatrix} 5 \\ -4 \end{bmatrix}; \begin{bmatrix} 1 & 3 & 5 \\ 2 & -1 & -4 \end{bmatrix}$ **3.** $\begin{bmatrix} 1 & 3 & 0 \\ 2 & 1 & -1 \\ 1 & -1 & 1 \end{bmatrix}; \begin{bmatrix} 0 \\ 0 \\ 0 \end{bmatrix}; \begin{bmatrix} 1 & 3 & 0 & 0 \\ 2 & 1 & -1 & 0 \\ 1 & -1 & 1 & 0 \end{bmatrix}$

5. $\begin{bmatrix} 1 & 2 & 3 & 1 \\ 1 & 1 & -1 & 0 \\ 3 & 5 & 0 & 2 \end{bmatrix} \begin{bmatrix} 5 \\ 3 \\ -2 \end{bmatrix}; \begin{bmatrix} 1 & 2 & 3 & 1 & 5 \\ 1 & 1 & -1 & 0 & 3 \\ 3 & 5 & 0 & 2 & -2 \end{bmatrix}$ **7.** $\begin{bmatrix} 5 & 3 & 2 \\ 1 & -2 & 0 \\ 3 & 1 & 1 \\ 1 & 0 & 4 \end{bmatrix}; \begin{bmatrix} 0 \\ 0 \\ 1 \\ 2 \end{bmatrix}; \begin{bmatrix} 5 & 3 & 2 & 0 \\ 1 & -2 & 0 & 0 \\ 3 & 1 & 1 & 1 \\ 1 & 0 & 4 & 2 \end{bmatrix}$

9. $2x \quad = 2$ **11.** $x \quad = -8$ **13.** $x \quad = 18$
    $x + 3y = 1$ $y = 7$ $y = 25$
                            $z = 30$

15. x = number of shares of Datafix;
    y = number of shares of Rocktite
    $30x + 20y = 4000$
    $2x + y = 220$

17. x = number of tons of coal; y = number of thousands of cubic feet of gas
    $3x + 4.5y = 183$
    $60x + y = 990$

19. x = number of wood pens; y = number of silver pens; z = number of gold pens
    $x + 0.5y + 3z = 12,000$
    $2x + 2y + 2.5z = 9600$

21. x = number of skilled workers;
    y = number of semiskilled workers;
    z = number of supervisors
    $12x + 9y + 15z = 1560$
    $3x + 5y + z = 588$
    $x + y = 140$

23. A = pounds of ingredient A;
    B = pounds of ingredient B;
    C = pounds of ingredient C;
    D = pounds of ingredient D
    $2A + B + 3C + D = 300$
    $4A + 6B + 2C + 2D = 600$
    $A + B + C + D = 200$

25. s = acres of soybeans; w = acres of wheat; c = acres of corn; b = acres of barley
    $25s + 50w + 40c + 30b = 10,000,000$
    $5s + 6w + 8c + 3b = 50,000$

27. x = number of workers on first shift;
    y = number of workers on second shift;
    z = number of workers on third shift
    $x + y = 10$
    $y + z = 8$
    $x - z = 0$

29. A = miles of highway in Alma;
    B = miles of highway in Bedford;
    C = miles of highway in Cooksville
    $2A + 0.5B + C = 120$
    $A - 2B = 0$

31. x = number of four-ply tires;
    y = number of radial tires
    $x + y = 600$
    $40x + 50y = 28,500$
    $3x - y = 0$

33. x = gallons from 8% tank; y = gallons from 13% tank

$$x + \quad y = 2000$$
$$0.08x + 0.13y = \quad 200$$

35. x = liters from 21% solution; y = liters from 14% solution

$$x + \quad y = 2$$
$$0.21x + 0.14y = 0.36$$

2.2 Exercises

1. $\begin{bmatrix} 1 & 1 & 1 & | & 2 \\ 0 & -1 & 0 & | & -2 \\ 0 & 1 & 3 & | & 2 \end{bmatrix}$ **3.** $\begin{bmatrix} 1 & 0 & -4 & | & -3 \\ 0 & 1 & 1 & | & 2 \\ 0 & 0 & -2 & | & -2 \end{bmatrix}$ **5.** $\begin{bmatrix} 1 & 0 & | & 2 \\ 0 & 1 & | & 1 \end{bmatrix}$ **7.** $\begin{bmatrix} 1 & 0 & 1 & | & 0 \\ 0 & 1 & 0 & | & 0 \\ 0 & 0 & 1 & | & \frac{4}{3} \end{bmatrix}$ **9.** $x = 1; y = 1; z = 2$

11. $x = 2; y = 1; z = 4$ **13.** $x = -\frac{1}{2}; y = 1; z = 1$ **15.** No solution.   **17.** $x = \frac{20}{7}; y = \frac{12}{7}$ **19.** $x = \frac{1}{7}; y = -\frac{10}{7}$

21. $x = 1; y = 0; z = -3$ **23.** No solution.   **25.** $x = 0; y = 0; z = 0$

27. $w = \frac{56}{261} \approx 0.2146; x = \frac{-53}{174} \approx -0.3046; y = \frac{1357}{1044} \approx 1.2998; z = \frac{73}{348} \approx 0.2098$ **29.** $x \approx 0.11; y \approx 2.61; z \approx -5.31$

31. $w \approx -5.57; x \approx 7.42; y \approx 8.49; z \approx 5.02$ **33.** To meet the pollution allowance, the plant should burn 16 tons of coal and 30,000 cubic feet of gas each hour.   **35.** To meet budget, printing, and memory needs, the firm needs to purchase 3 Q6 models, 2 Q10 models, and 2 Q12 models.   **37.** The charity should run 8 newspaper ads, 12 radio ads, and 4 television ads to meet its objectives.   **39.** Brenda cannot staff the three shifts to meet the given conditions; the system of equations has no solution.   **41.** The store manager should purchase 150 four-ply tires and 450 radial tires to meet the given conditions.   **43.** Cut the string into a 32-inch piece and a 128-inch piece.   **45.** The company should make 60 saws of type A, no saws of type B, and 70 saws of type C each week in order to fully utilize its labor hours fully.   **47.** The consolidated income is $2.4685 million for company A, $2.98 million for company B, and $4.1815 million for company C.

2.3 Exercises

1. a. $x = \frac{7}{3} + \frac{7}{3}z; y = -\frac{1}{3} + \frac{5}{3}z; z$ = any number   **c.** Many solutions are possible. Set z equal to any number and solve for x and y.

3. a. $x = \frac{1}{7} + \frac{2}{7}z; y = -\frac{2}{7} + \frac{3}{7}z; z$ = any number   **c.** Many solutions are possible. Set z equal to any number and solve for x and y.

5. a. $w = \frac{1}{5}y + z; x = 2 - \frac{3}{5}y + z; y$ = any number; z = any number.   **c.** Many solutions are possible. Set y and z equal to any number and solve for w and x.   **7. a.** $x_1 = \frac{15}{2} - 3x_4; x_2 = -8 + 3x_4; x_3 = \frac{51}{4} - 3x_4; x_4$ = any number.   **c.** Many solutions are possible. Set x_4 equal to any number and solve for $x_1, x_2,$ and x_3.   **9. a.** $x = 3 - y - z; y$ = any number; z = any number.   **c.** Many solutions are possible. Set y and z equal to any number and solve for x.   **11. a.** $x = -\frac{7}{5}z; y = \frac{3}{5}z; z$ = any number.   **c.** Many solutions are possible. Set z equal to any number and solve for x and y.   **13. a.** $x = z; y = -z; z$ = any number.   **c.** Many solutions are possible. Set z equal to any number and solve for x and y.   **15. a.** $x = -z; y = 0; z$ = any number.   **c.** Many solutions are possible. Set z equal to any number and solve for x.
17. No solution.   **19.** $w = 1 - 2x + 3y - z; x$ = any number; y = any number; z = any number.   **21.** No solution.
23. $x \approx 3.7097 - 0.7742z; y \approx 5.7742 - 1.5355z; z$ = any number.   **25.** No solution.   **27.** $w \approx -0.3505; x \approx 1.9576; y \approx 0; z \approx 2.7352$.
29. $x = 2z; y = 0; z$ = any number.   **31.** $x = -z; y = -2z; z$ = any number.   **33. a.** $x = 0; y = 0; u = 10; v = 5$.   **b.** $x = 5; y = 0;$ $u = 0; \quad v = 0$.   **35. a.** $x = 0; \quad y = 0; \quad u = -6; \quad v = 4$ **b.** $x = 2; \quad y = 0; \quad u \approx -4; \quad v = 0$.
37. The city of Burlington should allocate $500,000 to streets. The amount z allocated to parks can be any dollar value from $0 to $500,000. The amount allocated to urban renewal should be $500,000 - z$.   **39.** The number z of television ads can be any number from 0 to 10. The number of newspaper ads should be $30 - 3z$. The number of radio ads should be $2z$.   **41.** The number z of QT60 printers that should be manufactured in a week can be any number from 0 to 47.5. The number of QT20 printers manufactured in a week should be $60 + z$. The number of QT40 printers manufactured in a week should be $95 - 2z$.   **43.** The number z of orders that should be shipped from warehouse 2 to store 2 can be any number from 0 to 15. The number of orders shipped from warehouse 1 to store 1 should be $2z$. The number of orders shipped from warehouse 1 to store 2 should be $18 - z$. The number of orders shipped from warehouse 2 to store 2 should be $30 - 3z$. Many possibilities for supplying the requested orders exist.

Chapter 2 Summary Exercises

1. $\begin{bmatrix} 3 & 0 & -1 \\ 0 & 1 & 0 \\ 6 & 0 & -2 \end{bmatrix}$; $\begin{bmatrix} 0 \\ 7 \\ 0 \end{bmatrix}$; $\left[\begin{array}{ccc|c} 3 & 0 & -1 & 0 \\ 0 & 1 & 0 & 7 \\ 6 & 0 & -2 & 0 \end{array}\right]$ **3.** $\left[\begin{array}{ccc|c} 1 & 0 & 0 & 1 \\ 0 & 1 & 0 & -1 \\ 0 & 0 & 1 & 1 \end{array}\right]$. The system has exactly one solution.   **5.** $\left[\begin{array}{ccc|c} 1 & 0 & \frac{1}{3} & \frac{2}{3} \\ 0 & 1 & \frac{1}{3} & -\frac{1}{3} \\ 0 & 0 & 0 & 0 \end{array}\right]$. The system has infinitely many solutions.   **7.** No solution.   **9.** $x = 4$; $y = -3$; $z = 2$.   **11.** $x = 3$; $y = -5$;   **13.** $x = 3$; $y = 5$; $z = 1$.
15. The student spends 12 hours studying history, 4 hours studying English, 6 hours studying psychology, and 14 hours studying accounting each week.

17. $x + 2y - 3z = 5$
$\quad\;\; y + z = 4; x = 12; y = 1; z = 3.$
$\qquad\quad z = 3$

19. $2x + y - 3z = 16$
$\quad\;\; 3y + z = 12; x = 6; y = 4; z = 0.$
$\qquad\quad 4z = 0$

21. Neither student is correct. A system of linear equations with more equations than unknowns can have exactly one solution, infinitely many solutions, or no solution.   **23.** $x \approx 4.3$; $y \approx -0.6$; $z \approx -1.6$ **25.** $y = -\dfrac{3}{5}x + 13$ **27.** $x = -1$; $y = 3$

Sample Test Items for Chapter 2

1. $\begin{bmatrix} 1 & 0 & 2 \\ -3 & 5 & -8 \\ 1 & 1 & 0 \end{bmatrix}$ **2.** $\left[\begin{array}{cc|c} 3 & -1 & 5 \\ -5 & 17 & -1.2 \end{array}\right]$

3. Let N represent the number of nickels, D represent the number of dimes, and Q represent the number of quarters. Then
$$N + D + Q = 213$$
$$5N + 10D + 25Q = 2115$$
$$D - 2Q = 0$$

4. Let S represent the number of student tickets sold, and let A represent the number of adult tickets sold. Then
$$S + A = 9073$$
$$12.50S + 16.75A = 129{,}673$$

5. Let m represent the number of recipes of meatballs, c represent the number of recipes of crab cakes, and d represent the number of recipes of dumplings that Steven needs to prepare. Then
$$20m + 10c + 30d = 380$$
$$2c - d = 0$$
$$10m + 15c + 8d = 174$$

6. Only matrices (c) and (e) are in reduced row-echelon form.   **7.** $\left[\begin{array}{ccc|c} 1 & 1 & 3 & -6 \\ 0 & 2 & 4 & 10 \\ 0 & -7 & -14 & 30 \end{array}\right]$ **8.** $x = 2$; $y = -3$; $z = 7$.

9. $x = -1$, $y = 5$.   **10.** The parking meter contained 108 nickels, 70 dimes, and 35 quarters.   **11.** Many solutions are possible. If $z = 0$, then $x = 3$ and $y = -1$. If $z = 4$, then $x = 1$ and $y = 0$. It $z = -4$, then $x = 5$ and $y = -2$.   **12.** $x = 3 + z$; $y = 2 - z$; $z =$ any number.
13. No solution.   **14.** There were 5247 students and 3826 adults attending the performance.   **15.** Steven should prepare 5 recipes of the meatballs, 4 recipes of the crab cakes, and 8 recipes of the Chinese dumplings.

CHAPTER 3

3.1 Exercises

1. $t = 3$; $x = 9$.   **3.** $x =$ any number.   **5.** $\begin{bmatrix} 0 & -10 \\ -3 & 5 \end{bmatrix}$ **7.** $\begin{bmatrix} \frac{8}{3} & \frac{2}{3} \\ \frac{1}{3} & \frac{1}{3} \end{bmatrix}$ **9.** This operation is undefined.

11. $\begin{bmatrix} -40 & 60 & -100 \\ 0 & 0 & 40 \end{bmatrix}$ **13.** $\begin{bmatrix} a+e & b+f \\ c+g & d+h \end{bmatrix} = \begin{bmatrix} e+a & f+b \\ g+c & h+d \end{bmatrix}$ **15. a.** $\begin{array}{c} P_1 \\ P_2 \end{array}\begin{bmatrix} 30 & 50 & 90 \\ 50 & 60 & 100 \end{bmatrix}$ **b.** $\begin{array}{c} \text{First} \\ \text{Tourist} \\ \text{Economy} \end{array}\begin{bmatrix} 30 & 50 \\ 50 & 60 \\ 90 & 100 \end{bmatrix}$

17. a. $\begin{bmatrix} 300 \\ 287 \\ 250 \\ 240 \end{bmatrix}$ **b.** $\begin{bmatrix} 300 & 287 & 250 & 240 \end{bmatrix}$ **19.** $\begin{array}{c} \\ A \\ B \\ C \\ D \\ E \end{array}\begin{array}{c} \begin{array}{ccccc} A & B & C & D & E \end{array} \\ \begin{bmatrix} 0 & 1 & 1 & 1 & 0 \\ 0 & 0 & 0 & 1 & 0 \\ 0 & 0 & 0 & 1 & 0 \\ 0 & 0 & 0 & 0 & 1 \\ 0 & 0 & 0 & 0 & 0 \end{bmatrix}\end{array}$ **21.** $\begin{array}{c} \\ B \\ W \\ T \\ C \end{array}\begin{array}{c} \begin{array}{cccc} B & W & T & C \end{array} \\ \begin{bmatrix} 0 & 1 & 1 & 0 \\ 0 & 0 & 1 & 0 \\ 0 & 0 & 0 & 1 \\ 1 & 1 & 0 & 0 \end{bmatrix}\end{array}$ **23.** $X = 4B + \dfrac{20}{3}T$

25. $B = \dfrac{1}{6}A + \dfrac{1}{3}C - \dfrac{2}{3}D$ **27.** $T = \dfrac{1}{4}A - \dfrac{1}{2}C$ **29.** $\begin{bmatrix} \frac{1}{2} & -\frac{3}{2} & 1 \\ -\frac{7}{2} & -\frac{5}{2} & -\frac{5}{2} \end{bmatrix}$ **31.** $\begin{bmatrix} -\frac{4}{3} & -\frac{7}{3} \\ -1 & -1 \\ -\frac{2}{3} & 2 \end{bmatrix}$ **33.** Yes, $x = 3$ solves the equation.

35. $60B = \begin{bmatrix} 3.6 & 1.2 \\ 1.8 & 3 \end{bmatrix}$ **37.** $\begin{bmatrix} 1 & 6 \\ 16 & 13 \end{bmatrix}$ **39. a.** $0.08A$ **b.** $\begin{bmatrix} 2400 & 2000 \\ 4000 & 3200 \end{bmatrix}$ **c.** $A + 0.08A = 1.08A$ **d.** $\begin{bmatrix} 32,400 & 27,000 \\ 54,000 & 43,200 \end{bmatrix}$

41. $C + 0.05C = 1.05C = \begin{bmatrix} 4725 & 16,800 \\ 5250 & 9870 \end{bmatrix}$ **43.** $1.05^3C = \begin{bmatrix} 5209.31 & 18,522.00 \\ 5788.13 & 10,881.68 \end{bmatrix}$ **45.** $\begin{bmatrix} 14.3 & 28.6 \\ 37.5 & 87.5 \\ 66.7 & 116.7 \end{bmatrix}$ **47.** $B^T = \begin{bmatrix} 1 & 3 \\ 2 & -5 \end{bmatrix}$

49. $D^T = \begin{bmatrix} 1 & 2 & -3 & 4 \\ 0 & -1 & 2 & 1 \end{bmatrix}$

3.2 Exercises

1. $\begin{bmatrix} 1 & 2 \\ 3 & -1 \end{bmatrix}\begin{bmatrix} x \\ y \end{bmatrix} = \begin{bmatrix} 7 \\ 6 \end{bmatrix}$ **3.** $\begin{bmatrix} 1 & 1 \\ 3 & -1 \\ 1 & 5 \end{bmatrix}\begin{bmatrix} x \\ y \end{bmatrix} = \begin{bmatrix} 2 \\ 6 \\ -1 \end{bmatrix}$

5. $\begin{aligned} 2x \quad\quad &= 3 \\ x + 5y &= 5 \\ 4x + 6y &= 2 \end{aligned}$ **7.** $\begin{aligned} x \quad\quad + z \quad\quad + v &= -2 \\ 2x \quad\quad + 3z + 5w + v &= 3 \\ -x + 2y + 2z + 2w \quad\quad &= 6 \end{aligned}$

9. $\begin{bmatrix} 8 & 9 \\ -3 & -6 \end{bmatrix}$ **11.** This operation is not possible.   **13.** $\begin{bmatrix} 8 & 6 & 13 & 9 \\ -3 & 3 & 3 & 15 \end{bmatrix}$ **15.** $\begin{bmatrix} 3 & 0 \\ 4 & 7 \end{bmatrix}$ **17.** $\begin{bmatrix} 3 & 0 \\ 8 & 11 \end{bmatrix}$ **19.** $\begin{bmatrix} 7 & 0 \\ 0 & 7 \end{bmatrix}$

21. $\begin{bmatrix} 10 & 0 \\ 6 & 16 \end{bmatrix}$ **23.** The matrix product $(A - B)(A + B) = A^2 - BA + AB - B^2$. Since BA is not necessarily equal to AB, this expression cannot, in general, be simplified.

25. $(AB)C = \begin{bmatrix} 14 & 22 \\ 57 & 84 \end{bmatrix} = A(BC)$ **27.** $A(B + C) = \begin{bmatrix} 11 & 13 \\ 29 & 39 \end{bmatrix} = AB + AC$ **29.** Many examples of such matrices A, B, and C exist. Let $A = \begin{bmatrix} 1 & 0 & 0 \\ 0 & 0 & 0 \\ 0 & 0 & 0 \end{bmatrix}$, $B = \begin{bmatrix} 0 & 1 & 0 \\ 0 & 0 & 0 \\ 0 & 0 & 0 \end{bmatrix}$ and $C = \begin{bmatrix} 0 & 0 & 0 \\ 0 & 1 & 0 \\ 0 & 0 & 0 \end{bmatrix}$. Then $AB = \mathbf{0}$ and $AC = \mathbf{0}$, but $B \neq C$.

31. a., b., c. $\begin{bmatrix} 20 & 10 \\ 20 & 20 \end{bmatrix}$ **33. a., b., c.** $\begin{bmatrix} 10k & 5k \\ 10k & 10k \end{bmatrix}$ **35.** $(A + B)X$ **37.** $A(X + 2Y)$ **39.** $(3A + 4C)B$

41. $A(2X - 3Y)$ **43.** $B = [-100, 6]$; $BA = [-540, -920]$ **45. a.** $B = [1 \quad 1.5 \quad 0.73]$; $BA = [9.5 \quad 875.73 \quad 15.46]$

b. $C = \begin{bmatrix} -3 \\ -1.5 \\ 2 \end{bmatrix}$; $AC = \begin{bmatrix} -1247 \\ -44 \\ 2.5 \end{bmatrix}$ **47. a.** $B = \begin{matrix} \text{Jan} \\ \text{June} \end{matrix}\begin{bmatrix} 3 & 5 & 2 \\ 2 & 4 & 6 \end{bmatrix}$; $BA = \begin{bmatrix} 7600 & 9400 \\ 12,400 & 12,400 \end{bmatrix}$. The entry in the first-row, first-column

position indicates that 7600 men were influenced by ads in January. The entry in the first-row, second-column position indicates that 9400 women were influenced by ads in January. The entry in the second-row, first-column position indicates that 12,400 men were influenced by ads in June. The

entry in the second-row, second-column position indicates that 12,400 women were influenced by ads in June.   **b.** $C = \begin{matrix} \text{Men} \\ \text{Women} \end{matrix}\begin{bmatrix} 10 & 12 & 15 \\ 15 & 18 & 22 \end{bmatrix}$;

$AC = \begin{bmatrix} 22,000 & 26,400 & 32,500 \\ 17,000 & 20,400 & 25,100 \\ 33,000 & 39,600 & 48,900 \end{bmatrix}$. The rows indicate the amount of money spent due to newspaper ads, radio ads, and television ads, respectively.

The columns represent the amount of money spent during each of the three months. For instance, the entry in the second-row, third-column, tells

us the amount of money spent as a result of radio ads during the third month of the advertising campaign.   **49.** $B = [200 \quad 300]$; $C = \begin{bmatrix} 12 \\ 20 \\ 1 \end{bmatrix}$;

$BAC = [529]$. The total cost of filling this order is \$529.   **51.** $A = \begin{matrix} L \\ R \end{matrix}\begin{matrix} \;L & \;\;R \\ \begin{bmatrix} 0.10 & 0.90 \\ 0.40 & 0.60 \end{bmatrix} \end{matrix}$; $B = \begin{matrix} L & R \\ [48 & 52] \end{matrix}$. The number that should turn left or right

on the following day can be found by multiplying $BA = [25.76 \quad 74.4]$. The operation $BA^2 = [32.32 \quad 67.68]$ gives the number that should turn left or right on the third day. The operation $BA^3 = [30.304 \quad 69.696]$ gives the number of mice that should turn left or right, respectively, on the fourth day.

53. a. $B = \begin{bmatrix} 500{,}000 \\ 1{,}000{,}000 \end{bmatrix}$

b. $CB = \begin{matrix} \text{Belgium} \\ \text{France} \\ \text{Japan} \\ \text{Mexico} \\ \text{U.S.} \end{matrix} \begin{bmatrix} 45{,}000 \\ 49{,}250 \\ 27{,}500 \\ 28{,}000 \\ 27{,}000 \end{bmatrix}$.

The entries in this matrix represent the total construction cost allotted to architects and structural engineers in the respective countries.

55. a. $M = \begin{bmatrix} 0 & 1 & 1 \\ 0 & 0 & 1 \\ 0 & 1 & 0 \end{bmatrix}$

b. $M^2 = \begin{bmatrix} 0 & 1 & 1 \\ 0 & 1 & 0 \\ 0 & 0 & 1 \end{bmatrix}$.

The entry in the second row, second column, indicates that, in two stages, one can fly from City B back to City B. Comparing this with the diagram, we see that one can fly from City B to City C and then from City C back to City B. After only two stages, it is impossible to return to City A.

c. $M + M^2 = \begin{bmatrix} 0 & 2 & 2 \\ 0 & 1 & 1 \\ 0 & 1 & 1 \end{bmatrix}$.

This matrix indicates the total number of paths that can be followed to get from one city to another in two or fewer stages or less.

57. $A(B + C) = \begin{bmatrix} a_{11}b_{11} + a_{11}c_{11} + a_{12}b_{21} + a_{12}c_{21} & a_{11}b_{12} + a_{11}c_{12} + a_{12}b_{22} + a_{12}c_{22} \\ a_{21}b_{11} + a_{21}c_{11} + a_{22}b_{21} + a_{22}c_{21} & a_{21}b_{12} + a_{21}c_{12} + a_{22}b_{22} + a_{22}c_{22} \end{bmatrix} = AB + AC$

3.3 Exercises

1. Since $AB = BA = I$, it follows that $B = A^{-1}$.   **3.** Since $AB \neq I$, B is not the inverse of A.   **5.** $x = 1; \; y = -7$.

7. $x = 0.125, y = -0.175$.   **9.** $\begin{bmatrix} -5 & 3 \\ 2 & -1 \end{bmatrix}$ **11.** The matrix has no inverse.   **13.** The matrix has no inverse.   **15.** $\begin{bmatrix} 2 & \frac{3}{2} \\ 1 & \frac{1}{2} \end{bmatrix}$

17. $\begin{bmatrix} 0 & -\frac{1}{2} & 1 \\ -\frac{1}{2} & \frac{1}{4} & \frac{1}{2} \\ \frac{1}{2} & \frac{1}{4} & -\frac{1}{2} \end{bmatrix}$ **19.** The matrix has no inverse.   **21.** $\begin{bmatrix} 1 & 0 & -1 \\ 0 & -1 & 1 \\ -1 & 1 & 1 \end{bmatrix}$ **23.** $\begin{bmatrix} \frac{2}{3} & 0 & \frac{1}{3} & -\frac{1}{3} \\ -1 & \frac{1}{2} & 0 & 0 \\ \frac{1}{3} & 0 & -\frac{1}{3} & \frac{1}{3} \\ -\frac{4}{3} & 0 & \frac{1}{3} & \frac{2}{3} \end{bmatrix}$

25. a. $I - A = \begin{bmatrix} -1 & -3 \\ -1 & 5 \end{bmatrix}$ **b.** $\begin{bmatrix} -\frac{5}{8} & -\frac{3}{8} \\ -\frac{1}{8} & \frac{1}{8} \end{bmatrix}$ **27.** $\begin{bmatrix} 1 & 3 \\ 2 & 4 \end{bmatrix}\begin{bmatrix} x \\ y \end{bmatrix} = \begin{bmatrix} 7 \\ -3 \end{bmatrix}$ Solution: $x = -18.5; y = 8.5$.

29. $\begin{bmatrix} 5 & 2 \\ 4 & 2 \end{bmatrix}\begin{bmatrix} x \\ y \end{bmatrix} = \begin{bmatrix} 4 \\ -3 \end{bmatrix}$ Solution: $x = 7; y = -15.5$.

31. $\begin{bmatrix} 1 & 0 & 2 \\ 0 & 2 & 2 \\ 1 & 1 & 1 \end{bmatrix}\begin{bmatrix} x \\ y \\ z \end{bmatrix} = \begin{bmatrix} 1 \\ 3 \\ 5 \end{bmatrix}$ Solution: $x = 3.5; y = 2.75; z = -1.25$.

33. $\begin{bmatrix} 2 & 1 & 2 \\ 1 & -1 & 1 \\ 1 & 1 & 0 \end{bmatrix}\begin{bmatrix} x \\ y \\ z \end{bmatrix} = \begin{bmatrix} -5 \\ -5 \\ 2 \end{bmatrix}$ Solution: $x = \frac{1}{3}; y = \frac{5}{3}; z = -\frac{11}{3}$.

35. $\begin{bmatrix} 3 & 3 & 1 \\ 1 & -1 & 1 \\ 3 & 0 & 1 \end{bmatrix}\begin{bmatrix} x \\ y \\ z \end{bmatrix} = \begin{bmatrix} \frac{8}{3} \\ 1 \\ -1 \end{bmatrix}$ Solution: $x \approx -1.61; y \approx 1.22; z \approx 3.83$.

37. $x = 0; y = 0; z = 0$.   **39.** $w \approx 0.61; x = 3; y \approx 4.61; z \approx 1.6$

41. a. Let D represent the number of shares of Datafix and let R represent the number of shares of Rocktite. Then

$$30D + 20R = 4000$$
$$2D + 1R = 220.$$

b. $\begin{bmatrix} 30 & 20 \\ 2 & 1 \end{bmatrix} \begin{bmatrix} D \\ R \end{bmatrix} = \begin{bmatrix} 4000 \\ 220 \end{bmatrix}$

c. Emily purchased 40 shares of Datafix and 140 shares of Rocktite.

45. The police department can have 7.6 rookies and 9.6 sergeants under the new conditions.

47. $\begin{bmatrix} \frac{1}{3} & 0 \\ 0 & 1 \end{bmatrix}$
49. $\begin{bmatrix} \frac{1}{3} & 0 & 0 \\ 0 & -1 & 0 \\ 0 & 0 & \frac{1}{5} \end{bmatrix}$
51. $\begin{bmatrix} \frac{1}{3} & 0 & 0 & 0 \\ 0 & -\frac{1}{6} & 0 & 0 \\ 0 & 0 & \frac{1}{2} & 0 \\ 0 & 0 & 0 & \frac{1}{5} \end{bmatrix}$

59. a., b. $\begin{bmatrix} 4 & -7 \\ -2.5 & 4.5 \end{bmatrix}$

c. Given two $n \times n$ matrices A and B, $(AB)^{-1} = B^{-1}A^{-1}$.

d. $((AB)C)^{-1} = C^{-1}(AB)^{-1} = C^{-1}B^{-1}A^{-1}$

63. a. The matrix A must have an inverse.

b. No.

c. No.

d. Yes.

43. a. Let x represent the number of skilled workers, y the number of semiskilled workers, and z the number of supervisors. Then

$$12x + 9y + 15z = 1560$$
$$3x + 5y + \quad z = 588$$
$$x + \quad y \quad\quad = 140.$$

b. $\begin{bmatrix} 12 & 9 & 15 \\ 3 & 5 & 1 \\ 1 & 1 & 0 \end{bmatrix} \begin{bmatrix} x \\ y \\ z \end{bmatrix} = \begin{bmatrix} 1560 \\ 588 \\ 140 \end{bmatrix}$

c. The department needs 60 skilled workers, 80 semiskilled workers, and 8 supervisors.

53. $A = CB^{-1}$ **55.** $X = B(I - A)^{-1}$ **57.** $A = C(B + Z)^{-1}$

61. a., b. $\begin{bmatrix} 2 & -1 \\ -3 & 2 \end{bmatrix}$

c. Given an $n \times n$ matrix A, $(A^T)^{-1} = (A^{-1})^T$.

3.4 Exercises

1. The letters A through Z correspond to the numbers 1 through 26, respectively; a space corresponds to the number 30, a period corresponds to the number 40, and an exclamation point corresponds to the number 50.

a. $C = \begin{bmatrix} 9 & 30 \\ 8 & 1 \\ 22 & 5 \\ 30 & 1 \\ 30 & 10 \\ 15 & 2 \\ 40 & 30 \end{bmatrix}$

b. $CA = \begin{bmatrix} -21 & 48 \\ 7 & 17 \\ 17 & 49 \\ 29 & 61 \\ 20 & 70 \\ 13 & 32 \\ 10 & 110 \end{bmatrix}$

c. $A^{-1} = \begin{bmatrix} \frac{1}{3} & -\frac{2}{3} \\ \frac{1}{3} & \frac{1}{3} \end{bmatrix}$; $(CA)A^{-1} = C$

3. Use the same character–number correspondence as in Exercise 1.

a. $C = \begin{bmatrix} 14 & 15 \\ 30 & 23 \\ 1 & 25 \\ 40 & 30 \end{bmatrix}$

b. $CA = \begin{bmatrix} 29 & 44 \\ 53 & 76 \\ 26 & 51 \\ 70 & 100 \end{bmatrix}$

c. $A^{-1} = \begin{bmatrix} 2 & -1 \\ -1 & 1 \end{bmatrix}$; $(CA)A^{-1} = C$

5. a. $A^{-1} = \begin{bmatrix} 3 & -1 & 0 \\ -\frac{1}{2} & 0 & \frac{1}{2} \\ -2 & 1 & 0 \end{bmatrix}$

b. $(CA)A^{-1} = \begin{bmatrix} 7 & 18 & 5 \\ 1 & 20 & 30 \\ 4 & 1 & 20 \\ 5 & 40 & 30 \end{bmatrix}$; GREAT DATE.

7. a. $\begin{bmatrix} \frac{1}{2} & 0 & 0 \\ 0 & 1 & 0 \\ 0 & 0 & \frac{1}{3} \end{bmatrix}$

b. $(CA)A^{-1} = \begin{bmatrix} 7 & 12 & 15 \\ 2 & 1 & 12 \\ 30 & 23 & 1 \\ 18 & 13 & 9 \\ 14 & 7 & 30 \\ 3 & 15 & 14 \\ 20 & 9 & 14 \\ 21 & 5 & 19 \\ 40 & 30 & 30 \end{bmatrix}$; GLOBAL WARMING CONTINUES.

9. a. Each output unit of steel requires 0.2 unit of steel and 0.15 unit of electronics. Each output unit of electronics requires 0.10 unit of steel and 0.01 unit of electronics.

b. $(I - T)^{-1} \approx \begin{bmatrix} 1.0364 & 0.1047 \\ 0.1570 & 1.0260 \end{bmatrix}$

c. $X \approx \begin{bmatrix} 601.9682 \\ 899.2881 \end{bmatrix}$

d. Approximately 101.9682 units of steel and 99.2882 units of electronics are consumed internally.

11. a. Each output unit of manufacturing requires 0.10 unit of manufacturing and 0.15 unit of electronics. Each output unit of electronics requires 0.20 unit of manufacturing and 0.05 unit of electronics.

b. $(I - T)^{-1} \approx \begin{bmatrix} 1.1515 & 0.2424 \\ 0.1818 & 1.0909 \end{bmatrix}$

c. $X \approx \begin{bmatrix} 351.5152 \\ 581.8182 \end{bmatrix}$

d. Approximately 151.5152 units of manufacturing and 81.8182 units of electronics are consumed internally.

13. a. Each output unit of agriculture requires 0.02 unit of agriculture, 0.20 unit of manufacturing, and 0.15 unit of electronics. Each output unit of manufacturing requires 0.01 unit of agriculture, 0.10 unit of manufacturing, and 0.08 unit of electronics. Each output unit of electronics requires 0.01 unit of agriculture, 0.10 unit of manufacturing, and 0.05 unit of electronics.

b. $(I - T)^{-1} \approx \begin{bmatrix} 1.0248 & 0.0125 & 0.0121 \\ 0.2480 & 1.1246 & 0.1210 \\ 0.1827 & 0.0967 & 1.0647 \end{bmatrix}$

c. $X \approx \begin{bmatrix} 549.0623 \\ 1490.6231 \\ 2317.4834 \end{bmatrix}$

d. Approximately 49.0623 units of agriculture, 490.6231 units of manufacturing, and 317.4834 units of electronics are consumed internally.

15. a. $P = \begin{bmatrix} 0 & 2 & 3 \\ 0 & 0 & 1 \\ 0 & 0 & 0 \end{bmatrix}$

b. Each part of type P_2 requires 2 parts of type P_1 to manufacture. Each part of type P_3 requires 3 parts of type P_1 and 1 part of type P_2 to manufacture.

c. $X = \begin{bmatrix} 900 \\ 200 \\ 100 \end{bmatrix}$

17. a. $P = \begin{bmatrix} 0 & 3 & 2 & 0 \\ 0 & 0 & 0 & 1 \\ 0 & 1 & 0 & 2 \\ 0 & 0 & 0 & 0 \end{bmatrix}$

b. Each part of type P_2 requires three parts of type P_1 and one part of type P_3 to manufacture. Each part of type P_3 requires two parts of type P_1 to manufacture. Each part of type P_4 requires one part of type P_2 and two parts of type P_3 to manufacture.

c. $X = \begin{bmatrix} 1250 \\ 130 \\ 330 \\ 50 \end{bmatrix}$

19. a. $P = \begin{bmatrix} 0 & 2 & 0 & 0 \\ 0 & 0 & 2 & 0 \\ 0 & 0 & 0 & 3 \\ 0 & 0 & 0 & 0 \end{bmatrix}$

b. Each part of type P_2 requires 2 parts of type P_1 to manufacture. Each part of type P_3 requires 2 parts of type P_2 to manufacture. Each part of type P_4 requires 3 parts of type P_3 to manufacture.

c. $X = \begin{bmatrix} 2170 \\ 960 \\ 330 \\ 60 \end{bmatrix}$

21.

Chapter 3 Summary Exercises

1. $\begin{bmatrix} 12 & -5 \\ 9 & -1 \end{bmatrix}$ **3.** $\begin{bmatrix} 1 & 0 & 3 \\ -2 & 4 & -1 \end{bmatrix}$ **5.** $\begin{bmatrix} -4 & 2 & 1 \\ 12 & -4 & 0 \\ 3 & 1 & 3 \end{bmatrix}$ **7.** This operation is impossible. **9.** This operation is impossible.

11. $\begin{bmatrix} \frac{1}{9} & \frac{1}{9} \\ 0 & -1 \end{bmatrix}$ **13.** $x = 9$; $w = -24$; $t = 8.5$. **15.** $\begin{bmatrix} \frac{1}{3} & -\frac{1}{6} & \frac{1}{2} \\ -\frac{1}{3} & \frac{1}{6} & \frac{1}{2} \\ \frac{1}{3} & \frac{1}{3} & -1 \end{bmatrix}$ **17.** $\begin{array}{c} w \\ x \\ y \\ z \end{array}\begin{bmatrix} 0 & 1 & 0 & 1 \\ 1 & 0 & 1 & 0 \\ 1 & 1 & 0 & 1 \\ 1 & 0 & 0 & 0 \end{bmatrix}$ **19.** $CA = \begin{bmatrix} 15 & 52 \\ 16 & 78 \\ 1 & 11 \\ 20 & 74 \\ 13 & 54 \\ 50 & 169 \end{bmatrix}$

21. $(CA)A^{-1} = C$ **23.** $x = 1$; $y = 2$; $z = 1$. **25.** $\begin{bmatrix} \frac{5}{6} & -\frac{2}{3} \\ -\frac{1}{6} & \frac{1}{3} \end{bmatrix}$ **27.** $x = -7$

29. $P(x) = 20x - 2100$; the break-even point is $(105, 8400)$. **31.** Let t represent the number of additional hours of study. Then $f(x) = 72 + 5t$ is a linear function that gives your test score, where $0 \le t \le 5.6$.

Sample Test Items for Chapter 3

1. $r = 6$; $t = 6$; $w = 0$; $x = 3$; $y = 9$. **2.** $\begin{bmatrix} 3 & -2 & 3 \\ -6 & -3 & -6 \\ -7 & -8 & -1 \end{bmatrix}$ **3.** This operation is impossible. **4.** $\begin{bmatrix} 2 & -1 & 0 \\ 0 & 1 & 0 \\ 3 & 0 & 4 \end{bmatrix}$ **5.** $\begin{bmatrix} 30 & 36 & 42 \\ 66 & 81 & 96 \\ 102 & 126 & 150 \end{bmatrix}$

6. $\begin{bmatrix} 23 & 28 & 33 \\ 3 & 3 & 3 \\ 28 & 32 & 36 \end{bmatrix}$ **7.** This operation is impossible. **8.** $\begin{array}{c} w \\ x \\ y \end{array}\begin{bmatrix} 1 & 1 & 1 \\ 0 & 1 & 1 \\ 0 & 1 & 1 \end{bmatrix}$ **9.** $\begin{aligned} 2x \quad\;\; + z &= 4 \\ 3x - 2y + 5z &= 17 \end{aligned}$

10. $B = \begin{bmatrix} 175 & 250 \end{bmatrix}$; $BA = \begin{bmatrix} 4625 & 2850 \end{bmatrix}$. **11.** The matrices are not inverses of each other; $AB \ne I$.

12. $\begin{bmatrix} 3 & -7 \\ -2 & 5 \end{bmatrix}$ **13.** $\begin{bmatrix} 1 & 89 & -31 \\ 0 & -20 & 7 \\ 0 & 3 & -1 \end{bmatrix}$ **14.** $X = (Y - B)A^{-1}$ **15.** $x = 5$; $y = 13$; $z = 5$. **16.** $P = \begin{bmatrix} 0 & 1 & 2 \\ 0 & 0 & 3 \\ 0 & 0 & 0 \end{bmatrix}$

17. Each part of type P_2 requires one part of type P_1 to manufacture. Each part of type P_3 requires 2 parts of type P_1 and 3 parts of type P_2 to manufacture. **18.** $(I - P)^{-1} = \begin{bmatrix} 1 & 1 & 5 \\ 0 & 1 & 3 \\ 0 & 0 & 1 \end{bmatrix}$ **19.** $X = \begin{bmatrix} 1800 \\ 1100 \\ 300 \end{bmatrix}$ **20.** The production process consumes 1700 parts of type P_1, 900 parts of type P_2, and no parts of type P_3.

CHAPTER 4

4.1 Exercises

1. a.

	Silver	Gold	Limits
Grinder	1	3	30(60)
Bonder	3	4	50(60)
Number Made	x	y	
Profit	5	7	

b. Maximize $P = 5x + 7y$,
   subject to $x + 3y \leq 1800$
   $3x + 4y \leq 3000$
   $x \geq 0, y \geq 0.$

3. a.

	Financial	Scientific	Limits
Semiconductor	10	20	3200
Off–On Switch	1	1	300
Need to Make	1	0	100
Number Made	x	y	
Production Steps	10	12	

b. Minimize $P = 10x + 12y$,
   subject to $10x + 12y \geq 3200$
   $x + y \geq 300$
   $x \geq 100$
   $x \geq 0, y \geq 0.$

5. Minimize $F = 2x + 3y$
   subject to $4x + 6y \geq 10$
   $5x + 2y \geq 10$
   $x \geq 0, y \geq 0$

7. Maximize $L = 10{,}000x + 2500y + 800z + 600w$
   subject to $1000x + 800y + 400z + 300w \leq 20{,}000$
   $x \geq 5$
   $y \geq 3$
   $x \leq 8$
   $y \leq 5$
   $z \leq 3$
   $w \leq 10$
   $x \geq 0, y \geq 0, z \geq 0, w \geq 0$

9. Minimize $E = 5x + 3y + 4z$
   subject to $2x + 3y + 3z \geq 12$
   $2x + y + z \geq 6$
   $x \geq 0, y \geq 0, z \geq 0$

11. Maximize $P = 0.12x + 0.07y$
   subject to $x + y \leq 30{,}000$
   $2x - y \leq 0$
   $x \geq 0, y \geq 0$
   (x and y in dollars; divide by
   1000 to get the number of bonds.)

13. Maximize $P = 350x + 525y$
   subject to $x + y \leq 500$
   $200x + 350y \leq 25{,}000$
   $x \geq 0, y \geq 0$

15. Minimize $C = 12{,}000x + 10{,}000y$
   subject to $50x + 40y \geq 500$
   $60x + 30y \geq 1000$
   $100x + 80y \geq 1500$
   $x \geq 0, y \geq 0$

17. Maximize $R = 420x + 550y$
   subject to $x + y \geq 300$
   $x \geq 80$
   $y \geq 120$

19. Minimize $F = 2x + 5y$
   subject to $5x + 7y \leq 105$
   $80x + 110y \geq 400$
   $x - 3y \geq 0$
   $x \geq 0, y \geq 0$

4.2 Exercises

1. a. Yes **b.** Yes **c.** No   **3. a.** No **b.** Yes **c.** No

5. a. $3x + y = 9$
   b. $(0, 0)$ makes inequality true;
   shade toward the origin.
   c.
   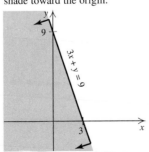

7. a. $2x + 3y = 6$
   b. $(0, 0)$ makes inequality true;
   shade toward the origin
   c.

9. a. $-2x + y = 8$
   b. $(0, 0)$ makes inequality false;
   shade away from the origin.
   c.
   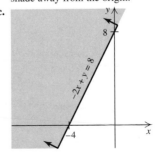

11. a. $x = 3$
   b. $(0, 0)$ makes inequality false; shade away from the origin
   c.

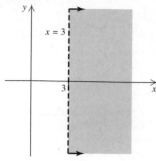

13. a. $3x - 2y = 12$
   b. $(0, 0)$ makes inequality true; shade toward the origin
   c.

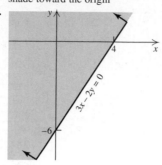

15. a. $x - y = 0$
   b. $(1, 0)$ makes inequality false; shade away from $(1, 0)$
   c.

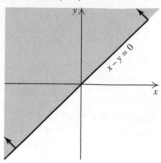

17. a. L: $x + 2y = 6$
   M: $x \quad\;\; = 0$
   N: $\quad\;\; y = 0$
   b.

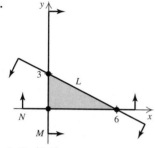

   c. L, M: $(0, 3)$
   M, N: $(0, 0)$
   L, N: $(6, 0)$

19. a. L: $x = y$
   M: $x = 2y$
   N: $x = 3$
   b.

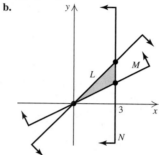

   c. L, M: $(0, 0)$
   M, N: $(3, \frac{3}{2})$
   L, N: $(3, 3)$

21. a. L: $x + 3y = 9$
   M: $3x + \;\; y = 11$
   N: $x \qquad\;\; = 0$
   O: $\qquad\;\; y = 0$
   b.

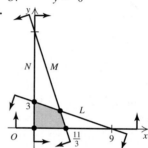

   c. N, L: $(0, 3)$
   N, O: $(0, 0)$
   M, O: $(\frac{11}{3}, 0)$
   L, M: $(3, 2)$

23. a. L: $x + y = 8$
   M: $2x + y = 10$
   N: $4x + y = 16$
   O: $x \qquad = 0$
   P: $\qquad y = 0$
   b.

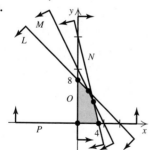

   c. L, O: $(0, 8)$
   O, P: $(0, 0)$
   N, P: $(4, 0)$
   M, N: $(3, 4)$
   L, M: $(2, 6)$

25.

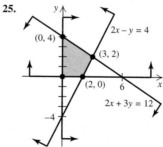

$(0, 0)$ is in the feasible region.
$(1, 3)$ is in the feasible region.

27.

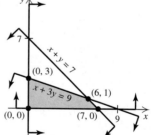

$(0, 0)$ is in the feasible region.
$(1, 3)$ is not in the feasible region.

29.

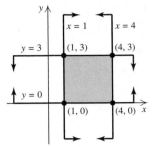

(0, 0) is not in the feasible region.
(1, 3) is in the feasible region.

31.

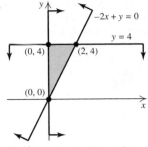

(0, 0) is in the feasible region.
(1, 3) is in the feasible region.

33.

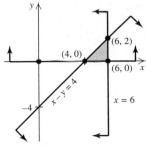

(0, 0) is not in the feasible region.
(1, 3) is not in the feasible region.

35. a. Maximize $S = 3000x + 2500y$
subject to $120x + 100y \leq 18{,}000$
$x \qquad\quad \geq \quad 50$
$\qquad\quad y \geq \quad 60$

b.

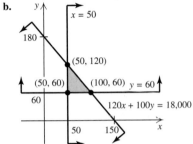

37. a. Minimize $C = 80x + 110y$
subject to $30x + 40y \geq 1200$
$90x + 50y \geq 2970$
$x \geq 0, y \geq 0$

b.

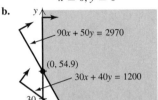

39. a. Minimize $W = 0.2x + 0.25y$
subject to $x \qquad\quad \geq \quad 1000$
$\qquad y \geq \quad 1200$
$2x + 3y \leq 14{,}000$

b.

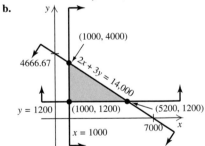

41. a. Maximize $R = 30x + 80y$
subject to $x + 3y \leq 1800$
$3x + 4y \leq 3000$
$x \geq 0, y \geq 0$

b.

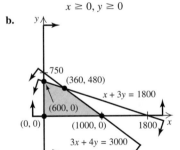

43. a. Maximize $R = 50x + 60y$
subject to $10x + 20y \geq 320$
$x - 2y \geq \quad 0$
$x \qquad\quad \leq \quad 50$
$x \geq 0, y \geq 0$

b.

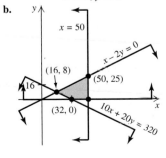

45. a. Maximize $T = x + y$
subject to $30x + 35y \leq 4800$
$20x + 45y \leq \quad 600$
$x \geq 0, y \geq 0$

b.

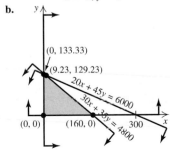

4.3 Exercises

1. a.

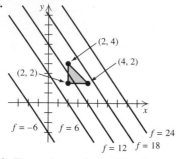

b. The maximum of f is 16 and occurs at $(4, 2)$; the minimum is 10 and occurs at $(2, 2)$. The larger c, the further upward the lines move.

c.

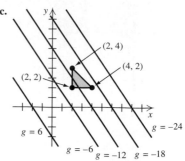

d. The maximum of g is -10 and occurs at $(2, 2)$; the minimum of g is -16 an occurs at $(4, 2)$. The larger c, the further downward the lines move.

3. a., b.

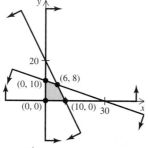

c.

Corner points	$P = 4x + 4y$
$(0, 10)$	40
$(0, 0)$	0
$(10, 0)$	40
$(6, 8)$	56

The maximum value of P is 56 and occurs when $x = 6$ and $y = 8$.

5. a., b.

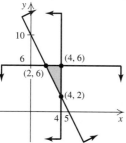

c.

Corner points	$f = x + 2y$
$(2, 6)$	14
$(4, 2)$	8
$(4, 6)$	16

The maximum value of f is 16 and occurs when $x = 4$ and $y = 6$.

7. a., b.

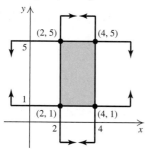

c.

Corner points	$C = x - 2y$
$(2, 5)$	-8
$(2, 1)$	0
$(4, 1)$	2
$(4, 5)$	-6

The minimum value of C is -8 and occurs when $x = 2$ and $y = 5$.

9. a., b.

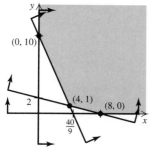

c.

Corner points	$C = 2x + 3y$
$(0, 10)$	30
$(4, 1)$	11
$(8, 0)$	16

The minimum value of C is 11 and occurs when $x = 4$ and $y = 1$.

11. a., b.

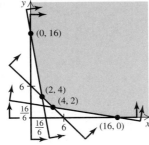

c.

Corner points	$C = 5x + 3y$
$(0, 16)$	48
$(2, 4)$	22
$(4, 2)$	26
$(16, 0)$	80

The minimum value of C is 22 and occurs when $x = 2$ and $y = 4$.

13. a., b.

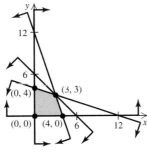

c.

Corner points	$P = 8x + y$
$(0, 4)$	4
$(0, 0)$	0
$(4, 0)$	32
$(3, 3)$	27

The maximum value of P is 32 and occurs when $x = 4$ and $y = 0$.

15. a., b.

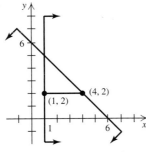

c.

Corner points	$f = 2x - 5y$
$(1, 2)$	-8
$(4, 2)$	-2

The minimum value of f is -8 and occurs when $x = 1$ and $y = 2$.

17. a. No **b.** Yes, at the point $(1, 2)$, with a minimum value of 7 for f.   **19.** Net profit is maximized at 14 computers and 16 fax machines. **21.** Cost is minimized if 1200 four-ply and 2800 radial tires are produced.   **23.** The amount of protein is maximized with 6 pounds of barley and 3 pounds of corn.   **25.** Profit is maximized when 60 student desks and 40 secretary desks are produced and sold.   **27.** Flight costs are minimized when approximately 17 type-A planes and no type-B planes are used.   **29.** Profit is maximized when 132 standard models and 168 deluxe models are produced daily.   **31.** Investment return is maximized when the club invests in 10 junk bonds and 20 premium bonds.   **33.** The number of experiments is maximized when 200 mice and 25 rats are purchased.

Chapter 4 Review

1. a. No **b.** Yes **c.** Yes

3.

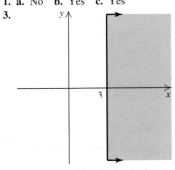

5.

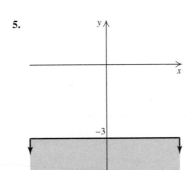

7. $\{(10, 10), (40, 35), (30, 45), (20, 45)\}$

9. 350   **11.** $\{(0, 0), (4, 0), (10/3, 5/3), (0, 5/2)\}$ **13.** 0   **15.** C has a minimum value of 4 at $(0, 2)$.   **17.** $y = 4x - 13$

19. $(1, 1, 1)$ **21.** $\begin{pmatrix} 1 & 2 & 3 \\ 3 & -4 & 2 \\ 0 & 1 & -2 \end{pmatrix}$

Sample Test Items for Chapter 4

1.

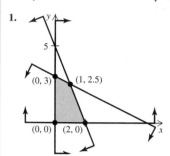

2.

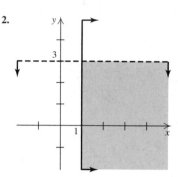

3. $\{(0, 0), (4, 0), (3, 2), (0, 13/5)\}$

4. $\{(0, 50), (25, 0)\}$ **5.** f has a maximum value of approximately 12.49 at about the point $(2.80, 1.38)$.

6. g does not reach a minimum in this linear program.

7. Maximize $g = 3x + 5y$

subject to
$$x \geq 10$$
$$x \leq 30$$
$$y \geq 6$$
$$y \leq 20$$
$$x + y \leq 36$$

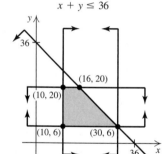

A maximum score of 148 can be achieved by answering 16 true–false questions and 20 multiple-choice questions correctly.

8. A maximum commission of \$1440 is earned from selling 96 unfinished bookcases a week.

CHAPTER 5

5.1 Exercises

1. In standard form   **3.** In standard form   **5.** Not in standard form   **7.** Not in standard form   **9.** 0   **11.** 9   **13.** Above the line $x + 3y = 9$ **15.** 5   **17.** $0 \leq s_2 \leq 8$ **19.** $x = 0, y = 2, s_1 = 7, s_2 = 0$. Yes, a corner point   **21.** $x = 0, y = 5, s_1 = 0, s_2 = -2$. No, not a corner point   **23.** $x = 0, y = 2, z = 1, s_1 = 0, s_2 = 0, s_3 = 5$. Yes, a corner point

25. $\begin{aligned} x + y + s_1 &= 9 \\ -x + 2y + s_2 &= 12 \end{aligned}$

27. $\begin{bmatrix} ① & 1 & 1 & 0 & 9 \\ -1 & 2 & 0 & 1 & 12 \end{bmatrix}$; a pivot on the circled element gives $\begin{bmatrix} 1 & 1 & 1 & 0 & 9 \\ 0 & 3 & 1 & 1 & 21 \end{bmatrix}$ with basic feasible solution $x = 9, y = 0, s_1 = 0$, and $s_2 = 21$. A pivot on the -1 in the x-column gives $\begin{bmatrix} 0 & 3 & 1 & 1 & 21 \\ 1 & -2 & 0 & -1 & -12 \end{bmatrix}$. With the nonbasic variables y and s_2 set equal to zero, $x = -12$ and $s_1 = 21$, which is not a point in the feasible region.

29. $\begin{bmatrix} ① & 1 & 1 & 0 & 9 \\ -1 & 2 & 0 & 1 & 12 \end{bmatrix} \rightarrow \begin{bmatrix} 1 & 1 & 1 & 0 & 9 \\ 0 & ③ & 1 & 1 & 21 \end{bmatrix} \rightarrow \begin{bmatrix} 1 & 0 & \frac{2}{3} & -\frac{1}{3} & 2 \\ 0 & 1 & \frac{1}{3} & \frac{1}{3} & 7 \end{bmatrix}$ The last matrix has basic feasible solution $x = 2$, $y = 7$, $s_1 = 0$,
and $s_2 = 0$.

31. $x + y + s_1 \qquad = 10$
$2x - 2y \qquad + s_2 = 8$

33. $\begin{bmatrix} 1 & 1 & 1 & 0 & 10 \\ ② & -2 & 0 & 1 & 8 \end{bmatrix}$; a pivot on the circled element gives $\begin{bmatrix} 0 & 2 & 1 & -\frac{1}{2} & 6 \\ 1 & -1 & 0 & \frac{1}{2} & 4 \end{bmatrix}$ with basic feasible solution $x = 4$, $y = 0$, $s_1 = 6$,
and $s_2 = 0$. A pivot on the 1 in the x-column gives $\begin{bmatrix} 1 & 1 & 1 & 0 & 10 \\ 0 & -4 & -2 & 1 & -12 \end{bmatrix}$. With the nonbasic variables y and s_1 set equal to zero, $x = 10$
and $s_2 = -12$, which is not a point in the feasible region.

35. $\begin{bmatrix} 1 & 1 & 1 & 0 & 10 \\ ② & -2 & 0 & 1 & 8 \end{bmatrix} \rightarrow \begin{bmatrix} 0 & ⊖2 & 1 & -\frac{1}{2} & 6 \\ 1 & -1 & 0 & \frac{1}{2} & 4 \end{bmatrix} \rightarrow \begin{bmatrix} 0 & 1 & \frac{1}{2} & -\frac{1}{4} & 3 \\ 1 & 0 & \frac{1}{2} & \frac{1}{4} & 7 \end{bmatrix}$ The last matrix has basic feasible solution $x = 7$, $y = 3$,
$s_1 = 0$, and $s_2 = 0$.

37. $x + y + s_1 \qquad = 8$
$x + 2y \qquad + s_2 = 10$

39. $\begin{bmatrix} ① & 1 & 1 & 0 & 8 \\ 1 & 2 & 0 & 1 & 10 \end{bmatrix}$; a pivot on the circled element gives $\begin{bmatrix} 1 & 1 & 1 & 0 & 8 \\ 0 & 1 & -1 & 1 & 2 \end{bmatrix}$ with basic feasible solution $x = 8$, $y = 0$, $s_1 = 0$, and
$s_2 = 2$. A pivot on the other 1 in the x-column gives $\begin{bmatrix} 0 & -1 & 1 & -1 & -2 \\ 1 & 2 & 0 & 1 & 10 \end{bmatrix}$. With the nonbasic variables y and s_2 set equal to zero, $x = 10$
and $s_1 = -2$, which is not a point in the feasible region.

41. $\begin{bmatrix} ① & 1 & 1 & 0 & 8 \\ 1 & 2 & 0 & 1 & 10 \end{bmatrix} \rightarrow \begin{bmatrix} 1 & 1 & 1 & 0 & 8 \\ 0 & ① & -1 & 1 & 2 \end{bmatrix} \rightarrow \begin{bmatrix} 1 & 0 & 2 & -1 & 6 \\ 0 & 1 & -1 & 1 & 2 \end{bmatrix}$ The last matrix has basic feasible solution $x = 6$, $y = 2$,
$s_1 = 0$, and $s_2 = 0$.

43. The corner points are $(0, 8)$, $(0, 0)$, $(4, 0)$, and $(2, 6)$. Clockwise:

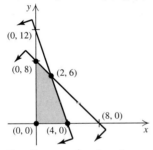

$\begin{bmatrix} 1 & ① & 1 & 0 & 8 \\ 3 & 1 & 0 & 1 & 12 \end{bmatrix} \rightarrow \begin{bmatrix} 1 & 1 & 1 & 0 & 8 \\ ② & 0 & -1 & 1 & 4 \end{bmatrix} \rightarrow \begin{bmatrix} 0 & 1 & ③\over② & -\frac{1}{2} & 6 \\ 1 & 0 & -\frac{1}{2} & \frac{1}{2} & 2 \end{bmatrix} \rightarrow \begin{bmatrix} 0 & \frac{2}{3} & 1 & -\frac{1}{3} & 4 \\ 1 & \frac{1}{3} & 0 & ①\over③ & 4 \end{bmatrix} \rightarrow \begin{bmatrix} 1 & 1 & 1 & 0 & 8 \\ 3 & 1 & 0 & 1 & 12 \end{bmatrix}$

Counterclockwise:

$\begin{bmatrix} 1 & 1 & 1 & 0 & 8 \\ ③ & 1 & 0 & 1 & 12 \end{bmatrix} \rightarrow \begin{bmatrix} 0 & ②\over③ & 1 & -\frac{1}{3} & 4 \\ 1 & \frac{1}{3} & 0 & \frac{1}{3} & 4 \end{bmatrix} \rightarrow \begin{bmatrix} 0 & 1 & \frac{3}{2} & -\frac{1}{2} & 6 \\ 1 & 0 & -\frac{1}{2} & ①\over② & 2 \end{bmatrix} \rightarrow \begin{bmatrix} 1 & 1 & ① & 0 & 8 \\ 2 & 0 & -1 & 1 & 4 \end{bmatrix} \rightarrow \begin{bmatrix} 1 & 1 & 1 & 0 & 8 \\ 3 & 1 & 0 & 1 & 12 \end{bmatrix}$

45. a. $x + y \le 6$
$7x + 3y \le 21$
$x \ge 0, y \ge 0$

b.

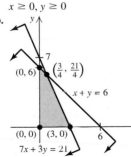

c. $\begin{bmatrix} 1 & 1 & 1 & 0 & 6 \\ ⑦ & 3 & 0 & 1 & 21 \end{bmatrix} \rightarrow \begin{bmatrix} 0 & \frac{4}{7} & 1 & -\frac{1}{7} & 3 \\ 1 & \frac{3}{7} & 0 & \frac{1}{7} & 3 \end{bmatrix}$

d. $\begin{bmatrix} 1 & ① & 1 & 0 & 6 \\ 7 & 3 & 0 & 1 & 21 \end{bmatrix} \rightarrow \begin{bmatrix} 1 & 1 & 1 & 0 & 6 \\ 4 & 0 & -3 & 1 & 3 \end{bmatrix}$

e. $\begin{bmatrix} 1 & ① & 1 & 0 & 6 \\ 7 & 3 & 0 & 1 & 21 \end{bmatrix} \rightarrow \begin{bmatrix} 1 & 1 & 1 & 0 & 6 \\ ④ & 0 & -3 & 1 & 3 \end{bmatrix} \rightarrow \begin{bmatrix} 0 & 1 & \frac{7}{4} & -\frac{1}{4} & \frac{21}{4} \\ 1 & 0 & -\frac{3}{4} & \frac{1}{4} & \frac{3}{4} \end{bmatrix}$

5.2 Exercises

1. a. Yes **b.** $f = 20$; $x = 5$, $y = 8$, $s_1 = 0$, $s_2 = 0$ **c.** No **3. a.** Yes **b.** $f = \frac{80}{5}$; $x = \frac{8}{5}$, $y = 0$, $s_1 = 0$, $s_2 = 16$ **c.** Yes; $s_2 = 16$, or 16 units of surplus in the second constraint. **5. a.** No **b.** $f = 25$; $x = 6$, $y = 0$, $s_1 = 8$, $s_2 = 2$, $s_3 = 0$ **c.** Yes; $s_1 = 8$ and $s_2 = 2$, or 8 units of surplus in the second constraint and 2 units of surplus in the third constraint.

7. a.

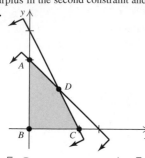

b. A: $x = 0$, $y = 7$, $s_1 = 3$, $s_2 = 0$, $f = 42$
B: $x = 0$, $y = 0$, $s_1 = 10$, $s_2 = 7$, $f = 0$
C: $x = 5$, $y = 0$, $s_1 = 0$, $s_2 = 2$, $f = 40$
D: $x = 3$, $y = 4$, $s_1 = 0$, $s_2 = 0$, $f = 48$

c.
$$
\begin{bmatrix}
② & 1 & 1 & 0 & 0 & 10 \\
1 & 1 & 0 & 1 & 0 & 7 \\
-8 & -6 & 0 & 0 & 1 & 0
\end{bmatrix}
\rightarrow
\begin{bmatrix}
1 & \frac{1}{2} & \frac{1}{2} & 0 & 0 & 5 \\
0 & ⓛ\tfrac{1}{2} & -\frac{1}{2} & 1 & 0 & 2 \\
0 & -2 & 4 & 0 & 1 & 40
\end{bmatrix}
\rightarrow
\begin{bmatrix}
1 & 0 & 1 & -1 & 0 & 3 \\
0 & 1 & -1 & 2 & 0 & 4 \\
0 & 0 & 2 & 4 & 1 & 48
\end{bmatrix}
$$
point B point C point D $x = 3$, $y = 4$, $f = 48$,

9. a.

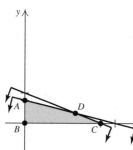

b. A: $x = 0$, $y = 3$, $s_1 = 5$, $s_2 = 0$, $f = 12$
B: $x = 0$, $y = 0$, $s_1 = 20$, $s_2 = 12$, $f = 0$
C: $x = 10$, $y = 0$, $s_1 = 0$, $s_2 = 2$, $f = 10$
D: $x = \frac{20}{3}$, $y = \frac{4}{3}$, $s_1 = 0$, $s_2 = 0$, $f = 12$

c.
$$
\begin{bmatrix}
2 & 5 & 1 & 0 & 0 & 20 \\
1 & ④ & 0 & 1 & 0 & 12 \\
-1 & -4 & 0 & 0 & 1 & 0
\end{bmatrix}
\rightarrow
\begin{bmatrix}
\frac{3}{4} & 0 & 1 & -\frac{5}{4} & 0 & 5 \\
\frac{1}{4} & 1 & 0 & \frac{1}{4} & 0 & 3 \\
0 & 0 & 0 & 1 & 1 & 12
\end{bmatrix}
$$
point B point A

11. a.

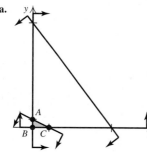

b. A: $x = 0$, $y = 3$, $s_1 = 0$, $s_2 = 111$, $f = 3$
B: $x = 0$, $y = 0$, $s_1 = 6$, $s_2 = 120$, $f = 0$
C: $x = 6$, $y = 0$, $s_1 = 0$, $s_2 = 96$, $f = 30$

c.
$$
\begin{bmatrix}
① & 2 & 1 & 0 & 0 & 6 \\
4 & 3 & 0 & 1 & 0 & 120 \\
-5 & -1 & 0 & 0 & 1 & 0
\end{bmatrix}
\rightarrow
\begin{bmatrix}
1 & 2 & 1 & 0 & 0 & 6 \\
0 & -5 & -4 & 1 & 0 & 96 \\
0 & 9 & 5 & 0 & 1 & 30
\end{bmatrix}
$$
point B point C

13. a.

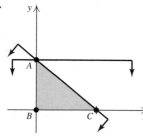

b. A: $x = 0$, $y = 5$, $s_1 = 0$, $s_2 = 0$, $f = 5$
B: $x = 0$, $y = 0$, $s_1 = 5$, $s_2 = 5$, $f = 0$
C: $x = 5$, $y = 0$, $s_1 = 0$, $s_2 = 5$, $f = 10$

c.
$$
\begin{bmatrix}
① & 1 & 1 & 0 & 0 & 5 \\
0 & 1 & 0 & 1 & 0 & 5 \\
-2 & -1 & 0 & 0 & 1 & 0
\end{bmatrix}
\rightarrow
\begin{bmatrix}
1 & 1 & 1 & 0 & 0 & 5 \\
0 & 1 & 0 & 1 & 0 & 5 \\
0 & 1 & 2 & 0 & 1 & 10
\end{bmatrix}
$$
point B point C

15. a.

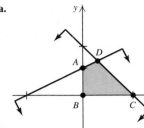

b. A: $x = 0$, $y = 5$, $s_1 = 4$, $s_2 = 0$, $f = 15$
B: $x = 0$, $y = 0$, $s_1 = 9$, $s_2 = 10$, $f = 0$
C: $x = 9$, $y = 0$, $s_1 = 0$, $s_2 = 19$, $f = 27$
D: $x = \frac{8}{3}$, $y = \frac{19}{3}$, $s_1 - 0$, $s_2 = 0$, $f = 27$

c.
$$\begin{bmatrix} ① & 1 & 1 & 0 & 0 & 9 \\ -1 & 2 & 0 & 1 & 0 & 10 \\ -3 & -3 & 0 & 0 & 1 & 0 \end{bmatrix} \rightarrow \begin{bmatrix} 1 & 1 & 1 & 0 & 0 & 9 \\ 0 & 3 & 1 & 1 & 0 & 19 \\ 0 & 0 & 3 & 0 & 1 & 27 \end{bmatrix}$$
point B point C

17. a.

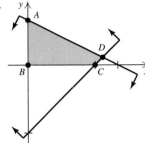

b. A: $x = 0$, $y = 4$, $s_1 = 10$, $s_2 = 0$, $f = -4$
B: $x = 0$, $y = 0$, $s_1 = 6$, $s_2 = 8$, $f = 0$
C: $x = 6$, $y = 0$, $s_1 = 0$, $s_2 = 2$, $f = -12$
D: $x = \frac{20}{3}$, $y = \frac{2}{3}$, $s_1 = 0$, $s_2 = 0$, $f = -14$

c.
$$\begin{bmatrix} 1 & -1 & 1 & 0 & 0 & 6 \\ 1 & 2 & 0 & 1 & 0 & 8 \\ 2 & 1 & 0 & 0 & 1 & 0 \end{bmatrix}$$
point B

19. a.

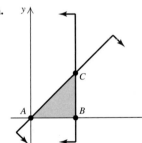

b. A: $x = 0$, $y = 0$, $s_1 = 0$, $s_2 = 2$, $f = 0$
B: $x = 2$, $y = 0$, $s_1 = 2$, $s_2 = 0$, $f = 6$
C: $x = 2$, $y = 2$, $s_1 = 0$, $s_2 = 0$, $f = 8$

c.
$$\begin{bmatrix} -1 & 1 & 1 & 0 & 0 & 0 \\ ① & 0 & 0 & 1 & 0 & 2 \\ -3 & -1 & 0 & 0 & 1 & 0 \end{bmatrix} \rightarrow \begin{bmatrix} 0 & ① & 1 & 1 & 0 & 2 \\ 1 & 0 & 0 & 1 & 0 & 2 \\ 0 & -1 & 0 & 3 & 1 & 6 \end{bmatrix} \rightarrow \begin{bmatrix} 0 & 1 & 1 & 1 & 0 & 2 \\ 1 & 0 & 0 & 1 & 0 & 2 \\ 0 & 0 & 1 & 4 & 1 & 8 \end{bmatrix}$$
point A point B point C

21. a.

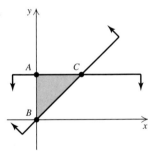

b. A: $x = 0$, $y = 3$, $s_1 = 3$, $s_2 = 0$, $f = 6$
B: $x = 0$, $y = 0$, $s_1 = 0$, $s_2 = 3$, $f = 0$
C: $x = 3$, $y = 3$, $s_1 = 0$, $s_2 = 0$, $f = 21$

c.
$$\begin{bmatrix} ① & -1 & 1 & 0 & 0 & 0 \\ 0 & 1 & 0 & 1 & 0 & 3 \\ -5 & -2 & 0 & 0 & 1 & 0 \end{bmatrix} \rightarrow \begin{bmatrix} 1 & -1 & 1 & 0 & 0 & 0 \\ 0 & ① & 0 & 1 & 0 & 3 \\ 0 & -7 & 5 & 0 & 1 & 0 \end{bmatrix} \rightarrow \begin{bmatrix} 1 & 0 & 1 & 1 & 0 & 3 \\ 0 & 1 & 0 & 1 & 0 & 3 \\ 0 & 0 & 5 & 7 & 1 & 21 \end{bmatrix}$$
point B point B point C

23. $x = \frac{24}{7}$, $y = \frac{36}{7}$, $f = \frac{144}{7}$ **25.** $x = 0$, $y = 0$, $z = 15$, $f = 45$ **27.** $x = 20$, $y = 0$, $z = 0$, $f = 100$ **29.** $x = 7$, $y = \frac{3}{2}$, $z = 0$, $P = 17$ or $x = 0$, $y = \frac{17}{2}$, $z = 0$, $P = 17$ **31.** $x = 8$, $y = 20$, $z = 0$, $f = 76$ **33.** No solution   **35.** $x = 2$, $y = 8$, $z = 0$, $w = 0$, $f = 40$ **37.** $x_1 = 1$, $x_2 = 0$, $x_3 = \frac{3}{4}$, $x_4 = 0$, $x_5 = 0$, $f = \frac{11}{4}$ **39.** A maximum profit of $5200 is earned by producing and selling 20 3-speed bikes and 60 10-speed bikes.   **41.** A maximum of 240 points is earned from correctly answering 0 true–false questions and 30 open-ended questions.   **43.** A maximum of 40 computers can be bought if all 40 are brand X.   **45.** A maximum of 100,000 people are reached when no radio ads, 110 newspaper ads, and 8 TV ads are run.   **47.** A maximum profit of $1280 per week is earned when 0 superdeluxe, 20 deluxe, and 80 standard-model toasters are produced each week.   **49.** A maximum attendance of 40,000 is reached by hiring five top stars, no faded stars, and no locals. This will result in a surplus of $1500 in advertising.   **51.** A maximum revenue of $205,000 is earned from assembling 410 QP20 models and none of the other two models.

5.3 Exercises

1. a.

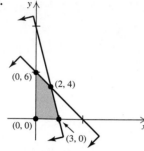

b. $\begin{bmatrix} 1 & 1 & 1 & 0 & 0 & | & 6 \\ ④ & 1 & 0 & 1 & 0 & | & 12 \\ \hline -2 & -1 & 0 & 0 & 1 & | & 0 \end{bmatrix} \rightarrow \begin{bmatrix} 0 & ③⁄₄ & 1 & -\frac{1}{4} & 0 & | & 3 \\ 1 & \frac{1}{4} & 0 & \frac{1}{4} & 0 & | & 3 \\ \hline 0 & -\frac{1}{2} & 0 & \frac{1}{2} & 1 & | & 6 \end{bmatrix} \rightarrow \begin{bmatrix} 0 & 1 & \frac{4}{3} & -\frac{1}{3} & 0 & | & 4 \\ 1 & 0 & -\frac{1}{3} & \frac{1}{3} & 0 & | & 2 \\ \hline 0 & 0 & \frac{2}{3} & \frac{1}{3} & 1 & | & 8 \end{bmatrix}$

$x = 0, y = 0, f = 0$ $x = 3, y = 0, f = 6$ $x = 2, y = 4, f = 8$

c.

Corner	$(0, 6)$	$(0, 0)$	$(3, 0)$	$(2, 4)$
f	6	0	6	8

3. a.

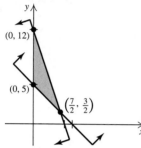

b. $\begin{bmatrix} -1 & -1 & 1 & 0 & 0 & | & -5 \\ 3 & ① & 0 & 1 & 0 & | & 12 \\ \hline -4 & -2 & 0 & 0 & 1 & | & 0 \end{bmatrix} \rightarrow \begin{bmatrix} 2 & 0 & 1 & 1 & 0 & | & 7 \\ 3 & 1 & 0 & 1 & 0 & | & 12 \\ \hline 2 & 0 & 0 & 2 & 1 & | & 24 \end{bmatrix}$

$x = 0, y = 0, f = 0$ $x = 0, y = 12, f = 24$

c.

Corner	$(0, 12)$	$(0, 5)$	$\left(\frac{7}{2}, \frac{3}{2}\right)$
f	24	10	17

5. a.

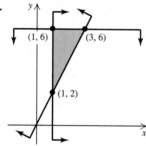

b. $\begin{bmatrix} ② & -1 & 1 & 0 & 0 & 0 & | & 0 \\ -1 & 0 & 0 & 1 & 0 & 0 & | & -1 \\ 0 & 1 & 0 & 0 & 1 & 0 & | & 6 \\ \hline -2 & -3 & 0 & 0 & 0 & 1 & | & 0 \end{bmatrix} \rightarrow \begin{bmatrix} 1 & -\frac{1}{2} & \frac{1}{2} & 0 & 0 & 0 & | & 0 \\ 0 & -\frac{1}{2} & \frac{1}{2} & 1 & 0 & 0 & | & -1 \\ 0 & ① & 0 & 0 & 1 & 0 & | & 6 \\ \hline 0 & -4 & 1 & 0 & 0 & 1 & | & 0 \end{bmatrix} \rightarrow \begin{bmatrix} 1 & 0 & \frac{1}{2} & 0 & \frac{1}{2} & 0 & | & 3 \\ 0 & 0 & \frac{1}{2} & 1 & \frac{1}{2} & 0 & | & 2 \\ 0 & 1 & 0 & 0 & 1 & 0 & | & 6 \\ \hline 0 & 0 & 1 & 0 & 4 & 1 & | & 24 \end{bmatrix}$

$x = 0, y = 0, f = 0$ $x = 0, y = 0, f = 0$ $x = 3, y = 6, f = 24$

c.

Corner	$(1, 6)$	$(1, 2)$	$(3, 6)$
f	20	8	24

7. a.

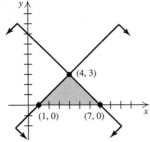

b. $\begin{bmatrix} -1 & 1 & 1 & 0 & 0 & | & -1 \\ ① & 1 & 0 & 1 & 0 & | & 7 \\ \hline 4 & -5 & 0 & 0 & 1 & | & 0 \end{bmatrix} \rightarrow \begin{bmatrix} 0 & ② & 1 & 1 & 0 & | & 6 \\ 1 & 1 & 0 & 1 & 0 & | & 7 \\ \hline 0 & -9 & 0 & -4 & 1 & | & -28 \end{bmatrix} \rightarrow \begin{bmatrix} 0 & 1 & \frac{1}{2} & \frac{1}{2} & 0 & | & 3 \\ 1 & 0 & -\frac{1}{2} & \frac{1}{2} & 0 & | & 4 \\ \hline 0 & 0 & \frac{9}{2} & \frac{1}{2} & 1 & | & -1 \end{bmatrix}$

$x = 0, y = 0, f = 0$ $x = 7, y = 0, f = -28$ $x = 4, y = 3, f = -1$

c.

Corner	$(1, 0)$	$(7, 0)$	$(4, 3)$
f	-4	-28	-1

9. a.

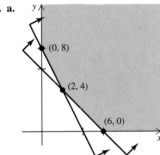

b. $\begin{bmatrix} -1 & -1 & 1 & 0 & 0 & | & -6 \\ -2 & ⊖1 & 0 & 1 & 0 & | & -8 \\ \hline 1 & 3 & 0 & 0 & 1 & | & 0 \end{bmatrix} \rightarrow \begin{bmatrix} ① & 0 & 1 & -1 & 0 & | & 2 \\ 2 & 1 & 0 & -1 & 0 & | & 8 \\ \hline -5 & 0 & 0 & 3 & 1 & | & -24 \end{bmatrix} \rightarrow$

$x = 0, y = 0, f = 0$ $x = 0, y = 8, f = 24$

$\begin{bmatrix} 1 & 0 & 1 & -1 & 0 & | & 2 \\ 0 & 1 & -2 & ① & 0 & | & 4 \\ \hline 0 & 0 & 5 & -2 & 1 & | & -14 \end{bmatrix} \rightarrow \begin{bmatrix} 1 & 1 & -1 & 0 & 0 & | & 6 \\ 0 & 1 & -2 & 1 & 0 & | & 4 \\ \hline 0 & 2 & 1 & 0 & 1 & | & -6 \end{bmatrix}$

$x = 2, y = 4, f = 14$ $x = 6, y = 0, f = 6$

c.

Corner	$(0, 8)$	$(2, 4)$	$(6, 0)$
f	24	14	6

11. a.

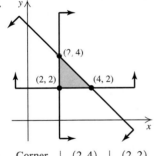

b. $\begin{bmatrix} ① & 1 & 1 & 0 & 0 & 0 & | & 6 \\ -1 & 0 & 0 & 1 & 0 & 0 & | & -2 \\ 0 & -1 & 0 & 0 & 1 & 0 & | & -2 \\ \hline 1 & 1 & 0 & 0 & 0 & 1 & | & 0 \end{bmatrix} \rightarrow \begin{bmatrix} 1 & 1 & 1 & 0 & 0 & 0 & | & 6 \\ 0 & ① & 1 & 1 & 0 & 0 & | & 4 \\ 0 & -1 & 0 & 0 & 1 & 0 & | & -2 \\ \hline 0 & 0 & -1 & 0 & 0 & 1 & | & -6 \end{bmatrix} \rightarrow$

$x = 0, y = 0, f = 0$ $x = 6, y = 0, f = 6$

$\begin{bmatrix} 1 & 0 & 0 & -1 & 0 & 0 & | & 2 \\ 0 & 1 & 1 & 1 & 0 & 0 & | & 4 \\ 0 & 0 & ① & 1 & 1 & 0 & | & 2 \\ \hline 0 & 0 & -1 & 0 & 0 & 1 & | & -6 \end{bmatrix} \rightarrow \begin{bmatrix} 1 & 0 & 0 & -1 & 0 & 0 & | & 2 \\ 0 & 1 & 0 & 0 & -1 & 0 & | & 2 \\ 0 & 0 & 1 & 1 & 1 & 0 & | & 2 \\ \hline 0 & 0 & 0 & 1 & 1 & 1 & | & -4 \end{bmatrix}$

$x = 2, y = 4, f = 6$ $x = 2, y = 2, f = 4$

c.

Corner	$(2, 4)$	$(2, 2)$	$(4, 2)$
f	6	4	6

13. a.

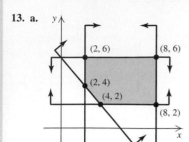

b.

$$\begin{bmatrix} -1 & -1 & 1 & 0 & 0 & 0 & 0 & 0 & -6 \\ -1 & 0 & 0 & 1 & 0 & 0 & 0 & 0 & -2 \\ 1 & 0 & 0 & 0 & 1 & 0 & 0 & 0 & 8 \\ 0 & -1 & 0 & 0 & 0 & 1 & 0 & 0 & -2 \\ 0 & ① & 0 & 0 & 0 & 0 & 1 & 0 & 6 \\ 1 & -2 & 0 & 0 & 0 & 0 & 0 & 1 & 0 \end{bmatrix} \rightarrow \begin{bmatrix} -1 & 0 & 1 & 0 & 0 & 0 & 1 & 0 & 0 \\ -1 & 0 & 0 & 1 & 0 & 0 & 0 & 0 & -2 \\ ① & 0 & 0 & 0 & 1 & 0 & 0 & 0 & 8 \\ 0 & 0 & 0 & 0 & 0 & 1 & 1 & 0 & 4 \\ 0 & 1 & 0 & 0 & 0 & 0 & 1 & 0 & 6 \\ 1 & 0 & 0 & 0 & 0 & 0 & 2 & 1 & 12 \end{bmatrix}$$

$x = 0, y = 0, C = 0$ $x = 0, y = 6, C = -12$

$$\begin{bmatrix} 0 & 0 & 1 & 0 & 1 & 0 & 1 & 0 & 8 \\ 0 & 0 & 0 & 1 & ① & 0 & 0 & 0 & 6 \\ 1 & 0 & 0 & 0 & 1 & 0 & 0 & 0 & 8 \\ 0 & 0 & 0 & 0 & 0 & 1 & 1 & 0 & 4 \\ 0 & 1 & 0 & 0 & 0 & 0 & 1 & 0 & 6 \\ 0 & 0 & 0 & 0 & -1 & 0 & 2 & 1 & 4 \end{bmatrix} \rightarrow \begin{bmatrix} 0 & 0 & 1 & -1 & 0 & 0 & 1 & 0 & 2 \\ 0 & 0 & 0 & 1 & 1 & 0 & 0 & 0 & 6 \\ 1 & 0 & 0 & -1 & 0 & 0 & 0 & 0 & 2 \\ 0 & 0 & 0 & 0 & 0 & 1 & 1 & 0 & 4 \\ 0 & 1 & 0 & 0 & 0 & 0 & 1 & 0 & 6 \\ 0 & 0 & 0 & 1 & 0 & 0 & 2 & 1 & 10 \end{bmatrix}$$

$x = 8, y = 6, C = -4$ $x = 2, y = 6, C = -10$

c.

Corner	(2, 6)	(2, 4)	(4, 2)	(8, 2)	(8, 6)
C	-10	-6	0	4	-4

15. a.

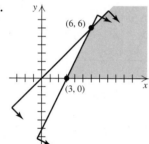

b.

$$\begin{bmatrix} -1 & 1 & 1 & 0 & 0 & 0 \\ ⊖2 & 1 & 0 & 1 & 0 & -6 \\ 2 & 1 & 0 & 0 & 1 & -10 \end{bmatrix} \rightarrow \begin{bmatrix} 0 & \frac{1}{2} & 1 & -\frac{1}{2} & 0 & 3 \\ 1 & -\frac{1}{2} & 0 & -\frac{1}{2} & 0 & 3 \\ 0 & 2 & 0 & 1 & 1 & -16 \end{bmatrix}$$

$x = 0, y = 0, f = 10$ $x = 3, y = 0, f = 16$

c.

Corner	(3, 0)	(6, 6)
f	16	28

17. $x = 4, y = 4, z = 2, C = 80$ **19.** $x = 4, y = 0, z = 2, f = 24$ **21.** $x = 1, y = \frac{9}{2}, z = 0, P = 21$
23. $x = 0, y = 3, z = 0, C = 9$ **25.** $x = 0, y = 3, z = \frac{9}{2}, f = \frac{21}{2}$ **27.** $x = \frac{3}{2}, y = 0, z = \frac{3}{2}, C = \frac{3}{2}$
29. $x_1 = 6, x_2 = 2, x_3 = 0, f = 26$
31. A maximum revenue of $154,600 per month is earned when 80 Daybrites and 220 Noglare televisions are stocked and sold monthly.
33. The candidate can minimize his time to 27.25 hours by using multiple strategies, one of which is giving no civic speeches and spending 3 hours on the phone and 24.25 hours at the shopping areas.   **35.** A maximum effectiveness of 24.5 can be achieved by taking 3 doses per day of brand A, 5 doses per day of Brand B, and 2 doses per day of Brand C.   **37.** A maximum revenue of $112,000 is earned by stocking 40 fax machines, 40 computers, and 20 CD players.   **39.** A minimal cost of $152,000 can be had by using multiple production schedules, one of which is a daily production of 1000 radial, 4200 snow, and 100 off-road tires.   **41.** A maximum of 9 units of protein can be achieved by a mixture of 6 pounds of barley and 3 pounds of corn.   **43.** A minimum total shipping cost of $440 is achieved by the following shipping plan: El Paso sends 60 units to Reynosa and 40 units to Tijuana, while Mexico City sends 40 units to Tijuana and 60 units to Nogales.   **45.** A minimum total shipping cost of $52 can be achieved through using multiple shipping plans, one of which is as follows: Hyderabad sends 20 units to New Delhi and 8 units to Mumbai, while Calcutta sends 7 units to Mumbai and 10 units to Kanpur.

47. a.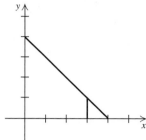

b. f has the maximum value of 11 at $(3, 1)$.

c. Again, f has the maximum value of 11 at $(3, 1)$.

5.4 Exercises

1. a. $\begin{bmatrix} 1 & 1 & 4 \\ 3 & 1 & 8 \\ 2 & 1 & 0 \end{bmatrix}$ **b.** $\begin{bmatrix} 1 & 3 & 2 \\ 1 & 1 & 1 \\ 4 & 8 & 0 \end{bmatrix}$ **c.** Maximize $d = 4u + 8v$
subject to $u + 3v \le 2$
$u + v \le 1$
$u \ge 0, v \ge 0$

Corner	A	B	C
	$(0, 8)$	$(2, 2)$	$(4, 0)$
Min $f = 2x + y$	8	6	8

Corner	P	Q	R	S
	$(0, \frac{2}{3})$	$(0, 0)$	$(1, 0)$	$(\frac{1}{2}, \frac{1}{2})$
Max $d = 4u + 8v$	$\frac{16}{3}$	0	4	6

d.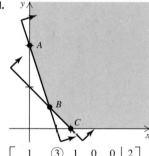

e. $\begin{bmatrix} 1 & ③ & 1 & 0 & 0 & 2 \\ 1 & 1 & 0 & 1 & 0 & 1 \\ -4 & -8 & 0 & 0 & 1 & 0 \end{bmatrix} \rightarrow \begin{bmatrix} \frac{1}{3} & 1 & \frac{1}{3} & 0 & 0 & \frac{2}{3} \\ ②\frac{2}{3} & 0 & -\frac{1}{3} & 1 & 0 & \frac{1}{3} \\ -\frac{4}{3} & 0 & \frac{8}{3} & 0 & 1 & \frac{16}{3} \end{bmatrix} \rightarrow \begin{bmatrix} 0 & 1 & \frac{1}{2} & -\frac{1}{2} & 0 & \frac{1}{2} \\ 1 & 0 & -\frac{1}{2} & \frac{3}{2} & 0 & \frac{1}{2} \\ 0 & 0 & 2 & 2 & 1 & 6 \end{bmatrix}$

$u = \frac{1}{2}, v = \frac{1}{2}, d = 6; x = 2, y = 2, f = 6$

3. a. $\begin{bmatrix} 1 & 1 & | & 10 \\ 2 & 1 & | & 12 \\ 3 & 1 & | & 0 \end{bmatrix}$ **b.** $\begin{bmatrix} 1 & 2 & | & 3 \\ 1 & 1 & | & 1 \\ 10 & 12 & | & 0 \end{bmatrix}$ **c.** Maximize $d = 10u + 12v$
subject to $u + 2v \le 3$
$\qquad\qquad u + v \le 1$
$\qquad\qquad u \ge 0, v \ge 0$

d.

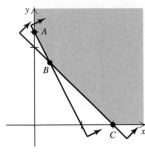
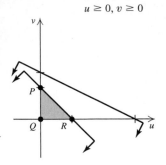

	A	B	C			P	Q	R
Corner	$(0, 12)$	$(2, 8)$	$(10, 0)$		Corner	$(0, 1)$	$(0, 0)$	$(1, 0)$
Min $C = 3x + y$	12	14	30		Max $d = 10u + 12v$	12	0	10

e. $\begin{bmatrix} 1 & 2 & 1 & 0 & 0 & | & 3 \\ 1 & ① & 0 & 1 & 0 & | & 1 \\ -10 & -12 & 0 & 0 & 1 & | & 0 \end{bmatrix} \rightarrow \begin{bmatrix} -1 & 0 & 1 & -2 & 0 & | & 1 \\ 1 & 1 & 0 & 1 & 0 & | & 1 \\ 2 & 0 & 0 & 12 & 1 & | & 12 \end{bmatrix}$ $u = 0, v = 1, d = 12$
$x = 0, y = 12, C = 12$

5. a. $\begin{bmatrix} 4 & 1 & | & 18 \\ -1 & -2 & | & -8 \\ 2 & -3 & | & 0 \end{bmatrix}$ **b.** $\begin{bmatrix} 4 & -1 & | & 2 \\ 1 & -2 & | & -3 \\ 18 & -8 & | & 0 \end{bmatrix}$ **c.** Maximize $d = 18u - 8v$
subject to $4u - v \le 2$
$\qquad\qquad u - 2v \le -3$
$\qquad\qquad u \ge 0, v \ge 0$

d.

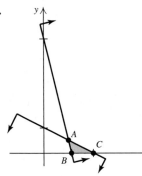

	A	B	C
Corner	$(4, 2)$	$(\frac{9}{2}, 0)$	$(8, 0)$
Min $f = 2x - 3y$	2	9	16

			Boundaries	
	P	Q	$u = \frac{1}{4}v + \frac{1}{2}$	$u = 0$
Corner	$(0, \frac{3}{2})$	$(1, 2)$		
Max $d = 18u - 8v$	-12	2	$9 - \frac{7}{2}v \to -\infty$	$-8v \to -\infty$

e. $\begin{bmatrix} 4 & -1 & 1 & 0 & 0 & | & 2 \\ 1 & ⟨-2⟩ & 0 & 1 & 0 & | & -3 \\ -18 & 8 & 0 & 0 & 1 & | & 0 \end{bmatrix} \rightarrow \begin{bmatrix} ⟨\frac{7}{2}⟩ & 0 & 1 & -\frac{1}{2} & 0 & | & \frac{7}{2} \\ -\frac{1}{2} & 1 & 0 & -\frac{1}{2} & 0 & | & \frac{3}{2} \\ -14 & 0 & 0 & 4 & 1 & | & -12 \end{bmatrix} \rightarrow \begin{bmatrix} 1 & 0 & \frac{2}{7} & -\frac{1}{7} & 0 & | & 1 \\ 0 & 1 & \frac{1}{7} & -\frac{4}{7} & 0 & | & 2 \\ 0 & 0 & 4 & 2 & 1 & | & 2 \end{bmatrix}$ $u = 1, v = 2, d = 2$
$x = 4, y = 2, f = 2$

7. a. $\begin{bmatrix} 1 & 1 & | & 8 \\ 1 & 0 & | & 1 \\ -2 & 1 & | & 0 \end{bmatrix}$ **b.** $\begin{bmatrix} 1 & 1 & | & -2 \\ 1 & 0 & | & 1 \\ 8 & 1 & | & 0 \end{bmatrix}$ **c.** Minimize $d = 8u + v$
subject to $u + v \geq -2$
$u \quad\geq 1$
$u \geq 0, v \geq 0$

d.

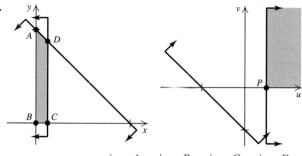

	A	B	C	D		P	Boundaries	
Corner	$(0,8)$	$(0,0)$	$(1,0)$	$(1,7)$	Corner	$(1,0)$	$v = 0$	$u = 1$
Max $f = -2x + y$	8	0	-2	5	Min $d = 8u + v$	8	$8u \to \infty$	$v + 8 \to \infty$

e. $\begin{bmatrix} -1 & -1 & 1 & 0 & 0 & | & 2 \\ \boxed{-1} & 0 & 0 & -1 & 0 & | & -1 \\ 8 & 1 & 0 & 0 & 1 & | & 0 \end{bmatrix} \to \begin{bmatrix} 0 & -1 & 1 & -1 & 0 & | & 3 \\ 1 & 0 & 0 & -1 & 0 & | & 1 \\ 0 & 1 & 0 & 8 & 1 & | & -8 \end{bmatrix}$ $\begin{array}{l} u = 1, v = 0, d = 8 \\ x = 0, y = 8, f = 8 \end{array}$

9. Minimum value of f is 8 when $x = 8$ and $y = 0$.   **11.** Minimum value of f is $\frac{52}{5}$ when $x = \frac{9}{5}, y = \frac{21}{5}$, and $z = \frac{2}{5}$.   **13.** Minimum value of f is 8 when $x = 8, y = 0$, and $z = 0$.   **15.** Maximum value of f is 20 when $x = 10, y = 0$, and $z = 0$.   **17.** A minimum total fat of 5 mg is obtained by using 20/11 of a serving of Food A and 5/11 of Food B.   **19.** The company can make the final mixing at a minimum of $16 by using 2 barrels of material A and 2 barrels of material B.   **21.** The ordering cost is at a minimum of $3300 when 175 VHS tapes and 75 DVD's are ordered.   **23.** Weekly cost is at a minimum of $32,300 when weekly production is held at 160 standard, 50 deluxe, and 90 superdeluxe microwaves.

Chapter 5 Review

1. The 2 in the second column is the smallest-quotient pivot.   **3.** Yes   **5.** $\begin{array}{l} x + y + s_1 \quad - 5 \\ 3x + 2y \quad + s_2 = 12 \end{array}$ **7.** $\begin{bmatrix} 1 & 1 & 1 & 0 & 0 & | & 5 \\ \boxed{3} & 2 & 0 & 1 & 0 & | & 12 \\ -4 & -3 & 0 & 0 & 1 & | & 0 \end{bmatrix}$

9. C has a minimum value of 12 at $(0, 12)$.   **11.** The club can have a maximum profit of $378.75 if it makes 120 batches of peanut butter cookies and 75 batches of pecan sandies.   **13.** Results vary according to imagination.   **15.** $x = 0, y = 3, z = 0, C = 9$

17. $A^{-1} = \begin{bmatrix} 1 & -1 \\ 0 & \frac{1}{2} \end{bmatrix}$. "You get an A"   **19.** $x = 2, y = -3, z = 6, w = 1$

Sample Test Items for Chapter 5

1. $\begin{array}{l} 2x + y \leq 8 \\ x + 5y \leq 10 \\ x \geq 0, y \geq 0 \end{array}$ **2.** $\begin{bmatrix} \boxed{2} & 1 & 1 & 0 & | & 8 \\ 1 & 5 & 0 & 1 & | & 10 \end{bmatrix} \to \begin{bmatrix} 1 & \frac{1}{2} & \frac{1}{2} & 0 & | & 4 \\ 0 & \frac{9}{2} & -\frac{1}{2} & 1 & | & 6 \end{bmatrix}$ **3.** No   **4.** $f = 35; x = 2, y = 0, s_1 = 8, s_2 = 4, s_3 = 0$

5. s_1; the surplus is 8 units. s_2; the surplus is 4 units.   **6.** $A: x = 0, y = 4, s_1 = 2, s_2 = 0$
$B: x = 0, y = 0, s_1 = 6, s_2 = 4$
$C: x = 6, y = 0, s_1 = 0, s_2 = 4$
$D: x = 2, y = 4, s_1 = 0, s_2 = 0$

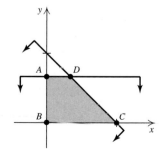

7. $\begin{bmatrix} \text{①} & 1 & 1 & 0 & 0 & | & 6 \\ 0 & 1 & 0 & 1 & 0 & | & 4 \\ -2 & -1 & 0 & 0 & 1 & | & 0 \end{bmatrix} \rightarrow \begin{bmatrix} 1 & 1 & 1 & 0 & 0 & | & 6 \\ 0 & 1 & 0 & 1 & 0 & | & 4 \\ 0 & 1 & 2 & 0 & 1 & | & 12 \end{bmatrix}$

Solution: $f = 12$ when $x = 6$ and $y = 0$.

8. Maximize $d = 3u + 4v + w$
subject to $u + v + w \le 2$
$u + 2v \le 3$
$u \ge 0, v \ge 0, w \ge 0$

9. $\begin{bmatrix} 1 & 1 & 1 & 1 & 0 & 0 & | & 2 \\ 1 & \text{②} & 0 & 0 & 1 & 0 & | & 3 \\ -3 & -4 & -1 & 0 & 0 & 1 & | & 0 \end{bmatrix} \rightarrow \begin{bmatrix} \text{(½)} & 0 & 1 & 1 & -\frac{1}{2} & 0 & | & \frac{1}{2} \\ \frac{1}{2} & 1 & 0 & 0 & \frac{1}{2} & 0 & | & \frac{3}{2} \\ 1 & 0 & -1 & 0 & 2 & 1 & | & 6 \end{bmatrix} \rightarrow \begin{bmatrix} 1 & 0 & 2 & 2 & -1 & 0 & | & 1 \\ 0 & 1 & -1 & -1 & 1 & 0 & | & 1 \\ 0 & 0 & 1 & 2 & 1 & 1 & | & 7 \end{bmatrix}$

Solution: $u = 1, v = 1, w = 0, d = 7,$
$x = 2, y = 1, f = 7$

10.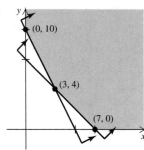

(0, 10)

(3, 4)

(7, 0)

11. $\begin{bmatrix} -1 & -1 & 1 & 0 & 0 & | & -7 \\ -2 & \text{(−1)} & 0 & 1 & 0 & | & -10 \\ 3 & 1 & 0 & 0 & 1 & | & 0 \end{bmatrix} \rightarrow \begin{bmatrix} 1 & 0 & 1 & -1 & 0 & | & 3 \\ 2 & 1 & 0 & -1 & 0 & | & 10 \\ 1 & 0 & 0 & 1 & 1 & | & -10 \end{bmatrix}$

Solution: $x = 0, y = 10, f = 10$

12.

Corner points	(0, 10)	(3, 4)	(7, 0)	$x = 0$	$y = 0$
$f = 3x + y$	10	13	21	$y \to \infty$	$3x \to \infty$

13. $x = 0, y = 6, z = 0, f = 18$

14. Total salary cost is at a minimum of $1.15 million when 25 professors and 10 lecturers are hired.

CHAPTER 6

Note: The answers to the exercises in this chapter were obtained by rounding only the final answer to two decimal places. All numerical solutions were obtained with MS Excel. Your answers may be slightly different if you rounded during any intermediate steps or if you used another technological tool.

6.1 Exercises

1. $2697.20 **3.** $6130.43 **5.** $6720.00 **7. a.** $1472.00 **b.** $1839.68 **c.** $1848.58 **9. a.** $3.21 **b.** $3.26 **c.** $3.32

11.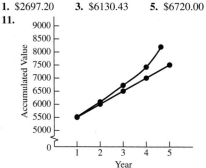

13. 9.5958% **15.** 11.7209% **17.** 19.1745%

19. Approximately 27 interest periods. **21.** Approximately 22 interest periods. **23.** Approximately 15 interest periods. **25.** $2791.97
27. $9196.71 **29.** $1420.41 **31.** $9948.94 **33.** $7730.75 **35.** 7.7834% **37.** Approximately 14 years. **39.** Approximately
6.4 years. **41.** $3.83 **43.** $60.83 **45.** Approximately fourteen years. **47.** $40,722.37 **49.** $29,549.11 **51. a.** $432,000
b. $897,795.29 **c.** $4,124,771.20

6.2 Exercises

1. $21,951.35   **3.** $15,413.30   **5.** $264,527.63   **7.** $3914.71   **9.** $5365.01   **11.** $3274.74   **13.** $15,162.14   **15.** $171.96
17. $95,454.20   **19.** The investment in part (a) has a future value of $81,990.98; the investment in part (b) has a future value of $291,450.92.
The investment in part (b) earns more.   **21. a.** $234,660.41   **b.** $165,644.10   **c.** $114,401.52   **d.** $76,355.43   **23.** $64,491.25   **25.**
$6258.36   **27.** $30,722.84   **29.** $2054.59   **31.** $3889.80 was transferred to the certificate of deposit; $4683.86 was transferred into the
90-day certificate of deposit. The final accumulation is $4761.53.

6.3 Exercises

1. 6.66%   **3.** 4.04%   **5.** 17.23%   **7.** 9.42%   **9. a.** 5.06%   **b.** 4.89%   **c.** 4.70%   **11. a.** 6.48%   **b.** 6.66%   **c.** 6.8%
13. a. 8.16%   **b.** 8.2%   **c.** 6.17%   **15.** 12.68%   **17.** $681.08   **19.** $881.92   **21.** $266.33   **23.** $792.13   **25.** $87.85
27. $313.36   **29.** $19.39   **31.** $3886.42   **33. a.** $147.21   **b.** $5972.22   **35. a.** $2649.11   **b.** $41,297.40   **37. a.** $1022.23
b. $306,667.53   **39. a.** $961.29   **b.** $403,739.94   **41. a.** $939.54   **b.** $507,353.14

43.

Pmt No.	Payment	Interest	Amt. Applied to Principal	Balance
1	$1286.49	$963.54	$322.95	$124,677.05
2	$1286.49	$961.05	$325.44	$124,351.61
3	$1286.49	$958.54	$327.95	$124,023.67
4	$1286.49	$956.02	$330.47	$123,693.19
5	$1286.49	$953.47	$333.02	$123,360.17
6	$1286.49	$950.90	$335.59	$123,024.58
7	$1286.49	$948.31	$338.18	$122,686.40
8	$1286.49	$945.71	$340.78	$122,345.62
9	$1286.49	$943.08	$343.41	$122,002.21
10	$1286.49	$940.43	$346.06	$121,656.16
11	$1286.49	$937.77	$348.72	$121,307.43
12	$1286.49	$935.08	$351.41	$120,956.02

45.

Pmt No.	Payment	Interest	Amt. Applied to Principal	Balance
1	−$1190.96	$828.13	$362.83	$124,637.17
2	−$1190.96	$825.72	$365.24	$124,271.93
3	−$1190.96	$823.30	$367.66	$123,904.27
4	−$1190.96	$820.87	$370.09	$123,534.17
5	−$1190.96	$818.41	$372.55	$123,161.63
6	−$1190.96	$815.95	$375.01	$122,786.61
7	−$1190.96	$813.46	$377.50	$122,409.12
8	−$1190.96	$810.96	$380.00	$122,029.12
9	−$1190.96	$808.44	$382.52	$121,646.60
10	−$1190.96	$805.91	$385.05	$121,261.55
11	−$1190.96	$803.36	$387.60	$120,873.95
12	−$1190.96	$800.79	$390.17	$120,483.78

47.

Pmt No.	Payment	Interest	Amt. Applied to Principal	Balance
1	$1209.04	$854.17	$354.88	$124,645.12
2	$1209.04	$851.74	$357.30	$124,287.82
3	$1209.04	$849.30	$359.74	$123,928.08
4	$1209.04	$846.84	$362.20	$123,565.88
5	$1209.04	$844.37	$364.68	$123,201.21
6	$1209.04	$841.87	$367.17	$122,834.04
7	$1209.04	$839.37	$369.68	$122,464.36
8	$1209.04	$836.84	$372.20	$122,092.16
9	$1209.04	$834.30	$374.75	$121,717.41
10	$1209.04	$831.74	$377.31	$121,340.11
11	$1209.04	$829.16	$379.88	$120,960.22
12	$1209.04	$826.56	$382.48	$120,577.74

49.

Pmt No.	Payment	Interest	Amt. Applied to Principal	Balance
1	$972.29	$494.79	$477.50	$124,522.50
2	$972.29	$492.90	$479.39	$124,043.11
3	$972.29	$491.00	$481.29	$123,561.83
4	$972.29	$489.10	$483.19	$123,078.64
5	$972.29	$487.19	$485.10	$122,593.53
6	$972.29	$485.27	$487.02	$122,106.51
7	$972.29	$483.34	$488.95	$121,617.56
8	$972.29	$481.40	$490.89	$121,126.67
9	$972.29	$479.46	$492.83	$120,633.84
10	$972.29	$477.51	$494.78	$120,139.06
11	$972.29	$475.55	$496.74	$119,642.32
12	$972.29	$473.58	$498.71	$119,143.61

Chapter 6 Summary Exercises

1. $2420   **3.** $2462.88   **5.** 35 periods   **7.** $2039.69   **9.** $2112.05   **11.** $337,006.30   **13.** $2662.09   **15.** 6.14%
17. 10.25%   **19.** $430.30   **21. a.** APR = 8%   **b.** APR = 8.06%   **c.** APR = 8.03%   **d.** APR = 7.98%   **e.** APR = 7.90%

f. APR = 7.79%   **23.** $\begin{bmatrix} 4 & -\frac{3}{2} & -3 \\ -\frac{1}{2} & \frac{1}{4} & 0 \\ -1 & \frac{1}{2} & 1 \end{bmatrix}$ **25.** $y = 3x + 2$

Sample Test Items for Chapter 6

1. $1560.00   **2.** $4533.21   **3.** $8492.79   **4.** $5333.333   **5.** Approximately 27 periods.   **6.** Approximately 22 periods.
7. $8670.31   **8.** $13,838.67   **9.** Approximately 8.7 years.   **10.** $26,452.76   **11.** $81,136.16   **12.** $5809.48   **13.** $6716.35
14. The present value of the total payments ($120 × 36 = $4320) received by the bank is $3429.14.   **15.** 7.19%   **16.** 9.42%
17. $1029.14   **18.** Approximately 12.4 years.

CHAPTER 7

7.1 Exercises

1. The student is a female or a sophomore.   **3.** The student is not a female or the student is a sophomore.   **5.** If the student is a female, then she is a sophomore.   **7.** The hat is red or the belt is black.   **9.** The hat is not red or the belt is black.   **11.** If the hat is red, then the belt is black.   **13.** The card is not a five or the card is a diamond.   **15.** The student will not do her homework or her test grades will improve.
17. $p \wedge q$ **19.** $p \wedge q$ **21.** $p \wedge q$ **23.** $\sim p \vee \sim q$ **25.** Exactly four of the students are taking mathematics or five are taking mathematics.   **27.** Exactly three shirts have flaws, or two shirts have flaws, or one shirt has a flaw, or no shirt has a flaw.   **29.** The car is not red

and it does not have a CD player.   **31.** The blouse is blue or the belt is not black.   **33.** Neither of the two marbles is red.   **35.** At most one of the two students graduated with honors.   **37.** All four marbles are green.

7.2 Exercises

1.

p	q	$\sim q$	$p \vee \sim q$
T	T	F	**T**
T	F	T	**T**
F	T	F	**F**
F	F	T	**T**

3.

p	q	$\sim p$	$\sim q$	$\sim p \Rightarrow \sim q$
T	T	F	F	**T**
T	F	F	T	**T**
F	T	T	F	**F**
F	F	T	T	**T**

5.

p	q	$\sim p$	$\sim p \vee q$	$\sim(\sim p \vee q)$
T	T	F	T	**F**
T	F	F	F	**T**
F	T	T	T	**F**
F	F	T	T	**F**

7.

p	q	$p \vee q$	$\sim(p \vee q)$	$\sim p$	$\sim q$	$\sim p \wedge \sim q$
T	T	T	**F**	F	F	**F**
T	F	T	**F**	F	T	**F**
F	T	T	**F**	T	F	**F**
F	F	F	**T**	T	T	**T**

9.

p	q	r	$q \vee r$	$p \wedge (q \vee r)$	$p \wedge q$	$p \wedge r$	$(p \wedge q) \vee (p \wedge r)$
T	T	T	T	**T**	T	T	**T**
T	T	F	T	**T**	T	F	**T**
T	F	T	T	**T**	F	T	**T**
T	F	F	F	**F**	F	F	**F**
F	T	T	T	**F**	F	F	**F**
F	T	F	T	**F**	F	F	**F**
F	F	T	T	**F**	F	F	**F**
F	F	F	F	**F**	F	F	**F**

11.

p	q	$p \wedge q$	$q \wedge p$	$\sim p$	$\sim(q \wedge p)$	$\sim p \Rightarrow \sim(q \wedge p)$
T	T	**T**	T	F	F	**T**
T	F	**F**	F	F	T	**T**
F	T	**F**	F	T	T	**T**
F	F	**F**	F	T	T	**T**

The statements are not equivalent.

13.

p	q	$\sim p$	$\sim p \vee q$	$q \wedge (\sim p \vee q)$	$\sim(\sim p \vee q)$
T	T	F	T	**T**	F
T	F	F	F	**F**	T
F	T	T	T	**T**	F
F	F	T	T	**F**	F

The statements are not equivalent.

15. F   **17.** T   **19.** F   **21.** T   **23.** $\sim(p \wedge q) = \sim p \vee \sim q$
The die does not show an odd number or the die shows a five.
25. $\sim(p \vee \sim q) = \sim p \wedge q$
The die does not show an odd number and it does not show a five.
27. $\sim(\sim p \vee \sim q) = p \wedge q$
The die shows an odd number and it does not show a five.

7.3 Exercises

1. $\{0, 4, 5, 6, 7\}$ **3.** U **5.** $\{5\}$ **7.** $\{0, 4, 5\}$ **9.** $\{2, 3\}$ **11.** $\{a, b, e, f\}$ **13.** $\{c\}$ **15.** $\varnothing$ **17.** $\{a, b, e, f\}$
19. A **21.**   **23.**   **25.**

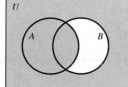

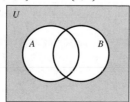

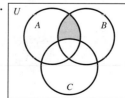

27.

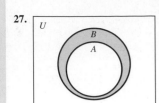

29. True   **31.** True   **33.** True   **35.** The number is odd and divisible by 3.   **37.** The number is odd or is not divisible by 3.   **39.** The number is odd, but not divisible by 3; the number is odd and not divisible by 3.   **41.** The person is at most 40 years of age.   **43.** The person wears glasses and is over 40 years of age.   **45.** The person wears glasses and is not over 40 years of age.

47.

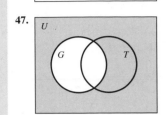

G'

49.

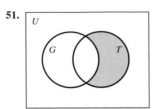

$G - T = G \cap T'$

51.

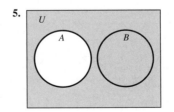

$G' \cap T = T - G$

53.

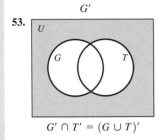

$G' \cap T' = (G \cup T)'$

55.

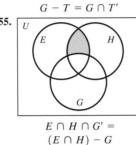

$E \cap H \cap G' =$
$(E \cap H) - G$

57.

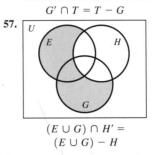

$(E \cup G) \cap H' =$
$(E \cup G) - H$

59.

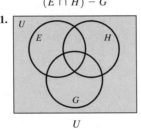

$(E \cup G \cup H)'$

61.

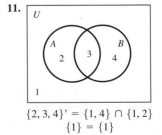

U

7.4 Exercises

1.

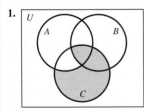

3.

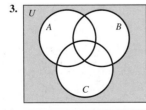

5.

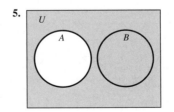

7.

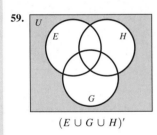

9.

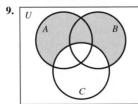

11.

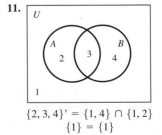

$\{2, 3, 4\}' = \{1, 4\} \cap \{1, 2\}$
$\{1\} = \{1\}$

13.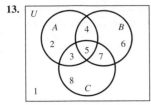

$\{6, 7, 8\} - \{4, 5, 6, 7\} =$
$\{1, 8\} \cap \{3, 5, 7, 8\}$
$\{8\} = \{8\}$

15. Equal   **17.** Not equal   **19.** Not equal   **21.** 17   **23.** 5   **25.** 4

27. a.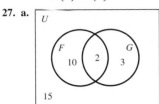

b. $n(F \cap G') = 10$ **c.** $n(F \cup G) = 15$ **d.** $n(F \cup G) = 15$ **e.** $n[(F - G) \cup (G - F)] = 13$
f. $n((F \cup G)') = 15$ **g.** $n(F \cup G) = 12 + 5 - 2 = 15$

29. a. 59 **b.** 41 **c.** 21 **d.** 7   **31. a.** 80 **b.** 30 **c.** 30 **d.** 6.67%   **33. a.** 39 **b.** 12 **c.** 27 **d.** 77 **e.** 54
35. a. 105 **b.** 123 **c.** 50 **d.** 18   **37. a.** 3 **b.** 8 **c.** 7   **39.** No   **41. a.** 75 **b.** 40

7.5 Exercises

1. 6   **3.** 11   **5.** 16; *HHHH, HHHT, HHTH, HTHH, THHH, HHTT, HTTH, TTHH, THHT, THTH, HTHT, TTTH, TTHT, THTT, HTTT,*
TTTT   **7.** 12, 33, 32, 31, 30, 23, 22, 21, 20, 13, 12, 11, 10   **9.** 3; $\{(E_1, L_1), (E_2, L_3), (E_3, L_2)\}$, $\{(E_1, L_3), (E_2, L_2), (E_3, L_1)\}$, $\{(E_1, L_2), (E_2, L_1), (E_3, L_3)\}$ **11.** 12; $S_1F_1I_1$, $S_1F_1I_2$, $S_1F_2I_1$, $S_1F_2I_2$, $S_2F_1I_1$, $S_2F_1I_2$, $S_2F_2I_1$, $S_2F_2I_2$, $S_3F_1I_1$, $S_3F_1I_2$, $S_3F_2I_1$, $S_3F_2I_2$
13. a. 1,000,000,000 **b.** 900,000,000 **c.** 810,000,000   **15. a.** 40,320 **b.** 0 **c.** 4320   **17. a.** 40,320 **b.** 5040   **19. a.** 10,000
b. 4536 **c.** 1120   **21.** 4; 32; 128   **23.** 3; 5   **25.** 1024   **27.** 6   **29. a.** 6,760,000 **b.** 6,500,000 **c.** 340,704   **31. a.** 37,152
b. 36,592 **c.** 29,040   **33.** 657,720   **35.** 1560   **37.** 1680   **39. a.** 6 **b.** 6   **41. a.** 100 **b.** 1900   **43. a.** 336 **b.** 512

7.6 Exercises

1. 720   **3.** 40,320   **5.** 1680   **7.** 792   **9.** 2,598,960   **11.** 2   **13.** 1680   **15.** 6   **17.** 5040   **19.** 1   **21.** 24   **23.** 1
25. a. $P(8, 4)$ **b.** 1680   **27. a.** $P(7, 7)$ **b.** 5040   **29.** 120   **31.** 336   **33.** 120   **35.** 24   **37.** 1260   **39.** 792
41. 60,060   **43.** 2520   **45.** 252   **47.** There are $26 \cdot 26 = 676$ possible pairs of initials, so any set of people larger than this must contain
repeated pairs of initials.

7.7 Exercises

1. 15   **3.** 84   **5.** 105   **7.** 194,580   **9.** 3432   **11. a.** Both are 10. **b.** Both are 36. **c.** Both are 56. **d.** $C(n, r) = C(n, n - r)$
13. $C(6, 0) = 1$; $C(6, 1) = 6$; $C(6, 2) = 15$; $C(6, 3) = 20$; $C(6, 4) = 15$; $C(6, 5) = 6$; $C(6, 6) = 1$ **15.** 10   **17.** 22,100   **19.** 5005
21. a. 455 **b.** 10 **c.** 525   **23. a.** 35 **b.** 21 **c.** 56   **25. a.** 5 **b.** 210 **c.** 210 **d.** 10   **27. a.** 1287 **b.** 5148 **c.** 11,761,893
d. 2574   **29. a.** 10 **b.** 30 **c.** 40   **31. a.** 14,950 **b.** 1 **c.** 69,667   **33. a.** 50 **b.** 990 **c.** 1155   **35.** 330   **37. a.** 134,044
b. 20 **c.** 1410   **39. a.** 108 **b.** 126 **c.** 6 **d.** 111   **41. a.** 2 **b.** 36   **43.** To open the lock, you must know both the numbers and
their order of entry.

Chapter 7 Summary Exercises

1.

p	q	$q \wedge p$	$p \vee (q \wedge p)$	$p \vee q$	$(p \vee q) \wedge q$
T	T	T	T	T	T
T	F	F	T	T	F
F	T	F	F	T	T
F	F	F	F	F	F

3. The current machine does break down and we do not buy a new one. **5.** $\{5, 6\}$ **7.** $\{1, 2, 4\}$ **9.** $\{2, 4, 5, 6\}$ **11.** $\varnothing$
13. $\{2, 3, 5, 6\}$ **15.** 128 **17.** 7,334,887,680 **19.** 61,124,064 **21.** 1,362,649,145 **23.** Some Fords are not red and not blue.

25. 4 **27.** 1820 **29.** 69,156,564,288,396,165,360 **31.** $C = 0.65x + 5.65$ **33.** $AB = \begin{bmatrix} -1 & 5 \\ -15 & 11 \end{bmatrix}$, $BA = \begin{bmatrix} 2 & -6 & 0 \\ 7 & 9 & -5 \\ 1 & 3 & -1 \end{bmatrix}$

Sample Test Items for Chapter 7

1. The first ball drawn is not red or the second ball drawn is blue. **2.** The first ball drawn is not red and the second ball drawn is not blue.
3. The first ball drawn is red and the second ball drawn is blue. **4.** If the first ball drawn is red, then the second ball drawn is blue. **5.** F
6. F **7.** T **8.** F **9.** T **10.** F **11.** T **12.** $\{4\}$ **13.** $\{2\}$ **14.** $\{\varnothing, 3\}$
15.

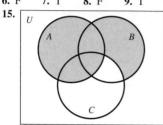

16. 0 **17.** 47 **18.** 900 **19.** 211,926 **20.** 60 **21.** 126
22. 70 **23.** 36 **24.** 384 **25.** 1023

CHAPTER 8

8.1 Exercises

1. a. $S = \{A, B, C, D\}$ **b.** $\dfrac{1}{4}$ **c.** $\dfrac{3}{4}$ **3. a.** $S = \{MM, MF, FM, FF\}$, where M denotes a male child and F denotes a female child. **b.** $\dfrac{1}{4}$
c. $\dfrac{3}{4}$ **d.** $\dfrac{1}{2}$ **5. a.** $S = \{MMM, MFM, MMF, MFF, FFF, FMF, FFM, FMM\}$, where M denotes a male child and F denotes a female
child. **b.** $\dfrac{1}{8}$ **c.** $\dfrac{7}{8}$ **d.** $\dfrac{3}{8}$ **7. a.** $S = \{TN, TR, NT, NR, RT, RN\}$, where T represents Tigist, N represents Noora, and R represents Ran.

In each outcome, the first letter represents the selection for president and the second represents the selection for secretary. **b.** $\dfrac{1}{6}$ **c.** $\dfrac{4}{6} = \dfrac{2}{3}$

d. $\dfrac{2}{6} = \dfrac{1}{3}$ **9. a.** $S = \{1, 2, 3, 4\}$ **b.** $\dfrac{1}{4}$ **c.** $\dfrac{2}{4} = \dfrac{1}{2}$ **d.** $\dfrac{1}{4}$ **11. a.** $S = \{H1, H2, H3, H4, T1, T2, T3, T4\}$ **b.** $\dfrac{1}{8}$ **c.** $\dfrac{2}{8} = \dfrac{1}{4}$

d. $\dfrac{2}{8} = \dfrac{1}{4}$ **13. a.** $S = \{1, 2, 3, 4, 5, 6, 7, 8, 9\}$ **b.** $\dfrac{1}{9}$ **c.** $\dfrac{2}{9}$ **d.** $\dfrac{5}{9}$ **15. a.** $S = \{(1, 2), (1, 3), (2, 1), (2, 3), (3, 1), (3, 2)\}$ **b.** $\dfrac{1}{6}$

c. $\dfrac{2}{6} = \dfrac{1}{3}$ **17. a.** $\dfrac{6}{36} = \dfrac{1}{6}$ **b.** $\dfrac{2}{36} = \dfrac{1}{18}$ **c.** $\dfrac{10}{36} = \dfrac{5}{18}$ **19. a.** 0 **b.** $\dfrac{1}{36}$ **c.** $\dfrac{6}{36} = \dfrac{1}{6}$ **21. a.** 0 **b.** $\dfrac{3}{216} = \dfrac{1}{72}$ **c.** $\dfrac{12}{216} = \dfrac{1}{18}$

23. a. 52 **b.** $\dfrac{13}{52} = \dfrac{1}{4}$ **c.** $\dfrac{4}{52} = \dfrac{1}{13}$ **25. a.** $\dfrac{C(13, 5)}{C(52, 5)} \approx 0.0004952$ **b.** $4 \cdot \dfrac{C(13, 5)}{C(52, 5)} \approx 0.001981$ **c.** $\dfrac{C(39, 5)}{C(52, 5)} \approx 0.2215$

27. a. $\dfrac{C(16, 6)}{C(52, 6)} \approx 0.0003933$ **b.** 0 **c.** $\dfrac{C(48, 6)}{C(52, 6)} \approx 0.6028$ **29. a.** $\dfrac{C(13, 2) \cdot C(39, 3)}{C(52, 5)} \approx 0.2743$ **b.** $\dfrac{C(4, 2) \cdot C(48, 3)}{C(52, 5)} \approx 0.0399$

c. $\dfrac{C(39, 3) \cdot C(13, 2)}{C(52, 5)} \approx 0.2743$ **31. a.** $\dfrac{C(190, 10)}{C(200, 10)} \approx 0.5915$ **b.** $\dfrac{C(10, 2) \cdot C(190, 8)}{C(200, 10)} \approx 0.0727$ **c.** $\dfrac{C(10, 5) \cdot C(190, 5)}{C(200, 10)} \approx 0.00002196$

33. a. $\dfrac{4 \cdot 3 \cdot 3 \cdot 2 \cdot 1}{P(5, 5)} = 0.6$ **b.** $\dfrac{2 \cdot 3 \cdot 2 \cdot 1 \cdot 1}{P(5, 5)} = 0.1$ **c.** $\dfrac{3 \cdot 2 \cdot 2 \cdot 1 \cdot 1}{P(5, 5)} = 0.1$ **35. a.** $\dfrac{1 \cdot 26 \cdot 26 \cdot 10 \cdot 10 \cdot 1}{26^3 \cdot 10^3} = \dfrac{1}{260}$

b. $\dfrac{26 \cdot 25 \cdot 24 \cdot 10 \cdot 9 \cdot 8}{26^3 \cdot 10^3} = \dfrac{11,232,000}{17,576,000} \approx 0.6391$ **c.** $\dfrac{25 \cdot 25 \cdot 25 \cdot 9 \cdot 9 \cdot 9}{26^3 \cdot 10^3} = \dfrac{11,390,625}{17,576,000} \approx 0.6481$ **37. a.** $\dfrac{C(60, 6)}{C(80, 6)} \approx 0.1666$

b. $\dfrac{C(20, 3) \cdot C(60, 3)}{C(80, 6)} \approx 0.1298$ **c.** $\dfrac{C(20, 6)}{C(80, 6)} \approx 0.0001290$ **39. a.** $\dfrac{P(7, 4)}{P(10, 4)} \approx 0.1667$ **b.** $\dfrac{3 \cdot 2 \cdot 7 \cdot 6}{P(10, 4)} = 0.05$ **c.** 0

41. a. $\dfrac{P(5,3)}{P(8,3)} \approx 0.1786$ **b.** $\dfrac{P(3,3)}{P(8,3)} \approx 0.01786$ **c.** $\dfrac{3 \cdot 5 \cdot 4}{P(8,3)} \approx 0.1786$ **43. a.** $\dfrac{C(6,4)}{C(10,4)} \approx 0.0714$ **b.** $\dfrac{C(4,2) \cdot C(6,2)}{C(10,4)} \approx 0.4286$

45. a. $\dfrac{C(6,5)}{C(10,5)} \approx 0.02381$ **b.** $\dfrac{C(4,2) \cdot C(6,3)}{C(10,5)} \approx 0.4762$ **c.** $\dfrac{C(6,3) \cdot C(4,2)}{C(10,5)} \approx 0.4762$

8.2 Exercises

1. $P(d) = \dfrac{1}{6}$ **3.** $P(a) = \dfrac{1}{2}$ **5.** The sum of the probabilities of the outcomes is greater than 1. To be a valid probability assignment, the sum should be exactly equal to 1.   **7.** The sum of the probabilities of the outcomes is less than 1. To be a valid probability assignment, the sum should be exactly equal to 1.   **9. a.** $S = \{A, B, C\}$; $P(A) = \dfrac{1}{4}$; $P(B) = \dfrac{1}{4}$; $P(C) = \dfrac{1}{2}$ **b.** $\dfrac{3}{4}$ **c.** $\dfrac{3}{4}$ **11.** $S = \{s, f\}$, where s denotes success and f denotes failure. $P(s) = \dfrac{1}{6}$; $P(f) = \dfrac{5}{6}$ **13. a.** $S = \{R, G, B\}$, where R indicates that the ball is red, G indicates that the ball is green, and B indicates that the ball is blue. $P(R) = \dfrac{6}{15}$; $P(G) = \dfrac{2}{15}$; $P(B) = \dfrac{7}{15}$. **b.** $\dfrac{8}{15}$ **c.** $\dfrac{9}{15}$ **15.** $S = \{s, f\}$, where s denotes success and f denotes failure. $P(s) = \dfrac{6}{36} = \dfrac{1}{6}$; $P(f) = \dfrac{30}{36} = \dfrac{5}{6}$.   **17.** $S = \{s, f\}$, where s denotes success and f denotes failure. $P(s) = \dfrac{13}{52} = \dfrac{1}{4}$; $P(f) = \dfrac{39}{52} = \dfrac{3}{4}$.   **19.** $S = \{s, f\}$, where s denotes success and f denotes failure. $P(s) = \dfrac{1}{4}$; $P(f) = \dfrac{3}{4}$.   **21. a.** $S = \{0, 1, 2, 3\}$; $P(0) = \dfrac{1}{8}$; $P(1) = \dfrac{3}{8}$; $P(2) = \dfrac{3}{8}$; $P(3) = \dfrac{1}{8}$ **b.** $\dfrac{1}{2}$ **c.** $\dfrac{7}{8}$ **d.**

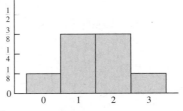

23. 0.076   **25.** 0.759

27. a. $S = \{1, 2, 3, 4, 5, 6\}$; the probability of each outcome is $\dfrac{1}{6}$
b. $\dfrac{3}{6} = \dfrac{1}{2}$ **c.** $\dfrac{2}{6} = \dfrac{1}{3}$
d.

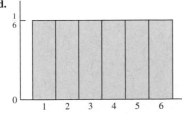

29. a. $S = \{0, 1, 2\}$; $P(0) = 0.3$; $P(1) = 0.6$; $P(2) = 0.1$ **b.** 0.9
c.

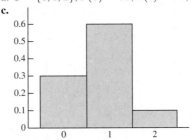

31. a. $S = \{0, 1, 2\}$; $P(0) = 0.81$; $P(1) = 0.18$; $P(2) = 0.01$
b. 0.99 **c.**

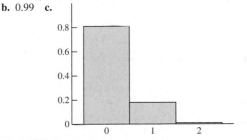

33. a. $S = \{0, 1, 2\}$; $P(0) \approx 0.8507$; $P(1) \approx 0.1448$; $P(2) \approx 0.0045$
b. 0.1493 **c.**

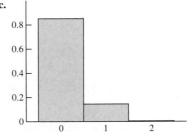

35. a. $S = \{0, 1, 2\}$; $P(0) \approx 0.4667$; $P(1) \approx 0.4667$; $P(2) \approx 0.0667$
b. 0.5334 **c.**

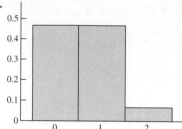

37. a.

Number of Radios Owned	Relative Frequency
3	0.25
4	0.40
5	0.35

b. **c.** 0.75

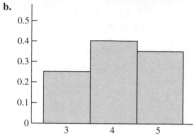

39. a.

Number of Beans in Pod	Relative Frequency
2	$\dfrac{2}{18}$
3	$\dfrac{5}{18}$
4	$\dfrac{8}{18}$
5	$\dfrac{3}{18}$

41. a.

Number of Heads	Assigned Probability
0	0.29
1	0.51
2	0.20

b. 0.71

b.

c. $\dfrac{16}{18} = \dfrac{8}{9}$

43. a.

Prediction	Assigned Probability
Recession	$\dfrac{64}{120}$
Slow Growth	$\dfrac{42}{120}$
Sharp Growth	$\dfrac{14}{120}$

b. $\dfrac{56}{120} \approx 0.4667$

45. a.

Test Result	Assigned Probability
Usable	0.92
Not Usable	0.08

b. 0.92

47. a. $\dfrac{16}{450} = \dfrac{8}{225}$ **b.** $\dfrac{110}{450} = \dfrac{11}{45}$ **c.** $\dfrac{166}{450} = \dfrac{83}{225}$ **d.** $\dfrac{144}{450} = \dfrac{8}{25}$ **49. a.** $\dfrac{182}{250} = \dfrac{91}{125}$ **b.** $\dfrac{110}{250} = \dfrac{11}{25}$ **c.** $\dfrac{30}{250} = \dfrac{3}{25}$ **d.** $\dfrac{120}{250} = \dfrac{12}{25}$

51. a. $\dfrac{4}{200} = \dfrac{1}{50}$ **b.** $\dfrac{12}{200} = \dfrac{3}{50}$ **c.** $\dfrac{20}{200} = \dfrac{1}{10}$ **d.** $\dfrac{113}{200}$ **53. a.** 1:12 **b.** 1:3 **c.** 1:1 **d.** 3:1 **55. a.** 22:703 **b.** 1:4324

c. 4324:1 **57. a.** 1:4 **b.** 7:3 **c.** 4:1 **59.** $\dfrac{1}{101} \approx 0.009901$

8.3 Exercises

1. a. $-\$0.50$ **b.** You should expect to lose \$750. **3. a.** $\dfrac{2}{11}$ spaces **b.** You should expect to advance 8 spaces. **5.** One boy.

7. 9.25 **9.** $-\$0.71$ **11.** $-\$0.25$ **13.** \$0.22 **15.** \$0 **17.** $-\$0.25$ **19.** \$48 **21. a.** \$100 **b.** \$200,000 **23.** \$4500 **25.** 3.6 **27.** $-\$0.05$ **29.** The campus location would have expected earnings of \$5.50 per customer, while the business location would have expected earnings of \$7.30 per customer. If 2000 customers were to visit the location in a day, the business location would have a larger expected earnings (\$14,600) than the campus location (\$11,000). **31.** The expected number of years of college attended is 3.15. **33.** Garret's expected earnings are approximately \$3.99. **35.** Your expected earnings from a single lottery ticket are approximately $-\$2.99$.

37. a.

Number of Red Balls	Probability
0	0.3571
1	0.5357
2	0.1071

b.

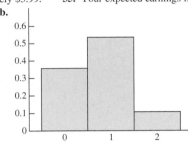

c. The expected number of red balls is approximately 0.7499.

39. The expected number of kings is approximately 0.1538. **41.** The expected number of red marbles is 1.25. **43.** The expected number of red balls is approximately 0.1667; the expected number of green balls is approximately 0.3333. **45.** The expected value of X is 0.4412. **47.** The expected value of X is 30. **49.** The expected value of X is 7.

Chapter 8 Summary Exercises

1. 312 **3.** 2:11 **5.** 0.2519 **7.** $S = \{s, f\}$, where s denotes success and f denotes failure. **9.** The expected number of girls in a three-child family is 1.5. **11.** The expected number of DVD's rented per person each month is approximately 3.04. **13.** The company should expect to earn \$800 per policy. **15.** If the probability of an event E occurring is 1, then the event will *definitely* occur. For example, consider the experiment of tossing a single coin once, and let E be the event that either heads or tails results. Then the probability of this event is 1. **17.** The expected value of a random variable is the average of the values of the random variable if the underlying experiment is repeated indefinitely.

19. The correct answer is (d). **21.** The correct answer is (c). **23.** $\begin{bmatrix} -5 & 0 & 10 \\ 4 & -2 & -20 \\ -6 & 0 & 25 \end{bmatrix}$

25. Since $P(\text{all the same sex}) = \dfrac{2}{16} = \dfrac{1}{8}$, $P(\text{three are of the same sex}) = \dfrac{8}{16} = \dfrac{1}{2}$, and $P(\text{two are of each sex}) = \dfrac{6}{16} = \dfrac{3}{8}$, the most likely outcome is that three of the four children will be of the same sex.

Sample Test Items for Chapter 8

1. 56 **2.** $\dfrac{8}{24} = \dfrac{1}{3}$ **3.** 3:5 **4.** $\dfrac{9}{14}$ **5.** $\dfrac{5}{36}$ **6.** $\dfrac{1}{36}$ **7.** $\dfrac{30}{36} = \dfrac{5}{6}$ **8.** $\dfrac{C(48, 3)}{C(52, 3)} \approx 0.7826$ **9.** $\dfrac{C(13, 2) \cdot C(39, 1)}{C(52, 3)} \approx 0.1376$

10. $4 \cdot \dfrac{C(13, 3)}{C(52, 3)} \approx 0.01294$ **11.** $\dfrac{C(3, 2) \cdot C(5, 1)}{C(8, 3)} \approx 0.2679$ **12.** $\dfrac{C(3, 1) \cdot C(5, 2)}{C(8, 3)} + \dfrac{C(3, 2) \cdot C(5, 1)}{C(8, 3)} + \dfrac{C(3, 3)}{C(8, 3)} \approx 0.8214$

13.

14. $\dfrac{10}{32} = \dfrac{5}{16}$ **15.** $\dfrac{16}{32} = \dfrac{1}{2}$ **16.** The expected number of girls in a five-child family is 2.5. **17.** The expected number of jacks is approximately 0.1538. **18.** The expected number of correct guesses is 38.

CHAPTER 9

Section 9.1 Exercises

1. a. 0.8   **b.** 0.5   **c.** 0.5   **3. a.** 0.9   **b.** 0.3   **c.** 0.7   **5. a.** 0.3   **b.** 0.2   **c.** 0.7   **7. a.** 0.6   **b.** 0.9   **c.** 0.8   **9.** 0.60

11. $\dfrac{30}{80} = \dfrac{3}{8}$ **13. a.** $\dfrac{86}{180} = \dfrac{43}{90}$ **b.** $\dfrac{93}{180}$ **c.** $\dfrac{13}{180}$ **15. a.** Yes, if $P(B) = 0.3$.   **b.** No; since $A \cap B \subseteq A$, $P(A)$ cannot be smaller than

$P(A \cap B)$.   **c.** No; this assignment would force $P(A)$ to be 1.1, which is not valid.   **17.** $\dfrac{155}{165} = \dfrac{31}{33}$ **19.** $\dfrac{140}{165} = \dfrac{28}{33}$

21. $\dfrac{5523}{32{,}166} \approx 0.1717$ **23.** $\dfrac{27{,}499}{32{,}166} \approx 0.8549$ **25.** $\dfrac{11}{20}$ **27. a.** $\dfrac{8}{52} = \dfrac{2}{13}$ **b.** $\dfrac{16}{52} = \dfrac{4}{13}$ **c.** $\dfrac{1}{52}$ **d.** $\dfrac{16}{52} = \dfrac{4}{13}$ **29. a.** 0.5714

b. 0.7143   **c.** 0.7143   **31. a.** $\dfrac{48}{120} = \dfrac{2}{5}$ **b.** $\dfrac{24}{120} = \dfrac{1}{5}$ **33. a.** $P((1 \text{ head and 1 tail}) \text{ or } (2 \text{ heads})) = \dfrac{3}{4}$ **b.** $1 - P(\text{no heads}) = \dfrac{3}{4}$

35. a. $P(\text{sum of 3 or 4 or 5 or 6 or 7 or 8 or 9 or 10 or 11 or 12}) = \dfrac{35}{36}$ **b.** $1 - P(\text{sum of 2}) = \dfrac{35}{36}$.

37. a. $P((1 \text{ king and 2 others}) \text{ or } (2 \text{ kings and 1 other}) \text{ or } (3 \text{ kings})) \approx 0.2174$ **b.** $1 - P(\text{no kings}) \approx 0.2174$

39. a. $P(\text{draw a 3 or 4 or 5 or 6}) = \dfrac{4}{6} = \dfrac{2}{3}$ **b.** $1 - P(\text{draw a 1 or 2}) = 1 - \dfrac{2}{6} = \dfrac{4}{6} = \dfrac{2}{3}$.   **41. a.** $\dfrac{6}{36} = \dfrac{1}{6}$ **b.** $\dfrac{9}{36} = \dfrac{1}{4}$ **c.** 1   **d.** $\dfrac{35}{36}$

e. $\dfrac{33}{36} = \dfrac{11}{18}$ **f.** 0   **43. a.** $\dfrac{5}{36}$ **b.** $\dfrac{26}{36} = \dfrac{13}{18}$ **c.** $\dfrac{23}{36}$ **45. a.** 0.5865   **b.** 0.0003620   **c.** 0.02534   **47. a.** 0.4048   **b.** 0.8810

c. 0.1190   **49.** 0.6779   **51. a.** $\dfrac{4}{8} = \dfrac{1}{2}$ **b.** $\dfrac{6}{8} = \dfrac{3}{4}$ **c.** $\dfrac{2}{8} = \dfrac{1}{4}$ **d.** $\dfrac{7}{8}$ **53.** 1

Section 9.2 Exercises

1. a. $S = \{1, 2, 3, 4, 5, 6\}$ **b.** $S^* = \{1, 3, 5\}$; $P(3 | \text{odd}) = \dfrac{1}{3}$

c. $P(3 | \text{odd}) = \dfrac{1/6}{3/6} = \dfrac{1}{3}$

3. a. $S = \{1, 2, 3, 4, 5, 6\}$ **b.** $S^* = \{1, 3, 5\}$; $P(2 | \text{odd}) = 0$

c. $P(2 | \text{odd}) = \dfrac{0}{3/6} = 0$

5. a. $S = \{HHH, HTH, HHT, HTT, TTT, THT, TTH, THH\}$
b. $S^* = \{HTH, HHT, HTT, TTT, THT, TTH, THH\}$;

$P(\text{exactly two tails} | \text{at least one tail}) = \dfrac{3}{7}$

c. $P(\text{exactly 2 tails} | \text{at least 1 tail}) = \dfrac{3/8}{7/8} = \dfrac{3}{7}$

7. a. $S = \{HHH, HTH, HHT, HTT, TTT, THT, TTH, THH\}$
b. $S^* = \{HTT, THT, TTH\}$;

$P(\text{second is heads} | \text{exactly 1 head}) = \dfrac{1}{3}$

c. $P(\text{second is heads} | \text{exactly 1 head}) = \dfrac{1/8}{3/8} = \dfrac{1}{3}$

9. $\dfrac{4}{51}$ **11.** $\dfrac{4}{51}$ **13.** $\dfrac{3}{51}$ **15.** $\dfrac{13}{52} = \dfrac{1}{4}$ **17. a.** $\dfrac{9}{14}$ **b.** $\dfrac{5}{14}$ **c.** $\dfrac{5}{14}$ **19.** $\dfrac{30}{70} = \dfrac{3}{7}$ **21.** $\dfrac{12}{52} = \dfrac{3}{13}$ **23.** $\dfrac{38}{90} = \dfrac{19}{45}$

25. $\dfrac{20}{40} = \dfrac{1}{2}$ **27.** $\dfrac{23}{40}$ **29.** $\dfrac{8}{20} = \dfrac{2}{5}$ **31.** $\dfrac{6}{12} = \dfrac{1}{2}$ **33.** $\dfrac{15}{25} = \dfrac{3}{5}$ **35.** $\dfrac{20}{44} = \dfrac{5}{11}$ **37.** $\dfrac{17}{25}$ **39.** $\dfrac{18}{100} = \dfrac{9}{50}$ **41.** $\dfrac{6}{36} = \dfrac{1}{6}$

43. $\dfrac{5}{6}$ **45.** $\dfrac{1}{6}$ **47. a.** $\dfrac{4}{12} = \dfrac{1}{3}$ **b.** $\dfrac{1}{2}$ **c.** 0   **d.** $\dfrac{2}{6} = \dfrac{1}{3}$ **49.** $\dfrac{3}{7}$ **51.** $\dfrac{8}{21}$ **53.** $\dfrac{20}{28} = \dfrac{5}{7}$ **55.** $\dfrac{2}{6} = \dfrac{1}{3}$ **57.** $\dfrac{3}{7}$ **59.** $\dfrac{4}{7}$

61. $\dfrac{2}{6} = \dfrac{1}{3}$

63.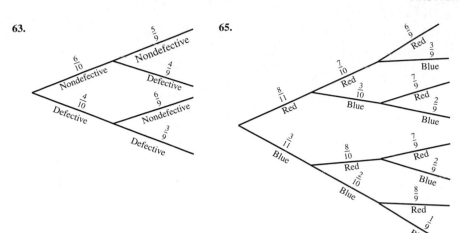

65.

67.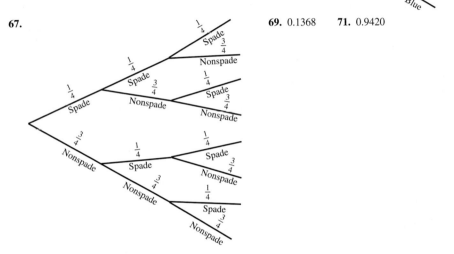

69. 0.1368   **71.** 0.9420

Section 9.3 Exercises

1. $\dfrac{12}{51} = \dfrac{4}{17}$ **3.** $\dfrac{507}{2652} \approx 0.1912$ **5.** $\dfrac{12}{2652} \approx .004525$ **7.** $\dfrac{192}{6840} \approx 0.02807$ **9.** $\dfrac{24}{6840} \approx 0.003509$ **11.** $\dfrac{1}{32}$ **13.** $\dfrac{1}{1024}$

15. $\dfrac{9}{1024} \approx 0.0088$ **17.** 0.1360   **19. a.** $\dfrac{4}{6} = \dfrac{2}{3}$ **b.** $\dfrac{1}{3}$ **21. a.** $\dfrac{1}{5}$ **b.** $\dfrac{4}{5}$ **23. a.** 0.4 **b.** 0.03 **c.** 0.67 **d.** 0.63 **e.** 0.33

25. a. $\dfrac{80}{504} = \dfrac{10}{63}$ **b.** $\dfrac{60}{504} = \dfrac{5}{42}$ **c.** $\dfrac{24}{504} = \dfrac{1}{21}$ **27. a.** $\dfrac{80}{1728} = \dfrac{5}{108}$ **b.** $\dfrac{60}{1728} = \dfrac{5}{144}$ **c.** $\dfrac{27}{1728} = \dfrac{1}{64}$ **29. a.** $\dfrac{25}{1296}$ **b.** $\dfrac{5}{1296}$

c. $\dfrac{8}{1296} = \dfrac{1}{162}$ **31. a.** $\dfrac{20}{1296} = \dfrac{5}{324}$ **b.** $\dfrac{36}{1296} = \dfrac{1}{36}$ **c.** $\dfrac{128}{1296} = \dfrac{8}{81}$ **33.** 0.0009   **35.** 0.0591   **37.** 0.0476   **39.** 0.8281

41. 0.008   **43.** 0.000003992   **45.** If two components are installed, the probability that at least one of them will function properly is 0.999996.

47. 0.996   **49.** $\dfrac{1}{128}$ **51.** $\dfrac{1}{8}$ **53.** 0.99999999   **55.** $\dfrac{1}{220}$ **57.** $\dfrac{136}{220} = \dfrac{34}{55}$ **59.** $\dfrac{1}{8}$ **61.** $0.25^{10} \approx 0.0000009537$

63. 0.9437   **65.** 0.936   **67.** 0.000008   **69.** 0.001   **71.** 0.243

Section 9.4 Exercises

1. 0.12   **3.** 0.9   **5.** 0.6   **7.** 0.2   **9.** 0.1905   **11.** 0.4054   **13.** $\dfrac{5}{13}$ **15.** $\dfrac{6}{504} = \dfrac{1}{84}$ **17.** $\dfrac{336}{504} = \dfrac{2}{3}$ **19.** $\dfrac{1}{9}$ **21.** $\dfrac{3}{7}$

23. 0.6143   **25.** 0.45   **27.** $\dfrac{7}{18}$ **29.** 0.05   **31.** 0.9524   **33.** 0.60   **35.** 0.68; the odds are 17:8.   **37.** 0.01449   **39.** 0.0425

41. 0.99   **43.** 0.9757   **45.** 0.40   **47.** 0.004   **49.** 0.25   **51.** 0.0125   **53.** 0.625   **55.** 0.04   **57.** 0.012   **59.** 0.25

61. 0.7143   **63.** $\dfrac{2}{6} = \dfrac{1}{3}$

Section 9.5 Exercises

1. 0.2780 **3.** 0.1636 **5.** 0.1300 **7.** 0.0571 **9.** 0.9415 **11.** 0.1419 **13.** 0.9921% **15.** 85.08% **17.** 6.8719%
19. 0.2164 **21.** 0.0551 **23.** 0.2734 **25.** 0.0007716 **27.** 0.4882 **29.** 0.008393 **31.** 0.9916 **33.** 0.0004158 **35.** 0.2461
37. 0.9893 **39.** 0.0009766 **41.** 0.8864 **43.** 0.3874 **45.** 0.9872 **47.** 0.0343 **49.** 0.00004096

51. a.

Successes	Probability
0	0.01024
1	0.0768
2	0.2304
3	0.3456
4	0.2592
5	0.07776

b. The expected number of successful sales calls is 3.
c. The salesperson can expect 120 successes out of 200 sales calls and 480 successes out of 800 sales calls.

53. a.

Successes	Probability
0	0.000125
1	0.007125
2	0.135375
3	0.857375

b. The expected number of field goals made in three attempts within the 30-yard range is 2.85.
c. The expected number of field goals made in 50 attempts within the 30-yard range is 47.5; in 200 attempts, the expected number is 190.

55. a. Kwasi should expect 6.5 of the next 10 calls to be for Amma. **b.** 0.9885 **c.** Thirty-five of the next 100 calls should be for someone other than Amma. **57.** 0.0046 **59.** 0.2099 **61.** 0.7558 **63.** 0.0889; 0.5556 **65.** 0.1207 **67.** 0.1 **69.** 0.015625 **71.** 0.3
73. 0.35 **75.** 0.25

Chapter 9 Summary Exercises

1. 0.2 **3.** 0.33 **5.** 0.77 **7.** 0.7 **9.** 0.6 **11.** $\dfrac{4}{7}$ **13.** $\dfrac{5}{7}$ **15.** $\dfrac{3}{5}$ **17.** 0.625 **19.** $\dfrac{1}{25}$ **21.** $\dfrac{1}{40}$ **23.** 0.2162

25. 0.5822 **27.** The correct answer is (c). **31.** $x = 5, y = -2, z = 1$ **33.** $\begin{bmatrix} 10 & -3 & -3 \\ 17 & -7 & -11 \end{bmatrix}$

Sample Test Items for Chapter 9

1. $\dfrac{2}{7}$ **2.** $\dfrac{5}{7}$ **3.** $\dfrac{4}{7}$ **4.** $\dfrac{3}{7}$ **5.** $\dfrac{140}{260} = \dfrac{7}{13}$ **6.** $\dfrac{150}{260} = \dfrac{15}{26}$ **7.** $\dfrac{86}{260} = \dfrac{43}{130}$ **8.** $\dfrac{204}{260} = \dfrac{51}{65}$ **9.** $\dfrac{86}{150} = \dfrac{43}{75}$ **10.** $\dfrac{8}{120} = \dfrac{1}{15}$
11. $\dfrac{196}{260} = \dfrac{49}{65}$ **12.** 0.0130 **13.** 0 **14.** $\dfrac{16}{52} = \dfrac{4}{13}$ **15.** 0.0004 **16.** 0.2016 **17.** 0.4213 **18.** 0.39 **19.** 0.6393
20. 0.375 **21.** $\dfrac{1}{2}$ **22.** 0.96 **23.** 0.4615 **24.** 0.2852 **25.** 0.6769

CHAPTER 10

10.1 Exercises

Note: Your tables and histograms may vary from those shown here if you used different class intervals when you organized your data.
1. The data appear to be normally distributed.

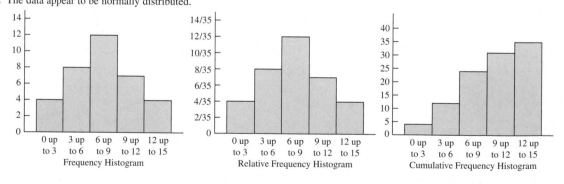

Frequency Histogram   Relative Frequency Histogram   Cumulative Frequency Histogram

3. The data appear to be skewed.

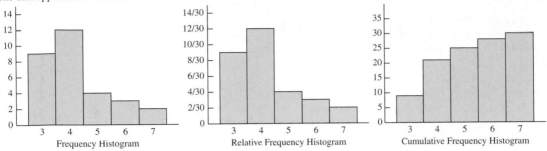

5. The data appear to be normally distributed.

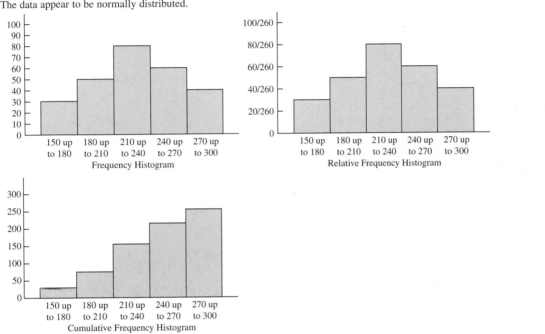

7. The data appear to be slightly skewed.

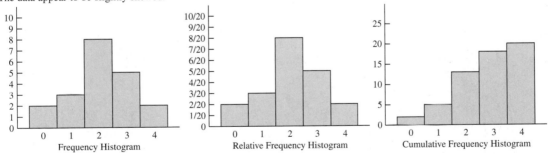

9. Fourteen people spent three or more minutes waiting in line.   **11.** 77.42% of the people spent less than three minutes waiting in line.

13.

Class	Frequency	Relative Frequency	Cumulative Frequency
10 up to 11	2	0.0455	2
11 up to 12	4	0.0909	6
12 up to 13	11	0.2500	17
13 up to 14	16	0.3636	33
14 up to 15	11	0.2500	44
15 up to 16	6	0.1364	50

15. About 88% of the batteries have a life of 12 or more hours.

17.

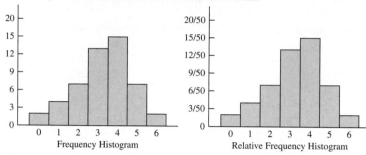

19. 48% of the pods have four or more beans in them.   **21.** The odds that a pod will have three or more beans in it are 37:13.

23. The data appear to be approximately normally distributed.

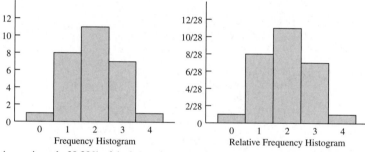

25. Approximately 39.29% of the homes have exactly two can openers.   **27.** The odds that a home will have three or more can openers are 2:5.

29.

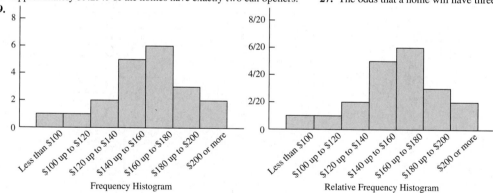

31.

Class	Frequency	Relative Frequency
1 up to 2	4	0.1333
2 up to 3	5	0.1667
3 up to 4	6	0.2000
4 up to 5	5	0.1667
5 up to 6	3	0.1000
6 up to 7	4	0.1333
7 or more	3	0.1000

33.

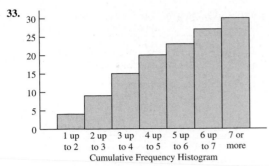

Cumulative Frequency Histogram

35.

Class	Frequency	Relative Frequency
3 up to 5	3	0.1250
5 up to 7	3	0.1250
7 up to 9	2	0.0833
9 up to 11	7	0.2917
11 up to 13	4	0.1667
13 up to 15	4	0.1667
15 or higher	1	0.0417

37. The probability that the cab company will keep a car 11 months or longer is 0.375.

39.

Class	Frequency
Under $200	2
$200 up to $250	1
$250 up to $300	2
$300 up to $350	4
$350 up to $400	3
$400 up to $450	3
$450 up to $500	2
$500 or more	3

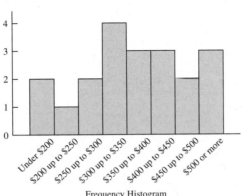

Frequency Histogram

41. a.

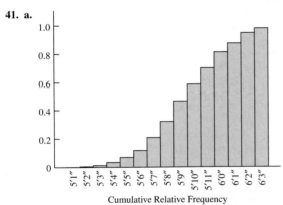

Cumulative Relative Frequency

c.

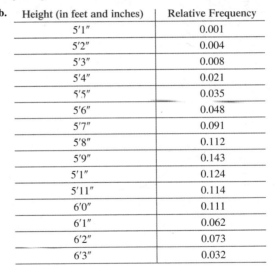

Relative Frequency

b.

Height (in feet and inches)	Relative Frequency
5'1"	0.001
5'2"	0.004
5'3"	0.008
5'4"	0.021
5'5"	0.035
5'6"	0.048
5'7"	0.091
5'8"	0.112
5'9"	0.143
5'1"	0.124
5'11"	0.114
6'0"	0.111
6'1"	0.062
6'2"	0.073
6'3"	0.032

43. a.

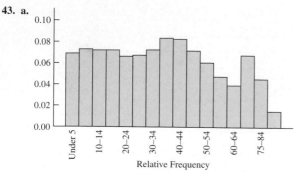

Relative Frequency

Cumulative Relative Frequency

b.

Age (in years)	Cumulative Relative Frequency
Under 5	0.069
5–9	0.142
10–14	0.214
15–19	0.286
20–24	0.352
25–29	0.419
30–34	0.491
35–39	0.574
40–44	0.656
45–49	0.727
50–54	0.787
55–59	0.834
60–64	0.873
65–74	0.94
75–84	0.985
85 and older	1

10.2 Exercises

1. a. There is no mode. **b.** $\bar{x} = 20.4$ **c.** The median is 18.   **3. a.** The mode is 102. **b.** $\bar{x} = 135$ **c.** The median is 107.
5. a. There is no mode. **b.** $\bar{x} = 0$ **c.** The median is 0.   **7.** The median, $22,000, is the best measure of central tendency for describing this data set.   **9.** The median, 20, is the best measure of central tendency for describing this data set.   **11.** The mode or the median would be best for describing this data set.   **13.** The mean, 22.5, is the best measure of central tendency for describing this data set.   **15.** 6
17. 2.875   **19.** −3   **21.** The mean, 76.875, is the measure of central tendency being used in this claim. The median, 30, would give a more accurate description of the data set.   **23.** The median is 365. This would give a more accurate description of the scores, since the mean is affected by the unusually high score of 990.   **25.** The mean and median are both 8.5.   **27.** The mean price for which they sold all 1000 mascots was $5.25.   **29.** The mean of the two selling prices is $4.75.   **31.** The mean purchase price per six-pack was $1.70.   **33.** The mean of the two purchase prices is $1.69.   **35.** The median purchase price per unit is $150.   **37.** The mean price per unit is approximately $54.91.
39. The mean of the selling prices is $71.   **41.** The mean quiz score was approximately 7.95.   **43.** The mean number of automobiles per household is approximately 2.30.   **45.** The median number of automobiles per household is 2.   **47.** The median number of peas per pod was 4.   **49.** The mean is approximately $102.5.   **51.** The mean is approximately 19.2 and the median is approximately 18.   **53.** The mean age is approximately 75.   **55.** The student's class average is 84.8.   **57.** You must make at least a 96 on the final exam to make an A in the course.   **59.**

Contribution (in dollars)	Appx. Frequency
$1 to $100	603.6
$101 to $200	339.9
$201 to $300	269.2
$301 to $400	217.5
$401 to $500	165.9
$501 to $600	127.8
$601 to $700	100.6
$701 to $999	165.9
$1000 or more	728.7

61. The median of the U.S. household charitable contributions in 1998 was approximately $250.   **63.** The median age of U.S. citizens in 1999 was approximately 37.

10.3 Exercises

1. a. $\bar{x} = 4$ **b.** $-2, -1, 0, 1, 2$ **c.** 0   **d.** 10   **e.** $s^2 = 2.5$ **f.** $s \approx 1.5811$ **3. a.** $\mu = 47.7$ **b.** $-35.7, -29.7, -27.7, -17.7, -15.7,$ $-14.7, 32.3, 34.3, 36.3, 38.3$ **c.** 0   **d.** 8704.1   **e.** $\sigma^2 = 870.41$ **f.** $\sigma \approx 29.5027$ **5. a.** $\bar{x} = 53.1$ **b.** $-3.1, -2.1, -1.1, -0.6, -0.1, 0.4,$ $0.9, 1.4, 1.9, 2.4$ **c.** 0   **d.** 27.9   **e.** $s^2 = 3.1$ **f.** $s \approx 1.7607$ **7.** The data given in **(b)** will have the largest standard deviation, since the values are more dispersed than those in **(a)**.   **9.** The data given in **(a)** will have a larger standard deviation, since the values are more dispersed than those in **(b)**.   **11. a.** $\mu = \$9.60$; $\sigma \approx \$1.75$ **b.** $\mu = \$9.60$; $\sigma \approx \$4.59$ **13. a.** $\bar{x} = \$1276.31$; $s \approx \$567.26$ **b.** Eight of the 10 claims fall within one standard deviation of the mean.   **15. a.** $\mu = 79.375$; $\sigma \approx 8.2906$ **b.** Six of the eight scores fall within one standard deviation of the mean.   **17. a.** $\mu = 76.25$; $\sigma \approx 6.4759$ **b.** All of the eight heart rates fall within two standard deviations of the mean.   **19. a.** $\bar{x} = 45.2°F$ **b.** $s^2 = 88.7(°F)^2$ **c.** $s \approx 9.4181°F$ **21. a.** $\mu = 315.6$ tourists; the median is 345 tourists.   **b.** $\sigma^2 = 6494.64$ tourists^2 **c.** $\sigma \approx 80.5893$ tourists   **23. a.** $\bar{x} = 78.5$ blooms; the median is 80.5 blooms   **b.** $s^2 = 52.5$ blooms^2 **c.** $s \approx 7.2457$ **25.** 0.75   **27.** 0.25   **29.** 0.75   **31. a.** $\mu \approx 75.9$ **b.** $\sigma \approx 8.7287$ **33. a.** $\bar{x} \approx 26.6$ mpg   **b.** $s \approx 2.0603$ mpg   **35.** $s^2 \approx 618,481.1029$; $s \approx \$786.44$ (*Note:* 2000 was used for the estimated mean of the last class interval.)   **37.** $E(X) = \$0.625$; $\sigma(X) \approx \$3.35$ **39.** $E(X) = \$0$; $\sigma(X) \approx \$7,000$

10.4 Exercises

1. The curve marked B has the larger standard deviation.   **3.** The curve changes concavity at $x = 18$ and at $x = 22$.   **5. a.** 23   **b.** 15.5   **c.** 17.6   **7.** 47.5%   **9.** 50%   **11.** 0.15%   **13.** 2.5%   **15.** 84%   **17.** 68%   **19.** 84%   **21.** 0.9772   **23.** 0.0107   **25.** 0.4880   **27.** 0.0189   **29.** Approximately 0.5.   **31. a.** $P(x > 90) = P(z > 2)$ **b.** 0.0228   **33. a.** $P(x < 62.8) = P(z < -3.44)$ **b.** Approximately 0.   **35. a.** $P(x < 93.67) = P(z < 2.734)$ **b.** Approximately 0.9968.   **37. a.** $P(x > 76.5) = P(z > 0.7)$ **b.** 0.2420   **39.** 275 grams.   **41.** Xiou's score is 1.7 standard deviations above Professor Feldman's mean; Crystal's score is 0.65 standard deviation above Professor Schein's mean. Therefore, Xiou receives a B and Crystal receives a C.   **43. a.** 15.87%   **b.** 2.28%   **c.** Of the 200 batteries, approximately 4.56 can be expected to last less than 46 months.   **d.** Approximately 9.18% of the batteries will be returned under warranty within 48 months.   **45. a.** About 1587 of the lightbulbs can be expected to last at least 775 hours.   **b.** Approximately 228 of the lightbulbs can be expected to last less than 700 hours.   **c.** Approximately 9544 of the lightbulbs will last between 700 and 800 hours.   **47. a.** Approximately 38.75% of the students will spend more than 5 minutes complaining.   **b.** 0.0764   **49. a.** 0.2483   **b.** Sharokh will arrive at his first class on time approximately 5.59% of the time.   **51. a.** 0.6915   **b.** 0.9104   **53. a.** Approximately 21,190 of the checking accounts have balances greater than $12,000.   **b.** Approximately 97.72% of the accounts have balances of more than $5,000.   **55. a.** Approximately 1.07% of the faculty will be eligible for the incentive program.   **b.** 0.1190   **57. a.** Approximately 86.43% of the tires will last between 23,000 and 28,000 miles.   **b.** Approximately 5.94% will be returned under warranty.   **59. a.** 0.8664   **b.** The percentage of domestic shorthair cats that weigh as much as or more than Ender is approximately 0.   **61.** $z = 1.53$ **63.** $z \approx -2.325$ **65.** $z \approx 3.09$ **67.** The warranty period should be approximately 9.75 years (9 years and 9 months).   **69.** The warranty period should be approximately 5.89 years (just over 5 years and 10 months).

Chapter 10 Summary Exercises

1. $2   **3.** $20.40   **5.** 0.0968   **7.** 0.7175   **9.** $s \approx 3.0586$ **11.** Approximately 0.3830.

13.

Number of Repair Calls	Frequency	Relative Frequency
0	5	0.25
1	7	0.35
2	4	0.20
3	3	0.15
4	1	0.05

15. Twenty-percent of the households in the Somerset neighborhood called a professional for home repairs in the past year.   **17.** $\sigma^2 = 1.34$; $\sigma \approx 1.1576$ **19.** $\bar{x} = 5.9625$; $\sigma = 1.4439$ **21.** $x = -7, y = 3, z = 4$ **23.** $(3, 7), \left(\dfrac{20}{3}, \dfrac{10}{3}\right), \left(\dfrac{18}{13}, \dfrac{49}{13}\right), \left(\dfrac{42}{17}, \dfrac{21}{17}\right)$

Sample Test Items for Chapter 10

1.

Number of Secretaries	Relative Frequency
1	0.2
3	0.3
5	0.5

2. The mean number of secretaries per firm is 3.6.   **3.** The median number of secretaries per firm is 4.   **4.** The probability is 0.5.   **5.** The median number of books carried is 2.   **6.** The mode is 1 book.   **7.** The mean is 3 books.   **8.** The standard deviation is 2.9155 books.   **9.** Five of the seven salaries are within one standard deviation of the mean.

10.

X	p
0	0.4135
1	0.4359
2	0.1376
3	0.0129

11. The mean is 72.6, the median is 77, and the mode is 82. If the data represent a *sample*, then the standard deviation of this sample is approximately 20.9083. If the data represent a *population*, then the standard deviation is approximately 19.8353. **12.** Her expected profit is $10,200. **13.** Approximately 0.2660. **14.** Approximately 616.375.

15.

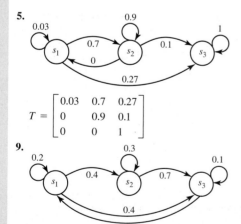

Relative Frequency Histogram

16. With $75,000 as the mean of the last class interval, the mean household income among those with Internet access in the year 2000 was approximately $60,503.50. **17.** With $75,000 as the mean of the last class interval, the median household income among those with Internet access in the year 2000 was approximately $75,000. **18.** With $75,000 as the mean of the last class interval, the standard deviation of household income among those with Internet access in the year 2000 was approximately $21,594.33. **19.** 0.9389

CHAPTER 11

Section 11.1 Exercises

1.

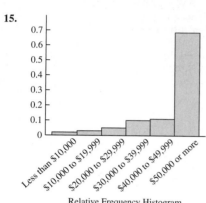

$$T = \begin{bmatrix} 0.6 & 0.4 \\ 0.7 & 0.3 \end{bmatrix}$$

3.

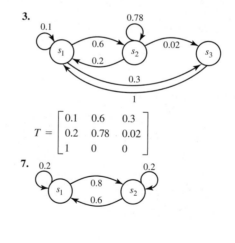

$$T = \begin{bmatrix} 0.1 & 0.6 & 0.3 \\ 0.2 & 0.78 & 0.02 \\ 1 & 0 & 0 \end{bmatrix}$$

5.

$$T = \begin{bmatrix} 0.03 & 0.7 & 0.27 \\ 0 & 0.9 & 0.1 \\ 0 & 0 & 1 \end{bmatrix}$$

7.

9.

11.

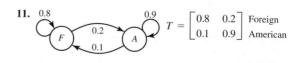

$$T = \begin{bmatrix} 0.8 & 0.2 \\ 0.1 & 0.9 \end{bmatrix} \begin{matrix} \text{Foreign} \\ \text{American} \end{matrix}$$

13. 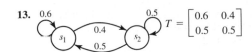 $T = \begin{bmatrix} 0.6 & 0.4 \\ 0.5 & 0.5 \end{bmatrix}$

15.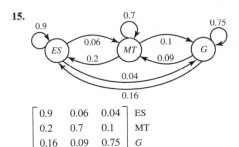

$$\begin{bmatrix} 0.9 & 0.06 & 0.04 \\ 0.2 & 0.7 & 0.1 \\ 0.16 & 0.09 & 0.75 \end{bmatrix} \begin{matrix} ES \\ MT \\ G \end{matrix}$$

17. This matrix cannot be a transition matrix, since the entries in row 2 add up to 1.1, which is greater than 1.

19. This matrix can be a transition matrix, since the entries in each row add up to 1.

21. a. $\begin{bmatrix} 0.80 & 0.20 \\ 0.40 & 0.60 \end{bmatrix} \begin{matrix} NumberKrunch \\ QuickDigit \end{matrix}$ **b., c.** The probability that a person now using a NumberKrunch computer will switch to a QuickDigit computer two purchases later is 0.28.

d. The probability that a NumberKrunch user will purchase another NumberKrunch computer two purchases later is 0.72.

23. a.  $\begin{bmatrix} 0.6 & 0.4 \\ 0.6 & 0.4 \end{bmatrix} \begin{matrix} Aligned \\ Misaligned \end{matrix}$

b., c. The probability that, if a probe is aligned now, it will still be aligned 2 minutes from now, is 0.6.

d. The probability that, if a probe is misaligned now, it will still be misaligned 3 minutes from now, is 0.4.

25. a.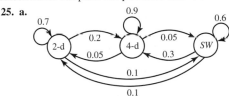

$$\begin{bmatrix} 0.7 & 0.2 & 0.1 \\ 0.05 & 0.9 & 0.05 \\ 0.1 & 0.3 & 0.6 \end{bmatrix} \begin{matrix} 2-d \\ 4-d \\ SW \end{matrix}$$

b., c. The probability that a person leasing a two-door sedan now will be leasing a station wagon three choices later is 0.1525.

27. If Denita misses her first free throw, then she will miss her third free throw with probability 0.248.

29. Row 1 of this transition matrix indicates that State 1 is an absorbing state; once the state is entered, it can never be left. This applies to all powers of the transition matrix.

Section 11.2 Exercises

1. $P_1 = \begin{bmatrix} 0.44 & 0.56 \end{bmatrix}$ **3.** $P_2 = \begin{bmatrix} 0.52 & 0.48 \end{bmatrix}$ **5.** $P_1 = \begin{bmatrix} 0.31 & 0.36 & 0.33 \end{bmatrix}$ **7.** $P_2 = \begin{bmatrix} 0.34725 & 0.266 & 0.38675 \end{bmatrix}$
9. $P_3 = \begin{bmatrix} 0.744 & 0.256 \end{bmatrix}$ **11. a.** $P_1 = \begin{bmatrix} 0.48 & 0.52 \end{bmatrix}$ **b.** $P_2 = \begin{bmatrix} 0.392 & 0.608 \end{bmatrix}$ **13. a.** $P_1 = \begin{bmatrix} 0.24 & 0.42 & 0.34 \end{bmatrix}$
b. $P_2 = \begin{bmatrix} 0.274 & 0.32 & 0.406 \end{bmatrix}$ **15.** On their third automobile purchase, 42.48% will buy a foreign make and 57.52% will buy an American make. **17.** In two years, approximately 45.12% of the State University students will be business majors, 34.116% will be engineering majors, and 20.724% will be mathematics majors.

19. a. 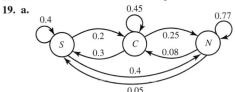 **b.** $\begin{bmatrix} 0.40 & 0.20 & 0.40 \\ 0.30 & 0.45 & 0.25 \\ 0.05 & 0.08 & 0.87 \end{bmatrix} \begin{matrix} sulfasalazine \\ Cipro® \\ Neither \end{matrix}$ **c.** After three generations, approximately 15.59% of the bacteria will be immune to sulfasalazine, 16.83% will be immune to Cipro®, and 67.57% will not be immune to either antibiotic.

21. $T^{-1} = \begin{bmatrix} 3.5 & -2.5 \\ -1.5 & 2.5 \end{bmatrix}; P_0 = \begin{bmatrix} 0.5 & 0.5 \end{bmatrix}$ **23.** $T^{-1} = \begin{bmatrix} -\frac{9}{67} & \frac{76}{67} \\ \frac{91}{67} & -\frac{24}{67} \end{bmatrix}; P_0 \approx \begin{bmatrix} 0.6119 & 0.3881 \end{bmatrix}$

25. $T^{-1} = \begin{bmatrix} 0 & \frac{10}{9} & -\frac{1}{9} \\ 0 & 0 & 1 \\ \frac{5}{2} & -\frac{5}{9} & -\frac{17}{9} \end{bmatrix}; P_0 = \begin{bmatrix} 0.7500 & 0.1111 & 0.1389 \end{bmatrix}$ **27.** $P_1 = \begin{bmatrix} 0.2 & 0.8 \end{bmatrix}; P_2 = \begin{bmatrix} 0.44 & 0.56 \end{bmatrix}; P_3 = \begin{bmatrix} 0.368 & 0.632 \end{bmatrix};$

$P_4 = \begin{bmatrix} 0.3896 & 0.6104 \end{bmatrix}; P_5 \approx \begin{bmatrix} 0.3831 & 0.6169 \end{bmatrix}; P_6 \approx \begin{bmatrix} 0.3851 & 0.6149 \end{bmatrix}; P_7 \approx \begin{bmatrix} 0.3845 & 0.6155 \end{bmatrix}; P_8 \approx \begin{bmatrix} 0.3847 & 0.6153 \end{bmatrix};$
$P_9 \approx \begin{bmatrix} 0.3846 & 0.6154 \end{bmatrix}.$ To four-decimal-place accuracy, P_n is approximately equal to P_9 for all $n \geq 10$. **29.** $P_1 = \begin{bmatrix} 0.38 & 0.62 \end{bmatrix};$
$P_2 = \begin{bmatrix} 0.386 & 0.614 \end{bmatrix}; P_3 = \begin{bmatrix} 0.3842 & 0.6158 \end{bmatrix}; P_4 \approx \begin{bmatrix} 0.3847 & 0.6153 \end{bmatrix}; P_5 \approx \begin{bmatrix} 0.3846 & 0.6154 \end{bmatrix}.$ To four-decimal-place accuracy, P_n is

approximately equal to P_5 for all $n \geq 5$. **31.** All state vectors are equal to P_0. **33.** $P_1 = [0.5 \quad 0.25 \quad 0.25]$; $P_2 = [0.625 \quad 0.25 \quad 0.125]$; $P_3 = [0.75 \quad 0.125 \quad 0.125]$; $P_4 = [0.8125 \quad 0.125 \quad 0.0625]$; $P_5 = [0.875 \quad 0.0625 \quad 0.0625]$; $P_6 = [0.90625 \quad 0.0625 \quad 0.03125]$; $P_7 = [0.9375 \quad 0.03125 \quad 0.03125]$; $P_8 = [0.953125 \quad 0.03125 \quad 0.015625]$; $P_9 = [0.96875 \quad 0.015625 \quad 0.015625]$; $P_{10} \approx [0.9766 \quad 0.0156 \quad 0.0078]$; $P_{11} \approx [0.9844 \quad 0.0078 \quad 0.0078]$; $P_{12} \approx [0.9883 \quad 0.0078 \quad 0.0039]$; $P_{13} \approx [0.9922 \quad 0.0039 \quad 0.0039]$; $P_{14} \approx [0.9941 \quad 0.0039 \quad 0.0020]$; $P_{15} \approx [0.9961 \quad 0.0020 \quad 0.0020]$. **35.** $P_1 = [0 \quad 1 \quad 0]$; $P_2 = [0.5 \quad 0 \quad 0.5]$; $P_3 = [0.5 \quad 0.5 \quad 0]$; $P_4 = [0.75 \quad 0 \quad 0.25]$; $P_5 = [0.75 \quad 0.25 \quad 0]$; $P_6 = [0.875 \quad 0 \quad 0.125]$; $P_7 = [0.875 \quad 0.125 \quad 0]$; $P_8 = [0.9375 \quad 0 \quad 0.0625]$; $P_9 = [0.9375 \quad 0.0625 \quad 0]$; $P_{10} \approx [0.9688 \quad 0 \quad 0.0312]$; $P_{10} \approx [0.9688 \quad 0.0312 \quad 0]$; $P_{12} \approx [0.9844 \quad 0 \quad 0.0156]$; $P_{13} \approx [0.9844 \quad 0.0156 \quad 0]$; $P_{14} \approx [0.9922 \quad 0 \quad 0.0078]$; $P_{15} \approx [0.9922 \quad 0.0078 \quad 0]$.

Section 11.3 Exercises

1. Since all entries in T are positive, the matrix is regular. **3.** Since all entries in T^2 are positive, the matrix is regular. **5.** Since the first row of T^n will contain 0's for all n, the matrix is not regular. **7.** $L \approx \begin{bmatrix} 0.4167 & 0.5833 \\ 0.4167 & 0.5833 \end{bmatrix}$ **9.** $L \approx \begin{bmatrix} 0.6667 & 0.3333 \\ 0.6667 & 0.3333 \end{bmatrix}$

11. $L \approx \begin{bmatrix} 0.7143 & 0.2857 \\ 0.7143 & 0.2857 \end{bmatrix}$ **13.** $S = [0.25 \quad 0.75]$; $L = \begin{bmatrix} 0.25 & 0.75 \\ 0.25 & 0.75 \end{bmatrix}$ **15.** $S = [\frac{4}{7} \quad \frac{3}{7}]$; $L = \begin{bmatrix} \frac{4}{7} & \frac{3}{7} \\ \frac{4}{7} & \frac{3}{7} \end{bmatrix}$

17. $S = [\frac{21}{41} \quad \frac{20}{41}]$; $L = \begin{bmatrix} \frac{21}{41} & \frac{20}{41} \\ \frac{21}{41} & \frac{20}{41} \end{bmatrix}$ **19.** $S = [\frac{1}{3} \quad \frac{4}{15} \quad \frac{2}{5}]$; $L = \begin{bmatrix} \frac{1}{3} & \frac{4}{15} & \frac{2}{5} \\ \frac{1}{3} & \frac{4}{15} & \frac{2}{5} \\ \frac{1}{3} & \frac{4}{15} & \frac{2}{5} \end{bmatrix}$

21. $S \approx [0.3885 \quad 0.4101 \quad 0.2014]$; $L \approx \begin{bmatrix} 0.3885 & 0.4101 & 0.2014 \\ 0.3885 & 0.4101 & 0.2014 \\ 0.3885 & 0.4101 & 0.2014 \end{bmatrix}$ **23.** Eventually, $\frac{4}{7}$ will buy American cars and $\frac{3}{7}$ will buy foreign cars.

25. In the long run, 32% of the voters in Mt. Vernon will vote Democrat, 60% will vote Republican, and 8% will vote Independent.

27. a. $S = [\frac{20}{41} \quad \frac{16}{41} \quad \frac{5}{41}]$ **b.** $L = \begin{bmatrix} \frac{20}{41} & \frac{16}{41} & \frac{5}{41} \\ \frac{20}{41} & \frac{16}{41} & \frac{5}{41} \\ \frac{20}{41} & \frac{16}{41} & \frac{5}{41} \end{bmatrix}$. Eventually, approximately 48.78% of the population will produce right-handed offspring, approximately 39.02% will produce left-handed offspring, and approximately 12.20% will produce ambidextrous offspring.

29. a.

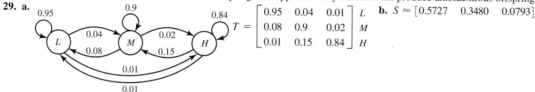

$T = \begin{bmatrix} 0.95 & 0.04 & 0.01 \\ 0.08 & 0.9 & 0.02 \\ 0.01 & 0.15 & 0.84 \end{bmatrix} \begin{matrix} L \\ M \\ H \end{matrix}$ **b.** $S \approx [0.5727 \quad 0.3480 \quad 0.0793]$

c. $L \approx \begin{bmatrix} 0.5727 & 0.3480 & 0.0793 \\ 0.5727 & 0.3480 & 0.0793 \\ 0.5727 & 0.3480 & 0.0793 \end{bmatrix}$ Eventually, 57.27% of the company's customers will be classified as low risk, 34.80% will be classified as medium risk, and 7.93% will be classified as high risk.

Chapter Summary Exercises

1.

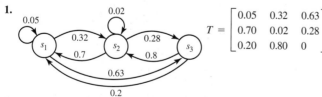

$T = \begin{bmatrix} 0.05 & 0.32 & 0.63 \\ 0.70 & 0.02 & 0.28 \\ 0.20 & 0.80 & 0 \end{bmatrix}$

3. $P_1 = [0.26 \quad 0.47 \quad 0.27]$ **5.** $L = \begin{bmatrix} 0.4375 & 0.5625 \\ 0.4375 & 0.5625 \end{bmatrix}$ **7.** $L = \begin{bmatrix} \frac{6}{11} & \frac{5}{11} \\ \frac{6}{11} & \frac{5}{11} \end{bmatrix}$ **9.** This matrix cannot be a transition matrix, since it has a negative entry in the second row, second column. **11.** If the archer misses the bull's-eye on the first shot, the probability that she will hit it on the third shot is approximately 0.7957.   **13.** $L \approx \begin{bmatrix} 0.2576 & 0.4697 & 0.2727 \\ 0.2576 & 0.4697 & 0.2727 \\ 0.2576 & 0.4697 & 0.2727 \end{bmatrix}$ **15.** $x = \frac{21}{16},\ y = -\frac{1}{16};\ z = \frac{45}{16}$

17.

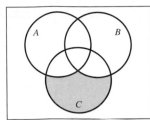

19. The maximum value of f is 85 when $x = 0$ and $y = 17$.

Sample Test Items

1. 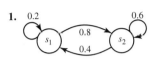   **2.** $T = \begin{bmatrix} 0.2 & 0.8 \\ 0.4 & 0.6 \end{bmatrix}$ **3.** 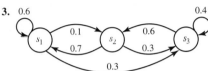   **4.** $T = \begin{bmatrix} 0.6 & 0.1 & 0.3 \\ 0.7 & 0 & 0.3 \\ 0 & 0.6 & 0.4 \end{bmatrix}$

5. 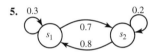   **6.** 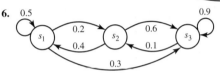   **7.**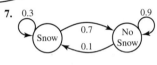

8. $T = \begin{bmatrix} 0.3 & 0.7 \\ 0.1 & 0.9 \end{bmatrix} \begin{matrix} \text{Snow} \\ \text{No Snow} \end{matrix}$ **9.** $P_1 = \begin{bmatrix} 0.26 & 0.57 & 0.17 \end{bmatrix}$ **10.** $P_2 = \begin{bmatrix} 0.206 & 0.597 & 0.197 \end{bmatrix}$ **11.** This transition matrix is not regular, since T^n will have a 0 in the first row for all powers of n.   **12.** Because T^2 has only positive entries, the transition matrix is regular.

13. $L \approx \begin{bmatrix} 0.5263 & 0.4737 \\ 0.5263 & 0.4737 \end{bmatrix}$ **14.** $L \approx \begin{bmatrix} 0.5714 & 0.4286 \\ 0.5714 & 0.4286 \end{bmatrix}$ **15.** $S \approx \begin{bmatrix} 0.3291 & 0.6709 \end{bmatrix};\ L \approx \begin{bmatrix} 0.3291 & 0.6709 \\ 0.3291 & 0.6709 \end{bmatrix}$

16. $S = \begin{bmatrix} \frac{7}{17} & \frac{10}{17} \end{bmatrix};\ L = \begin{bmatrix} \frac{7}{17} & \frac{10}{17} \\ \frac{7}{17} & \frac{10}{17} \end{bmatrix}$

CHAPTER 12

Exercises 12.1

1. Is strictly determined; 2; row player; row player's best strategy: Row 2; column player's best strategy: Column 2   **3.** Not strictly determined   **5.** Is strictly determined; 2; row player; row player's best strategy: Row 2; column player's best strategy: Column 1   **7.** Is strictly determined; 1; row player; row player's best strategy: Row 1; column player's best strategy: Column 1   **9.** Not strictly determined   **11.** The value of x must be greater than zero and less than three.

13.

Economy		
	Favorable	Recession
Remodel	100,000	−40,000
Don't remodel	−50,000	−45,000

The saddle point is −40,000. The store should remodel.

15. a.

		Fabless	
		Sports	Senior
	Sports	250	400
Nutone	Senior	350	300
	Drug	375	400

b. The saddle point is 375.
c. Nutone should start a drug therapy program and Fabless should start a sports medicine program.

17. a.

		Hector		
		R, 5	B, 7	B, 8
Juan	R, 4	1	−4	−4
	B, 9	−5	2	1

b. Not strictly determined

19. a.

		Computer hardware	
		Tallahassee	Austin
Computer software	Tallahassee	0	500,000
	Austin	−250,000	50,000

b. The game is strictly determined.

c. Both companies should locate in Tallahassee.

Exercises 12.2

1. 1   **3.** $\dfrac{4}{9}$ **5.** $\dfrac{7}{6}$ **7.** $\dfrac{5}{9}$ **9.** $-\dfrac{9}{88}$ **11.** Q_2 is the better column player strategy.   **13.** Q_2 is the better column player strategy.

15. Q_1 is the better column player strategy.

17. $P = \begin{bmatrix} 0 & 1 \end{bmatrix}$, $Q = \begin{bmatrix} 1 \\ 0 \end{bmatrix}$; $\begin{bmatrix} 0 & 1 \end{bmatrix} A \begin{bmatrix} q_1 \\ q_2 \end{bmatrix} = 4q_1 + 6q_2 \geq 4$; $\begin{bmatrix} p_1 & p_2 \end{bmatrix} A \begin{bmatrix} 1 \\ 0 \end{bmatrix} = 3p_1 + 4p_2 \leq 4$.

19. $P = \begin{bmatrix} 1 & 0 \end{bmatrix}$, $Q = \begin{bmatrix} 0 \\ 1 \end{bmatrix}$; $\begin{bmatrix} 1 & 0 \end{bmatrix} A \begin{bmatrix} q_1 \\ q_2 \end{bmatrix} = 4q_1 - 7q_2 \geq -7$; $\begin{bmatrix} p_1 & p_2 \end{bmatrix} A \begin{bmatrix} 0 \\ 1 \end{bmatrix} = -7p_1 - 8p_2 \leq -7$.

21. a.

		Ilke	
		1	2
Tanya	1	5	−5
	2	−5	5

b. $\frac{5}{9}$; favors Tanya

23. a.

		Jill	
		P	D
Jack	P	1	−10
	D	−1	10

b. −10   **c.** $-\dfrac{9}{2}$ **d.** The one in (b).

Exercises 12.3

1. a. This game is strictly determined.   **b.** 1   **3. a.** Row player: $\begin{bmatrix} \frac{9}{14} & \frac{5}{14} \end{bmatrix}$; column player: $\begin{bmatrix} \frac{4}{7} \\ \frac{3}{7} \end{bmatrix}$ **b.** $\dfrac{1}{7}$ **5. a.** This game is strictly determined.

b. 4   **7. a.** Row player: $\begin{bmatrix} \frac{6}{13} & \frac{7}{13} \end{bmatrix}$; column player: $\begin{bmatrix} \frac{8}{13} \\ \frac{5}{13} \end{bmatrix}$ **b.** $\dfrac{22}{13}$ **9. a.** Row player: $\begin{bmatrix} \frac{9}{16} & 0 & \frac{7}{16} \end{bmatrix}$; column player: $\begin{bmatrix} \frac{9}{16} \\ \frac{7}{16} \end{bmatrix}$ **b.** $\dfrac{31}{16}$ **11.** Row

player: $\begin{bmatrix} \frac{2}{5} & \frac{3}{5} \end{bmatrix}$; column player: $\begin{bmatrix} \frac{1}{2} \\ \frac{1}{2} \end{bmatrix}$; $v = 1$ **13.** Row player: $\begin{bmatrix} 1 & 0 \end{bmatrix}$; column player: $\begin{bmatrix} \frac{4}{7} \\ \frac{3}{7} \end{bmatrix}$; $v = 1$ (Alternatively, column players should always

play their second option.)   **15.** Row player: $\begin{bmatrix} \frac{1}{2} & \frac{1}{2} & 0 \end{bmatrix}$; column player: $\begin{bmatrix} \frac{3}{8} \\ \frac{5}{8} \end{bmatrix}$; $v = 3$ **17.** Row player: $\begin{bmatrix} \frac{2}{3} & \frac{1}{3} & 0 \end{bmatrix}$; column player: $\begin{bmatrix} \frac{1}{6} \\ \frac{5}{6} \\ 0 \end{bmatrix}$; $v = \frac{8}{3}$

19. Row player: $\begin{bmatrix} \frac{2}{3} & \frac{1}{3} & 0 \end{bmatrix}$ column player: $\begin{bmatrix} \frac{1}{3} \\ \frac{2}{3} \\ 0 \end{bmatrix}$; $v = \frac{1}{3}$ **21.** Row player: $\begin{bmatrix} \frac{1}{2} & \frac{1}{2} \end{bmatrix}$; column player: $\begin{bmatrix} \frac{1}{10} \\ \frac{9}{10} \\ 0 \end{bmatrix}$; $v = \frac{1}{2}$ **23. a.** Conceal coin in the

right hand $\frac{8}{13}$ of the time and in the left hand $\frac{5}{13}$ of the time.   **b.** Guess right hand $\frac{5}{13}$ of the time and left hand $\frac{8}{13}$ of the time.   **c.** Match

25. a. M-Mart should feature clothing $\frac{19}{31}$ of the time and sporting goods $\frac{12}{31}$ of the time. Q-Mart should feature electronics $\frac{11}{31}$ of the time and kitchen-ware $\frac{20}{31}$ of the time.   **b.** M-Mart   **27.** Bob should play a nickel $\frac{2}{3}$ of the time and a dime $\frac{1}{3}$ of the time. John should play each coin $\frac{1}{2}$ of the time.

Chapter 12 Summary

1. 9   **3.** 2   **5.** −1   **7.** $\dfrac{4}{3}$ **9.** 0   **11.** 4   **13.** 2   **15.** 60%   **17.** It gives the value of the game, and its position determines

the strategy of both players.   **19.** It is the same concept as the expected value in the probability sense. If it is positive, the game favors the row player. If it is negative, the game favors the column player. If it is zero, the game is fair.   **21.** $y = 8x + 160$

23.
$$\begin{array}{c} & \begin{array}{cccc} F & R & S & H \end{array} \\ \begin{array}{c} F \\ R \\ S \\ H \end{array} & \left[\begin{array}{cccc} 0 & 1 & 1 & 1 \\ 0 & 0 & 1 & 1 \\ 1 & 1 & 0 & 0 \\ 1 & 0 & 1 & 0 \end{array}\right] \end{array}$$

25.
Mean: 7
Median: 8
Mode: 10
Range: 9

Sample Test Items for Chapter 12

1. Strictly determined; saddle point: 3; favors row player.   **2.** Not strictly determined   **3.** Strictly determined; saddle point: -2; favors column player   **4.** None   **5.** Between -2 and 0

6.
$$\begin{array}{c} & \begin{array}{cc} C & M \end{array} \\ \begin{array}{c} C \\ M \end{array} & \left[\begin{array}{cc} 0 & 0.05 \\ -0.05 & 0 \end{array}\right] \end{array}$$

7. Both should open campus branches. The value is 0.   **8.** $\dfrac{3}{2}$ **9.** $\dfrac{3}{2}$ **10.** $\dfrac{14}{9}$ **11.** Q_1 **12.** Q_2

13. Q_1 **14.** Value: $\dfrac{29}{7}$; row strategy: $\left[\begin{array}{cc} \frac{1}{7} & \frac{6}{7} \end{array}\right]$; column strategy: $\left[\begin{array}{c} \frac{6}{7} \\ \frac{1}{7} \end{array}\right]$ **15.** Strictly determined; value: 3; both players should use their second option.   **16.** Strictly determined; value: 6; row player should use second option, while column player should use first option.

17. Value: 3.62; row strategy: $\left[\begin{array}{cc} \frac{8}{21} & \frac{13}{21} \end{array}\right]$; column strategy: $\left[\begin{array}{c} \frac{10}{21} \\ \frac{11}{21} \end{array}\right]$.   **18.** Value: 3.88; row strategy: $\left[\begin{array}{ccc} \frac{9}{16} & 0 & \frac{7}{16} \end{array}\right]$; column strategy: $\left[\begin{array}{c} \frac{9}{16} \\ 0 \\ \frac{7}{16} \end{array}\right]$.

19. Strictly determined; value: 7; row player should use option 1, column player should use option 3.   **20.** Both players should play their dime approximately 68% of the time.   **21.** The game favors Mia.

APPENDIX I

1. a. No.   **b.** Yes.   **c.** Yes.   **d.** No.   **2. a.** No.   **b.** Yes.   **c.** Yes.   **d.** No.   **3. a.** No.   **b.** No.   **c.** Yes.   **d.** No.   **4. a.** No.   **b.** No.   **c.** No.   **d.** Yes.   **5. a.** Yes.   **b.** Yes.   **c.** Yes.   **d.** No.

6.    **7.** 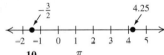   **8.**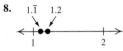

9.    **10.**

11. -8 **12.** 2   **13.** $-6 + 3x$ **14.** $-4x + 2$ **15.** 14   **16.** -2 **17.** $x = 3$; conditional.   **18.** $x = -1$; conditional.
19. $x = -6$; conditional.   **20.** $x = -2400$ **21.** This equation is consistent; any value of x is a solution of the equation.   **22.** No solution; this equation is inconsistent.   **23.** $x = 0$; conditional.   **24.** No solution; this equation is inconsistent.   **25.** This equation is consistent; any value of x is a solution of the equation.   **26.** 3   **27.** $\dfrac{1}{2}$ **28.** $\dfrac{8}{27}$ **29.** $\dfrac{9}{64}$ **30.** $\dfrac{64}{81}$ **31.** $x = \dfrac{\ln 3}{\ln 2}$ **32.** $x = \dfrac{\ln 2}{\ln 3}$

33. $x = \dfrac{\ln 3}{\ln 4}$ **34.** $x \approx 161.7472$ **35.** $x \approx 2$ **36.** 728   **37.** $\dfrac{6141}{1024}$ **38.** 1.1111   **39.** 202.0202   **40.** 307.6008

Index

T

Acknowledgments

Literary Credits

467 Reprinted with permission from PARADE and Marilyn vos Savant, © 1990.

Photo Credits

Title page ©PhotoDisc
9 ©Richard T. Nowitz/Corbis
10 Courtesy of Sara Anderson
22 Courtesy of Beth Anderson
23 ©Hulton-Deutsch Collections/Corbis
32 Courtesy of Beth Anderson
33 ©Corbis RF
48 ©Cowart's Construction Company
49 ©Digital Vision
62 ©Digital Vision
64 ©Corbis RF
67 ©PhotoDisc
69 ©Corbis RF
76 ©PhotoDisc
77 ©PhotoDisc
93 ©PhotoDisc Blue
94 ©Digital Vision
108 ©AFP/Corbis
104 ©AP Photo/Ed Reinke
123 Courtesy of Beth Anderson
125 ©Comstock Klips
137 ©PhotoDisc
138 ©Digital Vision
159 ©PhotoDisc Red
168 Courtesy of Beth Anderson
169 ©PhotoDisc
179 ©PhotoDisc
180 ©Ed Kashi/Corbis
193 ©PhotoDisc
227 ©Getty Images/Image Bank, Peter Frey
227 ©The Kobal Collection
242 ©PhotoDisc Blue
255 ©Cutlerrepaving.com

259 ©PhotoDisc
261 ©PhotoDisc
284 ©Digital Vision
295 ©PhotoDisc
297 Courtesy of Beth Anderson
301 ©Courtesy of Kinston, NC, City Council
336 Courtesy of Beth Anderson
346 Courtesy of Beth Anderson
372 Courtesy of Beth Anderson
395 Courtesy of Beth Anderson
398 ©PhotoDisc
415 public domain
428 public domain
449 ©Duomo/Corbis
450 ©AFP/Corbis
458 ©PhotoDisc Blue
469 Courtesy of Beth Anderson
471 ©PhotoDisc Blue
483 ©PhotoDisc
495 Courtesy of University of Delaware
497 ©AP Photo/The Indianapolis Star, Rich Miller
506 Courtesy of Beth Anderson
506 ©Digital Stock RF
519 public domain
521 Courtesy of Beth Anderson
527 ©Digital Vision
536 ©Dwayne Newton/Photo Edit
537 public domain
538 ©PhotoDisc
538 ©University of Tennessee Lady Vols
543 ©PhotoDisc
543 Courtesy of Beth Anderson
555 Courtesy of Beth Anderson
557 ©AP Photo/Jim Cole
564 ©Corbin Motorcycle Accessories
582 ©Bonnie Kamin/Photo Edit
583 Courtesy of Beth Anderson
586 ©PhotoDisc Red
571 ©Tom Stewart/CORBIS